IFMBE Proceedings 98

The IFMBE Proceedings Book Series is an official publication of *the International Federation for Medical and Biological Engineering* (IFMBE). The series gathers the proceedings of various international conferences, which are either organized or endorsed by the Federation. Books published in this series report on cutting-edge findings and provide an informative survey on the most challenging topics and advances in the fields of medicine, biology, clinical engineering, and biophysics.

The series aims at disseminating high quality scientific information, encouraging both basic and applied research, and promoting world-wide collaboration between researchers and practitioners in the field of Medical and Biological Engineering.

Topics include, but are not limited to:

- Diagnostic Imaging, Image Processing, Biomedical Signal Processing
- Modeling and Simulation, Biomechanics
- Biomaterials, Cellular and Tissue Engineering
- Information and Communication in Medicine, Telemedicine and e-Health
- Instrumentation and Clinical Engineering
- Surgery, Minimal Invasive Interventions, Endoscopy and Image Guided Therapy
- Audiology, Ophthalmology, Emergency and Dental Medicine Applications
- Radiology, Radiation Oncology and Biological Effects of Radiation
- Drug Delivery and Pharmaceutical Engineering
- Neuroengineering, and Artificial Intelligence in Healthcare

IFMBE proceedings are indexed by SCOPUS, EI Compendex, Japanese Science and Technology Agency (JST), SCImago. They are also submitted for consideration by WoS.

Proposals can be submitted by contacting the Springer responsible editor shown on the series webpage (see "Contacts"), or by getting in touch with the series editor Ratko Magjarevic.

Jefferson Luiz Brum Marques ·
Cesar Ramos Rodrigues ·
Daniela Ota Hisayasu Suzuki ·
José Marino Neto ·
Renato García Ojeda
Editors

IX Latin American Congress on Biomedical Engineering and XXVIII Brazilian Congress on Biomedical Engineering

Proceedings of CLAIB and CBEB 2022, October 24–28, 2022, Florianópolis, Brazil—Volume 1: Bioengineering and Biomaterials

Springer

Editors
Jefferson Luiz Brum Marques
Institute of Biomedical Engineering
Department of Electrical and Electronic Engineering
Federal University of Santa Catarina
Florianópolis, Brazil

Daniela Ota Hisayasu Suzuki
Institute of Biomedical Engineering
Department of Electrical and Electronic Engineering
Federal University of Santa Catarina
Florianópolis, Brazil

Renato García Ojeda
Institute of Biomedical Engineering
Department of Electrical and Electronic Engineering
Federal University of Santa Catarina
Florianópolis, Brazil

Cesar Ramos Rodrigues
Institute of Biomedical Engineering
Department of Electrical and Electronic Engineering
Federal University of Santa Catarina
Florianópolis, Brazil

José Marino Neto
Institute of Biomedical Engineering
Federal University of Santa Catarina
Florianópolis, Brazil

ISSN 1680-0737 ISSN 1433-9277 (electronic)
IFMBE Proceedings
ISBN 978-3-031-49400-0 ISBN 978-3-031-49401-7 (eBook)
https://doi.org/10.1007/978-3-031-49401-7

This Springer imprint is published by the registered company Springer Nature Switzerland AG
The registered company address is: Gewerbestrasse 11, 6330 Cham, Switzerland

Paper in this product is recyclable.

Preface

The IX Latin American Congress on Biomedical Engineering and XXVIII Brazilian Congress on Biomedical Engineering (CLAIB&CBEB 2022) took place simultaneously on October 24–28, 2022, in Florianópolis-SC, Brazil, and were organised by the Institute of Biomedical Engineering of The Federal University of Santa Catarina (IEB-UFSC), the Regional Council of Biomedical Engineering for Latin America (CORAL) and the Brazilian Biomedical Engineering Society (SBEB). These events were held remotely for the most part, with a small set of conferences taking place in person on the premises of IEB-UFSC (Florianópolis, Brazil). They included 11 hands-on technical workshops for students, 26 keynote speakers and symposia, and 40 oral and poster presentation sessions attended by about a thousand participants, including undergraduate and graduate students, academic researchers, and public and private sector agents.

We are proud to present in this book a selection of papers presented at this event by researchers from all over the world, reporting recent and innovative findings and technological outcomes in the many areas of interest of biomedical engineering. These papers represent nearly 50% of those original contributions presented at the CLAIB&CBEB 2022. Their academic quality has been warranted by careful peer review coordinated by an expert scientific committee of leading Latin American senior researchers in biomedical engineering. The content is organised into four volumes and eleven chapters, covering the most relevant areas of scientific and technological developments within the broad spectrum of biomedical engineering interests. We are sure that the contributions presented in this book give a deep overview of the leading edge in your expertise and other areas.

On behalf of Scientific and Organising Committees, we thank authors, academic reviewers and sponsoring societies such as CORAL, SBEB, UFSC, FAPESC and IEB-UFSC for their contributions. Moreover, we encourage readers to enjoy this amazing piece of scientific literature as a breadth of knowledge in the biomedical engineering field.

Organisation

CLAIB&CBEB 2022 was organised by the Regional Council of Biomedical Engineering for Latin America (CORAL) and the Brazilian Biomedical Engineering Society (SBEB) in cooperation with the International Federation for Medical and Biological Engineering (IFMBE).

Committees

The Organising Committee of the IX Latin American Congress of Biomedical Engineering (CLAIB 2022) and the XXVIII Brazilian Congress of Biomedical Engineering (CBEB 2022) was composed of the following members:

Organising Committee

President

Renato Garcia Ojeda

Vice-President

Jefferson Luiz Brum Marques

Scientific Committee

Jefferson Luiz Brum Marques

Programme Committee

José Marino-Neto

Special Events Committee

Cesar Ramos Rodrigues

Finance and Disclosure Committee

Daniela Ota Hisaysu Suzuki

Follow-Up Committee

Helio Schechtman	President of the Brazilian Society of Biomedical Engineering (SBEB)
Elliot Vernet	President Regional Council of Biomedical Engineering for Latin America (CORAL)
Shankar Krishnan	Presidente International Federation of Medical and Biological Engineering (IFMBE)

Part Editors

Biomedical Robotics, Assistive Technologies and Health Informatics	*Teodiano Freire Bastos Filho (UFES-Brazil)* *Luis Eduardo Rodriguez Cheu (CSEJG-Colombia)* *Renato Garcia Ojeda (UFSC-Brazil)*
Biomedical Image and Signal Processing	*Márcio Holsbach Costa (UFSC-Brazil)* *Virginia Laura Ballarin (UNMdP-Argentina)* *Jurandir Nadal (UFRJ-Brazil)*
Biomedical Optics and Systems and Technologies for Therapy and Diagnosis	*Renato Amaro Zângaro (UAM-Brazil)* *Regiane Albertini de Carvalho (UNIFESP-Brazil)*
Biomedical Devices and Instrumentation	*Percy Nohama (PUC-PR-Brazil)* *Guilherme Nunes Nogueira (PUC-PR-Brazil)*
Bioengineering, Modelling and Simulation, Bioinformatics and Computational Biology	*Antonio Carlos Guimarães de Almeida (UFSJ-Brazil)* *Antônio Márcio Rodrigues (UFSJ-Brazil)* *Luis Enrique Bergues Cabrales (UO-Cuba)*
Biomaterials, Tissue Engineering and Artificial Organs	*Idágene Aparecida Cestari (USP)* *Sônia Maria Malmonge (UFABC-Brazil)*
Biosensors, Bioinstrumentation and Micro-nanotechnologies	*Alcimar Barbosa Soares (UFU-Brazil)*

Biomechanics, Neural Engineering and Rehabilitation	*Adriano de Oliveira Andrade (UFU-Brazil)*
Clinical Engineering and Health Technology Assessment	*Fabiola Margarita Martínez-Licona (UAM-Mexico)* *Roberto Macoto Ichinose (UFRJ-Brazil)*
Health Technology Innovation and Development	*Antonio Adilton Oliveira Carneiro (USP-Brazil)*
Metrology and Quality of Healthcare Technologies	*Carlos Rubén Dell'Aquila (UNSJ-Argentina)* *Rodrigo Pereira Barretto da Costa Félix (InMetro-Brazil)*
Special Topics	*Adson Ferreira da Rocha (UnB-Brazil)* *Renato Garcia Ojeda (UFSC-Brazil)*

Support Committee

Alexandre Holzbach Junior
Bruna Fanchin
Clara Teresa de Souza Ramos
Daniella de Lourdes Luna Santana de Andrade
Daniel Ribeiro de Moraes
Esteferson Quadros
Felipe Rettore Andreis
Flavio Maurício Garcia Pezzolla
Guilberth Alves de Matos
Henrique Rezer Mosquér
Israel Santana
Jéssica Rodrigues da Silva
Jonas Martins Maciel
Jone Follmann
Juliano Martins
Lucas Bertinetti Lopes
Maicon Francisco
Marcelo Dérick Oliveira das Chagas
Maria Angelica Martins
Mariana Ribeiro Brandão
Mateus Andre Favretto
Matheus Gama Costa
Paulo Miguel Rodrigues da Rocha
Pedro Paulo Santos Gomes da Silva
Rafael Glatz
Rafael Schantone Silva
Raul Guedert

Sandra Cossul
Susana Bartnikowsky
Victor Hugo de Freitas Morales
Vinícius Rodrigues Zanon

Scientific Reviewers

Adilmar Coelho Dantas	Universidade Federal de Uberlândia, BR
Adriana Gabriela Scandurra	Universidad Nacional de Mar del Plata, AR
Adriana Kauati	State University of Western Paraná, BR
Adriane Parraga	Universidade Estadual do Rio Grande do Sul, BR
Adriano Pereira	Universidade Federal de Uberlândia, BR
Adriano Péricles Rodrigues	Universidade Federal de Goiás, BR
Agustina Bouchet	Universidad de Oviedo, ES
Agustina Garces Correa	Universidad Nacional de San Juan, AR
Alberto López Delis	Universidad de Oriente, CU
Aldira Domínguez	Universidade de Brasília, BR
Alejandro Espinel Hernández	Universidad do Oriente, CU
Alexander Alexeis Suárez-León	Universidad de Oriente, CU
Alexandre Balbinot	Universidade Federal do Rio Grande do Sul, BR
Alexandre Luís Cardoso Bissoli	Instituto Nacional da Propriedade Industrial, BR
Alexandre Romariz	Universidade de Brasilia, BR
Alfredo de Oliveira Assis	Universidade Federal de Goiás, BR
Alvaro David Orjuela Cañon	Universidad del Rosario, AR
Ana B. Pimentel-Aguilar	Instituto Nacional de Enfermedades Respiratorias, MX
Ana Karoline Almeida da Silva	Universidade de Brasília, BR
André Adami	Universidade de Caxias do Sul, BR
André Dantas	Instituto Santos Dumont, BR
Andre Lazzaretti	Universidade Tecnológica Federal do Paraná, BR
AndrÉ LuÍs Fonseca Furtado	Instituto Federal do Sudeste de Minas Gerais, BR
Andrés Alberto Ramírez Duque	Universidad El Bosque, CO
Andrés Felipe Ruíz Olaya	Universidad Antonio Nariño, CO
Angela Abreu Rosa de Sá	Universidade Federal de Uberlândia, BR
Angela Salinet	Universidade de São Paulo, BR
Antonio Mauricio Miranda de Sá	Universidade Federal de Rio de Janeiro, BR
Aparecido Carvalho	Centro Universitário Salesiano São Paulo, BR
Argenis Adrian Soutelo Jimenez	Universidad de Oriente, CO
Ariana Moura Cabral	Universidade Federal de Uberlândia, BR
Ariel Braidot	Universidad Nacional de Entre Rios, AR
Arnaldo Fim Neto	Universidade Federal do ABC, BR

Arquímedes Montoya Pedrón	Hospital General Docente Juan Bruno Zayas Alfonso, CU
Arthur Carvalho Pires	Universidade Federal de São João Del-Rei, BR
Aura Ximena González Cely	Universidad Pedagógica y Tecnológica, CO
Axayacatl Morales	Universidad Autónoma Metropolitana Iztapalapa, MX
Beatriz Fernandes	Pontificia Universidadd Catolica de Parana, BR
Beatriz Janeth Galeano Upegui	Universidad Pontificia Bolivariana, CO
Belkys Amador	Universidad Austral de Chile, CL
Bruno Bispo	Universidade Federal de Santa Catarina, BR
Caio C. E. De Abreu	Universidade do Estado de Mato Grosso, BR
Carla Scorza	Universidade Federal de São Paulo, BR
Carlos Alirio Lozano Ortiz	Universidade Federal de Rio de Janeiro, BR
Carlos Araujo	Instituto Federal do Paraná, BR
Carlos Castillo	Fundación Universitaria del Área Andina, CO
Carlos Danilo Miranda Regis	Instituto Federal da Paraiba, BR
Carlos Galvão Pinheiro Júnior	Universidade Federal de Goiás, BR
Carlos Maciel	Universidade de São Paulo, BR
Carlos Magno Medeiros Queiroz	Instituto Federal do Triângulo Mineiro, BR
Carlos Valadao	Instituto Federal do Espírito Santo, BR
Carolina Tabernig	Universidad Nacional de Entre Ríos, AR
Cesar Ferreira Amorim	Universidade Federal de Uberlândia, BR
Cesar Rodrigues	Universidade Federal de Santa Catarina, BR
César Teixeira	University of Coimbra, PT
Cicero Hildenberg Lima De Oliveira	Universidade Tecnológica Federal do Paraná, BR
Claudia Lescano	Universidad Nacional de San Juan, AR
Claudia Mirian de Godoy Marques	Universidae do Estado de Santa Catarina, BR
Cleber Zanchettin	Universidad Nacional de San Juan, AR
Cleison Daniel Silva	Universidade Federal do Pará, BR
Cristian Blanco	Universidad Antonio Nariño, CO
Cristian David Guerrero Mendez	Universidad Antonio Nariño, CO
Cristiano Jacques Miosso	Universidade de Brasília, BR
Cristina Shimoda Uechi	MCTIC, BR
Dafne Mendes Soares	Universidade Federal de São João del-Rei, BR
Damien Depannemaecker	Universidade Federal de São Paulo, BR
Daniel Cavalieri	Instituto Federal do Espírito Santo, BR
Daniela del Carmen Gonzalez	Universidad Nacional de San Juan, AR
Daniela O. H. Suzuki	Universidade Federal de Santa Catarina, BR
Daniella De Lourdes L. S. Andrade	Universidade Federal de Santa Catarina, BR
Danilo Melges	Universidade Federal de Minas Gerais, BR

Danilo Silva	Universidade Federal de Santa Catarina, BR
David Sérgio Adães de Gouvêa	Universidade Federal de Juiz de Fora, BR
Débora de Fátima Camillo Ribeiro	Pontificia Universidade Católica do Paraná, BR
Delmo Benedito Silva	Universidade Federal de São João Del-Rei, BR
Denis Delisle Rodriguez	Instituto Santos Dumont, BR
Diana Gutierrez	Universidad Manuela Beltran, CO
Diego Beltramone	Universidad Nacional de Córdoba, AR
Diego Sebastián Comas	National University of Singapore, SG
Dr. Luis Carlos Hernandez Barraza	Instituto Santos Dumont, BR
Éberte Freitas	Universidade Federal do Espírito Santo, BR
Edgard Morya	Instituto Santos Dumont, BR
Edras Pacola	Universidade Positivo, BR
Eduardo Antonio Fragoso Dias	Universidade Federal do Espírito Santo, BR
Eduardo M. Scheeren	Pontifícia Universidade Católica do Paraná, BR
Eduardo Naves	Universidade Federal de Uberlândia, BR
Elgison Da Luz Dos Santos	Centro Universitário Internacional Uninter, BR
Eliete Maria De Oliveira Caldeira	Universidade Federal do Espírito Santo, BR
Elisa Perez	Universidad Nacional de San Juan, AR
Elisangela Manffra	Pontifícia Universidade Católica do Paraná, BR
Elisângela Oliveira Carneiro	Universidade Estadual de Feira de Santana, BR
Emanuel Tello	Universidad Nacional de San Juan, AR
Enrique J. Marañón	Universidad de Oriente, CU
Enrique Mario Avila Perona	Universidad Nacional de San Juan, AR
Eric Laciar Leber	Universidad Nacional de San Juan, AR
Ernesto Suaste	CINVESTAV Sección Biolectrónica, MX
Esmirna Cascaret Fonseca	Centro Provincial de Medicina Deportiva, CU
Esteban Lanzarotti	Universidad de Buenos Aires, AR
Eugenio Orosco	Universidade Nacional de San Juan, AR
Euler Garcia	Universidade de Brasília, BR
Evandro Salles	Universidade Federal do Espírito Santo, BR
Fabiana Bertoni	Universidade Estadual de Feira de Santana, BR
Fabiana da Silveira Bianchi Perez	Faculdades Alfredo Nasser, BR
Fabio Viegas Caixeta	Universidade de Brasília, BR
Fabricio Lima Brasil	Instituto Santos Dumont, BR
Fabricio Neves Mendonca	Instituto federal de Minas Gerais, BR
Fabrício Noveletto	Universidade do Estado de Santa Catarina, BR
Fátima L. S. Nunes	Universidade de São Paulo, BR
Felipe Rettore Andreis	HST Aalborg University, DK
Fellipe Allevato Martins da Silva	Universidade Federal do Rio de Janeiro, BR
Fernando Henrique Magalhães	Universidade de São Paulo, BR
Fernando Jorge Muñoz Zapata	Universidad Nacional de San Juan, AR

Fernando Valdés-Pérez	Universidade do Oriente, CU
Flavio Buiochi	Universidade de Sao Paulo, BR
Floriano Salvaterra	Universidade Federal de Rio de Janeiro, BR
Francisco Assis de Oliveira Nascimento	Universidade de Brasília, BR
Franco Simini	Universidad de la Republica de Uruguay, UY
Frieda Saicla Barros	Universidade Tecnológica Federal do Paraná, BR
Gabriel Bruno M. Fernandes	Universidade Federal de Santa Catarina, BR
Gabriel Domingo Vilallonga	Universidade Nacional de Catamarca, AR
Gabriel Eduardo Cañadas Fragapane	Universidad Nacional de San Juan, AR
Gabriel Motta Ribeiro	Universidade Federal de Rio de Janeiro, BR
Gilcélio Silveira	Universidade Federal de São João del-Rei, BR
Glauco Cardozo	Instituto Federal de Santa Catarina, BR
Glécia Luz	Universidade de Brasília, BR
Gonzalo Quiroga	Universidad Nacional de San Juan, AR
Guilherme Augusto Gomes De Villa	Universidade Federal de Goiás, BR
Guillermo Abras	Universidad Nacional de Mar del Plata, AR
Gustavo Meschino	Universidad Nacional de Mar del Plata, AR
Gustavo Vivas	Empresa Brasileira de Serviços Hospitalares, BR
Harlei Leite	Universidade Federal do Espírito Santo, BR
Henrique Takachi Moriya	Universidade de Sao Paulo, BR
Hugo Líbero	Instituto Federal de Goiás, BR
Humberto Gamba	Universidade Tecnológica Federal do Paraná, BR
Humberto Romano	INTECNUS Foundation, AR
Isabela Alves Marques	Universidade Federal de Uberlândia, BR
Isabela Miziara	Universidade Federal do Pará, BR
Ismar Cestari	Instituto do Coração Universidade de São Paulo, BR
Jair Trapé Goulart	Universidade de Brasília, BR
Javier Camacho	Universidade EIA, CO
Jerusa Marchi	Universidade Federal de Santa Catarina, BR
Jessica Lara	Universidade Federal de São João Del-Rei, BR
João Henrique Kleinschmidt	Universidade Federal do ABC, BR
João Lameu	Universidae Federal do ABC, BR
Joao Marques	University of Saint Joseph, US
João Otávio Bandeira Diniz	Instituto Federal do Maranhão, BR
João Panceri	Instituto Federal do Espírito Santo, BR
Joaquim Cezar Felipe	Universidade de São Paulo, BR
Joaquim Mendes	Universidade do Porto, PT
Joaquin Azpiroz Leehan	UAM Iztapalapa, MX

Jorge Takenaga	Universidad de Monterrey, MX
Jose Carlos Cunha	Pontificia Universidade Católica do Paraná, BR
José Di Paolo	Universidad Nacional de Entre Ríos, AR
José Felicio Da Silva	Universidade de Brasília, BR
José Foggiatto	Universidade Tecnológica Federal do Paraná, BR
José María Flores	Universidad Nacional de Entre Rios, AR
José Wilson Bassani	Universidade de Campinas, BR
Juan Carlos Iturrieta Gimeno	Universidad Nacional de San Juan, AR
Juan Carlos Perfetto	Universidad de Buenos Aires, AR
Juan Enrique Palomares Ruiz	Instituto Tecnológico Superior de Cajeme, MX
Juan Graffigna	Universidad Nacional de San Juan, AR
Juan I. Montijano	University of Zaragoza, ES
Juan Ignacio Pastore	Universidad Nacional de Mar del Plata, AR
Juan Pablo Tello Portillo	Universidad del Norte, CO
Juan Pastore	Universidad Nacional de Mar del Plata, AR
Juan Vorobioff	Comisión Nacional de Energía Atómica CNEA, AR
Juliana Santa Ardisson	Universidade Federal do Espírito Santo, BR
Juliano Costa Machado	Instituto Federal Sul-Rio-Grandense, BR
Julio Alberto Rojas Vargas	Universidade do Oriente, CU
Julio Cesar Nievola	Pontifícia Universidade Católica do Paraná, BR
Karla Aparecida Ferreira	Universidade Federal de São João Del-Rei, BR
Katia Prus	Universidade Tecnológica Federal do Paraná, BR
Kelison Tadeu Ribeiro	Instituto Federal do Espírito Santo, BR
Lacordaire Kemel Pimenta Cury	Instituto Federal de Educação, Ciência e Tecnologia Goiano, BR
Leandro Bueno	Instituto Federal do Espírito Santo, BR
Lena Perez Font	Universidad de Oriente, CU
Leonardo Abdala Elias	Universidade de Campinas, BR
Leonardo Abrantes	Universidade Federal de Rio de Janeiro, BR
Leonardo Batista	Universidade Federal da Paraíba, BR
Leonardo Ramirez	Universidad Militar Nueva Granada, CO
Leticia Silva	Universidade Federal do Espírito Santo, BR
Lorena Orosco	Universidad Nacional de San Juan, AR
Lourenço Madruga Barbosa	Universidade Tecnológica Federal do Paraná, BR
Lucas Côgo Lampier	Universidade Federal do Espírito Santo, BR
Lucenildo Cerqueira	Universidade Federal de Rio de Janeiro, BR
Luciana Menezes Xavier de Souza	Universidade Federal de Santa Catarina, BR
Luciana Roberta Peixoto	Universidade de Brasília, BR
Luciane Aparecida Pascucci de Souza	Universidade Federal do Triângulo Mineiro, BR
Luciano Menegaldo	Universidade Federal de Rio de Janeiro, BR

Luis Eduardo Maggi	Universidade Federal do Acre, BR
Luis Jiménez-Angeles	Universidade Nacional Autônoma do México, MX
Luis Miguel Zamudio	Universidade do Oriente, CU
Luis Pulenta	Universidad Nacional de San Juan, AR
Luis Roberto Barriere	Universidad Don Bosco, SV
Luis Sifuentes	Universidad Cuauhtémoc Aguascalientes, MX
Luis Vilcahuaman	Pontificia Universidad Catolica del Peru, PE
Maikel Noriega Alemán	Universidade do Oriente, CU
Maira Ranciaro	Pontifícia Universidade Católica do Paraná, BR
Malki-çedheq Benjamim C. Silva	Universidade Federal de Pernambuco, BR
Marcella Carneiro	Universidade de Brasília, BR
Marcelo Lencina	National Technological University, AR
Marcio Nogueira de Souza	Universidade Federal de Rio de Janeiro, BR
Marcio Rodrigues	Instituto Federal do Sudeste de Minas Gerais, BR
Marco Antonio Garcia	Universidade Federal de Juiz de Fora, BR
Marcos Hara	Instituto Federal do Paraná, BR
Marcus Vieira Fraga	Universidade de Goiás, BR
Maria Claudia F. Castro	Centro Universitário da FEI, BR
Mariana de Mello Gusso	Pontifícia Universidade Católica do Paraná, BR
Mariela Azul Gonzalez	Universidad Nacional de Mar del Plata, AR
Mário Fabrício Fleury Rosa	Universidade de Brasília, BR
Marly Guimaraes Fernandes Costa	Universidade Federal do Amazonas, BR
Martha Ortiz	Universidad Autonoma de Mexico—Iztapalapa, MX
Martha Ribeiro	Universidade de São Paulo, BR
Martha Zequera	Pontificia Universidad Javeriana, CO
Martín Guzzo	Universidad Nacional de San Juan, AR
Maryangel Jhoseline Medina	ComisiónNacional de Energía Atómica, AR
Mateus Andre Favretto	Universidade do Estado de Santa Catarina, BR
Mauren Abreu de Souza	Pontificia Universidade Católica do Paraná, BR
Mauricio Cagy	Universidade Federal do Rio de Janeiro, BR
Mauricio Kugler	Nagoya Institute of Technology, JP
Mauricio Tavares	Contronic Tecnologia para Diagnósticos, BR
Mayra Aarao	Universidade Federal de São João Del-Rei, BR
Mohamed Ahmed	Universidade Federal de São João Del-Rei, BR
Monica Teresita Miralles	Universidad de Buenos Aires, AR
Munir Gariba	Universidade de São Paulo, BR
Murilo Contó	Boston Scientific, BR
Nahuel M. Olaiz	Universidad de Buenos Aires, AR
Nicolas Jacobo Valencia Jimenez	Universidad Santiago de Cali, CO
Nilton Correia Da Silva	Universidade de Brasília, BR

Olavo Luppi Silva	Universidade Federal do ABC, BR
Pablo Caicedo	Universidade Autônoma da Colômbia, CO
Pablo Daniel Cruces	Universidad de Buenos Aires, AR
Pablo Diez	Universidad Nacional de San Juan, AR
Pablo Turjanski	Universidad de Buenos Aires, AR
Patrick Marques Ciarelli	Universidade Federal do Espírito Santo, BR
Paula Brandao Furlan	Universidade Federal de Rio de Janeiro, BR
Paula Hembecker	Pontifícia Universidade Católica do Paraná, BR
Paulo Ambrosio	Universidade Estadual de Santa Cruz, BR
Paulo Broniera Junior	Federação das Indústrias Estado de São Paulo, BR
Paulo Eduardo Narcizo de Souza	Universidade de Brasília, BR
Paulo Gubert	Universidade Católica de Pelotas, BR
Paulo Jose Abatti	Universidade Tecnológica Federal do Paraná, BR
Paulo Roberto Sanches	Hospital de Clínicas de Porto Alegre, BR
Paulo Rogerio Scalassara	Universidade Tecnológica Federal do Paraná, BR
Pedro Bertemes Filho	Universidade do Estado de Santa Catarina, BR
Pedro Gewehr	Universidade Tecnológica Federal do Paraná, BR
Pedro Moisés de Sousa	Universidade Federal de Uberlândia, BR
Pedro Xavier de Oliveira	Universidade de Campinas, BR
Rafael Sanchotene Silva	Universidade Federal de Santa Catarina, BR
Raimes Moraes	Universidade Federal de Santa Catarina, BR
Raul Guedert	Universidade Federal de Santa Catarina, BR
Renata Coelho Borges	Universidade Tecnológica Federal do Paraná, BR
Renato de Araujo	Universidade Federal de Pernambuco, BR
Renato Naville Watanabe	Universidade Federal do ABC, BR
Ricardo Borsoi	University of Lorraine, FR
Ricardo Ruggeri	Autoridad Regulatoria Nuclear, AR
Ricardo Suyama	Universidade Federal do ABC, BR
Roberto Sagaró Zamora	Universidad de Oriente, CU
Rocio Belén Buenamaizón	Universidad Nacional de San Juan, AR
Rodolfo Dias Correia	Hospital das Clínicas de Ribeirão Preto, BR
Rodrigo Gomide	Instituto Federal Goiano, BR
Rodrigo Pinto Lemos	Universidade de Goiás, BR
Rodrigo Santos	Universidade Federal de Juiz de Fora, BR
Rodrigo Varejão Andreão	Instituto Federal do Espírito Santo, BR
Roger Gomes Tavares de Mello	Universidade Federal do Rio de Janeiro, BR
Ronni Amorim	Universidade de Brasília, BR
Rosa Itzel Flores	Universidad Autonoma de Mexico—Iztapalapa, MX
Rosangela Jakubiak	Universidade Tecnológica Federal do Paraná, BR
Rossana Rivas Tarazona	Pontificia Universidad Catolica del Peru, PE
Rubén Acevedo	Universidad Nacional de Entre Ríos, AR

Sandra Cossul	Universidade Federal de Santa Catarina, BR
Sandra Rocha Nava	Universidad La Salle, CO
Sara Rosa de Sousa Andrade	Faculdades Estácio de Sá, BR
Sérgio Francisco Pichorim	Universidade Tecnológica Federal do Paraná, BR
Sergio Furuie	Universidade de São Paulo, BR
Sergio Ioshii	Pontificia Universidade Católica do Paraná, BR
Sheila da Luz Schreider	Universidade Federal do Espírito Santo, BR
Shirley Quintero	Escuela de Ingenieros, CO
Silas Moreira de Lima	Universidade Federal de São João Del-Rei, BR
Silvia Rodrigo	Universidad Nacional de San Juan, AR
Suélia Rodrigues Fleury	Universidade de Brasília, BR
Suelia Rosa	Universidade de Brasília, BR
Sylvia Faria	Universidade de Brasília, BR
Sylvia Pires	Universidade Federal de São João Del-Rei, BR
Tatiane Ramalho	Universidade Federal de São Paulo, BR
Tenysson W. Lemos	Universidade de São Paulo, BR
Teodiano Bastos Filho	Universidade Federal do Espírito Santo, BR
Thaína Aparecida Azevedo Tosta	Universidade Federal de São Paulo, BR
Thais Winkert	Universidade Federal de Rio de Janeiro, BR
Thomaz Botelho	Instituto Federal do Espírito Santo, BR
Vanessa Pereira Gomes	Instituto de Pesquisas Tecnológicas, BR
Vesna Zeljkovic	Lincoln University, UK
Victor Diego Cupertino Costa	Universidade Federal de São João Del-Rei, BR
Victor Hugo C. de Albuquerque	Universidade Federal do Ceará, BR
Victor Luciano Carmona Viglianco	Universidad Nacional de San Juan, AR
Wally Auf der Strasse	Colégio Militar de Curitiba, BR
Wellington Pinheiro Dos Santos	Universidade Federal de Pernambuco, BR
Wemerson Parreira	Universidade do Vale do Itajaí, BR
Willian Bispo	Universidade Estadual de Londrina, BR
Yann Morère	Université de Lorraine, FR
Yolanda Torres Perez	Universidad Pedagógica y Tecnológica, CO

Sponsoring Institutions

CORAL: Consejo Regional de Ingeniería Biomédica para América Latina
www.coralbiomedica.org

SBEB: Sociedade Brasileira de Engenharia Biomédica
www.sbeb.org.br

IFMBE: International Federation of Medical and Biological Engineering
www.ifmbe.org

UFSC: Universidade Federal de Santa Catarina
www.ufsc.br

IEB-UFSC: Instituto de Engenharia Biomédica da UFSC
www.ieb.ufsc.br

FAPESC: Fundação de Amparo à Pesquisa e Inovação do Estado de Santa Catarina
www.fapesc.sc.gov.br

Contents

Biomaterials, Tissue Engineering and Artificial Organs

Bioengineering, Modelling and Simulation, Bioinformatics and Computational Biology

Mechanical Design and Modeling of COMBIOVENT Pulmonary Ventilator

Miguel Gómez Florido(✉), Juan Carlos Hernández Rodriguez, Pedro L. González Acosta, Yoel Vigil Lorenzo, and Luis José Pena Provedo

COMBIOMED-Digital Medical Technologies, 1704 202 St., Havana, Cuba
mgomez@icid.cu

Abstract. This paper presents the mechanical design of a Pulmonary Ventilator that allows the control, selection and monitoring of parameters necessary to provide life support to patients with acute respiratory failure. 3D models of the ventilation block are described. The sequence of iterations of modeling allowed perfecting the pieces before their manufacture. The simulation of the injection process of the front plastic piece of the equipment is analyzed. Finally, the design of the trolley and the tests to which they were subjected according to the IEC standards are examined. As a final conclusion, the equipment complies with the general safety requirements demanded in the standards for medical equipment.

Keywords: Ventilator · Mechanical Design · Simulation in 3D Models · Medical Equipment · Testing for Medical Trolleys

1 Introduction

COMBIOVENT is a pulmonary ventilator used in intensive care of adults, is medical equipment intended to treat patients who have difficulty or are unable to breathe on their own [1]. The patient receives air through a mask or a hose inserted into the trachea. Its operation is based on supplying an air flow that may or may not be enriched with oxygen, with a normalized frequency and volume, which is inhaled by the lungs and then exhaled.

This assisted breathing is controlled internally by means of an electro-pneumatic circuit that works according to defined parameters. The intensive care physician selects different breathing modes through a user interface with a friendly and intuitive design for easy communication. Likewise, other breathing modes can be established by the doctor with various parameters that respond to the patient's condition. Finally, the alarms are set to notify the medical group in case the patient registers parameters outside the permissible ranges [2].

The aim of this article is to show the final results of iterations to the mechanical design according to the industrial design guidelines, in order to achieved an equipment in the state of the art, aesthetically and as compact as possible, with easy access to its internal components and with technological feasibility [3].

J. L. B. Marques et al. (Eds.): CLAIB 2022/CBEB 2022, IFMBE Proceedings 98, pp. 3–14, 2024.
https://doi.org/10.1007/978-3-031-49401-7_1

It is analyzed the simulating of the filling time and the formation of sink marks, during the injection of the plastic piece that is located in front of the equipment.

Finally, the results of different tests to the ventilator trolley according to the international standards of the IEC related to medical equipment are exposed [4].

2 Materials and Methods

2.1 Theoretical Background

COMBIOVENT's design has the most common ventilation modes (assisted, controlled and pressure support modes); automatic control of the oxygen concentration and its visualization; capnography and oximetry monitoring. It also provides the option of using proximal and expiratory flow sensors for greater ventilation precision, as well as proximal pressure measurement.

Two circuits (inspiration and expiration) were located, separated from each other and both controlled by electronic boards that measure and regulate the flow and pressure parameters.

The inspiration circuit (Fig. 1) is composed by an inlet for high-pressure compressed oxygen that passes through a filter and a regulator. An air blower generates a flow which is mixed with this oxygen. This mixture passes through pressure and flow sensors and later for a distributor where hoses are connected carrying the mixture to the electronic boards sensors. At the end of the circuit, the hoses to the patient are connected [5].

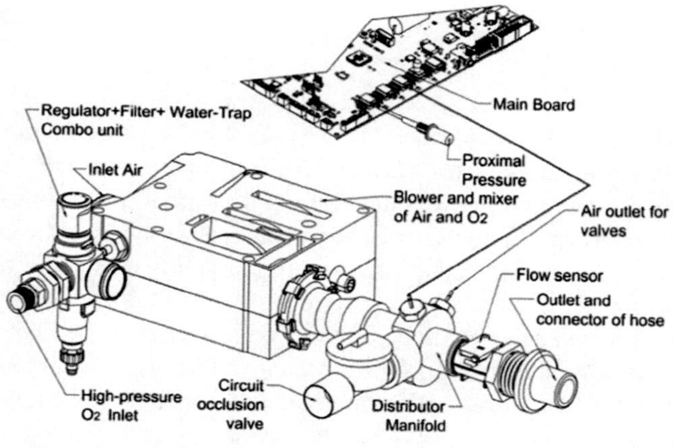

Fig. 1. Inspiration Circuit and its connection with main board.

The expiratory circuit handles residual carbon monoxide that has returned from the patient's expanded lung tissues. The flow of this air is also monitored by the equipment. A set of mechanical pieces was built with the aim of being able to easily decouple this entire circuit from the ventilator and to sanitize it. This assembly allows the attachment of an expiratory valve to control expiratory flow and positive end-expiratory pressure (PEEP) of the patient.

2.2 Mechanical Design Methodology

The mechanical design of the ventilator had several stages. In the first, conceptualization, it was decided to divide its elements into three blocks. The ventilation block contains the electronic boards and the electro-pneumatic controls where the oxygen inspiration circuit hoses and CO_2 expiration circuit hoses are connected to the patient. The other block is a touch screen used to operate the equipment. The decision to separate the screen and the electro-pneumatic system allows using any standard touch screen model. The last block is the trolley for the ventilator movement and where the two previous blocks are supported.

The second stage of the mechanical design was carried out based on the electro-pneumatic circuit proposed by the entire design team. The valves, sensors, regulators, with their mechanical and electrical parameters, were defined. Finally, their entire mechanical assembly was simulated, performing several iterations that ultimately led to a compact and technologically feasible design to manufacture.

2.3 Materials and Devices Used

All mechanical pieces were manufactured using 3000 Series Aluminum Alloy. This alloy has moderate strength, good workability, and good corrosion resistance. Sheets metals were manufactured with Aluminum 3105 and machined pieces with Aluminum 3005. Internal pieces were subjected to an anodizing process in order to increase resistance to corrosion, maintain its metallic appearance and create an external surface three times harder. External sheet metals pieces were covered with White Electrostatic Paint for its easy sanitization.

Pneumatic and the electrical modules were integrated into the mechanical assembling to achieve a functional system. The main modules or devices used were the following:

- Electronic Boards
- Touch screen
- Blower and Mixer of O_2 and Air
- Hoses and Couplings
- Flow sensors
- Valves ON/OFF and Proportional Solenoid Valves
- Pressure Regulators and Filters.
- Fastenal elements
- Manifold.

2.4 Modeling of the Mechanical Design of the Ventilation Block

Inside the inspiration circuit, there is one element with custom mechanical design: the distributor manifold. Initially, it had one inlet and four outlets for the air that come from the blower. The inlet is attached to the blower by means of clips. One of the outputs had a conical shape, to get a quick adjustment of the elements that are connected to it. The other three are threaded radial outlets. In the initial design (Fig. 2) the threaded outlets are in a fairly large thickness area. During the initial operation tests, reduction of total

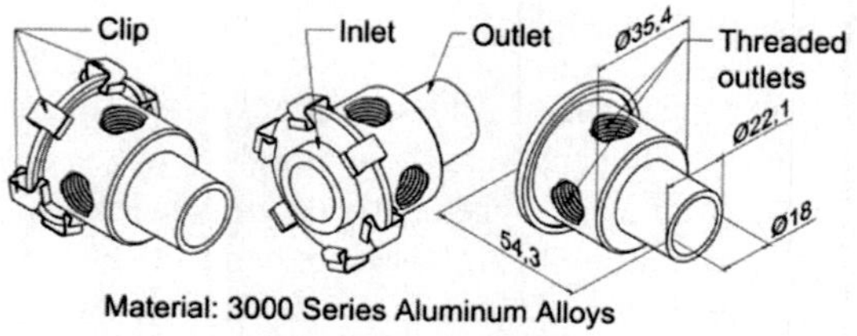

Fig. 2. Distributor manifold initial design

space of the piece was needed to add one outlet and also to change the way the piece is fastened to the circuit.

After doing several iterations of this design, the final result (Fig. 3) was reached eliminating the fastening clips, giving to the inlet the same shape as the conical outlet. The angle between the threaded radial outlets was reduced to 72 degrees (before was 90 degrees). A new outlet and a new piece were added, joining them by means of an interference fit and achieving a larger internal diameter of the outlet that allows a greater air flow inner. As a result, in a smaller space, six holes of different diameters were concentrated through which the air flows in different directions [6].

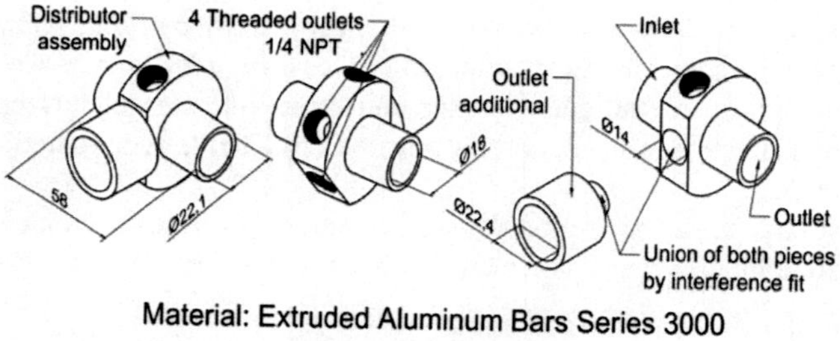

Fig. 3. Distributor manifold final design.

The mechanical element of the expiration circuit that had to be designed was the expiration elbow with an angle between both outlets of 90 degrees (Fig. 4). It connects the hoses coming from the patient with the expiratory pneumatic valve.

In the initial design, it was made up of two pieces with angles at their ends milled at 45 degrees and joined both pieces by welding. As the pieces are made of aluminum, the welding process becomes very expensive and productivity is very low. Under these circumstances it was necessary to make a new iteration of the design and change the way of the joint in both pieces.

This was solved by milling holes at 90 degrees and inserting one part into the cavity of the other with an interference fit joint.

The easy coupling and uncoupling of the elbow to the furniture plate was obtained with a nut that can be easily manipulated by hand. Finally, once the set is assembled, it is positioned in the equipment by means of a threaded knob.

Both pneumatic systems (inspiration and expiration) are controlled by a set of electronic boards (Fig. 5). The main board is where all the pressure sensors are located, acquire the electrical signals from the flow sensors, and the valves are controlled. It is also the central device where the others boards are interconnected.

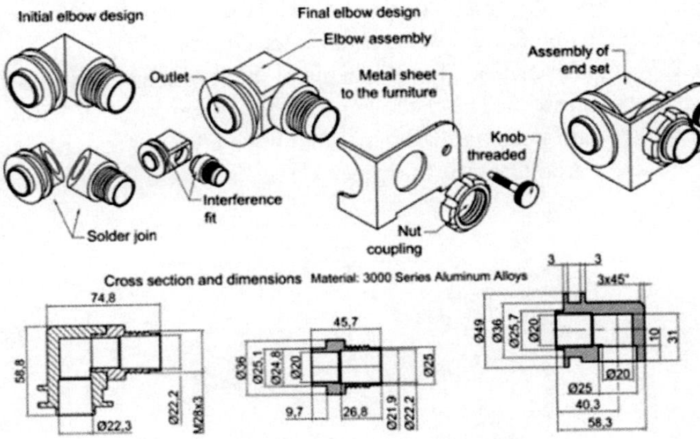

Fig. 4. Design of the expiration elbow set.

The second is the Raspberry Pi motherboard that sends and receives the signal from the monitor, USB and Network ports, among other functions. The third is the oximetry board that collects the values of this parameter from the patient and transmits it to the main board. The fourth is the blower control board, and finally, the power board, which is responsible for supplying the working voltages for the equipment.

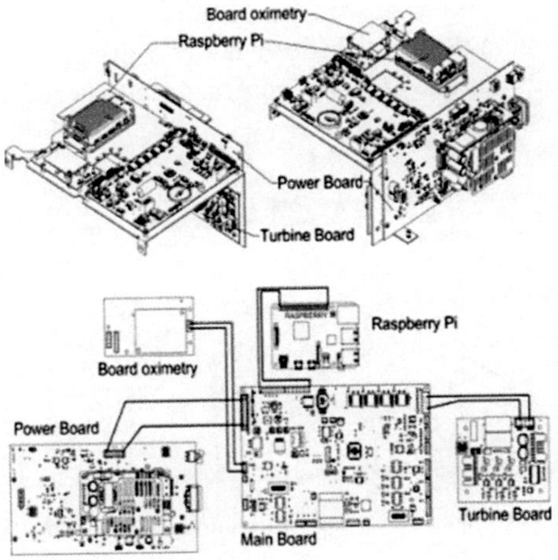

Fig. 5. Distribution of boards on the metal sheet and its connections.

Constructively and ergonomically, the decision was to separate the pneumatic from the electronic system. In first place, the interconnection cables between the boards must be few and short. Second, in case of a short circuit, the electronic boards and the pneumatic devices must be isolated, due to there may be a high oxygen concentration inside the equipment and any spark produces a fire. Third, to facilitate access to the pneumatic system by simply disassembling a piece that holds all the boards. Fourth, the plate serves as a shield between the pneumatic part and the electronic part [7].

All boards (Fig. 6) are positioned over of a metal sheet which has riveted metal inserts where the boards are screwed. The piece has external dimensions of 276.4 × 201.6 × 110.4 mm. In the case of areas on the boards that have high voltage, plastic posts with plastic screw are used. Additionally, this piece serves as a base for the battery pack.

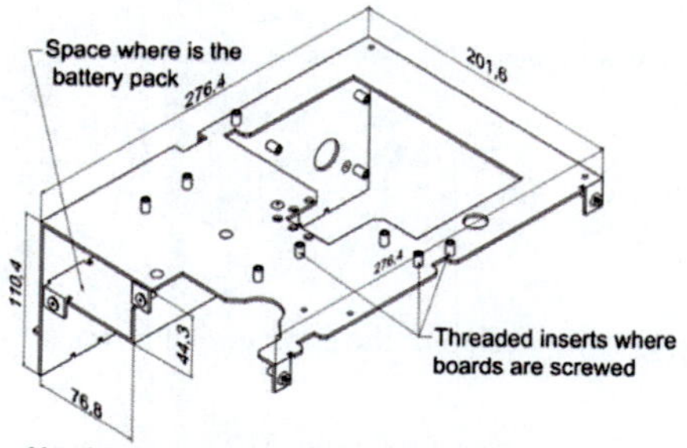

Fig. 6. Design of the support metal sheet for boards and battery.

A metal sheet supports all the elements of the ventilator (Fig. 7). It serves as the main base with enough rigidity so it does not suffer any type of deformation. It has 3 mm thickness, made of aluminum. It has bends on the sides to increase rigidity and inserts of riveted metal nuts in the places where different components are to be positioned, besides many punched holes where the entire set of ventilator pieces is assembled. The design of this piece allows everything to be easily accessed from the sides without disassemble other elements.

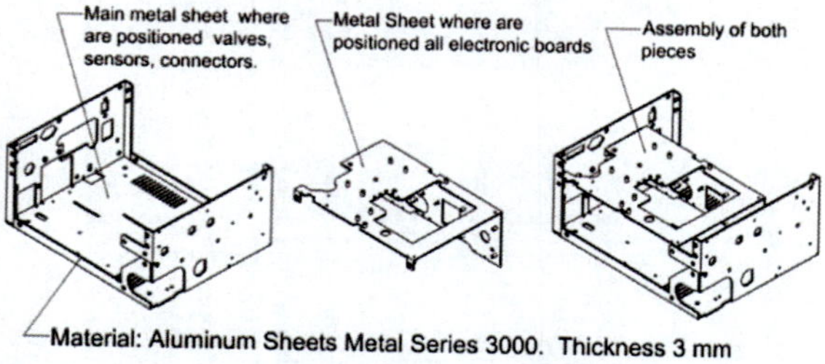

Fig. 7. Main base design and its assembly with the board base.

The assembly of the entire pneumatic circuit begins on the main base (Fig. 8), checking that there are no air leaks between the joints of different sensors and valves. To achieve this, it was necessary to manufacture pieces with a narrow tolerance range. It was also increased the sealing by using Teflon in the joints.

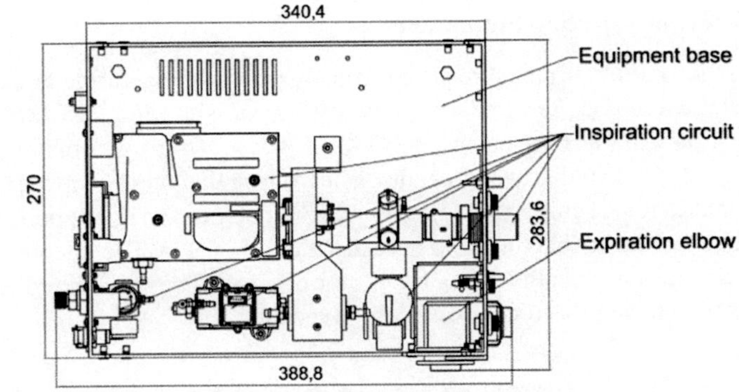

Fig. 8. Top view of equipment with pneumatic circuits.

Then (Fig. 9) the electronic boards are added to the assembly, interconnecting the main board sensors to the pneumatic system.

The front exterior side includes all the elements where the hoses, electrodes and LEDs are connected. In the rear exterior side are located the oxygen inlet, power switch, ground post, network connectors, USB port, windows for the air filter and battery cover.

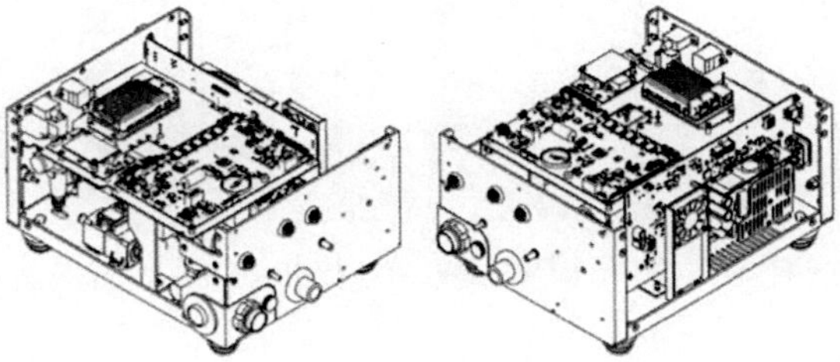

Fig. 9. Electro-pneumatic assembly.

With the addition of the set of pieces that support the touch screen monitor, the ventilation block is completed as shown in Fig. 10.

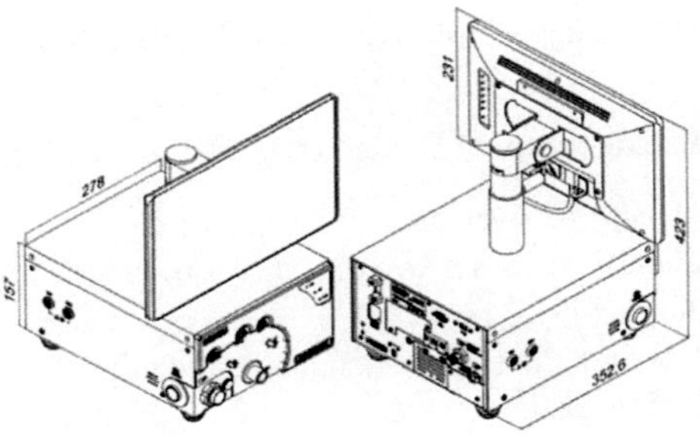

Fig. 10. Views of the ventilation block and the touch screen monitor.

2.5 Plastic Front Injection Simulation

The plastic front of the equipment required an analysis of manufacturing difficulties from the design stage. For this, its simulation was carried out with MOLDFLOW ADVISER [8]. The piece has external dimensions of 277.4 × 156.8 × 10 mm. One of its complexities is that the liquid plastic will travel great distances during the injection process inside the mold. This makes it very important to define the correct location of the points where the liquid will be injected so that the mold is easily filled (Fig. 11). The injection points are arranged in the piece so that the filling time is the shortest. The areas in red color are the critical ones but the values do not exceed three seconds.

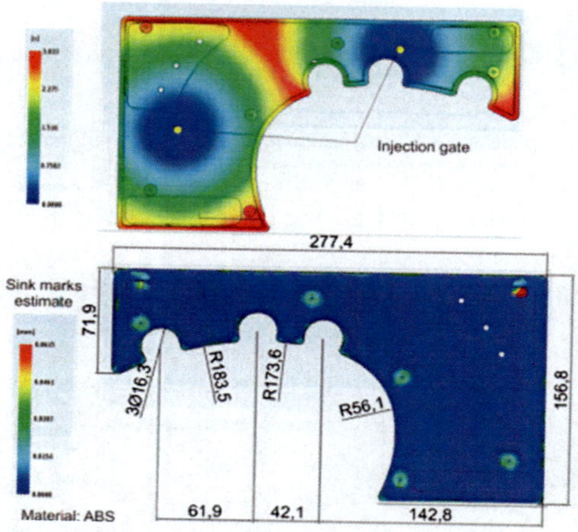

Fig. 11. Simulation with MOLDFLOW ADVISER of the plastic piece's injection.

The points where sink marks occur are also simulated. The areas with light blue colors show that the values obtained are in the order of 0.0154 mm, which are not appreciable at sight. In the red zone it is the highest value that reaches 0.0615 mm but it is in a low relief that will later be covered by a label.

Other parameters were simulated too, such as pressure drop inside the mold, temperature variation, cooling time, material cost, among others. Finally, the produced piece did not present defects due to the design during its manufacture.

2.6 Simulation of the Mechanical Design of the Medical Trolley

The ventilation block has to be over on a trolley that serves as a support during its use and movement. Additionally, it has an arm that serves to put the hoses in a fixed position once the patient is intubated [9].

To accomplish the design, the requirements demanded by the IEC 60601–1 standard in the section of mechanical hazards were consulted. In addition, a study was carried out that included interviews with doctors and technicians, visits to hospitals, bibliographical consultations and research of the main equipment manufacturers. The following ergonomic requirements for the trolley were identified:

- Its height must ensure that the touchscreen center is 1.4 m from the ground.
- It must have four wheels with brakes for easy movement and turns.
- It must have a standardized support for the placement of a dehumidifier.
- It must be able to stay, with the brakes applied, on surfaces inclined at 10°, without losing your balance.
- It must be able to pass a 20 mm drop without the trolley losing stability or something falling off.
- It must not suffer any kind of deformation or become unstable when applying inadequate force during use.
- It must have an arm installed to support the hoses.

The design of the trolley is composed by a base, a column, a support tray for the ventilation block, a manipulator and an arm [10].

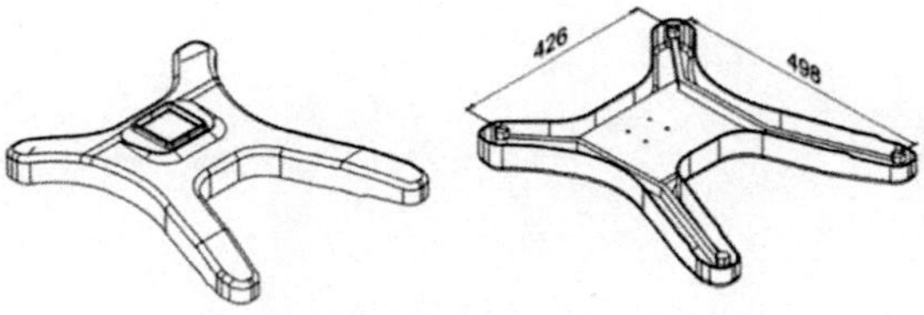

Fig. 12. Base of the trolley

The base of the trolley (Fig. 12) was made of steel sheet with thickness 2 mm. It has bends to achieve great rigidity and welded on the inside so that it does not suffer deformation easily, since it is the area where the greatest efforts are exerted. It also increases weight in this area, causing the center of gravity of the trolley to be as low as possible and therefore the stability is much greater. This piece has a design where the cross section, goes from a wider area towards the thinner. For this reason it can have greater resistance in the area where the greatest efforts are produced. The wheels have brakes and are placed in their ends.

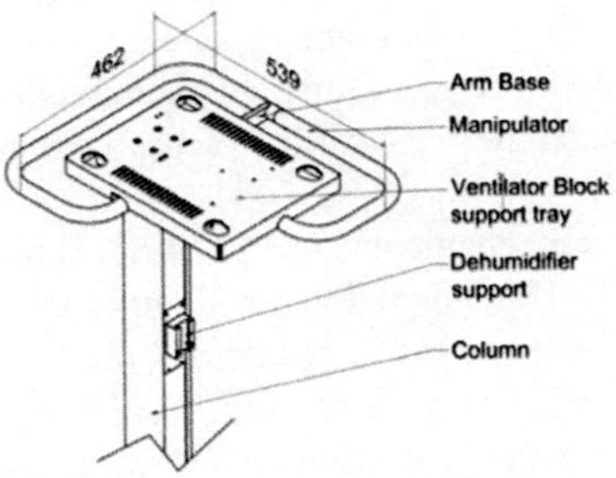

Fig. 13. Top of the trolley.

In the other trolley elements (Fig. 13) it is good to point out that the handle is made from a bent tube arranged in a way that the trolley can be manipulated from any position.

In the ventilation block support tray there are four holes to fix the equipment. The column slenderness is achieved by means of a sheet metal bend.

3 Results and Discussion

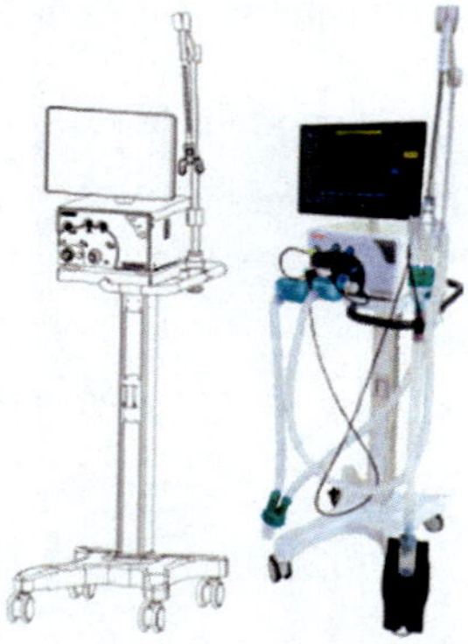

Fig. 14. Final virtual model (left), COMBIOVENT (right).

COMBIOVENT was evaluated according to the ISO and IEC standards [11–15].

First, parametric tests were carried out using simulators for five days of continuous work. The electronic, pneumatic, and mechanical parameters were verified. In parallel the risk analysis was carried out too. Second test was carried out with pigs, since these animals have an anatomy and lung capacity similar to that of humans. It was performed through an invasive procedure, anesthetizing them and intubating them. It was verified that the volume of air supplied by the ventilator was correct and the device had no faults during the entire procedure. The response of the program with the breathing modes and their alarms was adequate. The third was parametric tests, performed for ten days of continuous work. Ventilator was able to adequately control the breath parameters. Validation of ventilator for use in in humans is pending.

From the virtual final model (Fig. 14 left), the manufacture of all pieces was carried out. It is highlighted that machined and milled pieces were manufactured with narrow tolerances.

The entire manufactured mechanical set (Fig. 14 right) underwent the tests required according to the IEC60601-1 standard by the Regulatory Authority for Medicines, Equipment and Medical Devices of the Republic of Cuba.

In this article we will refer only to the results of tests carried out according to sections 9 and 15 that address the protection against mechanical hazards of equipment and systems.

The instability in transport position test of section 9.4.2.1, where the assembly does not lose its balance when it is positioned on an inclined plane with an angle of 10 degrees above the horizontal plane, was performed successfully.

The force for propulsion test of section 9.4.2.4.2 was verified moving the assembly along a flat and hard surface with a force of less than 200 N.

The movement test over an threshold (section 9.4.2.4.3) performed, verified that the assembly is capable of passing obstacles 20 mm high at a speed of approximately 0.5 m/s without tipping over it.

The three instability in transport tests of section 9.4.3.1 were verified. First, the brakes are activated and deactivated only by an external action. Second, the trolley has locking devices. Both of then was checked by inspection. Third, the assembly was placed in several positions (including the most critical one) on the incline and the brakes on were applied in each position and the equipment did not lose or move their position.

The rough handling test of section 15.3.5 was verified by means of three tests. First the assembly at a speed of 0.4 m/s in upward movement collided with an 40 mm hardwood step, the second one was to make fall the assembly over a same step at the same speed that first test, the third, crashing the assembly against the door frame was also carried out at the same speed. It did not suffer damages in any test.

Ventilation module has reduced dimensions, but separation between its different parts was respected, avoiding overheating due to there is enough space for air circulation.

Different iterations to the design and its modeling, made possible to eliminate errors, reducing material costs, manufacturing times and increasing quality in the final pieces.

4 Conclusions

Outlined objectives in the conceptualization stage were fulfilled. A pulmonary ventilator at the level of the state of the art was designed. Modular design allows for easy component replacement.

The modeling of all the mechanical parts, circuits, boards, pneumatic components and valves is, without a doubt, a powerful tool that cannot be omitted.

In the final design of the distributor manifold piece, much smaller external dimensions were achieved and a greater number of outlets were concentrated, one of them with a larger internal diameter so that a greater flow of air could pass during its operation.

The large enough internal diameters without obstacles of the expiration elbow achieved allow easy sanitize.

The simulation of the injection process of the plastic piece for the front of the equipment managed to control and eliminates errors and defects in the quality of the piece, allowing its correct manufacture.

The results obtained from the tests show that the equipment satisfies all the specifications of the standards for medical equipment and has shown excellent reliability.

Acknowledgements. We wish to thank the entire team that participated in the design, development and manufacture of COMBIOVENT and especially those whose names do not appear in this article.

Conflict of Interest The authors of this study declare that we have no conflicts of interest.

References

1. Grace, K.: The ventilator: selection of mechanical ventilators. In: Critical Care Clinics, vol. 14, no. 4 (1988)
2. Farias, J.A., Alía, I., Retta, A., Olazarri, F., Fernández, A., Esteban, A.: An evaluation of intubation failure predictors in mechanically ventilated infants and children. Intensive Care Med. 752–757 (2002)
3. Tobin, M.J.: Principles and practice of mechanical ventilation. In: Baum's Textbook of Pulmonary Diseases, 2nd edn. McGraw-Hill, Inc., USA (2006)
4. Gómez, M., Folgueras, J., Randulfe, J.: Empleo de técnicas de avanzada para el diseño rápido y eficiente de equipos biomédicos: La modelación 3D. In: Bioingeniería y Física Médica Cubana, vol. 7, no. 2, pp. 11–18. Mayo-Agosto (2006)
5. El-Khatib, M., Jamaleddine, G.: A new oxygenation index for reflecting intrapulmonary shunting in patients undergoing open-heart surgery. Chest (2004)
6. Gómez, M., Folgeras, J., Martínez, V., Portela, A.: Adaptador Intercambiable para desfibrilación en pacientes pediátricos. In: Bioingeniería y Física Médica Cubana, vo. l8, no. 1, pp. 25–31. (2007)
7. CEI: Compatibilidad electromagnética. Parte 4: Técnicas de ensayo y de medida. Sección 4: Ensayos de inmunidad a los transitorios eléctricos rápidos en ráfagas: CEI/IEC 61000-4-4 (1995)
8. Wiklund, M.E.: Eleven keys to designing error-resistant medical equipment. Medical Device & Diagnostic Industry, Mayo (2002)
9. Autodesk: Simulation Moldflow®, Reference Manual (2015)
10. Gómez, M., Randulfe, J., Fernández, I.: Diseño de un carro médico para electrocardiógrafos. In: VIII Congreso de la Sociedad Cubana de Bioingeniería – Bioingeniería'2009
11. International Electro Technical Commission: Medical electrical equipment: general requirements for basic safety and essential performance. IEC60601-1 (2005)
12. Particular requirements for basic safety and essential performance of critical care ventilators. ISO 80601-2-12
13. Particular requirements for basic safety and essential performance of ventilatory in sufficiency. ISO 80601-2-80
14. Particular requirements for basic safety and essential performance of ventilatory support equipment for ventilatory impairment. ISO 80601-2-79
15. Lung ventilators and related equipment—Vocabulary and semantics. ISO 19223

A General NAR Model for Seizure Identification Across Different Patients

João A. F. Brogin[1(✉)], Jean Faber[2], and Douglas D. Bueno[3]

[1] Department of Mechanical Engineering, São Paulo State University, Ilha, Solteira, SP 15385-000, Brazil
ferres.brogin@unesp.br

[2] Department of Neurology and Neurosurgery, Federal University of São Paulo, São Paulo, SP 04039-032, Brazil

[3] Department of Mathematics, São Paulo State University, Ilha, Solteira, SP 15385-000, Brazil

Abstract. Data-driven and dynamic models have been proposed to better understand and describe epileptiform activities. However, there is still a need for more comprehensive ones suitable for individual seizures from patients but also general across them, taking into account common features. Thus, this work proposes an alternative nonlinear autoregressive model to address this issue. Essentially, the regression coefficients obtained from individual models are taken as random variables, and a new one is constructed based on their respective median values. The results show that the proposed approach is effective in describing all of the individual seizures with a single model, yielding low estimation errors. This finding is useful when large datasets are not available, and can be adopted in the context of seizure prediction and attenuation in a similar way to that of transfer learning applications.

Keywords: epilepsy · autoregressive modeling · system identification

1 Introduction

Epilepsy affects millions of people worldwide every year and may severely compromise patients' quality of life [1,2]. From a physical point of view, it is usually associated to the progressive hypersynchrony of neuron spiking over time until seizure onset [2]. However, due to the lack of a fuller understanding about the precise mechanisms that trigger seizures [3], mathematical models aiming to describe their main features and signatures have been proposed in the literature as theoretical tools to investigate, for instance, how they are generated and spread [4–6].

Some of these models are based on coupled differential equations defined in terms of state variables which are related to biophysical factors and parameters, such as membrane potentials and current [4,6]. Their main advantage is that several computational simulations can be carried out by implementing them

J. L. B. Marques et al. (Eds.): CLAIB 2022/CBEB 2022, IFMBE Proceedings 98, pp. 15–22, 2024.
https://doi.org/10.1007/978-3-031-49401-7_2

and varying their parameters, such that experience and knowledge about the system's behavior can be attained. This procedure can avoid the indiscriminate use of animal models as well as anticipate some experimental observations [4].

Despite providing insight into seizure triggering and propagation, such models tend to be relatively abstract because of theoretical assumptions. In this sense, data-driven models may be of high relevance in the context of epilepsy as they are directly obtained from real recordings. Autoregressive models (AR) are well-established techniques in data-driven system identification approaches, presenting attractive features in terms of time series description [7–11]: they are predictors by design; there are many variations available today, as nonlinear autoregressive models (NAR), linear moving average and integrated moving average ones (ARMA and ARIMA) as well as their nonlinear versions (NARMA and NARIMA). Although AR models that incorporate inputs into their formulation exist (ARX, NARX and so forth, where X stands for eXogenous input), the classic modeling assumes only the output is available, as is often the case in practice for epilepsy.

Considering these characteristics, AR modeling has already been used for seizure description and prediction [12–15]. However, there is still a need for more general models which are not only suitable for describing individual activity from patients, but also general across subjects. In this sense, the present work introduces an alternative approach to constructing NAR models that adhere to individual seizures and the ones coming from different patients. The main assumption in this case is that, although each epileptiform signature is unique to a subject, there are common features across patients. These features are in turn reflected by the magnitude of the NAR coefficients.

Essentially, the proposed approach consists of analyzing a public dataset [16] and performing several consecutive regressions so as to obtain random variables for each of the NAR coefficients. The results show that it is possible to use median coefficients to build a model that summarizes all the remaining seizures. This finding is useful in applications where a sufficiently large number of recordings is not available, as well as in the context of seizure prediction and attenuation using NAR modeling.

2 Materials and Methods

2.1 Nonlinear Autoregressive (NAR) Models

NAR models can be defined in terms of linear and nonlinear contributions [14, 15]:

$$\hat{y}(n) = \underbrace{-\sum_{i=1}^{r} \phi_i y(n-i)}_{\text{Linear}} \underbrace{- \sum_{j=1}^{r}\sum_{k=1}^{r} \phi_{ij} y(n-j)y(n-k) + \ldots + \phi_0 + \varepsilon_n}_{\text{Nonlinear}} \quad (1)$$

where $n = 1, ..., N$ is a discrete-time series (i.e., the samples of the signal), ϕ_i, $i = 1, ..., r$, are the coefficients of the model, $y(n-i)$, $i = 1, ..., r$, are past observations of variable $\hat{y}(n)$, which is the output of the system. The first coefficient ϕ_0 is the mean value of the time series, and ε_n is the error associated to the n discrete-time instant. These coefficients are sensitive to the inherent dynamics present in the signals [17] and may indicate differences or similarities in behavior if they occur over time, which is why they are used in this work as features. To stress and better define the nonlinearities of the model, the following notation is used:

$$\begin{aligned}\Phi_i = [\phi_i + \sum_{j=i}^{r} \phi_{ij} y(n-j) + \sum_{j=i}^{r} \phi_{ijj} y^2(n-j) + \\ + \frac{1}{2} \sum_{j=i}^{r} \sum_{k=i}^{r} \phi_{ijk} y(n-j) y(n-k) + P^q] y(n-i), \quad j \neq k\end{aligned} \tag{2}$$

where $i = 1, ..., r$, P^q represents terms of higher order ($q \geq 3$). For the sake of understanding, ϕ_1 indicates the coefficient of the linear regressor for the first delayed time series $y(n-1)$, ϕ_{11} is the coefficient of the nonlinear regressor for the first delayed time series $y(n-1)^2$, and so forth, all of them associated to the $y(n-i)$ delayed series. The Akaike Information Criterion (AIC) is adopted to choose or evaluate the order of the model [18,19]:

$$\text{AIC} = \ln\left(\frac{\text{SSE}}{N}\right) N + 2r \tag{3}$$

where $\text{SSE} = \sum_{i=1}^{N} [y(n) - \hat{y}(n)]^2$ is the sum of squared errors. Essentially, this criterion attempts to balance the accuracy and complexity of the model to obtain the best trade-off between them at a reasonable order.

2.2 Available Dataset

The SWEC-ETHZ iEEG Database[1], which is public and free of charge for research and education purposes, presents one of the datasets containing short-term intracranial electroencephalograms (iEEGs) of 100 seizures coming from 16 patients[2]. All patients presented pharmaco-resistant epilepsies and were evaluated for surgery at the Sleep-Wake-Epilepsy-Center of the University Department of Neurology at the Inselspital Bern, after written informed consent was provided [16]. Each patient presents a varying number of seizures and electrodes, all of which digitally filtered between 0.5 and 150 Hz using a fourth-order Butterworth filter prior to analysis, and stored with a 512 Hz sampling frequency [16]. In this work, the ictal condition is used to construct the NAR models, whereas pre- and post-ictal conditions are neglected.

[1] SWEC: Sleep-Wake-Epilepsy-Center; ETHZ: Eidgenössische Technische Hochschule Zürich.

[2] The pre-ictal stage lasts up to the first three minutes of recording; the post-ictal stage, in turn, starts at the last three minutes of recording; at last, the ictal stage presents a variable number of time/samples [16].

2.3 Statistical Analysis

The number of available electrodes and seizures is not the same across the patients. For convenience, only the first seizure of each subject was selected, but including all of the channels. In this way, it is possible to assure a significantly high number of samples (a total of 1016 time windows from 16 patients) while accounting for representations of all patients at the same time. Histograms are used to depict the resulting distribution obtained after successive estimations of each coefficient. Medians are adopted to summarize the distributions, which implies that no assumptions regarding normality are made. To assess goodness of fit, two metrics are employed: the Pearson correlation and relative root mean squared error (RRMSE)[3] between the signal and the model.

3 Results

Figure 1 depicts an example of epileptiform signature and its reconstructed version. A second-order NAR model with six terms was considered. That is, the nonlinearity assumed is parabolic, and the number of terms resulting from the expansion of the polynomial $[\phi_i + \sum_{j=i}^{r} \phi_{ij} y(n-j)]y(n-i)$, $i = 1, 2$, is six. However, note that the AIC suggests that only the first two terms of the NAR model should be used, which implies that the effect of the nonlinearities in this case is not strong. This is endorsed by the resulting equation, where coefficients of the order 10^{-4} are related to the second-degree terms, whereas coefficients of the order 10^0 to 10^{-1} are associated to the first-degree terms.

Figure 2 presents the histograms for the random variables of each ϕ_i. The distributions tend to be assymetric, but there is a clear tendency of the values concentrating around their medians. This provides evidence of a common pattern across patients, represented by the NAR coefficients: a stable $\phi_0 \sim -0.03$ indicates that all iEEG have a similar mean value; besides, their linear components are well pronounced ($\phi_1 \sim 1.81$ and $\phi_2 \sim 0.83$) while the nonlinear ones ($\phi_{11} \sim -6.1.10^{-5}$, $\phi_{22} \sim -1.13.10^{-4}$ and $\phi_{12} \sim 1.68.10^{-4}$) less strong.

To better analyze how strong the influence of the nonlinearities in the model is, the first channel and seizure of the first patient was selected and new regressions were performed varying the number of lags to match the number of terms, i.e., $i = 1, ..., 6$, so that the last resulting model is purely the traditional linear AR ($\sum_i^6 \phi_i y(n-i)$, $i = 1, ..., 6$). Then, the RRMSE was calculated for all cases, resulting in Table 1. As seen, the error tends to decrease as the model becomes more linear, also suggesting that the addition of delays can provide a more refined description of the seizure. In other words, a precise identification of seizures required dependency on early stages of the electrophysiological activity. It is worth mentioning that the order of the polynomial has also been varied but no significant improvement was verified.

---

[3] The RRMSE is normalized by the difference between the maximum and minimum values of the time series in this work.

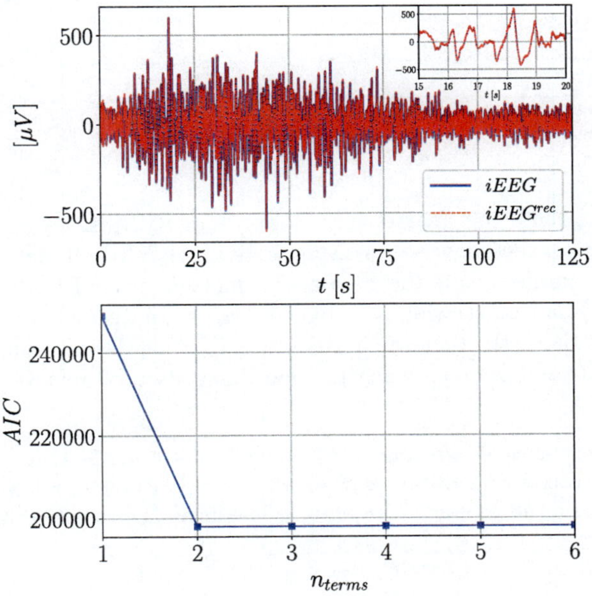

Fig. 1. NAR model reconstruction: – indicates the epileptiform activity obtained from the first seizure and first electrode of the first patient; - - indicates the reconstructed 2nd-order NAR model with 6 terms: $\hat{y}(n) = 0.0319 + 1.7363y(n-1) - 0.7404y(n-2) - 0.0005y(n-1)^2 - 0.0005y(n-2)^2 + 0.0009y(n-1)y(n-2)$. -□- indicates the Akaike Information Criterion. Coefficients are rounded up to the 4th significant figure

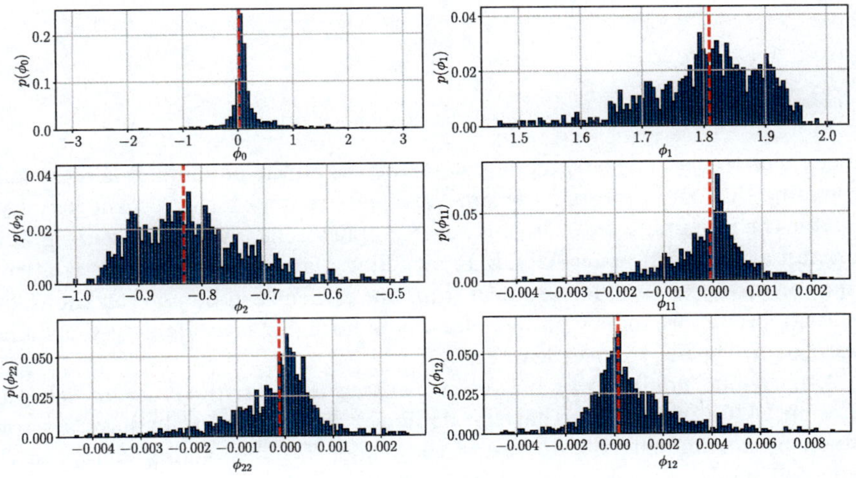

Fig. 2. Histograms obtained from the NAR coefficients after successive identifications: - - indicates the median values of the parameters

Table 1. Effect of the addition of lags to the NAR model considering the number of lags and the RRMSE

Lags	2	3	4	5	6
RRMSE	0.3973%	0.3130%	0.2299%	0.1894%	0.1508%

Table 2 presents the Pearson correlation and RRMSE between the seizures and their reconstructed versions using individual NAR models and a general NAR model, composed of the median coefficients from Fig. 2. For all cases, both RRMSE and correlation are slightly higher for the model built using the medians, but they still remain at a considerably low level, indicating that the median model provides a sufficiently good description of all seizures analyzed.

Table 2. Performance of individual NAR models for selected patients and the general one: ρ_1 and ρ_2 correspond to the Pearson correlation for each of them, respectively. P stands for patient and Sz stands for seizure. The mean is calculated using all patients

P/Sz	ρ_1	$RRMSE_1$ (%)	ρ_2	$RRMSE_2$ (%)
#01/#01	0.99892	0.3973	0.99890	0.4197
#02/#01	0.99896	0.4958	0.99891	0.5188
#15/#01	0.99719	0.6467	0.99712	0.6557
#16/#01	0.99511	0.9176	0.99504	0.9250
Mean	0.99789	0.5559	0.99781	0.5839

4 Discussion

Since there seems to be a common pattern across all the patients, it is reasonable to assume that any processing or analysis made to one signal/patient may hold true for the remaining ones. In this sense, applications involving seizure prediction and attenuation using AR/NAR modeling may benefit from this contribution. For instance, a model obtained from one patient can be used as knowledge to design predictors to new patients for whom no signal is available yet, a similar approach to that of transfer learning [20].

A parabolic nonlinearity has been assessed along with an order selection criterion (AIC), suggesting that the nonlinear regressors do not have a strong influence on enhancing the quality of the model. Besides, adding delays to the estimated model implies in a better goodness of fit, which indicates that the epileptiform activity is strongly state-dependent. This is consistent with the fact of seizures being assessed from the point of view of chaotic dynamics, which is sensitive to initial conditions [2,4]. Still, it is a good practice to use NAR models to generalize the approach to any type of signal and to mitigate stationarity issues [19].

At last, AR/NAR modeling can be used in future work to build state space models [18] from which state observers and controllers can be designed and integrated to study seizure reconstruction and attenuation [21], respectively. The same concept of transfer leaning can be adopted by identifying a model from one patient and using it to design controllers applied to other ones. Although there is an inherent fluctuation of the AR coefficients, which may lead to moderate outcomes, the general model proposed in this work is regarded as a helpful alternative for addressing seizures.

5 Conclusions

This work addressed a relevant issue in the context of epileptiform activity description: providing data-driven models suitable to individual seizures and a general one that can be applied to all of the seizures analyzed. This is particularly relevant for practical situations in which a sufficiently or arbitrarily large number of electrophysiological recordings is not available, and can be used in future applications of seizure prediction and attenuation using AR/NAR modeling.

Conflict of interest

The authors declare that they have no conflict of interest.

Acknowledgements. The first author thanks to Coordination for the Improvement of Higher Education Personnel (CAPES) for the financial support, Grant Number 88887.481049/2020-00.

References

1. Fisher, R., Schachter, S.: The postictal state: a neglected entity in the management of epilepsy. Epilepsy Behav. **1**, 52–59 (2000). https://doi.org/10.1006/ebeh.2000.0023
2. Iasemidis, L.: Epileptic seizure prediction and control. IEEE T Bio-Med. Eng. **50**, 549–558 (2003). https://doi.org/10.1109/TBME.2003.810705
3. Vezzani, A., French, J., Bartfai, T., Baram, T.Z.: The role of inflammation in epilepsy. Nat. Rev. Neurol. **7**, 31–40 (2011). https://doi.org/10.1038/nrneurol.2010.178
4. Jirsa, V.K., Stacey, W.C., Quilichini, P.P., Ivanov, A.I., Bernard, C.: On the nature of seizure dynamics. Brain **137**, 2210–2230 (2014). https://doi.org/10.1093/brain/awu133
5. Proix, T., Bartolomei, F., Guye, M., Jirsa, V.K.: Individual brain structure and modelling predict seizure propagation. Brain **140**, 641–654 (2017). https://doi.org/10.1093/brain/awx004
6. Chizhov, A.V., Zefirov, A.V., Amakhin, D.V., Smirnova, E.Y., Zaitsev, A.V.: Minimal model of interictal and ictal discharges "Epileptor-2." PLoS Comput. Biol. **14**, 1–25 (2018). https://doi.org/10.1371/journal.pcbi.1006186

7. Lennart, L.: System Identification. Birkhäuser, Boston (1998) https://doi.org/10.1007/978-1-4612-1768-8_11
8. Landau, I., Karimi, A.: A recursive algorithm for ARMAX model identification in closed loop. IEEE T Autom. Control **44**, 840–843 (1999). https://doi.org/10.1109/9.754830
9. Baldacchino, T., Anderson, S., Kadirkamanathan, V.: Computational system identification for Bayesian NARMAX modelling. Automatica **49**, 2641–2651 (2013). https://doi.org/10.1016/j.automatica.2013.05.023
10. Liu, H., Zhu, L., Pan, Z., Bai, F., Liu, Y., Liu, Y., Patel, M., Farantatos, E., Bhatt, N.: ARMAX-based transfer function model identification using wide-area measurement for adaptive and coordinated damping control. IEEE T Smart Grid **8**, 1105–1115 (2017). https://doi.org/10.1109/TSG.2015.2470648
11. Retes, P., Aguirre, L.: NARMAX model identification using a randomised approach. Int. J. Model. Identification Control **31**, 205–216 (2019). https://doi.org/10.1504/IJMIC.2019.098779
12. Chisci, L., Mavino, A., Perferi, G., Sciandrone, M., Anile, C., Colicchio, G., Fuggetta, F.: Real-time epileptic seizure prediction using AR models and support vector machines. IEEE T Bio-Med. Eng. **57**, 1124–1132 (2010). https://doi.org/10.1109/TBME.2009.2038990
13. Yu, P.N., Naiini, S.A., Heck, C.N., Liu, C.Y., Song, D., Berger, T.W.: A sparse Laguerre-Volterra autoregressive model for seizure prediction in temporal lobe epilepsy. In: 38th Annual International Conference of the IEEE Engineering in Medicine and Biology Society (EMBC), pp. 1664–1667. IEEE, Orlando, FL, USA (2016). https://doi.org/10.1109/EMBC.2016.7591034
14. Schiff, N.D., Victor, J.D., Canel, A., Labar, D.R.: Characteristic nonlinearities of the 3/s ictal electroencephalogram identified by nonlinear autoregressive analysis. Biol. Cybern. **72**, 519–526 (1995). https://doi.org/10.1007/BF00199894
15. Schiff, N.D., Victor, J.D., Canel, A.: Nonlinear autoregressive analysis of the 3/s ictal electroencephalogram: Implications for underlying dynamics. Biol. Cybern. **72**, 527–532 (1995). https://doi.org/10.1007/BF00199895
16. Burrello, A., Cavigelli, L., Schindler, K., Benini, L., Rahimi, A.: Laelaps: An energy-efficient seizure detection algorithm from long-term human iEEG recordings without false alarms. In: Design, Automation & Test in Europe Conference & Exhibition (DATE), pp. 752–757. Florence, Italy (2019). https://doi.org/10.23919/DATE.2019.8715186
17. Entezami, A., Shariatmadar, H.: An unsupervised learning approach by novel damage indices in structural health monitoring for damage localization and quantification. Struct. Health Monitor **17**, 325–345 (2018). https://doi.org/10.1177/1475921717693572
18. Durbin, J, Koopman, S.: Time series analysis by state space methods. OUP, Oxford (2012). https://doi.org/10.1093/acprof:oso/9780199641178.001.0001
19. Fox, J.: Applied Regression Analysis and Generalized Linear Models. Sage Publications Inc., Thousand Oaks, CA, USA (2015)
20. Taylor, M., Kacprzyk, J.: Transfer in Reinforcement Learning Domains. Springer, Berlin Heidelberg (2019). https://doi.org/10.1007/978-3-642-01882-4
21. Brogin, J., Faber, J., Bueno, D.: Estimating the parameters of the Epileptor model for epileptic seizure suppression. Neuroinformatics 1–23 (2022). https://doi.org/10.1007/s12021-022-09583-6

Solving the Cylindrical Cattaneo-Vernotte Bioheat Equation

Alessandro Avancini, Vinicius Rispoli(✉), and Ronni Amorim

Engineering College at Gama, University of Brasilia, Brasília, Brazil
vrispoli@unb.br

Abstract. The method of radio-frequency ablation is a widely accepted alternative to treat hepatic cancerous tumors. It is a minimal invasive technique that is capable of eradicating tumor cells and provides excellent recover for the subject. By constructing a theoretical model for radio-frequency ablation process with an internally cooled needle-like electrodes we can study the propagation of heat across the length of the tissue. Also allowing to visualize the evolution of the system and its behavior over time. In this paper we will approach the Cattaneo-Vernotte bioheat model. Unlike the gold standard Pennes Bioheat equation, the Cattaneo-Vernotte equation takes into account the effects of thermal relaxation on human tissues. Our goal is to deduce the analytical solution for the cylindrical coordinate Cattaneo-Vernotte bioheat equation on a unlimited domain. The obtained solution can be used to study and understand the heat transfer on the hepatic human tissue.

Keywords: radiofrequency ablation · mathematical modelling · Laplace transform · residue theorem

1 Introduction

The radio-frequency ablation with internally cooled needle-like electrode is a procedure applied for a number of conditions, such as tumors, autonomously functioning thyroid nodules and arrhythmia. Since it has a wide range of applications, it becomes extremely important to comprehend the effects on the tissue that has been applied to and its surroundings, as well as the parameters involved in the phenomena and how they interact with one another. It consists of a metallic electrode with an internal liquid cooling system that cools the electrode surface, and then the device is applied to the skin at the top of the damaged tissue, in order to eliminate the anomalous cells [1–5].

To achieve the level of understanding required for a careful safety analysis, a mathematical model of the procedure is needed to study the influence of the many variables involved in the propagation of heat within the hepatic tissue caused by the radio-frequency ablation. Is this paper we work on the analytical solution for the Cattaneo-Vernotte equation. The model developed in this study considers cylindrical symmetry for the needle-like electrodes, no restrictions or

J. L. B. Marques et al. (Eds.): CLAIB 2022/CBEB 2022, IFMBE Proceedings 98, pp. 23–34, 2024.
https://doi.org/10.1007/978-3-031-49401-7_3

limits regarding the extension of the spatial domain of solution and accounts for the thermal relaxation in order to generalize the solution.

The model worked in this paper is an evolution of the problem treated by Molina *et al.* [1], where is added a second-order derivative term in the partial differential equation that models the heat distribution on the human tissue. This modification brings, as expected, a different analytical solution and a few difficulties at the solutions steps that were not faced by Molina *et al.* [1] in their original paper. Moreover, the problem addressed here can be considered more realistic than others found in the recent literature that treats similar problem, as many of them deal with simplified versions of the Cattaneo-Vernotte problem in limited domains, or disregarding external heat sources and/or in Cartesian domains [6–8]. It is also important to note this kind of problem is also solved numerically, which, in general, allows solutions in more complex domains [9,10]. These approaches are complementary, since their joint use can enrich the analysis of this type of problem.

Here, the desired Cattaneo-Vernotte associated analytical solution is obtained using the Laplace transform, second order nonhomohegenous ordinary differential equations theory and also a combination between the Bromwich integral and the Cauchy residue theorem. In the following sections are presented the governing equations and the steps used to obtain the analytical solution.

2 Governing Equations

The temperature distribution on a biological tissue is usually described through the Pennes bioheat equation:

$$\rho c \frac{\partial T}{\partial t} = \nabla \cdot (k(T) \nabla T) - \omega_b \rho_b c_b (T - T_b) + S \tag{1}$$

where ρ, c, $k(T)$ represents the density, specific heat and thermal conductivity of the tissue, respectively. On the other hand, ρ_b, c_b, ω_b and T_b are constants representing the blood density, specific heat, perfusion coefficient and body temperature, in that order. Finally, S represents the heat sources acting on the tissue.

A modified version of the Pennes bioheat equation that takes into account the thermal relaxation of the tissue is given by the Cattaneo-Vernotte equation:

$$\rho c \left(\tau \frac{\partial^2 T}{\partial t^2} + \frac{\partial T}{\partial t} \right) = \nabla \cdot (k(T) \nabla T) - \omega_b \rho_b c_b (T - T_b) + S \tag{2}$$

where τ is an dimensional constant that regulates the tissue thermal relaxation. Here we are considering the materials to have isotropic characteristics, as well as a constant thermal conductivity $k(T) = k$. Cylindrical symmetry will also be assumed (Fig. 1), in order to make the analytical solution feasible to be obtained. The heat source term (S) is given by the electrical power density, which dissipated into the tissue through time and is given by the expression $S(r) = j_0^2 r_0^2 / \sigma r^2$.

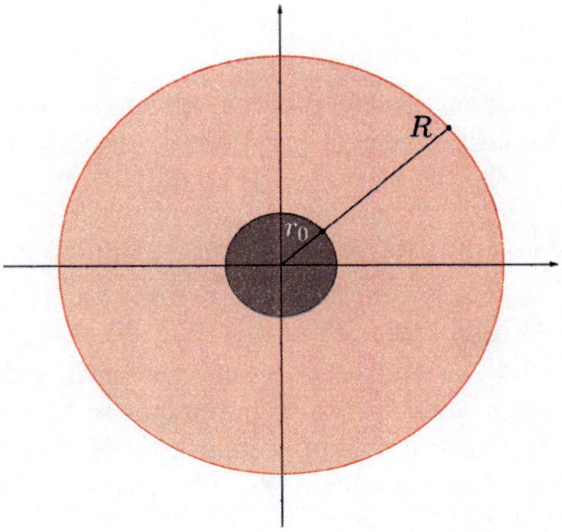

Fig. 1. Cylindrical symmetry assumed. Here r_0 is the electrode radius with external radius $R \to \infty$

Taking all those assumptions into consideration, and rewriting the expression in polar coordinates, the Eq. (2) becomes:

$$\rho c\left(\tau\frac{\partial^2 T}{\partial t^2}+\frac{\partial T}{\partial t}\right)=k\left(\frac{\partial^2 T}{\partial r^2}+\frac{1}{r}\frac{\partial T}{\partial r}\right)-\omega_b\rho_b c_b(T-T_b)+\frac{j_0^2 r_0^2}{\sigma r^2} \tag{3}$$

with the following initial and boundary conditions:

$$\begin{aligned} T(r,0)=T_b,\quad T_t(r,0)=0,\quad \forall\, r>r_0 \\ \lim_{r\to\infty} T(r,t)=T_b,\quad T(r_0,t)=T_C,\quad \forall\, t>0 \end{aligned} \tag{4}$$

where $T=T(r,t)$, r_0 is the electrode radius, j_0 is the electrical current density, σ is the electrical conductivity and T_C is the electrode refrigeration fixed temperature.

Finally, defining the dimensionless variables

$$\bar r=\frac{r}{r_0},\quad \bar t=\frac{k}{c\rho r_0^2}t \text{ and } \bar T(\bar r,\bar t)=\frac{\sigma k}{j_0^2 r_0^2}\left[T\left(r_0\bar r,\frac{c\rho r_0^2}{k}\bar t\right)-T_b\right] \tag{5}$$

we can change the initial-boundary value problem to dimensionless variables to obtain:

$$\Lambda\frac{\partial^2\bar T}{\partial\bar t^2}+\frac{\partial\bar T}{\partial\bar t}=\frac{\partial^2\bar T}{\partial\bar r^2}+\frac{1}{\bar r}\frac{\partial\bar T}{\partial\bar r}+\frac{1}{\bar r^2}-\Psi\bar T, \tag{6}$$

where $\Lambda=\tau k/\rho c r_0^2$ and $\Psi=r_0^2\rho_b c_b\omega_b/k$ are dimensionless constants, with the following initial and boundary conditions:

$$\begin{aligned} \bar T(\bar r,0)=0,\ \bar T_{\bar t}(\bar r,0)=0,\ \forall\,\bar r>1 \\ \lim_{\bar r\to\infty}\bar T(\bar r,\bar t)=0,\ \bar T(1,\bar t)=\frac{\sigma k(T_C-T_b)}{j_0^2 r_0^2}:=-\Gamma,\forall\,\bar t>0. \end{aligned} \tag{7}$$

3 On Bessel Functions

In order to solve the dimensionless problem given by Eqs. (6)–(7), we will use known properties of modified Bessel functions with complex argument. All the relations presented in this section can be found in more details on the references [11].

First, we start by recalling the Bessel functions of first and second kind of zero order $J_0(z)$ and $Y_0(z)$. These functions are defined as the linearly independent solutions of the Bessel differential equation $z^2y'' + zy' + z^2y = 0$. We note that $J_0(z)$ is analytic for all $z \in \mathbb{C}$ and $Y_0(z)$ is analytic for all z in its principal branch $\mathbb{C} \setminus (-\infty, 0]$. On the other hand, it is also important to recall the modified Bessel functions $I_0(z)$ and $K_0(z)$ of first and second kind of zero order as the linearly independent solutions of the modified Bessel equation $z^2y'' + zy' - z^2y = 0$. Again, we observe that $I_0(z)$ is analytic for all $z \in \mathbb{C}$ and $K_0(z)$ is analytic for all z in its principal branch $\mathbb{C} \setminus (-\infty, 0]$.

Bessel functions and modified Bessel functions have a lot of interesting properties and relations, but only a few are relevant for us in our deduction. One is related to the Wronskian $\mathcal{W}$ between the modified Bessel functions $I_0(z)$ and $K_0(z)$ that is given by

$$\mathcal{W}(I_0, K_0) = -\frac{1}{z} \tag{8}$$

And the others are the following non-trivial identities that will be required to the analytical solution calculation:

$$I_0(z) = J_0(zi), \ \forall\, z \in \mathbb{C} \tag{9}$$

$$K_0(iz) = -\frac{i\pi}{2}\left(J_0(z) - iY_0(z)\right), \ |\arg(z)| \leq \pi/2 \tag{10}$$

$$K_0(-iz) = \frac{i\pi}{2}\left(J_0(z) + iY_0(z)\right), \ |\arg(z)| \leq \pi/2 \tag{11}$$

4 Solution to the Initial-Boundary Value Problem

We start our process of finding the analytical solution by applying the Laplace transform on Eq. (6) with respect to the temporal variable $\bar{t}$:

$$\bar{r}^2\frac{\partial U}{\partial \bar{r}^2} + \bar{r}\frac{\partial U}{\partial \bar{r}} - \left(\Lambda s^2 + s + \Psi\right)\bar{r}^2 U = -\frac{1}{s}, \tag{12}$$

where

$$U(\bar{r}, s) = \mathcal{L}\{\bar{T}(\bar{r}, \bar{t})\}(\bar{r}, s) = \int_0^\infty e^{-s\bar{t}}\bar{T}(\bar{r}, \bar{t})d\bar{t}. \tag{13}$$

Also applying the Laplace transform on the boundary conditions defined in Eq. (7) we obtain:

$$\lim_{\bar{r}\to\infty} U(\bar{r}, s) = 0 \ \text{ and } \ U(1, s) = -\frac{\Gamma}{s} \tag{14}$$

Now we note that Eq. (12) is a nonhomogeneous modified Bessel equation of zero order. Its homogeneous solution is a linear combination of I_0 and K_0. So, the expected solution for the Eq. (12) should be in the following form:

$$U(\bar{r}, s) = C_1(\bar{r}, s) I_0\left(\bar{r}\sqrt{\Lambda s^2 + s + \Psi}\right) + C_2(\bar{r}, s) K_0\left(\bar{r}\sqrt{\Lambda s^2 + s + \Psi}\right) \quad (15)$$

In order to find the functions $C_1(\bar{r}, s)$ and $C_2(\bar{r}, s)$ we use the method of variation of parameters, recalling that Wronskian between I_0 and K_0 is given by Eq. (8), to reach on the following equations:

$$\frac{\partial}{\partial \bar{r}} C_1(\bar{r}, s) = -\frac{1}{s} \frac{K_0\left(\bar{r}\sqrt{\Lambda s^2 + s + \Psi}\right)}{\bar{r}} \quad (16)$$

$$\frac{\partial}{\partial \bar{r}} C_2(\bar{r}, s) = \frac{1}{s} \frac{I_0\left(\bar{r}\sqrt{\Lambda s^2 + s + \Psi}\right)}{\bar{r}} \quad (17)$$

Taking the integral:

$$C_1(\bar{r}, s) = -\frac{1}{s} \int_1^{\bar{r}} \frac{K_0\left(v\sqrt{\Lambda s^2 + s + \Psi}\right)}{v} dv + M_1(s) \quad (18)$$

$$C_2(\bar{r}, s) = \frac{1}{s} \int_1^{\bar{r}} \frac{I_0\left(v\sqrt{\Lambda s^2 + s + \Psi}\right)}{v} dv + M_2(s) \quad (19)$$

where M_1 and M_2 are functions independent on $\bar{r}$, chosen in such a way the boundary conditions in Eq. (14) are respected. And finally, we obtain the solution of the Eq. (12) in the $(\bar{r}, s)$-space:

$$\begin{aligned} U(\bar{r}, s) &= \frac{1}{s} I_0\left(\bar{r}\alpha(s)\right) \int_{\bar{r}}^{\infty} \frac{K_0\left(v\alpha(s)\right)}{v} dv \\ &+ \frac{1}{s} K_0\left(\bar{r}\alpha(s)\right) \int_1^{\bar{r}} \frac{I_0\left(v\alpha(s)\right)}{v} dv - \Gamma \frac{K_0\left(\bar{r}\alpha(s)\right)}{sK_0\left(\alpha(s)\right)} \\ &- I_0\left(\alpha(s)\right) \frac{K_0\left(\bar{r}\alpha(s)\right)}{sK_0\left(\alpha(s)\right)} \int_1^{\infty} \frac{K_0\left(v\alpha(s)\right)}{v} dv, \end{aligned} \quad (20)$$

where $\alpha(s) = \sqrt{\Lambda s^2 + s + \Psi}$.

Rewriting the Eq. (20) in a more convenient way we get:

$$\begin{aligned} U(\bar{r}, s) &= \int_{\bar{r}}^{\infty} \frac{1}{v} \frac{I_0\left(\bar{r}\alpha(s)\right) K_0\left(v\alpha(s)\right)}{s} dv \\ &+ \int_1^{\bar{r}} \frac{1}{v} \frac{K_0\left(\bar{r}\alpha(s)\right) I_0\left(v\alpha(s)\right)}{s} dv - \Gamma \frac{K_0\left(\bar{r}\alpha(s)\right)}{sK_0\left(\alpha(s)\right)} \\ &- \int_1^{\infty} \frac{1}{v} \frac{I_0\left(\alpha(s)\right) K_0\left(\bar{r}\alpha(s)\right) K_0\left(v\alpha(s)\right)}{sK_0\left(\alpha(s)\right)} dv \end{aligned} \quad (21)$$

And now we can write the solution Eq. (21) in terms of the inverse Laplace transform as follows:

$$U(\bar{r},s) = \int_{\bar{r}}^{\infty} \frac{\mathcal{L}^{-1}\{f_1\}(\bar{r},\bar{t},v)}{v} dv + \int_{1}^{\bar{r}} \frac{\mathcal{L}^{-1}\{f_2\}(\bar{r},\bar{t},v)}{v} dv$$
$$- B\mathcal{L}^{-1}\{f_3\}(\bar{r},\bar{t}) - \int_{1}^{\infty} \frac{\mathcal{L}^{-1}\{f_4\}(\bar{r},\bar{t},v)}{v} dv \qquad (22)$$

To obtain the inverse Laplace transform $\mathcal{L}^{-1}\{U(\bar{r},\bar{t})\}$, we shall proceed in several steps. We will use the Bromwich integral and the residue theorem in order to calculate all the inverses. Recalling that Bromwich's integral for $F(s)$, a function defined for $s \in \mathbb{C}$, is given by:

$$\mathcal{L}^{-1}\{F(s)\} = \lim_{\rho\to\infty} \frac{1}{2\pi i} \int_{\gamma-i\rho}^{\gamma+i\rho} e^{s\bar{t}} F(s) ds. \qquad (23)$$

First, we note that $\alpha(s) = \sqrt{\Lambda s^2 + s + \Psi}$, as a function of the complex variable s, has two branch points that happens on the roots of the polynomial $p(s) = \Lambda s^2 + s + \Psi$, i.e., the branch points are $b_{1,2} = -\frac{1\mp\sqrt{1-4\Lambda\Psi}}{2\Lambda}$. This contrasts with the model worked by Molina *et al.* [1]. It is also important to note that, for human hepatic tissue (see Ref.[12]), Λ and Ψ are positive constants such that $\Lambda\Psi \ll \frac{1}{4}$, then $0 < 1 - 4\Lambda\Psi < 1$ and the branch points lie on the negative real axis. So a modified Bromwich's contour, as shown in Fig. 2, will be necessary for the calculations.

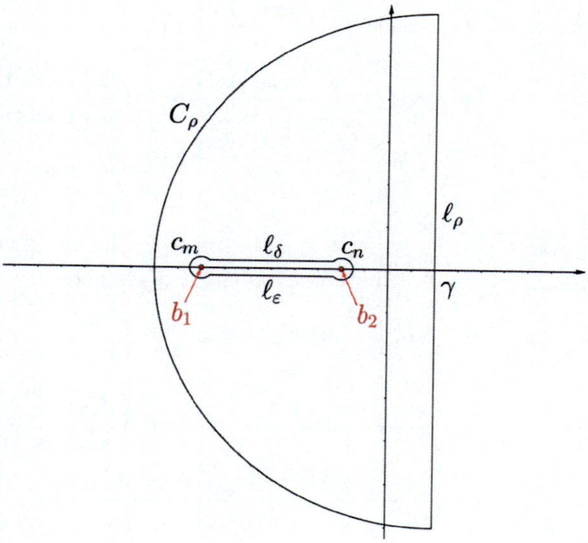

Fig. 2. Integration contour

We will define the domain $\mathbf{D} = \mathbb{C}\backslash[b_1, b_2]$ as a set of points to be used in following integrations, since $\alpha(s)$ is single valued and continuous in $\mathbf{D}$. Then, for a given function $F(s)$ the integral in Eq. (23) will be calculated based on the following line integral:

$$\begin{aligned}\oint_C e^{st}F(s)ds &= \int_{C_\rho} + \int_{\ell_\rho} + \int_{c_m} + \int_{\ell_\delta} + \int_{c_n} + \int_{\ell_\varepsilon} \\ &= 2\pi i \sum_k \text{Res}\left(e^{st}F(s), s_k\right), \end{aligned} \tag{24}$$

where s_k is a singularity on the left of the ℓ_ρ line.

Since the modified Bessel functions $I_0(z)$ and $K_0(z)$ have asymptotical exponential order as $|z| \to \infty$, it is possible to show under certain conditions on t that $\int_{C_\rho} \to 0$ as $\rho \to \infty$ (we note that the treatment of this limit is different and more intricate in contrast to Molina *et al.* [1] paper). Moreover, as $I_0(z) \sim 1$ and $K_0(z) \sim -\ln(z)$, as $|z| \to 0$ we can also verify that $\int_{c_m}$, $\int_{c_n} \to 0$ as $m, n \to 0$. Thus, we can state that the inverse Laplace transform in this case is given by:

$$\begin{aligned}\mathcal{L}^{-1}\{F(s)\} = \sum_k \text{Res}\left(e^{s\bar{t}}F(s), s_k\right) - \frac{1}{2\pi i}\int_{\ell_\delta} e^{s\bar{t}}F(s)ds \\ - \frac{1}{2\pi i}\int_{\ell_\varepsilon} e^{s\bar{t}}F(s)ds.\end{aligned} \tag{25}$$

So, our attention will be focused on the inner segments ℓ_δ and ℓ_ε. Although in Fig. 2 lines ℓ_δ and ℓ_ε appear to be far from the real axis, they should be understood as if they were sufficiently close (or over) the axis.

4.1 Computation of $\mathcal{L}^{-1}\{f_1\}(\bar{r}, \bar{t}, v)$

We start out inverse Laplace transform calculations with the first integrand in Eq. (22), i.e., we need to calculate:

$$\begin{aligned}F_1(\bar{r}, \bar{t}, v) &= \mathcal{L}^{-1}\{f_1\}(\bar{r}, \bar{t}, v) \\ &= \mathcal{L}^{-1}\left\{\frac{I_0(\bar{r}\alpha(s))K_0(v\alpha(s))}{s}\right\}(\bar{r}, \bar{t}, v),\end{aligned} \tag{26}$$

where $v \geq \bar{r}$.

First, we note that $\int_{C_\rho} \to 0$ as $\rho \to \infty$ when $\bar{t} > v + \bar{r}$. Now, it is immediate that the function $e^{s\bar{t}}\frac{I_0(\bar{r}\alpha(s))K_0(v\alpha(s))}{s}$ has only a simple pole on $s = 0$, then the residue at this pole is:

$$\begin{aligned}\text{Res}(e^{s\bar{t}}f_1, 0) &= \lim_{s\to 0} s\frac{e^{s\bar{t}}I_0(\bar{r}\alpha(s))K_0(v\alpha(s))}{s} \\ &= I_0\left(\bar{r}\sqrt{\Psi}\right)K_0\left(v\sqrt{\Psi}\right)\end{aligned} \tag{27}$$

Before we proceed, it is immediate to observe that $\Lambda s^2 + s + \Psi < 0$, for all $s \in (b_1, b_2)$, then we will define the real function $\beta(s) = \sqrt{-\Lambda s^2 - s - \Psi}$, for all $s \in (b_1, b_2)$. Now, using that the inverse given by Eq. (25) and calculating the contour integrals for ℓ_δ and ℓ_ε we obtain:

$$F_1(\bar{r}, \bar{t}, v) = I_0\left(\bar{r}\sqrt{\Psi}\right) K_0\left(v\sqrt{\Psi}\right) - \frac{1}{2\pi i}\int_{b_1}^{b_2} \frac{e^{s\bar{t}}}{s} I_0(\bar{r}\beta(s)i)K_0(v\beta(s)i)ds \quad (28)$$
$$+ \frac{1}{2\pi i}\int_{b_1}^{b_2} \frac{e^{s\bar{t}}}{s} I_0(-\bar{r}\beta(s)i)K_0(-v\beta(s)i)ds$$

Taking into consideration the relations given on Eqs. (9), (10) and (11), we can write Eq. (28) as:

$$\begin{aligned} F_1(\bar{r}, \bar{t}, v) &= I_0\left(\bar{r}\sqrt{\Psi}\right) K_0\left(v\sqrt{\Psi}\right) \\ &+ \frac{1}{4}\int_{b_1}^{b_2} \frac{e^{s\bar{t}}}{s} J_0(\bar{r}\beta(s))\left(J_0(v\beta(s)) - iY_0(v\beta(s))\right) ds \\ &+ \frac{1}{4}\int_{b_1}^{b_2} \frac{e^{s\bar{t}}}{s} J_0(\bar{r}\beta(s))\left(J_0(v\beta(s)) + iY_0(v\beta(s))\right) ds \end{aligned} \quad (29)$$

And finally, for $\bar{t} > v + \bar{r}$:

$$F_1(\bar{r}, \bar{t}, v) = I_0\left(\bar{r}\sqrt{\Psi}\right) K_0\left(v\sqrt{\Psi}\right) + \frac{1}{2}\int_{b_1}^{b_2} \frac{e^{s\bar{t}}}{s} J_0(\bar{r}\beta(s))J_0(v\beta(s))ds. \quad (30)$$

4.2 Computation of $\mathcal{L}^{-1}\{f_2\}(\bar{r}, \bar{t}, v)$

Now we must consider the second integrand in Eq. (22) which is given by:

$$\begin{aligned} F_2(\bar{r}, \bar{t}, v) &= \mathcal{L}^{-1}\{f_2\}(\bar{r}, \bar{t}, v) \\ &= \mathcal{L}^{-1}\left\{\frac{I_0(v\alpha(s))K_0(\bar{r}\alpha(s))}{s}\right\}(\bar{r}, \bar{t}, v) \end{aligned} \quad (31)$$

where $v \leq \bar{r}$. It is immediate to note, as a consequence of the previous result, that for $\bar{t} > v + \bar{r}$:

$$F_2(\bar{r}, \bar{t}, v) = F_1(v, \bar{t}, \bar{r}). \quad (32)$$

4.3 Computation of $\mathcal{L}^{-1}\{f_3\}(\bar{r}, \bar{t}, v)$

Now, from the third part of the right hand side of Eq. (22), we need to determine:

$$\begin{aligned} F_3(\bar{r}, \bar{t}) &= \mathcal{L}^{-1}\{f_3\}(\bar{r}, \bar{t}, v) \\ &= \mathcal{L}^{-1}\left\{\frac{K_0\left(\bar{r}\alpha(s)\right)}{sK_0\left(\alpha(s)\right)}\right\}(\bar{r}, \bar{t}). \end{aligned} \quad (33)$$

In order to calculate this last inverse we will proceed to an equivalent form, which is given as follows:

$$\begin{aligned}\mathcal{L}^{-1}\left\{\frac{K_0\left(\bar{r}\alpha(s)\right)}{sK_0\left(\alpha(s)\right)}\right\} &= \frac{d}{d\bar{t}}\mathcal{L}^{-1}\left\{\frac{K_0\left(\bar{r}\alpha(s)\right)}{s^2K_0\left(\alpha(s)\right)}\right\}\\ &= \frac{d}{d\bar{t}}\left\{\text{Res}\left(e^{s\bar{t}}\frac{K_0\left(\bar{r}\alpha(s)\right)}{s^2K_0\left(\alpha(s)\right)},0\right)+\frac{1}{2\pi i}\left(\int_{\ell_\delta}+\int_{\ell_\varepsilon}\right)\right\}\end{aligned} \tag{34}$$

noting that $\int_{C_\rho}\to 0$ as $\rho\to\infty$ when $\bar{t}>0$. Since $\frac{K_0(\bar{r}\alpha(s))}{s^2K_0(\alpha(s))}$ has a second order pole at $s=0$, the residue can be obtained by the following limit:

$$\begin{aligned}\text{Res}\left(e^{s\bar{t}}\frac{K_0\left(\bar{r}\alpha(s)\right)}{s^2K_0\left(\alpha(s)\right)},0\right) &= \lim_{s\to 0}\frac{d}{ds}\left(\frac{s^2e^{st}K_0(\bar{r}\alpha(s))}{s^2K_0(\alpha(s))}\right)\\ &= \frac{\bar{t}K_0(\bar{r}\sqrt{\Psi})}{K_0(\sqrt{\Psi})}+\frac{K_1(\sqrt{\Psi})K_0(\bar{r}\sqrt{\Psi})-\bar{r}K_1(\bar{r}\sqrt{\Psi})K_0(\sqrt{\Psi})}{2\sqrt{\Psi}K_0(\sqrt{\Psi})^2}\end{aligned} \tag{35}$$

Now, deriving Eq. (35) in relation to $\bar{t}$, we obtain:

$$\frac{d}{d\bar{t}}\text{Res}\left(e^{s\bar{t}}\frac{K_0\left(\bar{r}\alpha(s)\right)}{s^2K_0\left(\alpha(s)\right)},0\right)=\frac{K_0(\bar{r}\sqrt{\Psi})}{K_0(\sqrt{\Psi})} \tag{36}$$

Calculating the line integrals in ℓ_δ and ℓ_ε we get:

$$\int_{\ell_\delta}+\int_{\ell_\varepsilon}=\int_{b_1}^{b_2}\frac{e^{s\bar{t}}K_0(\bar{r}\beta(s)i)}{s^2K_0(\beta(s)i)}ds-\int_{b_1}^{b_2}\frac{e^{s\bar{t}}K_0(-\bar{r}\beta(s)i)}{s^2K_0(-\beta(s)i)}ds \tag{37}$$

Using again the relations given by Eq. (10) and Eq. (11) then Eq. (37) can be written as:

$$\begin{aligned}\int_{\ell_\delta}+\int_{\ell_\varepsilon} &= \int_{b_1}^{b_2}\frac{e^{s\bar{t}}}{s^2}\frac{J_0(\bar{r}\beta(s))-iY_0(\bar{r}\beta(s))}{J_0(\beta(s))-iY_0(\beta(s))}ds\\ &\quad-\int_{b_1}^{b_2}\frac{e^{s\bar{t}}}{s^2}\frac{J_0(\bar{r}\beta(s))+iY_0(\bar{r}\beta(s))}{J_0(\beta(s))+iY_0(\beta(s))}ds\\ &= 2i\int_{b_1}^{b_2}\frac{e^{s\bar{t}}}{s^2}\frac{J_0(\bar{r}\beta)Y_0(\beta)-Y_0(\bar{r}\beta)J_0(\beta)}{J_0^2(\beta)+Y_0^2(\beta)}ds\end{aligned} \tag{38}$$

Deriving Eq. (38) in relation to $\bar{t}$, we obtain:

$$\frac{d}{d\bar{t}}\left(\int_{\ell_\delta}+\int_{\ell_\varepsilon}\right)=2i\int_{b_1}^{b_2}\frac{e^{s\bar{t}}}{s}\frac{J_0(\bar{r}\beta)Y_0(\beta)-Y_0(\bar{r}\beta)J_0(\beta)}{J_0^2(\beta)+Y_0^2(\beta)}ds$$

which, finally, gives us the complete answer:

$$\begin{aligned}F_3(\bar{r},\bar{t}) &= \mathcal{L}^{-1}\left\{\frac{K_0\left(\bar{r}\alpha(s)\right)}{sK_0\left(\alpha(s)\right)}\right\}(\bar{r},s)\\ &= \frac{K_0(\bar{r}\sqrt{\Psi})}{K_0(\sqrt{\Psi})}-\frac{1}{\pi}\int_{b_1}^{b_2}\frac{e^{s\bar{t}}}{s}\frac{J_0(\bar{r}\beta)Y_0(\beta)-Y_0(\bar{r}\beta)J_0(\beta)}{J_0^2(\beta)+Y_0^2(\beta)}ds.\end{aligned} \tag{39}$$

4.4 Computation of $\mathcal{L}^{-1}\{f_4\}(\bar{r}, \bar{t}, v)$

At last, we proceed to find the fourth inverse on Eq. (22) which is:

$$\begin{aligned} F_4(\bar{r},\bar{t},v) &= \mathcal{L}^{-1}\{f_4\}(\bar{r},\bar{t},v) \\ &= \mathcal{L}^{-1}\left\{\frac{K_0(\bar{r}\alpha(s))}{sK_0(\alpha(s))}\cdot I_0(\alpha(s))K_0(v\alpha(s))\right\}. \end{aligned} \tag{40}$$

We start recalling that convolution theorem states:

$$F_4(\bar{r},\bar{t},v) = \mathcal{L}^{-1}\left\{\frac{K_0\left(\bar{r}\alpha\right)}{sK_0\left(\alpha\right)}\right\} * \mathcal{L}^{-1}\left\{K_0(v\alpha)I_0(\alpha)\right\}. \tag{41}$$

Then, using our previous results we can write:

$$\begin{aligned} F_4(\bar{r},\bar{t},v) &= F_3(\bar{r},\bar{t}) * \frac{d}{d\bar{t}}\mathcal{L}^{-1}\left\{\frac{K_0(v\alpha(s))I_0(\alpha(s))}{s}\right\} \\ &= F_3(\bar{r},\bar{t}) * \frac{d}{d\bar{t}}F_1(\bar{r},\bar{t},v) \\ &= \frac{1}{2}F_3(\bar{r},\bar{t}) * \int_{b_1}^{b_2} e^{s\bar{t}}J_0(\beta(s))J_0(v\beta(s))q(\bar{t})ds \\ &= \frac{1}{2}\int_0^{\bar{t}} F_3(\bar{r},w)\left(\int_{b_1}^{b_2} e^{s(\bar{t}-w)}J_0(\beta)J_0(v\beta)q(\bar{t})ds\right)dw, \end{aligned} \tag{42}$$

with $q(\bar{t}) = u(t - \bar{r} - 1)$ and $u(\cdot)$ is the unit step function.

4.5 Complete Answer

Gathering all our previous results, working some algebraic manipulations and performing some extensive analysis, the complete solution of the dimensionless initial-boundary value problem can be written as the following:

$$\begin{aligned} \bar{T}(\bar{r},\bar{t}) = &\int_{\bar{r}}^{\infty}\frac{1}{v}\left\{I_0\left(\bar{r}\sqrt{\Psi}\right)K_0\left(v\sqrt{\Psi}\right)p(\bar{t})\right. \\ &\left. + \frac{1}{2}\int_{b_1}^{b_2}\frac{e^{s\bar{t}}}{s}J_0(\bar{r}\beta(s))J_0(v\beta(s))p(\bar{t})ds\right\}dv \\ &+ \int_1^{\bar{r}}\frac{1}{v}\left\{I_0\left(v\sqrt{\Psi}\right)K_0\left(\bar{r}\sqrt{\Psi}\right)p(\bar{t})\right. \\ &\left. + \frac{1}{2}\int_{b_1}^{b_2}\frac{e^{s\bar{t}}}{s}J_0(v\beta(s))J_0(\bar{r}\beta(s))p(\bar{t})ds\right\}dv \\ &- \Gamma\frac{K_0(\bar{r}\sqrt{\Psi})}{K_0(\sqrt{\Psi})} \\ &+ \frac{\Gamma}{\pi}\int_{b_1}^{b_2}\frac{e^{s\bar{t}}}{s}\frac{J_0(\bar{r}\beta(s))Y_0(\beta(s)) - Y_0(\bar{r}\beta(s))J_0(\beta(s))}{J_0^2(\beta(s)) + Y_0^2(\beta(s))}ds \end{aligned} \tag{43}$$

$$- \int_{1}^{\infty} \frac{1}{2v} \left\{ \int_{0}^{\bar{t}} F_3(\bar{r}, w) \times \right.$$

$$\left. \times \left(\int_{b_1}^{b_2} e^{s(\bar{t}-w)} J_0(v\beta(s)) J_0(\beta(s)) q(\bar{t}) ds \right) dw \right\} dv$$

where $p(\bar{t}) = u(\bar{t} - v - \bar{r})$ and $u(\cdot)$ is the unit step function.

5 Conclusion & Discussion

It was shown that an analytical solution to the problem of heat dissipation in human tissue considering the thermal relaxation term is possible to be achieved, even if some assumptions about the theoretical model must be made at the beginning of the solving process. Knowing that said solution can be obtained not numerically but analytically hints at possible directions that future solutions may pursue when trying to find expressions to define the systems they are describing.

An important aspect related to the usage of this solution is its utility aiding the refinement of the radio-frequency ablation with internally cooled needle-like electrode process, with more understanding of the phenomena underlying the procedure. This way, technicians have more tools while designing the components and ease to determine the effects of their creation in the human body, which ultimately is the most crucial aspect regarding a medical procedure.

Finally, we note that important aspects related to the analytical solution presented were not treated in this paper and will be carried out as a future work, such as: (a) plots to elucidate the behavior of the calculated solution; (b) an analysis of the thermal relaxation parameter τ in human tissue; (c) comparisons between the presented model solution with *ex vivo* experiments and other models; and (d) the detailed proof why the line integrals $\int_{C_\rho}$, $\int_{c_m}$ and $\int_{c_n}$ vanishes.

Acknowledgments. The authors would like to thank the University of Brasília for the financial support.

References

1. Molina, J.A., Rivera, M.J., Berjano, E.: Analytical model based on a cylindrical geometry to study RF ablation with needle-like internally cooled electrode. Math. Probl. Eng. (2012). https://doi.org/10.1155/2012/834807
2. Baek, J.H., Moon, W.J., Kim, Y.S., et al.: Radiofrequency ablation for the treatment of autonomously functioning thyroid nodules. World J. Surg. **33**, 1971–1977 (2009). https://doi.org/10.1007/s00268-009-0130-3
3. Berjano, E.J.: Theoretical modeling for radiofrequency ablation: state-of-the-art and challenges for the future. BioMed. Eng. OnLine. **5** (2006). https://doi.org/10.1186/1475-925X-5-24

4. Burdío, F., Berjano, E.J., Navarro, A., et al.: RF tumor ablation with internally cooled electrodes and saline infusion: what is the optimal location of the saline infusion?. BioMed. Eng. OnLine. **6** (2017). https://doi.org/10.1186/1475-925X-6-30
5. Haemmerich, D., Chachati, L., Wright, A.S., Mahvi, D.M., Lee, F.T., Webster, J.G.: Hepatic radiofrequency ablation with internally cooled probes: effect of coolant temperature on lesion size. IEEE Trans. Biomed. Eng. **50**, 493–500 (2003). https://doi.org/10.1109/TBME.2003.809488
6. Askarizadeh, H., Ahmadikia, H.: Analytical study on the transient heating of a two-dimensional skin tissue using parabolic and hyperbolic bioheat transfer equations. Appl. Math. Model. **39**, 3704–3720 (2015). https://doi.org/10.1016/j.apm.2014.12.003
7. Choi, J. H., Yoon, S-H., Park, S.G., Choi, S-H.: Analytical solution of the Cattaneo-Vernotte equation (non-Fourier heat conduction). J. Korean Soc. of Marine Eng. **40**, 389–396 (2016). https://doi.org/10.5916/jkosme.2016.40.5.389
8. Lin, S-M., Li, C-Y.: Analytical solutions of non-Fourier bio-heat conductions for skin subjected to pulsed laser heating. Int. J. Therm. Sci. **110**, 146–158 (2016). https://doi.org/10.1016/j.ijthermalsci.2016.06.034
9. Ciesielski, M., Duda, M., Mochnacki, B.: Comparison of bio-heat transfer numerical models based on the Pennes and Cattaneo-Vernotte equations. J. Appl. Math. Comput. Mech. **15**, 33–38 (2016) https://doi.org/10.17512/jamcm.2016.4.04
10. Soares, D. Jr., Pinto, L., Mansur, W.: Solution of hyperbolic bioheat models by a fully-adaptive truly-explicit time-marching method. Structures **22** (2022)
11. Olver, F.W.J., Lozier, D.W., Boisvert, R.F., Clark, C.W.: NIST Handbook of Mathematical Functions. Cambridge University Press (2010)
12. Audigier, C., Mansi, T., Delingette, H., Rapaka, S. et al.: Efficient lattice Boltzmann solver for patient-specific radiofrequency ablation of Hepatic Tumors. IEEE Trans. Med. Imaging. **34**, 1576–1589 (2015). https://doi.org/10.1109/TMI.2015.2406575

Mathematical Modeling and Computer Simulation of ATP Metabolism in the Excitation-Contraction Coupling Phenomenon in the Rat Ventricular Myocyte

G. M. S. Motta[1,2](✉), M. L. A. Camargo[1,2], J. V. O. Mello[1,2], D. G. Goroso[1,2], and R. R. Silva[1,2]

[1] Research and Technology Center, University of Mogi das Cruzes, Mogi das Cruzes, Brazil
gabriel.motta@hotmail.com

[2] LabNECC/Center for Biomedical Engineering, University of Campinas, Campinas, Brazil

Abstract. Several mathematical and computational models have already been developed to describe the complex process of cardiac excitation-contraction (EC) coupling. The model of Mullins and Bondarenko (2013) is cited as an example, which describes detailed mechanisms ranging from the function of individual ion channels to other important transporters, and satisfactorily simulates the characteristics of Action Potential (AP) and contraction in the mouse ventricular myocyte. However, the mathematical modeling did not include the equations of energy production and consumption used during the EC process, so it cannot be used to simulate possible problems in the production and consumption of ATP (Adenosine Triphosphate), the main source of energy for cells. In this context, the aim of the present paper is to present the implementation of the model describing ATP metabolism proposed by Cortassa et al. (2006) in the electromechanical model of Mullins and Bondarenko (2013). From this new mathematical model, and using the MATLAB platform, the EnergyECLab tool was developed. As an application of the tool, "in silico" experiments were performed to obtain the time course of ATP and phosphocreatine (CrP) concentration and, as well as, the inhibition of Ca^{2+}-ATPase (SERCA) by Adenosine Diphosphate (ADP). The computational tool (EnergyECLab) developed to solve the system of Coupled Differential Equations (CDE's), has a friendly graphic interface and several didactic resources. As a result of the second "in silico" experiment, inhibition of SERCA by the ADP mechanism produced an increase in contraction force (30.7%) when compared to the control.

Keywords: Mathematical modeling · Cardiac electrophysiology · Ventricular myocyte · ATP metabolism

1 Introduction

According to the World Health Organization (WHO), about 17.5 million people die each year from cardiovascular disease [1]. This is the number one cause of death worldwide. In this context, it is necessary to use mathematical models applied to computational

J. L. B. Marques et al. (Eds.): CLAIB 2022/CBEB 2022, IFMBE Proceedings 98, pp. 35–43, 2024.
https://doi.org/10.1007/978-3-031-49401-7_4

methods for the understanding of the complex phenomenon of excitation-contraction (EC) coupling of the heart cell, which for more than 50 years have been able to identify, on a cellular scale, the causes of cardiac arrhythmias [2].

There are several models divided into electrical [3] and mechanical [4], and also intersection models that are called electromechanical [5]. However, much of it disregards the modeling of Adenosine Triphosphate (ATP) formation and consumption in the cardiac cell, which directly influences the generation of contraction force and cardiac Action Potential (AP) [6].

ATP is produced in the mitochondria by a 3-step process: Glycolysis, Krebs Cycle, and Oxidative Phosphorylation [7] and is used in important cellular structures, such as in the Ca^{2+}-ATPase Sarcoplasmic Reticulum (SR); in the Sodium-Potassium pump located in the sarcolemma; and in the mechanical process of contraction and relaxation of the cardiac cell [8].

Therefore, this paper has the objective of presenting the coupling of the ATP modeling of Cortassa et al. (2006) to the electromechanical model of Mullins & Bondarenko (2013). For this, a computational tool (EnergyECLab) was developed to solve the CDE's and display the results of the *"in silico"* experiments graphically. As an application of EnergyECLab, the influence of the ATP mechanism on contraction force and cytosolic Ca^{2+} concentration ($[Ca^{2+}]_i$) was tested.

2 Methods

2.1 Developing the Mathematical Model

To obtain the electromechanical model that takes into account ATP consumption and production, bidirectional coupling of the energy model of Cortassa et al. (2006) [8] into the electromechanical model of Mullins & Bondarenko (2013) [9] was performed.

In Mullins & Bondarenko's (2013) model [9], which describes the electromechanical mathematical modeling for the mouse ventricular myocyte, the AP is given by Eq. (1):

$$\begin{aligned} dV/dt = & -1\big(I_{cal} + I_{p(Ca)} + I_{NaCa} + I_{Cab} + I_{Na} + I_{Nab} + I_{NaK} + K_{to,f} \\ & + I_{Kto,f} + I_{K1} + I_{Ks} + I_{Kur} + I_{Kss} + I_{Kr} + I_{Cl,Ca} - I_{stim}\big)/C_m \end{aligned} \quad (1)$$

where C_m is the membrane capacitance in $\mu F/cm^2$, I_{CaL}, $I_{p(Ca)}$, I_{NaCa}, I_{Cab}, $I_{Cl,Ca}$, I_{Na}, I_{Nab} and I_{NaK} the currents, respectively of Ca^{2+} L-Type, of sarcolemma Ca^{2+} pump, of the sodium (Na^+)/Ca^{2+} exchanger, Ca^{2+} background, Ca^{2+} activated by chlorine (Cl^-), Na^+ fast, Na^+ background and of Na^+/potassium (K^+) pump in pA/pF. Whereas, $I_{Kto,f}$, $I_{Kto,s}$, I_{K1}, I_{Ks}, I_{Kur}, I_{Kss} and I_{Kr} indicate the main K^+ currents and I_{stim} is the external stimulus in pA/pF.

The calculation of the contraction force is described by Eq. (2) based on the model by Rice et al. (1999) [4]:

$$\begin{aligned} F_{contr} = & \ \alpha.(P1 + N1 + 2.(P2) + 3.(P3)) \\ & /(P1_{max} + 2.(P2_{max}) + 3.(P3_{max})) \end{aligned} \quad (2)$$

The contraction force F_{contr} is equated based on two non-permissive states ($N0$ and $N1$) and four permissive states permissivos of Tropomyosin ($P0$, $P1$, $P2$ and $P3$). Already, the constant α is a model fit factor [4].

The model of Cortassa et al. (2006) [8] describes ATP production and consumption during the EC coupling process. Equation (3) provides the ATP concentration during the entire process.

$$\begin{aligned} d[ATP]i/dt &= V_{ant}.(V_{mito}/V_{myo}) - V_{CK}^{mito} - V_{AM} - (1/2.J_{up}) \\ &\quad - (I_{p(Ca)} + I_{NaK}).\,A_{cap}/\,V_{myo}.\,F) \end{aligned} \tag{3}$$

Being, V_{ANT} e V_{CK}^{mito} parameters for the cytosolic reaction rate, F the Faraday constant, V_{myo} and V_{mito} are volume constants of the cytosol and mitochondria, A_{cap} capacitive area of the cell surface; $I_{p(Ca)}$ is the sarcolemma Ca^{2+} pump; I_{NaK} the current of Na^+/K^+ pump and J_{up} the Ca^{2+} flux of SERCA from the SR.

2.2 Development of the Computational Tool

The computational tool developed from the mathematical model, was implemented using the MATLAB platform version R2018a. The method used for solving the system of CDE's was the 4th order Runge-Kutta's with variable integration step contained in the ode15s routine.

2.3 Simulations Protocols

The simulations used the following protocol: stimulation pulse of 20 pA/pF with a width of 1 ms applied at the instant of 1000 ms and with a frequency of 0.25 Hz. The simulation time was 20000 ms, and in each simulation the final steady- state interval is presented. The initial sarcomere and cell lengths are given by 2.1 μm and 100 μm, respectively. Using these protocols the following *"in silico"* experiments were performed:

1. Time course of [ATP] and [CrP] related to the EC process.
2. Influence of SERCA Inhibition by ADP on contraction force and $[Ca^{2+}]_i$, using the parameters described in Table 1.

The following variables were used for analysis: the diastole (Dia), peak (Peak), amplitude (Δ) peak time (t_{peak}) and half time decline ($t_{0.5}$) values of the contraction force and Ca^{2+} transient.

Values of the main parameters used to simulate ATP consumption during the EC process and SERCA inhibition are shown in Table 1.

3 Results

3.1 Mathematical Model

The new mathematical model consists of 51 CDE's that describe ATP production and consumption during the EC process [8]. Below we describe the main equations that were innovative in the Mullins & Bondarenko model, in addition the main parameters can be found in Tables 2, 3 and 4 (2013) [9]:

Table 1. Parameters used in the simulation of SERCA inhibition

Symbol	Control	Test	Unit	Description
$K_{m,up}^{ATP}$	0.01	0.01	mM	ATP half saturation constant
$K_{i,up}$	0.14	0.14	mM	ADP inhibition constant
$K_{i,up'}$	5.1	0.051	mM	ADP inhibition constant

ATP consumption by the sarcolemma Ca^{2+} pump (Eq. 4).

$$I_{pCa} = I_{pCamax}[(1 + K_{m1pCa}^{ATP})/[ATP]_i.(1 + [ADP]_i/K_{ipCa}^{ADP}))^{-1} + (1 + Km2pCaATP/[ATP]_i)^{-1}] \tag{4}$$

Table 2. Parameters used in ATP consumption by the sarcolemma Ca^{2+} pump

Symbol	Value	Unit	Description
$I_{pCa_{max}}$	0.575	μA/μF	Maximum Ca^{2+} pump current
K_m^{pCa}	$5x10^{-4}$	mM	ATP half saturation constant for Ca^{2+} pump
$K_{m1_{pCa}}^{ATP}$	0.012	mM	First half ATP saturation constant for Ca^{2+} pump
$K_{m2_{pCa}}^{ATP}$	0.23	mM	Second half ATP saturation constant for the Ca^{2+} pump
$K_{i_{pCa}}^{ADP}$	1.0	mM	ADP inhibition constant for Ca^{2+} pump

ATP consumption of Na^+/K^+ pump (Eq. 5).

$$f_{NaK}^{ATP} = \left(1 + K_{NaK}^{1,ATP}/[ATP]_i.\left(1 + [ADP]_i/K_{NaK}^{i,ADP}\right)\right)^{-1} \tag{5}$$

ATP consumption in the generation of contraction force (Eq. 6)

$$VAM = V_{AM}^{max}\,(f_{01}.[P_0] + f_{12}.[P_1] + f_{23}.[P_2]/f_{01} + f_{12} + f_{23}).(1 + K_{M,AM}^{ATP}/[ATP]_i.[1 + [ADP]_i/K_{i,AM}]) \tag{6}$$

3.2 EnergyEClab Tool

The EnergyECLab tool was developed so as to present a friendly and intuitive graphical interface. EnergyECLab is structured in a main panel (Fig. 1) with a top menu containing the options START, CONTROL vs TEST, VIEW GRAPH, RESOURCE, RETURN and EXIT (Fig. 2). By clicking on the START tab, the user has the option to change

Table 3. Parameters used in modeling ATP consumption by Na+/K+ pump

Symbol	Value	Control	Unit	Unit	Description
$\bar{I}_{NaK}$	3.147	0.575	μA/μF	μA/μF	Maximum Na^+/K^+ pump current
$K_{m,Nai}$	10	5×10^{-4}	mM	mM	Half saturation of Na^+ for the Na^+/K^+ pump
$K_{m,Ko}$	1.5	0.012	mM	mM	Half saturation of K^+ for the Na^+/K^+ pump
$K_{NaK}^{1,ATP}$	$8x10^{-3}$	0	mM	mM	ATP half saturation constant for the Na^+/K^+ pump
$K_{NaK}^{i,ADP}$	0.1	1.0	mM	mM	ADP inhibition constant for the Na^+/K^+ pump

Table 4. Parameters used in modeling ATP consumption for contraction force generation

Symbol	Value	Unit	Description
V_{AM}^{max}	$7.2x10^{-3}$	mM/ms	Maximum rate of ATP hydrolysis by myofibrils (AM ATPase)
$K_{M,AM}^{ATP}$	0.03	mM	AM ATPase half-saturation constant
$K_{i,AM}$	0.26	mM	ADP inhibition constant for AM ATPase

parameters related to ATP consumption. Then the simulation protocols such as pulse width, stimulation frequency, cell length, sarcomere length, stimulus current, and plotting interval are entered. The results of each simulation can be viewed via graphs (VIEW GRAPH) that indicate ATP, ADP, CrP and Ca^{2+} concentrations as well as the time course of the contraction force. In CONTROL VS TEST, you have the option to select two simulations for comparison. In RESOURCE you have access to educational content such as videos on the functioning of important cellular structures. Finally, the last tabs RETURN and EXIT allow you to return to the main tab or quit the tool, respectively.

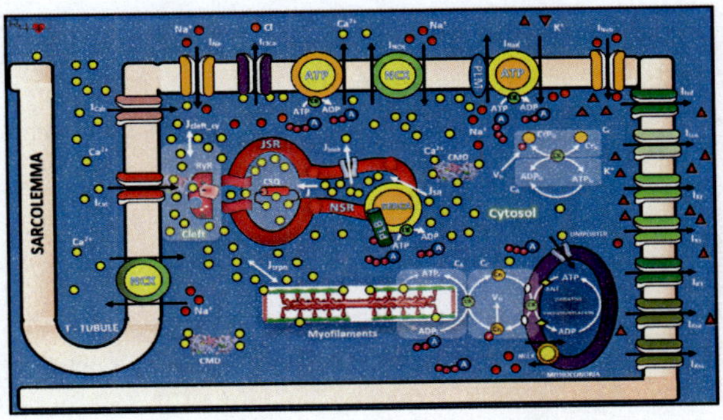

Fig. 1. EnergyECLab simulator main interface

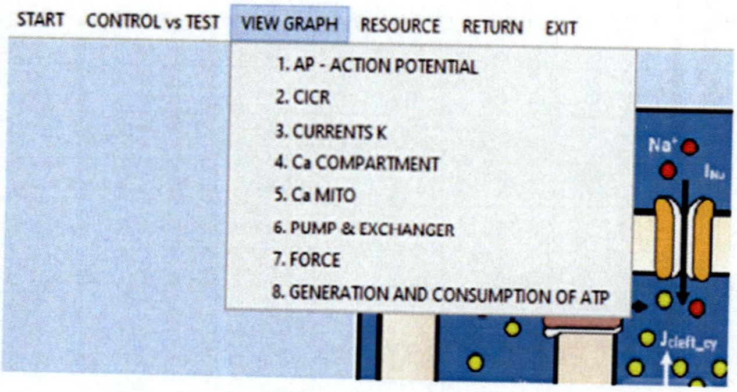

Fig. 2. EnergyECLab simulator options menu

3.3 In Silico Experiments

As an application of the tool the following simulations were performed:

"In silico" experiment 1 – The results of the simulations provide us as main output data: the time courses of ATP and CrP (Fig. 3).

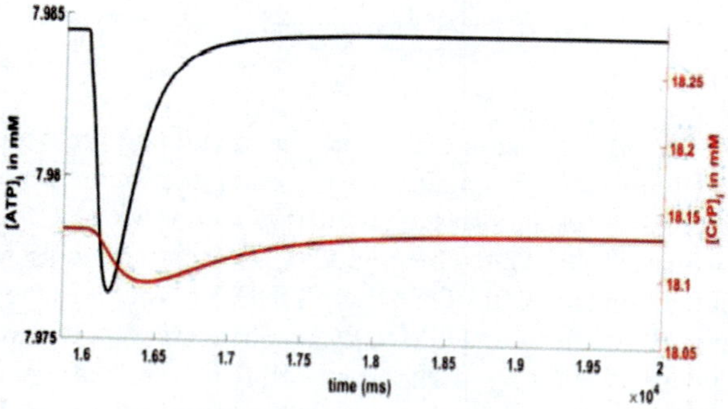

Fig. 3. Time course of ATP and CrP concentrations related to the EC process

"In silico" experiment 2 – Simulations whose output data are the graphs of the rate of Ca^{2+} uptake by the SR pump (Fig. 4) and the Ca^{2+} transient ($[Ca^{2+}]_i$) (Fig. 5) and contraction force (Fig. 6). Quantitative data regarding $[Ca^{2+}]_i$ and contraction force are shown in Tables 5 and 6, respectively.

4 Conclusions

The computational tool was able to satisfactorily reproduce the time course of ATP concentrations, $[Ca^{2+}]_i$ and contraction force, proving capable of integrating the models and enriching the investigation of ATP metabolism in the EC coupling phenomenon.

Further adjustments to the parameters will still be necessary, since the entire model must be adapted to mouse ventricular myocyte biophysical parameters. After the adjustments in these parameters it will also be possible to perform simulations involving the

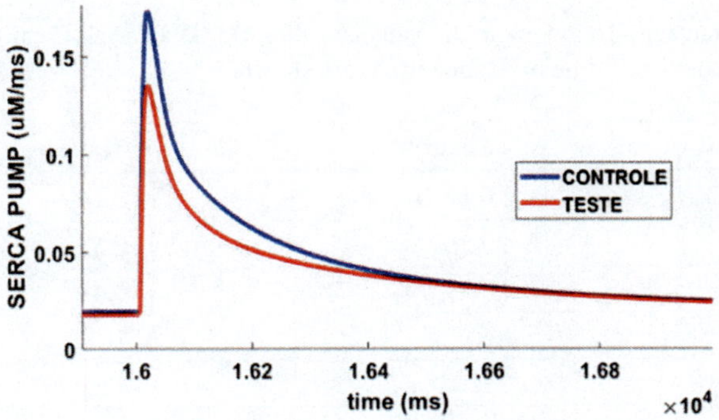

Fig. 4. Speed of Ca^{2+} uptake by SERCA: Control vs Test

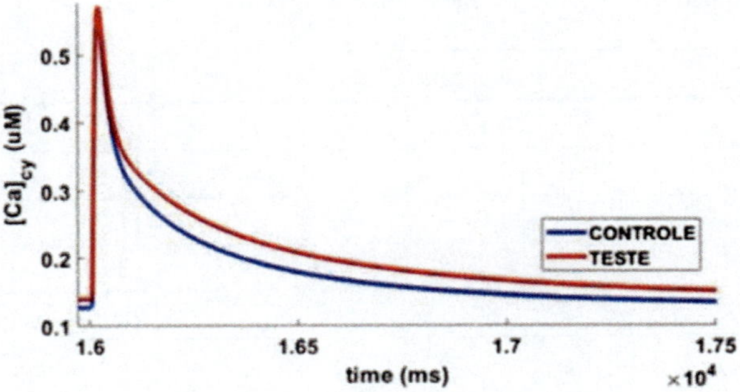

Fig. 5. Ca^{2+} transient: Control vs Test

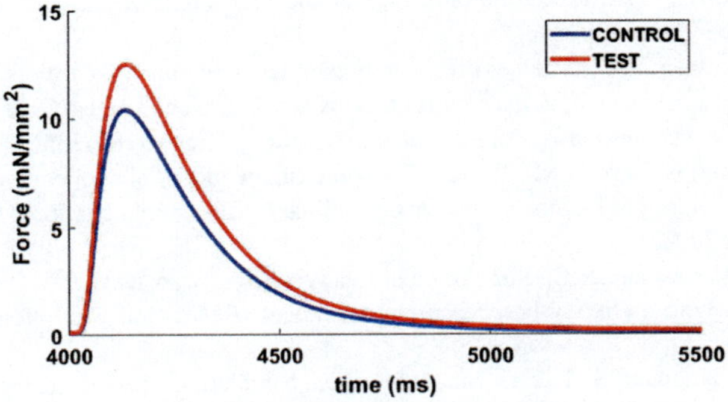

Fig. 6. Contraction force: Control vs Test

inhibition of the Na^{+}/K^{+} pump and its influence on the Ca^{2+} transient and time course of the contraction force. The EnergyECLab computational tool is available for download at the link: https://forms.gle/ENeRNMLMzDhQbMo78.

Table 5. Ca^{2+} transient. In this table the values of diastole (Dia), peak (Peak), amplitude (Δ), peak time (tpeak) and half time of decline (t0.5) are shown.

Symbol	Control	Test
Dia (μM)	0.13	0.14
Peak (μM)	0.56	0.57
Δ (μM)	0.43	0.43
t_{peak} (ms)	20	20
$t_{0.5}$ (ms)	120	180

Table 6. Contraction force. In this table the values of diastole (Dia), peak (Peak), amplitude (Δ), peak time (tpeak) and half time of decline (t0.5) are shown.

Variables	Control	Test
Dia (μM)	0.13	0.13
Peak (μM)	9.1	11.86
Δ (μM)	8.97	11.73
t_{peak} (ms)	130	130
$t_{0.5}$ (ms)	180	220

References

1. WHO at https://www.who.int/news-room/fact-sheets/detail/cardiovascular-diseases-(cvds). Last accessed 22 July 2022
2. Sutanto, H., Heijman, J.: Integrative computational modeling of cardiomyocyte calcium handling and cardiac arrhythmias: current status and future challenges. Cells **5**, 1–14 (2022)
3. Huxley, H.E.: The mechanism of muscular contraction. Science **164**, 1356–1365 (1969)
4. Rice, J., Wang, F., Bers, D.M., Tomb, P.P.: Approximate model of cooperative activation and crossbridge cycling in cardiac muscle using ordinary differential equations. Biophysical **95**, 2368–2390 (2008)
5. Negroni, J.A., Lascano, E.C.: Simulation of steady state and transient cardiac muscle response experiments with a Huxley-based contraction model. J Mol Cell Cardiology **45**, 300–312 (2008)
6. Satoshi, M., Nobuaki, S., Hikari, Jo., Akinori, N.: Simulation of ATP metabolism in cardiac excitation—contraction coupling. Prog. Biophys. Mol. Biol. **85**, 279–299 (2004)
7. Bonora, M., Patergnani, S., Rimessi, A., et al.: ATP synthesis and storage. Purinergic Signal **208**, 343–357 (2012)
8. Cortassa, S., Miguel, A., Brian O'Rourke, Robert, J., Hsiangler, T., Marbá, E., Raimond, L.: An integrated model of cardiac mitochondrial energy metabolism and calcium dynamics. Biophysical **84**, 2734–2755 (2006)
9. Bondarenko, V.E., Mullins, P.D.: A mathematical model of the mouse ventricular myocyte contraction. Pios One. N. **8**(5), 1–15 (2013)

10. Sakamoto, J., Tonomura, Y.: Order of release of ADP and Pi from phosphoenzyme with bound ADP of Ca^{2+}-dependent ATPase from sarcoplasmic reticulum and of Na^+, K^+-dependent ATPase studied by ADP-inhibition patterns. J. Biochem. **87**(6), 1721–1727 (1980)
11. Silva, R.R., Bissaco, M.A., Goroso, D.G.: MioLab, a rat cardiac contractile force simulator: applications to teaching cardiac cell physiology and biophysics. Comput. Methods Programs Biomed. **122**, 480–490 (2015)
12. Silva, R.R., Goroso, D.G., Bers, D.M., Puglisi, J.L.: MarkoLAB: a simulator to study ionic channel's stochastic behavior. Comput. Biol. Med. **87**, 258–270 (2017)
13. Silva, R.R., Filho, O.B., Bassani, J.W., Bassani, R.A.: The ForceLAB simulator: Application to the comparison of current models of cardiomyocyte contraction. Comput. Biol. Med. **131**, 1–10 (2021)

Non-invasive Mechanism Location of Atrial Arrythmias with Optical Flow from Electrocardiographic Imaging

Italo Sandoval[1](✉), Gabriel V. Costa[1], Camila R. Restivo[1], Maria S. Guillem[2], John A. Sims[1], and João Salinet[1]

[1] Center for Engineering, Modeling and Applied Social Sciences (CECS), HEartLab - Biomedical Engineering, Federal University of ABC, São Bernardo do Campo, Brazil
italo.sandoval@ufabc.edu.br

[2] ITACA Institute, Universitat Politècnica de València, Valencia, Spain

Abstract. The location of driver mechanisms in supra-ventricular tachyarrhythmias can be difficult to perform by human observers or in an automated manner, mainly when it comes to non-invasive methods. In this study, we investigated an automated methodology to locate two of the most common: Atrial Fibrillation (AF), driven by functional rotors; and atrial tachycardia (AT), driven by ectopic foci. In the proposed study, we used Farnebäck optical flow and calculated the divergence and curl of the generated vector field. First we validated the algorithm with signals from mathematical phantoms and then applied the method to AF and AT realistic in-silico models, comparing the results from the original simulated signals and the respective electrocardiographic imaging (ECGi), that is, after the originals went through forward solution followed by inverse solution. Results show that the present method is robust to noise, and outcomes are dependent of the predefined spatial resolution. In addition, the approach was able to locate both AT and AF mechanisms on the original atrial epicardial maps, and AF mechanism on the estimated atrial maps through the ECGi. However, AT mechanism location accuracy seems to be partially degraded by the ECGi.

Keywords: Atrial Arrhythmia · Optical Flow · Electrocardiographic Imaging · Curl · Divergent

1 Introduction

Atrial fibrillation (AF) and atrial tachycardia (AT) are among the most common supra-ventricular tachyarrhythmias [1]. These arrhythmias might differ in their underlying mechanisms: in AT, ectopic focus (EF) fires radially, generated by a set of cells in higher activation frequencies [1]; for AF, uncoordinated high frequency atrial activation is present, with consequent deterioration of mechanical function and, for a certain group of patients, the presence of functional rotors (FR) [2].

J. L. B. Marques et al. (Eds.): CLAIB 2022/CBEB 2022, IFMBE Proceedings 98, pp. 44–53, 2024.
https://doi.org/10.1007/978-3-031-49401-7_5

The most effective form of treatment for these arrhythmias are radiofrequency catheter ablation, which requires precise location of the mechanisms and is performed invasively, which increases the duration of the surgical procedure [3]. Therefore, we propose the development of methods, capable of locating mechanisms in a reliable and non-invasive manner. These could be used by professionals prior to ablation, enabling a reduction of intervention time.

1.1 Electrocardiographic Imaging

Electrocardiographic imaging (ECGi) is a diagnostic tool for reconstructing cardiac electrical activity from non-invasive signals [4,5]. ECGi depends on body surface potential mapping (BSPM), together with torso and heart geometry, as well as an implementation of a numerical solution for an inverse problem [4,5]. The forward problem of electrocardiography refers to the estimation of the body surface electrical potentials due to cardiac sources using theoretical equations of electromagnetism [6], allowing the estimation of BSPM signals using heart models. The forward problem can be solved by calculus based on the Laplace equation (Eq. 1) where Φ is the electric potential field, which is propagated to the body surface in passive conduction [7], and ∇^2 is the second order gradient (or Laplacian operator). In this approach (Eq. 2), the body potentials (Φ_T) are calculated by solving the Eq. 1 inside the body volume, by using the epicardium potentials (Φ_C) and a geometric relation between both (T = Torso, C = Epicardium) geometries, which generates a transference matrix A_{TC} [7].

$$\nabla^2\Phi = 0 \tag{1}$$

$$\Phi_T = A_{TC}\Phi_C \tag{2}$$

The inverse problem, on the other hand, consists of predicting the cause from its effects. In the field of electrocardiography, is the reconstruction of the epicardial potentials from the body surface potentials [8] [9], that is, reconstructing the epicardial potentials from the BSPM. Inverse problems are often ill-posed, since they do not have a unique solution, which implies that some method must be used to stabilize the solution. This stabilization can be performed by regularization methods [10]. The inverse solution is also based on Eq. 1 and is shown in Eq. 3, where ${A_{TC}}^+$ is the pseudo-inverse of A_{TC}, regularized by the Tikhonov method [11].

$$\Phi_C = {A_{TC}}^+\Phi_T \tag{3}$$

1.2 Farnebäck Optical Flow

Detection of FR and EF activity can be performed using many proposed methods, by using techniques based on phase analysis, local activation time and also motion detection algorithm. Optical flow is the distribution of apparent velocities of motion of brightness patterns in images [12]. Farnebäck Optical Flow (FOF) is a two-frame motion detection algorithm, based on a polynomial expansion transform [13]. The algorithm requires pairs of consecutive frames and results in a 2D motion vector field (MVF) that is a dense estimate of displacement, meaning that it is calculated for every pixel in the image. Here FOF algorithm is applied for characterize FR and EF activity, associated with MVF operations. Dense algorithms associated with MVF decomposition are known to have medical imaging applications, mainly in temporal evaluation of cardiac imaging [14][15]. In this work, we propose a methodology for detecting arrhythmic mechanisms such as EF and FR by using FOF and MVF analysis in simulated ECGi signals, initially validated with mathematical phantoms, and then tested in realistic computational models.

2 Methods

This study is composed of two steps: first we validate the proof of concept of our approach with mathematical phantoms. After validation, the analysis is extended to realistic models of the atria in AT and AF, comparing the results from the atrial eletrogram (AEGM) with the results obtained from its respective reconstructed AEGM (rAEGM, Fig. 1).

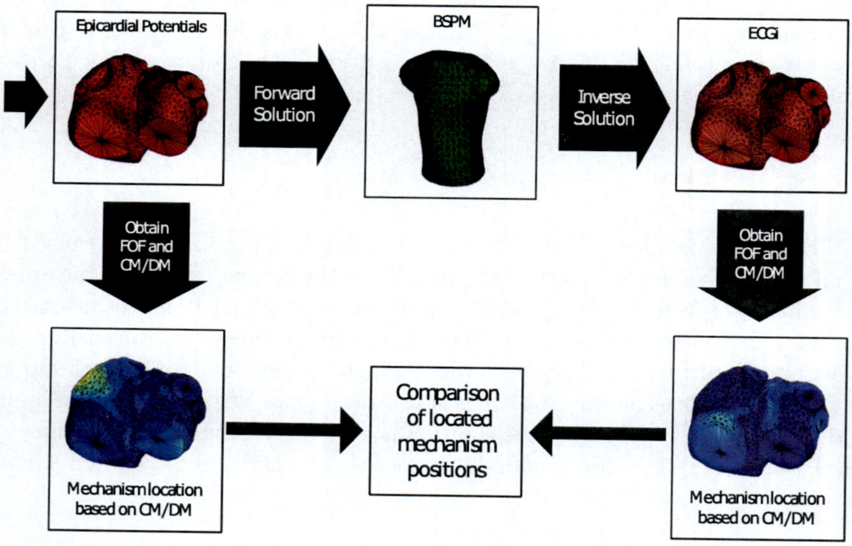

Fig. 1. Flowchart for the mechanism location method on ECGi. CM: curl map; DM: divergent map

2.1 Validation

The algorithm is initially tested with 12 mathematical phantoms (128x128 grid with 2000 time samples at $500Hz$), which were used for parameter tuning of the algorithm and assessing its noise robustness. The phantoms simulated EF (4 phantoms), FR (4 phantoms) and superposition of both mechanisms (4 phantoms).

The EF phantoms were created by using 4 steps: (1) the position of the mechanism center was determined; (2) the distance from each pixel to the center was calculated, then multiplied by a factor related to the spatial propagation of the signals, creating a phase map; (3) a temporal increment, related to the sampling frequency (fs) and the mechanism's underlying activation frequency (F), was added to the values for each time instant; and (4) the values were used in a reverse sawtooth function truncated to the 4th order of its respective Fourier series. Equation 4 summarizes the EF phantom (P_{EF}) creation.

$$P_{EF}(i,j,t) = SAW(A\sqrt{(i-i_0)^2 + (j-j_0)^2} + I(t)) \tag{4}$$

$$I(t) = \frac{t}{2fsF\pi} \tag{5}$$

$$SAW(x) = -\sum_{n=1}^{4} \frac{1}{n} sin(xn) \tag{6}$$

where A is a factor related to the conduction velocity, $I(t)$ is the temporal increment, (i_0, j_0) is the position of the mechanism and (i, j) is the position of the considered pixel.

The FR phantom was created as follow: (1) selecting two positions for the mechanism: one clockwise (CW) and another counter-clockwise (CCW); (2) separately, two maps were created by calculating the relative angle from each pixel to the mechanism position, creating a phase map; (3) the angle value referring to the relative CW and CCW mechanism position was subtracted from the CW map and added to the CCW map, aligning the maps; (4) the maps were then combined and, as in the EF phantom, a temporal increment, related to the sampling frequency and mechanism's underlying activation frequency, is added to the values; and (5) the values were used in a reverse sawtooth function. Equation 7 summarizes the creation of the rotor phantom (P_R).

$$P_R(i,j,t) = SAW(atan2(j-j_0, i-i_0) + I(t)) \tag{7}$$

Here, $atan2(a,b)$ is a function similar to $tan^{-1}(a/b)$, but returns values from $-\pi$ to $+\pi$ instead of $-\pi/2$ to $+\pi/2$, by considering the sign of a and b, and thereby returning different values for each of the four quadrants.

2.2 Realistic In-Silico Models

After validation in phantoms, the study proceeds by applying the proposed method on realistic AT and AF mathematical models of the atria. The models are composed of 284,578 nodes and 1,353,783 tetrahedrons, and simulate the behavior and geometry of the atria in arrhythmic conditions for both EF and FR mechanism (Table 1)[16].

Table 1. Characteristics of simulations

Simulation	Mechanism	Location
1	EF (AT)	Inferior vena cava
2	EF (AT)	Left superior pulmonary vein
3	EF (AT)	Right atrial appendage
4	EF (AT)	Right inferior pulmonary vein
5	Rotor (AF)	Right atrial appendage
6	Rotor (AF)	Right inferior pulmonary vein

From all simulations in Table 1, the atria geometry was simplified to a uniform triangular mesh, with 2048 nodes and its respective unipolar AEGM ($fs = 500$) was obtained. The AEGM is used in a forward solution for obtaining the BSPM signals referring to a torso mesh with 771 nodes. The BSPM signals are then used for obtaining the rAEGM, by using the inverse solution and Tikhonov regularization method [4].

For both rAEGM and AEGM, which are three-dimensional signals, a software was developed, which allows further bi-dimensional analysis such as FOF and MVF operations. In this process, for each node, a 35x35 grid is created with the interpolated signals from the node's neighbors. More specifically, for each node, its neighbors, a partial mesh, are projected into a plane, then an optimization based on the original mesh was made, reducing the spatial deformation originated by the projection. The signals from the projected partial mesh are then interpolated into the 35×35 grid with cubic splines, generating a two-dimensional signal for each of the 2048 nodes, here called partial map. In this stage, the partial map can be used in two-dimensional operations, such as convolution. After the two-dimensional analysis, the value for the central pixel of the new grid, generated after the 2D process, is attributed to the respective node, generating a new three-dimensional map.

2.3 Analysis

The FOF algorithm, available in the OpenCV library (v.4.5.4) for Python 3.8, is applied to consecutive time frames, obtaining the MVF for each time sample. MVF vectors are normalized and the temporal average calculated (mMVF, Fig. 2A, left). The curl and divergent potential maps are obtained using the following steps: (1) an x and an y oriented Sobel filter with size of 5×5 pixels is applied to each of the x and y components of the nMVF, respectively $nMVF_x$ and $nMVF_y$, to provide the maps for $\frac{\partial(nMVF_x)}{\partial x}$, $\frac{\partial(nMVF_y)}{\partial x}$, $\frac{\partial(nMVF_x)}{\partial y}$, $\frac{\partial(nMVF_y)}{\partial y}$. Then curl and divergent maps were obtained using Eqs. 8 and 9:

$$DivergentMap = \frac{\partial(nMVF_x)}{\partial x} + \frac{\partial(nMVF_y)}{\partial y} \tag{8}$$

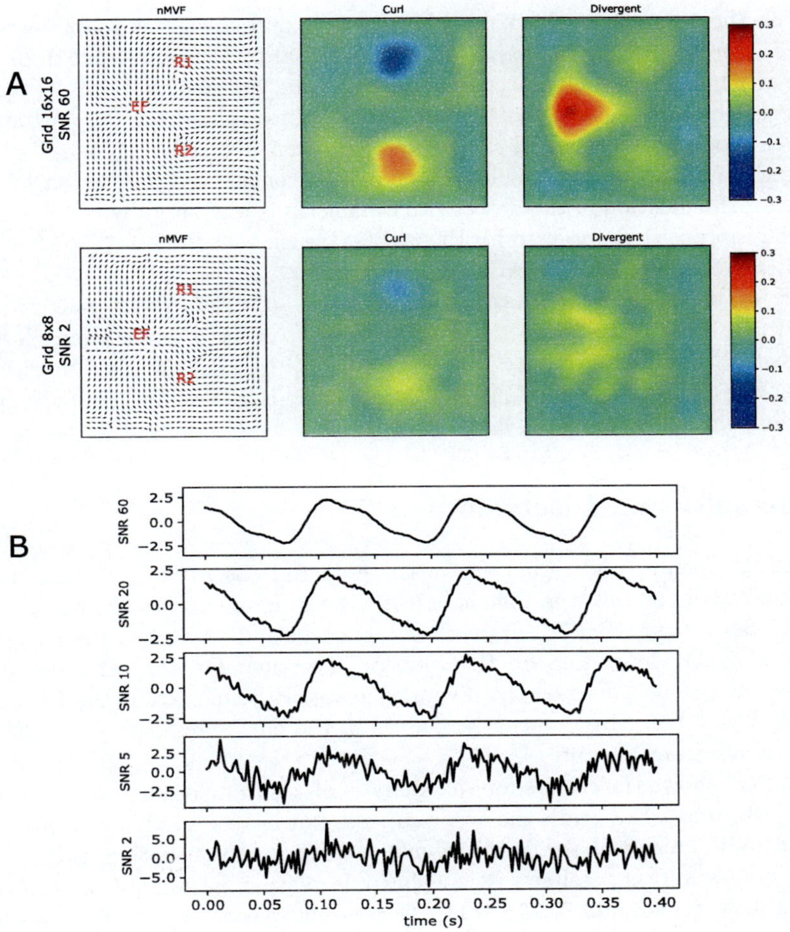

Fig. 2. Evaluation of noise robustness. A: signal for a FR phantom for different levels of SNR; B: Divergence and curl of the nMVF for a phantom of combined mechanisms, for two extremes of grid resolution and SNR, with the position of the mechanisms highlighted. R1: Clockwise rotor; R2: Counter-clockwise rotor; EF: Ectopic focus

$$CurlMap = \frac{\partial(nMVF_y)}{\partial x} - \frac{\partial(nMVF_x)}{\partial y} \tag{9}$$

For the phantoms, regions of interest were defined by applying a threshold to the curl and divergent maps, and the center of the regions was considered to be the location of the detected mechanisms, being FR the mechanism detected on the curl maps and EF on the divergent maps. The accuracy of detected mechanisms was measured considering as true positive a detection within a distance below 5% of the map size from the correct mechanism position.

For the realistic in-silico models, the curl and divergent maps were first obtained from the partial maps. Then the central pixel values from these maps were attributed to their respective nodes, allowing the generation of 3D curl and divergent maps. The node with maximum absolute value from the curl map, in the AF models, is considered as detected FR, then, similarly for the AT models, the maximum value from the divergent map is considered as detected EF. The error for the location of the detected mechanisms is measured by comparing the mechanism position obtained for the rAEGM, each with different signal to noise ratio (SNR) value, and the AEGM.

To evaluate the robustness of the proposed approach, the signals from the phantoms are sub-sampled in grids from 8x8 to 16x16, simulating electrode acquisition, and noise with a SNR from 60 to 2 (Fig. 2B) is applied to these signals. Noise is also applied to the BSPM simulations, before applying the inverse solution, with SNR of 60, 30 and 10.

3 Results and Discussion

Figure 2A (center and right) shows the curl and the divergent maps for the phantom with combined mechanisms with two different values of SNR and grid size. It can be seen that FR appears as higher absolute values in the curl maps, with its signal depending on the rotation direction. On the other hand, EF appears as higher values on the divergent maps. Physiologically, the FR can be CW or CCW (positive or negative values in the curl maps), while the EF can only radiate from a point.

Table 2 shows the results for the analysis of accuracy on phantoms. In some cases (10x10 and 8x8 grid), the accuracy does not have a maximum value in the higher SNR, which was verified to be due to false positives that disappear as the noise increases. It is possible to see a sharp decrease in accuracy from the 10x10 (0.75 ± 0.19) to 8x8 grid (0.32 ± 0.11). Decreasing spatial resolution causes maps to become fuzzier, limiting the location of mechanisms to greater than 5% error. Therefore, the results show reasonable noise robustness.

Table 2. Location accuracy of the algorithm for different levels of SNR and acquisition grids

	Grid			
SNR	16×16	12×12	10×10	8×8
60	1.000	0.878	0.750	0.350
20	1.000	0.878	0.806	0.354
10	1.000	0.867	0.879	0.361
5	1.000	0.750	0.889	0.403
2	0.917	0.625	0.417	0.125

The algorithm validation with the phantom showed good results, given an accuracy above 0.75 for the location of both mechanisms and great noise robustness, at least up to SNR of 5 and 10×10 grid. However, results were dependent on the spatial resolution of the maps, with poor accuracy for lower resolution grid (8×8).

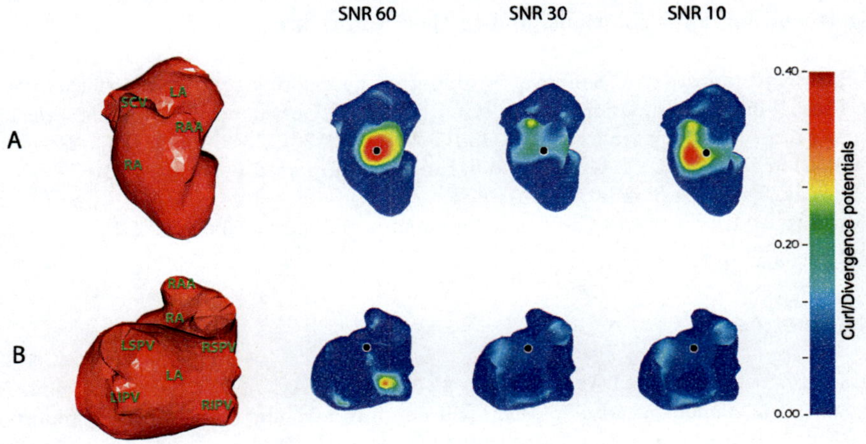

Fig. 3. Results for the analysis in realistic in-silico models. A: AF model and the 3D curl map for three different SNRs on rAEGM signals, with the correct position of the FR shown by the black dots; B: AT model and the 3D divergent map for three different SNRs on rAEGM signals with the correct position of the EF shown by the black dots. RA: right atrium; LA: left atrium; SCV: superior cava vein; RAA: right atrial appendage; LSPV: left superior pulmonary vein (PV); RSPV: right superior PV; LIPV: left inferior PV; RIPV: right inferior PV

Figure 3 shows the obtained maps for an AT and an AF model. For the AEGM, where the signals have a higher level of details, all detected mechanisms were located in the predetermined regions (Table 1). When comparing the located mechanism position for AEGM and rAEGM, the mean normalized error ($\pm$ standard deviation) for the rotor location was: 0.12 ± 0.11 for the SNR 60, 0.10 ± 0.01 for SNR 30 and 0.06 ± 0.02 for SNR 10. Despite of this, the error for the EF location was: 0.41 ± 0.14 for SNR 60, 0.50 ± 0.28 for SNR 30 and 0.55 ± 0.15 for SNR 10. These values show a clear difference between the error for FR and EF location in the rAEGM signals, being significantly higher for the EF. This difference could be due to a lack of robustness of the mechanism location algorithm, but perhaps due to a loss of the original signal characteristics in the computation of the forward and inverse solution. Further studies are desired to verify this hypothesis, and if confirmed, how the process affects the EF and how to adapt the proposed method.

4 Conclusion

This study proposes an approach for non-invasive location of two of the principal mechanism of atrial arrhythmias by using FOF and MVF analysis in rAEGM maps. The analysis shows that the proposed method could locate both EF and FR mechanisms presented on the AEGM maps, and FR on the rAEGM maps. Ectopic activity location accuracy seems to suffer with the forward and inverse electrocardiography solutions, and further studies are desired.

Acknowledgements. I. Sandoval is supported by Coordenação de Aperfeiçoamento de Pessoal de Nível Superior - Brasil (CAPES). G.V. Costa and C.R. Restivo are supported respectively by grants #2020/13017-2 and #2020/12841-3, São Paulo Research foundation (FAPESP), J. Salinet and J. Sims are supported by grant #2018/25606-2, São Paulo Research foundation (FAPESP).
Conflict of Interest MS Guillem is co-founder and shareholder of CORIFY CARE SL.

References

1. Page, R.L., Joglar, J.A., Caldwell, M.A., et al.: ACC/AHA/HRS guideline for the management of adult patients with supraventricular tachycardia: a report of the American College of Cardiology/American Heart Association Task Force on Clinical Practice Guidelines and the Heart Rhythm Society. J. Am. Coll. Cardiol. **67**, e27–e115 (2016). https://doi.org/10.1161/cir.0000000000000310
2. Hindricks, G., Potpara, T., Dagres, N., et al.: ESC Guidelines for the diagnosis and management of atrial fibrillation developed in collaboration with the European Association for Cardio-Thoracic Surgery (EACTS). Eur. Heart J. **42** (2021). https://doi.org/10.1093/eurheartj/ehaa612
3. Issa, Z.F., Miller, J.M., Zipes, D.P.: Clinical Arrhythmology and Electrophysiology: A Companion to Braunwald's Heart Disease. Elsevier Health Sciences (2018)
4. Salinet, J., Molero, R., Schlindwein, F.S., et al.: Electrocardiographic imaging for atrial fibrillation: a perspective from computer models and animal experiments to clinical value. Front. Physiol. **12**(2021). https://doi.org/10.3389/fphys.2021.653013
5. Coll-Font, J., Dhamala, J., Potyagaylo, D., et al.: The consortium for electrocardiographic imaging. Comput. Cardiol. **43** (2016)
6. Farina, D.: Forward and inverse problems of electrocardiography: clinical investigations. Ph.D. thesis, Universität Karlsruhe (2008)
7. Rudy, Y.: Noninvasive electrocardiographic imaging of arrhythmogenic substrates in humans. Circ. Res. **112**, 863–874 (2013). https://doi.org/10.1161/circresaha.112.279315
8. Brooks, D.H., MacLeod, R.S.: Electrical imaging of the heart. IEEE Signal Process. Mag. **14**, 24–42 (1997). https://doi.org/10.1109/79.560322
9. Gulrajani, R., Roberge, R., Savard, P.: The inverse problem of electrocardiography. Crit. Rev. Biomed. Eng. **16**, 171–214 (1989)
10. Hansen, C.: Rank-deficient and discrete ill-posed problems: numerical aspects of linear inversion. In: SIAM Monographs on Mathematical Modeling and Computation Society for Industrial and Applied Mathematics (1987)
11. Tikhonov, A.N.: On the solution of ill-posed problems and the method of regularization. Dokl. Akad. Nauk SSSR **151**, 501–504 (1963)

12. Horn, B.K.P., Schunck, B.G.: Determining optical flow. Artificial Inteligence, vol. 17, pp. 185–203 (1981) https://doi.org/10.1016/0004-3702(81)90024-2
13. Farnebäck, G.: Two-frame motion estimation based on polynomial expansion. In: Bigun, J., Gustavsson, T. (eds.) Image Analysis, pp. 363–370 (2003)
14. Guo, Q., Mandal, M.K., Liu, G., Kavanagh, K.M.: Cardiac video analysis using Hodge-Helmholtz field decomposition. Comput. Biol. Med. **36**, 1–20 (2006). https://doi.org/10.1016/j.compbiomed.2004.06.011
15. Sims, J.A., Giorgi, M.C., Oliveira, M.A., Meneghetti, J.C., Gutierrez, M.A.: Directional analysis of cardiac motion field from gated fluorodeoxyglucose PET images using the Discrete Helmholtz Hodge Decomposition. Comput. Med. Imaging Graph. **65**, 69–78 (2018). https://doi.org/10.1016/j.compmedimag.2017.06.004
16. Rodrigo, M., Climent, A.M., Liberos, A., et al.: Technical considerations on phase mapping for identification of atrial reentrant activity in direct-and inverse-computed electrograms. Circ: Arrhythmia Electrophysiol. **10**, e005008 (2017) https://doi.org/10.1161/circep.117.005008

Integrated Numerical Model for Hemodynamic Simulation of Human Atria Under Atrial Fibrillation by Coupling 3D Electrophysiological Activity

João Lameu[1(✉)], Jimena G. Siles[1,2], Italo Sandoval[1], Melissa McInnis[1], and João Salinet[1,2]

[1] Center for Engineering, Modeling and Applied Social Sciences (CECS), HEartLab - Biomedical Engineering, Federal University of ABC, São Bernardo do Campo, Brazil
joao.lameu@ufabc.edu.br

[2] Graduate Program in Biotechnoscience, Federal University of ABC, São Paulo, Brazil
https://heartlab.pesquisa.ufabc.edu.br

Abstract. Atrial Fibrillation (AF) is the most common cardiac arrhythmia, increasing the chance of ischemic events five-fold. AF is characterized by fast and abnormal electrical activity, leading to disturbance in atrial contraction, resulting in pro-thrombotic zones due to blood stagnation. Literature shows that hemodynamic modeling using Computational Fluid Dynamics (CFD) has been used to evaluate the hemodynamic outcomes of atria chambers under AF; however, with simplifications related to the electromechanical effects of AF. This study aims to present an integrated analysis by coupling electrophysiological information to a hemodynamic model. A patient undergoing an electrophysiology study was enrolled. Firstly, the patient underwent Magnetic Resonance Imaging (MRI) scan, followed by an electrophysiology study. From MRI images, atrial shell was segmented through a semiautomatic image processing pipeline. Then, multi-electrode catheters were used to acquire electrophysiology signals, allowing determination of local activation times in the 3D atria shell. The 3D mapping provided AF activation patterns used as boundary conditions in the hemodynamic model. The mechanical effects from AF activity were coupled into the blood flow using a momentum method, allowing the local oscillatory wall motion to be mimicked due to the AF activity and its propagation effects through the blood flow. Statistical analysis of fundamental hemodynamic field properties (pressure and wall shear stress, WSS) exhibited a higher dispersion when considering the momentum method. The hemodynamic field presented larger pro-thrombotic zones (WSS ≤ 0.1 Pa) when considering the wall effects of AF activity

Keywords: cardiac arrhythmia · thrombogenesis · electrophysiology · atrial fibrillation · computational fluid dynamics

J. L. B. Marques et al. (Eds.): CLAIB 2022/CBEB 2022, IFMBE Proceedings 98, pp. 54–64, 2024.
https://doi.org/10.1007/978-3-031-49401-7_6

1 Introduction

Atrial Fibrillation (AF) is the most common cardiac arrhythmia worldwide [1], being the main cause of ischemic events, myocardium infarction and venous thromboembolism [2]. It is characterized by fast and abnormal electrical activity in the atria, contributing to deteriorating mechanical function [3]. Different from other cardiac arrhythmias, AF presents irregular rhythms without clear repetition patterns of activation [3]. AF leads to disturbance of atrial contraction, reducing the total emptying blood volume to the ventricles. Also, persistent AF can result in an increase of atrial wall rigidity, resulting in pro-thrombotic zones due to blood stagnation [4], that is directly related to flow field indicators, such as wall shear stress (WSS). Low values of WSS are related to pro-thrombotic zones, indicating blood stagnation. An inferior threshold for thrombus deposition of WSS $\leq$ 0.1 Pa was reported [5]. In contrast, WSS $\geq$ 0.3 Pa were associated with thrombus destruction [5].

Hemodynamic modeling by Computational Fluid Dynamics (CFD) has been used to evaluate AF outcomes, however, with simplifications related to electromechanical effects of AF in the blood flow patterns, e.g. use of rigid walls [4,6]. The intrinsic nature of AF with abnormal activations in a relative high frequency (between 3 to 10 Hz), when compared to other atrial arrhythmias, is a major challenge in atrial blood flow modeling, since, until now, medical image techniques were not available with sufficient spatial-temporal resolution to provide dynamic details of oscillatory wall motion [7].

Few numerical studies studied the high frequency AF effects in wall motion [6,8]. Koizumi et al. [8] considered the AF by proposing a modified model with different left atrial (LA) volumes and absence of atrial contraction waves in the transmitral flow. Masci et al. [6] accounted the AF effects by a sinusoidal function modulated to AF frequency of 4 Hz and oscillation amplitude of 0.1 mm, due to convergence limitations of the numerical solution. This amplitude is considerably lower than clinical data obtained with Tissue Velocimetry Imaging (0.88 mm and 2.95 mm for persistent and paroxysmal AF patients, respectively) [9] and seems to lead to underestimation of fibrillatory effects in hemodynamics. In this context, the present study proposes an integrated analysis using imaging and electrophysiological data from an AF patient for *in silico* hemodynamic analysis. The 3D atrial geometry was obtained from MRI of an AF patient, subjected to electrophysiological study by catheter ablation. Electrograms (EGMs) were acquired simultaneously from multi-electrode catheters, then mapped in 3D atrial shell and then used as a boundary condition in the hemodynamic model as a 3D profile. The mechanical effects due to fibrillatory wall activity were coupled to the blood motion using a momentum method.

2 Materials and Methods

2.1 3D Atrial Surface Segmentation

MRI scans from one AF patient (each: 15 slices, 512 $\times$ 512 pixels, the distance between slices 11.0 mm) (Fig. 1-a-i.), available from the EDGAR database [10],

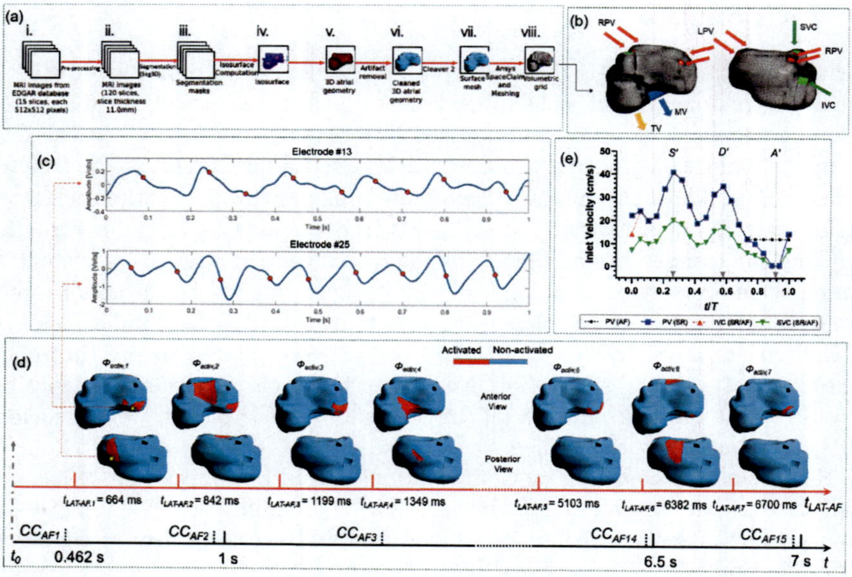

Fig. 1. (a) Pipeline for creation of the volumetric grid from MRI image [12]. (b) CFD mesh of left and right atria with details of boundary conditions (described in Section C): RPV and LPV - right and left pulmonary veins, SVC and IVC - superior and inferior vena cava, MV - mitral valve, TV - tricuspid valve. (c) Examples of EGM from two electrodes (13 and 25) illustrating the LATs in the red circle markers. (d) Details of timescales used in numerical model with anterior and posterior views of atria with 3D activation mapping (ϕ_{activ}) obtained from patient-specific electrophysiological data: t_{LAT-AF} denotes the starting time of a certain region activated, according to the LATs obtained from the 64 simultaneous EGMs, t is the physical time, CC_{AF1}, ..., CC_{AF15} indicates the completed cardiac cycles of blood flow in the atria, i.e. a heartbeat, defined as a complete cycle of pulsatile flow profile from (e). (e) Inlet velocity profiles as a function of normalized time (t/T) used as boundary condition for the hemodynamic model. T denotes a complete cardiac cycle of blood flow: $T = 462$ ms for AF conditions [15], $T = 960$ ms for sinus rhythm [16]. S', D' and A' are the systolic, diastolic and atrial contraction peaks: S' is due to the blood inflow from PVs combined with atrial relaxation and longitudinal motion of the mitral valve, D' is related to the rapid ventricular filling and A' is a negative peak due to atrial contraction, resulting in a slight backflow of atrial blood to the PV and it is not observed in AF cases [15,17]

were used in this study. Segmentation was performed in Seg3D software [11]. Preprocessing (Fig. 1-a-ii.): thickness of each slice was adjusted to 11 mm, as indicated by the DICOM headers. A Median filter with radius of 1 was applied, followed by a Gaussian Blur filter through a discrete Gaussian kernel (pixel value of 0.3). A resample was applied 3 times, yielding a layer with 120 slices. Atrial Segmentation (Fig. 1-a-iii.): the heart was delimited using the Implicit Model function. Following heart segmentation, the Paint Brush tool isolated the area of the atria. Connected component analysis was applied to connect the highlighted regions and exclude any unwanted noise. The isosurface was computed (Fig. 1-a-iv.), meshes generated, loaded in Matlab (R2015a, The MathWorks), for generating 3D atria maps. The 3D atrial geometry (Fig. 1-a-v.) was further processed to remove artifacts (Fig. 1-a-vi.) and improve smoothness, by following 4 iterations of: merging vertices closer than 2 and removing triangles with less than 2 edges in common with other triangles. For the first two iterations, simple smoothing also occurred. The 3D atria's surface meshes (Fig. 1-a-vii.) were generated with the software Cleaver2 [11] containing 1,114 faces and 17,760 vertices [12]. From the surface mesh, a volumetric grid was obtained by approximately 500,000 elements (Fig. 1-a-vii and b) by the Ansys SpaceClaim and Meshing (ANSYS, Inc.).

2.2 Atria Electrical Activity

After MRI scan, the AF patient underwent an electrophysiology study. A 64-electrode multipolar basket catheter (Constellation, Boston Scientific, Natick, MA) was used to acquire signals from the LA. These data were available at EDGAR database [10]. EGMs were acquired for 7 s (sampling frequency of 2034 Hz). The local activation times (LATs) were obtained with the maximum negative slope ($-dV/dt$) [13]. Once identified, different patterns of activations were used as an input into the hemodynamic model, denoted by the field ϕ_{activ}. Hemodynamic simulation was performed using the patient-specific endocardium activation to evaluate the AF effects in the blood flow field and the feasibility of clot formation.

2.3 In Silico Hemodynamic Modeling

The numerical model was based on CFD coupling electromechanical effects of AF. The Ansys Fluent (ANSYS, Inc.) code was used in the simulations. Due to the multi-physics nature of this model, it was divided in governing equations of fluid motion and coupling of electromechanical effects. Also, we defined two timescales for better comprehension, as illustrated in Fig. 1. t_{LAT-AF}: the time where the tissue underneath the electrical mapping electrode is activated, as obtained from EGMs and T the total cardiac cycle time related to the blood flow to complete a full cycle of the pulsatile flow profile depicted in Fig. 1-e. For instance, the initial time of t_{LAT-AF} was synchronized to the start of cardiac flow at physical time $t_0 = 0$. The mechanical effects of electrophysiological activity were considered to last up to the next activation time.

Governing Equations for Non-Newtonian Fluid
A transient model was proposed, considering non-Newtonian, incompressible blood flow under laminar regime conditions [6,8]. The mass and momentum conservations for fluid are given by Eqs. 1 and 2, respectively. The non-Newtonian behavior of blood was considered using a generelized Newtonian approach (Eq. 3) with the Carreau-Yasuda model (Eq. 4). The variables are listed in Table 1.

$$\nabla \cdot \mathbf{U} = 0 \tag{1}$$

Table 1. Variables for the numerical model

	Variable	SI Units
ρ	Blood specific mass	(kg m^{-3})
$\mathbf{U}$	Blood velocity vector	(m s^{-1})
p	Blood pressure	(kg m^{-1}s^{-2})
τ	Viscous stress tensor	(kg m^{-1}s^{-2})
$\mathbf{f}$	Momentum source	(kg m^{-2}s^{-2})
$\eta(\dot{\gamma})$	Apparent blood viscosity	(kg m^{-1}s^{-1})
η_∞	Apparent viscosity at infinite shear	(kg m^{-1}s^{-1})
η_0	Apparent viscosity at zero shear	(kg m^{-1}s^{-1})
λ	Fluid relaxation time	(s)
n	Power law index	Dimensionless
a	Yasuda index (shape parameter)	Dimensionless
U_{wall}	Wall velocity	(m s^{-1})
t^i	Current time	(s)
t^{i-1}	Previous time	(s)
d_{wall}	Wall displacement	(m)
A	Wall displacement amplitude	(m)
f_{AF}	AF frequency	(Hz)
ϕ	AF activity coefficient	Dimensionless
φ	Activity propagation coefficient	Dimensionless

$$\rho\left(\frac{\partial \mathbf{U}}{\partial t}\right) + \mathbf{U}\cdot\nabla\mathbf{U} = -\nabla p + \nabla\cdot\boldsymbol{\tau} + \mathbf{f} \tag{2}$$

$$\tau = \eta(\dot{\gamma})\left(\nabla\mathbf{U} + \left(\nabla\mathbf{U}^T\right)\right) \tag{3}$$

$$\eta(\dot{\gamma}) = \eta_\infty + (\eta_0 - \eta_\infty)\left[1 + (\lambda\dot{\gamma})^a\right]^{\frac{(n-1)}{a}} \tag{4}$$

The blood transport and rheological parameters used were: $\rho = 1055$ kg m^{-3}, $\eta_\infty = 0.00355$ kg m^{-1}s^{-1}, $\eta_0 = 0.16$ kg m^{-1}s^{-1}, $\lambda = 8.2$ s, $n = 0.2128$, $a = 0.64$ [14].

AF Electromechanical Effects in Hemodynamics

The mechanical effects of fibrillatory motion of atria walls were considered by a momentum-based method, i.e., the atria walls were considered rigid and an additional momentum source was included in the momentum balance of the fluid (**f** in Eq. 2). Basically, an inertial effect induced by the wall oscillatory motion was considered, as given by Eqs. 5–6. The wall velocity was considered to be an isotropic velocity field and determined from the wall displacement, d_{wall} (Eq. 6), given by a first-order sinusoidal function, with wall displacement amplitude A

= 0.88 mm and AF frequency f_{AF} = 6.76 Hz [9]. The model variables are listed in Table 1.

$$\mathbf{f} = \rho\phi\left(\frac{U_{wall} - \mathbf{U}}{\Delta t}\right) \tag{5}$$

$$U_{wall} = \frac{d_{wall}^{t_i} - d_{wall}^{t_{i-1}}}{\Delta t} \quad \text{with} \quad d_{wall} = A\sin\left(2\pi f_{AF}\, t\right) \tag{6}$$

$$\phi = \begin{cases} \phi_{activ} = 1.0 & \text{for AF active wall zones} \\ \phi_{activ} = 0 & \text{for non-active wall zones} \\ \varphi & \text{for fluid zones (Eq. 8)} \end{cases} \tag{7}$$

$$\frac{\partial\varphi}{\partial t} + \mathbf{U}\cdot\nabla\varphi = \phi_{activ} \tag{8}$$

The mechanical effects of wall oscillatory motion were considered to propagate the blood flow inwards by solving an additional transient-advective equation for φ (Eq. 8). The 3D activation profile from EGM data was taken as a source of φ. From Eq. 5, we notice that wall velocities are greater than blood velocities, resulting in local acceleration of the fluid. This is due to the physical effect of the solid wall pushing on the fluid.

Boundary Conditions and Numerical Details
Figure 1-b shows the boundary conditions used in simulations. Briefly, the blood flow was described by transient velocity profiles (Fig. 1-e) at pulmonary veins (PV) (considering symmetry for blood flow rate) [15] and at inferior and superior vena cava, based on literature data [17]. Both mitral and tricuspid valves were considered as fixed open areas, each with constant outlet pressure prescribed according to clinical data [7]. Despite the simplification of inlet and outlet geometries, the total cross-sectional areas of these boundary conditions were near to those observed clinically [8]. The no-slip condition was considered at the atrial walls. 3D profiles of AF activity obtained from EGM were prescribed at atria walls, allowing us to consider the local oscillatory effects of AF in the blood flow. For transient patterns, the 3D AF activity mapping were changed in the CFD setup according to the LATs. The numerical solution employed the SIMPLE algorithm with high order discretization schemes. A timestep of $1 \cdot 10^{-3}$ s was defined with root mean square (RMS) residual of $1 \cdot 10^{-4}$.

2.4 Case Studies

In order to evaluate the proposed model, case studies were performed under distinct conditions, as summarized in Table 2. The SR case was simulated to obtain the hemodynamic field under healthy conditions, defined here as reference. AF_0 presents a simplified model of AF since it neglects the third velocity pulse, related to the atrial contraction (A' in Fig. 1-e), without any mechanical consideration from atria walls.

Table 2. Description of the cases of study

Case	Description	Purpose
SR	Sinus rhythm, heart rate of 64 bpm ($T = 940$ ms) [15]	Reference case under healthy conditions
AF_0	AF case without wall activity. AF effects considered only by inlet PV velocities	Simplified AF modeling
AF_{1-3}	AF case with three hypothetical ϕ fields (Fig. 2-a),	To verify the momentum method effects in hemodynamic fields
AF_4	AF case with realistic ϕ field (Fig. 1) obtained from EGM in a patient-specific atria	To evaluate real fibrillatory activity in atrial hemodynamics

All AF cases were simulated with heart rate of 130 bpm ($T = 462$ ms) [16]

Cases AF_{1-3} present the proposed momentum method model implemented in atria hemodynamics. Five cardiac cycles of blood flow were simulated, the first two without any AF activity, and the subsequent three cardiac cycles with hypothetical ϕ fields as shown in Fig. 2-a. In this case, the AF electrical activity was considered to start at the beginning of the cardiac cycle of blood flow, with the mechanical effects lasting along the full cycle. Five cardiac cycles were simulated for SR, AF_0 and AF_{1-3} cases.

In the AF_4 case, realistic AF activation patterns, obtained from EGMs of a patient-specific atria (Fig. 1-d) were coupled as a boundary condition for the fluid dynamic solver. The local mechanical effects were considered to last up the next activation. The simulation was 7s long to encompass all activation patterns from EGM, resulting in about 15 cardiac flow cycles. The results of global hemodynamic indicators (WSS and atrial wall pressure) were taken from the last cardiac cycle of blood flow, to evaluate the effects of local wall oscillatory motion from AF activity in the hemodynamic field.

3 Results and Discussion

3.1 Model Validation

Statistical data variability was assessed, sampling each simulation timestep, discarding the first cardiac cycle due to initial numerical instabilities. The electromechanical effects in hemodynamic indicators were evaluated for area-weighted averages of WSS and pressure in LA walls (Fig. 2-b). Fig. 2-b shows a greater variability of the evaluated flow indicators (wall pressure and WSS) with the increment in active area of LA and also with the time advance, related to the propagation of oscillatory motion through the blood flow. In the AF_0 case (without wall activity and with AF effects considered only by inlet PV velocities), we found lower average of WSS and lower data dispersion of wall pressure and WSS. The increase in AF activity in LA walls (AF_4) resulted in higher dispersion of WSS and wall pressure results, demonstrating the effective influence of wall oscillatory motion, which is considered by the momentum method.

3.2 Electrophysiology AF Effects in Atria Hemodynamics

Figure 2-c shows 3D atrial isovolumes of the activation coefficient transported through the blood, as a propagation (advection) effect of the oscillatory motion from active walls for the AF_4 case (Fig. 1-d). As expected, the inward propagation of blood flow due to wall oscillatory motion was stronger after a longer activation time (e.g. after the 5th activation pattern, $t_{LAT-AF} \geq 5103$ ms), affecting a larger portion of blood flow in LA. Figure 2-d shows the WSS field for LA and right atrium (RA) walls for the time instants related to systolic, diastolic and atrial contraction peaks of blood inflow (S', D' and A' in Fig. 1-e).

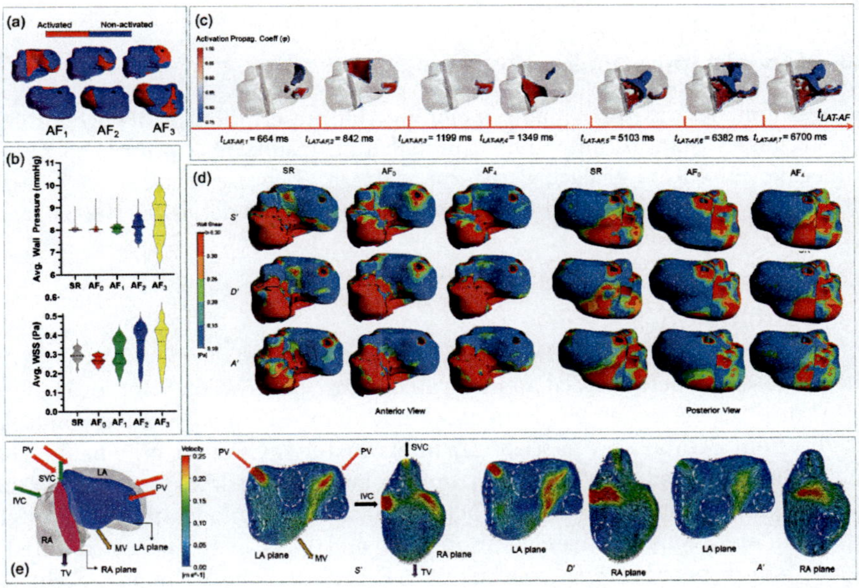

Fig. 2. (a) Hypothetical 3D AF activation fields prescribed at atria walls with % relative active area (ratio of AF activated area to total area of atria endocardium) of 23.28% (AF_1) 16.77% (AF_2) and 39.68% (AF_3). (b) Violin plots for area-weighted averages of wall pressure (mmHg) and WSS, at LA walls for SR, AF_0-AF_3 cases. (c) Isovolumes of φ (from 0.75 to 1.0), showing the mechanical effect of wall motion propagation into the blood flow for case AF_4. (d) Anterior and posterior view of atria with WSS contours for SR, AF_0 and AF_4 cases at S', D' and A' instants of cardiac flow. (e) Velocity vector fields for two transversal planes at LA and RA for AF_0 case at S', D' and A' with details of inlets (PVs, IVC, SVC) and outlets (MV and TV), white dashed lines highlight the larger recirculation zones with low blood velocity observed in LA

In Fig. 2-d we can observe higher WSS in RA. This is related to the RA anatomy, that favors an almost direct transport of the blood from IVC and SVC through the tricuspid valve, without major recirculation or stagnation zones, as observed in RA plane (see white dashed lines in Fig. 2-e). The WSS observed in RA are predominantly above 0.3 Pa for the three instants evaluated. These hemodynamic characteristics along with the major occurrence of AF activity and thrombus in LA is in accordance with literature. The LA exhibited low values of WSS even for the SR case. This is attributed to the recirculation zones observed in the upper portion of the LA, between the left and right PV inlets (white dashed lines in LA plane in Fig. 2-e). The blood flow jet is directed to the mitral valve, and then generates recirculation zones with low blood velocities. From the posterior view (Fig. 2-d), we can observe a larger area of low values of WSS for AF_0 and AF_4 cases. These results indicate a combination of the AF effects from inlet velocity profile (absence of atrial contraction peak) and oscillatory wall motion (only in AF_4 case), indicating low blood velocities, longer blood residence times and consequently pro-thrombotic environment.

3.3 Model Limitations

Some simplifications were made, including the prescription of 3D AF profiles in determined time instants. This simplification results in a prescribed period of mechanical effects of each activation pattern. The use of atrial geometries with higher spatial resolution to evaluate hemodynamic details in the left atria appendage is also important.

4 Conclusion

The proposed model provided an integrated approach to account for local effects of AF in the hemodynamics. Patient-specific data (3D atria geometry and AF activity from EGMs) were successfully integrated in a CFD solver. The momentum method coupled with local electrophysiological AF allowed the observation of mechanical effects inherent from active walls into the bloodstream. Transient data from area-weighted average of pressure and WSS in LA walls exhibited a higher data dispersion when considering AF activity. The AF model predicted larger pro-thrombotic zones (WSS $\leq$ 0.1 Pa) concerning the simplified AF and sinus rhythm models. The proposed integrated model could use patient-specific data for customization to perform comprehensive hemodynamics analysis.

Acknowledgements. The authors acknowledge the financial support from FAPESP (São Paulo Research Foundation, Process 18/25606-2). J.G. Siles is supported by FAPESP (grant n. 2020/03601-9). I. Sandoval and M. McInnis is supported by Coordenação de Aperfeiçoamento de Pessoal de Nível Superior - Brasil (CAPES). Special thanks to Prof. John Sims (CECS, UFABC) for the reviewing suggestions in the final manuscript.

Conflict of Interest The authors declare that they have no conflict of interest.

References

1. Rahman, F., Kwan, G.F., Benjamin, E.J.: Global epidemiology of atrial fibrillation. Nat. Rev. Cardiol. **11**, 639–654 (2014). https://doi.org/10.1038/nrcardio.2014.118
2. Duenas-Pamplona, J., Garcia, J.G., Sierra-Pallares, J., et al.: A comprehensive comparison of various patient-specific CFD models of the left atrium for atrial fibrillation patients. Comput. Biol. Med. **133**, 104423 (2021). https://doi.org/10.1016/j.compbiomed.2021.104423
3. Salinet, J., Molero, R., Schlindwein, F.S., et al.: Electrocardiographic imaging for atrial fibrillation: a perspective from computer models and animal experiments to clinical value. Front. Physiol. **12**, 653013 (2021). https://doi.org/10.3389/fphys.2021.653013
4. Zhang, L.T., Gay, M.: Characterizing left atrial appendage functions in sinus rhythm and atrial fibrillation using computational models. J. Biomech. **41**, 2515–2523 (2008). https://doi.org/10.1016/j.jbiomech.2008.05.012
5. Taylor, J.O., Witmer, K.P., Neuberger, T., et al.: In vitro quantification of time dependent thrombus size using magnetic resonance imaging and computational simulations of thrombus surface shear stresses. J. Biomech. Eng. **136**, 071012 (2014). https://doi.org/10.1115/1.4027613
6. Masci, A., Barone, L., Dede, L., et al.: The impact of left atrium appendage morphology on stroke risk assessment in atrial fibrillation: a computational fluid dynamics study. Front. Physiol. **9**, 1–11 (2019). https://doi.org/10.3389/fphys.2018.01938
7. Mill, J., Agudelo, V., Olivares, A.L., et al.: Sensitivity analysis of in- silico fluid simulations to predict thrombus formation after left atrial appendage occlusion. Mathematics **9**, 2304 (2021). https://doi.org/10.3390/math9182304
8. Koizumi, R., Funamoto, K., Hayase, T., et al.: Numerical analysis of hemodynamic changes in the left atrium due to atrial fibrillation. J. Biomech. **48**, 472–478 (2015). https://doi.org/10.1016/j.jbiomech.2014.12.025
9. Limantoro, I., de Vos, C.B., Delhaas, T., et al.: Clinical correlates of echocardiographic tissue velocity imaging abnormalities of the left atrial wall during atrial fibrillation. Europace **16**, 1546–1553 (2014). https://doi.org/10.1093/europace/euu047
10. EDGAR Database at www.ecg-imaging.org
11. NIH Center for Integrative Biomedical Computing, University of Utah Scientific Computing and Imaging Institute at www.sci.utah.edu/cibc-software
12. McInnis, M.: The forward and inverse solution during atrial fibrillation: from proof of concept to clinical application. Master's Dissertation. Federal University of ABC (UFABC) (2021)
13. Siles, J.G.: A robust conduction velocity pipeline method for guiding evaluation of the electrophysiological effect due to radiofrequency catheter ablation: outcomes from an isolated mice atria model. Master's Dissertation. Federal University of ABC (UFABC) (2020)
14. Boyd, J., Buick, J.M.: Analysis of the Casson and Carreau-Yasuda non-Newtonian blood models in steady and oscillatory flows using the lattice Boltzmann method. Phys. Fluids **19**, 093103. https://doi.org/10.1063/1.2772250
15. Fernandez-Perez, G.C., Duarte, R., Calle, M.C., et al.: Analysis of left ventricular diastolic function using magnetic. Radiologia **4**, 295–305 (2012). https://doi.org/10.1016/j.rx.2011.09.018

16. Muellerleile, K., Sultan, A., Groth, M.,et al.: Velocity encoded cardiovascular magnetic resonance to assess left atrial appendage emptying. In: J. Cardiovasc. Magnetic Resonance **14**, 39 (2012). https://doi.org/10.1186/1532-429X-14-39
17. Wexler, L., Bergel, D.H., Gabe, I.T., et al.: Velocity of blood flow in normal human Venae Cavae. Circ. Res. **23**, 349–359 (1968). https://doi.org/10.1161/01.RES.23.3.349

Identification of the Parameters of Double-Pendulum Intermittent Delay Feedback Control Model Using Optimization Method

Diego Gonzalez[1,2], Luis Aureliano Imbiriba[3], and Frederico Jandre[1](✉)

[1] Biomedical Engineering Program, Federal University of Rio de Janeiro, Rio de Janeiro, Brazil
jandre@peb.ufrj.br

[2] Federal Center for Technological Education, Celso Suckow da Fonseca, Nova Friburgo, Brazil

[3] School of Physical Education and Sports, Federal University of Rio de Janeiro, Rio de Janeiro, Brazil

Abstract. The identification of the parameters of postural control models may aid evaluating the balance control mechanism. A novel approach suggests that the central nervous system adopts an intermittent active feedback control to maintain upright posture, since this strategy may explain better the spontaneous sway. The aims of this work are to analyze the intermittent strategy of subjects in different sensorial conditions and to propose a method to identify the parameters of an intermittent delay feedback control model, applied to the double-pendulum biomechanical model. Four sway parameters (mean velocity, range, signal RMS and frequency band) were calculated from signals available in a public dataset containing body sway records of young and old subjects in trials with eyes open/closed on rigid/foam surface. An intermittence parameter ρ was obtained using system identification with a genetic algorithm optimization approach. The sway parameters were significantly different between eyes open/closed on rigid/foam surface ($p < 0.001$); the intermittence parameter ρ suggests that the subjects use more the active control when the trial becomes more challenging ($p < 10–9$), and the older subjects use more the active control than younger ($p < 0.01$). The optimization meth-od was able to identify parameters that could reproduce sway patterns that also distinguished between trials with eyes open/closed on rigid/foam surface (velocity $p = 0.00005$, range $p = 0.00005$, RMS $p = 0.00005$). The evaluation of further models, incorporating a different set of parameters, search intervals and optimization criteria, may improve system identification.

Keywords: Postural control · Body sway · Double-pendulum

1 Introduction

Balance is a complex issue that involves the interaction of different physiological subsystems, such as the central nervous system, musculoskeletal, and sensorial systems. Even simple tasks, such as upright standing, cannot be described by a single mechanism

J. L. B. Marques et al. (Eds.): CLAIB 2022/CBEB 2022, IFMBE Proceedings 98, pp. 65–75, 2024.
https://doi.org/10.1007/978-3-031-49401-7_7

[1, 2]. The effects of aging, diseases which affect sensory outputs, or any other factor that impairs these systems may degrade the subject's balance. Several techniques have been developed to objectively describe balance and postural control overall, both qualitatively and quantitatively [3]. Direct or indirect measurement of body sway in upright stance from the oscillations of center of mass (CoM) and center of pressure (CoP) positions, is one of the most usual approaches to assess balance, given that the stabilization of CoM is commonly considered as the object of postural control [2, 3].

However, due to the multifactorial composition of postural control mechanisms, changes in body sway may reflect variations in different subsystems, and the increase in sway amplitude or velocity, for example, do not necessarily indicate a balance impairment or postural control instability [4]. Also, factors that are not associated with sensory disturbances, diseases, or biomechanical changes, e.g. mental fatigue, may influence body sway and cannot be explained directly by physical causes [5]. One method to aid the comprehension of postural control is the development of models that identify the separated roles of the musculoskeletal system, the sensory systems and the neural controller acting to stabilize the position of the CoM. These models may target the assessment not only of the output variable (CoM position), but also of each subsystem that composes the whole postural control system [6, 7].

The most usual biomechanical models are the single and the double inverted pendulum [8, 9], whose parameters can be obtained from anthropometric data of the subject [10]. In quiet standing models, the internal reference of CoM is frequently considered as its value at the equilibrium point (normally when the angle or the position of the CoM are equal to zero) and there's some difficulty in directly obtaining the model parameters. Proportional-Integral-Derivative (PID) and Proportional-Derivative (PD) with continuous feedback time delay control models have been assessed [11], but a novel approach under investigation assumes an intermittent strategy of feedback control of stance [12–14]. This type of model may explain the body's spontaneous sway in quiet standing as a natural strategy to reduce resource expenditure, requiring lower levels of internal noise and control gains, which could also explain changes in postural sway when different sensorial tests are applied to the same subject. The use of an intermittent strategy could, in fact, even indicate a younger or a healthier postural control system [15].

The single inverted pendulum model has been investigated with simulation studies [12–14] and system identification from experimental data has been done with methods such as Bayesian inference [15] and optimization techniques associated with Kalman filters [16]. However, the double pendulum model has been less investigated using experimental data [17–19]. The use of intermittent strategies within the latter model was also less investigated in different sensorial conditions such as with closed eyes or with the challenge of a foam pad under the feet. The purposes of this work are to compare the use of intermittent strategy of postural control in the same subjects between different sensorial conditions and to evaluate a method to identify model parameters of the double inverted pendulum controlled with intermittent feedback using experimental data from real subjects.

2 Materials and Methods

2.1 Experimental Data

A public dataset [20] with body kinematics from an infrared motion capture system and ground reaction forces from a force plate, of 49 subjects, approved by a local ethics committee, was employed in the present study. The data was collected from 27 young and 22 old male and female subjects, and contain measurements of ground reactions forces, CoP displacement in anterior-posterior (AP) and medial-lateral (ML) directions, three-dimensional positions of body markers, CoM in AP, ML and vertical direction, and anthropometric data of the subjects. A protocol with 4 different conditions was used in that study: stance with eyes open (EO) and closed (EC) on a rigid force plate, stance with eyes open (EOC) and closed (ECF) on a force plate with a foam pad under the feet. Signals from 3 60-s trials were acquired for each subject in each condition, resulting in a total of 12 trials per subject, per condition.

2.2 Data Analysis

Sway parameters. Signals were filtered with a digital 4th order Butterworth lowpass filter at a cutoff frequency of 10 Hz. The first 10 s of each signal were discarded to remove transient effects of the filter. Signals were detrended towards zero mean. Four parameters were calculated from CoM displacement, in the anteroposterior (AP) direction: maximal range of the signal (Range), RMS value (RMS), mean velocity of displacement (Velocity) and the frequency band encompassing 80% of the power of the signal (F80). These parameters were calculated according to other studies [11, 21, 22] and are presented in Eqs. 1, 2, 3 and 4

$$Range = |x_{max} - x_{min}| \tag{1}$$

$$RMS = \frac{\sqrt{\sum x_i^2}}{N} \tag{2}$$

$$Velocity = \frac{\sum |x_i - x_{i-1}|}{T} \tag{3}$$

$$F80 = f_{80}| \int_0^{2\pi f_{80}} S_{xx}(f)df = 0.8 \int_0^{\infty} S_{xx}(f)df \tag{4}$$

where x_{max} and x_{min} are respectively the maximum and the minimum value of CoM at one trial, x_i is the i-th and x_{i-1} is the i-1-th samples of a N-samples CoM signal, T is the duration of the trial and $S_{xx}(f)$ is the power spectrum of the CoM displacement calculated as [23]. These parameters were calculated for all trials and simulations.

Intermittence parameter. The intermittent control paradigm defines that there are regions on the CoM phase plane ($CoMxCo\dot{M}$) where the PD feedback controller is activated (S_{on}) and others where it is inactivated (S_{off}). A parameter ρ was proposed to indicate the ratio $\rho = S_{on}/(S_{on} + S_{off})$ [15], with $\rho = 1$ meaning that the PD feedback controller is always active during the trial. This parameter was calculated for all trials of the 49 subjects, using the criterion found elsewhere [14–16, 19] to identify if the PD controller is activated.

If $CoM\left(\dot{CoM} - \alpha CoM\right) > 0$ and $CoM^2 + \dot{CoM}^2 > r^2$.

Then, active control "on".

Else, active control "off".

The values of r and α used were fixed for all tests and subjects and were based on the provided by [14], $\alpha = 0.4$ and $r = 0.004.h_{cm}$, where h_{cm} is the total CoM height above the ankle.

Statistical analyses. Mean and standard deviation of each parameter were calculated for the two groups, young and old subjects, and tested with nonparametric statistics due to small sample size and heteroscedasticity. Friedman test was used to compare the parameters between conditions for each group, with repeated measures, and the Wilcoxon signed rank test with Bonferroni adjustment ($\alpha = 0.008$) for post hoc comparison.

2.3 Double-Pendulum and Control Models

The double inverted pendulum model used in the present work comes from other studies [18, 19, 24]. The parameters weight and height were obtained from the dataset information, and the segment lengths were calculated from its three-dimensional markers data [18]. Moments of inertia and the mass of respective segments were calculated according to literature [10]. The 2-part model, comprising lower L (Legs) and upper segment HAT (Head-Arms-Trunk), is depicted in Fig. 1. The parameters are lower segment mass (m_L), center of mass height (h_L) and length (l_L) and upper segment mass (m_{HAT}), center of mass height (h_{HAT}) and length (l_{HAT}). The output variables of this model are the ankle joint angle θ_a and the hip joint angle θ_h. Other variables of interest are the angular velocity of the ankle (ω_a) and angular velocity of the hip (ω_h) and the moments of inertia of the lower and upper segments, respectively I_L and I_{HAT}.

Since the angular displacements θ_a and θ_h and their velocities ω_a and ω_h are small in these conditions near the equilibrium point ($\theta_a = \theta_h = \omega_a = \omega_h = 0$):

$$\sin(\theta) \approx \theta, \sin(\omega) \approx \omega, \cos(\theta) = \cos(\omega) \approx 1$$

The linearized model is therefore:

$$\boldsymbol{M\ddot{\theta}} + \boldsymbol{G\theta} = \boldsymbol{\tau} + \boldsymbol{\sigma} \tag{5}$$

where $\boldsymbol{\theta}$ is the angle vector $\boldsymbol{\theta} = [\theta_a \theta_h]^T$, M the inertia matrix and G contains the gravitational torques. These matrices depend on the model parameters and the acceleration of gravity ($g = 9,8\ \text{m/s}^2$).

$$M = \begin{bmatrix} I_L + I_{HAT} + m_{HAT} l_L^2 + 2m_{HAT} l_L h_{HAT} & I_{HAT} + m_{HAT} l_L h_{HAT} \\ I_{HAT} + m_{HAT} l_L h_{HAT} & I_{HAT} \end{bmatrix}$$

$$G = \begin{bmatrix} -g m_L h_L - g m_{HAT} - g m_{HAT} h_{HAT} & -g m_{HAT} h_{HAT} \\ -g m_{HAT} h_{HAT} & -g m_{HAT} h_{HAT} \end{bmatrix}$$

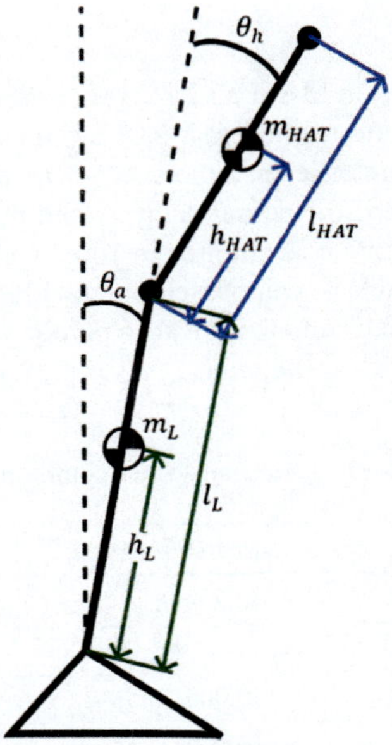

Fig. 1. Double-pendulum model with the variables m_L, h_L, l_L, m_{HAT}, h_{HAT}, l_{HAT}, θ_a and θ_h.

The torque vector comprises a passive and an active components $\boldsymbol{\tau} = \boldsymbol{\tau_p} + \boldsymbol{\tau_a}$ [18, 19]. The passive torque component $\boldsymbol{\tau_p}$ represents the stiffness and viscoelastic properties of musculoskeletal system; $\boldsymbol{\tau_a}$ represents the (possibly intermittent) delay feedback control torque. The vector $\sigma = [\sigma_1 \sigma_2]^T$ represents the internal noise. The state space model considering the state vector $\boldsymbol{x} = [\theta_a \theta_h \omega_a \omega_h]^T$ from Eq. 5 is

$$\dot{\boldsymbol{x}} = \begin{bmatrix} 0 & I \\ -M^{-1}G & 0 \end{bmatrix}\boldsymbol{x} + \begin{bmatrix} 0 \\ M^{-1} \end{bmatrix}(\boldsymbol{\tau} + \boldsymbol{\sigma})$$

$$\boldsymbol{\tau_p} = \begin{bmatrix} -K_a & 0 & -B_a & 0 \\ 0 & -K_h & 0 & -B_h \end{bmatrix}\boldsymbol{x}$$

$$\boldsymbol{\tau_a} = \begin{bmatrix} -P_a & 0 & -D_a & 0 \\ 0 & -P_h & 0 & -D_h \end{bmatrix}\boldsymbol{x_\Delta}$$

where K_a and K_h are respectively the ankle and hip stiffness, B_a and B_h the ankle and hip damping coefficients, P_a and P_h the proportional control gains and D_a and D_h the derivatives control gains. The neural delay in the feedback control is expressed as $\boldsymbol{\Delta}$ at the state vector with delay $\boldsymbol{x_\Delta}$. This active control is equivalent to the PD controller since it is equivalent to the feedback of the output and its first derivative. The model of the position of the CoM is:

$$CoM = [m_L h_L + m_{HAT}(l_L + h_{HAT}) m_{HAT}(l_L + h_{HAT}) 0 \ 0]\boldsymbol{x}$$

The criterion to activate (or inactivate) the PD controller is as described in Sub-section B. When the PD controller becomes inactivated, $\boldsymbol{\tau_p}$ becomes zero.

2.4 System Identification

The model was implemented in MATLab 2022 (MathWorks, USA) using the Simulink toolbox with the Runge-Kutta method at a 0.01 s time step, in order to simulate and identify: the value of the controller parameters P_a, D_a, the parameters that determine the intermittent condition r,α, the neural delay Δ and the variance of internal control noise σ_1 and σ_2 (both are white noise limited to 10 Hz). The intervals of search and the fixed variables, shown in Table 1, were based on previous studies [15, 16, 18, 19]. The values of P_h and D_h were set to zero to indicate a purely passive control of the hip joint. The passive parameters are $K_a = 0.8, mgh_{cm}$, $K_h = 2m_{HAT}gh_{HAT}$ and $B_a = B_h = 30$.

Table 1. Parameters of the control model

Parameter	Search intervals
P_a	$[0.2.mgh_{cm}\ 1.5.mgh_{cm}]$
D_a	[4 400]
σ_1	[0.0001 0.5]
σ_2	[0.0001 0.5]
r	[0.0001 0.01]
Δ	[0.15 0.30]
α	[0.3 0.8]

The parameters were identified by optimization using the Genetic Algorithm (GA), a derivative-free method supposed to overcome local minima. The algorithm was implemented using the GA toolbox with the specifications: population size = 75 individuals; crossover rate = 80%; maximum stall generations within 10^{-4} error tolerance = 75 generations. The goal of the algorithm is to minimize an objective function error E based on the sway parameters

$$E = \sqrt{E_{vel}^2 + E_{Range}^2 + E_{RMS}^2 + E_{F80}^2}$$

where E_{vel}^2, E_{Range}^2, E_{RMS}^2 and E_{F80}^2 are the relative error between the real and simulated sway parameters Velocity, Range, RMS and F80, respectively. The parameters were estimated for 9 young subjects in the 4 conditions EO, EC, EOF, ECF.

3 Results

3.1 Sway and Intermittence Parameters Comparison

The 4 sway parameters (Velocity, Range, RMS and F80) and the intermittence parameter ρ calculated for both groups (young and old subjects) are shown in Table 2. Friedman test results showed that there is a significant difference between the four conditions (EO,

EC, EOF and ECF) for all sway parameters ($p \leq 0.001$) and the intermittence parameter ρ ($p \leq 10–9$) in both groups. Only Velocity showed significant difference for all paired comparisons in post hoc tests. The other parameters presented significant differences only for some of the paired comparisons. The value of ρ was higher for older subjects in each condition ($p \leq 0.01$), suggesting that this group use more the active control during the tests than the younger subjects.

Table 2. Sway and intermittence parameters (mean ± s.d.)

Parameter	Group	EO	EC	EOF	ECF
Velocity	Young	2.25 ± 0.44	2.54 ± 0.58	5.17 ± 1.14	7.69 ± 1.88
(mm/s)	Old	2.70 ± 0.69	3.25 ± 0.77	7.41 ± 2.50	10.5 ± 2.7
Range	Young	19.7 ± 8.3	19.5 ± 6.2	38.8 ± 13.3	38.8 ± 11.1
(mm)	Old	18.6 ± 6.4	20.2 ± 7.0	39.8 ± 12.7	47.8 ± 16.7
RMS	Young	4.62 ± 2.23	4.36 ± 1.58	8.90 ± 3.44	7.95 ± 2.42
(mm)	Old	4.22 ± 1.90	4.43 ± 1.72	8.60 ± 3.16	9.17 ± 3.22
F80	Young	0.27 ± 0.07	0.28 ± 0.08	0.28 ± 0.08	0.34 ± 0.07
(Hz)	Old	0.29 ± 0.09	0.31 ± 0.09	0.31 ± 0.08	0.34 ± 0.07
ρ	Young	0.30 ± 0.06	0.32 ± 0.05	0.33 ± 0.06	0.39 ± 0.04
	Old	0.34 ± 0.07	0.35 ± 0.05	0.38 ± 0.04	0.42 ± 0.03

3.2 Identification of System Parameters

The values of the estimated model parameters are presented in Table 3. Since the gain P_a is estimated based on the value of mgh_{cm}, the fraction P_a/mgh_{cm} is presented.

Table 3. Model estimated parameters (mean ± s.d.)

Parameter	EO	EC	EOF	ECF
P_a/mgh_{cm}	1.42 ± 0.07	1.44 ± 0.05	1.40 ± 0.09	1.46 ± 0.02
D_a	70.3 ± 76.9	61.1 ± 42.2	59.3 ± 30.7	55.1 ± 28.6
σ_1	$(2 \pm 2).10^{-4}$	$(2 \pm 2).10^{-4}$	$(10 \pm 7).10^{-4}$	(32 ± 31).10–4
σ_2	$(2 \pm 2).10^{-4}$	$(1 \pm 1).10^{-4}$	$(6 \pm 5).10^{-4}$	$(5 \pm 4).10^{-4}$
r	$(3 \pm 1).10^{-3}$	$(4 \pm 1).10^{-3}$	$(7 \pm 2).10^{-3}$	$(8 \pm 2).10^{-3}$
Δ	0.23 ± 0.04	0.25 ± 0.04	0.25 ± 0.04	0.22 ± 0.03
α	0.40 ± 0.05	0.41 ± 0.13	0.43 ± 0.08	0.35 ± 0.06

The values of sway and intermittence parameter from the simulation with the estimated model and the values of error function E are presented in Table 4. Friedman test results showed that there is a significant difference between the four conditions (EO, EC, EOF and ECF) for Velocity ($p = 0.00005$), Range ($p = 0.00005$) and RMS ($p = 0.00005$). Figure 2 presents one comparison between the real and the simulated CoM oscillation from one subject in ECF condition. Figures 3 and 4 represents the phase plane of real and simulated signals from one subject in ECF condition.

Table 4. Sway and intermittent parameters of simulation (mean ± s.d.)

Parameter	EO	EC	EOF	ECF
Velocity (mm/s)	3.78 ± 0.93	3.98 ± 1.13	8.27 ± 2.04	9.49 ± 2.07
Range (mm)	11.1 ± 3.3	13.1 ± 3.7	27.4 ± 5.8	29.7 ± 5.3
RMS (mm)	2.30 ± 0.85	2.64 ± 0.96	5.59 ± 1.24	6.03 ± 1.02
F80 (Hz)	0.35 ± 0.17	0.35 ± 0.17	0.30 ± 0.18	0.35 ± 0.09
ρ	0.45 ± 0.02	0.45 ± 0.03	0.46 ± 0.03	0.46 ± 0.02
E	0.96 ± 0.33	0.66 ± 0.15	0.62 ± 0.20	0.23 ± 0.12

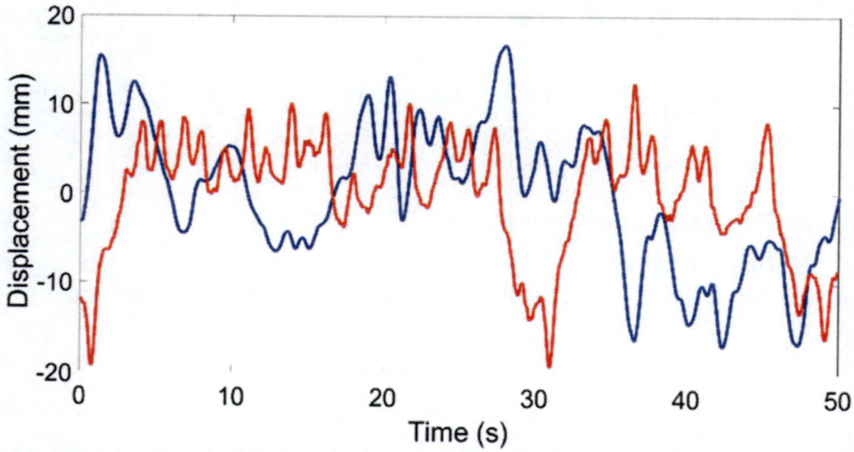

Fig. 2. Experimental CoM displacement in AP direction (blue) and simulated CoM in AP direction (red).

4 Discussion

The comparisons of the sway parameters Velocity, Range, RMS and F80 in AP direction showed that there is a statically significant difference in the CoM sway when varying the eye condition (open or closed) and the support surface (rigid or foam), for both young and old subjects. These results (Table 2) are in agreement with a recent study using other instruments at the same conditions [21], but also reinforce the idea that larger sway

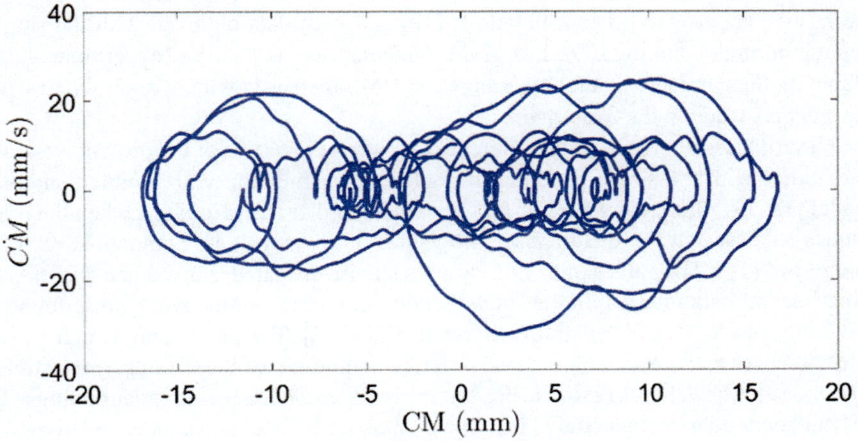

Fig. 3. Experimental CoM phase plane.

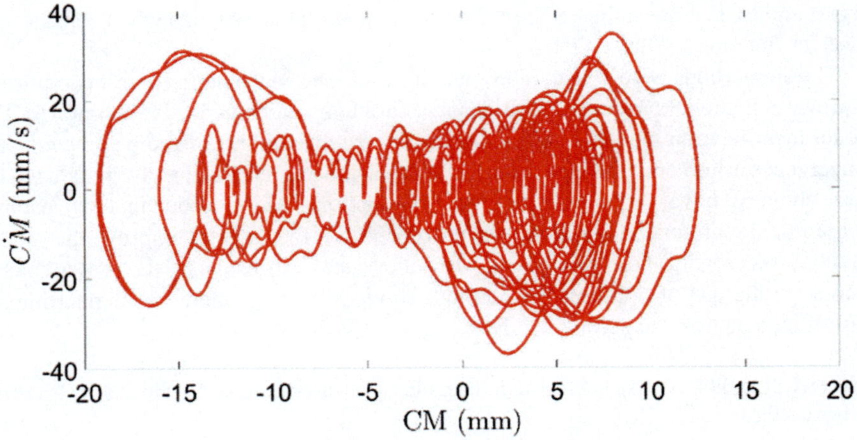

Fig. 4. Simulated CoM phase plane.

parameters do not necessarily mean worse balance, and that the control model analysis may provide additional information [4]. For the young subjects, RMS value was lower with eyes closed than with eyes open. Range and RMS values were lower for old than for young subjects with eyes open. The difference in ρ between conditions suggests that the subjects employed more active control in more challenging conditions (closed eyes and/or on a foam pad). Furthermore, the results support the assumption that the older subjects more than the younger ones may use the active control in the same conditions, which is in agreement with other findings [15].

The model parameters identified P_a, D_a, σ_1, σ_2, Δ, r and α, were able to generate CoM sway signals in AP direction with magnitudes similar to those of the experimental data using the double-pendulum intermittent delay control model (Table 4). These results are in agreement with other simulations [19] and with data obtained with the single-pendulum model and the EO trials of the same dataset [16]. As the experimental data (Table 2), the parameters Velocity, Range and RMS increased with eyes closed or when the subjects were on the foam pad.

Nevertheless, there are some differences between the results of the present work and the studies with the single-pendulum [15, 16] and simulation with double-pendulum model [18, 19]. First, the values of P_a found were higher than $1.mgh_{cm}$, when the other studies suggest that $P_a < 1.mgh_{cm}$. The values of D_a and Δ are compatible with the findings of [16]. The values of σ_1 and σ_2 were less investigated, since other studies with experimental data focused on the single-pendulum model, but the ankle noise found in this study resulted smaller than in other works [15, 16]. The parameters which govern intermittence, r and α, have the same order of magnitude of those in previous studies [14–16, 19]. The values of α in the present study are compatible with other findings [14, 15] but lower than in other study [16]. The ρ value from simulated data were larger than those in Table 4. Indeed, results from a previous work [15] found ρ values larger than those calculated here for young subjects, suggesting that α is not a fixed value, but may vary among subjects. For the simulated sway parameter, the EO condition showed the largest and ECF showed the smallest disagreement with the real data when considering the error function E (Table 4).

The simulations were able to reproduce some of the sway patterns in test conditions. However, the model presents limitations and further research is needed to improve the model identification processes. Changes in the search intervals of the parameters and convergence criteria may reduce the error between the experimental and simulated data, as the choice of novel variables in the objective function. The inclusion of passive parameters in the identification algorithm, as proposed by [16] may also improve the model accuracy. Even with these limitations, we believe that the results of the present study encourage the use of optimization methods to identify the parameters of intermittent control models using experimental data.

Acknowledgments. The authors wish to thank financial support of the Brazilian state agencies CNPq, CAPES and FAPERJ.

Conflict of Interest The authors declare that they have no conflict of interest.

References

1. Peterka, R.J.: Sensory integration for human balance control. Elsevier B.V. (2018)
2. Mancini, M., Horak, F.B.: The relevance of clinical balance assessment tools to differentiate balance deficits. Eur. J. Phys. Rehabil. Med.Rehabil. Med. **46**, 239–248 (2010)
3. Paillard, T., Noé, F.: Techniques and Methods for Testing the Postural Function in Healthy and Pathological Subjects. Biomed Res. Int. 2015 (2015)
4. Pasma, J.H., Engelhart, D., Schouten, A.C., van der Kooij, H., Maier, A.B., Meskers, C.G.M.: Impaired standing balance: the clinical need for closing the loop. Neuroscience **267**, 157–165 (2014)
5. Varas-Diaz, G., Kannan, L., Bhatt, T.: Effect of mental fatigue on postural sway in healthy older adults and stroke populations. Brain Sci. **10**, 1–17 (2020)

6. Peterka, R.J.: Sensorimotor integration in human postural control. J. Neurophysiol.Neurophysiol. **88**, 1097–1118 (2002)
7. Van Der Kooij, H., Van Asseldonk, E., Van Der Helm, F.C.T.: Comparison of different methods to identify and quantify balance control. J. Neurosci. Methods. **145**, 175–203 (2005)
8. Crétual, A.: Which biomechanical models are currently used in standing posture analysis? Neurophysiol. Clin. Clin. **45**, 285–295 (2015)
9. Winter, D.A., Patla, A.E., Prince, F., Ishac, M., Gielo-perczak, K.: Stiffness control of balance in quiet standing. J. Neurophysiol.Neurophysiol. **80**, 1211–1221 (1998)
10. Winter, D.A.: Biomechanics and Motor Control of Human Movement: Fourth Edition (2009)
11. Maurer, C., Peterka, R.J.: A new interpretation of spontaneous sway measures based on a simple model of human postural control. J. Neurophysiol.Neurophysiol. **93**, 189–200 (2005)
12. Bottaro, A., Yasutake, Y., Nomura, T., Casadio, M., Morasso, P.: Bounded stability of the quiet standing posture: an intermittent control model. Hum. Mov. Sci. **27**, 473–495 (2008)
13. Bottaro, A., Casadio, M., Morasso, P.G., Sanguineti, V.: Body sway during quiet standing: is it the residual chattering of an intermittent stabilization process? Hum. Mov. Sci. **24**, 588–615 (2005)
14. Asai, Y., Tasaka, Y., Nomura, K., Nomura, T., Casadio, M., Morasso, P.: A model of postural control in quiet standing: robust compensation of delay-induced instability using intermittent activation of feedback control. PLoS ONE **4**, 6169 (2009)
15. Suzuki, Y., Nakamura, A., Milosevic, M., Nomura, K., Tanahashi, T., Endo, T., Sakoda, S., Morasso, P., Nomura, T.: Postural instability via a loss of intermittent control in elderly and patients with Parkinson's disease: a model-based and data-driven approach. Chaos. 30
16. McKee, K.L., Neale, M.C.: Direct estimation of the parameters of a delayed, intermittent activation feedback model of postural sway during quiet standing. PLoS One. **14** (2019)
17. Suzuki, Y., Nomura, T., Casadio, M., Morasso, P.: Intermittent control with ankle, hip, and mixed strategies during quiet standing: a theoretical proposal based on a double inverted pendulum model. J. Theor. Biol. **310**, 55–79 (2012)
18. Suzuki, Y., Morimoto, H., Kiyono, K., Morasso, P.G., Nomura, T.: Dynamic determinants of the uncontrolled manifold during human quiet stance. Front. Hum. Neurosci.Neurosci. **10**, 618 (2016)
19. Morasso, P., Cherif, A., Zenzeri, J.: Quiet standing: the single inverted pendulum model is not so bad after all. PLoS ONE **14**, 1–20 (2019)
20. dos Santos, D.A., Fukuchi, C.A., Fukuchi, R.K., Duarte, M.: A data set with kinematic and ground reaction forces of human balance. PeerJ. 2017 (2017)
21. Gonzalez, D.R.G., Imbiriba, L.A., Jandre, F.C.: Comparison of body sway measured by a markerless low-cost motion sensor and by a force plate. Res. Biomed. Eng. **37**, 507–517 (2021)
22. Prieto, T.E., Myklebust, J.B., Hoffmann, R.G., Lovett, E.G., Myklebust, B.M.: Measures of postural steadiness: differences between healthy young and elderly adults. IEEE Trans. Biomed. Eng. **43**, 956–966 (1996)
23. Vieira, T.M.M., Oliveira, L.F., Nadal, J.: Estimation procedures affect the center of pressure frequency analysis. Brazilian J. Med. Biol. Res. **42**, 665–673 (2009)
24. Lee, J., Zhang, K., Hogan, N.: Identifying human postural dynamics and control from unperturbed balance. J NeuroEng. Rehabil. **18**, 54 (2020)

Check for updates

The Forward and Inverse Solution Pipeline for Robust Characterization of Supraventricular Cardiac Arrhythmias

G. V. Costa[1], C. R. Restivo[1], I. Sandoval[1], M. S. Guillem[2], and J. Salinet[1(✉)]

[1] Modelling and Applied Social Sciences Centre, HEartLab—Biomedical Engineering, Federal University of ABC/Engineering, São Bernardo do Campo, Brazil
joao.salinet@ufabc.com

[2] ITACA Institute, Universitat Politècnica de València, Valencia, Spain

Abstract. Supraventricular arrhythmias, including atrial tachycardia, flutter and fibrillation are common cardiac arrhythmias, and responsible for reduction of patients' quality of life in long term. The most common treatment is by radiofrequency catheter ablation, where a catheter with electrodes is introduced inside the atria chamber, and in contact with the endocardium wall, deliver a high-power energy, allowing cauterization of predefined group of cells, responsible for triggering and maintenance of the atrial arrhythmias. Either atrial epicardial electrograms, or atria torso surface electrocardiograms, can be estimated through the inverse and forward solutions of the electrocardiography. This study aimed to evaluate the robustness of both forward and inverse solutions, front of different levels of noise, 3D geometry changes due to the respiratory and cardiac cycle, and reduction of number of leads.

Keywords: Atrial arrhythmias · Computational modelling · Diagnostic · Forward solution · Inverse solution

1 Introduction

Atrial tachycardia (AT), flutter (AFL), and fibrillation (AF) are the most common supraventricular cardiac arrhythmias in clinical practice [1]. Patients with AT and AFL showed to evolve to AF, by changing electrophysiological and structural atrial cells properties [2]. Invasive electrophysiological studies where a catheter with elec-trodes is introduced inside the atria, allowing physicians to identify and target, with radiofrequency cauterization, regions responsible for the arrhythmias' maintenance [1]. This procedure, although necessary, increases the risk and duration of the surgical procedure.

Non-invasive arrhythmias characterization, with a high number of simultaneous electrocardiograms (ECGs), has been shown as an important auxiliary clinical tool for decision-based therapy, prior to invasive procedures. In this direction, recently, the non-invasive electrocardiographic imaging technique, has been used in clinical prac-tice with

J. L. B. Marques et al. (Eds.): CLAIB 2022/CBEB 2022, IFMBE Proceedings 98, pp. 76–85, 2024.
https://doi.org/10.1007/978-3-031-49401-7_8

great expectation, aiming to improve catheter ablation therapy outcomes [3, 4]. In this method, atrial epicardial electrograms (I-AEGs) is estimated non-invasively, through a high density of body potentials and 3D geometries of torso/atria [4]. On the other way around, atrial non-invasive electrocardiograms (I-AECGs) can be estimated, through the forward problem of electrocardiography, by considering the 3D geometry of the torso and atria, and atria signals (AEGs) [3].

This study aimed to propose a pipeline of techniques to reduce disparity in SVA frequency and phase biomarkers, due to signals noise, reduced number of leads layout and 3D geometries modification.Please note that the first paragraph of a section or subsection is not indented. The first paragraphs that follows a table, figure, equation etc. does not have an indent, either.

2 Materials and Methods

2.1 Database

In this study, 13 simulations representing SVA, originated by three different mechanisms are used: AT (4 simulations), conducted by ectopic focus; FLA (4 simulations), conducted by re-entrant macro circuit; and AF (5 simulations), conducted by functional rotors. The frequency of sampling (fs) of the signals is 500 Hz. The locality of the occurrence of the mechanisms and their respective activation frequencies (Fac) are present below.

Table 1. Characteristics of arrhythmia models

Model	Arrythmia	Location	Fac (Hz)	Length (s)
1	TA	Inferior vena cava	4.0	3
2		Left superior PV	4.0	3
3		Right atrium Appendix	4.0	3
4		Right inferior PV	4.0	3
5	FLA	Counterclockwise rotation	4.0	5
6		Clockwise rotation	4.0	3
7		Inferior vena cava	4.0	3
8		Pulmonary veins	4.0	1
9	FA	RA	5.4	4
10		Appendix	5.8	4
11		Right superior PV	6.8	4
12		Left inferior PV	5.5	4
13		Right inferior PV	5.4	4

A standard 3D geometry of the torso and atria composed with 771 and 2048 vertices, respectively, are used for estimation of the torso potentials, by the forward electrocardiographic solution, and epicardial atria electrograms, by the inverse solution of the electrocardiographic.

2.2 Forward and Inverse Solutions

The description below presents an overview from the physics of both forward and inverse solutions. For further details, please see the supplemental material of a recent multicenter bibliography review study [2].

The forward problem of electrocardiography refers to the calculation of potentials on the body surface due to cardiac sources, using the theoretical equations of electromagnetism [1]. The physics of the forward electrocardiographic solution is ruled by the Laplace equation (Eq. 1), where Φ (mV/m) is the electric field generated by the excitation of the atria and is propagated to the surface of the body passing through a volume with passive electrical conduction, and ∇^2 is the Laplacian operator.

$$\nabla^2\Phi = 0 \tag{1}$$

In this method, the potentials in the torso (Φ_T) are calculated by solving the Laplace equation within the torso volume, using the electrical potentials acquired in the atria (Φ_A) and the geometric relationship between the 3D surfaces of the torso and atria (A_{TA} matrix) (Eq. 2).

$$\Phi_T = \Phi_A A_{TA} \tag{2}$$

The solution of Eq. 2 is a well-posed problem, where solution is unique, stable, and accurate, and depends on the input data. In this study, 771 atrial torso signals (I-AECGs) from each mathematical model are estimated, considering the standard 3D torso and atria surface, and 2048 atria signals (AEGs) undergoing the atrial arrhythmias (Table 1).

The inverse solution is a medical tool that allows the estimate of the atria epicardial electrical activity through the atria non-invasive signals of the torso and 3D geometries of the atria and torso [2]. It is an ill-posed mathematical problem solution of Eq. 2, meaning that small errors (i) in the measurement of the signals of the torso (i.e., noise and/or position of the electrodes) and (ii) in the 3D geometries of the torso/atria can result in large errors in the estimation of the atria epicardial potentials (Φ_A). To suppress this error, we used the zero-order Tikhonov Regularization method, where Φ_A is delimited in their or derivatives of the potentials of the epicardium in space, time or both, limiting these parameters to be within the electrophysiological and electric field limits of the atria.

In this method, the obtained solution is smooth, compatible with observation data, for a certain noise levels. The search for the regular solution is an additional information, which transforms the ill-posed problem into a well-posed. Through the regularization, it is possible to find the solution of the inverse problem of electrocardiography represented in the following equation:

$$\Phi_A = \Phi_T A_{TA}^{+} \tag{3}$$

where A_{TA}^{+} is the regularized inverse matrix obtained through the Tikhonov Regularization Method, also known as pseudo-inverse. In this study, for each arrhythmia model, 2048 epicardial atrial electrograms (I-AEGs) are estimated, through the inverse solution, based in the standard 3D torso and atria surfaces, and 771 non-invasive atria signals (AECGs) obtained in the torso undergoing the atrial arrhythmias.

2.3 Impact of Noise on the Estimation of the Potentials on Torso/Atria

For each model, the AECGs in each of the 771 vertices of the 3D geometry of the torso (for the forward solution), and the AEGs in each of the 2048 vertices of the 3D geometry of the atria (for the inverse solution) are contaminated with white Gaussian noise at different levels of noise signal ratio (SNRs: 60, 30, 10, 5, 1 and −3 dB), with the objective of simulating the effects of several random processes, that may occur in measurements of electrical signals [5]. After applying the noise, signals are filtered out by a Butterworth band-pass filter, of 4th order, in three distinct cut-off passbands frequencies: (i) 0.5 to 30 Hz, (ii) 3 to 30 Hz, and (iii) around the highest dominant frequency ($H_{DF} \pm 1$ Hz) obtained from the signals, through the discrete Fourier transform. Followed the filtering of AECGs and AEGs, the inverse and forward solutions are applied respectively, allowing estimation of the I-AECGs on the torso (forward solution), and estimation of the I-AEGs on the atria epicardium through the inverse solution [2].

2.4 Impact of 3D Torso/atria Morphological Changes on the Estimation of the Potentials on the Torso/Atria

Torso: to evaluate the impact of respiratory dynamics, different stages of the respiratory cycle are considered, according to the increase and decrease in air volume inside the torso: (i) beginning of inspiration, in which the torso is filled in 1/3 of its volume; (ii) torso filled with 2/3 of its volume; (iii) fully expanded torso representing maximum inspiration.

Atria: the alteration of the geometry of the atria during the cardiac cycle is performed as follows: (i) geometric representation of the atrial systole phase (contraction of the atria) where there is a reinforcement in the ventricular filling, causing the volume of the ventricles to increase approximately 25%; (ii) onset of atrial diastole, and (iii) end of atrial diastole.

The morphological modifications are developed in Python 3, with spatial alteration in the horizontal plane, applying different coefficients of deformation in the axial, left-right axis, and in the anterior-posterior axis, to the torso, and stretching the atria in the direction of the base to the apex, for the atria.

2.5 Impact of the Reduction of the Electrodes

To approximate the results obtained in this study to the clinical environment, we analyzed the impact of reducing the number of electrodes, as follows [5].

Torso: from 771 electrodes to (i) 16, (ii) 32, (iii) 62, and (iv) 131 electrodes.
Atria: from 2048 electrodes to (i) 256, (ii) 128, (iii) 64, and (iv) 32 electrodes.

For the calculation of the unknown potentials of the vertices, two types of interpolation are applied: linear interpolation and Laplacian interpolation [6, 7]. In this case, the objective was to verify the stability of the calculated biomarkers through the forward solution (i.e., estimation of I-AECGs) and inverse solution (i.e., estimation of I-AEGs), with the use of two different methods of interpolation.

2.6 Biomarkers

The above analysis, introduction of noise, changes in 3D torso/atria morphology, and reduction of number of leads, are applied into the signals prior the forward and inverse solutions. After that, I-AECGs and I-AEGs are estimated accordingly, by the respective forward and inverse solutions, and then 3D frequency and phase maps are calculated [5], as follows. For each type of map, its intrinsic biomarkers are extracted, allowing evaluation of their values, according to the above introductory influences (noise, spatial deformation, and change of leads layout). Details of the noise influence, leads layout and explanation of some of the extracted biomarkers, please see [5–7].

Frequency: The spectrum of each I-AECG and I-AEG was obtained with the application of the Fourier Rapid Transform (FFT) and a padding zero of 5 times. Subsequently, the dominant frequency (DF) is obtained from the spectrum, as the highest power frequency. The highest frequency in 3D of the DFs is called H_{DF} [5]. Seven biomarkers obtained through frequency maps are calculated [6, 7]: number of distinct areas in the 3D of H_{DF} atriums, mean size of H_{DFs}, total area of H_{DFs}, ratio between mean DF and H_{DF}, Interquartile Range (IQR) from the DF of the 3D vertices, and the average and IQR of the Regularization Index (RI).

Phase: The Hilbert transform is used to generate an analytical signal, from which the instantaneous phase of the estimated I-AECGs/ I-AEGs can be calculated as the inverse tangent of the ratio between the imaginary part and the real part of the analytical signal [5]. Phase analysis allows the quantification of the dynamics of excitation profiles contributing to the identification of singularity points (PSs, i.e., points where the phase around them progresses in a complete cycle of $-\pi$ to $+\pi$) [2]. Sequential propagation of PSs defines a filament, being a connection of the PSs along the subsequent phase maps [5]. Through phase maps, filament distributions and heatmaps (HMs) [5], six biomarkers are extracted [6, 7], as follow: number of filaments per second, average filament duration in percentage, spatial displacement of filaments; number of regions of PSs, the size of the agglutinated PSs areas, and the average density of the PS.

2.7 Statistics

Statistical analysis of the distribution was performed by the Kolmogorov-Smirnov statistical method. The variables with positive asymmetry were normalized using the base logarithm 10 of the variable. The analysis of bilateral variance (two-way ANOVA) was performed to evaluate the outcomes of the type of arrhythmia and morphological alterations of the torso and/or atrium, and layout alteration. In the case of a significant interaction effect, Bonferroni's Post Hoc procedure was performed to identify the source of the difference. $P < 0.05$ values were considered statistically significant.

In order to graphically represent the impact of geometric changes in the calculation of biomarkers, a graphic feature known as word cloud is used, which allows visual illustrating the most frequent terms of a given text/context. In this case, the more the biomarker is impacted due to geometric changes, the larger and stronger it will be illustrated on the chart. For this, the variation between the phase and frequency biomarkers estimated according to the geometric change and the biomarkers estimated with the reference geometries, method applied for each variation used and SVA is calculated. Moreover, for each arrhythmia, an analysis of the impact of the change in geometries is made in all situations applied in the methodology through ANOVA.

3 Results and Discussion

3.1 Forward Solution

Influence of the signal's level noise. The estimation of I-AECGs during the complex rhythm of AF, in different noise ranges, limits the effectiveness of the forward solution, compared to regular high frequency rhythms (AT and FLA). For example, for a SNR of 60 dB, all I-AECGs estimated during AF presented a relative error higher than 30%, independently of the applied bandpass filtering strategy. In contrast, for both AT and FLA the error was $< 10\%$. 3D DF maps from the torso, obtained through the forward solution applied to the AEGs filtered out around HDF $\pm$ 1 Hz, preserved the intrinsic activation frequency of each arrhythmia, diagnosed noninvasively, even at high noise levels (-3dB; percentage of points of the torso with error $< 10\%$: AT/AFL $= 100\%$ and FA $= 78.8\%$). For the 3D phase maps and its biomarkers (number of filaments, average filament propagation, and spatial displacement of the filaments), when the same strategy of filtering is applied (i.e. AEGs filtered around HDF $\pm$ 1 Hz), the biomarkers estimated at SNR 60 dB had the best agreement with the estimated with other SNRs (from 30 dB to -3 dB), for all arrhythmias, face to the other two bandpass filters (0.5–30 Hz and 3–30 Hz).

Influence of the 3D morphological torso/atria modifications. The 3D torso morphology variability seen during the respiratory cycle showed to modify the 3D DF torso maps, by introducing fluctuations on the size and locations of the higher frequency areas, for AF and AT. These fluctuations were more evident during the torso morphology at the end of the inspiration.

Influence of the atria number of electrodes. The biomarkers from the torso frequency and phase maps are obtained from the I-AECGs estimated through the forward solution, considering the highest number AEGs (i.e. 2048 AEGs), seen in each vertex of the atria 3D shell. As we remove some AEGs, the respective unknown potentials are estimated by the two proposed interpolation methods (Linear and Laplacian). Decreasing the number of electrodes means, in practice, removing more and more AEGs, on the 3D atrial shell vertices, prior the forward solution (i.e., I-AECGs). With that in mind, the interpolation strategy starts to play an important role, and should not introduce nonphysiologically events. It is commonly applied to the commercial equipment. As expected, our results showed that the phase biomarkers were more sensitivity to different leads layouts when compared with the frequency biomarkers. Table 1 illustrates a typical example, where the biomarker number of filaments per second, associated in somehow with reentry, were obtained from torso phase maps, with the forward solution applied to the original number of AEGs presented on the atrial shell vertices. The same biomarker is computed also considering 128 AEGs (i.e., electrodes) and the remaining 1,920 (2,048–128) 3D atrial shell vertices were interpolated by both linear and Laplacian methods. In summary, Table 2 shows that reduction of electrodes do interfere, and the interpolation strategy could make the scenario even worst, hampering non-invasive arrhythmias characterization by the clinicians. AF is most affected since its underlaying mechanism in this study is micro reentry. TA outcomes is more related to threshold-base decision rather than electrophysiological meaning, but previous study has also presented a certain level of reentrance on the torso for TA [5].

Table 2. The variability of the number of filaments per second according to the reduction the number of AEGs from 2,048 to 128 and type of interpolation.

Arrhythmia	Original	Linear	Laplacian
AF	6.4 ± 1.1	2.3 ± 1.5	1.1 ± 0.6
FLA	0	0	0.2 ± 0.3
TA	0.4 ± 0.8	0.8 ± 0.9	0.7 ± 1.3

3.2 Inverse Solution

Influence of the signal's level noise. Similar to presented for the forward solution, the bandpass filter with the cut-off frequency centered around HDF, showed to reduce the disparity between biomarkers extracted from the atria phase maps, in different SNRs, and arrhythmias. For example, in AF, the number of filaments per second obtained by the above preprocessing bandpass filter, reduced the biomarker variability across the SNRS in 13%, face to the other two bandpass filters (0.5–30 Hz and 3–30 Hz). Similarly, the spatial displacement of the phase singularity points on the atria, did presented a variability reduction at around 38%. This biomarker is associated with the spatial localization of the reentry mechanism core on the atria, responsible to the arrhythmia maintenance.

Influence of the 3D morphological torso/atria modifications. In general, it was observed that the cardiorespiratory did not contribute to significantly fluctuations of phase and frequency biomarkers extracted from I-AEGs maps, in all three arrhythmias. But in the cases where modifications were pronounced, it showed to be associated with AF. Some can argue that it could be intrinsically associated with its irregularity nature. Furthermore, phase biomarkers were not identified in the two most regular atria arrhythmias (FLA and AT), when the inverse solution was applied with morphological changes of the 3D geometries.

Influence of the torso number of electrodes. It was observed that as higher the number of electrodes in the torso layouts, the smaller the difference of the biomarkers calculated, face to the estimated with original number of signals, present in each vertex of the 3D torso geometry shell. For example, the number of filaments per second during AF, in the atria, extracted from the I-AEGs phase maps (inverse solution applied to the 771 ECGs), was about is 0,4. As the number of leads decrease, and unknown torso potentials are dependent on interpolation, to make feasible the inverse solution, the outcomes changes, accordingly, Linear interpolation: 131 torso electrodes (1,1 ± 1,0 filaments/second); 62 (1,6 ± 1,8); 32(1,3 ± 0,7); 16(1,1 ± 1,9) vs. Laplacian interpolation: 131 (0,6 ± 0,5); 62 (0,9 ± 1,0); 32 (2,2 ± 3,1) and 16(2,7 ± 1,6). In this example, with 131 electrodes, disparity reduced with Laplacian interpolation rather linear. When the analysis was extended to other biomarkers, it is also possible to observe that Laplacian interpolation still performed better. Currently, Cardioinsight Mapping Vest (Medtronic, InC.) has a vest with embedded 252 torso electrodes. Recently, a system with 62 electrodes has been developed by the research group of this study, with collaboration of investigators from InCor, HCFMUSP. It is believed that further studies on evaluation of different interpolation methods, would allow atria characterization under AF, even with a lower number of electrodes (i.e., 62).

Figures 1 and 2 highlight the most affected biomarkers in AF, according to the Laplacian interpolation method and 3D atria changes of morphology, respectively. The higher the font size, the most affected is the biomarker. Moreover, as much dark is the font in the same color, most affected is the biomarkers. In Fig. 1, representing the effect of Laplacian interpolation in the biomarkers, obtained from the forward (A) and inverse solution (B). Phase biomarkers are varying in shades of red and frequency in blue. For the forward solution, the phase biomarkers are the most affected, in special, the number of regions of PSs and filament rate, both related to the propagation of the micro reentrant activity in AF. Figure 1B shows that in the inverse solution, Laplacian interpolation also modify biomarkers extracted from frequency maps, reflecting the high frequency areas in the atria (localization, size and organization).

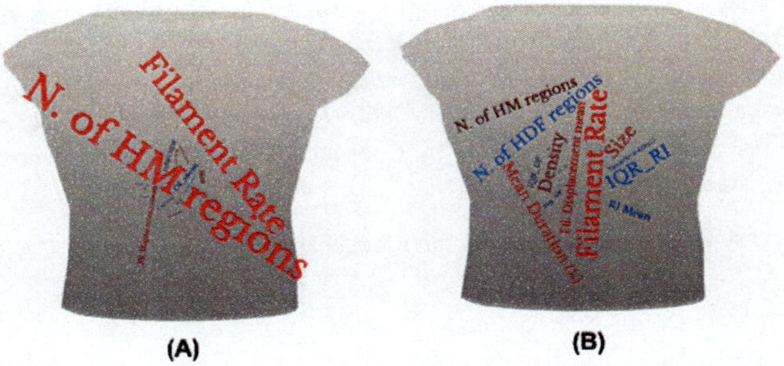

Fig. 1. Effect of Laplacian interpolation on the extracted biomarkers for the forward (A) and inverse solution (B).

In Fig. 2, representing the impact of atria morphological changes, due its systole and diastole, phase biomarkers vary in yellow tones, and frequency in blue. In both forward (A) and inverse (B) solution, phase biomarkers are the most affected. In Fig. 2A, the filament rate and the area covered by the PS regions, are the two most affected, and for Fig. 2B, also the filament rate, and number of regions of PSs. These biomarkers are related to the AF micro reentry characterization, in terms of localization, propagation and number of micro reentries.

(A) (B)

Fig. 2. Sensitivity of the biomarkers extracted from the forward (A) and (B) inverse solution maps, after the 3D atria morphologic changes.

4 Conclusion

The pre-processing strategy may distort the interpretation of TA, FLA and AF. The pre-processing based on the application of the bandpass filter centered on the H_{DF} prior to the implementation of the forward and inverse solutions, allowed to preserve the characteristics of the activation frequency even for a complex rhythm such as AF, demonstrating the need for a customized pre-processing strategy. Furthermore, for the application of the forward and inverse solution, the morphological changes in the geometries of the torso and atria throughout the cardiorespiratory cycle should be considered and, if it requires interpolation, the Laplacian interpolation should be used. This pipeline strategy allows the customization of the method reducing the disparity of results in the evaluation of atrial arrhythmias.

Acknowledgments. G.V.C., C.R.R and J.S are supported respectively by grants #2020/13017–2, #2020/12841–3 and #2018/25606–2, São Paulo Research Foundation (FAPESP). M.S.G. received funding from the European Union's Horizon research and Innovation programme under the Marie Skłodowska-Curie grant agreement no. 860974 and the Generalitat Valenciana Grants (ACIF/2020/265). I. Sandoval is supported by Coordenação de Aperfeiçoamento de Pessoal de Nível Superior—Brasil (CAPES).

Conflict of Interest MS Guillem is co-founder and shareholder of CORIFY CARE SL.

References

1. Page, R.L., Joglar, J.A., Caldwell, M.A., et al.: 2015 acc/aha/hrs guideline for the management of adult patients with supraventricular tachycardia: a report of the american college of cardiology/American heart association task force on clinical practice guidelines and the heart rhythm society. J. Am. Coll. Cardiol.Cardiol. **67**, e27–e115 (2016)
2. Salinet, J., Molero, R., Schlindwein, F.S., et al.: Electrocardiographic imaging for atrial fibrillation: a perspective from computer models and animal experiments to clinical value. Front. Physiol. **12**, 653013 (2021)
3. Farina, D. Forward and inverse problems of electrocardiography: clinical investigations. PhD thesis, Universit¨at Karlsruhe 2008

4. Yoram, R.: Noninvasive electrocardiographic imaging of arrhythmogenic substrates in humans. Circ. Res. **112**, 863–874 (2013)
5. Marques, V.G., Rodrigo, M., Guillem, M.S., et al.: Characterization of atrial arrhythmias in body surface potential mapping: a computational study. Comput. Biol. Med.. Biol. Med. **127**, 103904 (2020)
6. Restivo, C.R., Costa, G.V., Sandoval, I., Guillem, M.S., et al.: Characterization of atrial Arrhythmias through the forward problem of electrocardiography: preliminary analysis. In: 17th International Symposium on Medical Information Processing and Analysis (SIPAIM) and 10th Symposium on Medical Instrumentation and Imaging (SIIM) 2021, Campinas
7. Costa, G.V., Restivo, C.R., Sandoval, I., et al.: Characterization of atrial arrhythmias using the eCGi: preliminary analysis. In: 17th International Symposium on Medical Information Processing and Analysis (SIPAIM) and 10th Symposium on Medical Instrumentation and Imaging (SIIM) 2021, Campinas

Panoramic Optical Mapping Setup in a Langendorff Perfused Rabbit Heart for the Study of Atrial Fibrillation

Jimena Siles[1,2], Ilija Uzelac[3], Italo Sandoval[2], Vinicius Silva[2], Giovanni Weber[2], Rhuan Correia[2], and João Salinet[2](✉)

[1] Graduate Program in Biotechnoscience, Federal University of ABC, São Paulo, Brazil

[2] HEartLab, Federal University of ABC, São Paulo, Brazil
joao.salinet@ufabc.edu.br

[3] Georgia Institute of Technology, School of Physics, Georgia, USA

Abstract. Atrial fibrillation (AF) is the most common sustained cardiac arrhythmia in clinical practice, affecting up to 2% of the world's population and becoming an important public health problem. Basic and clinical research has significantly improved the diagnosis and treatment of AF. Current commercial electrical mapping systems are prone to errors and it is challenging to characterize and locate the AF arrhythmogenic sources correctly for certain types of AF [1]. Due to its non-invasive and high spatiotemporal resolution, optical mapping has helped researchers to improve current knowledge of AF. This study presents preliminary results from three rabbit's heart experiments, obtained with a Langendorff perfusion setup assembled to a panoramic optical mapping setup, with three optical cameras and high-power LEDs, automatically managed by an in-house developed user graphical unit interface (GUI). Fluorescence obtained after dye excitation was recorded using three special cameras, and data was processed to present the normalized optical action potential (OAP) maps. A 3D heart surface reconstruction equipment records a series of images around the whole heart to obtain a 3D mesh and project the normalized OAP maps.

Keywords: Atrial fibrillation · Isolated heart perfusion · Panoramic optical mapping · Animal experiment · Mechanisms

1 Introduction

Atrial fibrillation (AF) is the most frequent sustained cardiac arrhythmia, affecting up to 2% of the world population [2]. It is estimated that AF is responsible for 33% of all hospital admissions with an average annual cost of $5,450 per patient [3,4]. This arrhythmia is characterized by the collapse of organized atria cardiac electrical activity [2], decreasing cardiac output and increasing the chances of occurrence of thromboembolic phenomena, heart failure, and sudden arrhythmic death [3,5,6].

J. L. B. Marques et al. (Eds.): CLAIB 2022/CBEB 2022, IFMBE Proceedings 98, pp. 86–94, 2024.
https://doi.org/10.1007/978-3-031-49401-7_9

The mechanisms responsible for AF trigger and maintenance could differ from patient to patient, and multiples can coexist simultaneously, or interspersed with each other in the same patient, favoring the complex heart rhythm [7]. In intracavitary mapping catheter with built-in-electrodes, factors such far field effects, electrical mapping system type, duration of the recorded signals, and the number of electrodes can influence the identification of AF mechanisms [8]. Non-invasive electroanatomic mapping systems such as body surface potential mapping (BSPM) and electrocardiographic imaging (ECGi) are used for the characterization of AF sources. BSPM allows torso potential acquisition and representation in a 3D shell. ECGi allows the prediction of epicardial atria electrical activity through the BSPM signals and 3D geometries of the torso and atria [9]. It has been applied with great expectation in the clinic due to its versatility and higher spatial resolution comparable to intracardiac mapping [9].

The technologies above have played an important role in improving the current knowledge of AF mechanisms and treatments. Due to AF complexity [10,11] and a significant prone to errors in current commercial systems (inability to provide information about repolarization characteristics of electrically active cells, presence of far-field information, and sensitivity to inter-electrode spacing and wavefront orientation [12]) for certain groups patients it is challenging to characterize and locate the AF arrhythmogenic sources correctly [1]. In recent decades, important advances in the understanding of AF and development of therapeutic strategies, have been achieved through the advent of the high spatial resolution, obtained by optical mapping, considered the gold standard in mapping cardiac electrical activity [8,13]. In vivo and in vitro studies allowed the identification of AF mechanisms, later observed in patients. In these systems, voltage-sensitive dyes alter its emission spectrum proportionally to voltage changes of transmembrane myocardial cells, allowing a non-invasive and high spatiotemporal resolution mapping of the heart electrical activity, contributing to a reliable characterization of the complex patterns of AF.

A video camera with high sensitivity and low noise acts as an optical sensor for high-speed visualization of photons emitted by the dye attached to the cell membrane. There are many advantages of optical mapping in the face of electrical mapping by electrodes and, among them, stand out: the fluorescent signal is more proportional to the potential of transmembrane than the electrogram by the electrode (extracellular measurement of the potential), it is also less dependent on the direction of wavefront propagation compared to bipolar electrodes or large number of unipolar electrodes, its spatiotemporal resolution is high, and optical signals are more immune to contamination of field electrical activity far from adjacent areas. Optical mapping can generate images of the heart from various points of view, using more than one optical camera spaced between each other around the heart, called panoramic optical mapping [14], which the main objective is to obtain fluorescent images of the electrophysiological activity of the entire epicardial surface of the heart, allowing the study of the spatiotemporal propagation of electrical activity responsible for the maintenance of AF.

In this study is proposed a Langendorff perfusion setup for perfusion of rabbit's heart, assembled to a panoramic optical mapping setup, with three optical cameras and high-power LEDs, automatically managed by an in-house developed user graphical unit interface (GUI). The GUI allows also reconstruction of the 3D heart surface, for development of 3D electrophysiology standard maps. After the tests, three experiments with ex-vivo rabbit's hearts were carried out.

2 Methods

The research protocol described below has been approved by the local Committee on Ethics in the Use of Animals (CEUA), process number no. 3947230519.

2.1 Protocol and Langendorff Perfusion System

Prior to the experiment, the following solutions were prepared: (1) 2 liters of cardioplegia solution (in mM, 110 NaCL, 16 KCL, 10 $NaHCO_3$, 16 $MgCL_2.6H_2O$, 1.2 $CaCl_2.2H_20$), and (2) 4 liters of Tyrode solution (in mM 130 NaCL, 4 KCL, 24 $NaHCO_3$, 1.2 NaH_2PO_4, 1 $MgCl_2.6H_2O$, 1.8 $CaCl_2.2H_20$, 10 Glucose $C_6H_{12}0_6$). Contraction motion was suppressed by (±)-Blebbistatin at Tyrode's concentration of 1.7 μM, prepared as a stock solution dissolved in DMSO (2.5 μM) at the ratio of 5 mg/ml. V_m dye Di-4-ANBDQPQ (JPW-6003) (Potentiometric dyes) was previously prepared as a stock solution dissolved in ethanol at the ratio of 20 mg/ml. We used 0.25 mg of the V_m dye per heart.

The experiments were carried out with three male New Zealand rabbits (3.8, 3.8, and 3.5 kg). In each experiment, the animal was lightly anesthetized by an infusion of acepromazine (0.95 ml), injected intramuscularly. After 30 min, ketamine (1.52 ml) and xylazine (0.95 ml) were infused. Five minutes before the euthanasia, heparin 500 U/kg was administered to prevent clots from forming in blood vessels. The euthanasia was performed by thoracotomy, under deep anesthesia, for the heart extraction. Briefly, thoracic cavity was accessed and quickly opened. Heart was extracted by cutting at the upper end of ascending aorta [14]. After removal, heart was washed and dipped with cold cardioplegia solution, to induce bradycardia, by decreasing energy consumption [15]. After lung, trachea, and fat removal, a silicone tubbing (7 cm long and 3 mm in diameter) was used to cannulate the aorta and connect the heart to the Langendorff-perfusion setup. Figure 1A presents the experimental setup with the Langendorff and the panoramic optical mapping system developed. The heart is located in the center of the panoramic optical mapping setup, inside of an acrylic tank (Fig. 1B).

The Langendorff system pressure ranged between 50 to 60 mmHg. A peristaltic pump (Gilson, Minipuls Evolution) pumped Tyrode solution at around 15 ml/min, and adjusted to keep pressure constant. The Tyrode solution was gassed with carbogen (95% O_2 and 5% CO_2) to keep pH stable around 7.4. After, the solution was passed through a heat exchanges connected to a thermal circulatory pump inside a digital water bath (7Lab, Brazil) filled with distilled water

A

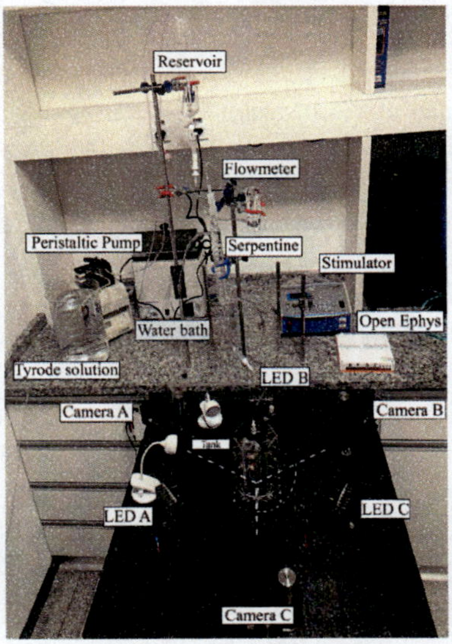

B

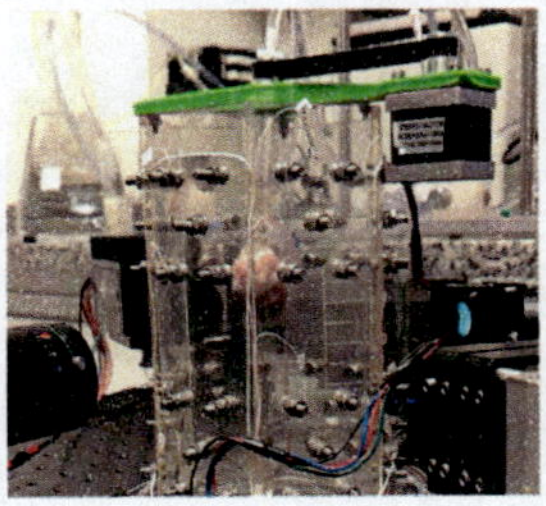

Fig. 1. A. The experimental setup with the Langendorff and the panoramic optical mapping system; B. Heart inside the acrylic tank

at 37°C. Figure 1A shows the Langendorff system's components. After, the cannulated heart was connected to the setup, and the system was cycled for 15 min, for equilibrium. After that, (±)-Blebbistatin stock solution was added to the Tyrode solution (1.71 μM) followed by the voltage-sensitive dye Di-4ANBDQPQ (0.25 mg).

A stimulator (EFF 341, Insight Equipamentos-Pesquisa e Ensino, Brazil) was used for AF induction. The positive and negative electrodes from the stimulator were mounted on a support, allowing contact with the right atrium of the heart. The AF induction was performed through a sequence of pulses, ranging between 7 to 8 ms, and intervals between stimuli ranging from 130 to 350 ms, initially. The interval between stimuli was decremented until fibrillation. Fluorescence recordings from the experiments were analyzed during sinus rhythm, pacing stimulation, and AF.

2.2 Panoramic Optical Mapping System

Excitation Light Source. After staining the heart with the Di-4-ANBDQPQ, heart was excited using a custom assembled apparatus. Figure 2A presents the excitation system consisting of three high-power deep-red LEDs (650 nm, Luminus Devices Inc., USA) placed on a heatsink. Each LED light was collimated (Aspheric Condenser Lens, Thorlabs) and band-pass filtered (650/40 nm Thorlabs, 5.5 V and 5A) before reaching the heart epicardia. The LED intensity was controlled with a custom in-house LabVIEW GUI (version 21, National Instruments), controlling the DC power supply (Programmable DC Power Supply, Tektronix Keithley).

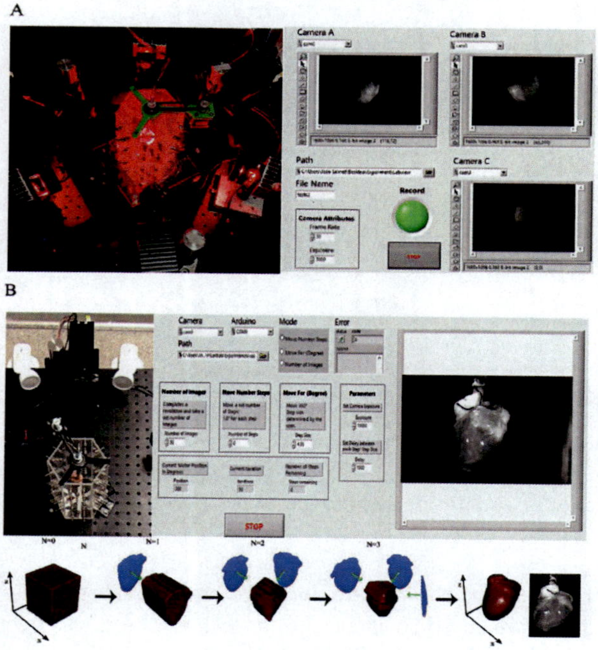

Fig. 2. A. Panoramic optical mapping system under acquisition of the heart electrical activity (left) and the in-house LabVIEW GUI for simultaneous recording of the three optical cameras (right); B. 3D heart epicardial surface reconstruction system (left) and the in-house LabVIEW GUI for capturing heart images for 3D heart surface reconstruction (right). At the bottom, the pipeline used to generate the 3D heart surface reconstruction.

Fluorescence Recording. The emitted fluorescence was passed through a long pass filter of 715 nm cut-on wavelength (BrightLine Long / Short pass Single-edge Filter, Semrock, USA) on each of the three cameras (Emergent technologies) with a C-mount lens (Fujinon). Each camera was separated 120°

(Fig. 1A white dotted line) between them. A custom in-house made LabVIEW GUI (version 21, National Instruments) triggers the cameras and allows acquisition profile customization, with default at acquisition of fluorescence images at 30 frames/s, at a resolution of 1024×1024 pixels (Fig. 2A right).

2.3 3D Heart Surface Reconstruction System

Figure 2B (Left) shows the strategy for the 3D heart surface reconstruction. This system has a customized three-pointed star-shaped adapter printed out with a 3D printer, and the adapter was placed on the top of the translucent (transparent and low autofluorescence) hexagonal acrylic tank (180×57x4 mm). Positioned at a fixed distance from the tank, one optical camera records the sequence of heart images while it is 360°C rotated. Two white LEDs were used to produce uniform light on the heart, for later silhouette segmentation and 3D mesh generation. The in-house LabVIEW GUI (Fig. 2B right) allows the user to control degree of rotation on clockwise, or counterclockwise directions. This system is composed by an Arduino Uno, bipolar stepper motor driver (DRV8825), and a stepper motor (JK35HS34-1004) connected to a belt in the central part of the star-shaped device. The toolbox driver software Vision Acquisition Software IMAQdx (version 21, National Instrument) was used to acquire, display, and save the images. It was followed by a silhouette identification in each acquired image, using image processing and analysis in Java (ImageJ). Then, each region of interest (ROI) was projected over the volume along its respective direction (0°, 120° and 240°, respectively), each voxel outside the ROI was removed. After, the volume contour was obtained by verifying each voxel neighbor, and these were used to get a triangular mesh using the Poisson method [16]. Finally, the mesh was simplified by merging close vertices until reach a number of vertices between 2500 and 3000 (Fig. 2B bottom).

2.4 Data Processing

Sequence of recorded images representing fluorescence intensity from each camera was separately processed in MATLAB (2021, Mathworks). For each signal trace data was filtered with spatio-temporal Gaussian smoothing filter of 3×3 spatial filter kernel size, and 1×5 temporal filter kernel size. In the last step, each signal trace was normalized and OAP maps were generated. For 3D representation, the normalized OAP maps were overlaid on the developed 3D epicardia heart surface meshes. Projection was obtained as follow: (1) for each camera (A, B, and C), the generated 3D mesh was rotated in its respective direction (0°, 120° and 240°, respectively); then (2) the visible nodes were projected on to the map, and finely adjust was manually made through the interface; (3) the image values was interpolated to each projected node by using the nearest pixel value; and (4) nodes with more than one value assigned had their value considered as the weighted average of these values, the weights being the inverse of the distance between this node and the region without overlapping cameras, generating a gradual transition between the regions.

3 Results

Figure 3A presents one heart image with its 2D normalized OAP map, for each of the three cameras of the panoramic optical mapping system.

The maps were normalized between 0 and 1, where no activation is represented by 0, and the highest seen potential as 1. Figure 3B presents the respective 3D surface generated, throughout the pipeline of the imaging processing applied

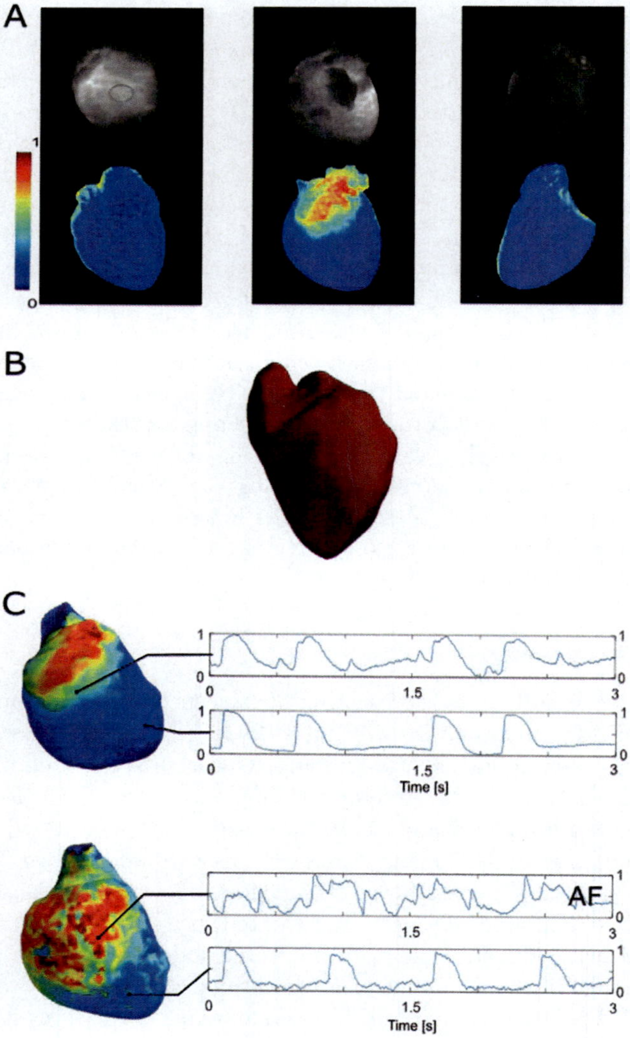

Fig. 3. A. The real heart image and the 2D normalized OAP map for cameras A, B, and C. B. Generated 3D surface C. 3D surface projected with atrial and ventricular OAP in sinus rhythm and AF

to the acquired 50 images (.tiff), steps of 7.2°, during 360° rotation. Figure 3C shows the normalized maps projected over the 3D surface generation under sinus rhythm and induced AF (the frequency of the stimulus was 6.9 Hz, and the pulse width 7 ms).

4 Conclusion

An experimental setup for ex-vivo heart studies, including AF, with acquisition of the epicardium electrical activity from an innovative panoramic optical mapping was proposed and the preliminary results presented. Inspired by an interdisciplinary character between the areas of engineering and health, this study involves important areas of knowledge such as signal and image processing, hardware, and animal experimentation. The setup will be used simultaneously with electrical mapping techniques (i.e. invasive, BSPM and ECGi) to customize electrophysiology maps and biomarkers during the initiation and maintenance of AF. The outcomes would potentially improve current commercial systems used to guide AF catheter ablation.

Acknowledgements. This study is supported by grant #2018/25606-2, São Paulo Research Foundation (FAPESP). J.G. Siles is supported by grant #2020/03601-9, FAPESP. I. Sandoval is supported by Coordenação de Aperfeiçoamento de Pessoal de Nível Superior - Brasil (CAPES). The authors would like to thank Alexandre Alves de Sousa Nascimento and Cayo Antônio Soares de Almeida for helping with the euthanasia and heart extraction.

Conflict of Interest. The authors declare no conflict of interest.

References

1. Malki, G., Zlochiver, S.: Cardiac spiral wave drifting due to spatial temperature gradients-a numerical study. Med. Eng. Phys. **61**, 69–80 (2018)
2. Magalhães, L.P., Figueiredo, M.J.O., Cintra, F.D., Saad, E.B., Kuniyoshi, R.R., Teixeira, R.A., Lorga, A.M., D'Avila, A.,de Paola, A.A.V., Kalil, C.A.: Diretrizes brasileiras de fibrilação atrial.Arquivos Brasileiros de Cardiologia **106**, 1–22 (2016)
3. Bollman, A., Lombardi, F.: Electrocardiology of atrial fibrillation. In: IEEE Engineering in Medicine and Biology Magazine, IEEE, vol. 25(6), 15–23 (2006)
4. Wodchis, W.P., et al.: A review of the cost of atrial fibrillation. Value in Health **15**(2), 240–248 (2012)
5. Lilly, L. S.: Pathophysiology of heart disease: a collaborative project of medical students and faculty. Lippincott Williams & Wilkins (2012)
6. January, C.T., Wann, L.S., Calkins, H., et al.: AHA/ACC/HRS focused update of the 2014 AHA/ACC/HRS guide- line for the management of patients with atrial fibrillation: a report of the American College of Cardiology/American Heart Association Task Force on Clinical Practice Guidelines and the Heart Rhythm Society. J. Am. Coll. Cardiol. **74**, 104–132 (2019)

7. Zipes, D.P., Jalife, J., Stevenson, W.G..: Cardiac Electrophysiology: From Cell to Bedside E-Book. Elsevier Health Sciences (2017)
8. Berenfeld, O., Ennis, S., Hwang, E., et al.: Time-and frequency domain analyses of atrial fibrillation activation rate: the optical mapping reference. Heart Rhythm **8**, 1758–1765 (2011)
9. Salinet, J., Molero, R., Schlindwein Fernando S., et al.: Electrocardiographic imaging for atrial fibrillation: a perspective from computer models and animal experiments to clinical value. Front. Phys. **12**, 653013 (2021)
10. Allessie, M., Groot, N.: CrossTalk opposing view: rotors have not been demonstrated to be the drivers of atrial fibrillation. J. Physiol. **592**, 3167 (2014)
11. Zaman, J.A.B., Baykaner, T., Schricker, A.A. , Krummen, D.E., Narayan, S.M.: Mechanistic targets for the ablation of atrial fibrillation. Glob. Cardiol. Sci. Pract. (2017)
12. Cantwell, C.D., Roney, C.H., Ng, F., S., Siggers, J. H., Sherwin, S. J., Peters, N. S.: Techniques for automated local activation time annotation and conduction velocity estimation in cardiac mapping. Comput. Biol. Med. **65**, 229–242 (2015)
13. Laughner, J.I., Ng, F.S., Sulkin, M.S., Arthur, R.M., Efimov, I.R.: Processing and analysis of cardiac optical mapping data obtained with potentiometric dyes. Am. J. Physiol. Heart Circ. Physiol. **303**, H753–H765 (2012)
14. Lou, Q., Li, W., Efimov, I.R.: Multiparametric optical mapping of the Langendorff-perfused rabbit heart. JoVE (J. Visual. Exper.) e3160 (2011)
15. Dhein, S.: The langendorff Heart in Practical Methods in Cardiovascular Research. Springer, pp. 155–172 (2005)
16. Kazhdan, M., Bolitho, M., Hoppe, H.: Poisson surface reconstruction. In: Proceedings of the fourth Euro Graphics Symposium on Geometry Processing, vol. 7 (2006)

Simulation of Electromyographic Signals Based on Point Distribution Model

Adriano de Oliveira Andrade[1(✉)], Ariana Moura Cabral[1], Marcus Fraga Vieira[2], and Adriano Alves Pereira[1]

[1] Federal University of Uberlândia, Faculty of Electrical Engineering, Centre for Innovation and Technology Assessment in Health, Uberlândia, Brazil
adriano@ufu.br

[2] Federal University of Goiás, Bioengineering and Biomechanics Laboratory, Goiânia, Brazil

Abstract. A central issue in EMG signal modeling is the simulation of its basic units designated as motor unit action potentials (MUAPs). In this paper we show how the Point Distribution Model (PDM) may be employed for generation of synthetic MUAPs based on experimental data. Our results show that the PDM allows for a simple way of controlling the distortion of synthetic MUAPs. This distortion takes into account underlying statistical properties of experimental data. Furthermore, we illustrate how these artificial MUAPs can be used for generation of electromyographic signals via a data-driven model. A comparison between the power spectra of synthetic and actual EMG signals shows that this model may generate signals that mimic the reality.

Keywords: Electromyography · Point distribution model · Motor unit action potential · Biomedical signal modelling

1 Introduction

Since the mid 1970s, a considerable amount of papers dealing with the use and development of EMG models have been published. Some of them are described and available on the SENIAM (Surface EMG for the Non-Invasive Assessment of Muscles) CD-ROM [1].

Rosenfalck [2] describes the simulation of intra and extracellular potential fields of active muscle fibres. A mathematical analysis of the spread of action currents within muscle fibres and through the volume conductor embedding the fibres is considered in this work. An interesting feature of this model is that intra-cellular action potentials of muscle fibres are estimated without regarding the membrane theories described by the Hodgkin-Huxley equations. This leads to a considerable simplification in the final model and in its mathematical expressions.

Models like the one described in [2] have focused primarily on the spike produced by the propagating wave of excitation in the muscle fibres. More sophisticated approaches have included many other features. For instance, in [3] a

J. L. B. Marques et al. (Eds.): CLAIB 2022/CBEB 2022, IFMBE Proceedings 98, pp. 95–106, 2024.
https://doi.org/10.1007/978-3-031-49401-7_10

system describing the relationship between latencies and shapes of motor unit action potentials with the motor unit architecture was introduced.

If the aim is to investigate the effects of electrode dimension and spacing, innervation zone location, propagation of the motor unit action potential in a non-homogeneous medium, the thickness of the subcutaneous tissue layer, and many other physiological factors that influence on the final form of the composite EMG signal then the models described in [4] may be suitable.

A model that takes into account the volume conduction and anatomy of the limb was introduced in [5]. This model uses the basic approach for generation of muscle fibre action potentials described in [2] but includes some innovative features, for instance, tissue capacitance (very often considered as a purely resistive element) and fibre curvature. In [6] this signal simulator was extended in order to account for insulation and shielding of fibres. The authors highlighted that those parameters may be relevant for applications regarding the control of multifunctional prostheses.

For studies that are focused on the analysis of the amplitude of the EMG signal, for example, those investigating the relation between EMG amplitude and force, the modelling of the signal as a band-limited Gaussian random process may be employed [7].

In some investigations, for instance, those concerning the generation of artificial signals for evaluation of EMG decomposition systems, the main interest is not on the simulation of the properties of the volume conductor but instead in controlling characteristics that primarily affect the decomposition process. Some of those important features are highlighted in [8]: (i) the shape similarity between MUAPs of different motor units; (ii) the shape changes of MUAPs belonging to the same motor unit during time.

Different strategies may be followed in order to generate signals that take into account the characteristics above. In [9] experimental MUAPs were distorted by random phase shifts computed via the Discrete Fourier Transform. In [10] MUAPs were simulated as a summation of three terms of a Gaussian probability function. The random selection of the parameters of the Gaussian functions allowed for MUAP distortion.

In [11] MUAPs obtained from experimental wire EMG were expanded in Associated Hermite functions so that a mathematical description of their morphology was obtained. By varying the expansion coefficients it was possible to create other MUAP shapes with some common characteristics with the original ones.

There are many strategies for simulating MUAPs, but most overlook the statistical properties of available experimental data. In this paper, we show how to generate artificial motor unit action potentials using the Point Distribution Model (PDM). This simulation strategy satisfies those important features highlighted in [8] and takes into account statistical properties of experimental data. Furthermore, we illustrate how artificial MUAPs obtained from the PDM can be used for generation of electromyographic signals by means of a data-driven model.

2 Methodology

2.1 The Point Distribution Model

A common problem that researchers of Computer Vision and Image Processing fields face is the one of creating new instances of an image given a set of known examples. The new image patterns have to share common information with the examples in an image library but at the same time they should be unique. The PDM has been successfully applied for solving this problem [12]. The PDM of a random vector $\mathbf{X}$ is defined in Eq. 1.

$$\hat{\mathbf{x}} = \bar{\mathbf{x}} + \mathbf{P}\mathbf{b} \tag{1}$$

where $\hat{\mathbf{x}}$ is a new instance of $\mathbf{X}$, $\bar{\mathbf{x}}$ is the mean of $\mathbf{X}$, $\mathbf{P}$ is an array of eigenvectors obtained from $\mathbf{X}$ and $\mathbf{b}$ is a vector that indicates how much variation is exhibited with respect to each of the eigenvectors. For instance, $\mathbf{X}$ could be a group of experimental MUAPs originated from the same motor unit.

$\mathbf{P}$ is normally obtained from an eigen-decomposition of the covariance matrix of $\mathbf{X}$. The eigenvectors with larger eigenvalues correspond to the directions of larger variation in the underlying data, thus, the PDM can be seen as a statistical model of the global shape variation of the training set $\mathbf{X}$.

The vector $\mathbf{b}$ is unique for each new instance $\hat{\mathbf{x}}$. In practice it is related to the eigenvalues of $\mathbf{P}$ and each one of its ith components, b_i, is randomly generated as a scalar limited to a given interval dependent on the eigenvalues of $\mathbf{P}$.

The new instances of $\mathbf{X}$, $\hat{\mathbf{x}}$, share a common information $\bar{\mathbf{x}}$, and they are unique because of the random nature of $\mathbf{b}$. This means that this model satisfies the requirements for MUAP generation.

Selection of Parameters for the PDM In practice the vector $\mathbf{P}$ is formed by a few eigenvectors that account for most part of data variance. In this work, for instance, eigenvectors that account for more than 90% of the data variance are used. This was a practical choice that produced satisfactory results, that is, those eigenvectors retained enough information for describing different groups of MUAPs.

A typical interval for generation of $\mathbf{b}$ in image processing is $-k\sqrt{\lambda_i} \leq b_i \leq k\sqrt{\lambda_i}$, where k is a user-defined constant and λ_i is the ith eigenvalue related to the ith eigenvector in $\mathbf{P}$. This interval defines the distance between a new instance $\hat{\mathbf{x}}$ and the most typical pattern in the data set represented by $\bar{\mathbf{x}}$. The smaller the value of k, the closer a new example $\hat{\mathbf{x}}$ is to the centre $\bar{\mathbf{x}}$.

2.2 The EMG Simulator

Figure 1 depicts the main steps required for EMG signal synthesization. Note that this simulator is totally data-driven. The synthesization of signals requires four basic steps labeled in the diagram. First, the number of active motor units is defined. This parameter does not change during simulation, meaning that the recruitment of new motor units is not allowed. In practice such condition may be obtained during isometric and isotonic contractions. This input is then used by the *Sampler* for a random and exclusive selection of different groups of experimental MUAPs from a MUAP library. Each group of motor unit action potentials belongs to a different motor unit.

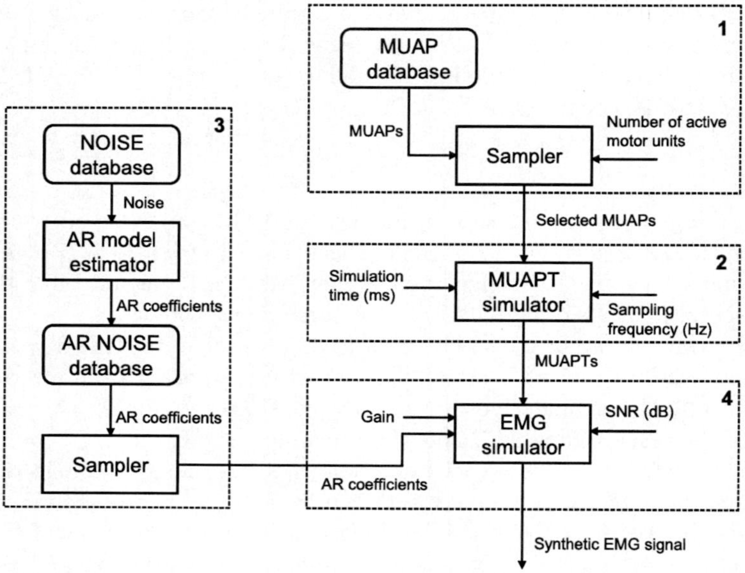

Fig. 1. Block diagram of the EMG simulator.

In the second step, the selected groups of MUAPs, the simulation time and the sampling frequency are employed as input variables to the MUAPT simulator, which generates the firing time distribution of different motor units and position MUAPs at their occurrence time. If *Sampling frequency* is different from the original sampling frequency used for digitization of experimental MUAPs then they are resampled by using interpolation (e.g. by means of splines).

The positioning of motor unit action potentials at their occurrence time is performed by an operation of convolution between a delta function, with a unit pulse positioned at the occurrence time of a given MUAP, and the MUAP waveform.

The definition of MU firing statistics consists in selecting a probability distribution function that best represent the behaviour of MUs in practice and also in the choice of adequate parameters for that function. Many experimental studies have shown that an adequate choice is to employ a Gaussian distribution function for simulation of MU firing statistics [13].

Practical investigations concerning the firing of MUs of the *First Dorsal Interesseous* (the muscle in study in this research) of health subjects have shown that typical mean firing frequencies may vary from 5 Hz to 20 Hz for muscle contractions inferior to 40% of the maximum voluntary contraction [14,15]. A typical coefficient of variation, that is, the ratio between the standard deviation and the mean of the distribution may range from 0.1 to 0.25. This is used for simulation of MU firing statistics. Each active motor unit has an unique Gaussian distribution with parameters randomly defined.

It is also in the second stage that as many MUAP patterns as it is necessary are created. This means that these new examples of MUAPs (generated by means of PDM) are those that will be used for signal simulation and not the original samples stored in the MUAP library. The exact number of MUAP patterns to be generated is a function of the simulation time.

In the third stage, an autoregressive model of noise obtained from practical EMG signals is selected from a library (*AR NOISE database*). Again the *Sampler* is responsible for an exclusive and random selection of one of the available models. Different autoregressive models of noise are calculated by the *AR model estimator* from data available in *NOISE database*.

The coefficients of the autoregressive models in *AR NOISE database* are reported in Table 1. These coefficients were estimated using the Matlab function *ar* with the option *ls*, or the least squares method for estimation of AR coefficients. This algorithm is described in details in [16]. The order of the model (4) was estimated according to the work of Snider and Bonds [17].

Table 1. Autoregressive models available in the *AR NOISE database*. These coefficients were obtained from experimental signals and may be employed for generation of synthetic noise.

Data set	Autoregressive coefficients
1	[2.0176 −1.5844 0.5438 −0.0385]
2	[2.2534 −2.0640 0.9718 −0.1923]
3	[2.4261 −2.3979 1.2280 −0.2638]
4	[2.1937 −1.9037 0.7807 −0.0979]
5	[2.3060 −2.2431 1.2272 −0.3112]
6	[2.3398 −2.2147 1.0791 −0.2297]
7	[2.2534 −2.1438 1.0548 −0.2119]
8	[2.3675 −2.2030 1.0512 −0.2294]
9	[2.2287 −2.0010 0.8949 −0.1403]
10	[2.1388 −1.9062 0.9030 −0.2049]

Finally, in the fourth step, the motor unit action potential trains are combined in order to generate the artificial EMG signal. This signal is then amplified by an input variable (*Gain*) and noise corrupted for generation of the composite EMG signal with a given signal-to-noise ratio defined by *SNR*.

2.3 Experimental Protocol

Needle and surface electromyographic signals were collected from the First Dorsal Interosseous of 15 young and healthy subjects during isometric and weak contraction. A force sensor was used to constrain the level of muscle contraction. A weak contraction was equivalent to the subject holding a 100-gram load.

A pair of standard passive surface electrodes (MedTech Systems Ltd, Ag/AgCl, diameter of 1 mm, inter-electrode distance of 1 mm, type pellet) were used for signal detection. The electrodes were positioned in the direction of muscle fibres. The skin was properly cleaned with alcohol and abraded.

A concentric needle electrode (TECA - Oxford Instruments - model X53153, length = 25 mm, diameter = 0.3 mm, recording area = 0.03 mm^2) was used for detection of needle EMG signals. The electrode was positioned in the direction of muscle fibres and the skin was properly cleaned with alcohol before electrode insertion. The depth of insertion was approximately 0.5 mm.

The signal was differentially amplified via a commercial amplifier (CMRR $>$ 80 dB, input impedance $> 10^{15}\Omega$, noise level $< 1.2\mu V$) with gain set to 10000, pass-band filtered (10 Hz - 2 kHz), and digitized by a 16-bit A/D converter (PC-card DAS 16/16 Measurement Computing) at 10 kHz.

Ethical approval for this research has been granted by the Berkshire Local Research Ethics Committee (Reference number 33/04) and also by the Ethical Committee of the University of Reading, UK. Before data collection, a consent form was signed by each subject.

The SNR was set to 20 dB. For both synthetic and experimental signals the sampling frequency was set to 10 kHz, the gain set to 10,000 and the simulation time to 30 s.

3 Results

In this study a MUAP library formed by 844 examples of MUAPs was created. All MUAPs were obtained from experimental EMG signals collected from the First Dorsal Interosseous of those 15 different subjects. From each subject, one MUAP group, which was composed by MUAPs classified as members of the same motor unit, was available. The MUAPS were obtained by using the decomposition method described in [18]. Thus, in total, 15 different groups of MUAPs were available for study. Thirteen of them were surface EMG signals whereas 2 were needle EMG signals. They are shown in Fig. 2.

3.1 Generation of MUAPs from a PDM

Figure 3 shows how PDM generates MUAPs. Left-top figure shows the original MUAP data set with 58 samples. 800 MUAPs were generated. Various k simulations are presented.

In the second row of Fig. 3 the MUAP template or average is shown. Note that similar waveforms are obtained for experimental and synthetic signals. This was expected as the term $\bar{\mathbf{x}}$ should be constant in the PDM.

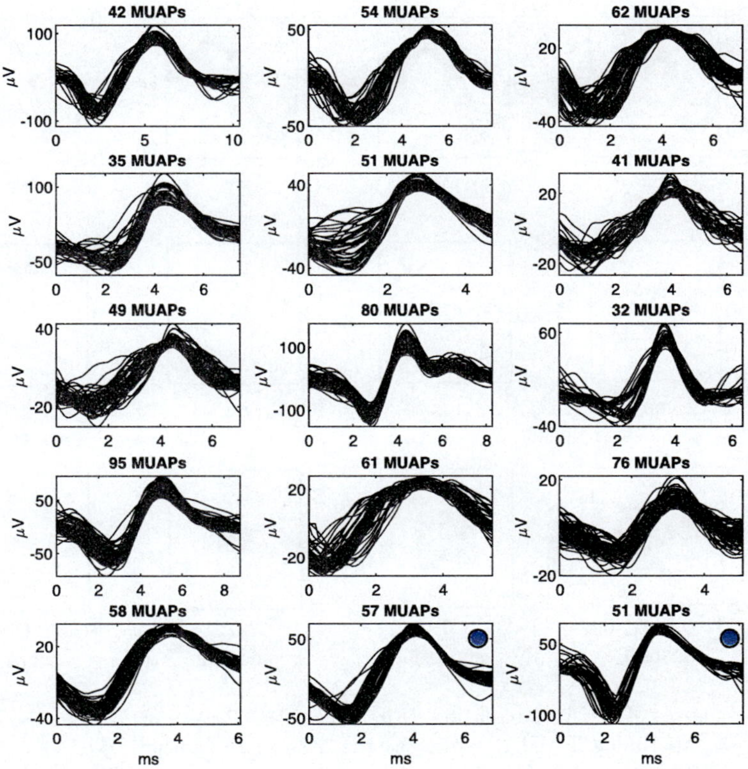

Fig. 2. Groups of experimental MUAPs available in the MUAP library. The two groups marked with a circle are from needle electromyography whereas the others are from surface electromyography.

A projection of MUAP waveforms is presented in the third row of Fig. 3. The first (score 1) and second (score 2) scores obtained from Principal Component Analysis (PCA) are employed for data visualization. This allows for an easier interpretation of the effect of the constant k on data generation. As the values of k reduce new examples are generated closer to data centres (circles in the figure). This simulation also shows that data generation is constrained to an interval, which is defined by the data themselves (dependence on eigenvalues) and can be externally controlled by adjustment of a pre-defined constant k.

3.2 Generation of Synthetic EMG Signals

Figure 4 shows two synthetic EMG signals, EMG1 and EMG2, and an experimental EMG signal. Both EMG1 and EMG2 have five active motor units and were simulated considering a PDM with $k = 0.075$.

For illustration, Fig. 5 depicts the activities of individual motor units, generated via PDM, that compose EMG1.

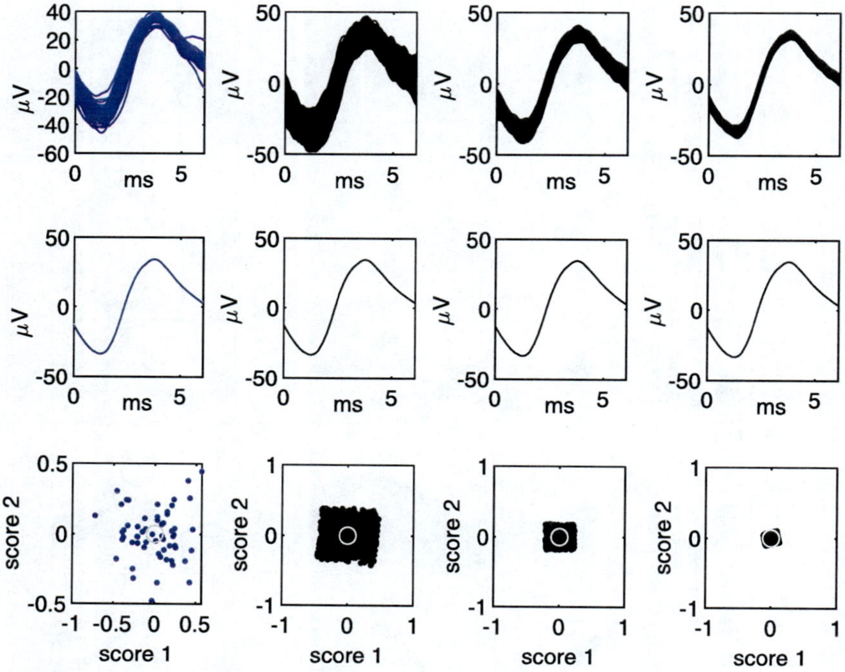

Fig. 3. Example of MUAP simulation. (First row) MUAP waveforms. The first column is a set of experimental MUAPs whereas the others are synthetic MUAPs obtained from a PDM with $k = 0.3$, $k = 0.15$ and $k = 0.075$. (Second row) MUAP template or average. (Third row) Scatter plot of first and second scores obtained from the principal components of MUAPs. The data centre is highlighted with a circle.

The distribution of the firing frequency of those motor units is shown in Fig. 6. This represents the distribution of the time intervals between the peaks of MUAPs.

Figure 7 depicts a comparison between the power spectra of the experimental EMG signal with the simulated ones. The Welch method was employed to estimate the PSD. An overlapping (50%) Hanning window of 512 points was used. Note that the experimental EMG is not correlated with the synthetic signals, meaning that information from this experimental EMG was not used in the model.

4 Discussion

This work introduced PDM to simulate MUAPs. The PDM takes into account underlying statistical properties of experimental data set to generate new instances of MUAPs. The generation of MUAPs can be externally controlled by the user by means of a constant, k, which controls how distant new examples

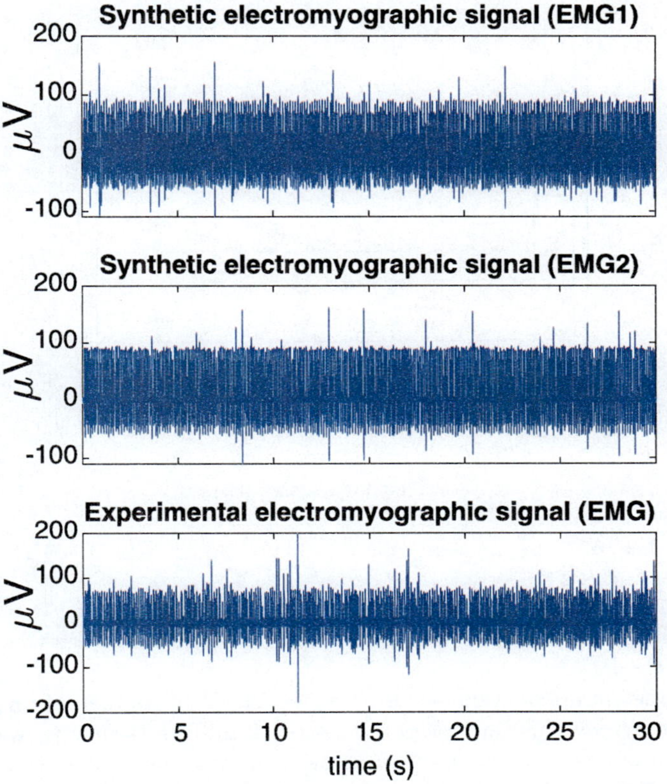

Fig. 4. Synthetic (EMG1 and EMG2) and experimental (EMG) electromyographic signals.

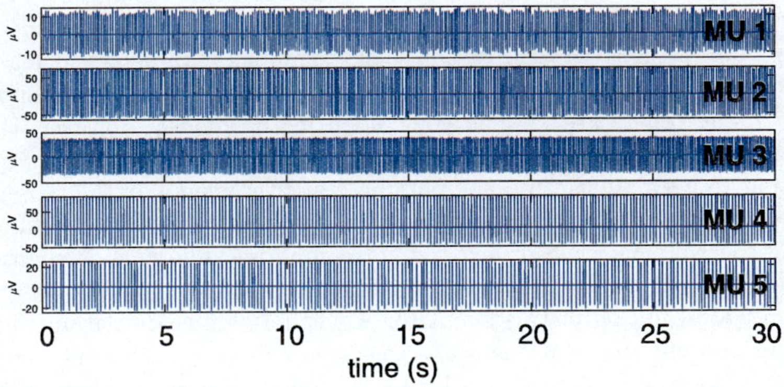

Fig. 5. Activities of single motor units of EMG1. Note that those MUAPs were simulated by means of PDM.

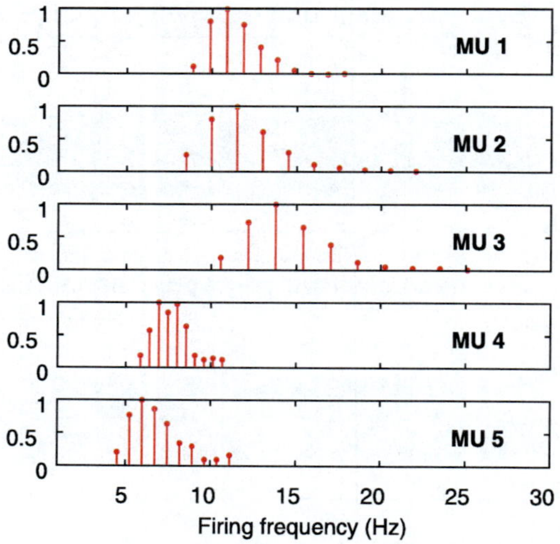

Fig. 6. Firing frequency distribution of motor units of EMG1.

are from the mean of the data set. The results in Fig. 3 illustrates this important feature of the PDM.

An important application of synthetic MUAPs is in the generation of artificial EMG signals. For this, an EMG simulator, totally data-driven, was devised. This simulator takes into account some basic parameters for simulation of EMG signals, e.g., the number of active motor units and their firing properties, noise generation based on AR models obtained from actual noise and the generation of MUAPs based on the PDM. The comparison between the power spectra of experimental and synthetic signals showed that the EMG simulator devised in this work generate signals that may mimic the reality.

As experimental MUAPs are the base of the model, parameters such as influence of electrode material and dimension, signal attenuation due to adipose tissue and many others are implicitly included in the model. However, there is no explicit control over those variables. Furthermore, additional tests should be carried out to evaluate the effect of parameters of the model on the power spectrum. A quantitative comparison between power spectra could be obtained by applying the Kolmogorov-Smirnov test for comparing different distributions of two samples.

A natural evolution of this simulator would be to include parameters that takes into account the relationship between motor unit recruitment and force. Another important feature would be to allow for simulation of multichannel signals. For this purpose MUAP databases collected simultaneously from close detection surfaces could be employed.

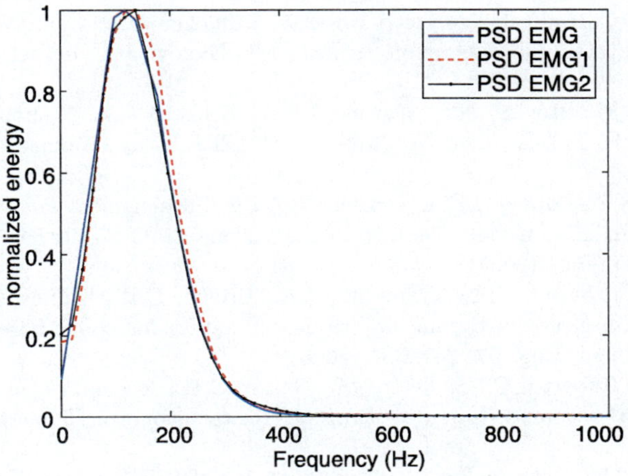

Fig. 7. Comparison between power spectra of synthetic (EMG1 and EMG2) and experimental (EMG) signals

5 Conclusion

This study introduced a method to simulate motor unit action potentials based on the point model distribution. The simulated potentials can be used in more complex models aiming to generate synthetic EMG signals. The comparison of synthetic data with experimental data showed a large similarity between the power spectra of synthetic and experimental signals.

Acknowledgements. This work was supported by the National Council for Scientific and Technological Development (CNPq), Coordination for the Improvement of Higher Education Personnel (Programs CAPES/DFATD-88887.159028/2017-00, CAPES/ COFECUB - 88881.370894/2019-01 and CAPES-PRINT-UFU), and the Foundation for Research Support of the State of Minas Gerais. A.O. Andrade, M. F. Vieira and A. A. Pereira are fellows of CNPq (304818/2018-6 and 305223/2014-3; 304533/2020-3; 309525/2021-7).

Conflict of Interest. The authors declare that they have no conflict of interest.

References

1. Hermens Hermie, J., Bart, F.: Roessingh research and development, The SENIAM CD-rom. Netherlands 90-75452-14-4 (1999)
2. Poul, R.: Intra- and extracellular potential fields of active nerve and muscle fibres: a physico-mathematical analysis of different models. Acta Physiologica Scandinavica. Supplementum **321**, 1–166 (1969)
3. McGill, K.C., Lateva, Z.C., Shaojun, X.: A model of the muscle action potential for describing the leading edge, terminal wave, and slow after wave. IEEE Trans. Biomed. Eng. **48**, 1357–1365 (2001)

4. Dario, F., Corrado, C.: Merletti Roberto. Influence of anatomical, physical, and detection-system parameters on surface EMG Biological Cybernetics. **86**, 445–456 (2002)
5. Lowery M.M., Stoykov N.S., Dewald J.P.A., Kuiken, T.A.: Volume conduction in an anatomically based surface EMG model. IEEE Trans. Biomed. Eng. **51**, 2138–2147 (2004)
6. Stoykov, N.S., Lowery, M.M., Kuiken, T.A.: A finite-element analysis of the effect of muscle insulation and shielding on the surface EMG Sig. IEEE Trans. Biomed. Eng. **52**, 117–121 (2005)
7. Prakash, P., Salini, C.A., Tranquilli, J.A., Brown, D.R., Clancy, E.A.: Adaptive whitening in electromyogram amplitude estimation for epoch-based. Appl. IEEE Trans. Biomed. Eng. **52**, 331–334 (2005)
8. Dario, F., Roberto, C., Roberto, M., Baare, O.H.: Evaluation of intra-muscular EMG signal decomposition algorithms. J. Electromyography Kinesiol. **11**, 175–187 (2001)
9. Hassoun M.H., Chuanming, W., Spitzer, A.R.: NNERVE: neural network extraction of repetitive vectors for electromyography. II. Performance analysis. IEEE Trans. Biomed. Eng. **41**, 1053–1061 (1994)
10. Fang, J., Agarwal, G.C., Shahani, B.T.: Decomposition of multiunit electromyographic signals. IEEE Trans. Biomed. Eng. **46**, 685–697 (1999)
11. Farina, D., Crosetti, A., Merletti, R.: A model for the generation of synthetic intramuscular EMG signals to test decomposition algorithms. IEEE Trans. Biomed. Eng. **48**, 66–77 (2001)
12. Tim, M.: Computer Vision and Image Processing. Palgrave Macmillanfirst ed., New York (2004)
13. Stashuk, D., Qu, Y.: Robust method for estimating motor unit firing-pattern statistics. Med. Biol. Eng. Comput. **34**, 50–57 (1996)
14. Basmajian John V., De Luca Carlo J.: Muscles Alive. Williams & Wilkins (1985)
15. Yoshihisa, M., Kazuto, A., Masa, N., Akio, K., Naoichi, C.: Motor unit firing behavior in slow and fast contractions of the first dorsal interosseous muscle of healthy men. In: Electroencephalography and Clinical Neurophysiology/Electromyography and Motor Control, vol. 97, pp. 290–295 (1995)
16. Simon, H.: Adaptive Filter Theory. Prentice Hallthird ed. (1996)
17. Snider, R.K., Bonds, A.B.: Classification of non-stationary neural signals. J. Neurosc. Methods. **84**, 155–166 (1998)
18. Andrade, A.O., Nasuto, S.J.: Kyberd Peter. Extraction of motor unit action potentials from electromyographic signals through generative topographic mapping. J. Franklin Inst. **344**, 154–179 (2007)

Bioinspired PID Controller Based on Izhikevich Neurons Optimized by Differential Evolution for Neuromorphic Implementations

Júlia Nepomuceno Mello[1](✉), Mariane Rodrigues Garcia[1], Alcimar Barbosa Soares[1], and Frederico Caetano Jandre[2]

[1] Faculdade de Engenharia Elétrica, Universidade Federal de Uberlândia, Uberlândia, Brasil
julianepomucenomello@gmail.com

[2] Programa de Engenharia Biomédica, Universidade Federal do Rio de Janeiro/COPPE, Rio de Janeiro, Brasil

Abstract. The well-known proportional-integral-derivative (PID) controllers are challenged by variations in the process dynamics, inadequate tuning and other difficulties associated with the process itself. Other controller structures, such as the biologically inspired spike neural networks (SNN), were developed and essayed against such challenges. The goal of the present work is to develop a bioinspired PID controller with a small SNN using the neuron model proposed by Izhikevich, and optimized by a Differential Evolution algorithm for neuromorphic implementations. The controller architecture is based on 3 neurons, each responding with a proportional, derivative or integrative output. The parameters of each neuron were optimized with either a step signal or a sinusoidal waveform of variable amplitude and frequency as reference. The results show that the SNN controller performed similarly to a traditional PID controller for the two applications. The best result found is represented by a mean square error of 1.605 for the step signal. Improvements might be achieved by further optimization steps, an aim for future research. Neuromorphic control applications, such as in the control of prostheses, appear as potential applications.

Keywords: PID controller · Izhikevich neurons · Neuromorphic controller · Bioinspired control

1 Introduction

Controllers, such as the proportional integral derivative (PID), are widely studied in the literature being based on the initial idea of controlling the outputs of a system based on the difference between the desired and measured value of some variable of the system [1–3]. Specifically PID controllers are already well validated in the literature as the most used controller, due to its simplicity of the algorithm, robustness of response and use in industrial processes [1]. However, such controllers still have challenges associated with variations in the process dynamics, control difficulties associated with non-linearities, disturbances and noise, control strategies incompatible with the needs and objectives

J. L. B. Marques et al. (Eds.): CLAIB 2022/CBEB 2022, IFMBE Proceedings 98, pp. 107–117, 2024.
https://doi.org/10.1007/978-3-031-49401-7_11

of the process, implementation errors, in addition to inadequate tuning and different parameterizations [1, 3].

Thus, other strategies can be used to perform process control, such as fuzzy controller [4], artificial neural networks [5] and spike neural networks (SNN) [4, 6, 7]. SNN-based controllers are inspired by biological control structures, which are represented by a neural structure that relates a stimulus received through sensor neurons to a movement generated from the output of motor neurons such as in the patelar reflex [6]. These SNN-based controllers can be made with a variable number of neurons and, consequently, with different complexity and computational cost. Applications of these controllers are presented in biological systems [4, 6] and also in other engineering control issues such as DC motors [4], damped pendulum [8] and Unmanned Aerial Vehicles [9].

A large controller was presented by Pérez Fernández et al., 2018 who built an SNN, with 3 layers of 48 Izhikevich neurons each, to control one human arm model (the Hill's arm model) with flexion and extension movements allowed by an elbow joint [6]. Previously, the same group had controlled the same arm model using a two-layer network (3 input and 2 output Izhikevich neurons) with a Differential Evolution to optimize the control system parameters [4]. The same SSN control model was used to the control of a DC motor [4] and in a traction control system [7].

Controllers with an even smaller number of neurons have already been used in order to approximate their responses to those of classical controllers, such as proportional-derivative (PD) controllers [2, 8]. The work of Dorothy, 2015 presents a PD-like controller with only two mutually excitatory Izhikevich neurons, conceived of as being antagonistic motoneurons, such as in antagonistic muscles [8]. Meanwhile, the work developed by Webb, Davies and Lester, 2011 presented the approach of building a PID controller using three spiking neurons. In this sense, each neuron approaches one of the controller terms, based on the spike frequency adaptation behavior of biological neurons [2]. In this context, using the Izhikevich model, they were able to approximate 3 neurons for the proportional, derivative and integrative functions, without, however, integrating such responses to generate a complete PID controller itself [2].

The objective of this study was to develop a bioinspired PID controller using the Izhikevich neuron model for neuromorphic implementations. Aiming at optimizing the parameters of the model, a Differential Evolution algorithm (DEA) was used to seek to minimize the error associated with a simulated PID controller when submitted to a target input (reference signal). The next section presents the proposed methodology, detailing the controller developed and the optimization process used. Next, the results of the simulations performed with different reference signals are presented. Finally, the applications for this type of system are detailed, as well as their advantages and disadvantages.

2 Materials and Methods

2.1 Neuron Model

Our bioinspired controller was built around the Izhikevich neuron model. As shown above, this model has already been tested in control applications [2, 4, 6–8]. It has several advantageous features, in relation to other models of neurons, due to its ease of implementation and good balance between accuracy and computational cost [4]. In addition, this model is able to reproduce, by changes in its parameters, variable spiking rate patterns of biological neurons [10, 11]. The model is composed of a system of two second-order differential equations (solved by Euler's method), where the membrane potential and the model recovery variable are described as [10, 11]:

$$\frac{dv}{dt} = 0.04v^2 + 5v + 140 - u + I(t) \tag{1}$$

$$\frac{du(t)}{dt} = a[bv(t) - u(t)] \tag{2}$$

where u and v represent, respectively, the membrane potential and the recovery variable responsible for cell membrane repolarization; I is the input synaptic current ant t is time (in milliseconds). The parameters of the model are a, the time scale factor of the recovery variable (higher values result in in slower recovery) and b, the sensitivity of the recovery variable to fluctuations in membrane potential [10, 11].

Whenever the membrane potential v exceeds the 30 mV threshold, a new spike is generated. After triggering the action potential or spike, the restoration of the membrane potential and the recovery variable occurs as described in the equation below [10, 11].

$$I\,fv \geq 30\,mV\ then \begin{cases} v \leftarrow c \\ u \leftarrow u + d \end{cases} \tag{3}$$

where c is the resting membrane potential value (in mV) and d is the value of the recovery variable after the spike. Typical values used for a regular spike response consist of a = 0.02 s-1, b = 0.2, c = −65 mV and d = 2 [10, 12]. In this context, the firing rate of a regular spiking excitable neuron is a function of both the input current (proportional term) and the recent rate of change of the stimulus (derivative term); the relative weights of the two terms depend on the level of adaptation that the neuron exhibits [2]. This proportional-derivative nature is the basis of the proposed neuromorphic controller.

2.2 Bioinspired PID Controller

Considering the adaptation properties described for neurons, we seek to approximate 3 neurons to represent the functions of a PID controller. In order to model neurons with proportional, derivative and integrative input-output transfer, the following limitations proposed by [2] were considered. The proportional neuron is a regular spiking neuron with no adaptation behavior ($d = 0$). The derivative unit includes a positive spike frequency adaptation behavior ($d > 0$) as in biological neurons, such that the neuron discounts previous stimuli from the current stimulus. Finally, the integrative neuron has a negative spike frequency adaptation behavior (d < 0) to create a positive feedback loop, that is, every time the neuron spikes the input current required to elicit a spike decreases [2].

The controller is thus composed of the 3 neurons described combined in a parallel system. In addition, the error signal needs to be presented to the controller and a neural coding must be defined in order to be able to determine the value of the input or output of the neuron based on the recent spike sequence [2]. Neural coding commonly used in SNN are based on time between spikes or firing rate [6]. In the present work, the neural coding is based on an input current injected in each neuron, representing the error signal, and an output value corresponding to the mean firing rate of the system. The output code is based on the traditional biological model by which most of the information in a spike train is carried in its firing rate [2]. For this purpose, a 100-ms moving average window, with step of 5 ms, was used, holding the output value constant until the next window, in order to verify the average firing rate of the output of the 3 neurons combined [2].

Finally, for the construction of the neuromorphic PID controller, it is necessary to define the parameters of each neuron, seeking the desired responses for them [8, 10, 11]. The best set of Izhikevich's parameters that describe the desired behaviors is not yet well established in the literature. These can also vary according to the application, hence the need for an optimization stage. In the present work, DEA was used to optimize the Izhikevich parameters for each neuron in the proposed PID controller based on an identity transfer function. A diagram of the proposed architecture is presented in Fig. 1.

Differential Evolution is a method belonging to the group of Evolutionary Algorithms, as well as Genetic Algorithms [13]. This method was chosen because it has several advantages, such as simplicity of implementation compared to Genetic Algorithm, for example. In addition, this methodology has few parameters to be determined and optimized in the search for the best solution [14]. Finally, it still has a good performance, with high convergence speed and robustness. Its principle is based on the concept of population, mutation and crossover used by Genetic Algorithms, but with a different approach [13, 14].

The algorithm initializes a random population covering the parameters space, in which each chromosome/individual has the dimension of the number of variables of the problem to be solved and represents one possible solution [13, 14]. In the search of the solution optimization, each chromosome (group of bits) will act as a target vector that will be compared with a mutant vector in the search for the best adapted individual, for example, with the lowest fitness function (cost function) in a minimization problem [13, 14]. The selection occurs by the greedy criterion, in which the trial and target vectors are compared and the fittest is selected. This process occurs for all individuals in the population, thus yielding a new generation. The simplest algorithm termination condition is associated with the maximum number of generations [13, 14]. The model's parameters include F, a real and constant factor that controls the amplification of the differential variation and CR, the crossover constant which establishes the probability of crossover [14].

In the proposed work, the parameters of the Izhikevich neuron were considered, that is, a, b, c and d for each neuron, thus totaling 12 bits, one tor each parameter of each neuron. The chosen fitness function was the mean squared error (MSE) of the desired signal and the decoded output signal of the PID neurons. Two 3000 ms signals at a sampling rate of 0.1 MHz were tested: (i) a step signal and (ii) a sinusoidal waveform of variable amplitude and frequency. The training signal was chosen in order to expose the system to different frequencies and amplitudes looking for a good frequency and amplitude response. Also, the addition of constant zones seeks to improves the output of the system when a steady input is applied by diminishing the response oscillations [4].

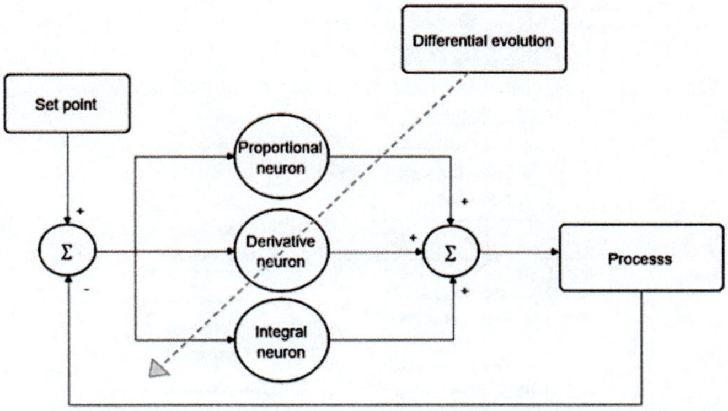

Fig. 1. Bioinspired PID controller architecture based on 3 Izhikevich neurons optimized by a Differential Evolution algorithm (DEA). The diagram shows the 3-neuron parallel structure that is optimized by the DEA. The set point was considered as a simulated signal with non-constant amplitude and frequency. The error function is the difference between the reference signal and the decoded output of the PID neurons (identity transfer) and is used as the neurons input current.

The initial ranges of the Izhikevich parameters a, b, c and d were, respectively, 0.001 to 0.3; 0.001 to 0.7; −40 to −80; and 0.001 to 14 (with positive values for the derivative neuron and negative values for the integral neuron). A population of 20 individuals was randomly initialized within the established ranges. The maximum number of generations was set to 60, F was set to 0.5 and the CR to 0.9. The ED/best/2/bin variation was employed for the mutation and crossover, due to the good performance of this variation in initial tests. This variation takes the fittest individual as the target vector, uses two difference vectors in the mutation stage and is based in the binary crossover method [13]. Figure 2 presents the architecture of the proposed optimization algorithm.

3 Results

First, the optimization of the neuron parameters was performed considering a sinusoidal waveform of variable amplitude and frequency as a reference signal [4]. The response of the fittest individual (smallest MSE value) can be seen in Fig. 3, as well as the optimized PID controller response to the sinusoidal signal. The final error of the neuromorphic PID system was 2.406. Such MSE can be considered high, especially for applications that need greater accuracy. This result points to the need for a longer optimization process, as discussed later.

Table 1 shows the optimized Izhikevich parameters considering the simulation presented in Fig. 3. The set of parameters, especially a and b, are outside the range of parameters originally proposed by Izhikevich for simulating the dynamics of biological cortical neurons [10, 11]. Such a result is expected, since a biologically compatible response is not sought, but rather the approximation of the behavior of such neurons to a PID controller, which is composed of the joint response of such neurons.

The step response of the optimized PID controller had a MSE of 2.448, similar to the error found during the training with the sinusoidal signal. The PID response can be seen

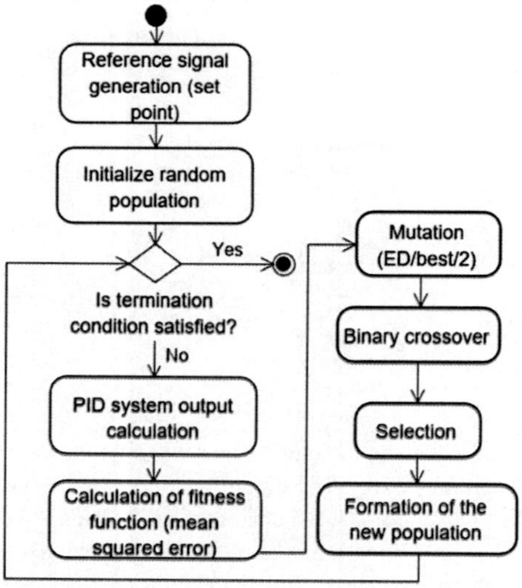

Fig. 2. Differential Evolution Architecture used to optimize the parameters of the Izhikevich neurons. Each individual in the population is represented by 12 bits (one for each parameter of each neuron). The fitness function compares the output of the PID controller with the reference signal for each individual in the population. The DEA variation ED/best/2/bin was used in the mutation and crossing of individuals. The optimization process is repeated until the defined number of generations is reached.

Fig. 4, in which it is possible to observe that the controller has a fast response, however with an overshoot considering the reference amplitude. The curve (Fig. 4) shows that the controller has a rise time of around 36.68 ms, however with a prominent overshoot of about 20.78%. A more detailed description of the optimized controller response metrics can be seen in Table 2.

In a second simulation, seeking to analyze the response to a step considering the optimization of the system for this reference signal, the Differential Evolution was performed with the step as the set point signal. The parameters found by using the step signal as reference are shown in Table 3, also outside the range originally proposed by Izhikevich, especially a and d, as expected due to the desired behaviors for the neurons modeled.

Figure 5 presents the evolution of the fitness function and the optimized bioinspired controller response for the step reference. The MSE was of 1.605. The response is closer to the desired compared with the other set of parameters, with smaller rise and fall times. Even though small fluctuations can be observed in the output of the PID controller, these are within the range of 5% error.

The system has a rise time of around 43.25 ms, and overshoot of 26,93%. The controller output metrics can be seen in Table 4. It should be noted that, for both simulations, a reduced number of generations was used due to the computational cost. It can be seen from Figs. 3 and 5 that the algorithm had not yet reached a plateau for a large number of generations, being an indication that the optimization process may have reached a local minimum. Thus, it is believed that a larger number of generations could allow an even

(A)

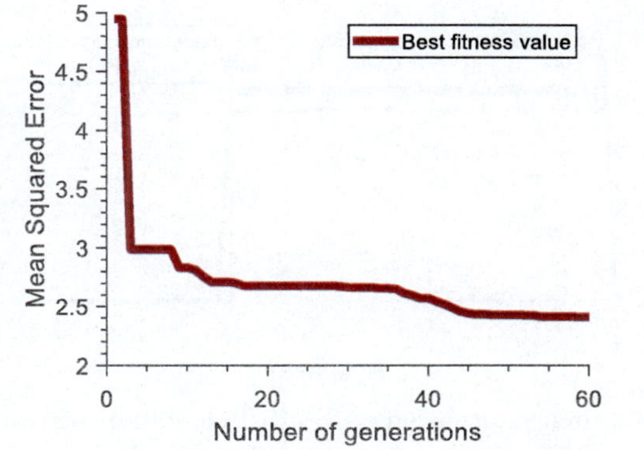

(B)

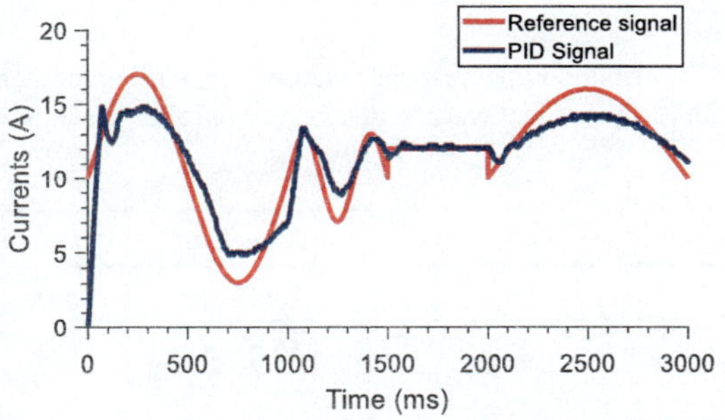

Fig. 3. (A) Evolution of the fitness function over generations and (B) Optimized PID controller response to a sinusoidal input of variable amplitude and frequency (reference signal/set point).

Table 1. Optimized Izhikeich parameters for the PID controller considering the sinusoidal waveform.

Neurons	Parameters			
	a	b	c	d
Proportional	0.109	$9.914 * 10^{-4}$	– 40.321	0.000
Derivative	$2.350*10^{-4}$	0.136	–61.530	4.911
Integral	0.250	0.222	– 62.673	– 9.810

greater reduction in the final error of the control system. Another limitation of the work is the use of identity transfer, so a different process can impose greater challenges to the proposed model. However, the results presented are promising and indicate the good ability of the proposed system to approximate bioinspired neurons in a PID controller.

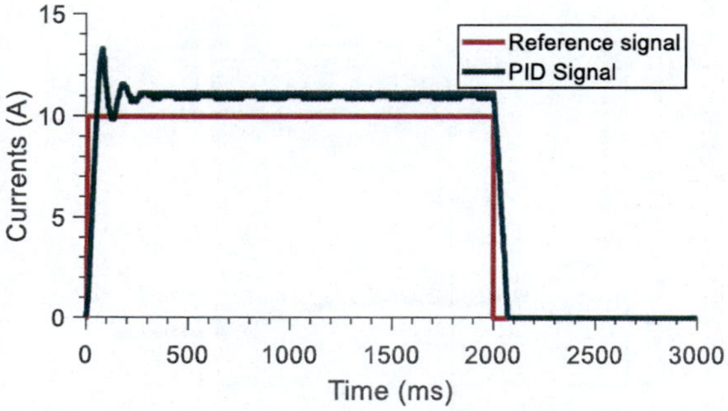

Fig. 4. Optimized PID controller response to a step signal, considering the optimization with the sinusoidal waveform signal (controller based on the parameters of Table 1).

Table 2. Optimized PID controller output response metrics considering the step as the test signal and the optimization with the sinusoidal waveform signal (controller based on the parameters of Table 1).

Metric	Value
MSE	2.448
Rise time (ms)	36.68
Settling time (ms)	276.5
Peak time (ms)	82.54
Peak overshoot (A)	13.29
Overshoot	20.78%
Peak undershoot (A)	9.86

Table 3. Optimized Izhikeich parameters for the PID controller considering the step signal.

Neurons	Parameters			
	a	b	c	d
Proportional	0.231	0.219	– 58.918	0.000
Derivative	0.084	0.167	– 8.938	8.407
Integral	0.374	0.264	– 96.436	– 3.543

4 Discussion

The ability of the Izhikevich neurons to reproduce spike frequency adaptation is associated with the slow variable u [8]. Such behavior is crucial to reproduce the derivative and integral performance necessary for the PID controller. Considering the Izhikevich parameters, such characteristics are achieved by having the d parameter as the weight given

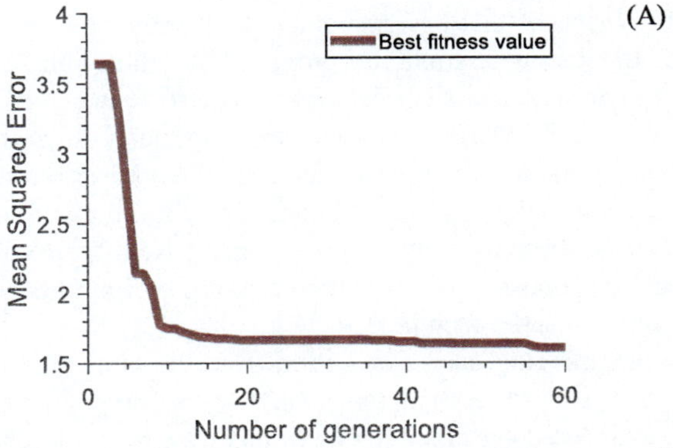

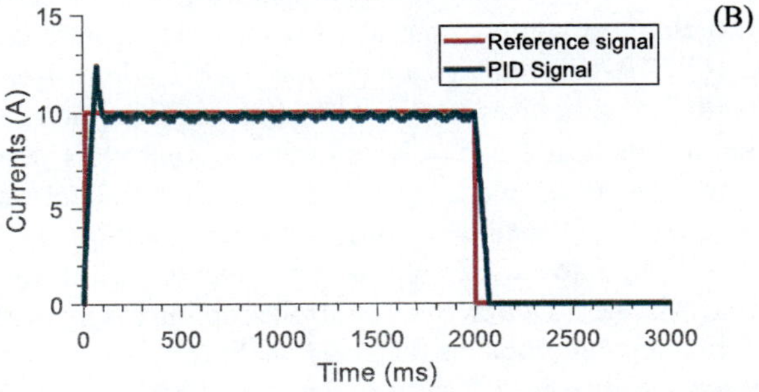

Fig. 5. (A) Evolution of the fitness function over generations and (B) Optimized PID controller response to a step signal (reference signal). Note that fluctuations are in the range of 5% error.

Table 4. Optimized PID controller output response metrics considering the step signal as the test and training signal (controller based on the parameters of Table 3).

Metric	Value
MSE	1.605
Rise time (ms)	43.25
Settling time (ms)	135.5
Peak time (ms)	71.83
Peak overshoot (A)	12.43
Overshoot	26.83%
Peak undershoot (A)	9.57

to stimuli within the window, with positive values associated with the usual derivative behavior of biological neurons and negative values associated with the integral behavior. Also, changes to parameters a change the window size over which the differentiation and integration are approximated [2]. Thus, it can be observed that, during the simulations, different window sizes and weights for the stimuli were considered for the construction of the PID based on neurons that best responded to each input signal presented. Such results are outside the parameters initially proposed by Izhikevich 2003 for the simulation of biological neurons, and such behavior was also found in the work of Webb, Davies and Lester, 2011 [2]. This emphasizes the importance of a good algorithm to optimize the system parameters in order to seek a result with less error. In the proposed work, the results are still below the desired and it is believed that a longer simulation time and number of generations would allow escaping from the local minima found. Tuning of classical PID controllers depend on the choice of figures of merit as well, and some established results can be seen in [15].

Neuromorphic engineering takes inspiration from the biological organization and operation of nervous systems to rethink the technology of computing, sensing and actuating in practical engineered system [16, 17]. In this way, neuromorphic control draws inspiration from biology to address control issues and are represented by adaptable neural structures [6]. These technologies have many applications, from modeling and understanding biological processes to the development of cognitive systems, neural prosthetics and similar applications [17]. Neuromorphic controllers based on SNN have already proven themselves to be able to control a mathematical model of the arm (flexion and extension movements of the elbow), which leads us to believe in exciting possibilities in the use of such systems in the control of prostheses [4, 6]. Thus, it is expected to apply in future works the 3-neuron SNN controller optimized by the DEA to control real systems and engineering problems that can benefit from a neuromorphic solution, such as the control of prostheses [17] and slip detection systems [18], for example.

5 Conclusions

A simple 3-neuron SNN based on biological systems was described in the present work to approximate a PID controller response. Furthermore, a supervised learning based on DEA has been developed to obtain the parameters of the Izhikevich neurons that best adapt to the system to be controlled. The response of the proposed SNN-based PID controller was analyzed considering two input signals and encouraging results were found to the utilization of such system. The proposed controller was able to track the presented references (sinusoidal signal and step), with a response similar to traditional PID controllers even when training was performed with one reference signal and the test with another. Nonetheless, future studies should be carried out for a better understanding of each neuron and parameter used in the neural network, as well as improvements in the optimization system. In addition, it is expected to use this system in the future in neuromorphic control applications, such as in the control of prostheses.

Acknowledgments. Acknowledgment to CNPq, CAPES and FAPEMIG for the partial financial support and to the Federal University of Uberlândia (UFU) and BIOLAB-FEELT-UFU for providing the physical structure.

Conflict of Interest The authors declare that they have no conflict of interest.

References

1. Borase, R.P., Maghade, D.K., Sondkar, S.Y., Pawar, S.N.: A review of PID control, tuning methods and applications. Int. J. Dynam. Cont. **9**(2), 818–827 (2020)
2. Webb, A., Davies, S., & Lester, D.: Spiking neural PID controllers. Lecture Notes in Computer Science (Including Subseries Lecture Notes in Artificial Intelligence and Lecture Notes in Bioinformatics), 7064 LNCS(PART 3), 259–267 (2011)
3. Faccin, F.: Abordagem Inovadora no Projeto de Controladores PID. Universidade Federal do Rio Grande do Sul (2004)
4. Pérez, J., Cabrera, J.A., Castillo, J.J.: Bio-inspired spiking neural network for nonlinear systems control. Neural Netw. **104**, 15–25 (2018)
5. Mie, S., Okuyama, Y., & Saito, H.: Simplified quadcopter simulation model for spike-based hardware PID controller using SystemC-AMS. In: 12th International Symposium on Embedded Multicore/Many-Core Systems-on-Chip on Proceedings IEEE, MCSoC 1, pp. 23–27 (2018)
6. Pérez Fernández, J., Alcázar Vargas, M., Velasco García, J.M., Cabrera Carrillo, J.A., Castillo Aguilar, J.J.: A biological-like controller using improved spiking neural networks. Neurocomputing **463**, 237–250 (2021)
7. Pérez, J., Cabrera, J.A., Castillo, J.J.: A traction control system based on co-evolutionary learning in spiking neural network (SNN). In: 14th International symposium on advanced vehicle control (AVEC' 18) on proceedings (2018)
8. Dorothy, M.R.: Neuroinspired control strategies with applications to flapping flight. University of Illinois at Urbana-Champaign (2015)
9. Stagsted, R., Vitale, A., Binz, J., Renner, A., Bonde Larsen, Leon Sandamirskaya, Y.: Towards neuromorphic control: a spiking neural network based PID controller for UAV. Robot: Sci Syst (2020)
10. Izhikevich, E.M.: Simple model of spiking neurons. IEEE Trans. Neural Netw. **14**(6), 1569–1572 (2003)
11. Izhikevich, E.M.: Which model to use for cortical spiking neurons? IEEE Trans. Neural Netw. **15**(5), 1063–1070 (2004)
12. Skocik, M.J., Long, L.N.: On the capabilities and computational costs of neuron models. IEEE Trans. Neural Netw. Learn. Syst. **25**(8), 1474–1483 (2014)
13. Engelbrecht, A.P.: Computational intelligence: an introduction (2nd ed). John Wiley & Sons (2007)
14. Storn, R., Price, K.: Differential evolution—a simple and efficient heuristic for global optimization over continuous spaces. J. Global Optim. **11**(4), 341–359 (1997)
15. Ogata, K.: Modern control engineering (vol. 5). Prentice hall, Upper Saddle River, NJ
16. Ribar, L., Sepulchre, R.: Neuromorphic Control: designing multiscale mixed-feedback systems. IEEE Control. Syst. **41**(6), 34–63 (2021)
17. Furber, S.: Large-scale neuromorphic computing systems. J. Neural Eng. **13**(5), 1–14 (2016)
18. Nakagawa-silva, A., Thakor, N. V, Cabibihan, J., & Soares, A. B.: A bio-inspired slip detection and reflex-like suppression method for robotic manipulators. IEEE Sens J PP(c), 1 (2019)

Genetic Algorithms in Machine Learning Applied to Computer Vision: Facial Emotion Recognition

Amanda Rosa F. Jorge(✉), Marcio J. da Cunha, and Alcimar B. Soares

Electrical Engineering Faculty, Federal University of Uberlândia, Uberlândia, Brazil
amandarosafj@live.com

Abstract. In the literature, the recognition of human facial emotions has been highly improved. However, the great challenge is to perform the training of efficient and fast algorithms due to the large volume of input data. This article presents a new approach to emotion classification. Instead of using information vectors of all pixels extracted from the image, the pre-sent study works with 7 Euclidean distances considering key points of the face: corners of the mouth, eyes and tip of the nose, in addition to the mean and variance of the facial region, totaling 9 input attributes. Thus, we aim verify by neurogenetic algorithms if it is possible to use one-dimensional data as information carrier to facial emotion classification. Two classification algorithms were performed separately: Support Vector Machine and Multi-Layer Perceptron. Then, a genetic algorithm was applied to select the more appropriate input at-tributes for each architecture. By this we analyze which variables were really relevant for the classification. The database of facial emotions used was the Warsaw Set of Emotional Facial Expression Pictures (WSEFEP) containing 210 images of 30 individuals. Altogether, the database presents images of 7 facial emotions: angry, disgusted, happy, neutral, sad, surprised, and afraid. The best result was found for the MLP architecture with 6 input attributes. The global average accuracy found was 51%, being 88% and 83% for the prediction of happy and surprised emotions, respectively.

Keywords: Emotion Recognition · Computer Vision · Machine Learning · Genetic Algorithm

1 Introduction

Facial expressions are a non-verbal form of communication. The premise for facial expressions is, in fact, to show what we feel in social environment. In order for a facial expression to be understood, the recipient must have the ability to recognize the expression, interpret its meaning, understand its motivation and respond to it. This whole process in which this non-verbal communication takes place is the social cognition [1, 2].

Considering human behavior facial expressions are similar among different individuals explained by biological and social aspects. It can interfere with human communication over time. These topics are interesting questions to be studied and computationally

J. L. B. Marques et al. (Eds.): CLAIB 2022/CBEB 2022, IFMBE Proceedings 98, pp. 118–128, 2024.
https://doi.org/10.1007/978-3-031-49401-7_12

modeled, as well as their correlation [3, 4]. According to studies by Paul Ekman, there are a total of 7 basic facial expressions recognized by the vast majority of people, regardless of their origin, educational level or culture. The following expressions are joy, sadness, anger, fear, disgust, surprise and contempt [5].

The benefits presented from projects and developments in algorithmic solutions capable of interpreting facial emotions from human faces are new possibilities for human-computer interactions, text writing, as in robotics, games, digital marketing, intelligent assistance, and rehabilitation systems, among many others [6]. The sooner the development and improvement of algorithms for facial interpretation occurs, the more efficient the role of technology will be in understanding natural areas of the human sciences such as psychology, neuroscience, human cognition, and learning [7, 8].

In the past decades, deep learning is one of the branches of artificial intelligence and is being widely applied in the field of computer vision [9]. The area of computer vision encompasses processes such as: Image Classification, Object Detection, Image Segmentation, Semantic Segmentation, Image Reconstruction, among others. Thus, a wide variety of application fields can be followed from pre-processing to the data training algorithm acquired by computer vision [10, 11].

In general, before the image data pass through processing phase it is proper extract some features [11, 12]. We can classify these features in two groups: global and local features. Examples of global features are color, histograms, texture, edges and, in case of some filter application, a specific global descriptor extracted [12].

On the other hand, the local features are strictly dependent to the regions of interests from detection or salient regions invariant to illumination and viewpoint changes. These local structures extracted from local feature descriptors are considered keypoints [12]. The Fig. 1 shows an illustration of dependent behavior and amount of data related to global or local image features. The main point before extracting image information is a considered attention to the pre-processing stage: volume of extracted input data, relevant features or even resizing images which can lead to a loss of information [13].

Canal et al. (2022) presents a systematic review of the state-of-the-art in recent years regarding the interpretation and recognition of human emotions from facial expressions. This review presents the main techniques used and the main characteristics of the methods found [14].

In essence, it presents as "classical" approaches those algorithms based on a face detection procedure followed by some recognizer capable of interpreting the feature vector and making some decision according to a fixed number of output classes. These approaches usually are Support Vector Machines (SVM), Naive Bayes, k-Nearest Neighbors, Fuzzy, among others. On the other spectrum of problem solvers, they present approaches based on neural networks, as NNB approaches. In this way, the input layer contains a discretization of the facial image or more commonly vectors of features extracted using some feature descriptor method [14].

CNN is on the rise lately in many areas, especially those involving image processing and pattern recognition, such as those used in medical image classification, object labeling and facial emotion recognition [3, 4, 7, 15–18]. Meantime, the CNNs are not considered in this paper due to the fact that our input data is one-dimensional.

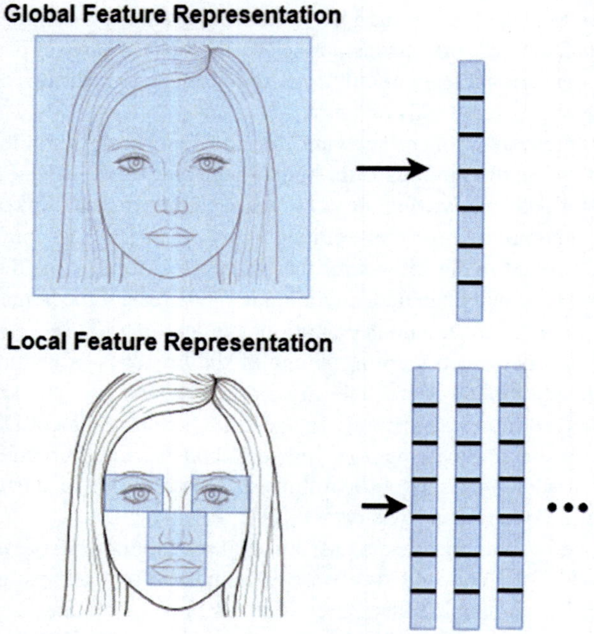

Fig. 1. Global and local feature vectors representations from images.

Another computational tool that has been used in the area of machine learning technologies is the Genetic Algorithm (GA) [19, 20]. Genetic Algorithms are globally optimized logical algorithms that are based on genetic and natural selection mechanisms. For a binary genetic algorithm, the population is generated in binary form. Each gene on a chromosome is represented by a number of bits 0 and 1. These algorithms are commonly used for separation of blind sources of biological electrical potentials, estimation of torque and power of movement, also for the gravitational point balance approach of prostheses, among other areas [21, 22].

This study considered the keypoints of the face such as: eyes, right and left corners of the mouth and tip of the nose, the average and global variance of the pixels of the face image. The aim is to verify by neurogenetic algorithms if it is possible to use one-dimensional data to carrier information and be used as input variables for facial emotion classification in computer vision field.

2 Methodology

2.1 Database

For data processing, the private database, Warsaw Set of Emotional Facial Expression Pictures (WSEFEP) is used, containing 210 images of 30 individuals. It must have a request to obtain access to database. The photos collected are facial images with static individuals performing different expressions, one for each expression performed. As shown in Fig. 2, the categories of emotions present in the dataset are: Happy, Sad, Neutral, Surprised, Afraid, Angry and Disgusted.

Fig. 2. Illustration of the categories of emotions analyzed from top left to bottom right: Happy, Sad, Neutral, Surprised, Afraid, Angry and Disgusted.

2.2 Data Pre-processing

The algorithms were developed in Python language. The images from each database were grouped into seven classification classes from 0 to 6. The qualitative targets were changed to numbers: 0-angry, 1-disgusted, 2-happy, 3-neutral, 4-sad, 5-surprised, 6-afraid. The pre-processing algorithm then goes through the directories with the images and creates spreadsheets with the analytical values of the data in addition to the emotion category. The open-source library RetinaFace is used to detect and cut the face, in addition to collecting the coordinates of keypoints such as eyes, nose and mouth in the facial region [23].

The input parameters to feed the training algorithms are statistical values of the cropped face image, mean and variance and the Euclidean distances calculated from the face coordinates. These coordinates consist of the right eye (OD), left eye (LE), nose (N), right corner of the mouth (BD) and left corner of the mouth (BE). From this, the distances presented below and visualized in Fig. 3 are calculated:

- Right eye-nose (N-OD)
- Right eye-right corner of mouth (OD-BD)
- Left eye-nose (N-OE)
- Left eye-left corner of mouth (OE-BE)
- Nose-right corner of mouth (N-BD)
- Nose-left corner of mouth (N-BE)
- Mouth right corner-mouth left corner (BD-BE)

2.3 Data Analysis

According to the study Canal et al. (2022), a review of the architectures of emotion recognition algorithms was carried out, dividing between two groups: classical and neural network-based approaches [14]. Except from CNNs, the most common methods

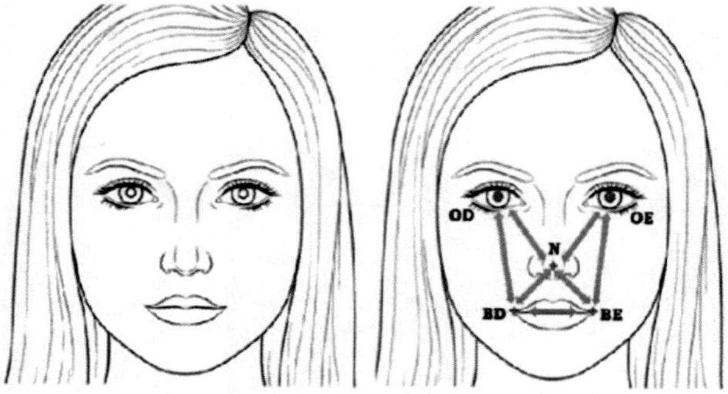

Fig. 3. Facial keypoints: Right Eye (OD), Left Eye (OE), Nose (N), Right Mouth (BD) and Left Mouth (BE).

for analyzing one-dimensional data from each category are Support Vector Machine (SVM) and Multi-Layer Perceptron (MLP). For this reason, both architectures were chosen and used in this study.

The Genetic Algorithm was inserted in the analysis as a selector of input attributes for machine learning. Depending on the level of positive or negative correlation of the considered attributes, the accuracy of the classifier can be improved or impaired.

Thus, a population of 50 individuals with 9-bit binary chromosomes was applied to the input data. A total of 20 elite individuals were selected and 10 tournaments 3–3 were held to choose the best individual. The fitness function considered was the accuracy presented by the algorithm, MLP or SVM. Further, for a wider range of individuals, 25 crossovers and mutations were performed, respectively, with 80% and 50% probabilities of occurrence.

For both the SVM and the MLP, the test group was considered 25% of the total. For the SVM algorithm, the sigmoid kernel was used. For the MLP algorithm, 4 hidden layers with 30, 30, 30 and 13 neurons were used, respectively. With the maximum iteration value set to 3500.

3 Results

Initially, the SVM algorithm showed an improvement in the accuracy value of 9% from the application of the genetic algorithm in the selection of attributes. According to Figs. 4 and 5, the confusion matrix of the classifier with no selection and with GA selection is verified, respectively. Considering all attributes in the training, the classifier reached 36%. With the selection of attributes by the AG, the training reached 45%. The best chromosome for the SVM was the following vector [1,0,1,0,1,0,1,0,1]. The attributes selected resulted in average, OD-BD, N-BE, N-OE, BE-BD.

According to Fig. 4, the classification of the SVM algorithm without the application of the genetic algorithm, there was confusion in the classification between angry, disgusted, and happy emotions. Regarding the correct answers, the best classified emotions reached 50% of correct answers in neutral and sad. The afraid emotion was highly confused with the surprised emotion.

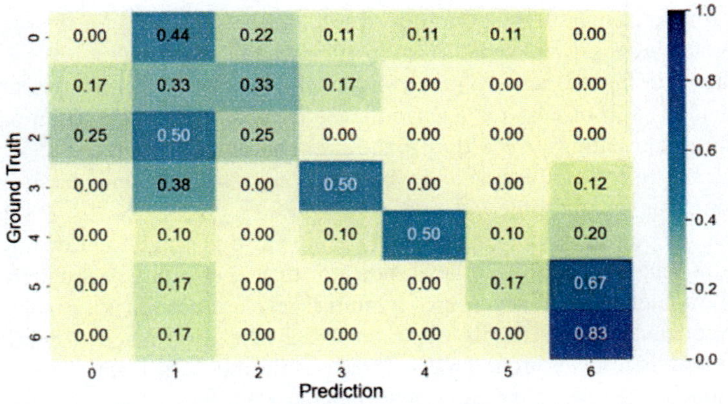

Fig. 4. Confusion matrix by SVM algorithm considering all input variables reached 36% average accuracy.

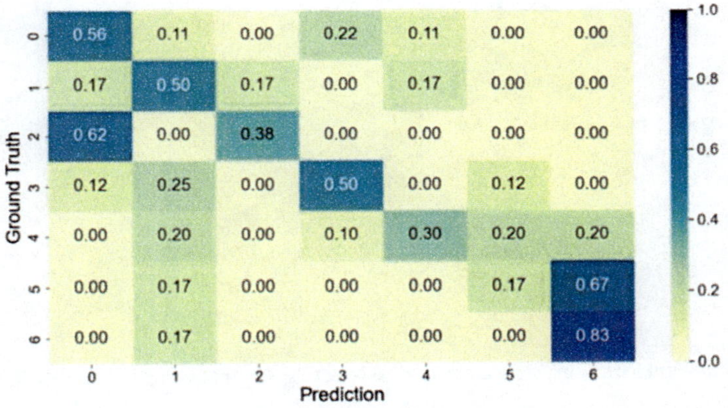

Fig. 5. Confusion matrix by SVM algorithm with GA application reached 45% average accuracy.

In the classification of the SVM algorithm with the application of the genetic algorithm, the angry label was highly confused with happy. Despite this fact, there was an increase in the accuracy for angry, disgusted and happy. In both cases, with and without ag, the most correct class was afraid, according to the confusion matrix shown in Fig. 5.

For the second classifier analyzed, the MLP algorithm, without the application of GA, showed an improvement in the accuracy value of 45%, according to Fig. 6. With the analysis of all input attributes considered, there were misclassifications between angry, disgusted and happy emotions, similar to SVM.

The best chromosome for the MLP was the following vector [0,0,1,1,0,1,1,1,1]. With the application of AG, in the MLP, six input variables from pictures were selected in total: OD-BD, OE-BE, N-BD, N-OD, N-OE and BD-BE. According to confusion matrix in Fig. 7, the average accuracy of the algorithm reached 51% accuracy among the 7 emotion classes. After selecting input attributes, the best ranked emotions were:

happy and surprised. However, there were still misclassifications for the prediction of emotions such as angry and sad.

As shown in Fig. 8, the ROC curve of the MLP classifier for the different emotions is presented. The ROC curve is a performance measure for classification problems at various threshold settings. According to the emotions considered, it can be seen that the afraid and surprised emotions were the best adapted to learning and classification. The worst rating was found for the sad emotion.

A ROC curve can be drawn per class, but it is also possible to draw a ROC curve considering each element of the label indicator matrix as a binary prediction (micro-average). Another evaluation measure for multi-label classification is the macro-average, which gives equal weight to the classification of each label. Both macro-media and micro-media present themselves around 67% of true positive rate. For further information and reproduction of the results, the code is available in a repository [24].

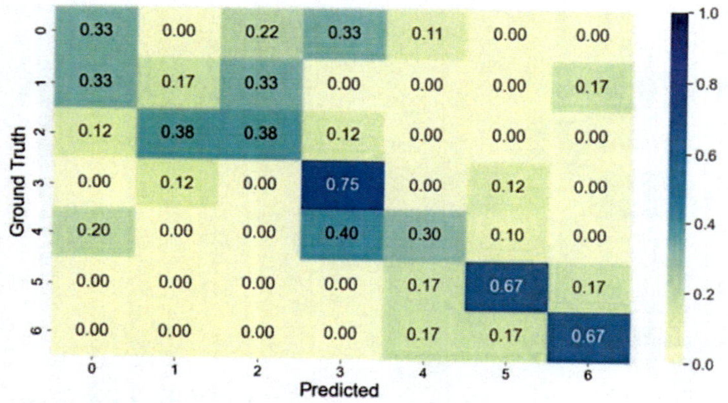

Fig. 6. Confusion matrix by MLP algorithm considering all input variables reached 45% average accuracy.

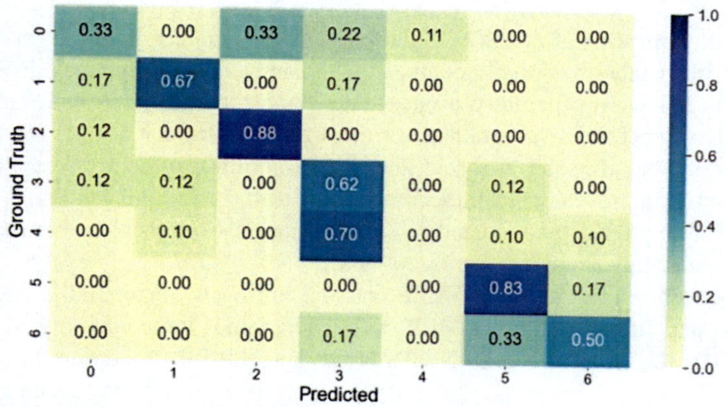

Fig. 7. Confusion matrix by MLP algorithm with GA application reached 51% average accuracy.

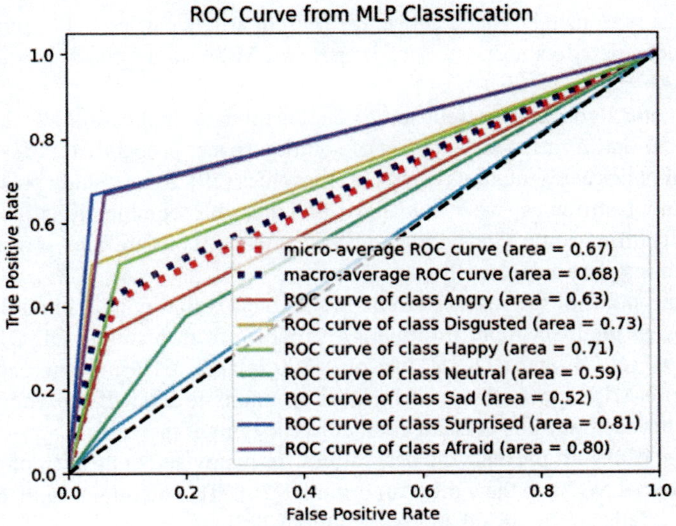

Fig. 8. ROC Curve by MLP with GA among all trained classes.

4 Discussion

The study by Canal et al. (2022) presents a literature review of several techniques for classifying human facial emotions. According to this study, classical techniques (SVM, Fuzzy Logic and others) and based on neural networks (CNN, MLP and others) are presented. For other facial emotion datasets, the highest accuracy was found in the study of [25] whose average system accuracy reached 96.76% with NNB approach, in 2017, with the CK + dataset. The CK + dataset features 593 video sequences (30 fps) available with 7 emotion classes [26].

From a classic approach, by Canal et. Al (2022), shows that the highest average accuracy among the classified classes, present in the literature, is 92.40% in 2018 by the study of [27]. The CK dataset features has 486 available video sequences with 7 emotion classes as well [28].

Regarding classifiers based on neural networks, we find CNN, MLP architectures, among others. CNNs use training with image resizing that can lead to loss of image information [13]. They use the parameter arrays of pixels and local variables such as textures and contours. On the other hand, the present study presents an analysis of facial keypoint coordinates and statistical variables of faces, mean and variance. This simplification of the input attributes, one-dimensional input data, is focused on reducing the volume of information analyzed while maintaining the desired efficiency among human emotions.

The benefits of using facial coordinates are that the information analyzed is not interfered with by different ambient lighting and is independent of the individual's skin tone. Also, the facial movement, which is an obstacle for other methods, as long as the facial coordinates are detected, the movement does not generate noise or major problems.

Based on the reasons mentioned above, the present article carried out training analysis by SVM, considered as a classical approach and MLP, an approach based on neural networks.

The genetic algorithms present some benefits on their application which are: parallelism, global optimization, a larger set of solution space, probabilistic based in nature selection, genetic representations using chromosomes [29]. As our study presented analysis of 9 input attributes, we would have 9!-1 possible combinations, excluding an entirely null chromosome. Considering the vast amount of possibilities, it would be very time consuming if tested manually.

The main point of the research to be highlighted is that with a small dataset, and considering, in total, six input information variables from pictures: OD-BD, OE-BE, N-BD, N-OD, N-OE and BD-BE. Just over 200 images, 51% average accuracy was obtained with MLP for 7 classes of facial expressions in total. If the volume dataset increases, this accuracy and classification efficiency might increase.

The state of the art presents the best ratings for happy and neutral emotions, reaching the mark of 90% of the rating algorithms [26]. The emotions with the biggest classification fails are for afraid and sad emotions [26].

In the present study, the classes that presented the best performance were happy and surprised, and those that had the highest percentage of error were angry and sad as presented in Figs. 7 and 8. Therefore, the results of the article are in accordance with the literature stating that our approach may be promising.

The limitation of the present study is the full front region face has to be detected in the image of the find the coordinates. Plus, advanced "humanly" emotions, for example contempt emotion, the framework training level increases and fine muscle details must be considered.

In future study, improve the MLP with genetic algorithm to find the best number of maximum interactions and number of neurons in each hidden layer size. And finally, add other input variables to be analyzed. It aimed to consider datasets with affective biomarkers for emotion recognition. For example, facial coordinates summed with body temperature.

5 Conclusion

With the increasing number of different applications for the classification of facial emotions by machine interfaces, in recent years new techniques and classification methods have been presented. The present study aims to verify by neurogenetic algorithms if it is possible to use one-dimensional data to carrier information for emotion classification. From the Euclidean distances between keypoints of the face: corners of the mouth, eyes and tip of the nose, we concluded that the MLP architecture presented an average accuracy of 51% considering the total of 7 classes.

From the application of AG in both SVM and MLP input variables, the best predicted classes were with MLP architecture for happy and surprised targets, in line with the literature. From 9 input attributes, the AG selected 6, these being: OD-BD, OE-BE, N-BD, N-OD, N-OE and BD-BE. While the dataset used has just over 200 images, it is believed that from training with a greater number of images of each label. The results of the article are in accordance with the literature, showing that approach with facial keypoint coordinates may be promising in facial emotion recognition field.

Acknowledgments. The authors are thankful to the Federal University of Uberlândia, and to the funding agencies Fapemig and CNPq.

Conflict of Interest The authors declare that they have no conflict of interest.

References

1. Li, J., Jin, K., Zhou, D., Kubota, N., Ju, Z.: Attention mechanism-based CNN for facial expression recognition. Neurocomputing **411**, 340–350 (2020)
2. Sajjad, M.S., Zahir, A., Ullah, Z., Akhtar, Muhammad, K.: Human behavior understanding in big multimedia data using CNN based facial expression recognition. Mobile Netw Appl **25**,1611–1621 (2020)
3. Kumar, A., Sharma, K., Sharma, A.: MEmoR: a multimodal emotion recognition using affective biomarkers for smart prediction of emotional health for people analytics in smart industries. Image Vis. Comput. **123**, 104483 (2022)
4. Saxena, S.: An intelligent facial expression recognition system with emotion intensity classification, 39–52 (2022)
5. Ekman, P.: An argument for basic emotions. Cognition and Emotion, **6** (1992)
6. Lopes, A.T., De Aguiar, E., Oliveira-Santos, T.: A facial expression recognition system using convolutional networks. In: Brazilian Symposium of Computer Graphic and Image Processing, pp. 273–280 (2015)
7. Kaushik, H., Kumar, T., Bhalla, K.: iSecureHome: a deep fusion framework for surveillance of smart homes using real-time emotion recognition. Appl. Soft Comput. **122** (2022)
8. Guo, Y.: Facial Expressions recognition with multi-region divided attention networks for smart education cloud applications, 119–128 (2022)
9. Chai, J., Zeng, H., Li, A., Ngai, E.W.T.: Deep learning in computer vision: a critical review of emerging techniques and application scenarios. Mach. Learn. with Appl. **6**(August), 100–134 (2021)
10. Nixon, M., Aguado, A.: Feature extraction and image processing for computer vision (2019)
11. Snyder, W., Qi, H.: Fundamentals of Computer Vision. Cambridge University Press, Cambridge (2017)
12. Awad, A.I., Hassaballah, M.: Image feature detectors and descriptors. Springer International Publishing AG, vol. 630 (2016)
13. Jerubbaal John Luke, M.B., Joseph, R., Cognitive, impact of image size on accuracy and generalization of convolutional neural networks. Int. J. Res. Analyt. Rev. (IJRAR), **6**, 70–80 (2019)
14. Canal, F.Z., Müller, T.R., Matias, J.C., Scotton, G.G., de Sa Junior, A.R., Pozzebon, E., Sobieranski, A.C.: A survey on facial emotion recognition techniques: a state-of-the-art literature review. Inf. Sci. **582**, 593–617 (2022)
15. Minaae, S.: Deep-emotion: facial expression recognition using attentional convolutional network (2021)
16. Thuseethan, S., Rajasegarar, S., Yearwood, J.: EmoSeC: emotion recognition from scene context. Neurocomputing **492**, 174–187 (2022)
17. Xu, C., Yan, C., Jiang, M., Alenezi, F., Alhudhaif, A., Alnaim, N., Polat, K., Wu, W.: A novel facial emotion recognition method for stress inference of facial nerve paralysis patients. Exp. Syst. Appl.**197**, 116705 (2022)

18. Zhang, J., Yu, H.: Improving the facial expression recognition and its interpretability via generating expression pattern-map. Pattern Recogn. **128** (2022)
19. Pal, A.R., Pratihar, D.K.: Estimation of joint torque and power consumption during sit-to-stand motion of human-being using a genetic algorithm. Proced. Comput. Sci. **96**, 1497–1506 (2016)
20. Leal-Naranjo, J.A., Ceccarelli, M., Miguel, C.R.T.S.: Mechanical design of a prosthetic human arm and its dynamic simulation. Adv. Intell. Syst. Comput. **540**, 482–490 (2017)
21. Ganesh, R., Kumar, D.: Blind source separation based classification scheme for myoelectric prosthesis hand. Intell. Mechatron. (2011)
22. Dadula, C.P., Dadios, E.P.: A genetic algorithm for blind source separation based on independent component analysis. In: 2014 International Conference on Humanoid, Nanotechnology, Information Technology, Communication and Control, Environment and Management, pp. 0–5 (2014)
23. Deng, J., Guo, J., Zhou, Y., Yu, J., Kotsia, I., Zafeiriou, S.: RetinaFace: single-stage dense face localisation in the wild. Cornell University (2019)
24. Amanda, R.F.J.: Machine learning emotion recognition, GitHub repository. Available at: https://github.com/AmandaRosa/Machine_Learning_Emotion_Recognition. Last accessed 01 Jan 2023
25. Lopes, A.T., de Aguiar, E. De Souza, A.F., Oliveira-Santos, T.: Facial expression recognition with convolutional neural networks: coping with few data and the training sample order. Pattern Recognit. **61**, 610–628 (2017)
26. Lucey, P., et al.: The extended Cohn-Kanade (CK+): A complete dataset for action unit and emotion-specified expression. IEEE Comput. Soc. Conf. Comput. Vis. Pattern Recogn. Works. **4**, 94–101 (2010)
27. Slimani, K., Kas, M., El Merabet, Y., Messoussi, R., Ruichek, Y.: Facial emotion recognition: a comparative analysis using 22 LBP variants. ACM Int. Conf. Proc. Ser. 88–94 (2018)
28. Kanade, T., Cohn, J., Yingli, T.: Comprehensive database for facial expression analysis. In: Proceeding fourth IEEE international conference automatic face and gesture recognition (Cat. No. PR00580) (2000)
29. Katoch, S., Kumar, V.: A review on genetic algorithm: past, present, and future. Multimedia Tools Appl **80**, 8091–8126 (2021)

ForceLAB Simulator Update: A Study on the Positive Cooperativity of Cross-Bridge Formation in Ventricular Myocyte

Matheus Leonardo Alves de Camargo[1(✉)], José Wilson Magalhães Bassani[2,3], Rosana Almada Bassani[2], and Robson Rodrigues da Silva[1,2]

[1] Research and Technology Center, University of Mogi das Cruzes, Mogi das Cruzes, Brazil
matheus.leonardo675@gmail.com
[2] LabNECC, Center for Biomedical Engineering, University of Campinas, Campinas, Brazil
[3] Department of Biomedical Engineering, School of Electrical and Computing Engineering, University of Campinas, Campinas, Brazil

Abstract. Cardiovascular diseases are one of the leading causes of death in the world. Understanding the mechanisms that control heart contraction is essential to deal with this problem. As auxiliary tools, mathematical models have been used to describe these phenomena at the cellular level. Electrical and mechanical models of ventricular myocytes have been implemented in computer simulators for analysis. One of them is ForceLAB, in which two of the main models of contraction force were implemented: Rice et al. (1999) (RWH) and Negroni & Lascano (2008) (NL). The objective of this paper is to present an updated version of the ForceLAB tool, to which additional features were added, among which the values of important variables, as one of the outputs. The new version of ForceLAB was implemented using App Designer for the development of the new interfaces. As an application, the influence of the positive crossbridge cooperativity on the unusual behavior of the Ca^{2+}-force relationship simulated with the RWH was tested. The waveforms of twitch transient of cytosolic Ca^{2+} concentration ($[Ca^{2+}]_i$), developed force and the force versus pCa curve (where pCa = $-\log\,[Ca^{2+}]_i$) were simulated. The simulation results showed that the non-paralel shift of the force versus pCa curve upon variation of myofilament Ca^{2+} was not corrected by the absence of the positive cooperativity mechanism did in the RWH model formulation.

Keywords: Mathematical model · Contraction · Cardiac myocyte · Crossbridge cooperativity · *In silico* experiments

1 Introduction

Cardiovascular diseases are the leading cause of death in the world. In 2019, they caused about 17.9 million deaths, accounting for 32% of global death count, 85% of which were from myocardial infarction and stroke [1].

Understanding the mechanisms of cardiac contraction is a crucial task to elucidate the genesis of several heart diseases. This understanding involves the description of the

J. L. B. Marques et al. (Eds.): CLAIB 2022/CBEB 2022, IFMBE Proceedings 98, pp. 129–138, 2024.
https://doi.org/10.1007/978-3-031-49401-7_13

dynamics of Ca^{2+} ions in the ventricular myocyte, which plays a key role in contraction development [2]. Ca^{2+} entry into the cell occurs through the L-type channel during the action potential (AP), triggering Ca^{2+}-induced Ca^{2+} release from the sarcoplasmic reticulum when Ca^{2+} binds to the Ca^{2+} channels of the organelle, the ryanodine receptors. The increase in cytosolic free Ca^{2+} concentration ($[Ca^{2+}]_i$) favors Ca^{2+} binding to troponin C (TnC), located in the thin filament of the sarcomeres, triggering the contraction process [3–5]. The Ca^{2+}-TnC binding starts a sequence of conformational changes in proteins of the myofibril thin filament, leading to actin-myosin interaction (crossbridge attachment). Different approaches have been used to model crossbridge formation, for example, taking into account the cooperativity between the formation of neighboring crossbridges and/or the size of the sarcomere [6].

Mathematical and computational modeling is an important tool for understanding biological phenomena [7]. It is possible to investigate, for instance, the mechanisms of cardiac arrhythmias by simulation of myocardial cell electrical activity [8–12]. For the description of the phenomenon of cardiac contraction in ventricular myocytes, two models have been widely used: the 1999 model of Rice et al. (the Rice, Winslow and Hunter model, RWH) [13] and the 2008 model of Negroni & Lascano (NL) [14]. These models were implemented in a simulator recently developed by Silva et al. [15], the ForceLAB. In that paper, the simulator functions and comparison of the two models were presented. The simulator was initially developed using the 2017a version of MATLAB, which has limitations regarding the development of applications. One of the pending issues of the tests performed in that study was the atypical result observed in the force versus pCa curve simulated with the RWH model when the affinity of the Ca^{2+}-TnC interaction was varied. One of the hypotheses raised was that this unusual behavior would be due to the positive cooperativity of crossbridge attachment implemented by the authors in the RWH model.

The aim of the present article was to report an update of the ForceLAB simulator, presenting the new functions and the main differences between the two versions of the tool. As an application of the new version of the simulator, the test of the previous hypothesis of the involvement of the positive crossbridge cooperativity in the unusual curve simulated with the RWH model was carried out.

2 Materials and Methods

A. **Rice et al. (1999) Model (RWH)**. The model to estimate the myocyte contraction force proposed by Rice et al. [13] (RWH) is based on the formation of crossbridges between myosin and actin. The binding of these proteins occurs after exposure of the myosin binding site in actin filament after tropomyosin displacement triggered by Ca^{2+}-TnC association. The RWH model describes 6 permissivity states for crossbridge formation: 2 non-permissive (N0 and N1), and 4 permissive (P0, P1, P2 and P3). All these states are described in the model as coupled differential equations:

$$N0 + N1 + P0 + P1 + P2 + P3 = 1 \tag{1}$$

The transition between the N0 and N1 states to the permissive states depends on the Ca^{2+}-TnC association, which is modeled with basis on the association and dissociation rate constants of these ligands, k_{on} and k_{off}, respectively. In the permissive states, the transition occurs via parameters that predict positive cooperativity of crossbridge attachment. The peak developed contraction force (F_{contr}), expressed in mN per mm^2 of cell cross sectional area, is formulated in the RWH model as:

$$F_{contr} = \alpha((P1 + N1 + 2(P2) + 3(P3))/(P1_{max} + 2(P2_{max}) + 3(P3_{max}))) \quad (2)$$

where α is a model fit factor.

Figure 1A illustrates the transition scheme among the states of crossbridge formation in the RWH model. In the study of Mullins & Bondarenko [16], the RWH model was coupled to the electrophysiological model of the mouse ventricular myocyte, previously developed by Bondarenko et al. [17]. These coupled models were used also in the present study.

B. **Negroni & Lascano 2008 Model (NL)**. In the NL model [14], the force of contraction is directly related to the Ca^{2+}-TnC interaction. This binding is described in 6 states of the troponin system (TS): (a) [TS]: Ca^{2+}-free state (b) [TSCa]: Ca^{2+} binds to TnC without crossbridge formation; (c) [TSCa~]: Ca^{2+} binds to TnC leading to the formation a weak crossbridge; (d) [TSCa*]: Ca^{2+} binds to TnC leading to the formation of a strong crossbridge; (e) [TS~]: maintainance of weak crossbridge interaction without Ca^{2+}-TnC binding; (f) [TS*]: maintainance of strong crossbridge interaction without Ca^{2+}-TnC binding. Figure 1B illustrates the transition among these states.

In the NL model, the contraction force (F, in mN/mm^2) is given by:

$$F = A_w([TSCa\sim] + [TS\sim])h_w + A_p([TSCa*] + [TS*])h_p \quad (3)$$

where A_w and A_p are elastic constants for the weak and strong crossbridge states, respectively, and h_w and h_p indicate the crossbridge elongation in these states. In this study, the NL model was also coupled to the electrophysiological model of Bondarenko et al. [17].

C. **The ForceLAB 2.0 Simulator**. To allow solving the system of 51 coupled differential equations and presenting the results of the simulations graphically, an upgrade was made in the previously published ForceLAB simulator, [15]. The new version was developed using MATLAB 2019a, with the interfaces created with the App Designer.
D. **Simulation Protocols.** Using the new version of the ForceLAB tool, sustained for values in response to variation in extra myofibrillar Ca^{2+} concentration and twitch Ca^{2+} transients and contractions were evoked by suprathreshold electric stimuli were simulated with the RWH and NL models using protocols previously employed [15]: (a) total simulation duration: 20 s; (b) duration of electric stimulation pulse: 1 ms; (c) electric stimulation frequency: 1 Hz; (d) stimulation current: 20 pA/pF; (e) initial sarcomere length: 2.1 μm; (f) cell length: 100 μm. For the RWH model, two simulations were performed, one including the positive cooperativity for crossbridge formation, and one excluding it. This change was accomplished by modifying the values assumed for the parameters f01, f12 and f23 that control the transition between the states P0 and P1, P1 and P2, and P2 and P3, respectively, as shown in Fig. 2.

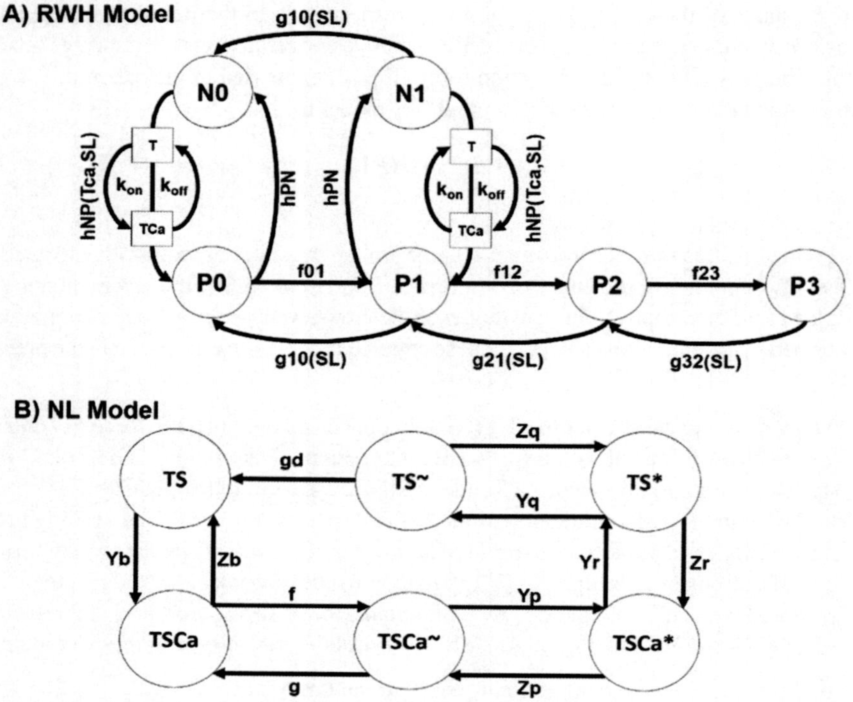

Fig. 1. State diagrams of the models implemented in ForceLAB. A) RWH model state diagram. B) State diagram of the NL model.

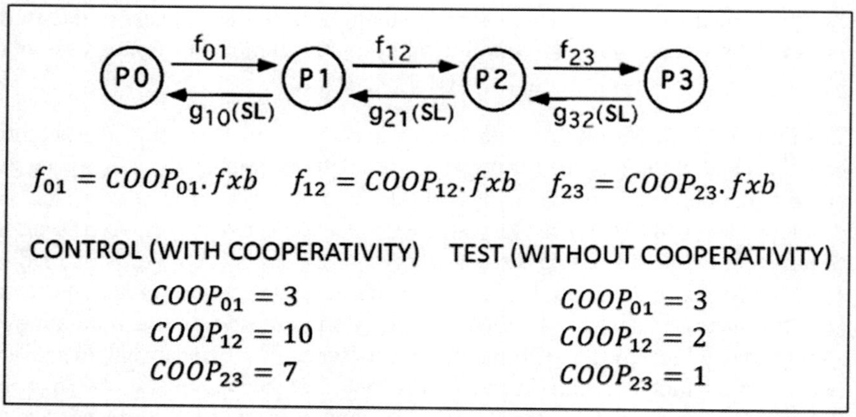

Fig. 2. Illustration of the steps that involved and assumed values of the parameters changed to test the influence of positive cooperativity of crossbridge formation on contraction simulated with the RWH model.

From the twitch $[Ca^{2+}]_i$ and force waveforms evoked in response to electric stimulation, the following variables were obtained directly in the ForceLAB simulator: the diastolic (Dia, i.e., just before stimulus delivery) and peak values; amplitude (Δ, i.e., difference of peak and diastolic values); and times to peak (t_{peak}) and half decay ($t_{0.5}$). The force versus pCa relationship was determined using the steady-state force corresponding to each value of pCa (i.e., the negative logarithm of the molar Ca^{2+} concentration).

The force versus pCa curve was simulated following the same changes in TnC affinity for Ca^{2+} as tested in [15], namely, increase and decrease of the apparent dissociation constant of Ca^{2+} at TnC (Kd) by dividing and multiplying kon by 3, respectively. Using the ForceLAB simulator, the values of pCa_{50} (pCa value that corresponds to 50% of the maximum contraction force recorded at saturating Ca^{2+} concentration) and the Hill coefficient N_H (slope constant of the curve) were calculated from the force versus pCa curve.

3 Results

A. **ForceLAB Simulator Update**. The main result of this work was the updated version of the ForceLAB simulator, in which the major functions of the original version [15] were recreated by adding new features. The main and implemented features of ForceLAB 2.0 were:

Start *Screen*: It starts the tool and gives access to the "ABOUT" section, where information on the tool, the collaborating institutions and the main authors is presented.

Simulation: The first tab shown when starting the tool (Fig. 3) offers the user the possibility to run simulations of the electrical model [16] coupled to the RWH and NL contraction models. The user can also change the simulation protocols, the general parameters of the electrical model and the main parameters of the two contraction force models.

View Graph: It makes it possible to graphically display up to 3 of the simulations performed in each of two graphs. The waveforms that can be plotted are: AP, $[Ca^{2+}]_i$, TnC-bound $[Ca^{2+}]$ ($[Ca^{2+}]_b$) and contraction force. One of the new features implemented was the automated calculation of important values extracted from the curves: Dia and peak values, Δ, t_{peak}, $t_{0.5}$ and AP duration at 20%, 50% and 90% of repolarization (APD20, APD50 and APD90, respectively). These values are shown in tables below the graphics, and can be exported in the ".xlsx" format.

Myofilament Sensitivity: In this part of the simulator, one can perform tests by changing Ca^{2+}-TnC k_{on} and k_{off}, and choose the simulation protocols. Simulations are performed for the two implemented models of contraction force, and the Ca^{2+} transient and contraction traces are plotted along with the calculated Dia, peak, Δ, t_{peak} and $t_{0.5}$ values, which can be exported as ".xlsx" files.

Force versus pCa: In this tab the user can vary the pCa and obtain the corresponding steady-state contraction force. The user can choose the contraction model and the value of K_d by changing k_{on} and/or k_{off} ($K_d = k_{off}/k_{on}$). Other available options are the initial

Fig. 3. ForceLAB simulation tab: Access to general and electrophysiologic parameters, and myofilament-related parameters of the RWH and NL models.

and final pCa values, pCa step, and general simulation protocols. The K_d, pCa_{50} and N_H values are calculated and can also be exported as ".xlsx" files.

View Data: Area where the user can access the exported data in the other tabs as a table. It is also possible to open the generated ".xlsx" files.

B. **Influence of Positive Cooperativity on Cross-Bridge Formation.** Simulations were performed to investigate if and how the positive cooperativity mechanism of cross-bridge formation, present in the RWH model, affects the simulation results. In Fig. 4, which shows the "View Graph" screen generated by the ForceLAB simulator itself, the comparison of the twitch Ca^{2+} transients and contractions simulated with the NL and RWH models, the latter with and without cooperativity, is illustrated. Tables 1 and 2 show the values of the twitch $[Ca^{2+}]_i$ and contraction, respectively, extracted from the traces depicted in Fig. 4.

Figure 5 shows the force versus pCa curves obtained with the RWH model without cooperativity considering the original Kd value (control) and the situations of increase and decrease in Kd by a factor of 3. The corresponding variables are shown in Table 3.

4 Discussion

The present update of ForceLAB aimed to use the new features of the MATLAB 2019 version, especially the App Designer. Regarding usability, the simulator has now its functions grouped in a single screen where the user accesses the resources by means of

Table 1. Twitch $[Ca^{2+}]_i$ variables obtained with the NL and RWH models, the latter also without including the positive cooperativity for crossbridge formation (*no coop*).

Variables	NL	RWH	RWH (no coop)
Dia (μM)	0.13	0.14	0.14
Peak (μM)	0.51	0.53	0.53
Δ (μM)	0.38	0.38	0.38
t_{peak} (ms)	31.09	18.03	17.90
$t_{0.5}$ (ms)	199.37	157.31	157.41

The values of diastolic (*Dia*) and peak $[Ca^{2+}]_i$, transient amplitude (Δ) and times to peak (t_{peak}), and to half-decay ($t_{0.5}$) are presented

Table 2. Contraction variables obtained with the NL and RWH models, the latter also without including the positive cooperativity for crossbridge formation (*no coop*).

Variables	NL	RWH	RWH (no coop)
Dia (mN/mm^2)	0.27	0.23	0.05
Peak (mN/mm^2)	9.01	9.10	4.74
Δ (mN/mm^2)	8.74	8.87	4.69
t_{peak} (ms)	149.12	127.50	87.84
$t_{0.5}$ (ms)	169	187	97

The values of diastolic (*Dia*) and peak force, developed tension (Δ) and times to peak (t_{peak}), and to half-decay/relaxation ($t_{0.5}$) are presented.

tabs. This change makes the simulator easier to use, as it provides a single environment for simulations. One of the main advantages over the previous version [15] is the automatic calculation of values of the main variables used for analysis, such as Dia and peak values, Δ, t_{peak} and $t_{0.5}$ of the twitch waveforms. Thus, the analysis of the simulation results has become much easier, and the function of data export as ".xlsx" files ensures the prompt saving and storage of the data.

The removal of the positive crossbridge cooperativity from the formulation of the RWH model did not affect the $[Ca^{2+}]_i$ values or the transient time-course during a twitch. However, diastolic tension was almost null and the developed force was decreased by almost 50%, with marked acceleration of contraction development and relaxation. Thus introduction of positive cooperativity, which reflects a physiological phenomenon [18], is required for a realistic simulation of a twitch contraction. Nevertheless, alterations were not observed in the force versus pCa relationship upon removal of the cooperativity from the RWH model formulation.

Symmetrical (increase and decrease by the same factor), moderate changes in K_d are expected to result in symmetrical and parallel shifts (to the right and the left, respectively) in this relationship, as observed with the NL model. However, this was not the case

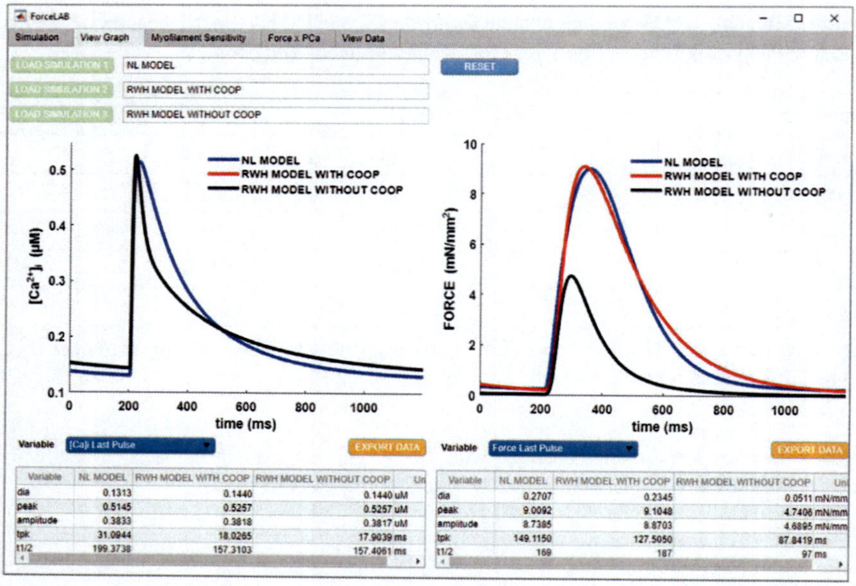

Fig. 4. Timecourse of twitch $[Ca^{2+}]_i$ transients and contractions shown in ForceLAB graph view tab. They correspond to the waveforms simulated with the NL and RWH models, the latter including or not positive cooperativity for crossbridge formation.

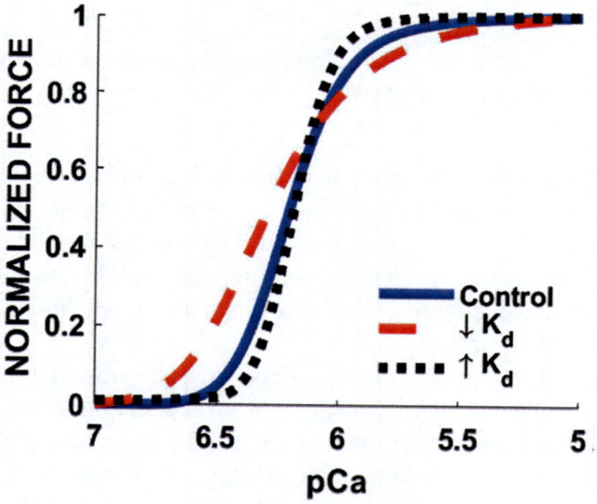

Fig. 5. Force versus pCa relationship simulated with the RWH model without the cooperativity mechanism using the original K_d value (control), and after it was increased (upward arrow) and decreased (downward arrow) by 3 times.

with the RWH model: decreasing K_d also decreased the slope of the force versus pCa curve, whereas the opposite was observed when K_d was enhanced [15]. Unfortunately,

Table 3. Parameters of the force versus pCa curves shown in Fig. 5, without the crossbridge positive cooperativity, using the original (control), increased and decreased values of the apparent dissociation constant (K_d) by a factor of 3.

Variables	Control	Kd↑	Kd↓
pCa50	6.21	6.19	6.29
NH	3.17	3.79	2.31
Kd (μM)	0.60	1.80	0.20

The values of the pCa at which force is half-maximal (pCa_{50}) and the Hill coefficient (N_H) and (K_d) are shown.

model instability prevented us to test a possible role of the myosin positive cooperativity mechanism in this unusual result at that time. With the 2.0 ForceLAB version, software limitations of the former version could be overcome, and the observed model instability could be eliminated, which has made it possible to test the previously raised hypothesis. Nevertheless, the present results (Table 3 and Fig. 5) show that variation of n_H resulting from variations in the K_d value for the Ca^{2+}-TnC interaction is still present, also with very similar n_H values as when the cooperativity was included in the model [15]. Thus, other aspect of the formulation should be responsible for the odd pattern of dependence of the slope of the force versus pCa relationship on the affinity of Ca^{2+} for TnC.

5 Conclusions

The new features of the updated ForceLAB simulator have made it more user-friendly, and make it possible to gather quantitative data as simulation proceeds, allowing their prompt recording and storage. ForceLAB version 2.0 is available for download after a brief registration at the following link: https://forms.gle/9nneEyX9NKDcjiah8.

With this improved version of the tool, it was possible to rule out the hypothesis that the formulation of the crossbridge positive cooperativity mechanism could be responsible for the non-parallel shifts in response to variations in the Ca^{2+}-TnC affinity in the force versus pCa relationship simulated with the RWH model. Thus, more studies are required to clarify the origin of this odd result.

Acknowledgments. This study was financed in part by the Coordenação de Aperfeiçoamento de Pessoal de Nível Superior - Brasil (CAPES) - Finance Code 88887.627770/2021-00, São Paulo Research Foundation (FAPESP; grant # 2016/18422-7), Brazilian Council for Scientific and Technological Development (CNPq; grant # 304010/2016-2), the Brazilian Health Ministry/FINEP (Conv. 0113021400, ref. 0441/12), and the University of Mogi das Cruzes.

Conflict of Interest The authors declare that they have no conflict of interest.

References

1. WHO: https://www.who.int/news-room/fact-sheets/detail/cardiovascular-diseases-(cvds)
2. Niederer, S.A., Campbell, K.S., Campbell, S.G.: A short history of the development of mathematical models of cardiac mechanics. J. Mol. Cell. Cardiol. **127**, 11–19 (2019). https://doi.org/10.1016/j.yjmcc.2018.11.015
3. Bers, D.M.: Cardiac excitation-contraction coupling. Nature **415**(6868), 198–205 (2002). https://doi.org/10.1038/415198a
4. Huxley, H.E.: The mechanism of muscular contraction. Science (New York, N.Y.) **164**(3886), 1356–1365 (1969). https://doi.org/10.1126/science.164.3886.1356
5. Fuchs, F., Briggs, F.N.: The site of calcium binding in relation to the activation of myofibrillar contraction. J. Gen. Physiol. **51**(5), 655–676 (1968). https://doi.org/10.1085/jgp.51.5.655
6. Winslow, R.L., Walker, M.A., Greenstein, J.L.: Modeling calcium regulation of contraction, energetics, signaling, and transcription in the cardiac myocyte. Wiley interdisciplinary reviews. Syst. Biol. Med. **8**(1), 37–67 (2016). https://doi.org/10.1002/wsbm.1322
7. Sutanto, H., Heijman, J.: Integrative computational modeling of cardiomyocyte calcium handling and cardiac arrhythmias: current status and future challenges. Cells **11**(7), 1090 (2022). https://doi.org/10.3390/cells11071090
8. Iyer, V., Mazhari, R., Winslow, R.L.: A computational model of the human left-ventricular epicardial myocyte. Biophys. J. **87**(3), 1507–1525 (2004). https://doi.org/10.1529/biophysj.104.043299
9. Hunter, P.J., McNaughton, P.A., Noble, D.: Analytical models of propagation in excitable cells. Prog. Biophys. Mol. Biol. **30**(2–3), 99–144 (1975). https://doi.org/10.1016/0079-6107(76)90007-9
10. Beeler, G.W., Reuter, H.: Reconstruction of the action potential of ventricular myocardial fibres. J. Physiol. **268**(1), 177–210 (1977). https://doi.org/10.1113/jphysiol.1977.sp011853
11. Yanagihara, K., Noma, A., Irisawa, H.: Reconstruction of sino-atrial node pacemaker potential based on the voltage clamp experiments. Jpn. J. Physiol. **30**(6), 841–857 (1980). https://doi.org/10.2170/jjphysiol.30.841
12. Plank, G., et al.: The openCARP simulation environment for cardiac electrophysiology. Comput. Methods Programs Biomed. **208**, 106223 (2021). https://doi.org/10.1016/j.cmpb.2021.106223
13. Rice, J.J., Winslow, R.L., Hunter, W.C.: Comparison of putative cooperative mechanisms in cardiac muscle: length dependence and dynamic responses. Am. J. Physiol. **276**(5), H1734–H1754 (1999). https://doi.org/10.1152/ajpheart.1999.276.5.H1734
14. Negroni, J.A., Lascano, E.C.: Simulation of steady state and transient cardiac muscle response experiments with a Huxley-based contraction model. J. Mol. Cell. Cardiol. **45**(2), 300–312 (2008). https://doi.org/10.1016/j.yjmcc.2008.04.012
15. Silva, R.R., Filho, O.B.S., Bassani, J.W.M., Bassani, R.A.: The ForceLAB simulator: application to the comparison of current models of cardiomyocyte contraction. Comput. Biol. Med. **131**, 104240 (2021). https://doi.org/10.1016/j.compbiomed.2021.104240
16. Mullins, P.D., Bondarenko, V.E.: A mathematical model of the mouse ventricular myocyte contraction. PLoS ONE **8**(5), e63141 (2013). https://doi.org/10.1371/journal.pone.0063141
17. Bondarenko, V.E., Szigeti, G.P., Bett, G.C., Kim, S.J., Rasmusson, R.L.: Computer model of action potential of mouse ventricular myocytes. Am. J. Physiol. Heart Circulatory Physiol. **287**(3), H1378–H1403 (2004). https://doi.org/10.1152/ajpheart.00185.2003
18. Moss, R.L., Razumova, M., Fitzsimons, D.P.: Myosin crossbridge activation of cardiac thin filaments: implications for myocardial function in health and disease. Circ. Res. **94**(10), 1290–1300 (2004). https://doi.org/10.1161/01.RES.0000127125.61647.4F

Atrial Fibrillation Mechanisms: A Contribution from Computational Modelling

M. E. T. Antunes[1], F. O. Campos[2], I. Sandoval[1], J. G. Siles[1], I. Uzelac[3], and J. Salinet[1](✉)

[1] HEartLab Biomedical Engineering, Modelling and Applied Social Sciences Center, Federal University of ABC/Engineering, São Bernardo do Campo, Brazil
joao.salinet@ufabc.com

[2] School of Biomedical Engineering and Imaging Sciences, King's College London, London, UK

[3] School of Physics, Georgia Institute of Technology, Georgia, USA

Abstract. Atrial Fibrillation (AF) is the most common cardiac arrhythmia worldwide; it is complex and can be multifactorial, and the underlying mechanisms are not yet completely understood, limiting treatment and reducing patients' quality of life. Computational modelling has been shown to play an important role at improving current knowledge of AF. This study aims to simulate a normal heart condition as a planar wavefront propagation, and the three main AF mechanisms: ectopic foci, rotors and multiple and continuous intra-atrial reentries. This can be done through the OpenCARP platform, using its python framework called carputils, and applying Courtemanche's human atrial action potential model. Action potential duration, local activation time and conduction velocity are estimated and compared, alongside the simulation images themselves. In summary, the mechanisms believed to underlie AF were reproduced here using 2D in-silico simulations performed in a standard computer. Action potential and cell-to-cell propagation variables had their roles identified in the mechanisms triggering and maintaining.

Keywords: Computational modelling · Atrial fibrillation · Mechanisms

1 Introduction

Atrial fibrillation (AF) is the most common cardiac arrhythmia worldwide [1], and might be associated with cardiac structural diseases and thromboembolic complications [2]. For the elderly population (i.e. above 80 years), the prevalence of developing AF increases as high as 10% [3], becoming a chronic non-infectious cardiovascular epidemic requiring a high budget for its management and treatment.

Multiple and continuous intra-atrial re-entries, ectopic and rotor mechanisms showed to play an important role at AF trigger and maintenance [1]. Briefly, multiple random reentries are associated with wavefront length (i.e. conduction velocity × refractory period) shortening to critical values, due to complex interactions between substrate,

J. L. B. Marques et al. (Eds.): CLAIB 2022/CBEB 2022, IFMBE Proceedings 98, pp. 139–146, 2024.
https://doi.org/10.1007/978-3-031-49401-7_14

triggers and modulators, causing a non-homogeneous atrial depolarization and repolarization. Ectopic activity is originated by an excitable group of cells, such as the pulmonary veins, triggering abnormal and faster concentric ripples from inner center to the outside, overriding the regular sinus atrial node wavefront activation, and rotor a macro-organization that occurs in spiral reentrant waves (rotors), self-perpetuating through the circular interaction of the wavefront with an unexcited group of cells in its cores. Rotors have shown to be stationary, or propagate and anchor in anatomical heterogeneities of the atria, whose fibrillation conduction is generated by the collision between the reentries and the areas with conduction block, resulting in fibrillary conduction [1].

The predominant underlying AF mechanism showed to differ from patient to patient, and in patients in which AF persisted for longer periods, multiple mechanisms can coexist, increasing AF complexity and limiting treatment rate successes [1]. Basic science, with the help of computer modelling of AF and its mechanisms, has been helping the field, with new insights and complementary information obtained from clinic and animal models [4]. This first year undergraduate research study aims to simulate the three above specified AF mechanisms, through a human atrial action potential (AP) model [5] implemented in a 2D sheet of cells.

2 Materials and Methods

The openCARP's Python framework, carputils, is used to develop the in-silico AF mechanisms models [6]. Simulation results were visualized using Meshalyzer, a tool available within the openCARP framework. All simulations are performed with an Acer laptop of 6 GB of RAM, i5-6200U CPU @ 2.40 GHz and with the Windows 10 operational system. Carputils is used in the Sublime Text, a text editor for coding with Python3 inside a Windows Subsystem for Linux (WSL) with Kali distribution.

2.1 Human Atrial Action Potential

The human atrial cells' AP is modelled with the Courtemanche model [5]. In this model, the cellular membrane is described as an electric circuit composed by a capacitor connected in parallel with multiple resistances and a direct current power source, representing the ionic channels embedded into the bilayer lipid plasmatic membrane. The cell transmembrane potential is defined in Eq. 1, where I_{ion} is the total ionic current and I_{st} is the stimulus current, both flowing across the membrane, whose total capacitance is represented by C_m, considered a constant in this case, with a fixed value of $1\,\mu F/cm^2$.

$$dV/dt = -(I_{ion} + I_{st})/C_m \tag{1}$$

The total ionic current is a contribution of individual ionic currents (Eq. 2). Table 1 summarizes the ionic currents of the implemented human atrial AP.

$$I_{ion} = I_{Na} + I_{K1} + I_{to} + I_{Kur} + I_{Kr} + I_{Ks} + I_{Ca,L} + I_{NaCa} + I_{NaK} + I_{pCa} + I_{bCa} + I_{bNa} \tag{2}$$

Table 1. Ionic currents considered in the Courtemanche model.

Symbol	Definition
I_{Na}	Fast inward Na^+ current
I_{K1}	Inward rectifier K^+ current
I_{to}	Transient outward K^+ current
I_{Kur}	Ultra-rapid delayed rectifier K^+ current
I_{Kr}	Rapid delayed rectifier K^+ current
I_{Ks}	Slow delayed rectifier K^+ current
$I_{Ca,L}$	L-type inward Ca^{2+} current
I_{NaCa}	Na^+/Ca^{2+} exchanger current
I_{NaK}	Na^+-K^+ pump current
I_{pCa}	Sarcoplasmic Ca^{2+} pump current
I_{bCa}	Background Ca^{2+} current
I_{bNa}	Background Na^+ current

I_{ion} is obtained considering the intracellular concentrations as follows: Calcium $[Ca^{2+}]_i = 0.18\ \mu M$, Sodium $[Na^+]_i = 11.2$ mM and Potassium $[K^+]_i = 139.0$ mM. Calcium concentration in the sarcoplasmic reticulum (i.e. release compartment) is 0.64 mM. The extracellular concentrations are considered constants (not variating with time), since the model applies the monodomain equations, those being differential equations to represent cell-to-cell stimulus propagation, considering only the intracellular space.

2.2 Planar Wavefront

A regular planar wavefront propagation is simulated within a 2D sheet of 1×1 cm of tissue with myocyte cells paced at a 1 Hz frequency. The stimulus repeats itself at a 1 Hz pace, lasts 2 ms and has an electric current density of 100 $\mu A/cm^2$. Intracellular longitudinal and transversal conductivities are set to 0.174 nS/pF and the extracellular to 0.625 nS/pF. The wavefront propagates from left to right, and the simulation length is 1200 ms.

2.3 Ectopic

Firstly, the above described planar wavefront is simulated, with stimulus duration of 2 ms and electric current density of 100 $\mu A/cm^2$. After that, 3 ectopic foci stimuli placed in distinct positions in the 2D cells' sheet mesh spontaneously and simultaneously depolarize at 6 Hz, and the simulation lasts 1000 ms. The first depolarization started 50 ms after the beginning of the planar wavefront. The ectopic focus closest to the left border has an electric current density of 130 $\mu A/cm^2$ and the remaining 100 $\mu A/cm^2$. Longitudinal and transversal conductivities are also set to 0.174 nS/pF intracellularly and 0.625 nS/pF extracellularly.

2.4 Rotor

A 2D sheet of cells with 2 × 2 cm is divided into two regions with different conductivities. The lower region has intracellular longitudinal and transversal conductivities set to 0.080 nS/pF, and the same parameters set to 0.521 nS/pF extracellularly, whereas the upper region has longitudinal and transversal conductivities of 0.270 nS/pF intracellular and of 0.705 nS/pF extracellular. The inward rectifier K^+ current (I_{K1}) was reduced by multiplying the original value by 0.65, resulting in 0.0513 pA/pF. The first stimulus is triggered at time = 0 ms and the second at time = 340 ms. Both stimuli are located in the upper left side of the tissue, and have an electric current density of 100 $\mu A/cm^2$. The simulation duration is 1000 ms.

2.5 Multiple Wavelets Mechanism

This mechanism presents a higher level of complexity, and its simulation is still under development. At this moment, we introduced anisotropy, testing different hypotheses of effective refractory period duration, and implementing the cross-field stimulation [7].

2.6 Quantitative Analysis

For each in-silico experiment, the local activation time (LAT), action potential duration (APD) and the conduction velocity (CV) are calculated. LATs are estimated during the depolarization phase (i.e. positive slope), as the time instant when transmembrane potential crosses the –10 mV threshold value. Local repolarization time (LRT) is defined as the time instant when transmembrane potential crosses the −70 mV, during the negative slope of the repolarization phase. APDs are computed as the difference between repolarization and activation times (Eq. 3) [8]. CV for planar wavefront is obtained by the single vector method (Eq. 4), through the Euclidean distance divided by LAT difference between two distinct regions [9]. Since the single vector method for tracking CV is not suitable for the rotor and ectopic activity, CV is then obtained by the circle method [10], as follows: firstly, the differences in the LATs across the endpoints of chords, passing through the center of the circle are calculated. From the LAT differences, an effective conduction speed can be calculated as a function of the chord's orientation angle. The minimum CV or the maximum LAT difference corresponds to the true conduction speed along the propagation direction.

$$\mathrm{APD} = \mathrm{LRT} - \mathrm{LAT} \tag{3}$$

$$\mathrm{CV} = \Delta d/(\mathrm{LAT_o} - \mathrm{LAT_i}) \tag{4}$$

For the ectopic foci simulation, every stimulus triggering generated a 1 × 1 cm matrix of cells. To calculate CV, a matrix of 40 × 40 cells is taken from the original 2D cells's sheet for each stimulus, and the radius used to track CV is of 5 cells (= 1 mm). The CV value is calculated for each stimulus (6 Hz), and the average is obtained. For the rotor simulation, the mean CV is also an average of CV obtained across the paced stimuli.

3 Results and Discussion

Figure 1 presents a typical AP from the atrial myocyte according to Courtemanche [5]. In this figure, it is possible to observe the five AP phases, ionic currents, APD and refractory period. Briefly, while resting, the cell potential is negative on the inside and positive on the outside. With a stimulus, sodium channels open and Na^+ enters the cell, increasing the membrane voltage rapidly, reaching the AP trigger threshold. It is followed by the inactivation of the sodium channels, and activation of both calcium (slow inward of Ca^{2+}) and potassium channels (outward of K^+ ions), characterizing the early repolarization phase. In the plateau, there is a balance between K^+ efflux and Ca^{2+} influx. Then, the influx of Ca^{2+} activates the release of a higher calcium concentration, stocked inside the sarcoplasmic reticulum. The increase of $[Ca^{2+}]$ inside the cell would trigger the complex contraction process. The last phase is characterized by the efflux of K^+, returning the transmembrane potential to its resting potential.

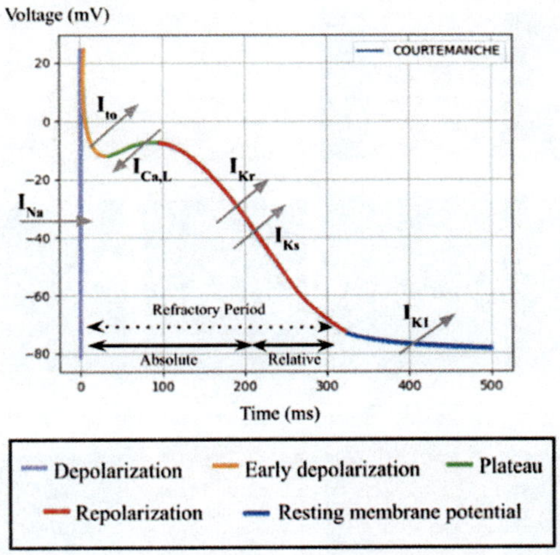

Fig. 1. Simulated AP of the human atrial myocytes according to Courtemanche's model. The gray arrows represent the current fluxes and each color represents one of the AP phases, according to the subtitle. The dotted arrow denotes the total refractory period, whereas the full black arrows indicate the absolute and relative phases of the refractory period.

In Fig. 2a, it is possible to see the propagation of a planar wavefront in a "healthy" 2D atrial sheet of cells. The darkest blue represents the resting potential, which is −81.2 mV, according to Courtemanche's model, and the red is the maximum value established as 35 mV. Figure 2b shows a planar wavefront triggered at 0 ms, and after 50 ms, three ectopic foci are depolarized simultaneously and recurrent every 167 ms (6 Hz pace). In the first two triggers of the ectopic stimuli, the tissue cells are still in the effective refractory period, so it is not possible to trigger another AP. In the third stimulus, however, the tissue is already repolarized, therefore out of the effective refractory period, and the ectopic foci triggers stimulation that propagates throughout the tissue. The rotor mechanism is represented in Fig. 2c, where the first stimulus depolarizes the tissue normally, but, since there are two regions with different APD, the upper half region becomes re-excitable before the lower half. When the second stimulus is applied, the part of the tissue that has not repolarized yet creates a unidirectional block, and the AP continues to travel along the upper half of the domain until the lower half of the domain becomes excitable, causing the reentry.

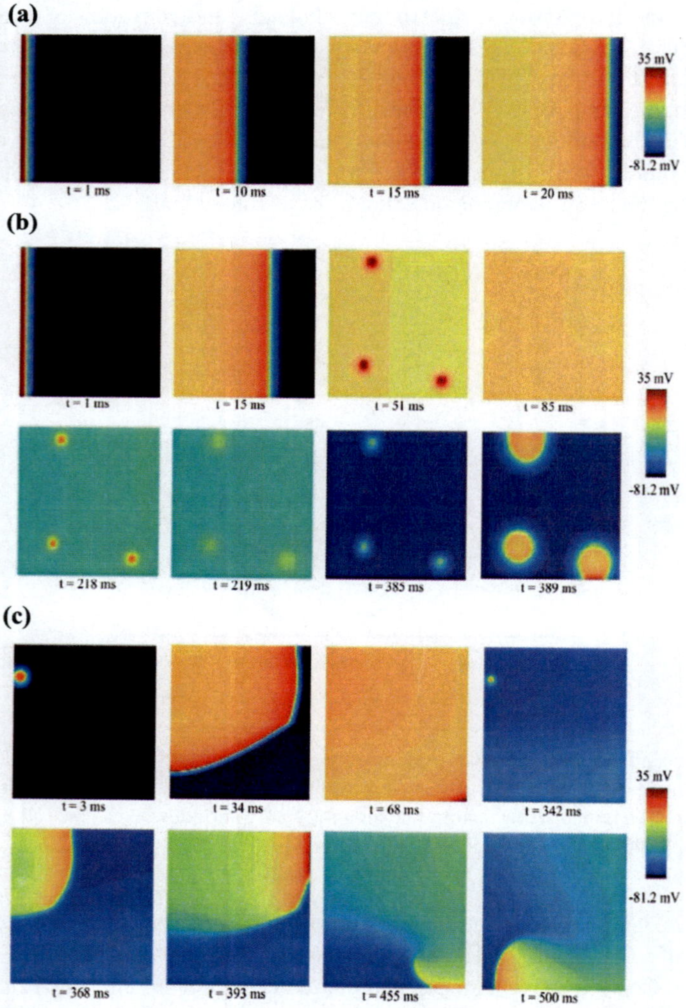

Fig. 2. The simulations of (a) normal propagation (planar wavefront); (b) ectopic mechanism; and (c) rotor mechanism). Simulations can be visualized with the following link: https://youtube.com/playlist?list=PLom9zID5FoVEjtFlT4BHLZs86Uv6VXrBR

The average APD, LAT and CV for planar wavefront are 299.47 ms, 1011.15 ms and 44.78 cm/s, respectively. For both ectopic and rotor mechanisms, the median APD are, respectively, 300.09 and 404.53 ms. The LAT and CV are 297.47 ms and 52.74 cm/s for the ectopic, and 444.91 and 39.77 cm/s for the rotor.

4 Conclusions

This study shows that implementation of simplistic 2D in-silico AF mechanisms is feasible, with a standard computer and minimum level of programming. Multiple wavelet mechanism models are still under development and its details will be included during the review process. Up to now, the understanding of the electrophysiology theory basis beyond AP formation, its variables, and how cell-to-cell propagation is performed were crucial for the development of the simulations.

Acknowledgements. The authors acknowledge the financial support from São Paulo Research Foundation (FAPESP), process 18/25606–2. M.E.T Antunes is performing 1-year undergraduate research (Edital: Edital 01/2021 - PIC/PIBIC/PIBITI/PIBIC-AF, process: 23006.015594/2021–31. I. Sandoval is supported by Coordenação de Aperfeiçoamento de Pessoal de Nível Superior - Brasil (CAPES). J.G. Siles is supported by São Paulo Research Foundation (FAPESP), grant #2020/03601–9.

Conflict of Interest The authors declare that they have no conflict of interest.

References

1. Salinet, J., Molero, R., Schlindwein, F.S., Karel, J., Rodrigo, M., Rojo-Álvarez, J.L., Berenfeld, O., Climent, A.M., Zenger, B., Vanheusden, F., Paredes, J.G.S., MacLeod, R., Atienza, F., Guillem, M.S., Cluitmans, M., Bonizzi, P.: Electrocardiographic imaging for atrial fibrillation: a perspective from computer models and animal experiments to clinical value. Front. Physiol. **12**, 653013 (2021)
2. Zimerman, L.I., Fenelon, G., Martinelli Filho, M., et al.: Sociedade Brasileira de Cardiologia. Diretrizes Brasileiras de Fibrilação Atrial. Arq. Bras. Cardiol. **92**(6 supl. 1), 1–39 (2009)
3. Karamichalakis, N., et al.: Managing atrial fibrillation in the very elderly patient: challenges and solutions. Vasc. Health Risk Manage. **11**, 555–562 (2015)
4. Dössel, O., Krueger, M.W., Weber, F.M., Wilhelms, M., Seemann, G.: Computational modeling of the human atrial anatomy and electrophysiology. Med. Biol. Eng. Comput. **50**(8), 773–799 (2012)
5. Courtemanche, M., Ramirez, R.J., Nattel, S.: Ionic mechanisms underlying human atrial action potential properties: insights from a mathematical model. Am. J. Physiol. **275**(1), H301–H321 (1998)
6. Plank, G., Loewe, A., Neic, A., Augustin, C., Huang, Y.L., Gsell, M.A.F., Karabelas, E., Nothstein, M., Prassl, A.J., Sánchez, J., Seemann, G., Vigmond, E.J.: The openCARP simulation environment for cardiac electrophysiology. Comput. Methods Prog. Biomed. **208**, 106223 (2021)
7. Williams, S.E., et al.: Left atrial effective conducting size predicts atrial fibrillation vulnerability in persistent but not paroxysmal atrial fibrillation. J. Cardiovasc. Electrophysiol. **30**(9), 1416–1427 (2019)
8. Local activation time | openCARP. Cardiac Electrophysiology Simulator | openCARP, https://opencarp.org/documentation/examples/02_ep_tissue/08_lats. Last accessed 01 Mar 2023

9. Lab 9: Conduction Velocity of Nerves—OpenWetWare. https://openwetware.org/wiki/Lab_9:_Conduction_Velocity_of_Nerves. Last accessed 01 Mar 2023
10. Siles, J., Salinet, J., Pollnow, S., Dössel, O., Uzelac, I.: Estimation of the Ablated area size based on local conduction velocity simulations and animal experiments. In: Computing in Cardiology (CinC). IEEE, Brno, Czech Republic (2021)

Analysis of Light Transmission in Skin Tissue Considering Scattering Condition in Beer-Lambert Type Models

J. Valladares-Pérez(✉), F. Cortés-Ortegón, A. García-Valenzuela, and C. Sánchez-Pérez

Instituto de Ciencias Aplicadas y Tecnología, Universidad Nacional Autónoma de México, Ciudad Universitaria, 04510 Mexico City, Mexico
jose.valladares@icat.unam.mx

Abstract. In this work, we present an analysis of light transmission of the Beer-Lambert Theory using a multilayer skin model for oximetry applications that considers a system of different layers of skin (dermis and epidermis) containing physical properties of whole blood as a suspension of erythrocytes in plasma. Erythrocytes are assumed with hemoglobin inside in its oxygenated and deoxygenated form according to the oxygen saturation. Due to the non-homogeneity of the skin, the Beer-Lambert law is extended for a suspension of erythrocytes in plasma considering: (i) its absorption cross-section and (ii) its absorption and scattering cross-sections. The obtained results proof that the scattering is stronger than the absorption in a spectral range from 350 to 1000 nm. Likewise, the scattering in this model is presented as forward scattering, giving a slight difference to the Beer-Lambert model of homogeneous medium respect to the extended models of inhomogeneous medium analyzed. These analytic models can be a simple way for interpreting experiments based in oximetry.

Keywords: Beer-Lambert law · Oximetry · Multilayer model · Scattering

1 Introduction

1.1 Pulse Oximetry

Pulse oximetry it's a non-invasive optical technique by which the percentage of oxygenated haemoglobin (HbO_2) is indirectly measured respect to total haemoglobin, this represents the percentage of oxygen that is being transported in the blood circulation throughout the body, this parameter is known as partial oxygen saturation (SpO_2) [1].

A conventional pulse oximeter measures oxygen saturation by the photoplethysmography (PPG) technique [2].

The device emits light in two wavelengths, one in the visible and the other in the infrared that when interacting mainly with HbO_2 and deoxygenated hemoglobin (Hb), they are attenuated according to the oxygenation of the tissues in the body as part of

J. L. B. Marques et al. (Eds.): CLAIB 2022/CBEB 2022, IFMBE Proceedings 98, pp. 147–157, 2024.
https://doi.org/10.1007/978-3-031-49401-7_15

the cardiac cycle, so, it is possible that from these signals, one can determine the partial oxygen saturation HbO_2.

The light transmission in oximetry is normally interpreted from the Beer-Lambert (BL) Law that was proposed for homogeneous mediums, in this the amount of transmitted light varies according to the amount of light absorbed and that it's proportional to the concentration of HbO_2 and total haemoglobin in the blood. Each Hb molecule absorbs a certain amount of light, whereas the grater the amount of Hb per unit area, the greater the amount of light that is absorbed. Likewise, the light absorbed is proportional to the length of the path it travels.

However, the biological tissue specifically the skin is a non-homogeneous medium, so the need arises to analyze the classic BL method with models that consider haemoglobin as a suspension of these in the whole blood; that is, considering its absorption and scattering. Thus, this work suggests an analysis of light transmission in a skin model from the extension of the BL model to consider haemoglobin in the blood as particles with absorption and scattering.

To study the transmitted light on a proposed skin model, analytic methods are of great interest because physical and optical properties are directly related to the calculations. There is also the possibility to use numeric methods like Monte Carlo simulations, this kind of work have been done in the past [3, 4], according to [5] this method has the advantage of being able to model complex structures and substrate to produce physical quantities. But it has the disadvantage of being computationally expensive and the simulations has to be run multiple times to have a better result. This give us a better reason to use analytic methods, because the geometry we are going to simulate is simple and the physical response can be calculated by using the equations available in the literature.

1.2 Theoretical Background

Beer-Lambert Law. BL law describes the attenuation of light propagating through a uniform medium with absorption ($BL_{\varepsilon a}$) [6]. The light intensity that travels though the medium ($I_{\varepsilon a}$) decreases exponentially with distance, this law is define bay the Eq. 1:

$$I_{\varepsilon_a} = I_0 e^{-\varepsilon_a(\lambda)cd}. \tag{1}$$

where $\varepsilon_a(\lambda)$ is the absorption coefficient of the medium, which depends on the wavelength λ, c is the concentration of the absorbent substance that is constant in the medium, and d is the length of the way.

BL law applied to a medium stratified in several layers considers the sum of the nth absorbances related to its volumetric ratio.

This law is of great interest for homogeneous mediums because it does not take into consideration physical processes, dispersion, or light scattering. To analyze the latest, it is essential to extend this formulations. On the other hand, the model of Bören-Huffman's book [7], can be applied to calculate the amount of light transmitted in a suspension on absorbent particles (Eq. 2) ($BL_{\sigma a}$):

$$I_{\sigma_a} = I_0 e^{-\rho\sigma_a(\lambda)d}. \tag{2}$$

where ρ is the number of particles per volume of the absorbent substance, σ_a is the absorption cross-section and d is the length of the way. We will refer to this equation as BL law for absorbent particles ($BL_{\sigma a}$).

On the other hand, we can consider the Twersky approximation [8], which is the base of the so called modified Beer-Lambert law (Eq. 3), that calculates the light intensity transmitted through particle suspensions that produces scattering ($BL_{\sigma a,\sigma s}$):

$$I_{\sigma a,\sigma s} \approx I_0 e^{-\gamma d}\Big[e^{\beta d} + q(\alpha)\Big(1 - e^{-\beta d}\Big)\Big], \tag{3}$$

where the arguments $\gamma \approx (\sigma_a/\upsilon)\omega$, $\beta \approx (\sigma_s/\upsilon)\omega(1-\omega)$ and $\omega \approx \rho\upsilon$, being σ_a y σ_s the absorption and scattering cross-sections of the particle respectively, for the volume fraction occupied by the particles ω, υ is the volume of one particle, and ρ is the numerical density of particles. The factor $q(\alpha)$ is the area of the absorption cross-section of the light source this is $q(\alpha) = \sigma(\alpha)/\sigma$ where σ is the integral of the scattering cross-section over the forward hemisphere ($\sigma \approx \sigma_t$) for large particles such as biological cells, σ_t is the complete scattering cross section), and $\sigma(\alpha)$ is the integral of the differential scattering cross section over a detection solid angle α around the direction of incidence of the light [8]. We refer to this equation as the BL law for absorbent particles ($BL_{\sigma a,\sigma s}$) where $0 < q(\alpha) < 1$.

To calculate the absorption and scattering cross-sections there are different approximations. This paper will discuss the differences between two of these, Mie Theory (MT), and the Anomalous Diffraction Approximation (ADA).

Mie Theory. The first complete theory on spherical scattering was developed by Gustav Mie in 1908, this theory describes the interaction between a flat electromagnetic wave and a homogeneous spherical dielectric particle, thus predicting the relative intensity of the scattered light as a function of the size of the particles (a) (generally, "Mie scattering" formulas are more useful in situations where the size of the scattering particles is comparable to the light's wavelength), the observation angle which is not limited to scattering angles less than 90° (forward/sideways direction), but they are also produced for scattering angles greater than 90° (backward direction) [9]. According to MT, more forward scattering occurs than that produced in any other direction and this scattering is proportional to the particle size. [10].

The equations necessary to perform the dispersion calculation with the MIE theory can be found in the book by Böhren and Huffman [7].

Anomalous Diffraction Approximation. Developed in 1957 by Van de Hulst [11], which represents the extinction efficiency (σ_{ext}), for spherical and non-spherical dielectric particles, which turns out to be a complementary approximation to MT. This method requires little computationally time, and the equation works best for large parameters ($a >> \lambda$) and for the ratio of the refractive index of the particle to the background near to one [12].

The assumptions of Van de Hulst's, describe that the scattered intensity is concentrated near the original propagation direction. The light that travels through the sphere suffers it a significant lag, although it does not suffer a significant refraction. This light will interfere constructively and destructively with the diffracted light such as that of a circular disc, whose surface is equal to the geometric section of the sphere.

The electric field in the shadow of the sphere is E $= e^{-ipsen\tau}$, where τ is the angle relative to the tangent of the sphere in the point where the ray collides with the sphere, and p = 2 × (m − 1) describes the lag of the beam that goes though the center of the sphere ($\tau = 90°$). Given a unit magnitude field outside of the geometric shadow, the scattered field adds the factor $e^{-ip} - 1$ to the original field in the forward direction [13].

In order to find the scatter amplitude $S(0)$, the Babinet principle is applied, since the scattered and the remaining fields are added to the original field, resulting in:

$$S(0) = -\frac{k_m^2}{2\pi}\int_{\infty}^{\infty}\left(e^{2i[k_p-k_m]Re\left(\sqrt{a^2-x^2-y^2}\right)} - 1\right)dxdy, \tag{4}$$

with $k_p - k_m = k(n_p - n_m)$ where k is the wavenumber in the void, n_p y n_m are the refractive index of the particle in the medium.

Finally, solving Eq. 4 we get to:

$$\sigma_{ext} = -\frac{4\pi}{k_m^2}Re\{S(0)\}. \tag{5}$$

The analysis of the transmitted light was divided into two subsections. The first one calculates the resulting transmitted angle considering the refractive indices of the different layers and the second uses the different BL models to determine the differences in absorption of the hemoglobin mixtures.

1.3 Light Transmission in Skin

To analyze the variations on the trajectory of light at an incidence different from the normal one, Snell law will be used (Eq. 6) and the Fresnel coefficients for calculating the transmittance at each interface for the TE polarization (Eqs. 7 and 8).

$$n_1 sen\theta_1 = n_2 sen\theta_2, \tag{6}$$

$$t_1 = \frac{2n_1\cos\theta_1}{n_1\cos\theta_1 + n_2\cos\theta_2}, \tag{7}$$

$$T_1 = \frac{n_2\cos\theta_2}{n_1\cos\theta_1}|t_1|^2, \tag{8}$$

where n_1 and n_2 are the refractive index of the incident and transmitted media, respectively θ_1 and θ_2 are the angles of incidence and transmission. For the polarization TE, t_1 is the transmission Fresnel coefficient and T_1 is the transmittance of the medium. The response for Eqs. 1–3 of the three BL models will be compared.

1.4 Skin Model

To analyze the transmitted light in tissue, a multilayer model was simulated for a simplified skin model. This is composed of 5 layers: epidermis, dermis and a layer of whole blood, as shown in Fig. 1a) In the layer of blood, plasma and erythrocytes were considered, the latter are assumed spherical and only with hemoglobin inside, either in its oxygenated HbO_2 or deoxygenated Hb form, in a concentration that depends on the partial oxygen saturation SpO_2. This is calculated from the ratio of HbO_2 and the total hemoglobin (H_{bt}):

$$SpO_2 = \frac{HbO_2}{HbO_2 + Hb} = \frac{HbO_2}{Hbt}. \tag{9}$$

The absorption coefficient of H_{bt} (ε_{Hbt}) is calculated as:

$$\varepsilon_{Hbt} = c_{HbO2}\varepsilon_{HbO_2} + c_{Hb}\varepsilon_{Hb}, \tag{10}$$

where c_{HbO2} and c_{Hb} are the concentrations of oxygenated and deoxygenated hemoglobin respectively and ε_{HbO2} and ε_{Hb} are the absorption coefficients for each of these [14].

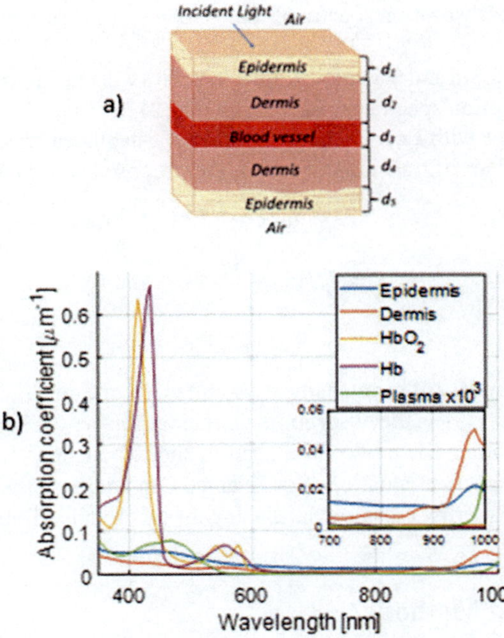

Fig. 1. Skin tissue model. a) Diagram of multilayered model and b) Absorption coefficient spectra for the elements considered.

To perform the simulation, it was necessary to know the refractive indices of all the media, as well as their thicknesses, this data is condensed in Table 1.

The thickness of the skin varies depending on the area of the body, therefore, values of 50 and 600 μm were chosen for the epidermis and dermis respectively. The blood vessel was considered with the thickness of an arteriole. To calculate the absorption cross-section, we consider the complex refractive index for the erythrocytes ($n_{er} = n_{er'} + in''_{er}$) [18].

$$n''_{er} = \log(10)\frac{\lambda}{4\pi M_h}\varepsilon_{a,HC}(\lambda)HC, \tag{11}$$

Table 1. Optogeometric characteristics of the skin model.

Media	Ref	n	Thickness (μm)
Epidermis	[6, 15]	1.43	27–150
Dermis	[15, 16]	1.43	600–3000
Plasma	[17]	1.35	—
Erythrocytes	[18, 19]	1.40	5
Blood Vessel	[20]	—	18

where HC is the hematocrit (0.4 for women and 0.45 for men), M_h is the molecular weight ($M_h = 64500$) [18] and $\varepsilon_{a,HC}(\lambda)$ is the absorption coefficient [14].

The values of the absorption spectra of the elements of the skin model are presented in Fig. 1b).

Values of hemoglobin and plasma absorption spectra were taken from [14] and dermis and epidermis absorption spectra for light skin tone [21].

Combining Eq. 11 with Eqs. 6–8, it is possible to establish a series of new formulas to describe the behavior of transmitted light in the proposed skin model, for the $BL_{\varepsilon a}$.

$$I_p = I_0 T_p e^{-BL_{\varepsilon ap}}, \tag{12}$$

$$I_{p+1} = I_p T_{p+1} e^{-BL_{\varepsilon ap+1}}, \text{ and} \tag{13}$$

$$I_T = I_{p+1} T_{p+2}, \tag{14}$$

where I_{p+1} corresponds to the intensity transmitted by the pth layer, I_T is the total transmitted light. $BL\varepsilon_{ap+1}$ is the absorption coefficient for the pth layer.

To evaluate the other BL modes, the term of the exponential I_p and I_{p+1} will be changed with the respective equation (Eqs. 2 and 3), in addition, the values calculated for absorption and scattering cross-sections using the ADA and MT approximations will be analyzed.

2 Materials and Methods

Equation 12 (a to c) were introduced into Matlab to calculate the transmitted light in the skin model. The values for all the variables used were taken from the references consulted in the *Skin model* section. The simulations were done using a personal PC.

The angle of incidence was taken at 0° according to the normal plane, and the light intensity was taken as 1 to normalize the calculations. Then, the cross-sections were calculated using the two theories. With all the equations programed, we run the simulation to calculate light transmitted through the skin model and then calculate the transmittance of the system.

We simulated two slightly different models, the first one considering the skin transparent so we can only analyze the contribution of blood for SpO_2 of 99 and 0%. In the second we add the absorption of the skin while also analyzing the SpO_2 concentrations in the same values as before.

3 Results

When the response of this system was analyzed by eliminating absorption of the epidermis and dermis to understand only the contribution of the blood when varying the ratio of HbO_2 and Hb, these results are presented in Fig. 2.

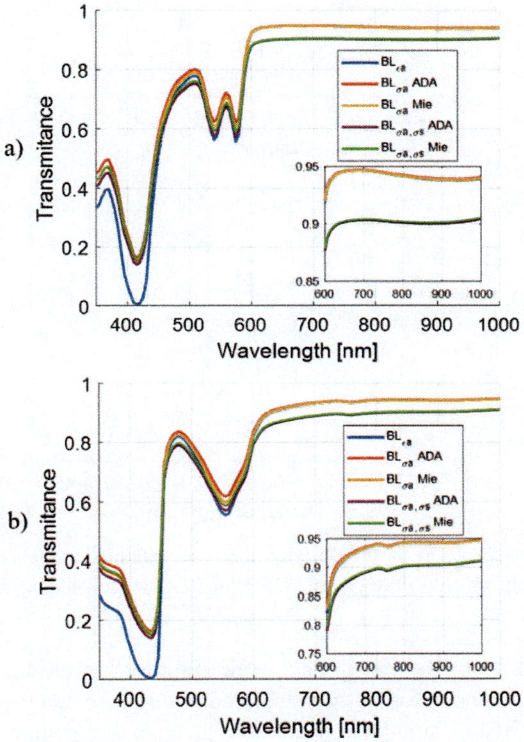

Fig. 2. Transmittance of the different Beer-Lambert models without the skin absorption for a) $SpO_2 = 99\%$ (superior) and b) $SpO_2 = 0\%$ (inferior). The model of $BL_{\sigma a,\ \sigma s}$ has a $q(\alpha) = 0.9$.

The graphs in Fig. 2 show that the transmittance spectrum is similar for the $BL_{\varepsilon a}$ model and the 4 extended models. From wavelengths above 600 nm the difference between them is minimal for the two values of SpO_2, in this range the $BL_{\sigma a,\sigma s}$ has a difference in transmittance for both of the calculated approximations. Below this wavelength range, the biggest difference can be seen between the models, although the one that presents the highest variation is $BL_{\varepsilon a}$ at 435 nm for the two SpO_2. This same behavior occur in the range between 500 and 600 nm, showing two minimums at 540 and 575 nm, in difference from the 0% of SpO_2 where it only has one minimum at 555 nm.

Then, we analyzed the models considering the skin absorption. Results are presented in Fig. 3.

It can be observed how the shape of the signal is modified, now presenting transmittance peaks at 645, 730 and 825 nm, all of them above 600 nm in wavelengths commonly

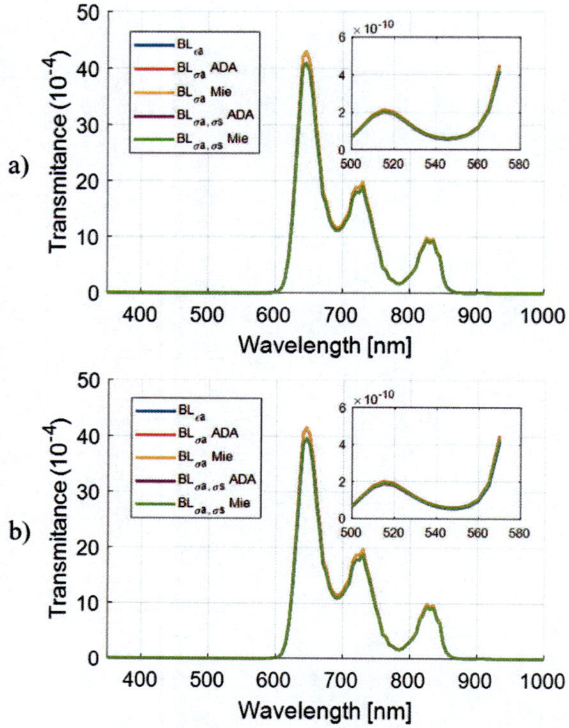

Fig. 3. Transmittance spectrum for the Beer-Lambert and extended models for: a) $SpO_2 = 99\%$ (superior) and b) $SpO_2 = 0\%$ (inferior), the models of $BL_{\sigma a,\sigma s}$ consider $q\ (\alpha) = 0.9$.

used in oximetry, however, the transmittance value falls by 4 magnitude orders compared to the model without absorption in the skin, in addition to the fact that in these case the variation between the two simulated SpO_2 are much smaller than in the previous case and therefore, the variation of the $BL_{\sigma a,\sigma s}$ model is also reduce in regard to the others.

The analysis of this model for the conditions with or without absorption in the skin, allows to visualize the variations between the different BL models, as well as apply them to a partial oxygen saturation value to observe changes in the transmittance spectra. When observing the model curves without skin absorption, these variations between models are more evident, and given a high $q(\alpha)$ value (0.9), which can mean that the light is spread forward, indicating that in the optical window for oximetry (>600 nm) do not represent a significant difference.

But when introducing the absorption of the skin, it can be observed that the differences between these become even smaller, so if it were possible to use a wavelength close to the peaks, the transmittance spectrum would allow to transmit more light and, therefore, generate more useful signals for oximetry.

The value of $q(\alpha)$ used is high, so the forward scattering is high which may explain why there were not significant differences for transmittance considering classic BL and the extended model with absorption and scattering is not substantial. The election of the

$q(\alpha)$ value was made considering that the size of the erythrocyte is very big respect to the two wavelengths VIS and IR considered in this analysis.

4 Discussion

Comparing this work with others that have done a similar research whether in the use of different BL law approximations to study the light propagation response on tissue [22, 23] or using it to study the response of the oxygenated blood like the ones discussed in [24]. In [22] they used the modified Beer-Lambert Law to study the propagation of light in skin with high bilirubin concentration and in [23] they studied the amplitude modulation of the PPG signal depending on the position of a reflection type oximetry emitter and receiver. The BL law used is a different more simple approximation of the one they were using, that one can lead to a confusion on how to interpret the parameters needed for the simulation, so we went for the one originally derived in [8]. While in [24] they reviewed different articles in which they used different wavelength sometimes both of them below the isosbestic point to calculate SpO_2, and those are in the range we saw while analyzing the model without the skin absorption.

5 Conclusions

In this paper we analyzed the light transmission in a skin model considering a plasma and erythrocytes suspension within an arteriole as part of a skin model. The Beer-Lambert model was considered in its classical form and a rethinking of this considering the absorption of erythrocytes; as well as an extended model that considers absorption and scattering.

In a skin model without absorption, in the wavelengths most used in oximetry (600 to 1000 nm), the models for homogeneous media ($BL_{\varepsilon a}$) and for particle suspension ($BL_{\sigma a}$) have a difference respect to the modified model ($BL_{\sigma a,\sigma s}$), as long as the $q(\alpha)$ value gets closer to 1, it will mean that the light spreads forward, and therefore, the modified model will approach the particle suspension one. For the opposite case, below 600 nm, the models present areas where they approach each other and others where they usually differ to a great extent, especially at the points of greatest absorption in the hemoglobin spectrum, approximately at 435 and 555 nm.

Another point to highlight is that the absorption generated by the skin when it is added to the model, strongly attenuates the resulting spectrum, and generates transmission maximums in spectral bands at 645, 730 and 825 nm, some of which are usually used in oximetry, however, in these cases, the differences between models become even higher.

Acknowledgments. The authors acknowledge financial support from UNAM-PAPIIT grant IV100320 and SECTEI grant SECTEI/080/2020. J. Valladares-Pérez and F. Cortés-Ortegón thank CONACYT for the PhD studies grant (CVU: 929080) and master studies grant (CVU: 1086326).

Conflict of Interest The authors declare that they have no conflict of interest.

References

1. López-Herranz, G.P.M.G.: Oximetría de pulso. Rev. Med. Hosp. Gen. Mex. **66**(3), 160–169 (2003)
2. Bencomo, S., Salas, B., Villazana, S.: Diseño y construcción de un oxímetro de pulso. Rev. Ingeniería UC **23**(2), 162–171 (2016)
3. Li, H., Zhang, C., Feng, X.: Monte Carlo simulation of light scattering in tissue for the design of skin-like optical devices. Biomed. Opt. Express **10**, 868–878 (2019)
4. Chatterjee, S., Kyriacou, P.A.: Monte Carlo analysis of optical interactions in reflectance and transmittance finger photoplethysmography. Sensors (Basel) **19**(4), 789 (2019). PMID: 30769957; PMCID: PMC6412556. https://doi.org/10.3390/s19040789
5. So-Ling, C., Li, L.: A multi-layered reflection model of natural human skin. In: Proceedings of Computer Graphics International 2001, pp. 249–256 (2001). https://doi.org/10.1109/CGI.2001.934681
6. Webster, J.G. (ed.): Design of Pulse Oximeters. Taylor & Francis, Medical Science Series (1997)
7. Böhren, C.F., Huffman, D.R.: Absorption and Scattering of Light by Small Particles. New York (1983)
8. Twersky, V.: Interface effects in multiple scattering by large, low-refracting, absorbing particles. J. Opt. Soc. Am. **60**(7), 908–914 (1970)
9. La teoría de MIE at https://www.fritsch.es/medicion-de-particulas/conocimientos-de-fritsch/la-teoria-de-mie/
10. Mie dispersion at https://rammb.cira.colostate.edu/wmovl/vrl/tutorials/euromet/courses/spanish/satmet/s2400/s2400005.htm
11. Hulst, H.C.: Light Scattering by Small Particles. New York (1957)
12. Thomas, M.E.: Improved-anomalous diffraction approximation for accurate extinction, scatter, and absorption efficiency calculations. In: Proceedings of SPIE 10750, Reflection, Scattering, and Diffraction from Surfaces (2018). https://doi.org/10.1117/12.2323110
13. Thovsen, K.B.: Evaluation of Mie Scatter Approximation Formulas for the Scattering of Infrared Light at Biological Cells (2013)
14. Baranoski, G.V.G.: Natural Phenomena Simulation Group at http://www.npsg.uwaterloo.ca/data/blood.php
15. Ding, H., Lu, J.Q., Wooden, W.A., Kragel, P.J., Hu, X.H.: Refractive indices of human skin tissues at eight wavelengths and estimated dispersion relations between 300 and 1600 nm. Phys. Med. Biol. **51**(6), 1479–1489 (2016)
16. Krishnaswamy, A., Baranoski, G.V.: A Study on Skin Optics. Natural Phenomena Simulation Group, School of Computer Science, University of Waterloo, Canada (2004)
17. Belgacem, O., et al.: Molecular mass determination of plasma-derived glycoproteins by ultraviolet matrix-assisted laser desorption/ionization time-of-flight mass spectrometry with internal calibration. J. Mass Spectrom. **37**, 1118–1130 (2002). https://doi.org/10.1002/jms.372
18. Vázquez, E.O., García, V.A.: Reflectivity and transmissivity of a surface covered by a disordered monolayer of large and tenuous particles: theory versus experiment. Appl. Opt. **56**, 7158–7166 (2017)

19. MANUAL MSD, Version for general public at https://www.msdmanuals.com/es-ar/hogar/multimedia/table/v12857021_es
20. Tu, T.Y., Chao, P.C.P.: Continuous blood pressure measurement based on a neural network scheme applied with a cuffless sensor. Microsyst. Technol. **24**, 4539–4549 (2018). https://doi.org/10.1007/s00542-018-3957-4
21. Finlayson, L., Barnard, I.R.M., McMillan, L., et al.: Depth penetration of light into skin as a function of wavelength from 200 to 1000 nm. Photochem. Photobiol. (2021). https://doi.org/10.1111/php.13550
22. Ong, P., Huong, A., Ngu, X., Mahmud, F., Philimon, S.: Modified lambert beer for bilirubin concentration and blood oxygen saturation prediction. Int. J. Adv. Intell. Inf. **5**, 113 (2019). https://doi.org/10.26555/ijain.v5i2.363
23. Lapitan, D.G., Tarasov, A.P.: Analytical assessment of the modulation depth of photoplethysmographic signal based on the modified Beer-Lambert law. In: 2019 IEEE 8th International Conference on Advanced Optoelectronics and Lasers (CAOL), pp. 103–106 (2019). https://doi.org/10.1109/CAOL46282.2019.9019552
24. Oshina, I., Spigulis, J.: Beer-Lambert law for optical tissue diagnostics: current state of the art and the main limitations. J. Biomed. Opt. **26**(10), 100901 (2021). PMID: 34713647; PMCID: PMC8553265. https://doi.org/10.1117/1.JBO.26.10.100901

Analysis of Low Cost 3D Printer Process Parameters and Their Influence on Manufacturing Functionality of Printed Parts

Alessandro Márcio Hakme da Silva[1(✉)], Giovani Biteli do Prado[2], Carlos Alberto Fortulan[3], Luis Gustavo Sabino[1], Alessandra Baptista[1], and Jonas de Carvalho[3]

[1] University Brazil, São Paulo, Brazil
alessandro.silva@ub.edu.br
[2] University of Araraquara, São Paulo, Brazil
[3] University of São Paulo, São Carlos, Brazil

Abstract. Additive Manufacturing (AM), also known as 3D printing or Rapid Prototyping (RP) is the additive production process that works by adding successive layers of material to create the desired three-dimensional model. There are several techniques that can be used in 3D printing, among them, Fusion and Deposition Modeling (FDM) stands out, as it is the most practical. Areas of application of AM can be more varied, reaching areas of great relevance and social impact such as engineering, medicine, among others. The major reason around of this technology is given by characteristics of speed of creation project and the quality of the final product when compared with the traditional means of production. In order to have the final quality always assured, it is necessary the parameters must be properly adjusted. In this study we investigated the influence of layer thickness and the mean orientation, mechanical tests were performed on a variety of samples, showing how the mechanical properties undergo large changes. It is possible to observe that the samples that obtained the best results were the highest layer thickness and the samples printed in the Z direction. Thus, it can be concluded that for the production of Scaffolds in polymeric biomaterials it is necessary to master each of the parameters involved in their production process, as well as to be able to control them.

Keywords: Additive manufacturing (AM) · Layer thickness · Mean direction · Mechanical parameters

1 Introduction

Additive Manufacturing (AM) or Rapid Prototyping (RP) is the given name to 3D print process using polymeric materials and/or biomaterials. The term Additive Manufacturing is employed in a variety of industries to describe the process to quickly create a system or piece representation [1].

J. L. B. Marques et al. (Eds.): CLAIB 2022/CBEB 2022, IFMBE Proceedings 98, pp. 158–165, 2024.
https://doi.org/10.1007/978-3-031-49401-7_16

According to Gibson et al. [2], the emphasis of the AM process is to build something rapidly, where a prototype or a basic model is produced, from which a final product is conceived. This type of process is commonly used in industries in order to accomplish the best results on products improvement, in the acquisition of innovative, efficient and lighter structures.

The AM comprises many methods of production, nonetheless the Fusion and Deposition Modeling (FDM) of great interest for testing due to its association to low cost 3D printers [3].

The FDM technique builds a 3D geometry by assembling individual layers of extruded thermoplastic filaments such as Acrylonitrile Butadiene Styrene (ABS) or Polylatic Acid (PLA), which have fusion temperatures low enough for use in 3D printers that work with this type of process [4].

According to Klift et al. [5], FDM is a complex process with a large number of parameters that directly influences on the product quality and material properties, and the combined set of these parameters are of difficult comprehension.

Print parameters like the Layer Thickness (LT), print rapidness and building orientation are parameters to be studied once they will directly impact the final product both in their aesthetic and mechanical properties.

The main goal of this paper was to perform mechanical and structural tests on rectangular printed parts, in order to acquire more data about the RP, because the parameters for a high quality print are poor discussed in literature and also varies from individual manufacturers, and the information related to low cost 3D printer's manufacturing process are scarce. At the end of the tests we expect to observe if similar results would be in agreement with the research [3, 6] where mechanical strength were higher on samples printed in Z direction than that printed in other directions, and also that the parts printed in the same direction have the same mechanical properties, however that the production parts containing thicker layers (higher layer heights) have a different mechanical resistance from that with thinner layers (lower layer heights).

This paper discusses the adjustment of a 3D printer process parameters for a 3D printer with manufacturing functionality to make it a viable platform for bioprinting process. The choice of controller is justified because it stands out by operating in order to eliminate the proportional error of the relationship between expected and observed magnitude, having variations in the system, ensuring the stability of the controlled parameters into the system [7].

Bioprinting, despite being similar to 3D printing regarding to its stages, is fundamentally different from the latter due to the materials used the parameters and their influence on manufacturing functionality. In bioprinting, layers of biological products are deposited, such as connective tissue cells to obtain total temperature to control inside the bioprinting capsule, in order not to compromise the process of complete adhesion of the biological material deposited which could lead to loss of its biocompatibility [8].

2 Materials and Methods

A 3D GTMAX H4 low cost printer using FDM technique with ABS as main material for printing was utilized to perform a complete study related to the print parameters at the Center for Engineering Applied to Health (CEAS) – School of Engineering of São Carlos - University of São Paulo.

It was formulated a tridimensional specimens measuring 3 × 5 × 10 mm using Inventor 3D software and created two sets of samples following the same parameters regarding to building orientation, but with different layer thickness between them.

Figure 1 demonstrates the three build orientation of samples in their respective axis, Vertical (Upright or Z), Lateral (On-Edge) and Surface (Flat).

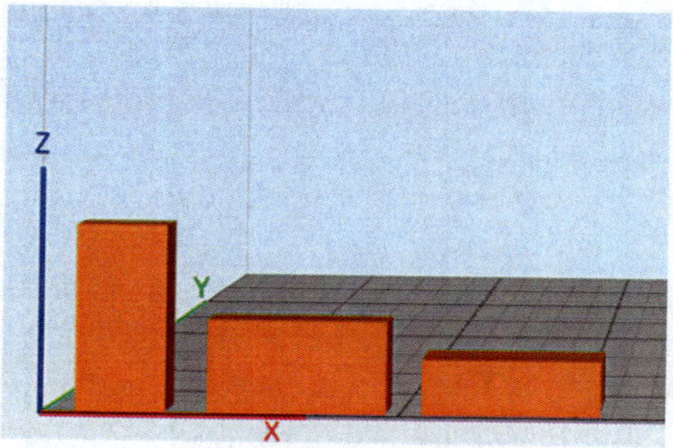

Fig. 1. Orientation for samples printing.

Data of each sample printed can be checked on Table 1. The build orientation and thickness layer were the main parameters approached in this paper.

All samples printed were subjected to compressive mechanical tests by using a MTS Bionix® Servohydraulic Test System from the Center for Engineering Applied to Health and the Mechanics Department at School of Engineering of São Carlos, São Paulo, Brazil.

Table 1. Print data of each sample.

Sample	Layer thickness	Build orientation	Number of layers	Time (min)	Standard printing speed (mm/min)
1	0.1	Upright (Z)	252	66	4200
2	0.1	On-edge	125	33	4200
3	0.1	Flat	74	20	4200
4	0.2	Upright (Z)	125	33	4200
5	0.2	On-edge	61	16	4200
6	0.2	Flat	36	10	4200

3 Results

3.1 Build Orientation

The first test was performed in order to assess how the build orientation directly influences on mechanical and structural properties of the printed parts. As showed on Table 1, the samples printed on vertical direction have a larger number of layers. It is also evident from the graphs in Figs. 2 and 3 and from Table 2 that samples printed in the Z direction supported a higher load comparing to the other specimens.

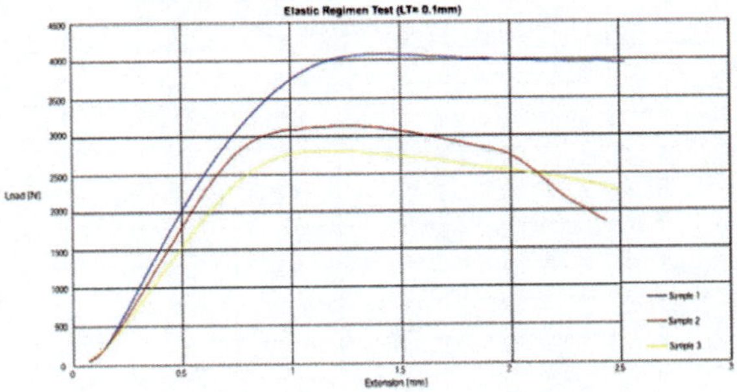

Fig. 2. Graph: elastic regimen test (LT: 0.1 mm)

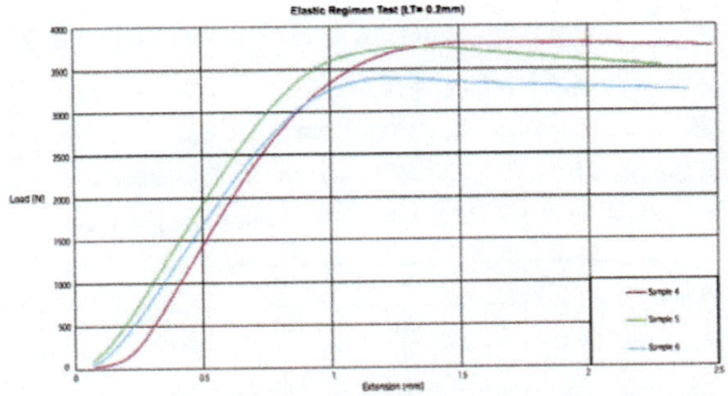

Fig. 3. Graph: elastic regimen test (LT: 0.2mm)

3.2 Layer Thickness

The second test was performed in samples printed in the same direction, but in different layer thickness in order to assess the impact of layer thickness. Table 2 shows that samples printed with thicker layers generally tends to bear a higher load as observed in the behavior of compression tests in the graphs of Figs. 4, 5 and 6.

Table 2. Compressive test results for each sample

Sample	Young Modulus (MPa)	Load (Flow) [N]	Number of layers	Flow strain	Max load [N]
1	1373.89	3110.4	3110.4	31.24	4081.3
2	1230.04	1230.04	2831.8	28.44	3127.9
3	1023.41	1023.41	2532.6	25.43	2795.3
4	1282.71	1282.71	3054.2	30.67	3798.6
5	1239.65	1239.65	3292.0	33.06	3759.5
6	1148.31	1148.16	2961.7	29.74	3394.1

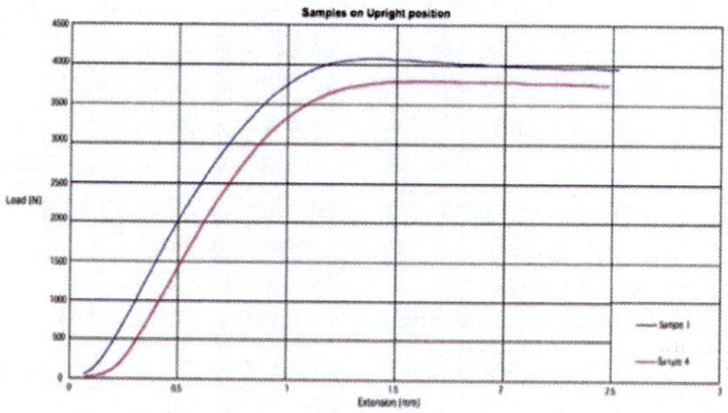

Fig. 4. Graph: samples in vertical orientation (Upright or Z)

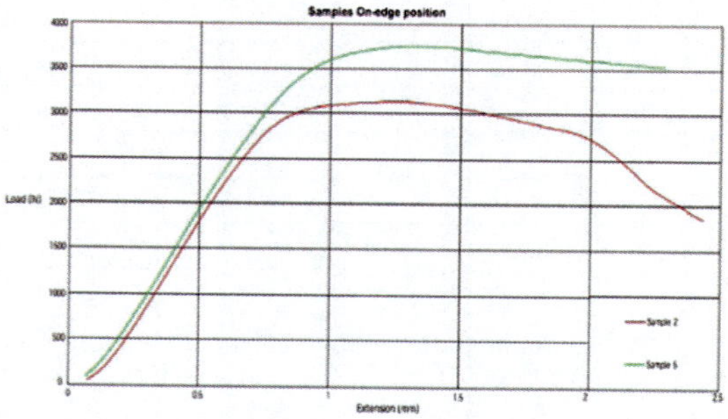

Fig. 5. Graph: samples in lateral orientation (On-edge)

4 Discussion

These results had already been presented in several researches like [6] for example, that indicated that many experimental studies investigated the three different building direction, also addressed in this paper, resulting in the same answer, that parts printed in the Z direction has a higher compressive strength compared to other ones (higher than 3500 N load supported and 25 mm deformation extension, as a stress-strain diagram

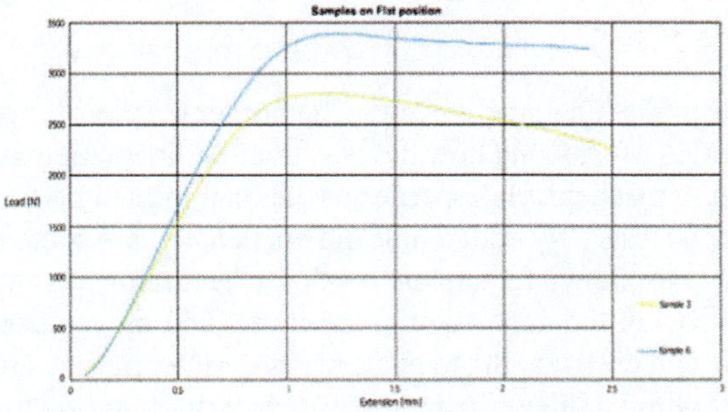

Fig. 6. Graph: samples in surface orientation (Flat)

which was used to determine qualities such as elastic limit, maximum deformation, yield point, yield strength, and, for the samples, compressive strength), supporting higher mechanical loads as in the maximum load supported.

The layer thickness is directly related to the number of layers required to print a part. As observed by Chacón et al. [3], this paper showed that the effect of layer thickness on mechanical properties was different for samples builder on vertical, lateral and surface orientation, due to build direction. However, the results disagreed from that from [3], in which the printed parts on vertical presented a higher strength when the layer thickness was increased, unlike the results of this paper that showed a higher strength in other orientations (lateral and surface) with the increasing of layer thickness.

According to Zhang et al. [9] highlight that 3D printing process provides advantages and disadvantages. Still, there are many technical challenges to overcome to find out optimal 3D printing parameters and process for bone engineering. In this way, the present study also shows the importance of parameter determination in the 3D printing pre and post process performing.

The test of different filament (orthogonal and isometric; aligned and staggered) configurations performed by Baptista and Guedes [10, 11] showed little impact on mechanical performance of the scaffolds. Also, besides observation of differences between aligned and staggered scaffolds, no significant statistical difference between different designs and configurations were reported. It suggests that parameters associated with deposition of materials in 3D printing processes need to be better understood.

A different and simpler approach to explain this result is to consider that increasing layer thickness leads to lesser layers for a given thickness, therefore decreasing the number of layer connections, enhancing the strength of the printed part. In the next stage of bioprinting and tissue engineering the key developments in the field of 3D printing technologies involves the incorporation of various biomaterials and cells in printing composite scaffolds mimicking physiologically relevant complex geometry and gradient porosity. Recently, the emerging trend of printing smart scaffolds which can respond to external stimulus such as temperature, pH and magnetic field, known as 4D printing is gaining immense popularity and can be considered as the future of 3D printing applications in the field of tissue engineering [12].

5 Conclusions

The analysis of the process parameters of the 3D printer that was proposed in this paper achieved its objectives of showing how the construction orientation and layer thickness directly influence the mechanical properties of the final printed parts.

Within these same tests only one sample did not behave as expected, which may have occurred due to several external factors such as the difference in the values of air humidity (internal environment) in the deposition of the layers and the temperature variation of the extruder nozzle and desk leading to unevenness during printing. However, this value that does not agree with the literature in any way disturbs or shows that there are errors in reading or that this printing is outside the standards of the other samples used.

Even so, for impressions from biomaterials (bioprinting) such as bio-glass with bovine hydroxyapatite (HA) injection (biomaterials for bone quality improvement in fracture regeneration) or other material injection scaffolds, tests should be performed with a wider range of parameters in order to obtain the best results, so that HA scaffolds can be produced to have the mechanical properties closest to the desired for each situation in the laboratories, controlling the environment and nozzle temperatures better, as well as moisture in the final print.

References

1. Volpato, N.: Manufatura aditiva: tecnologias e aplicações da impressão 3D. Editora Blucher (2017)
2. Gibson, I., Rosen, D., Stucker, B.: Additive Manufacturing Technologies: Rapid Prototyping to Direct Digital Manufacturing, pp. 1–459 (2010). https://doi.org/10.1007/978-1-4939-2113-3
3. Chacon, J., Camineiro, M.Á., Garcia-Plaza, E., et al.: Additive manufacturing of pla structures using fused deposition modelling: effect of process parameters on mechanical properties and their optimal selection. Mater. Des. **124**, 143–157 (2017). https://doi.org/10.1016/j.matdes.2017.03.065
4. Tymrak, B., Kreiger, M., Pearce, J.M.: Mechanical properties of components fabricated with open-source 3-d printers under realistic environmental conditions. Mater. Des. **58**, 242–246 (2014). https://doi.org/10.1016/j.matdes.2014.02.038
5. Klift, F.V.D., Koga, Y., Todoroki, A., et al.: 3D printing of continuous carbon fibre reinforced thermo-plastic (CFRTP) tensile test specimens. Open J. Compos. Mater. **6**(01), 18 (2015). https://doi.org/10.4236/ojcm.2016.61003
6. Mohamed, O.A., Masood, S.H., Bhowmik, J.L.: Optimization of fused deposition modeling process parameters: a review of current research and future prospects. Adv. Manuf. **3**(1), 42–53 (2015). https://doi.org/10.1007/s40436-014-0097-7
7. Franklin, G.F., Powel, J.D., EMANI-NAEINI: Sistemas de Controle para Engenharia, 6th ed. Bookman, Porto Alegre, Brasil (2013)
8. Oliveira, N.A., Roballo, K.C.S., Lisboa Neto, A.F.S., et al.: Bioimpressão e podução de mini-órgãos com células tronco. Pesq. Vet. Bras. **37**(9), 1032–1039 (2017). https://doi.org/10.1590/S0100-736X2017000900020
9. Zhang, L., Yang, G., Johnson, B.N., Jia, X.: Three-dimensional (3D) printed scaffold and material selection for bone repair. Acta Biomater. **84**, 16–33 (2019). https://doi.org/10.1016/j.actbio.2018.11.039

10. Baptista, R., Guedes, M.: Morphological and mechanical characterization of 3D printed PLA scaffolds with controlled porosity for trabecular bone tissue replacement. Mater. Sci. Eng. C **118**, 1–17 (2021). https://doi.org/10.1016/j.msec.2020.111528
11. Gang, F., et al.: 3D printing of PLLA/biomineral composite bone tissue engineering scaffolds. Materials (Basel). **15**(12), 4280 (2022). https://doi.org/10.3390/ma15124280
12. Anandhapadman, A., Venkateswaran, A., Jayaraman, H., Ghone, N.V.: Advances in 3D printing of composite scaffolds for the repairment of bone tissue associated defects. Biotechnol. Prog. **38**(3), e3234 (2022). https://doi.org/10.1002/btpr.3234

Short-Term Prediction of COVID-19 Deaths in Argentina

Frederico J. R. Pelogia(✉), Vitor S. T. Soares, and Henrique M. Paiva

Instituto de Ciência e Tecnologia (ICT), Universidade Federal de São Paulo (UNIFESP), São José dos Campos, Sao Paulo, SP, Brazil
fpelogia@unifesp.br

Abstract. This paper proposes a dynamic model for accurate short-term prediction of the coronavirus disease (COVID-19) death rates in Argentina. For this purpose, the daily number of deaths in three of the most densely populated cities in the country were observed: Buenos Aires, Rosario and Mendoza. This methodology is based on the Richards growth model, and parameters are numerically optimized using a sequential least squares programming (SLSQP) method, with parameters obtained in the fitting process being accumulated and directly applied to a daily version of the model. The model fit is analyzed by inspecting its predictions for the first three epidemiological waves in each of the studied cities. The results presented for the selected cities in Argentina confirm the model's accuracy in depicting death trends, despite the profound noise observed in data regarding the daily number of deaths.

Keywords: COVID-19 deaths · Argentina · Mathematical modeling · Epidemiology · Forecast

1 Introduction

COVID-19, officially declared as a global pandemic by the World Health Organization (WHO) on March 11, 2020, is a deadly and infectious disease which has deeply shaken society, politics, and economy internationally [2]. The Coronavirus, caused by the SARS-CoV-2 virus, has truly taken its toll on the world in the past 2 years, with over half a billion confirmed cases and more than 6 million total deceased reported to date [21].

Symptoms for COVID-19 have a common pattern, with fever, coughing, shortness of breath, diarrhea, and nausea occurring in most infected [4]; loss of smell or taste, inflammatory skin conditions, and even ocular symptoms are also experienced by many [10]. However, it has also been noted that the Coronavirus is occasionally capable of expressing more alarming, peculiar symptoms, potentially causing long-term damage to patients. According to Guo [3], a substantial portion of symptomatic individuals also report some instances of cognitive dysfunction for a period that lasted over 12 weeks from the identification of the

J. L. B. Marques et al. (Eds.): CLAIB 2022/CBEB 2022, IFMBE Proceedings 98, pp. 166–175, 2024.
https://doi.org/10.1007/978-3-031-49401-7_17

first symptoms. Amidst the disorders identified, the most consistent are memory loss, concentration difficulty, brain fog, forgetfulness, word-finding problems, and overall confusion, all of which are extremely detrimental to the patients ability to communicate and work.

Amongst the most effective protocols to prevent and slow down viral transmission, the WHO strongly recommends practicing social distancing, using properly fitted masks, and washing your hands or applying alcohol-based hand sanitizers frequently [1]. It is also critical to adequately follow immunization schedules, as data proves new doses significantly contribute to maintaining maximum antibody levels in the system [16].

In this context, mathematical modeling has become an invaluable tool to further prevent the spread of COVID-19. According to the WHO, the key to combating the pandemic is to be well informed about the disease and how the virus spreads [1]. Thus, mathematical models created by scientists and researchers worldwide have greatly assisted society and local governments as they implemented safety procedures, determined the efficacy of new policies, and evaluated the numerous indicators associated with the Coronavirus [7].

In general, most epidemiologic models nowadays take some inspiration from the seminal work of Kermack and McKendrick in 1927 [5] and subsequent studies, incorporating different new techniques and technologies to further improve epidemiologic models, and apply them in new studies.

During the current pandemic, numerous different methods were adopted to assess the situation, with a wide range of mathematical models applied around the world to observe and study the evolution of COVID-19 transmissions, manage hospital occupancy, and more.

In terms of mathematical and computational modeling, the current Coronavirus epidemic has been faced from all sorts of perspectives, with researchers applying methods such as fractional order differential equations [22], stochastic models [9], compartmentalized models [15], data-driven modeling [6] and many more. Researchers have also approached the issue by studying models employed previously in similar epidemic outbreaks like the MERS in 2012 [14] or the influenza pandemic in 20th century London [8], amongst many. Other alternatives involve incorporating a wide range of computational tools in order to enrich the modeling process, as many prevalent COVID-19 modeling approaches do, obtaining a more precise response. Namely, the use of machine learning and artificial intelligence, or AI, appears to be a powerful ally to researchers, as self-learning algorithms tend to elevate a models accuracy significantly [7].

Modern modelling approaches [13,14] aim to simplify the process so that the resulting model is not only specific, but also reliable and easily adaptable within the situations predicted. By categorizing the population in different classes (hospitalized, recovered, infectious), defining the main parameters involved in the diseases spread (transmissibility, recovery rates, death rates), and estimating through research and data analysis the adequate values for such parameters, researchers are able to create a model which closely resembles the actual spread of COVID-19 in different countries. Through intensive monitoring of the viral evolution in the studied population, parameter adjustments are also employed in order to guarantee the authenticity and accuracy of the mathematical model created.

As for the case of managing hospital occupancy, studies such as Oliveira [12] have been critical in directly saving lives and avoiding a collapse in the local healthcare system. By monitoring almost 15 million individuals in Bahia, Brazil, scientists provided policymakers with an extensive amount of information, which served as guidelines for enforcing stricter lockdown procedures or opting for flexibilization. According to the study, the most critical indicator tracked was intensive care unit (ICU) bed occupancy rates, as a high rate indicates a number of hospitals in the area may reach their full capacity, leading to a subsequent growth in mortality, both COVID and non-COVID related.

Each approach has its specific applications, strengths, and faults, all of which culminate in a diverse collection of tools that have aided society in combating COVID-19 to a great extent. There are, however, limitations that are common to many different methods of mathematical modeling, especially those involved in epidemiology. Namely, a factor which prevents models from being applied in many countries is simply the lack of data. As an example, the African continent has been a modeling challenge during the pandemic, especially in 2020. Studies indicate that, in numerous countries, a varying portion that can reach up to 10 percent of COVID-related data is missing, making it considerably tougher to design a precise mathematical model with adequate parameters [11].

This work is concerned with short-term predictions of COVID-19 deaths in Argentina. Since data are more useful in a regional context, three of the most populous cities in Argentina were selected to test the model: Buenos Aires, Rosario and Mendoza. The dataset containing both the daily and accumulated number of deaths by COVID-19 was obtained from [20]. The chosen database collects the data directly from the official source, but provides a more suitable structuring and standardization for practical use in this research.

2 Materials and Methods

To describe the cumulative number of deaths by COVID-19, the present paper uses the Richards growth model [17,18], characterized by the Ordinary Differential Equation (ODE)

$$\frac{dD}{dt}(t) = \frac{1}{\delta\nu} D(t) \left(1 - \left(\frac{D(t)}{A}\right)^{\nu}\right) \tag{1}$$

Here, D(t) represents the number of deaths that occurred from the first to the tth day of the pandemic. The term $(\delta\nu)^{-1}$ scales $D(t)$ with time and represents an intrinsic growth rate. The parameter ν controls the asymmetry of the model, giving it the ability to represent uneven acceleration and deceleration phases. The parameter A represents the value that the model reaches in its saturation regime, conceiving the possibility of being interpreted as the final number of deaths in the studied region.

The analytical solution of (1) is well known [13,17,18] and can be written as

$$D(t) = \frac{A}{\left(1 + \nu e^{-\frac{(t-t_p)}{\delta}}\right)^{\frac{1}{\nu}}} \tag{2}$$

Thus, the derivative of (2), which models the daily variation in the number of deaths, can also be obtained and is expressed as [13]

$$\frac{dD}{dt}(t) = \frac{A}{\delta} \frac{e^{-\frac{(t-t_p)}{\delta}}}{\left(1 + \nu e^{-\frac{(t-t_p)}{\delta}}\right)^{\frac{\nu+1}{\nu}}} \tag{3}$$

The parameter t_p, present in both (2) and (3), is the turning point of the model, meaning that it marks the day in which the number of deaths per day start decreasing instead of increasing.

Since the pandemic has been taking place for more than two years, several epidemiological waves have already occurred. The onset of a new wave is announced by a change from a deceleration phase to an acceleration phase. In order to properly model this multiple wave behavior, the present paper uses a model that consists of a sum of n different sigmoid functions, where n is the number of epidemiological waves and each function is characterized by (2). Therefore, the depicted model has $4n$ adjustable parameters.

For the purpose of finding the optimal set of parameters that better captures the system dynamics, it is possible to formulate an optimization problem that minimizes the Mean Squared Error (MSE) between the model and the actual data. This problem can be computationally solved by using a sequential least squares programming (SLSQP) method, implemented in the minimize function of the optimization toolbox of the SciPy Python library [19].

In this work, the optimization process described is used in a two step procedure to fit the model. First, the algorithm performs the fit using a symmetric sigmoid by setting $\nu = 1$. Then, the optimal point obtained is used as the initial condition for the fitting of an asymmetric sigmoid with the parameter ν free to be adjusted.

This logic is applied to each sigmoid (2) added to the model, incrementally. For each new wave, the model used to fit the data is composed by the sum of all the sigmoids, but with only the parameters of the new one being adjustable. The parameters of the other functions are fixed to their optimal values. Thus, the sum ensures that the model will be continuous and representing the pandemic behaviour up to the analysed point in time.

The parameters obtained in the fitting process of the cumulative data can be directly applied to a daily version of the model, which, in a similar manner, consists of a sum of n functions characterized by (3). Hence, this tends to approximate the dynamics of the daily number of deaths by COVID-19, which is one of the most relevant pieces of information to assess the severity of the pandemic in a given region.

3 Results

The present work analyzed the model fit and its predictions for the first three epidemiological waves in each studied city. The predictions were made by omitting the last two weeks of data in the optimization process and then comparing it to the values forecast by the model.

The results for the cities of Buenos Aires, Rosario and Mendoza are shown in Figs. 1, 2 and 3, respectively. Each figure contains three graphs with the comparison between the daily version of the model, as in (3), and the daily number of deaths up to the first three waves. The starting point of all graphs represent the first day with uninterrupted notification of deaths, which happened in March 2020 for all cities. The interval of interest, that includes the first three waves, ends in July 2021.

The optimal values of the parameters found for each sigmoid in the model are presented in Table 1, grouped by city. While these values do not correspond to actual epidemiological parameters, they are the ones that better capture the dynamics in the data for the growth model used.

Table 1. Optimal parameters of the sigmoids for each city

City–Sigmoid	**A**	t_p	ν	δ
Buenos Aires—1	5745	182	0,384	37,7
Buenos Aires—2	1326	359	0,343	25,9
Buenos Aires—3	4254	460	0,100	30,5
Rosario—1	2146	219	0,265	22,7
Rosario—2	415	328	0,100	21,8
Rosario—3	1736	453	0,407	25,3
Mendoza—1	1779	224	0,885	21,4
Mendoza—2	215	369	0,100	34,0
Mendoza—3	2482	451	0,100	32,6

The relative root mean squared error (RMSE) found was 0,31% for the city of Buenos Aires, 0,24% for the city of Rosario and 0,20% for the city of Mendoza.

4 Discussion

Observing the results, it is clear that the model fits well the data, accurately describing its trend, despite the noise effects present in the real data.

In addition, the model was also able to represent the multiple wave behavior, revealing that the composition of sigmoids was successful. Regarding the predictions of the number of deaths in the last two weeks of each wave, they were

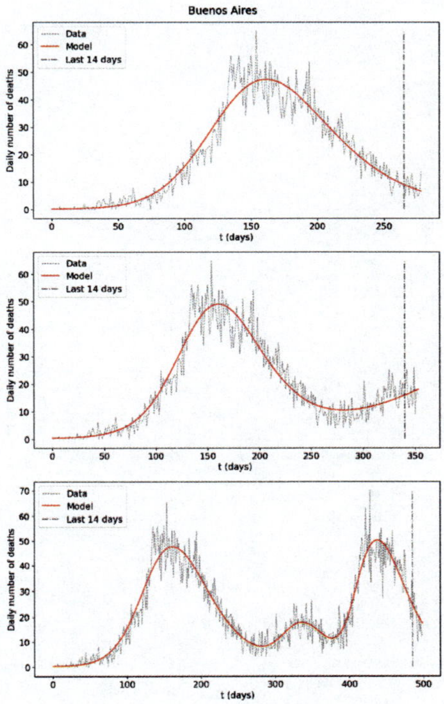

Fig. 1. Results for the first three epidemiological waves in the city of Buenos Aires

also capable of approximating the tendency of the real data in the period, for all cities studied.

From Fig. 1, it is possible to notice that in the city of Buenos Aires the first wave presented a peak of about 50 deaths per day. The second wave was smaller, with less than half of the daily deaths of the first one. As for the third one, it was as big as the first but slightly narrower.

With respect to the predictions shown in Fig. 1, it is possible to see that the model describes really well the mean behavior of the real data, for the three waves.

Fig. 2 shows us that the city of Rosario suffered from a peak of about 30 deaths per day, followed by a wide and small wave, peaking about 10 daily deaths. The third wave in the city was slightly bigger than its predecessor, having a peak of around 20 deaths per day.

The short-term predictions shown in Fig. 2 are excellent for the second and third wave, presenting essentially the same trend as the real data. As for the first wave, the model suggests a less steep deceleration for the first week, but this is balanced out by the fact that in the second week the number of deaths rises, getting closer to the forecast value.

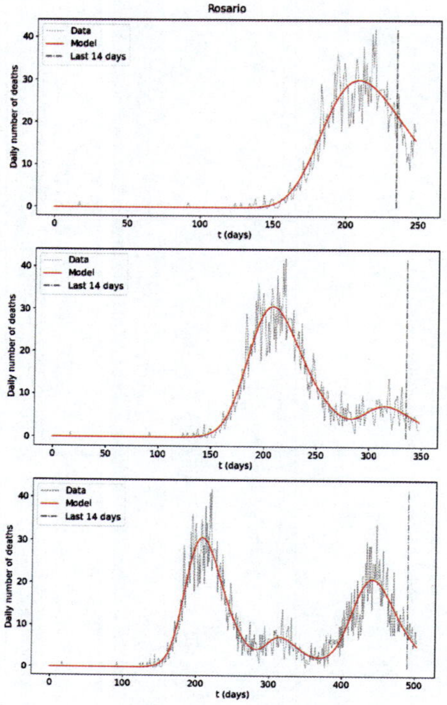

Fig. 2. Results for the first three epidemiological waves in the city of Rosario

As said before, a change from a deceleration phase to an acceleration phase indicates the onset of a new wave and, thus, the need to add another sigmoid to the model. Therefore, the choice of the transition points is crucial for the quality of the predictions. If a transition point is set to a larger value than it needs to be, the model might be insufficient to represent the dynamics in the interval it is supposed to work with. On the other hand, a transition point that is too low can overwhelm the subsequent wave.

The results for the city of Mendoza, shown in Fig. 3, reveal a first wave with a peak with about 20 daily deaths. After this, a very wide and subtle second wave occurred, peaking less than 5 deaths per day. The third wave presented a peak with around 5 additional deaths per day, when compared to the first one. The predictions for the city fitted really well with the real data for all of the first three waves.

In general, the two-week predictions of the model for the daily number of deaths could successfully describe the trend of the real data in the period. This is especially good when considering the low complexity of the model used, with only one sigmoid for each wave.

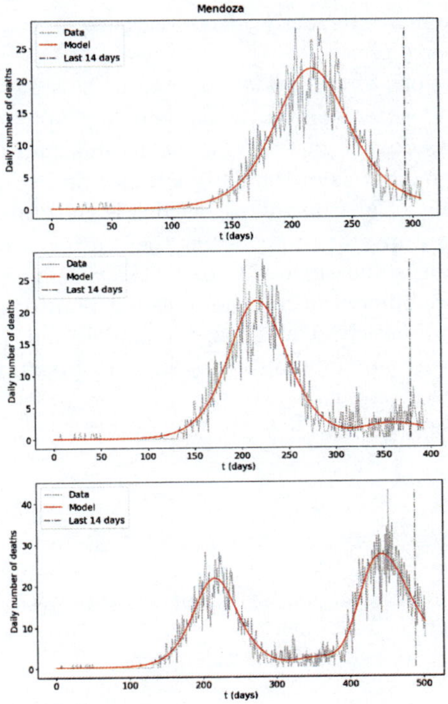

Fig. 3. Results for the first three epidemiological waves in the city of Mendoza

The choice of focusing on the daily number of deaths and its predictions is due to its relevance in terms of monitoring the pandemic. This tracking is especially important to support the decision making process of the local governments.

5 Conclusion

This work brought forth an efficient model for predicting the daily number of deaths in Argentina in short-term, derived from an analysis of COVID-19's epidemiological waves in three cities amongst the most populous in the country: Buenos Aires, Rosario and Mendoza.

In all three waves analyzed for each Argentinian city, the model implemented in this work was able to satisfactorily represent, in a two-week prediction, the real data observed throughout the 500 days studied, showing its efficacy.

Seeing as the daily number of COVID-related deaths is one of the most critical indicators to track during the pandemic, mainly due to its direct impact in public health, to accurately predict this factor in short-term is essential. Thus, in being capable of producing reliable data on the current repercussions of coronavirus disease in Argentina, the model depicted could provide information to help the nation's local governments to safely lift restrictions as the pandemic advances into the end of its acute phase.

One of the limitations of the model adopted by this paper is the lack of correspondence between the model and the epidemiological parameters. Unlike the compartmental approach, the main form of contribution of this model consists in the quantitative predictions of the daily number of deaths.

The desired output of all controlling actions is the stabilization of the pandemic. Still, it is also important to consider the impact of outer factors such as new variants of the virus and the evolution of the vaccination coverage.

Moreover, the development of reliable tools for monitoring the spread of diseases is extensible and non-specific. Thus, the models and methods employed in this work can contribute to modeling the dynamics of possible future epidemics, including other diseases.

Conflict of Interest

The authors declare that they have no conflict of interest.

Acknowledgments. This research has been supported by grant # 2020/14357-1, São Paulo Research Foundation (FAPESP).

References

1. Coronavirus disease (COVID-19) at www.who.int/health-topics/coronavirus
2. Cucinotta, D., Vanelli, M.: WHO declares COVID-19 a pandemic. Acta Bio Medica: Atenei Parmensis **91**(1), 157 (2020)
3. Guo, P., Benito Ballesteros, A., Yeung, S.P., Liu, R., Saha, A., Curtis, L., Kaser, M., Haggard, M.P., Cheke, L.G.: COVCOG 2: Cognitive and memory deficits in long COVID: a second publication from the COVID and cognition study. In: Frontiers in Aging Neuroscience, p. 204 (2022)
4. Kalra, R.S., Tomar, D., Meena, A.S., Kandimalla, R.: SARS-CoV-2, ACE2, and hydroxychloroquine: cardiovascular complications, therapeutics, and clinical readouts in the current settings. Pathogens **9**(7), 546 (2020)
5. Kermack, W.O., McKendrick, A.G.: A contribution to the mathematical theory of epidemics. In: Proceedings of the Royal Society of London. Series A, Containing Papers of a Mathematical and Physical Character, vol. 115, no. 772, pp. 700–721 (1927)
6. Kuhl, E.: Data-driven modeling of COVID-19-lessons learned. Extreme Mechan. Lett. **40**, 100–921 (2020)
7. Latif, S., Usman, M., Manzoor, S., Iqbal, W., Qadir, J., Tyson, G., Castro, I., Razi, A., Boulos, M.N.K., Weller, A., et al.: Leveraging data science to combat COVID-19: a comprehensive review. IEEE Trans. Artif. Intell. **1**(1), 85–103 (2020)
8. Lin, Q., Zhao, S., Gao, D., Lou, Y., Yang, S., Musa, S.S., Wang, M.H., Cai, Y., Wang, W., Yang, L., et al.: A conceptual model for the coronavirus disease 2019 (COVID-19) outbreak in Wuhan, China with individual reaction and governmental action. Int. J. Infect. Dis. **93**, 211–216 (2020)

9. Ndii, M.Z., Supriatna, A.K.: Stochastic mathematical models in epidemiology. Information **20**, 6185–6196 (2017)
10. Ng, S.L., Ong, Y.S., Khaw, K.Y., Teh, S.P., Tan, C.S., Ming, L.C., Chan, K.G., Lee, L.H., Goh, B.H.: Focused review: potential rare and atypical symptoms as indicator for targeted COVID-19 screening. Medicina **57**(2), 189 (2021)
11. Okeahalam, C., Williams, V., Otwombe, K.: Factors associated with COVID-19 infections and mortality in Africa: a cross-sectional study using publicly available data. BMJ Open **10**(11), e042,750 (2020)
12. Oliveira, J.F., Jorge, D.C., Veiga, R.V., Rodrigues, M.S., Torquato, M.F., da Silva, N.B., Fiaccone, R.L., Cardim, L.L., Pereira, F.A., de Castro, C.P., et al.: Mathematical modeling of COVID-19 in 14.8 million individuals in Bahia, Brazil. Nat. Commun. **12**(1), 1–13 (2021)
13. Paiva, H.M., Afonso, R.J.M., de Lima Alvarenga, F.M.S., de Andrade Velasquez, E., et al.: A computational tool for trend analysis and forecast of the COVID-19 pandemic. Appl. Soft Comput. **105**, 107,289 (2021)
14. Paiva, H.M., Afonso, R.J.M., de Oliveira, I.L., Garcia, G.F.: A data-driven model to describe and forecast the dynamics of COVID-19 transmission. PloS One **15**(7), e0236,386 (2020)
15. Pinto Neto, O., Reis, J.C., Brizzi, A.C.B., Zambrano, G.J., de Souza, J.M., Pedroso, W., de Mello Pedreiro, R.C., de Matos Brizzi, B., Abinader, E.O., Zângaro, R.A.: Compartmentalized mathematical model to predict future number of active cases and deaths of COVID-19. Res. Biomed. Eng. 1–14 (2020)
16. Regev-Yochay, G., Gonen, T., Gilboa, M., Mandelboim, M., Indenbaum, V., Amit, S., Meltzer, L., Asraf, K., Cohen, C., Fluss, R., et al. (2022). Efficacy of a fourth dose of Covid-19 mRNA vaccine against omicron. N. Engl. J. Med. **386**(14), 1377–1380 (2022)
17. Richards, F.J.: A flexible growth function for empirical use. J. Exp. Botany **10**(29), 290–300 (1959). www.jstor.org/stable/23686557
18. Smirnova, A., Chowell, G.: A primer on stable parameter estimation and forecasting in epidemiology by a problem-oriented regularized least squares algorithm. Infect. Dis. Model. **2**(2), 268–275 (2017)
19. Virtanen, P., Gommers, R., Oliphant, T.E., Haberland, M., Reddy, T., Cournapeau, D., Burovski, E., Peterson, P., Weckesser, W., Bright, J., van der Walt, S.J., Brett, M., Wilson, J., Millman, K.J., Mayorov, N., Nelson, A.R.J., Jones, E., Kern, R., Larson, E., Carey, C.J., Polat, İ., Feng, Y., Moore, E.W., VanderPlas, J., Laxalde, D., Perktold, J., Cimrman, R., Henriksen, I., Quintero, E.A., Harris, C.R., Archibald, A.M., Ribeiro, A.H., Pedregosa, F., van Mulbregt, P., SciPy 1.0 Contributors: SciPy 1.0: Fundamental Algorithms for Scientific Computing in Python. Nat. Methods **17**, 261–272 (2020). DOI: https://doi.org/10.1038/s41592-019-0686-2
20. Wahltinez, O., et al.: COVID-19 open-data: curating a fine-grained, global-scale data repository for SARS-CoV-2 (2020). www.goo.gle/covid-19-open-data
21. World Health Organization: WHO Coronavirus (COVID-19) Dashboard at www.covid19.who.int
22. Zhang, Z., Zeb, A., Egbelowo, O.F., Erturk, V.S.: Dynamics of a fractional order mathematical model for COVID-19 epidemic. Adv. Diff. Equ. **2020**(1), 1–16 (2020)

Evaluating the Mechanical Properties of Isolated Rat Cardiomyocytes Sarcolemma Using Atomic Force Microscopy

A. A. Almazloum[1(✉)], M. A. Cotta[2], J. W. M. Bassani[1,3], and R. A. Bassani[1,3]

[1] Department of Electronics and Biomedical Engineering, School of Electrical and Computer Engineering, University of Campinas (UNICAMP), Campinas, SP, Brazil
ahmad.almazloum@yahoo.com
[2] Department of Applied Physics, Institute of Physics "Gleb Wataghin", UNICAMP, Campinas, SP, Brazil
[3] Center for Biomedical Engineering, LabNECC, UNICAMP, Campinas, SP, Brazil

Abstract. Incidence of cardiac diseases, one of the leading causes of death worldwide, is still on the rise. As the cardiomyocyte is the contractile unit of the heart, knowledge on the details of its mechanical structure and functionality is helpful to the understanding of myocardial pathophysiology and development of therapeutic approaches. Atomic force microscopy (AFM) has been used not only for topographic surface imaging but also for direct assessment of the mechanical characteristics of the plasma membrane. In this work, a preliminary study was conducted focusing on factors affecting the elasticity (***E***) of the sarcolemma (SL) of isolated rat ventricular myocytes, which was determined using AFM. After collecting the topographic image, 16 × 16 force map was generated and post-analyzed using the Hertz model to estimate the Young's modulus (i.e., ***E***). The investigated factors influencing on ***E*** were position (32), cell (18), heart (12), and post-isolation storage period. Appreciable variability was observed for all factors, apparently due to differences in maximum forces, speeds and depths of indentation which could have a great influence on ***E***. Also, longer storage was associated with decreased ***E*** values (< 10 h: 10.96 ± 0.08kPa; n = 5902; > 12h; 6.48 ± 0.05kPa; n = 2012), probably due to cell deterioration with time. Accurate measurement of ***E*** is essential to detect alterations in cell function and effect of therapeutic drugs due to changes in SL mechanical properties. From the present results, it is possible to conclude that storage must be as short as possible, and that other alternatives should be sought for verifying cell viability and improving cell attachment.

Keywords: Cell mechanics · Cardiomyocyte · Force spectroscopy

1 Introduction

Elasticity, viscoelasticity, or stiffness are frequently used to evaluate the mechanical properties of the plasma membrane (sarcolemma, SL, in myocytes) of living cells, which greatly impacts cell function. Elasticity (represented by the Young's modulus,

J. L. B. Marques et al. (Eds.): CLAIB 2022/CBEB 2022, IFMBE Proceedings 98, pp. 176–187, 2024.
https://doi.org/10.1007/978-3-031-49401-7_18

E) is an intensive property of solid materials that describes the relationship between the mechanical stress (force) and the strain (deformation) [1]. Deviation of plasma membrane ***E*** values from the normal range may indicate defects (e.g., pathologies), and ***E*** determination may help assessing the effect of treatments, such as drugs, in living cells/tissues [2]. Many techniques have been used to estimate ***E*** of living materials, such as optical tweezers, micropipette aspiration, tracing force microscopy, and atomic force microscopy (AFM) [1, 2]. The latter offers high resolution of surface topography and interaction force in the sub-nanometer and pico-Newton ranges, respectively, which is appropriate for biological samples, especially myocytes [1–4].

AFM is commonly used for force spectroscopy, in which an external vertical force is applied to the sample via the tip, with a well-defined geometry and spring constant, mounted on a cantilever [1]. AFM has been used on several cell types to produce SL surface topography and force maps. To obtain high resolution topographic images, fixation (e.g., with glutaraldehyde) has been used to assure cell immobility. On the other hand, this chemical treatment has adverse effects on cell viability and may modify membrane mechanical properties [1, 2], thus compromising the reliability of ***E*** estimation.

Investigation of the mechanical properties of ventricular myocytes is relevant, considering that alterations in membrane elasticity may be accompanied by cell dysfunction. By allowing the generation of force maps, force spectroscopy has been used to determine SL ***E*** in several cell types, but very few studies were performed on cardiomyocytes, investigating alterations in SL ***E***, associated with aging and diabetes [3, 4].

In this study, we present some preliminary results and considerations on the estimation of ***E*** for (presumably) live isolated rat cardiomyocytes in a physiological liquid environment, following different periods of post-isolation storage in cardioplegic solution. Considering factors that may affect the reliability of the results, variability of the measurements was evaluated at different levels: among scanned positions in the same cell, among myocytes of the same heart, among hearts, and in fresh versus stored myocytes.

2 Material and Methods

2.1 Isolated Cardiomyocytes

Protocols of animal care and use were approved by the institutional Committee of Ethics in Animal Use (CEUA/IB/UNICAMP; process number 4429-1/2019). Myocytes were isolated from the left ventricle of adult (3–5-month-old) male Wistar rats (n = 12) by enzymatic digestion with collagenase I via coronary perfusion, as previously described by Penna and Bassani [5]. After digestion, the left ventricle was dissected and cells were mechanically dissociated in cardioplegic solution, with the following composition (in mM): 30 KCl, 10 KH_2PO_4, 1 $MgCl_2$.6H2O, 10 N-2 hy-droxyethylpiperazine-N”-2 ethane-sulfonic acid (HEPES), 11 glucose, 20 taurine, 70 glutamic acid, 1 mg/ml bovine serum albumin, pH 7.4.

Myocytes were kept in the cardioplegic solution at 4 °C until use. Before use, cells were left to sediment by gravity and the pellet was washed 2–3 times with Ca^{2+}-free Tyrode’s solution (composition in mM: 140 NaCl, 6 KCl, 2.5 $MgCl_2$, 10 HEPES, 11 glucose; pH 7.4) to remove the cardioplegic solution and debris. The cell suspension in the Ca^{2+}-free Tyrode's solution was added to the AFM liquid cell (chamber), of which the coverslip bottom was treated with collagen to facilitate cell adhesion. Ca^{2+} was removed from the extracellular medium to minimize spontaneous contractions, which is crucial to decrease noise of the force map recordings and to prevent the myocyte from adhering to the cantilever tip. It is noteworthy that ***E*** of myocyte SL does not appear to be affected by major decreases in extracellular Ca^{2+} concentration [4].

2.2 Atomic Force Microscopy System

Surface topography and force curve maps were produced with a 5500 Scanning Probe Microscope (Keysight Technologies, USA) which has a video camera system to align the laser on the cantilever. This provides good-quality optical images of the samples.

A soft silicon nitride probe MSCT (Bruker AFM Probes, CA, USA) was used. This probe has a triangular cantilever A (L = 175 μm, W = 22 μm, spring constant (k) = 0.07 N/m, resonance frequency (f_0) = 22 kHz), with a pyramidal tip with 20 ± 1.25° half-opening angle, 5 μm height and 10 nm radius at the extremity. However, because the nominal specifications of the cantilever have a wide range of tolerance, we used the thermal noise method (thermal tuning) [6], implemented in this microscope, to determine f_0 and k experimentally. The fact that both f_0 and k have a large variability comes from the uncertainty on the cantilever dimensions, particularly its thickness [7, 8]. That is why k calibration needs to be repeated each time that the probe is installed or even markedly drifts from its position during the measurements [7, 8].

The values of mean ± standard error of the mean (SEM) of f_0 and k obtained with the thermal noise method were, respectively, 23.3 ± 1.6 kHz and 0.051 ± 0.006 N/m (n = 12). The deflection sensitivity (*def*), which is another calibration for the laser system that includes the optical detectors, showed a larger variability (1.75 ± 0.16 μm/V; n = 12). This is due to *def* dependence on the system alignment: the laser incidence on the cantilever, its reflection received by the detector and even the transmission medium. Thus, the presence of particles in suspension in the solution may affect the *def* value. The calibration of *def* was carried out according to standardized protocols in the literature [7, 8], in a clean liquid cell containing the Ca^{2+}-free Tyrode's solution, but not cells, and considering the chamber bottom as an infinite rigid surface.

2.3 Experimental Protocol

Experiments were carried out at 25 ± 3 °C. After spreading collagen solution on the coverslip bottom of the liquid chamber, it was allowed to dry for 15 min. Then 500 μL of the myocyte suspension in Ca^{2+}-free Tyrode's solution were pipetted in the liquid cell, and cells were left to settle down and adhere to the chamber bottom for 30 min. Care was taken not to extend this sedimentation period because all measurements needed to be made within 1 h after the sample insertion in the AFM head, after which the preparation was discarded. Even though the cardiomyocyte oxygen consumption is lower in Ca^{2+}-free Tyrode, cells tend to deteriorate after a long time without perfusion, which introduces more variability to the measurements [2]. Cells were used between 1 and 10 (fresh group) and 12–24 h (stored group) after isolation.

After choosing a rod-shaped, striated myocyte with fairly straight borders and not in contact with other cells, the tip of the cantilever was moved upon the cell surface to produce a surface topographic image in the contact mode. Three transversal regions were delimited along the whole major axis of the cell. Avoiding the central region, which usually contains the nucleus, the tip of the cantilever was positioned upon either of the two remaining regions (extremities). The topographic image covered a 20 × 20 μm^2 area and the scanning speed varied between 0.3 and 1.0 lines/s according to the surface topography. The SL was scanned by applying a force of 0.86 ± 0.15 nN (n = 32), sufficient to track the myocyte surface. Then, a 16 × 16-matrix of force curves, known as force-volume map, was obtained after generating a good-quality topographic image and also some individual force curves to calibrate the ramp range. This ramp (i.e., the distance travelled by the cantilever in the approach and retraction phases, to generate the force curve), depended on the height variation measured in the topographic image. Because this variation was on the average 3.85 μm, the indentation speed was set as low as 2 μm/s with wide ramp to increase the possibility to find the contact point correctly. A limit of 5 nN was preselected as a precaution to avoid SL damage or getting the cell stuck on the tip, which might have required recalibrating or even exchanging the cantilever.

Empirically, the suitable force to be applied could be established as the minimum force necessary to produce a standard force curve (see Fig. 2), taking into consideration an adequate indentation distance. Indentations were limited to a maximum of 1.5 μm (~10% of the average myocyte height [9, 13]) and a minimum of 0.5 μm (necessary for an acceptable precision in the detection of the contact point in the force curve [10, 11]). It is important to mention that adhesion forces and hysteresis were kept to a minimum by reducing the approach/retraction velocity (2 μm/s in this study), which minimizes energy dissipation [2, 3] and thus allows a more precise $\boldsymbol{E}$ estimation.

After obtaining a 256 force curve maps for the first position, a different position (preferentially in other cell positions) was selected and the whole procedure was repeated. In some cases, it was not possible to obtain more than a couple of maps because of myocyte instability (lack of adequate adhesion to the chamber bottom) or lack of time, due to the limited cell viability interval for measurement acquisition. The absence of cell damage or morphological alteration during the measurement procedure was carefully monitored by the optical image to assure data reproducibility.

2.4 Data Analysis

All topographic images, force curves and maps were analyzed with AtomicJ [12] and Gwyddion [13] programs, which are free software. Firstly, the topographic images were pre-processed to improve contrast. Artifacts and surface flatness were corrected by aligning the rows using the polynomial method (degree 3) after removing the background (degree 2). Subsequently, mean and root mean square (RMS) roughness values were calculated for a 10 × 10 μm^2 central mask to reduce the effect of the scanner curvature (range = 90 μm). Then, a 3-dimensional projection was generated to allow a better representation of the surface topography. Secondly, force curves were pre-processed by applying a 2nd-degree Gaussian filter to remove noise and improve detection of the contact point [10, 11]. In this step, curves were verified individually for adherence to the standard configuration, otherwise they were discarded.

As shown in Fig. 2, the standard force curve has two phases: approach (magenta) and retraction (green). The former starts with a horizontal baseline (at least 0.5 μm) followed by a continuous ascending line with slope < 1 (vertical) reaching the setpoint (transition) without abrupt peaks or valleys. The retraction phase has the same form in the reversed order, but it has a valley at the end of the descending line before the baseline, caused by the adhesion of the sample to the tip.

The Hertz model for pyramidal tips modified by Sneddon [14] and later by Bilodeau [15] was selected to fit the approach curve. Following the procedure of automatic detection of the contact point [12], each individual force curve in each force map was checked for successful determination of the contact point. The criterium of contact point detection depended simply on the Eq. 1 of the model, which represents the points of an ideal curve resulting from exerting on the sample a vertical force F by an indenter with pyramidal geometry:

$$F = \frac{E^* \tan\theta}{\sqrt{2}}\delta^2 \tag{1}$$

where E^* is the reduced Young's modulus of the tip material, θ is the half-opening angle of the pyramid and δ is the indentation into the sample. For tips made of materials much stiffer than the sample, $E^* \cong \boldsymbol{E}/\left(1-\nu^2\right)$, where $\boldsymbol{E}$ is the Young's modulus and ν is the Poisson ratio of the sample ($\nu = 0.5$ for incompressible materials, such as the cytoplasm).

Knowing that the cantilever can be considered as a linear spring, Hook's law can be applied (Eq. 2):

$$F = k.(d - d_0) \tag{2}$$

where k is the spring constant expressed as N/m and $\Delta d = (d - d_0)$, i.e., the cantilever deflection starting from the contact point (z_0, d_0), which should be converted from V into nm by multiplying it by *def* [nm/V]. For the special case of very rigid materials, the displacement on the vertical axis equals the deflection of the cantilever, but for soft materials, this equality is not valid and indentation in the sample can be described by the Eq. 3, assuming that z is the movement of the piezo-electric motor:

$$\delta = (z - z_0) - (d - d_0) \tag{3}$$

Thus, combining Eqs. 1, 2 and 3, Δd can be calculated as [16]:

$$z - z_0 = (d - d_0) + \sqrt{\left(\sqrt{2}.\frac{k(d - d_0)\left(1 - \nu^2\right)}{E\tan\theta}\right)} \tag{4}$$

With k and d_0 obtained from the calibrations and the baseline, Eq. 4 was used to fit the approach curve, from which $\boldsymbol{E}$ and z_0 were retrieved. By comparing the experimentally determined curves with the model, it was possible to select for further analysis only the curves with coefficient of determination $\left(R^2\right)$ value ≥ 0.95.

Histograms, box-and-whisker plots, and maps were generated for each individual scanned position and pooling all positions. Depending on the histogram distribution, the mean (normal distribution) or the median (Lorentz distribution) was considered for ***E***. Additionally, 2D ***E*** maps were produced, in which each pixel represents the stable ***E*** value (20 × 20 μm^2 contact image with 256 pixels). Other maps, such as of contact point, deformation, and adhesion force, were also generated and checked for anomalies.

Data are presented as means ± SEM.

3 Results and Discussion

3.1 Images of Surface Topography

Tridimensional topographical surface images using the contact mode were obtained from 18 cells in at least one position. The surface topography was characterized by quantitative analysis. Figure 1 illustrates the surface topography of a myocyte within a 20 × 20 μm^2 area, in which micro-irregularities can be detected within 0.7 μm.

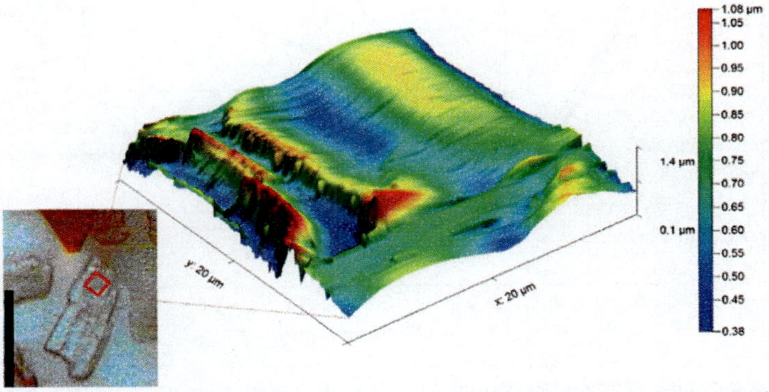

Fig. 1. Tridimensional surface topology of the sarcolemma of a rat ventricular myocyte, generated from AFM contact image, showing micro-irregularities up to 0.7 μm depth, where tips and valleys were colored in red and blue, respectively. Inset: optical microscopic image of the cell, where the red square indicates the region from which the force map was obtained and the black vertical bar measures 100 μm.

As expected, the RMS roughness was greatly reduced by preprocessing and artifact correction, from 823.34 ± 62 to 34.53 ± 8 nm (n = 32). Interestingly, fresh myocytes had greater roughness than stored ones (41.0 ± 10.0 nm; n = 24, vs. 14.4 ± 17.0 nm; n = 8 (p = 0.010; *t* test for heteroscedastic variances).

3.2 Image Force Mapping

Of the 8064 force curves obtained from 32 positions in 18 different myocytes isolated from 12 hearts, 98% were considered suitable for further analysis, according to the criteria previously described. Figure 2 shows examples of force-distance, force-indentation, and ***E***-indentation curves, and Fig. 3 shows the frequency-histogram of ***E*** values obtained from a single position with 16 × 16 force curves.

After analyzing each individual force map (scanned position) for all cells obtained from all hearts, a global histogram and box-and-whisker plot were created. The former offers an overall view of all force curves and enables fitting the data distribution, while individual box-and-whisker plots allows comparing multiple sources of variations: position (in the same cell), cells (from the same heart), and hearts.

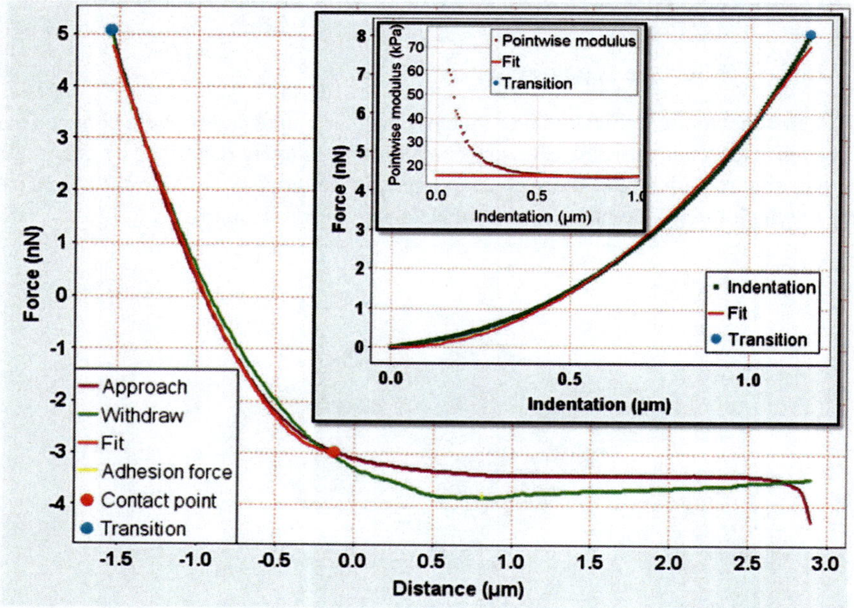

Fig. 2. Force-distance curve obtained by AFM spectroscopy in a cardiomyocyte, taken from a 16×16 force map after the generation of a $20 \times 20\ \mu m^2$ contact topographic image. The main graph shows the approach (magenta) and withdraw (green) phases and the fitted curve (red). Inset: proportion of the applied force within the indented distance, starting from the contact point. Inset in the inset: pointwise calculation of the Young's modulus (E) for each individual point in the indentation trace, stability of the E value when indentation $> 0.5\ \mu m$. The maximum indenting force in this curve was ≈ 5 nN.

Figure 4 illustrates a global histogram, fitted with the Lorentz distribution fit (in red) ($\boldsymbol{E} = 10.96 \pm 0.08$ kPa; n = 5902). Two peaks can be distinguished in the histogram, at ~4 kPa and ~13 kPa, which can possibly be attributable to scanning over the cell nucleus, as this organelle was not always centrally located, compared with other peripheral regions [17].

3.3 Impact of the Scanned Region on SL Elasticity

Figure 5 shows differences in the ***E*** values among 3 indented positions in the same myocyte, such as F6-1, F6-2 and F6-3; F4-1 and F4-2; F8-1 and F8-2, and F10-1 and F10-2, although not in all cells. This variation, also seen in other cell types, might be

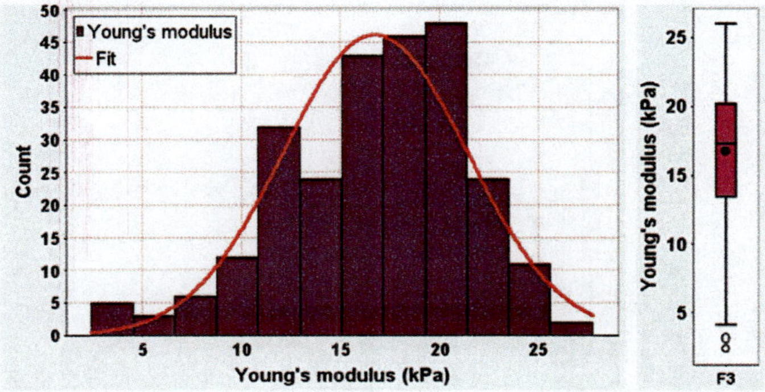

Fig. 3. Left: frequency histogram of the Young's module in cardiomyocytes, fitted with a normal distribution curve (red) (16.77 ± 0.29 kPa; n = 256). Right: box-and-whisker plot showing mean and median (central dot and line, respectively) and interquartile interval (box) of the Young's modulus obtained with AFM spectroscopy force mapping (16 × 16 maps) in a single fresh myocyte. The whiskers indicate the minimum e maximum values. Outliers are shown when their values exceed 1.5 times that of the interquartile range above (none here) or below the limits of the latter.

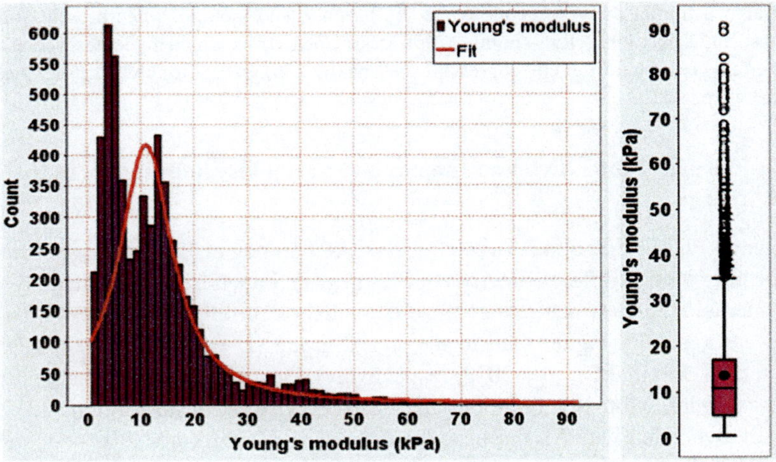

Fig. 4. Similar presentation as in Fig. 3, but with data from a larger sample (fresh group: 5904 force curves, 24 positions, 13 cells, 9 hearts). The curve fitted for Lorentz distribution of the histogram is shown in red (10.96 ± 0.08 kPa) and the respective box-and-whisker plot (right) for Young's modulus. For description of the box-and-whisker plot see the caption in Fig. 3.

explained to non-homogeneous spatial distribution of subsarcolemmal structures, such as cytoskeleton and cytoplasm [18]. As in skeletal myocytes, the SL of ventricular myocytes present a periodic structure characterized by invaginations (T tubules) at the line Z, where $\boldsymbol{E}$ is greater [19].

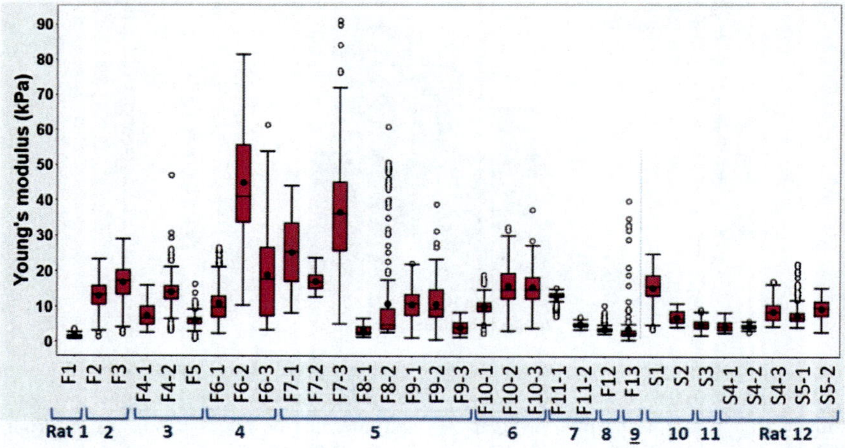

Fig. 5. Box-and-whisker plots of the sarcolemmal elasticity (Young's modulus) estimated by AFM spectroscopy force mapping in rat cardiomyocytes at different scanning positions (16 × 16 force maps: 7914 force curves, 32 positions, 18 cells, 12 hearts). Data are shown as described in the caption of Fig. 3, for cells from the fresh (F) and stored (S) groups (see Material and Methods). The 1st algorism after the group code is the cell number, and the 2nd, when present, indicates different scanned positions. "Rat" (bottom) indicates the heart from which myocytes were isolated. The boxes are computed from 209 curves at least (max = 256), except for cell F1, for which only 93 curves were used.

3.4 SL Elasticity: Inter-Cell and Inter-Heart Variability and Impact of the Time of Storage

As shown in Fig. 6, ***E*** varied between 3 and 40 kPa among cells. This large variability might be attributed to different degrees of SL integrity. Intact, freshly isolated myocytes have firmer and tenser structure than the older cells. Accordingly, ***E*** values in the fresh group were generally higher than in the stored group (***E*** = 10.96 ± 0.08 kPa; n = 5902 vs. 6.48 ± 0.05 kPa; n = 2012, respectively). Decrease in SL ***E*** was reported in fibroblasts under cytotoxic conditions that lead to cytoskeletal disruption [20]. Also, cell death associated with SL permeabilization (verified by intracellular staining with propidium iodine) is accompanied by a fall in ***E*** values [21].

Approach/withdraw speed and maximum indenting force were found to have substantial impact on ***E*** values. An indentation velocity of 0.6–6 μm/s is suitable to reduce hysteresis [3, 4], so 2 μm/s was used in this work. Another probable source of variability might be the presence of molecules anchored to the external SL face [7, 22]. The geometry of the indenting tip is another important factor to determine the detected features of the SL surface. Sharp tips, as used in this study, usually result in ***E*** overestimation when under high applied forces, compared with blunted, truncated or colloidal ones. However, previous studies with rodent cardiomyocytes [3, 4] reported higher mean ***E*** values than in the present report (~35–40 vs. ~11 kPa), although we have obtained values > 20 kPa in a number of cases.

The present results indicate that methodological improvement is necessary. For instance, cell viability was confirmed simply by myocyte appearance, a method which

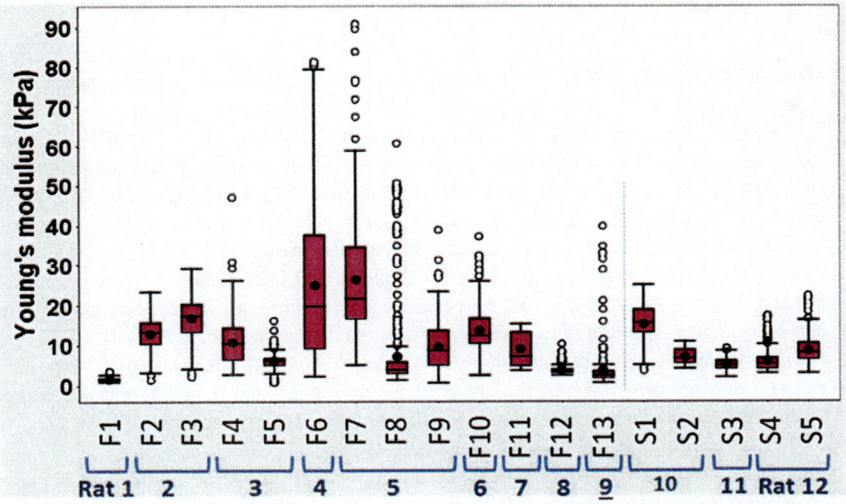

Fig. 6. Young's modulus of cardiomyocytes sarcolemma estimated by AFM spectroscopy force mapping. Data are the same as in Fig. 5, except those points of different scanned positions of the same cell were pooled.

is not the most adequate. Working with freshly isolated myocytes rather than cultured cells requires efficient means of cell immobilization. Adhesion to the collagen layer is not sufficiently strong and not spatially homogeneous, and a more suitable substrate should be used. Finally, cell storage for several hours should be avoided. Understanding how these problems may affect the collected data is important to set the limitations necessary for elaborating the experimental protocols for this kind of experiment with cardiomyocytes.

4 Conclusions

Several factors such as scanned position, indentation speed, maximum indenting force, indentation depth, tip's shape of the probe, and storage time for cells may influence ***E*** determination, taking into consideration the heterogeneous structure of this cellular type. These preliminary experiments performed under challenging conditions (i.e., loosely attached contractile cells with high degree of SL corrugation [9], absence of extracellular Ca^{2+}) are helpful to reveal crucial methodological aspects that are important for reliable measurements and result interpretation.

Acknowledgments. The technical support of Ms. Elizângela S. Oliveira and Mr. João H. Clerici is greatly appreciated. Financial support: Brazilian National Council for Scientific and Technological Development (CNPq, Proc. N. 304010/2016-2 and 429326/2018-1), São Paulo Research Foundation (FAPESP 2019/07616-3), Brazilian Ministry of Health/FINEP (grant# 01-13-0214-00) and Coordination for the Improvement of Higher Education Personnel (CAPES, Ph.D. scholarship for A.A. Almazloum).

Conflict of Interest The authors declare that they have no conflict of interest.

References

1. Moeendarbary, E., Harris, A.R.: Cell mechanics: principles, practices, and prospects. WIREs Syst. Biol. Med. **6**, 371–388 (2014). https://doi.org/10.1002/wsbm.1275
2. Lachaize, V., Formosa-Dague, C., Smolyakov, G. et al.: Atomic force microscopy: an innovative technology to explore cardiomyocyte cell surface in cardiac physio/pathophysiology. Lett. Appl. Nanobiosci. **4**, 321–324 (2015). https://hal.archives-ouvertes.fr/hal-01552834
3. Lieber, S.C., Aubry, N., Vatner, S.F., et al.: Aging increases stiffness of cardiac myocytes measured by atomic force microscopy nanoindentation. Am. J. Physiol. Heart Circ. Physiol. **287**, H645–H651 (2004). https://doi.org/10.1152/ajpheart.00564.2003
4. Benech, J.C., Benech, N., Damián, J.P., et al.: Diabetes increases stiffness of live cardiomyocytes measured by atomic force microscopy nanoindentation. Am. J. Physiol. Cell Physiol. **307**, C910–C919 (2013). https://doi.org/10.1152/ajpcell.00192.2013
5. Penna, L.B., Bassani, R.A.: Increased spontaneous activity and reduced inotropic response to catecholamines in ventricular myocytes from footshock-stressed rats. Stress **13**, 73–82 (2010). https://doi.org/10.3109/10253890902951778
6. Butt, H.-J., Jaschke, M.: Calculation of thermal noise in atomic force microscopy. Nanotechnology **6**, 1 (1995). https://doi.org/10.1088/0957-4484/6/1/001
7. Schillers, H.: Measuring the elastic properties of living cells. In: Santos, N., Carvalho, F. (eds.) Atomic Force Microscopy. Methods in Molecular Biology, vol. 1886. Humana Press, New York (2019)
8. Schillers, H., Rianna, C., Radmacher, M., et al.: Standardized nanomechanical atomic force microscopy procedure (SNAP) for measuring soft and biological samples. Sci. Rep. **7**, 5117 (2017). https://doi.org/10.1038/s41598-017-05383-0
9. Milan, H.F.M., Bassani, R.A., Santos, L.E.C., et al.: Accuracy of electromagnetic models to estimate cardiomyocyte membrane polarization. Med. Biol. Eng. Comput. **57**, 2617–2627 (2019). https://doi.org/10.1007/s11517-019-02054-2
10. Gavara, N., Chadwick, R.S.: Determination of the elastic moduli of thin samples and adherent cells using conical atomic force microscope tips. Nat. Nanotech. **7**, 733–736 (2012). https://doi.org/10.1038/nnano.2012.163
11. Rico, F., Roca-Cusachs, P., Navajas, D., et al.: Probing mechanical properties of living cells by atomic force microscopy with blunted pyramidal cantilever tips. Phys. Rev. E **72**, 021914 (2005). https://doi.org/10.1103/physreve.72.021914
12. Hermanowicz, P., Sama, M., Gabrys, H., et al.: AtomicJ: an open source software for analysis of force curves. Rev. Sci. Instrum. **85**, 063703 (2014). https://doi.org/10.1063/1.4881683
13. Nečas, D., Klapetek, P.: Gwyddion: an open-source software for SPM data analysis. Cent. Eur. J. Phys. **10**, 181–188 (2012). https://doi.org/10.2478/s11534-011-0096-2
14. Sneddon, I.N.: The relation between load and penetration in the axisymmetric Bousinesq problem for a punch of arbitrary profile. Int. J. Eng. Sci. **3**, 47–57 (1965). https://doi.org/10.1016/0020-7225(65)90019-4
15. Bilodeau, G.G.: Regular pyramid punch problem. J. Appl. Mech. **59**, 519–523 (1992). https://doi.org/10.1115/1.2893754
16. Radmacher, M.: Measuring the elastic properties of living cells by AFM. In: Jena, B.P., Hörber, J.K.H. (eds.) Methods in Cell Biology: Atomic Force Microscopy, pp. 67–90. Academic Press (2002)
17. Charras, G.T., Horton, M.A.: Determination of cellular strains by combined atomic force microscopy and finite element modeling. Biophys. J. **83**(2), P858-879 (2002). https://doi.org/10.1016/S0006-3495(02)75214-4
18. Liu, Y., Mollaeian, K., Ren, J., et al.: Effect of F-actin and microtubules on cellular mechanical behavior studied using atomic force microscope and an image recognition-based cytoskeleton

quantification approach. Int. J. Mol. Sci. **21**, 392 (2020). https://doi.org/10.1039/C1NR11303H

19. Ogneva, I., Levedev, D.V., Shenkman, B.S.: Transversal stiffness and Young's modulus of single fibers from rat soleus muscle probed by atomic force microscopy. Biophys. J. **98**, 418–424 (2010). https://doi.org/10.1016/j.bpj.2009.10.028
20. Pastrana, H.F., Cartagena-Rivera, A.X., Raman, A., et al.: Evaluation of the elastic Young's modulus and cytotoxicity variations in fibroblasts exposed to carbon-based nanomaterials. J. Nanobiotechnol. **17**, 32 (2019). https://doi.org/10.1186/s12951-019-0460-8
21. Nikolaev, N., Muller, T., Liu, Y., et al.: Changes in the stiffness of human mesenchymal stem cells with the progress of cell death as measured by atomic force microscopy. J. Biomech. **47**, 625–630 (2014). https://doi.org/10.1016/j.jbiomech.2013.12.004
22. Sokolov, I., Lyer, S., Venkatesh, S.R., et al.: Detection of surface brush on biological cells in vitro with atomic force microscopy. Appl. Phys. Lett. **91**, 023902 (2007). https://doi.org/10.1063/1.2757104

Check for updates

Calibration of Unicapillary Pneumotachograph Through Computational Fluid Dynamics (CFD) Approach

Vinicius Ferreira Peres dos Santos(✉), Felipe Fava de Lima, and Henrique Takachi Moriya

Biomedical Engineering Laboratory, University of São Paulo, Sao Paulo, Brazil
viniciusfperes@usp.br

Abstract. A pneumotachograph is a device applied to measure gas flow of the respiratory system and can be utilized in mechanical ventilators, spirometry equipments and in scientific researches on the respiratory system. These devices need to be properly characterized to provide reliable measurements. In general, pneumotachographs are characterized experimentally by the calibrated syringe strokes method. Computational Fluid Dynamics (CFD) has been used in different areas of science for the analysis of fluids both in laminar and turbulent regimes quickly and inexpensively. In this work the feasibility of using CFD to obtain calibration curves of a pneumotachograph was verified using the calibrated syringe strokes technique. In this first study it was possible to calculate calibration curves of a pneumotachograph by the novel proposed computational methodology with results similar to the obtained experimentally. The proposed method could be used as preliminary tool to predict the behaviour of each variable involved in this process. Finally, the calculation of volumetric error and standard deviation was realized to compare with experimental results by Minimal Mean Square method.

Keywords: Pneumotachograph · Calibration · Computational fluid dynamics · Minimal mean square · Low flow

1 Introduction

A pneumotachograph (PT) is a device applied to measure respiratory gas flow and can be utilized in mechanical ventilators, spirometry equipments, and instruments employed in scientific researches of the respiratory system. Different physical principles can be employed by a PT, such as the pressure drop in a mechanical restriction (Fleisch/Lilly, fixed orifice, and variable orifice PT), ultrasound, and heat transfer (hot wire PT). A review about PT for mechanical ventilation is available on [1].

Many studies to establish the correlation and right comprehension between normal functioning of the lung and lung diseases are conducted in small animals

J. L. B. Marques et al. (Eds.): CLAIB 2022/CBEB 2022, IFMBE Proceedings 98, pp. 188–198, 2024.
https://doi.org/10.1007/978-3-031-49401-7_19

(mice, rats, rabbits, etc.) [2]. In these studies, there is a need to accurately measure low gas flows and several technical issues must be evaluated to guarantee the reliability of the parameters under measurement [2].

Additionally, an accurate measurement of volume and flowrate is extremely important to avoid diseases due uncorrected ventilation parameters such as: barotrauma and volutrauma [3].

Giannella-Neto et al. developed and experimentally characterized an unicapillary PT for use in mice [4]. The developed PT geometry was based on Poiseuille's law assuming airflow regime with Reynolds number compatible laminar flow ($Re \leq 2000$). After construction, the calibration function of the PT (Flow x Pressure Drop) was obtained using the calibrated syringe strokes (CSS) method [4,5]. The calibration was performed in different conditions: 1) Inspiration without load and PT opened to atmospheric pressure, 2) Inspiration in a apparatus emulating compliance and resistance similar to a mice, and 3) Expiration without load. The authors also validated the calibration function in different flow ranges.

When compared with a calibration using constant airflow, the method using CSS is easier to perform and allows the characterization of the PT with airflow profiles closer to those observed in experimental conditions [5]. This method uses different airflow profiles applied manually by a calibrated syringe to the PT with known total volume . The resulting pressure drop waveforms generated on the PT are then processed using a multiple linear regression algorithm to obtain the calibration function (See [5]).

Computational Fluid Dynamic (CFD) has been utilized in several research fields [6]. CFD allows the digital simulation of non-stationary behavior of fluids in both laminar and turbulent flow conditions. Analyses can be performed quickly and inexpensively, as it does not depend on the construction and utilization of physical models.

CFD solves problems through numerical approximation of governing equations that describes fluid motion. These equations are composed by: continuity, momentum and energy equations, that originate the Navier Stokes equations. The following steps are used to solve numerically the problem: pre processing, solver and post processing. The differential equations are solved by a set of algebraic equations in a discrete domain [7].

To the best of our knowledge, there are no published studies that evaluate the feasibility of using CFD to obtain calibration functions of PT using CSS as similar approach. If feasible, the design, optimization and characterization of PT using CFD could generate faster and more reliable results since it emulates computationally the calibration method used in practice. In this initial study the PT developed by Giannella-Neto et al. [4] was analysed using CFD to check the feasibility of obtaining calibration functions of PT using similar curves of CSS method.

2 Materials and Methods

The three-dimensional (3D) geometry of the pneumotachograph proposed by Giannella et al. was reproduced using Computer Aided Design (CAD) software (Solidworks, 3DS, France) (Fig. 1).

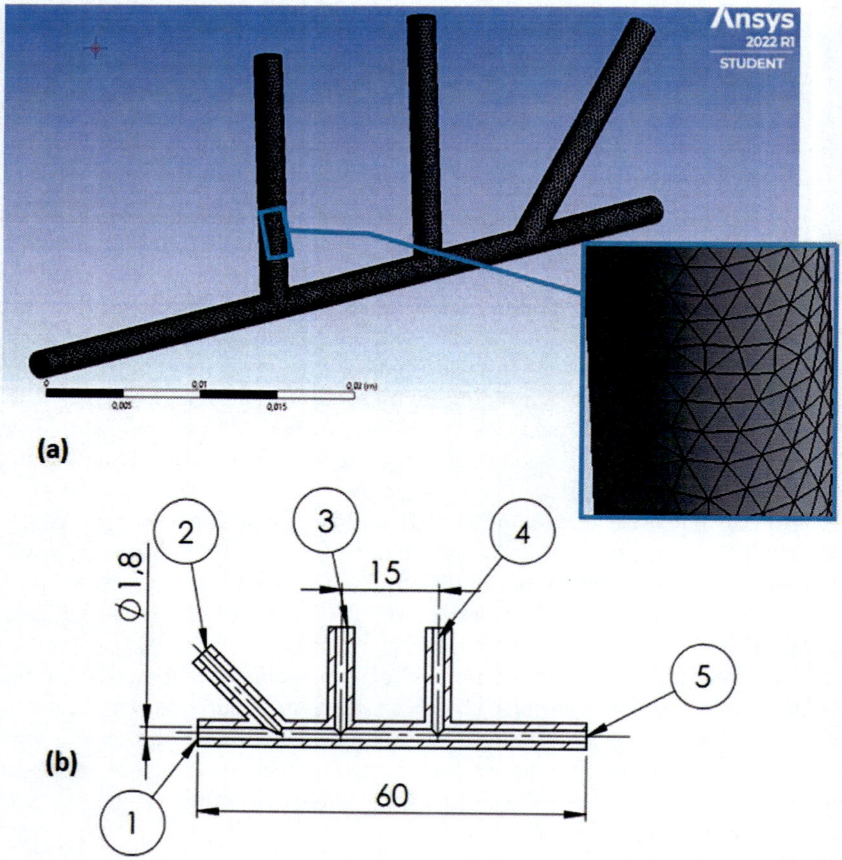

Fig. 1. (a) 3D model with geometry as proposed by Giannella-Neto et al. [4] and the generated mesh in the CFD software; (b) Cross section view with dimensions of the 3D model of the PT. All units in milimeters. 1—Inlet port, 2—Expiratory port, 3—Pressure tap 1, 4- Pressure tap 2, 5—Outlet port

The generated 3D model was exported to be utilized in the CFD software (Fluent, Ansys, USA). The Dirichlet (outlet surface) and Neumann (inlet surface) boundary conditions were defined according to reference surfaces and values setted up experimentally by Giannella-Neto et al. The outer face of the inlet port was configured to have normal airflow with intensities determined by predefined profiles (Fig. 2). The outer face of the outlet port was defined to have constant atmospheric pressure (1 ATM). Other outer faces were configured to have normal flow equal to zero (walls). An unstructured mesh was generated using tetrahedron elements with constant side length of 0.8 mm. This setup resulted in 512,000 discrete elements that was sufficient to provide acceptable results according to the skewness and aspect ratio criteria [6].

A total of 12 airflow profiles passing through the inlet port of the PT were simulated. Six profiles were used to obtain calibration functions and twelve to validate the calibration functions.

Six trapezoidal-shaped airflow profiles with maximum flow rates from 1 to 6 mL$\cdot$ s^{-1} were used to compute the calibration curves and in the validate step, the same profile was used when this range was performed (Fig. 2). These profiles reached their maximum at 250 ms and the time duration was adjusted to ensure a total applied air volume of 2 mL. Six ramp-shaped airflow curves with maximum airflow from 7 to 12 mL $\cdot$s^{-1} with total applied volume of 2 mL were also simulated to the validation process (Fig. 3). The PT was opened to the atmosphere and the results were obtained with sampling frequency of 218.2 Hz. The CFD solver was configured as listed on Table 1.

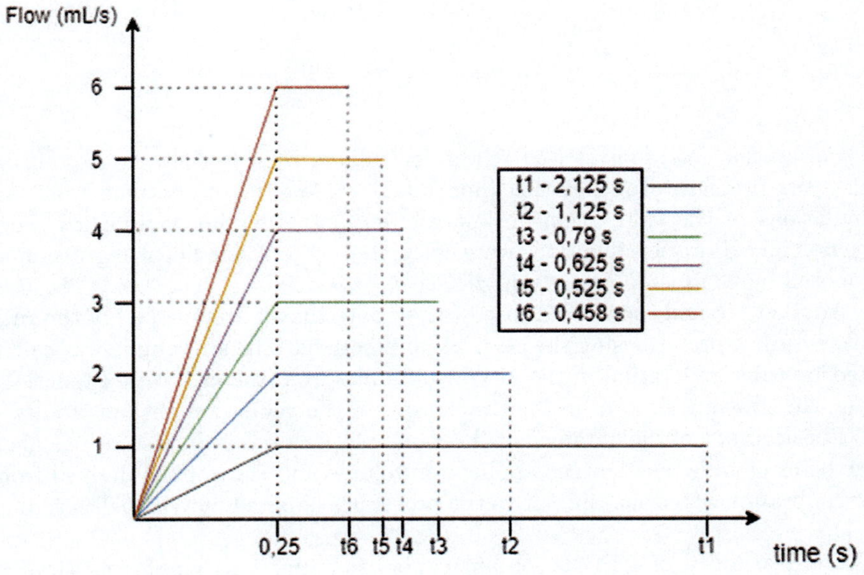

Fig. 2. Trapezoidal-shaped airflow profiles utilized in the calibration and also in the validation process

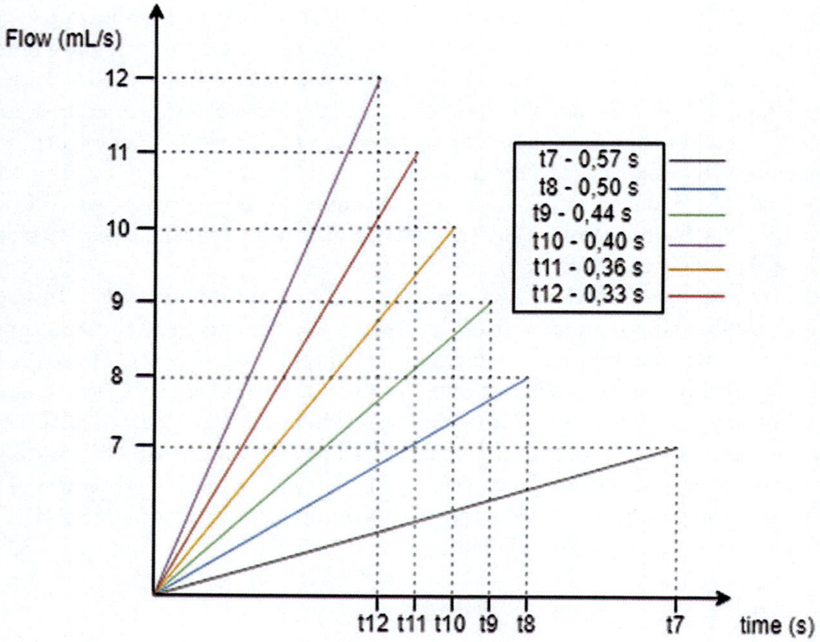

Fig. 3. Ramp-shaped airflow profiles used in the validation process

Calibration functions ($\dot{V}(\Delta P)$) were calculated by a Python 3.0 [8] routine using the pressure drop between taps 1 and 2 (see Fig. 1) obtained from the simulations of the calibration airflow profiles. 1st order linear and 3rd order polynomial calibration functions were calculated by multiple linear regression as proposed by Giannella-Neto et al. [4].

Another Python routine was used to compute the air volume (V) of the validation airflow profiles using the calibration functions. The air volume was calculated by time integration of the airflow obtained from the calibration functions using the pressure drop from the simulations of the validation airflow profiles.

The accuracy of the obtained calibration functions was evaluated by calculating the error between the true applied air volume and the volume derived from the calibration functions, similar to the procedure adopted by Giannella-Neto et al. [4].

The feasibility of CFD to obtain calibration functions employing similiar curves of CSS was verified by comparing the volume error obtained computationally with those obtained experimentally by Giannella-Neto et al. [4].

The described methods, starting from CAD modeling to the verification of the feasibility of CFD to obtain calibration functions, are schematized in Fig. 4.

Table 1. CFD Solver Configuration

Parameter	Method
State	Transient
Turbulence model	k-ϵ
Pressure-velocity coupling	Pressure
Gradient	Least square cell-based
Pressure	Second order upwind
Momentum	Second order upwind
Energy	Second order upwind
Volume fraction	Quick

3 Results

CFD simulations with different airflow profiles resulted pressure distributions on the PT as exemplified by Fig. 5. Equations 1 and 2 present the 1st order linear and 3rd order polynomial calibration functions, respectively, calculated by the Python routine utilizing the pressure difference curves between taps 1 and 2 when the PT was subject to the calibration airflow profiles.

$$\dot{V} = 89.39 \cdot \Delta P \tag{1}$$

$$\dot{V} = -6723.74 \cdot \Delta P^3 + 383.13 \cdot \Delta P^2 + 85.98 \cdot \Delta P \tag{2}$$

$\dot{V}$ represents the air flow (mL$\cdot$ s^{-1}) and ΔP the pressure drop (cmH_2O) between PT taps 3 and 4.

Fig. 6 graphically displays the calibration curves (Eqs. 1 and 2) for pressure drops from 0 cmH_2O to 0.2 cmH_2O and curves obtained experimentally by Giannella-Neto et. al. (curves plotted manually by web tool.

Table 2 presents the calculated air volumes by time integration of the airflow obtained from the calibration equations using pressure drops resulted from the simulations of the validation airflow profiles.

Table 3 provides the mean and the standard deviation error from the calculated air volumes by the calibration functions considering that the true applied volume was 2 mL.

4 Discussion

A qualitative comparison between the calibration curves obtained experimentally by Giannella-Neto et al. (Fig. 6) indicates that the computational technique produce similar results. A quantitative comparison between the experimental and computational calibrations functions was not possible because Giannella-Neto et al. [4] reported only the graphs of the calibration functions and not the coefficients of the equations per se. Therefore, it was possible to observe that both

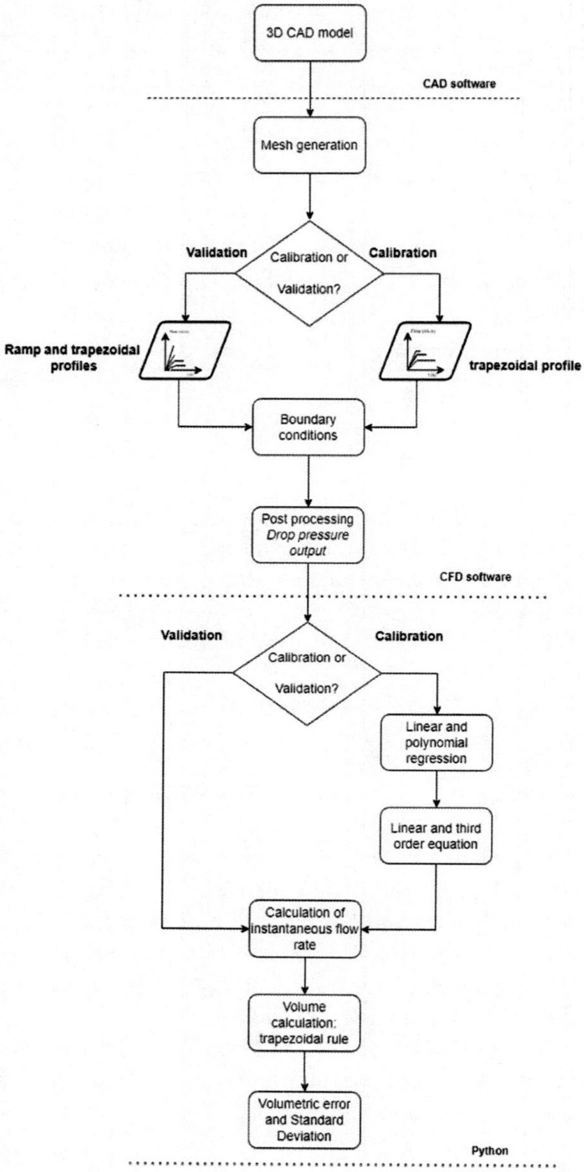

Fig. 4. Schematic of the methods, starting from CAD modeling to the verification of the feasibility of CFD to obtain calibration functions

process seems to present volumetric error of similar magnitude in the same range of calibration (low flow range) when the validation is performed for full range values (from 1 to 12 $mL \cdot s^{-1}$). In the study by Giannella-Neto et al. [4], under

Fig. 5. Pressure distribution on the PT for the calibration curve with maximum flow of 7 mL/s at the instant of time when airflow reached the peak

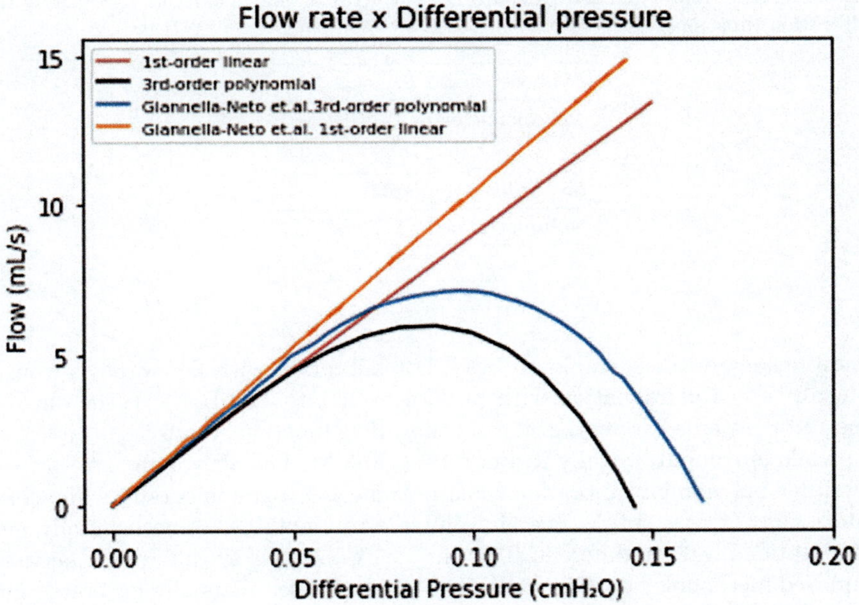

Fig. 6. Curves obtained by computational calibration method vs experimental by Giannella-Neto et al. (curves plotted manually by web tool)

Table 2. Calculated volumes by time integration of the airflow obtained from the calibration equations using the pressure drop resulted from simulations of the validation airflow profiles

Calculated volume by regression method			
Air flow rate reference (mL/s)	Volume by linear equation (mL)	Volume by third order equation (mL)	Profile
1	1.88	1.88	Trapezoidal
2	1.93	1,96	Trapezoidal
3	1.99	2.03	Trapezoidal
4	2.03	2,04	Trapezoidal
5	2.09	2.04	Trapezoidal
6	2.14	1.97	Trapezoidal
7	2.10	1.97	Ramp
8	2.15	1.89	Ramp
9	2.19	1.75	Ramp
10	2.24	1.56	Ramp
11	2.29	1.30	Ramp
12	2.33	0.94	Ramp

Table 3. Mean and Standard Deviation of the error of the calculated volumes by the calibration functions

1st order linear	
Volumetric error (%)	SD(±)
7.4	4.9
3rd order polynomial	
Volumetric error (%)	SD(±)
11.9	15.8

conditions close to the employed in CFD (calibration with flow profiles from 1 to 6 $\mathrm{mL}\cdot\mathrm{s}^{-1}$ and validation with profiles from 1 to 12 $\mathrm{mL}\cdot\mathrm{s}^{-1}$), the air volume error obtained from the linear calibration function was 6.8% (Table 1 of [4]) while in computationally it was 7.4% (Table 3). The air volume error of the 3rd order polynomial calibration function was 5.4% experimentally while computationally it was 11.9%. The air volume errors obtained experimentally and computationally were different. This may be generated by differences between employed methodologies. Giannella-Neto et al. [4] used manually generated airflow profiles from a calibrated syringe while in the CFD simulations the airflow profiles had a trapezoidal wave shape to emulate the syringe strokes. The manually generated profiles had stochastic behavior with excursions through different airflow intensities while those used in CFD were deterministic and remained most of the time at a single airflow value. This may contributed to the differences between the air volume errors.

In this first study it was possible to verify the feasibility of performing CFD simulations of a PT to obtain calibration curves but it has not yet been possible to verify the accuracy of the results. In future studies, calibration curves obtained from profiles closer to those that are manually generated by a calibrated syringe will be used in the characterization process. The simulation of all conditions presented by Giannella-Neto et al. [4] will also be performed enabling an analysis of the real accuracy of this novel methodology.

The proposed method could be used as preliminary tool to predict the behaviour of each variable involved in this process and then, it is possible to optimize some points of interest before experimental validation approach, for example: to evaluate the influence of geometry of a pneumotachograph, the response of a pneumotachograph at different air flow profile and to test different setups of patient circuits.

5 Conclusion

In this work the feasibility of using CFD for the characterization of using the CSS calibration methodology was investigated. It was possible to verify that CFD simulations of a PT can be applied to obtain calibration functions using similarity with CSS methodology. The qualitative comparison between the calibration curves obtained experimentally and computationally shown the feasibility of the proposed novel methodology. Future studies must be carried out to analyze the accuracy of the method.

Conflict of Interest

The authors declare that they have no conflict of interest.

Acknowledgments. This study was financed in part by the *Coordenacao de Aperfeicoamento de Pessoal de Nivel Superior—Brazil* (CAPES)—Finance Code 001 and the *Conselho Nacional de Pesquisa e Desenvolvimento Cientfico e Tecnologico*—Brazil (CNPq) (308280/2019-9 to HTM).

References

1. Jaffe, M.B.: Gas flow measurement. In: Capnography, pp. 397–406. Cambridge University Press, New York (2011)
2. Bates, J.H.T., Irvin, C.G.: Measuring lung function in mice: the phenotyping uncertainty principle. https://doi.org/10.1152/japplphysiol.00706.2002, www.jap.org
3. Schena, E., et al.: Flow measurement in mechanical ventilation: a review (2015). https://doi.org/10.1016/j.medengphy.2015.01.010
4. Giannella-Neto, A. et al.: Design and calibration of unicapillary pneumotachographs. J. Appl. Physiol. 84, 335–343 (1998). https://doi.org/10.1152/jappl.1998.84.1.335, www.jap.org

5. Tang, Y., et al.: Calibration of pneumotachographs using a calibrated syringe. J. Appl. Physiol. **95**(2), 571–576 (2003)
6. Date, A.W.: Introduction to Computational Fluid Dynamics. Cambridge University Press (2005)
7. Zawawi M.H., et al.: A Review: Fundamentals of Computational Fluid Dynamics (CFD), vol. 2030. American Institute of Physics Inc. (2018). https://doi.org/10.1063/1.5066893. ISBN: 9780735417526
8. Van Rossum, G., Drake, F.L.: Python 3 Reference Manual. CreateSpace, Scotts Valley (2009). ISBN: 1441412697

Use of Three Distinct Types of Cuneate Neurons in the Classification of Naturalistic Textures

Ana Clara Pereira Resende da Costa[1](✉), Vinícius de Carvalho Zanini[1], Pedro Faria de Bessa[1], Ailton Luiz Dias Siqueira-Junior[2], and Alcimar Barbosa Soares[1]

[1] Faculty of Electrical Engineering, Federal University of Uberlandia, Uberlandia 38408-100, Brazil
anaclaraapereira13@gmail.com

[2] Federal Institute of the Triângulo Mineiro, Ituiutaba 38305-200, Brazil

Abstract. The tactile sensation is encoded by mechanoreceptors in the skin that characterize information such as pressure and vibration. This information is processed by the somatosensory system and the first major processing step is in the cuneate nucleus. Based on the functionality of the cuneate nucleus, a model of the intercellular dynamics of this processing stage has recently been proposed. The validation of this model considered that there was only one type of cuneate neuron that received synaptic input from slow and fast adapting mechanoreceptors; however, physiologically there are several types of cuneate neurons with specific synaptic inputs. Considering this specificity, this paper hypothesizes that the use of different types of artificial cuneate neurons would allow better segregation and classification of tactile artificial data. To prove this hypothesis, we implemented a neuromorphic model with three types of bioinspired artificial cuneate neurons and evaluated its ability to classify eight naturalistic textures. Two indexes were extracted from the output of these tree cuneate neuron models and used as a feature vector for five different support vector machines (SVM). The results show that combining the output of the three neural models it was obtained an average accuracy in the texture's classification of 90.47%.

Keywords: Tactile sensations · Cuneate nucleus · Texture classification

1 Introduction

The feeling of touch is indispensable for the individual and spatial perception. The skin is the interface that allows interaction and information exchange between the person and the environment. This is done through approximately seventeen thousand primary afferents (mechanoreceptors) [1]. Primary afferents transduce sensations like pressure and vibration into different action potentials that are processed and interpreted in higher structures of the somatosensory system [1, 2].

The cuneate nucleus is the structure responsible for the first tactile processing of the mammals' upper limbs. It is composed of secondary neurons, known as cuneate neurons,

J. L. B. Marques et al. (Eds.): CLAIB 2022/CBEB 2022, IFMBE Proceedings 98, pp. 199–209, 2024.
https://doi.org/10.1007/978-3-031-49401-7_20

which receive synapses from the different primary afferents [3, 4]. Its neural circuits are responsible for enabling the segregation and classification of tactile information [4].

Inspired by the functionality of this structure, recently, Rongala et al. proposed a neural model of the intracellular dynamics of the cuneate nucleus able to classify different naturalistic textures under different conditions, proving its capacity for generalization and representation of artificial information of touch [5, 6]. This model can reproduce the intrinsic membrane dynamics and the highly specified distribution of the synaptic weights observed in vivo. It also implements the synaptic learning of the excitatory and inhibitory weights of primary afferents and interneurons, respectively [5]. The authors also considered that all implemented artificial cuneate neurons received inputs from slow and fast adapting primary afferents [5, 6].

However, studies have shown that there are several types of cuneate neurons. Unimodal neurons that receive afferents exclusively from a single type of class, slow adaptation (SA) or fast adaptation (FA). Bimodal and trimodal neurons receive combination of different classes [7, 8]. Thus, some cuneate neurons are more sensitive to indentation, and others to vibration, but there are secondary neurons that respond to both types of stimuli suggesting that part of the spatial and temporal selectivity of the primary somatosensory system is inherited from the cuneate nucleus [7].

Considering these functionalities, this paper hypothesizes that the use of different types of artificial cuneate neurons would allow better segregation and classification of tactile artificial data. In this work we implemented a neuromorphic model with three distinct types of bioinspired cuneate neurons: (i) receiving and processing signals from slowly adapting mechanoreceptors, (ii) from rapidly adapting mechanoreceptors, and (iii) from both mechanoreceptor models. We also evaluated the ability of these models to classify naturalistic textures.

2 Materials and Methods

It was implemented an artificial neural network biologically inspired into the somatosensory process to evaluate the use of three cuneate neurons in the segregation of tactile information. This network is composed of two different layers, the first one is composed of 32 primary afferents and the second one of 3 cuneate neurons.

Figure 1 presents the block diagram of the proposed model. Basically, the data of eight different naturalistic textures were obtained by a force sensor and used as the input of the neuromorphic neural network inspired in the somatosensory processing. The first layer is composed of primary afferents and the second one of the synaptic learning of 3 different cuneate neurons sending 3 different spikes trains.

To evaluate the model performance in texture classification, it was extracted from each spike train two features: the Average Spike Rate (ASR) for time in seconds and the Coefficient of Variation of the Intervals between Spikes (CV_ISI) which were used as input of a Support Vector Machine (SVM).

2.1 Data Base

The data used to evaluate this model are from a database developed by [9] with pressure sensing data acquisition from active palpation of a robot's hand of 8 different naturalistic

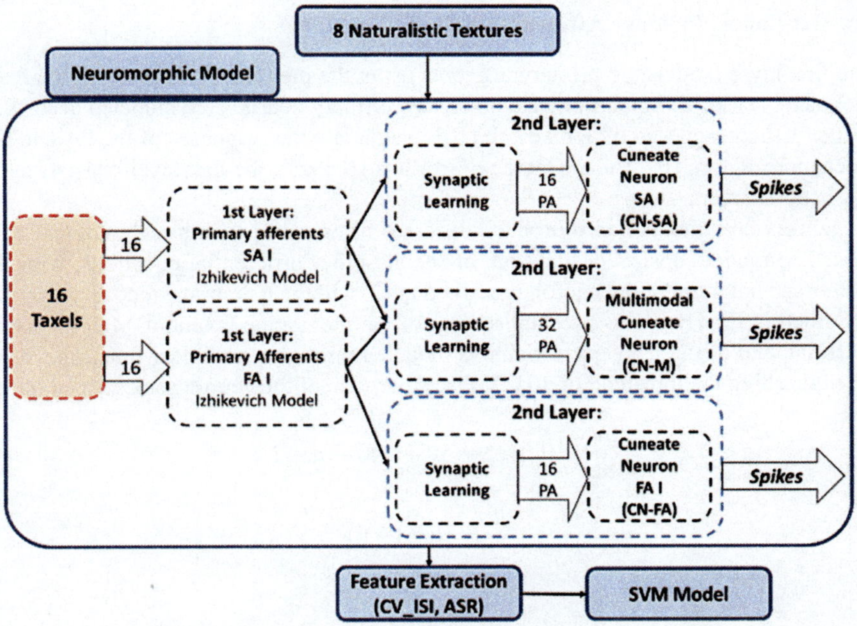

Fig. 1. Block diagram of the methodology implemented in the paper.

textures. The hand had a piezoresistive sensor with 16 pressure sensitive points (taxels) organized in a 4 × 4 grid. The data was sampled in a 1 kHz frequency, a constant palpation speed of 10 mm/s and a normal force of 1 N. The 8 textures were labeled as Floor Tile 1 (F1), Floor Tile 2 (F2), Floor Tile 3 (F3), Corrugated Paper (CP), Floor Rubber Tile (RT), Textile Rug (TR), Scotch Brite (SB), Styrofoam (ST) and for each one of them were made 20 trials.

2.2 Pre-processing of Data

After being collected from each of the 16 taxels, the signals were sent to a Butterworth low-pass filter of 4th order, which had a cutoff frequency of 30 Hz, then normalized in direct relation to the samples with higher and lower amplitudes, according to (1).

$$n_i = \frac{x_i - \upsilon_{max}}{\upsilon_{max} - \upsilon_{min}} \tag{1}$$

where n_i is the normalized value of a taxel at position i; x_i is the sample at position i; υ_{max} and υ_{min} are, respectively, the samples with the highest and lowest sampling amplitudes.

The signal was normalized in 6 s segments in the active palpation phase, with 6000 samples. This was used to generate the input current for the neurons, modeled to simulate the primary tactile afferents (mechanoreceptors), of the first layer.

2.3 1st Layer: Primary Afferents–Izhikevich's Model

The first layer models the primary afferents using the preprocessed tactile signal from each taxel sensor as input. For each taxel, two primary neurons are modeled according to the method proposed by Izhikevich [10] to simulate the responses of the FA and SA mechanoreceptors [10]. As the tactile sensor has 16 taxels, the first layer consists of 32 mechanoreceptors.

Izhikevich's model can reproduce different firing rate patterns of biological neurons. The model consists of a system of second-order differential equations, with the membrane potential of the neuron described in (2) and the membrane recovery variable described in (3) [10]. The re-establishment of the membrane potential, after an action potential, and of the recovery variable is modeled through a nonlinear function where, when reaching the threshold of 30 mV resets the state of the neuron as described in (4).

$$\frac{dv}{dt} = 0.04v^2 + 5v + 140 - u + I(t) \tag{2}$$

$$\frac{du}{dt} = a(bv - u) \tag{3}$$

$$\text{If } v \geq 30\,\text{mV, then } \begin{cases} v \leftarrow c \\ u \leftarrow u + d \end{cases} \tag{4}$$

v represents the membrane potential of the neuron, u the recovery variable, $I(t)$ the synaptic current (input current), t the time in a millisecond scale. The parameters ***a*, *b*, *c*** and ***d*** describe respectively the time scale of u; the sensitivity of the variable u to fluctuations of membrane potential; the value of membrane resting potential (in mV); the value of u redefinition after a shot.

From the parameters (***a*, *b*, *c*** and ***d***) it is possible to model different types of neurons [10]. Table 1 show the values of the parameters used in the implementation of the Izhikevich's model for each type of mechanoreceptor used in this work.

Table 1. Parameters of Izhikevich's model

Variables	a	b	c (mV)	d (mV)
FA	0.1 s^{-1}	0.2	– 65	2
SA	0.02 s^{-1}	0.2	– 65	8

Since the fast-adaptation receptors are sensitive only to transient phases of the stimuli, the derivative of the normalized signal associated with a gain of 3000 was used before being used as input current of the FA model to emulate this effect, as represented in (5). For the SA model the input current was the normalized signal with a gain of 20 for adjustment of the current scale, according to (6).

$$I(t)_{FA} = \frac{dVtaxel_n}{dt} \cdot G_{FA} \tag{5}$$

$$I(t)_{SA} = Vtaxel_n \cdot G_{SA} \tag{6}$$

$I(t)_{FA}$ and $I(t)_{SA}$ are the input currents of the Izhikevich Model to the fast and slow adaptation afferents respectively, $Vtaxel_n$ is the normalized signal of each taxel, G_{FA} is the gain for the fast adaptation neural model equal to 3000, and G_{SA} is the gain for the slow adaptation model equal to 20.

According to Izhikevich's model proposal, a spike (activation potential trigger) is generated as soon as the membrane potential reaches 30 mV. Thus, every time the neuron fires, the binary output of the neuromorphic model is equal to 1 and if it does not reach 30 mV, equals to 0. This binary signal generated for each afferent was used as input of the second neural layer composed of three models of cuneate neurons.

2.4 2nd Layer: Secondary Afferents - Cuneate Neurons

The second layer was biologically inspired by the tactile somatosensory process of the cuneate nucleus, where the first big somatosensory signals processing occurs. It was used 3 neurons in this stage: the first one received only the output of the 16 SA afferents, the second one only the 16 FA and the last one received all the signals for each texture.

To implement the 3 cuneate neurons, it was used the model proposed by [5]. This model is based on the Exponential Integrate and Fire (EIF) with a detailed activity modeling of the low threshold voltage-dependent calcium channels and the calcium-dependent potassium channels. The complete dynamic of the cuneate neuron's membrane potential is described in (7).

$$C_m \frac{dV_m}{dt} = I_L + I_{spike} + I_{ion} + I_{ext} + I_{syn} \tag{7}$$

where $w_{exc,i}$ is the excitatory synaptic weight; t^* is the time that occurs a spike; w_{inh} is the inhibitory weight; g_{max} is the maximum conductance, that is 23×10^{-8} S; τ is a decay time constant of 6.4 ms. The excitatory learning is described in (9), and the potentializing of the excitatory weights occurs when the calcium local activity of a synapse (mechanoreceptor with the cuneate neuron) is correlated with the cuneate neuron's total calcium activity, the opposite of the synapse will be depreciated.

$$\begin{aligned} I_{syn} = {} & g_{max} \sum_i w_{exc,i} \exp(-\tau(t - t^*))(E_{rev,exc} - V_m) \\ & + g_{max} w_{inh} \sum_i \exp(-\tau(t - t^*))(E_{rev,inh} - V_m) \end{aligned} \tag{8}$$

where $A_{Tot}^{Ca^{2+}}$ is the cuneate neuron's total calcium activity; $Avg_{A_{tot}^{Ca^{2+}}}$ is the mean of the 3 last values from the total calcium activity; Syn_{EQ} is the synaptic equilibrium; $A_{Loc}^{Ca^{2+}}$ is the calcium local activity from a synapse and K is a gain factor.

$$\Delta w_{exc,i} = \int_{t_0}^{t_{max}} \left\{ \left(A_{Tot}^{Ca^{2+}}(t) - (Avg_{A_{tot}^{Ca^{2+}}} * Syn_{EQ}) \right) * A_{Loc}^{Ca^{2+}}(t) \right\} * K * dt \tag{9}$$

The total inhibitory weight update (w_{inh}) is based on the calcium channels' spike rate, it is related to the total calcium activity and represents the cuneate neuron's inhibitory synapse weight. This way, the increase of the calcium activity will result in the increase of the inhibitory weight and the opposite is also true.

To determine the excitatory and inhibitory weights it was realized a 4-Fold training, all the 8 textures were presented 300 times using a one second sample randomly generated for the first cuneate neuron and followed by the others, creating 2400 training cycles. The initial weights were generated randomly with values between 0.001 and 0.5 following a normal distribution and the initial inhibitory weight was defined as 0.125 as proposed in [6], generating 3 weight distributions, one for each cuneate neuron.

2.5 Validation Index

For the validation of the model and comparison of the results two metrics were calculated, the ASR and the CV_ISI. The ASR, described in (10), shows the global rate of spikes, average frequency, of the neuromorphic model.

$$ASR = \frac{nSpikes}{T} \tag{10}$$

where nSpikes is the number of spikes present the interval T in seconds, for this paper, 6 s.

The CV_ISI gives an estimation of the time interval variation of consecutive spikes and has the formula described in (11).

$$CV\_ISI = \frac{\sigma_{is}}{\mu_{is}} \tag{11}$$

where σ_{is} is the standard deviation and μ_{is} is the mean of the vector that contains the time difference between two consecutive spikes, the spike interval of the model's binarized vector output.

It was evaluated for the three cuneate neurons if those indexes correspond to a characteristic vector capable of segregating those textures in a vectorial space, pointing that the information to detect different textures is present in their outputs. Those outputs were then applied, as seen in Fig. 2 in five different ways into a SVM classifiers, three of them with only one cuneate neuron in each (slow adaptation, fast adaptation, and multimodal), another one with the fast adaptation and the slow adaptation cuneate neurons and the last one with all 3 cuneate neurons. Their performance were validated using a leave-one-out cross validation with the 20 samples of each texture, making it possible to define which combination of neuron outputs gives a better texture classification.

3 Results

3.1 Synaptic Learning—2nd Layer

The initial synaptic excitatory weights distribution was randomly generated, using synaptic weights lower than 0.5 for each one of the cuneate neurons and applied in their 4 respective folds.

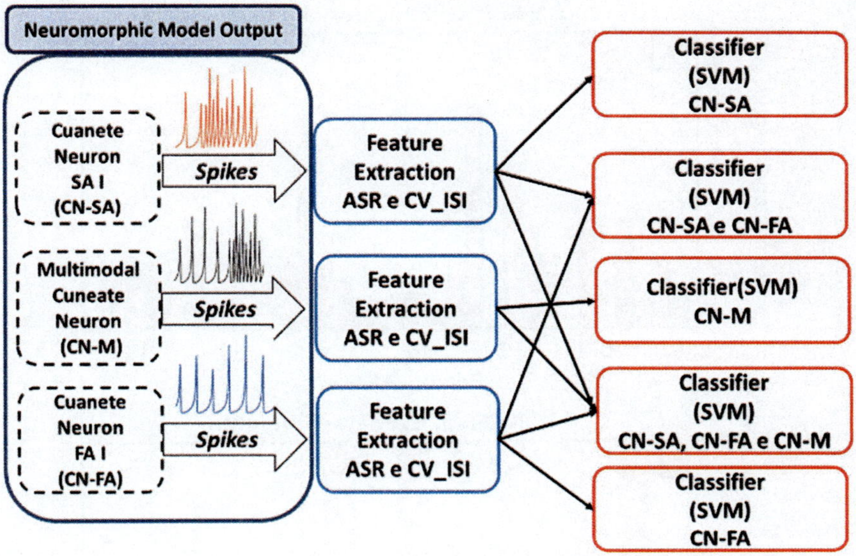

Fig. 2. Block diagram of the neural network validation.

Similar results in the 4 folds were found after training. Figure 3 show the results for the fold 4 of the 3 neurons combination. It is observed that the CN-SA had 3 afferents with significative synaptic weights, that obtained a value higher than 0.5, while the CN-M and CN-FA only had one significative synaptic weight.

For the inhibitory weights, the initial value used in all the cuneate neurons was 0.125, which resulted in a final inhibitory weight of 0.28 to the CN-SA, 0.16 to CN-FA and 0.2 to CN-M in the fold 4. Showing that CN-SA have the bigger inhibitory weight and the bigger number of significative afferent weights.

3.2 Neuromorphic Model Classification Performance

To evaluate the model performance, it was extracted the CV-ISI and the ASR of the spike vector output from the 8 textures of the 3 cuneate neurons.

Those characteristics where then used as input vectors of each cuneate neuron's SVM (CN-M, CN-SA, CN-FA). It was also used more two SVM models to evaluate the combined information carried from the spikes, one with the CN-FA and CN-SA inputs and the other one with the 3 of them. For all the SVM models it was used a leave-one-out cross validation with 20 samples of each texture. The results can be seen in Table 2 and in Fig. 4 it was generated a confusion matrix of the fold 4 from each SVM.

As seen in Table 2 the best result accomplished by a single cuneate neuron output was the SVM CN-SA with 81.41% Accuracy, the CN-FA with 65.15% and the CN-M with 69.38% accuracy. It was also seen that the better results came from the SVMs with more than one cuneate neuron data used, the SVM with the FA and the SA got a mean accuracy of 86.66% and the biggest one was the SVM that used the information of the three cuneate neurons, with a mean accuracy of 90.47%.

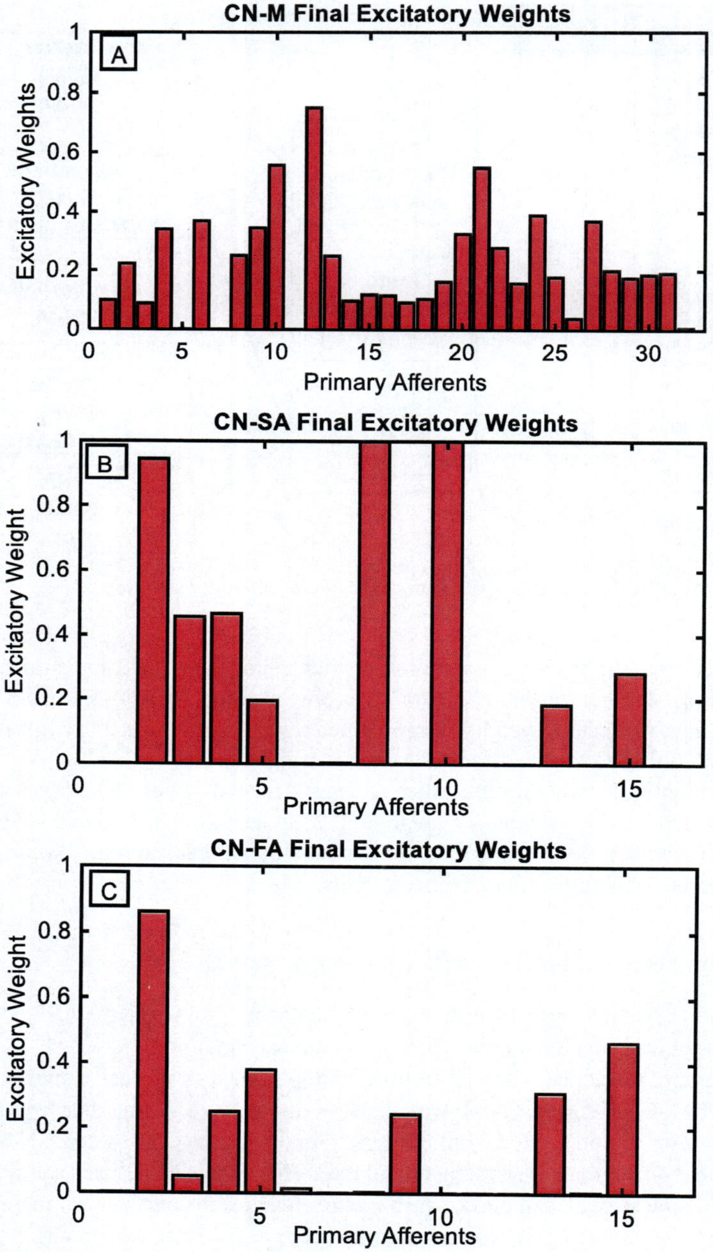

Fig. 3. Fold-4 final excitatory synaptic weights, A) multimodal cuneate neuron (CN-M), B) type 1 slowly adapting cuneate neuron (CN-SA), C) type 1 rapidly adapting cuneate neuron (CN-FA).

Table 2. SVM results

	Mean accuracy (%)				
SVM	Fold-1	Fold-2	Fold-3	Fold-4	$\mu \pm \sigma$
CN-SA	80.63	80.63	80.63	83.75	81.41 ± 1.56
CN-FA	64.38	64.38	67.50	64.38	65.15 ± 1.56
CN-M	70.63	68.75	68.13	70.00	69.38 ± 1.14
CN-SA and CN-FA	86.63	85.00	86.88	88.13	86.66 ± 1.28
CN-SA, CN-FA and CN-M	91.88	90.63	86.25	93.13	90.47 ± 2.99

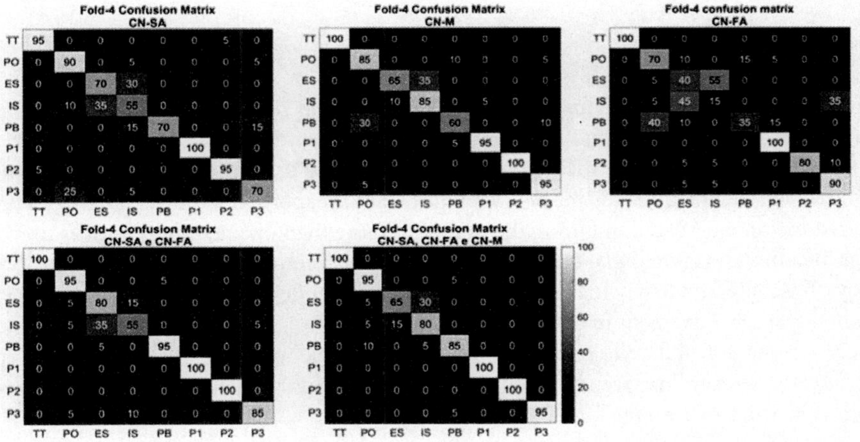

Fig. 4. Confusion matrices of the SVM classification results for the different textures

4 Results

From the models of second-order neurons implemented, the one that obtained the best performance in the classification of textures was the CN-SA, which received only SA I afferents, with average accuracy of 81.41%. The highest sensitivity in the classification and segregation of stimuli by CN-SA is in detriment of the afferents of slow adaptation having a high sensitivity to tactile signals due to the low speed of palpation of the collection (10 mm/s).

The accuracy found for CN-SA was higher than that found by [9] for the same data, these authors obtained an average accuracy of 73% using only one taxel modeled to simulate a primary afferent SA I. In addition, the biggest classification error found in [9] was for the textures: Sponge, Styrofoam, and Textile Carpet, while in our model the biggest classification error was for textures: Sponge, Styrofoam, and Rubber Floor. It should be noted that the Sponge and Styrofoam textures have similar physical characteristics, such as complacency and roughness, consequently the tactile signals of the same were similar. Which demonstrates that the model was able to perform the segregation and generalization of tactile stimuli.

In addition, the results obtained from the combination of the three cuneate neurons models obtained an average accuracy of 91.88%, and the best fold of 93.13%. These results are comparable to those found by Rongala et al. [6] when used a neural network of twenty-five cuneate neurons that obtained an average accuracy around 96% in the classification of naturalistic textures under different collection conditions with different palpation speeds. It is emphasized that in our model validation only one acquisition condition was used (speed of 10 mm/s) and that implementation of the second neural layer has a different methodology since all cuneate neurons modeled by Rongala et al. [6] received two models of primary afferents (SA I and FA I) and the authors also used noise addition in the primary afferent's spikes.

5 Conclusion

This work implements, evaluate, and compare different types of biologically inspired cuneate artificial neurons and the combination of these neurons in the naturalistic texture's segregation and classification. Our results showed that the CN-SA was the implemented neuron that obtained the best average accuracy, approximately 81.41% in texture classification, and that combining the output of more than one neuron improves texture classification, it is noted that the CN-SA and CN-FA together obtained an average accuracy of 86.66% and that using the three implemented models had an accuracy of 90.47% showing an improvement in classification.

From these results, we concluded that the implementation of more than one type of cuneate neuron improves the tactile data classification, which is in accordance with somatosensory processing.

Acknowledgment. Acknowledgment to CNPQ, CAPES, and FAPEMIG for the financial support and the Federal University of Uberlândia (UFU), the Federal Institute of the Triângulo Mineiro (IFTM) and the BIOLAB-FEELT-UFU for providing the physical structure.

Conflict of Interest. The authors declare that they have no conflict of interest.

References

1. Abraira, V.E., Ginty, D.D.: The sensory neurons of touch. Neuron **79**, 618–639 (2013)
2. Johansson, R.S., Flanagan, J.R.: Coding and use of tactile signals from the fingertips in object manipulation tasks. Nat. Rev. Neurosci. **10**, 345–359 (2009)
3. Bengtsson, F., Brasselet, R., Johansson, R.S., Arleo, A., Jörntell, H.: Integration of sensory quanta in cuneate nucleus neurons in vivo. PLoS One **8** (2013)
4. Jörntell, H., Bengtsson, F., Geborek, P., Spanne, A., Terekhov, A. V., Hayward, V.: Segregation of tactile input features in neurons of the cuneate nucleus. Neuron **83**, 1444–1452 (2014)
5. Rongala, U.B., Spanne, A., Mazzoni, A., Bengtsson, F., Oddo, C.M., Jörntell, H.: Intracellular dynamics in cuneate nucleus neurons support self-stabilizing learning of generalizable tactile representations. Front. Cell. Neurosci. **12**, 1–25 (2018)

6. Rongala, U.B., Mazzoni, A., Spanne, A., Jörntell, H., Oddo, C.M.: Cuneate spiking neural network learning to classify naturalistic texture stimuli under varying sensing conditions. Neural Netw. **123**, 273–287 (2020)
7. Suresh, A.K., Greenspon, C.M., He, Q., Rosenow, J.M., Miller, L.E., Bensmaia, S.J.: Sensory computations in the cuneate nucleus of macaques. **124**, 122333 (2021)
8. Douglas, P.R., Ferrington, D.G., Rowe, M.: Coding of information about tactile stimuli by neurones of the cuneate nucleus. J. Physiol. **285**, 493–513 (1978)
9. Gupta, A.K., Nakagawa, A., Lepora, N.F., Thakor, N.V.: Spatio-temporal encoding improves neuromorphic tactile texture classification. IEEE Sens. J. **XX**, 1–1 (2021)
10. Izhikevich, E.M.: Simple model of spiking neurons. IEEE Trans. Neural Netw. **14**, 1569–1572 (2003)

Check for updates

Computational Fluid Dynamics for Assessment of Airflow Patterns in the Airway Tree

Daniel Formariz Legendre[1,2](✉), Gustavo Sensaki Lucente[1], Vinicius Ruiz Uemoto[1,3], Helio Minamoto[4], Cristiane Yumi Koga Ito[5], Bruno Utiyama da Silva[2], Renata Valeri de Freitas[1], Paulo Francisco Guerreiro Cardoso[4], Henrique Takachi Moriya[3], and Fabio Gava Aoki[6]

[1] Adib Jatene Foundation, Dante Pazzanese Institute of Cardiology, São Paulo, Brazil
daniel@fajbio.com.br
[2] Department of Bioengineering, Sao Judas University, Sao Paulo, Brazil
[3] Biomedical Engineering Laboratory, University of Sao Paulo, Sao Paulo, Brazil
[4] Department of Thoracic Surgery, InCor-HCFMUSP, Sao Paulo, Brazil
[5] Institute of Science and Technology, UNESP, Sao Jose dos Campos, Brazil
[6] Institute of Science and Technology ICT, UNIFESP, Sao Jose dos Campos, Brazil

Abstract. Tracheal stenosis lesions have a negative impact on patients' quality of life. The incidence of post-intubation tracheal stenosis ranges from 6 to 20%, and after tracheostomy from 0.6 to 21%. Since the symptoms are only noticeable after the presence of a 70% reduction in the tracheal lumen, early diagnosis is not always possible. Computed tomography (CT) scan can be used to assess stenosis even before symptoms appear, however it is unable to provide dynamic data (e.g., airflow and pressure patterns) that would be important in early diagnosis, as well as for prognostic assessment and therapeutic planning. Computational Fluid Dynamics (CFD) is a non-invasive analysis that allows computer simulations of airflow using reconstructions based on CT scan images. We hypothesize that CFD simulation can be a feasible tool to assess the fluid-mechanical behavior within the tracheal stenosis. Results of the streamlines allowed us to understand the flow pattern present along the trachea. In addition, it was possible to identify the occurrence of vortex in the proximal region of the trachea that precedes the stenosis and in the left bronchus also affected by stenosis. The assessment of pressure gradient and WSS values may provide additional information for risk estimating to the tracheal wall.

Keywords: Tracheal stenosis · CT scan · Image segmentation and computational fluid dynamics

1 Introduction

The main functions of the trachea include the conduction of air from the external environment to the alveoli, filtering, heating and humidifying the air taken to the lungs, in order to protect the organism from microorganisms and particles dispersed in the air

J. L. B. Marques et al. (Eds.): CLAIB 2022/CBEB 2022, IFMBE Proceedings 98, pp. 210–222, 2024.
https://doi.org/10.1007/978-3-031-49401-7_21

during breathing [1]. The most common benign and malignant tracheal disorder is tracheal stenosis, which is a segmental narrowing of the trachea that can result in dyspnea, reduced exercise capacity and stridor. The degree of tracheal stenosis can range from mild to severe. Tracheal stenosis can be congenital, neoplastic, or secondary to benign lesions, among which post-intubation stenosis is the most frequent in our country [2–4].

Treatment may involve resection of the stenotic area and reconstruction of the airway continuity [5, 6]; however, surgical treatment is restricted to conditions that affect less than 50% of the tracheal length in adults and 30% in children.

In cases in which resection is not possible, either due to the extension of the stenosis or due to patient comorbidities, dilatations and placement of endoluminal prostheses aim to maintain airway permeability. In these situations, monitoring the diameter in the stenotic area can help in making a decision about the most appropriate time to perform more invasive procedures [7–9].

The CT (Computed Tomography) image allows the confirmation of the disease, even before the appearance of symptoms. However, diagnostic imaging by itself is unable to provide dynamic data (e.g. flow and pressure patterns) that are important in early diagnosis, management decision and prognostic assessment.

Computational Fluid Dynamics (CFD) is the method that uses numerical analysis to solve flow problems. It allows evaluating the behavior of the airflow in the trachea by providing information of clinical interest that can help in the evaluation of the evolution of the stenosis, as well as in the therapeutic decision-making and the effectiveness of the treatment.

Brouns et al. [10] studied how tracheal stenosis artificially inserted into a realistic upper and tracheal airway structure affects local pressure drops in with different degrees of tracheal constriction. They explore how the increase in pressure drop could be detected at an early stage by using computational fluid dynamics (CFD) that are consistent with the clinically observed sudden appearance of breathing impairment at a relatively advanced stage of tracheal obliteration. Also, they propose a simple equation to quantify pressure drops generated by different degrees of stenosis, providing that breathing flow and stenosis cross section can be reliably estimated.

Other authors also used CFD analysis to evaluate tracheal stenosis. Mylavarapu et al. [11] performed CFD study in the process of planning surgical treatment in order to select the most optimal one that shows most promise. Xu et al. [12] used to model the effect of airway geometry on internal pressure in the upper airway of three children with obstructive sleep apnea syndrome. It helps the surgeon during the decision-making process of a proposed surgery with a more scientific rationale, in faster and affordable ways.

The present work aims to non-invasively evaluate the detailed characteristics of the airflow during the respiratory cycle being calculated by means of transient computational fluid dynamics simulation. It is an important tool that can be used in order to understand the fluid-mechanical behavior within the tracheal stenosis that is not easily possible through evaluations in clinical practice.

The results will be presented in terms of spatial and temporal distribution of pressure fields, flow velocity and wall shear stress based on CT scan images.

2 Materials and Methods

2.1 CT Image Acquisition and Processing

Anonymized CT image (reconstructed into a 512 × 512 matrix with a resolution of 0.405 × 0.405 mm^2 and slice thickness of 0.8 mm) acquired in a clinical routine examination of one patient, 26 years old, female, with tracheal and left bronchus stenosis caused by tumor was selected out from image database for retrospective study. The study had the approval of the Medical Ethics Committee of InCor—Clinical Hospital: Faculty of Medicine—USP (CAAE: 52501821.4.0000.0068). Written informed consent was collected from this subject.

In the present study, the three-dimensional model of the stenotic airway tree (Fig. 1) was delimited from the height of the subglottis (an anatomical subregion of the larynx) to the branch just below the carina, to facilitate segmentation of the CT image and due to the high computational cost. In the segmented model, the trachea divides into the main bronchi (primary bronchi) at the carina, with the right bronchus wider, shorter and more vertical than the left bronchus. The right main bronchus bifurcates posteriorly into the right superior lobe bronchus.

The degree of constriction was evaluated by measuring the cross-sectional area of the healthy proximal region of the trachea compared to the region of greatest constriction [13]. Based on this definition, a stenosis ratio of 86.96% was found in the patient-specific model.

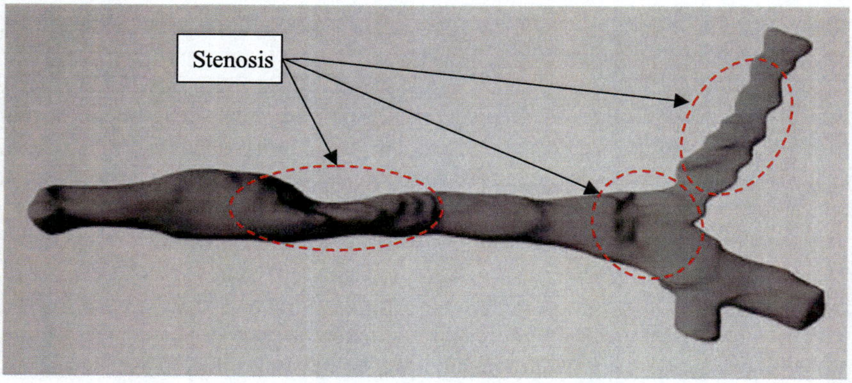

Fig. 1. Segmented trachea with stenosis extracted from CT images

The CT scan image in DICOM format was imported into the specialized computer platform for image analysis (3D Slicer) [14], with the objective of performing the segmentation and reconstruct the 3D tracheobronchial geometry, respecting the scale 1 to 1 to preserve the actual size of the model (Fig. 2).

In this work, image segmentation was performed manually by iterating through all the slices of a CT scan images and drawing a contour at the boundary, and in a semi-automatic way method using the grow from seeds and scissors tools. The segmented model was saved into surface mesh with a format of STL (Standard Tessellation Language) and imported into computer numerical simulation platform (Workbench 2022R1, ANSYS Inc., USA). The simulation was performed on an workstation (Intel Xeon W-2133, 32GB RAM and NVIDIA QUADRO P4000 graphic card).

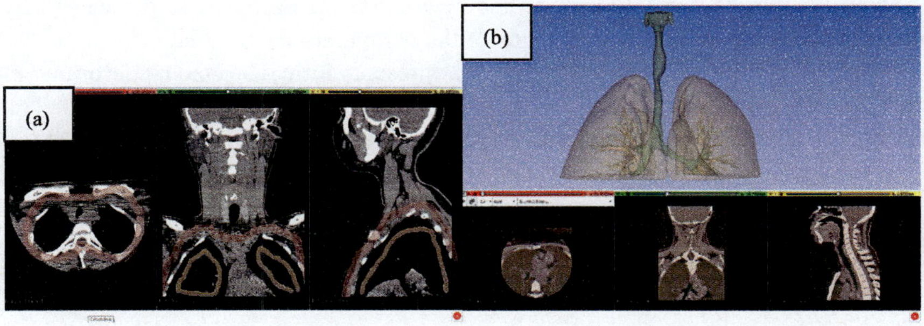

Fig. 2. Segmentation process (a) and creation of three-dimensional model of isolated airway tree (b).

2.2 CFD Procedures and Boundary Conditions

The SpaceClaim module was used to correct any problems in the triangle mesh that forms its geometry and to convert the faceted body into a solid model.

The geometry previously treated in the SpaceClaim module was imported into the Fluent environment, where the mesh was created.

The governing equations for flows, in any regime and with any value for the Reynolds number, are the Navier-Stokes (N-S) equations, which are nonlinear second-order partial differential equations, based on the principles of conservation of mass, momentum and energy.

Flow will generally be laminar for $Re \leq 2300$ and turbulent for larger values [15]. In the study, Re was determined to verify the expected behavior of the flow according to the diameter of the highest ring of the proximal trachea and the average inlet velocity of the flow at this cross section. The Reynolds number ranged from 0 to 2531 at inspiration and from 0 to 3050 at expiration. Since the Reynolds number was greater than 2500 a transient state and turbulent flow solver model was used in the simulation.

The SST k-ω (Shear-Stress Transport k-ω) model was used in Fluent environment, which is widely adopted in cases with boundary layer separation. This turbulence model has been used in several computational fluid dynamics studies [16, 17]. In Fluent environment, control equations were solved using a transient input-based solver. The SIMPLE algorithm (semi-implicit method) was adopted for the pressure-velocity coupling, which is the standard for transient simulations. In the spatial discretization of the moment, turbulent kinetic energy, turbulent dissipation rate and energy, "First Order Upwind" was used.

A residual of 1E−6 was adopted as the convergence criterion, a time step size of 0.005 s, a time step number of 1300 and the simulation results saved every 0.05 s at flow time.

The mesh refinement process is important to increase the accuracy of the solution. However, it may produce an excessive number of elements that causes a strong impact in terms of computational cost, demanding a very long time for simulation. The total number of mesh elements was 505,270 for the segmented geometry.

The mesh can be measured by geometric metrics for each cell in the mesh and it influences the accuracy and convergence of the computational simulation.

A good mesh is considered to be one that presents little distorted or deformed elements. This study used the element orthogonality as a conceptual metric to evaluate the mesh quality and its impact on obtaining an accurate solution. The minimum Orthogonal Quality is 0.500 for the whole model (Fig. 3).

A local refinement technique was applied as boundary layer to evaluate critical flow region such as near-wall boundaries. The computational model was composed of a poly-hexcore mesh elements, which means, a mesh with a polyhedral surface and hexahedral volume elements (Fig. 4a, b).

We assumed air to be a Newtonian fluid with a constant density (ρ) of 1.225 kg/m^3 and dynamic viscosity (μ) of 1.7984 $\times$ 10^{-5} kg/m.s. These properties were provided by the Fluent library for standard atmosphere [18, 19].

The respiratory profile applied at the trachea inlet was obtained by measuring the airflow inhaled and exhaled from the lungs at rest in supine position of a healthy volunteer (Fig. 5).

This breathing profile was applied through transient table file as inlet boundary condition.

The airflow rate at trachea was set to be negative for expiration and positive for inspiration, the pressure at the distal bronchi was set as the standard atmosphere and non-slip condition on the model wall was adopted as a boundary condition. A room temperature of 25 °C and a physiological temperature of 36.5 °C were also adopted for the study.

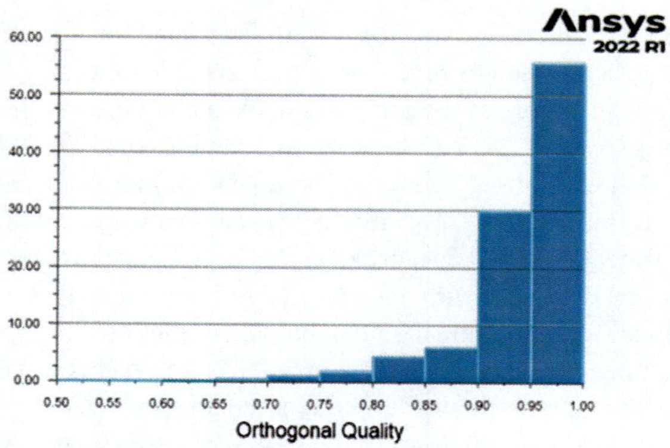

Fig. 3. Histogram of orthogonal mesh quality.

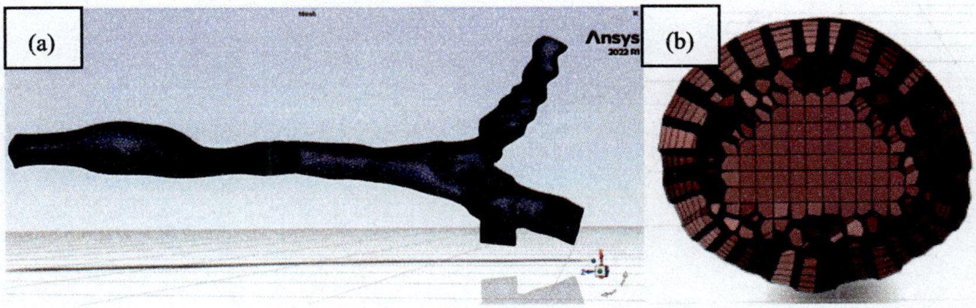

Fig. 4. Computational model mesh composed with a polyhedral surface (a) and hexahedral volume (b) elements.

3 Results

3.1 Airflow Behavior

Airflow behavior within the breathing cycle can be evaluated through transient CFD analysis.

The results were showed in terms of spatial and temporal distribution of pressure, flow velocity and wall shear stress.

All results were presented at times of 1.135 s (t1) and 4.54 s (t2) which represent, respectively, the peaks of inspiration and expiration velocities.

3.2 Streamlines, Flow Velocity, Wall Pressure and Wall Shear Stress

Streamlines showed at inspiration phase (time of 1.135 s) a uniform velocity pattern along the subglottic larynx and after the stenosis site up to the right bronchus (Fig. 6a, b).

However, it was possible to verify a secondary flow vortex in the proximal trachea upstream to the stenosis, caused by a sudden airway expansion from larynx to trachea, and also in the left main bronchus which exhibit constrictions and shape changes caused by stenosis.

Velocity increased with decreasing diameter of the trachea, with a maximum value in the region after stenosis.

Streamlines showed at expiration phase (time of 4.54 s) virtually the same behavior as in the inspiration phase (Fig. 7a, b), however, the flow went from bronchi towards the glottis. The image showed vortices in the same place as in the inspiration phase, but with different intensity. Also, it presented velocity peaks after shape and diameter changes in the region of stenosis (Fig. 7c).

Figure 8a, b showed the cross-sectional plane colored by the normal velocity through them at the instant of time 1.135 s (a) and 4.54 s (b). It can be correlated with the streamlines patterns through the airway tree. In both cases, it was possible to verify the airflow low velocities in the recirculation zones, as well as high velocities at the site of tightest stenosis due to the constriction influence.

Wall pressure decreased from the subglottic larynx inlet to bronchial outlets during the inspiration phase (Fig. 9) and, on the other hand, from the main bronchi to the

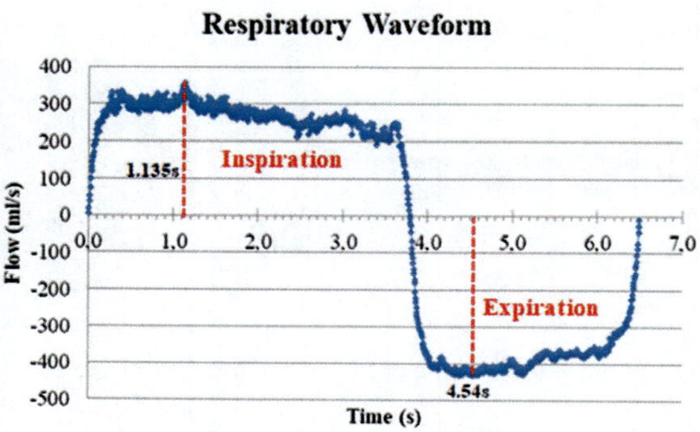

Fig. 5. Breathing profile in a respiration cycle at rest.

subglottic larynx during the expiration phase (Fig. 10). Pressure behaviors were different according to the instant time and airflow direction.

At inspiration phase, higher pressures appeared in the proximal region of the airway tree because the stenosis increased the flow resistance and hence inlet pressure. On the contrary, at expiration time, the pressure was higher at bronchi and decayed along the trachea with minimum value at larynx.

The tangential forces exerted to the tracheobronchial wall are characterized by the WSS (Wall Shear Stress). At low Reynolds with laminar flow condition, a small WSS is achieved due to the undisturbed boundary layer. However, in the airway tree model under CFD investigation, the airflow was accelerated due to the stenosis sites and the flow pattern was disturbed because of the complex geometry leading to turbulence condition which disrupts the boundary layer and results in higher WSS.

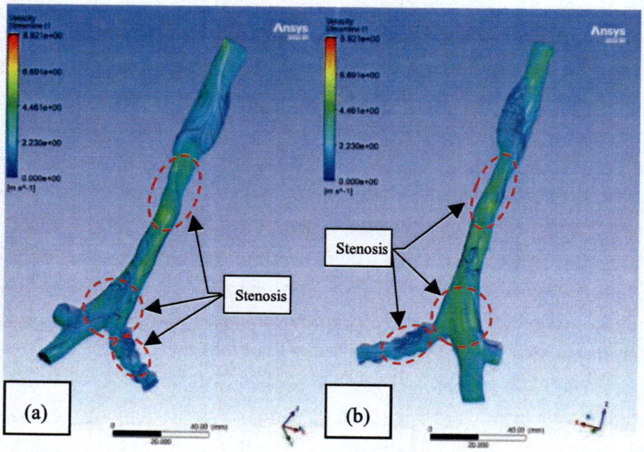

Fig. 6. Anterior (a) and posterior (b) view of the airway tree at the instant of time 1.135 s.

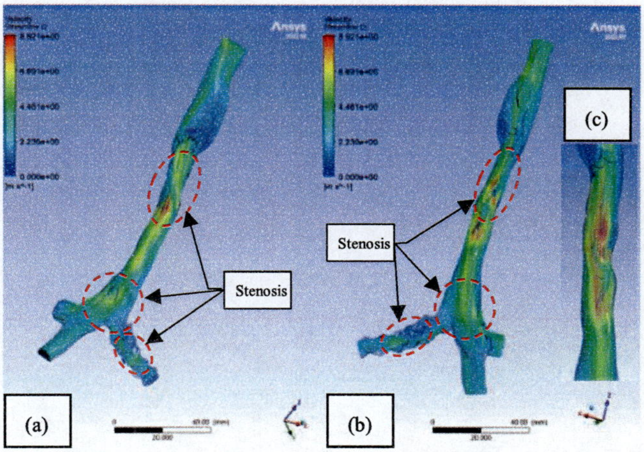

Fig. 7. Anterior (a) and posterior (b) view of the airway tree at the instant of time 4.54 s and a close view of the high velocity region (c).

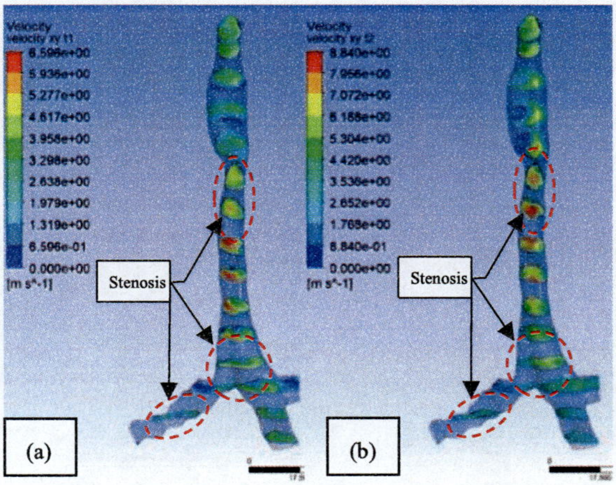

Fig. 8. Posterior view of the airway tree with cross sectional plane colored by the normal velocity through them at the instant of time 1.135 s (a) and 4.54 s (b).

Higher WSS were found at the stenosis sites regardless whether it was at the trachea or at the bronchi. It was also noted high WSS at the carina where the flow bifurcates into the bronchi (Figs. 11 and 12).

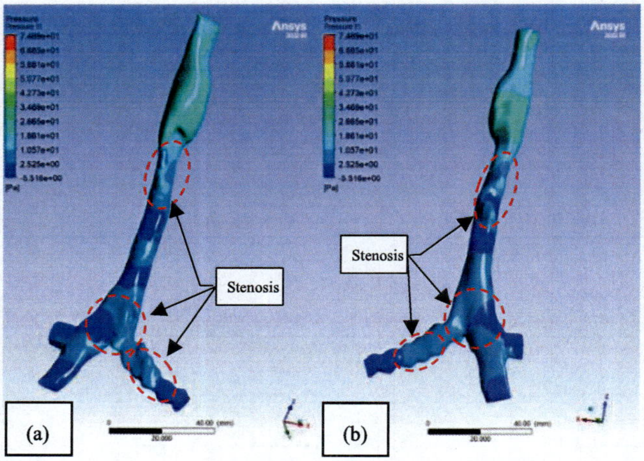

Fig. 9. Anterior (a) and posterior (b) view of the airway tree showing the wall pressure behavior during the inspiration phase.

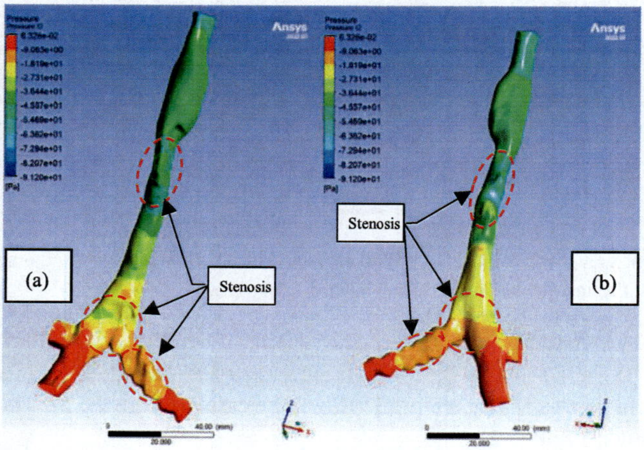

Fig. 10. Anterior (a) and posterior (b) view of the airway tree showing the wall pressure behavior during the expiration phase.

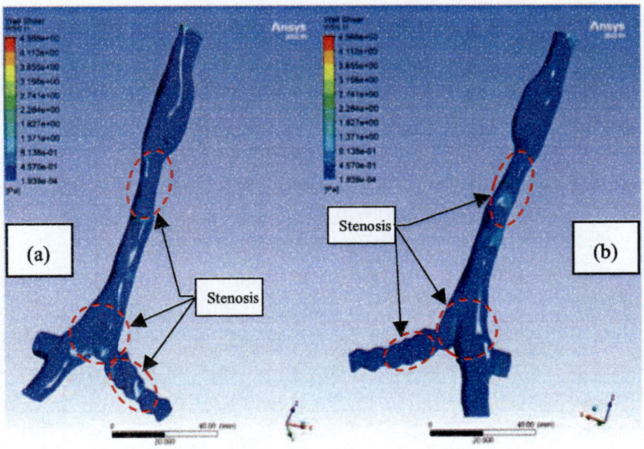

Fig. 11. Anterior (a) and posterior (b) view of the airway tree showing the wall shear stress behavior during the inspiration phase.

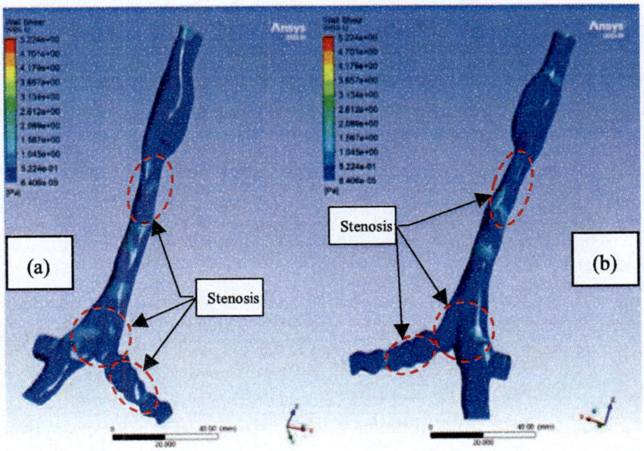

Fig. 12. Anterior (a) and posterior (b) view of the airway tree showing the wall shear stress behavior during the expiration phase.

4 Conclusions

The present work showed the feasibility of using computed tomography images of the airways tree obtained from patient with tracheal stenosis to perform a CFD study.

The use of computational fluid dynamics as an aid tool for medical evaluation allowed us to understand the spatial and temporal patterns of airflow in the tracheobronchial region of a patient suffering from tracheal stenosis due to the presence of a tumor.

This study considered a respiratory pulse acquired from a normal subject without stenosis in a rest condition. The model reconstructed to this work considered only the larynx, trachea and three main bronchi, although, including higher generation's bronchi might give more accurate information. Also, the tracheobronchial wall was assumed to be rigid. The results are based on CT image of one unique volunteer with tracheal stenosis and this prevents a comprehensive validation of the flow pattern in different kinds of tracheal stenosis.

However, there was a clear association between the pressure drop and stenotic site that could be understood as higher degree of respiratory resistance and effort in tracheal stenosis. The greater pressure gradient is associated with increased resistance caused by stenosis.

There is a relationship between the trajectory lines, velocity and pressure fields. Based on the computational results presented, it was possible to evaluate velocity and pressure field both temporally and spatially according to the respiratory cycle. Velocities fall in regions where vortices occur and increase in constriction sites, and it was related to geometric changes present in the trachea.

According to Qi et al. [20], large WSS variations in blood vessels may increase the risk of arteriosclerosis. In the same way, high wall shear stress at the stenotic airway could be related with the trauma of the mucosa and an inflammatory response. This WSS can affect the structural integrity of the tracheal wall.

Results of the streamlines allowed us to understand the flow pattern present along the trachea. In addition, it was possible to identify the occurrence of vortex in the proximal region of the trachea that precedes the stenosis and in the left bronchus also affected by stenosis.

It was also noted that, despite the presence of vortices, there was a realignment of the streamlines in most of the trachea and in the right bronchus, where the flow settles again without disturbances.

The assessment of pressure gradient and WSS values may provide additional information for risk estimating to the tracheal wall. In addition, it should be associated with clinical outcome in order to provide better understanding related to this disease.

On the whole, all works, including this one, aim to prove the robustness of CFD study in the analysis of respiratory flow patterns and also to be used as a powerful non-invasive tool to support surgeons during the decision-making process. In general, tracheal stenosis is asymptomatic until reaching a critical narrowing of the airway lumen, from 75% or more, and the use of CFD tool could prevent critical stage which requires immediate intervention.

Acknowledgment. This study was financed in part by the Coordenacao de Aperfeicoamento de Pessoal de Nivel Superior—Brazil (CAPES)—Finance Code 001 and the Conselho Nacional de Pesquisa e Desenvolvimento Científico e Tecnologico—Brazil (CNPq) (308280/2019-9 to HTM).

Conflict of Interest. We do not have any conflict of interest.

References

1. Furlow, P.W., Mathisen, D.J.: Surgical anatomy of the trachea. Ann. Cardiothorac. Surg. **7**(2), 255–260 (2018)
2. Leske, V. et al: Tracheobronchopathia osteochondroplastica: a study of 41 patients. Medicine (Baltimore), Philadelphia **80**(6), 378–390 (2001)
3. Marom, E.M., et al.: Diffuse abnormalities of the trachea and main bronchi. Am. J. Roentgenol. Leesburg **176**(3), 713–717 (2001)
4. Prince, J.S., et al.: Nonneoplastic lesions of the tracheobronchial wall: radiologic findings with bronchoscopic correlation. RadioGraphics, Oak Brook **22**(1), 215–230 (2002)
5. Wood, D.E., Thomas, C.R.: Mediastinal tumors. Springer, Berlim (1995)
6. Mandour, M., et al: Chronic subglottic and tracheal stenosis: endoscopic management vs. surgical reconstruction. Eur. Arch. Oto-Rhino-Laryngol. Leiden **260**(7), 374–380 (2003)
7. Dos Anjos, D.M., et al.: Endopróteses: opção terapêutica para estenoses traqueobrônquicas adquiridas. Revista de Medicina, São Paulo **86**(3), 174–184 (2007)
8. Grillo, H.C., et al.: Postintubulation tracheal stenosis: treatment and results. J. Thorac. Cardiovasc. Surg. Beverly **109**(3), 486–493 (1995)
9. Minamoto, H., Terra, R.M., Cardoso, P.F.G.: Estenoses Benignas da Via Aérea. Pulmão RJ, Rio de Janeiro **20**(02), 48–53 (2011)
10. Brouns, M., et al.: Tracheal stenosis: a flow dynamics study. J. Appl. Physiol. **102**(3), 1178–1184 (2007)
11. Mylavarapu, G., et al.: Planning human upper airway surgery using computational fluid dynamics. J. Biomech. **46**(12), 1979–1986 (2013)
12. Xu, C., et al.: Computational fluid dynamics modeling of the upper airway of children with obstructive sleep apnea syndrome in steady flow. J. Biomech. **39**(11), 2043–2054 (2006)
13. Bates, A.J., Cetto, R., Doorly, D.J., Schroter, R.C., Tolley, N.S., Comerford, A.: The effects of curvature and constriction on airflow and energy loss in pathological tracheas. Respir. Physiol. Neurobiol. **234**, 69–78 (2016)
14. Kikinis, R., Pieper, S. D., Vosburgh, K.: 3D slicer: a platform for subject-specific image analysis, visualization, and clinical support. In: Jolesz, F. (ed.) Intraoperative imaging and image-guided therapy. Springer, New York (2014). https://doi.org/10.1007/978-1-4614-7657-3_19
15. Pritchard, P. J., Mitchell, J. W.: Fox and McDonald's introduction to fluid mechanics (9th ed.). Wiley (2015)
16. Mimouni-Benabu, O., et al.: A preliminary study of computer assisted evaluation of congenital tracheal stenosis: a new tool for surgical decision-making. Int. J. Pediatr. Otorhinolaryngol. **76**(11), 1552–1557 (2012)
17. Gemcia, T., et al.: Computational model of airflow in upper 17 generations of human respiratory tract. J. Biomech. **41**, 2047–2054 (2008)
18. ANSYS, Inc.: ANSYS Fluent User's Guide, Release 2021R2 (2021)

19. Qi, S. et al.: Transient dynamics simulation of airflow in a CT-scanned human airway tree: more or fewer terminal bronchi? Comput. Math. Methods Med. Article ID 1969023, 14p (2017)
20. Qi, S., et al: Computational fluid dynamics simulation of airflow in the trachea and main bronchi for the subjects with left pulmonary artery sling. BioMed. Eng. OnLine **13**, 1–85 (2014)

Molecular Docking Prediction Between Anthocyanins Reported for Andean Berry (*Vaccinium meridionale Swartz*) with Proteins Involved in the Apoptosis

Myriam Liliana Agudelo-Quintero(✉), Sandra Arango Varela, and Jorge Alejandro Lopera-Rodríguez

Grupo de Investigación e Innovación Biomédica GI2B, Instituto Tecnológico Metropolitano, Medellin, Colombia
myriamagudelo305350@correo.itm.edu.co

Abstract. Anthocyanins are polyphenols contained in berries. There is evidence that act as possible modulators in cell death mechanisms like apoptosis. Cancer is a relevant scenario where modulate apoptosis can be relevant. For this reason, in this study it was evaluated the binding molecular affinity with apoptotic proteins of some anthocyanins from *Vaccinium meridionale Swartz* berry to identify potential antitumoral biological compounds. Cyanidin 3-glucoside, cyanidin 3-galactoside, and delphinidin 3-glucoside structures were downloaded from PubChem and crystalographic structures of apoptosis proteins were obtained from Protein Data Bank. Docking experiments *in silico* were implemented with structures prepared. Cyanidin 3-glucoside and cyanidin 3-galactoside showed the best affinity with FADD/fas protein while delphinidin 3-glucoside with procaspase 3 protein. These results of affinity suggest a potential proapoptotic activity related with these anthocyanins.

Keywords: Molecular affinity · Apoptosis modulators and anthocyanins

1 Introduction

Cell death is a fundamental biological process in the development of multicellular organisms and tissue homeostasis. Cells are removed during metamorphosis, embryogenesis, pathogenesis, and tissue renewal, when this process increases or decreases, it is associated with different diseases such as cancer [1]. Apoptosis is a regulated cell death mechanism that affects individual cells in living tissues without an inflammatory reaction associated. This occurs in response to cell damage or external stress, and during normal development and morphogenesis [2]. Apoptosis induction is an important mechanism to maintain the balance between survival and death in cells, to prevent cancer and other related diseases. In this context, it is considered one of the most important targets in the preventive and therapeutic approach of cancer [3].

During apoptosis process a signaling pathway occurs involving massive activation of caspases. Intrinsic or mitochondrial apoptosis occurs because of events in the intracellular environment, the critical point is the permeabilization of the mitochondrial outer

J. L. B. Marques et al. (Eds.): CLAIB 2022/CBEB 2022, IFMBE Proceedings 98, pp. 223–232, 2024.
https://doi.org/10.1007/978-3-031-49401-7_22

membrane. This process is controlled by a group of pro and antiapoptotic proteins of the Bcl-2 family [4]. Single domain BH3 proteins can induce apoptosis by direct activation of Bax and Bak or by inactivation of other antiapoptotic proteins of the Bcl-2 family. Bax and Bak can in turn be activated by the tumor suppressor protein p53. Bax and Bak activated induce the assembly of a pore that alters mitochondrial permeability and causes important rearrangements in the mitochondrial superstructure. This promotes the proapoptotic factors releasing from the intermembrane space of mitochondrial towards the cytosol. Factors such as cytochrome C bind to the adaptor protein Apaf-1 and procaspase 9, consolidating a molecular complex called apoptosome, which is responsible for the activation of procaspase 9 that catalyzes the proteolytic activation of procaspases 3 and 7. These are known as effector caspases, and they are responsible for carrying out the final stage of the apoptotic process [5].

Extrinsic apoptosis is triggered by disturbances in the extracellular environment. This type of cell death is regulated by various death receptors, whose activation depends of ligand binding. Among the death receptors that activate this pathway are the Fas surface receptors (also known as CD95 or APO-1), and some members of the TNF superfamily such as receptor 1A better known as TNFR1, the receptor 10A known as DR4 and 10B known as DR5 [6].

Apoptosis initially appears as a disturbance in the cytoskeleton, with subsequent chromatin condensation, nuclear fragmentation, and loss of cell membrane integrity, ending with the formation of apoptotic bodies that can be easily phagocytosed by macrophages, avoiding an inflammatory response [7].

On the other hand, anthocyanins are polyphenols with chemopreventive properties in some types of cancers such as breast, colon, prostate [8], liver [9], among others. These properties have been evaluated in multiple studies. It has been established that they can act through different biological mechanisms such as the activation of apoptosis in cancer cells by different pathways, avoiding the phases of tumor promotion/progression and disruption of the angiogenesis process [10].

In silico docking evaluation can predict interactions between important biological targets with ligands to study and development of chemical therapies for diseases like cancer.

In this study, the molecular binding affinity between anthocyanins present in *Vaccinium meridionale Swartz* with pro and antiapoptotic proteins was evaluated, to identify compounds that act as possible modulators of apoptosis and therefore with potential antitumor activity.

2 Materials and Methods

The targets presented in Table 1 are proteins involved in human apoptosis pathways. The crystallographic structures of the targets were downloaded from the Protein Data Bank (PDB) [11] and European Protein Data Bank (PDBe) [12]. The structures with the best resolution (< 3 Å) and R values (near to zero) were selected. The ligands presented in Table 2 were obtained from PubChem database [13]. The ligands are anthocyanins reported in Andean berry *Vaccinium meridionale Swartz* with apoptotic activity in colon adenocarcinoma cell [14–16].

The molecules preparation was implemented in Chimera 1.15 software [17]. Previously the 2D structure of ligands was prepared to obtain the lowest energy conformation with MMFF94 algorithm using Avogadro 1.2 software [18]. For the molecular docking preparation, solvents and ligands were removed from the protein structures. Next, hydrogens and electrical charges were added in proteins and anthocyanins ligands. The next changes were done: amino acid residues of selenomethionine (MSE) to methionine (MET) and nucleic acids bromo-UMP (5BU) to UMP (U), methylselenyl-dUMP (UMS) to UMP, and methylselenyldCMP (CSL) to CMP (C). The protonation states were considered for histidine residues using the AMBER ff14SBand AM1-BCC algorithms. Then, the molecular docking experiment was performed with Autodok Vina [19]. The docking box was positioned in all protein and reported active site in PDB [20–24].

Table 1. Targets

Protein	PDB code	Resolution (Å)
Bcl-2	2VM6	2.20
Caspase 3	3KJF	2.00
Cytochrome C	3ZCF	1.65
FADD	3EZQ	2.73
Pro-Caspase-3	2J30	1.40

Table 2. Ligands

Ligand	CID PubChem
Cyanidin 3-glucoside (C3glu)	44256700
Cyanidin 3- galactosidase (C3gala)	44256715
Delphinidin 3-glucoside (D3glu)	443650

3 Results

Molecular affinity values obtained of global docking is presented in Tables 3, 4 and 5 and addressed docking in Tables 6, 7 and 8.

The best molecular affinity value for all ligands was obtained to global docking. The best interaction is presented in Figs. 1, 2 and 3.

The best molecular affinity value for all ligands is presented in Figs. 4, 5 and 6.

The type to binding is presented in Fig. 7.

Table 3. Molecular affinity between cyanidin 3-glucoside of global docking.

Protein	C3glu	SD
Bcl-2	− 7.57	0.058
Caspase 3	− 7.50	0.000
Cytochrome c	− 7.10	0.100
FADD/Fas	− 8.03	0.208
Pro-caspase 3	− 7.33	0.289

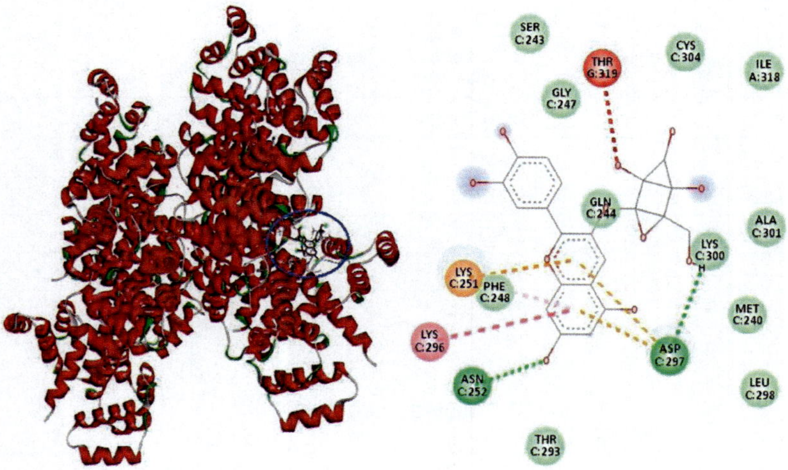

Fig. 1. Molecular interaction between Cyanidin 3-glucoside and FADD/Fas.

Table 4. Molecular affinity between cyanidin 3-galatosidase of global docking.

Protein	C3gala	SD
Bcl-2	− 7.77	0.058
Caspase 3	− 7.57	0.208
Cytochrome c	− 7.23	0.289
FADD/Fas	− 8.17	0.115
Pro-caspase 3	− 7.70	0.000

4 Discussion

In this study, the affinity of apoptosis proteins with anthocyanins present in the *Vaccinium meridionale Swartz* berry was evaluated through molecular docking tools. The anthocyanins included were cyanidin 3-glucoside, cyanidin 3-galactosidase, and delphinidin

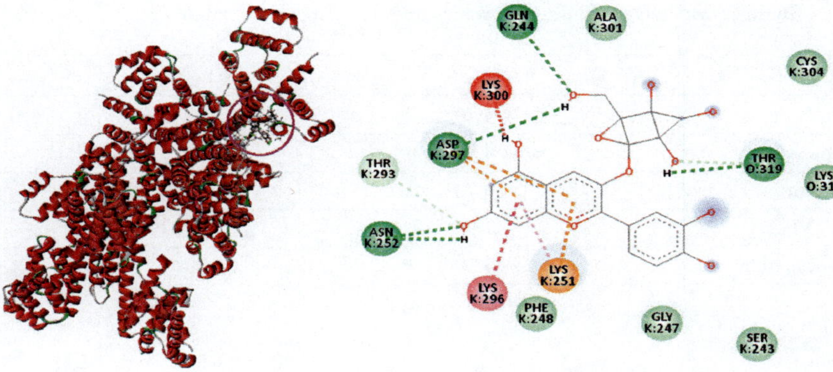

Fig. 2. Cyanidin 3-galactoside and FADD/Fas.

Table 5. Molecular affinity between delphinidin 3-glucoside of global docking.

Protein	D3glu	SD
Bcl-2	– 7.37	0.153
Caspase 3	– 8.10	0.173
Cytochrome c	– 7.67	0.231
FADD/Fas	– 7.90	0.173
Pro-caspase 3	– 8.20	0.000

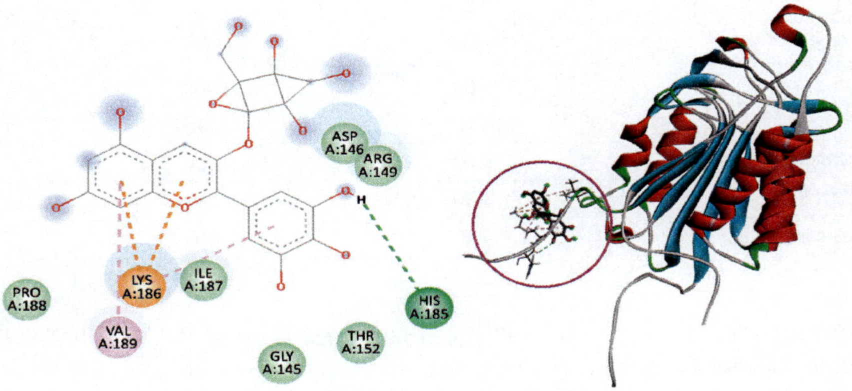

Fig. 3. Delphinidin 3-glucoside and procaspase 3.

3-glucoside. For both cyanidin 3-glucoside and cyanidin 3-galactosidase, the greater affinity was found with the FADD/Fas death protein complex, coinciding with the same interaction site and only differing with glutamine 244, lysine 300, and threonine 319. On the other hand, delphinidin 3-glucoside showed higher affinity with pro-caspase 3 with

Table 6. Molecular affinity between cyanidin 3-glucoside of addressed docking.

Protein	C3glu	SD
Bcl-2	− 5.70	0.173
Caspase 3	− 4.90	0.000
Cytochrome c	− 6.70	0.000
FADD/Fas	− 7.90	0.346
Pro-caspase 3	− 5.10	0.000

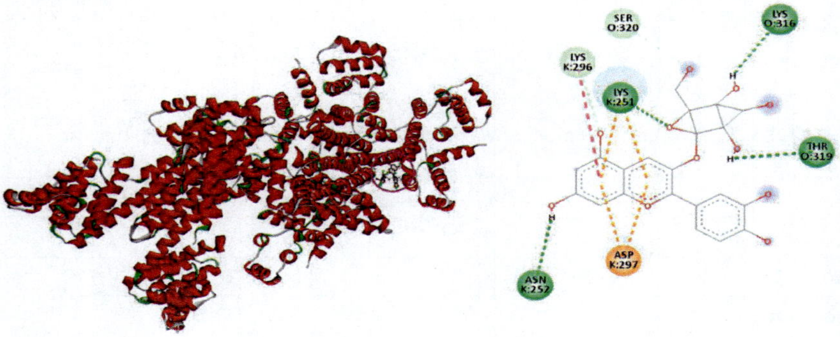

Fig. 4. Cyanidin glucoside and FADD/Fas.

Table 7. Molecular affinity between cyanidin 3-galatosidase of addressed docking.

Protein	C3gala	SD
Bcl-2	− 5.60	0.000
Caspase 3	− 4.70	0.115
Cytochrome c	− 8.80	0.000
FADD/Fas	− 8.30	0.058
Pro-caspase 3	− 5.40	0.000

amino acids histidine 185, lysine 186, isoleucine 187, and valine 189. Both FADD/Fas and pro-caspase 3 are involved in promoting apoptosis. Studies with molecular dynamic are necessary to determine activation or other activity, because this protein complex can initiate the apoptosis extrinsic pathway [23].

A relevant affinity with Bcl-2 was also identified, which may suggest that there is a potential proapoptotic role by interfering with the inhibition of apoptosis.

The effects of cyanidin 3-glucoside on apoptosis have been associated with the promotion of apoptosis in glioblastoma cells [25] and tumor growth inhibition, caspase 3 activation, chromatin condensation, and cell death in HS578T breast cancer cells [26].

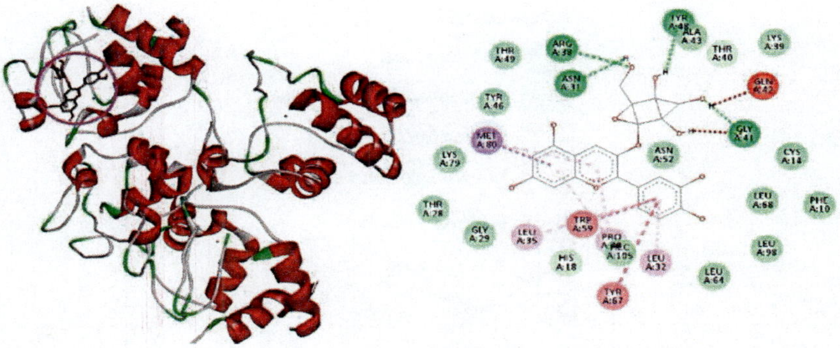

Fig. 5. Cyanidin 3-galactosidase and Cytochrome c.

Table 8. Molecular affinity between delphinidin 3-glucoside of addressed docking.

Protein	D3glu	SD
Bcl-2	– 5.67	0.153
Caspase 3	– 5.63	0.473
Cytochrome c	– 6.33	0.115
FADD/Fas	– 7.80	0.000
Pro-caspase 3	– 5.20	0.000

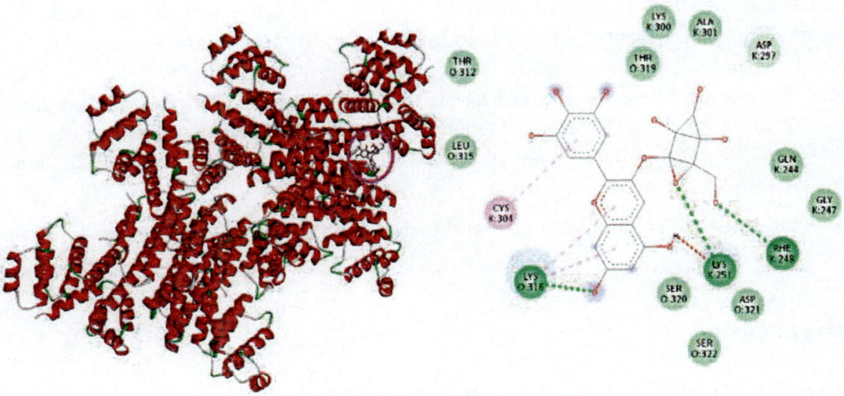

Fig. 6. Delphinidin 3-glucoside and FADD/Fas.

Although in the present study it was not possible to identify the effect of the interaction of cyanidin 3-glucoside on pro-caspase 3, previous studies have suggested that this role can be apoptosis activator [27]. However, additional studies are required to confirm this.

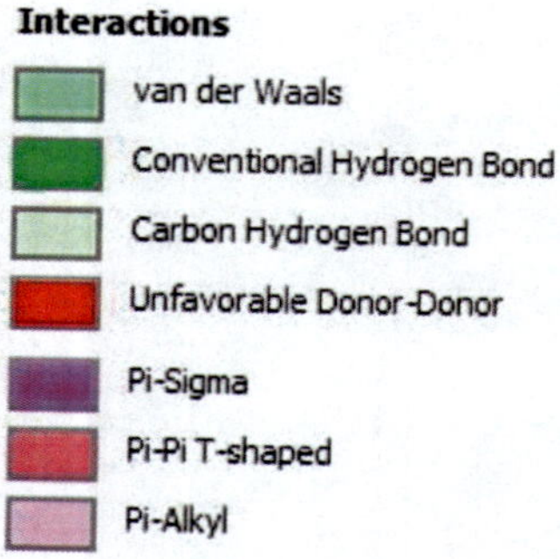

Fig. 7. Binding between protein residues and ligands.

In previous studies, anthocyanins have shown a proapoptotic role in Hep3B liver cancer cells by induction of apoptosis via mitochondrial pathway, leading to an increase caspase 3 activity, which is dose-dependent [28].

Thus, the present study presents computational evidence of the affinity of anthocyanins for proteins related with apoptosis, both anti and proapoptotic proteins. *In silico* molecular dynamics and *in vitro* affinity studies will have to be performed in the future to characterize the effects of the identified interactions.

5 Conclusions

The results suggest potential proapoptotic activity of the evaluated anthocyanins. It will be necessary *in silico* molecular dynamics studies to characterize the interactions and *in vitro* studies to evidence potential modulatory activity in apoptosis pathways.

Acknowledgment. The authors would like to express their acknowledgments to the Instituto Tecnológico Metropolitano for financial support. Young researchers and innovators program (code P20247) and Research Group "Grupo de Investigación e Innovación Biomédica GI2B" are also acknowledged.

Conflict of Interest The authors declare that they have no conflict of interest.

References

1. Arcy, M.S.D.: Cell death: a review of the major forms of apoptosis, necrosis and autophagy. Cell Biol. Int. **43**(6), 582–592 (2019). https://doi.org/10.1002/cbin.11137
2. Almalki, W.H., Aftab, R.M., Alqurashi, A.E., Toonsi, R.T.: Current updates on the role of apoptosis on human health and disease. Nat. Volatiles & Essent. Oils **8**(6), 3564–3574 (2021)
3. Jan, R., Chaudhry, G.-S.: Understanding apoptosis and apoptotic pathways targeted cancer therapeutics. Adv. Pharm. Bull. **9**(2), 205–218 (2019). https://doi.org/10.15171/jcvtr.2015.24
4. Yee, C., Yang, W., Hekimi, S.: The intrinsic apoptosis pathway mediates the pro-longevity response to mitochondrial ROS in *C. elegans*. Cell **157**(4), 897–909 (2014). https://doi.org/10.1016/j.cell.2014.02.055

5. Sayers, T.J.: Targeting the extrinsic apoptosis signaling pathway for cancer therapy. Cancer Immunol. Immunother. **60**(8), 1173–1180 (2011). https://doi.org/10.1007/s00262-011-1008-4
6. Ashkenazi, A.: Targeting the extrinsic apoptotic pathway in cancer: lessons learned and future directions find the latest version: targeting the extrinsic apoptotic pathway in cancer: lessons learned and future directions. J. Clin. Investig. **125**(2), 487–489 (2015). https://doi.org/10.1172/JCI80420.normal
7. Santagostino, S.F., Assenmacher, C., Tarrant, J.C., Adedeji, A.O., Radaelli, E.: Mechanisms of regulated cell death: current perspectives. Vet. Pathol. **58**(4), 596–623 (2021). https://doi.org/10.1177/03009858211005537
8. Motilva, M., Bars-cortina, D., Sakhawat, A., Pi, C.: Chemopreventive effects of anthocyanins on colorectal and breast cancer: a review. Semin. Cancer Biol. **81**(January 2021), 241–258 (2022)
9. Pathways, N.-B., et al.: Anthocyanins derived from vitis coignetiae pulliat contributes anti-cancer effects by suppressing. Molecules (Basel, Switzerland) **25**(5445), 1–12 (2020)
10. Lin, B., Gong, C., Song, H., Cui, Y.: Effects of anthocyanins on the prevention and treatment of cancer tables of links. Br. J. Pharmacol. **174**, 1226–1243 (2017). https://doi.org/10.1111/bph.13627
11. RCSB PDB: Homepage. https://www.rcsb.org/. Accessed 29 Jun 2021
12. PDBe home: EMBL-EBI. https://www.ebi.ac.uk/pdbe/. Accessed 30 May 2022
13. PubChem. https://pubchem.ncbi.nlm.nih.gov/. Accessed 29 Jun 2021
14. Carlos, D.A., Sandra, A., Fabián, C.-M., Benjamín, R., Maria, E.M.: Antiproliferative and pro-apoptotic effects of Andean berry juice (*Vaccinium meridionale* Swartz) on human colon adenocarcinoma SW480 cells. J. Med. Plants Res. **11**(24), 393–402 (2017). https://doi.org/10.5897/jmpr2017.6401
15. Arango-Varela, S.S., Torres-Camargo, D., Reyes-Dieck, C., Zapata-Londoño, M.B., Maldonado-Celis, M.E.: Aqueous extract of andean berry induces apoptosis in human colon cancer cells without mitochondrial damage. J. Berry Res. **11**(3), 377–393 (2021). https://doi.org/10.3233/JBR-200684
16. Arango-Varela, S.S., Luzardo-Ocampo, I., Maldonado-Celis, M.E., Campos-Vega, R.: Andean berry (*Vaccinium meridionale* Swartz) juice in combination with Aspirin modulated anti-inflammatory markers on LPS-stimulated RAW 264.7 macrophages. Food Res. Int. 137(July 2020). https://doi.org/10.1016/j.foodres.2020.109541
17. UCSF Chimera References. https://www.cgl.ucsf.edu/chimera/docs/morerefs.html. Accessed 20 Sep 2021
18. Avogadro—Free cross-platform molecular editor—Avogadro. https://avogadro.cc/. Accessed 20 Sep 2021
19. Trott, A.J.O.O.: AutoDock Vina: improving the speed and accuracy of docking with a new scoring function, efficient optimization and multithreading. J. Comput. Chem. **31** (2010). http://vina.scripps.edu/index.html. Accessed 19 Sep 2021
20. Dolores, M., Nyman, T., Welin, M., Lehtio, L., Flodin, S., Tre, L.: Completing the family portrait of the anti-apoptotic Bcl-2 proteins: crystal structure of human Bfl-1 in complex with Bim. FEBS Lett. **582**, 3590–3594 (2008). https://doi.org/10.1016/j.febslet.2008.09.028
21. Wang, Z., et al.: Biochimica et biophysica acta kinetic and structural characterization of caspase-3 and caspase-8 inhibition by a novel class of irreversible inhibitors. BBA—Proteins and Proteomics **1804**(9), 1817–1831 (2010). https://doi.org/10.1016/j.bbapap.2010.05.007
22. Kapralov, A.A., et al.: The hydrogen-peroxide-induced radical behaviour in human cytochrome c—phospholipid complexes: implications for the enhanced pro-apoptotic activity of the G41S mutant. Biochem. J. **452**, 441–452 (2013). https://doi.org/10.1042/BJ20130758

23. Scott, F.L., et al.: The Fas—FADD death domain complex structure unravels signalling by receptor clustering. Nature **457**(7232), 1019–1022 (2009). https://doi.org/10.1038/nature07606
24. Feeney, B., Pop, C., Swartz, P., Mattos, C., Clark, A.C.: Role of loop bundle hydrogen bonds in the maturation and activity of (Pro)caspase-3. Biochemistry 13249–13263 (2006)
25. Hosseini, M.M., et al.: Cytotoxic and apoptogenic effects of cyanidin-3-glucoside on the glioblastoma cell line. World Neurosurg. **108**, 94–100 (2017). https://doi.org/10.1016/J.WNEU.2017.08.133
26. Chen, P.N., Chu, S.C., Chiou, H.L., Chiang, C.L., Yang, S.F., Hsieh, Y.S.: Cyanidin 3-glucoside and peonidin 3-glucoside inhibit tumor cell growth and induce apoptosis in vitro and suppress tumor growth in vivo. Nutr. Cancer **53**(2), 232–243 (2005). https://doi.org/10.1207/S15327914NC5302_12
27. Boice, A., Bouchier-Hayes, L.: Targeting apoptotic caspases in cancer. Biochim. Biophys. Acta Mol. Cell Res **1867**(6), 118688 (2020). https://doi.org/10.1016/j.bbamcr.2020.118688
28. Shin, D.Y., et al.: Induction of apoptosis and inhibition of invasion in human hepatoma cells by anthocyanins from meoru. Ann. N. Y. Acad. Sci. **1171**, 137–148 (2009). https://doi.org/10.1111/J.1749-6632.2009.04689.X

Flux MRI: Accelerating with Aid of Physical Models

Matheus Dalvi[1], Marcus Vinicius Costa[2], and Vinicius Rispoli[2](✉)

[1] Department of Electrical Engineering, University of Brasília, Brasília, Brazil
[2] Engineering College at Gama, University of Brasília, Brasilia, Brazil
vrispoli@unb.br

Abstract. In the context of Magnetic Resonance Imaging (MRI), *in vivo* blood flow measurements can be obtained directly by Phase-Contrast (PC) technique. Another possibility lies in the approach by alternative flow estimation methods, such as Computational Fluid Dynamics (CFD) simulations. Despite being widely applied on exams, PC-MRI can result in excessively long scan times and may suffer from partial volume effects. Patient specific CFD simulations, on the other hand, require reduced scan times and provide high spatial and temporal resolution. At the same time, its accuracy can be hugely affected by model assumptions and simulated velocity fields can be inconsistent when compared to the subject blood flow. In order to reduce scan times, Compressive Sensing (CS) technique has been applied on flow MRI exams, such as PC-MRI, to allow blood flow to be measured from a small dataset in k-space. The main objective of this work is, under the CS framework, to propose a novel non-linear signal reconstruction method that can combine PC-MRI signal data with *a priori* information based on velocity images. Its feasibility is then verified, showing its capacity to integrate these distinct nature signals. Finally, it is expected, by applying the proposed approach, that patient specific blood flow measurements could be quantified by an even smaller set of measurements acquired in k-space. Then, drastically reducing examination and sampling times.

Keywords: Magnetic resonance · Compressive sensing · Phase contrast · Computational fluid dynamics

1 Introduction

The gold standard technique for flow quantification in magnetic resonance imaging is the phase contrast (PC) [1–3]. In this imaging technique, a bipolar gradient is adjusted to align with the flow of interest and quantify the average velocity, under a certain direction axis, for each voxel. Despite the numerous possibilities provided by this technique, some limitations still make its use unfeasible in certain cases. Due to the fact that each voxel is assigned an average of the distribution of velocities in its interior, this velocity is usually underestimated due

J. L. B. Marques et al. (Eds.): CLAIB 2022/CBEB 2022, IFMBE Proceedings 98, pp. 233–244, 2024.
https://doi.org/10.1007/978-3-031-49401-7_23

to partial volume effect [4,5]. Another downside related to the phase contrast technique is related to its long scan/sampling times.

On the other hand, Compressive Sensing (CS) has shown significant advances in the area of acceleration MR images, allowing, from a significantly reduced amount of samples, the reconstruction of images with equivalent or even superior quality to those of other techniques [6]. The CS theory is based on the fact that if it is known the domain where a given signal has a sparse representation, than the signal can be represented by a reduced number of coefficients. Therefore, it can be reconstructed with an amount of samples much lower than those required from Nyquist-Shannon theorem [7,8]. Initially presented as a technique for acquiring static images, has already been shown in the literature that CS can also be applied to dynamic images and flow quantification, providing rapid and reliable exams [9–11]. In order to enhance the images reconstructed with CS technique, the use of other types of information about the signal can help reduce even more the amount of samples needed for the reconstruction. Has been presented in the literature over the years, that the use of some kind of *a priori* information can improve image reconstruction [12–14].

An alternative approach to patient specific flow quantification is the use of Computational Fluid Dynamics (CFD) [15–17]. Studies comparing and relating CFD estimated flows to MRI measured flows have already been carried out in several vascular geometries, such as intracranial aneurysms [18], the thoracic aorta [19] and the carotid bifurcation, both in models (phantoms) [20] and *in vivo* [21]. Exams associated with CFD calculations require less time for data acquisition in patients, since flow patterns are predicted with only part of the information obtained by conventional MRI exams with velocity encoding. In recent years, several authors worked on integration techniques between measured flow, compressive sensing and the use physical models in the fields of flow MRI acquisition [22–26] and mechanical engineering [27–29]. Has been shown the that the combination of aforementioned techniques is feasible and, in the case of MRI, can improve the quality of MRI acquired flow images. Finally, it is relevant to note that lately authors are combining artificial intelligence techniques with CFD to assist and improve the reconstruction of MRI flow images [30,31].

In this regard, the main purpose of this work is, through a case study, to develop a method capable of improving flow MRI image reconstruction technique. This will be achieved through the combination of compressive sensing, applied to raw phase contrast flow MRI data, and the use of velocity *a priori* information. It is relevant to note that *a priori* information can be estimated from any technique, such as computational fluid dynamics. Then, is proposed a novel solver, in comparison to the presented literature, that integrates simultaneously both MRI and other kind of velocity data. The main difference of the proposed method, as opposed to those found in the literature, is in the way MRI measured velocity is compared to any source of *a priori* information. Given the nonlinear character of the presented method, the algorithms are solved in an alternate approach. Finally, it has been shown that it is possible to obtain velocity images that integrates data from sources of different nature. Thus, allowing the number of samples reduction necessary for the reconstruction by CS.

2 Methods

2.1 Reconstruction Problem

Phase contrast velocity encoding is reconstructed based on a pair of complex images (phase images), with measurements acquired after the application of two bipolar gradients [1–3]. The classical formulation for the CS problem proposes the reconstruction of such images from the optimization of two independent problems, based only on the measurements acquired by the scanner and on the *a priori* information about sparsity in a given domain [32,33]. Specifically, the original CS reconstruction problem is a NP-hard problem, generally of combinatorial complexity [7], and is not viable except for very low-dimensional cases. Thus, the original problem can be relaxed and a precise reconstruction can be achieved using the following non-linear constrained optimization problem:

$$\mathbf{x}^* = \operatorname{argmin}_{\mathbf{x}} \| \mathbf{Bx} \|_p^p \quad \text{s. t.} \quad \mathbf{Ax} = \mathbf{b}, \tag{1}$$

where $0 < p \leq 1$, $\mathbf{B} \in \mathbb{C}^{N \times N}$ is the sparsifying transform, $\mathbf{A} \in \mathbb{C}^{N \times N}$ is the acquisition process matrix, $\mathbf{x} \in \mathbb{C}^N$ is the desired image and $\mathbf{b} \in \mathbb{C}^N$ is the acquired signal and

$$\| \mathbf{x} \|_p = \left(\sum_{n=1}^{N} |x_n|^p \right)^{1/p} . \tag{2}$$

In order to add *a priori* velocity information to the CS model, this work proposes the reconstruction of the phase images through the minimization of the following objective function:

$$\begin{aligned} J(\mathbf{x_1}, \mathbf{x_2}) = {} & \|\mathbf{Ax_1} - \mathbf{b_1}\|_2^2 + \|\mathbf{Ax_2} - \mathbf{b_2}\|_2^2 \\ & + \|\mathbf{Bx_1}\|_p^p + \|\mathbf{Bx_2}\|_p^p \\ & + \lambda \, \|(\mathbf{x_1} \odot \bar{\mathbf{x}}_\mathbf{2} - \mathbf{C}) \odot \mathbf{m}\|_2^2 , \end{aligned} \tag{3}$$

where $\mathbf{x_1}$ and $\mathbf{x_2}$ corresponds to the phase images of each bipolar gradient, $\mathbf{b_1}$ and $\mathbf{b_2}$ the measurements acquired in k-space, $\mathbf{A}$ is the sensing matrix, $\mathbf{B}$ the matrix referring to the sparsifying transform, $\mathbf{C}$ the comparison image containing the a priori information about velocity and $\mathbf{m}$ a mask to select in the images only areas where there is flow information. The operator $\odot$ refers to pointwise matrix multiplication and $\bar{\mathbf{x}}_s$ is the complex conjugate of the image $\mathbf{x}_s$. Furthermore, the constant λ is the weight assigned to the *a priori* data associated norm. This formulation for the reconstruction problem, as well as the proposed solution, is loosely based on the approach used in [34,35].

The vector $\mathbf{C}$ is the key of this work, as it provides the means of comparison between MRI data and *a priori* information brought from the physical model. In this way, its modeling is essential to obtain the best possible results. Considering $\mathbf{v}$ a normalized vector containing velocity information from some source, for example CFD. Then, the vector $\mathbf{C}$ will be modeled in the form

$$\mathbf{C} = \rho e^{i\pi \mathbf{v}}, \tag{4}$$

where ρ is the magnitude of the vector $\mathbf{x_1} \odot \bar{\mathbf{x}}_2$ and i is the imaginary unit. This modeling is adequate, as the desired velocity image is obtained by the phase difference between optimal vectors $\mathbf{x_1}$ and $\mathbf{x_2}$ [34,35]. Also, since the magnitude between $\mathbf{x_1} \odot \bar{\mathbf{x}}_2$ and $\mathbf{C}$ is the same, then the optimizer will retain its minimization in the phase variable, which is the desired one.

2.2 Optimization

In the solution of the proposed optimization problem, an alternate approach [36] will be used. So, the minimization of the objective function (3) will be iteratively given by two dependent optimization problems. This approach will be used since the optimization is question is nonlinear, due to the term $\|(\mathbf{x_1} \odot \bar{\mathbf{x}}_2 - \mathbf{C}) \odot \mathbf{m}\|_2^2$ in Eq.(3). Each step of the iterative method will be responsible for minimizing, through the solutions of two linear systems, the norms of Eq. (3) depending on one of the images $\mathbf{x}_s$, with $s = 1, 2$, while the other image is fixed and used as input in the process of obtaining the image $\mathbf{x}_s$. The IRLS method, as used in [33], will be used to calculate the p norm.

The calculation of each partial image $\mathbf{x_1}^{k+1}$ and $\mathbf{x_2}^{k+1}$, where k is the iteration index, used to minimize the objective function $J(\mathbf{x_1}, \mathbf{x_2})$ will be given by the solutions of the following linear systems:

$$\begin{aligned}[\mathbf{A^H A} + \mathbf{B^H w_1 B} + \lambda \mathrm{diag}(\mathbf{m} \odot \mathbf{x_2}^k \odot \bar{\mathbf{x}}_2^k)]\mathbf{x_1^{k+1}} \\ = \lambda(\mathbf{m} \odot \mathbf{x_2}^k \odot \mathbf{C}) + \mathbf{A^H b_1}\end{aligned} \tag{5}$$

$$\begin{aligned}[\mathbf{A^H A} + \mathbf{B^H w_2 B} + \lambda \mathrm{diag}(\mathbf{m} \odot \mathbf{x_1}^k \odot \bar{\mathbf{x}}_1^k)]\mathbf{x_2^{k+1}} \\ = \lambda(\mathbf{m} \odot \mathbf{x_1}^k \odot \bar{\mathbf{C}}) + \mathbf{A^H b_2}\end{aligned} \tag{6}$$

where $\mathbf{w}_s$, with $s = 1, 2$, is the vector of weights referring to the IRLS method and the superscript $^{\mathbf{H}}$ represents the Hermitian conjugate (conjugated transpose). It is worth mentioning that all variables involved in the system are in vector form.

2.3 Data Acquisition

High-spatial-resolution four-dimensional PC data of a pulsatile carotid flow phantom (Phantoms by Design, Inc., Bothell, WA) were obtained using a 3DFT SPGR pulse sequence. The scan parameters were: $0.5 \times 0.5 \times 1$ mm^3 spatial resolution; field-of-view $39 \times 12.5 \times 32.5$ cm^3; TR 11.4 ms; flip angle 8.5°; temporal resolution 91.2 ms; VENC 60 cm/s; 40 min per scan; 9 NEX. The data were acquired on a GE Discovery MR750 3T system, with a 32-channel receive-only head coil array (Nova Medical, Inc., Wilmington, MA, USA). The phantom's pulse cycle was set to 60 bpm. The velocity map for each spatial axis—u_{pc}, v_{pc}, and w_{pc}—was reconstructed using data from all channels of the receive coil array. The lumen was segmented by manually outlining the vessel borders from a stack of 2D axial images, obtained from the reconstructed 3D volume.

2.4 k-Space Sampling

Since the dataset used in this work consists of phase contrast data already in the spatial domain, the first step consists of selecting a slice at an instant of the cardiac cycle and simulating the space k during the acquisition process. The selected section (Fig. 1)—was such that it allowed the quantification and visualization of flow in two axes (x and z) of the plane. For this, it was necessary to simulate the space k for three distinct phase images: reference, flow encoded on the x axis—echo x—and flow encoded on the z axis—echo z.

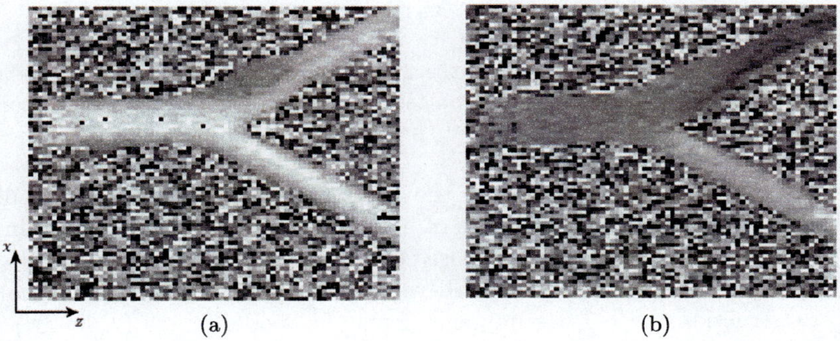

Fig. 1. Phase contrast images from the chosen slice. (a) Velocity on the z axis. (b) Velocity on the x axis. The lighter regions represents the velocity of 60 cm/s. The darkest region, −60 cm/s. Values defined by VENC

The k-space simulation for the three images was carried out by applying the two-dimensional Fourier transform (Fig. 2). In order to meet the inconsistency criterion and considering the hardware limitations of the scanner, we chose to sample the k-space according to a spiral trajectory with variable density, in order to concentrate the sampling in the center of the frequency domain and, at the same time, satisfying the randomness criterion. Figure 2 illustrates the spiral trajectory applied to the sampling process of the k-space.

Here the simulated k-space was sampled by approximately 22%. This value was chosen in order to be noticed distortions when comparing the CS reconstructed image to the reference image.

2.5 Experiments

This work will be restricted to verify the feasibility of the proposed minimizer as *a priori* velocity information is added to the optimization problem. That is, from the variation, in (3), of the parameter λ, it will be evaluated if, in fact, the increment of this value over the other weights brings the reconstructed velocity image closer to the one chosen for comparison. If, as λ is increased, there is an increase in the similarity between the reconstructed PC image and the one

(a) (b)

Fig. 2. Obtaining the k-space through the 2D Fourier transform. (a) k-space relative to the *echo z* image. (b) k-space spiral pattern sampling used in the experiments

adopted as a means of comparison, the functionality of the optimizer will be confirmed, confirming the hypothesis of inclusion of *a priori information* in the reconstruction process. For the proposed tests, values between 10^0 and 10^4 will be assigned to the parameter λ. Such values were defined in order to guarantee reconstructions with little influence from the model image ($\lambda = 10^0$) and with predominance ($\lambda = 10^4$) of this image over the complementary information (measures and sparseness). Based on experiments performed, the influence of the model image on the final reconstructed image was verified only for values of λ from 10^1. Thus, reconstructions with $\lambda = 10^0$ represent results equal to those obtained when such *a priori* information is disregarded.

In addition to the calculation as a function of the parameter λ, the reconstructions will be made using $p = 1$ and the sparsification of the signal of interest will be done using the discrete cosine transform (DCT). Such choice was based on preliminar tests with different sparsifying transforms without *a priori* information ($\lambda = 0$). Thus, DCT was chosen since it presented the best overall results when considered only the CS technique.

The norm $\|(\bar{\mathbf{x}}_1 \odot \mathbf{x}_2 - \mathbf{C}) \odot \mathbf{m}\|_2^2$, in Eq. (3), performs, in the optimizer, the role of approximating the MRI signal to the one used for comparison ($\mathbf{C}$). So, our proposed test is aiming, simultaneously, to demonstrate the influence of such a norm on the reconstructed image and validate the analytical solutions that iteratively minimize the objective function. In the experiments conducted here, we choose to adopt $\mathbf{C}$, for each λ, as the original image to be reconstructed. At first, there is no restriction on what type of velocity image that can be inserted on the objective function. The use of the reference phase contrast image should facilitate our analysis after the experiments and provide greater accuracy, when observed both objective and subjective quality metrics of the reconstructed signal. That is, unlike what would happen if, for example, a simulated image was used via CFD, the introduction of the reference image as a comparative model allows the optimizer to always have this as a reference, facilitating the identification of how the introduction of this extra information contributes to the reconstruction process.

All algorithms used were implemented in MATLAB (MathWorks, Inc., Natick, MA, USA). Furthermore, to calculate the solutions of the linear systems involved in the optimization process, the preconditioned conjugate gradient method, native to MATLAB, was used.

2.6 Quantitative Comparison

In order to quantitatively compare the reconstructed images with the reference, we will use two different metrics, peak signal-to-noise ratio (PSNR) and structural similarity index (SSIM).

First, given to images $\mathbf{x}$ and $\mathbf{y}$, both with dimensions $M \times N$, the PSNR is defined by [37]:

$$\mathrm{PSNR}(\mathbf{x},\mathbf{y})|_{dB} = 10 \cdot \log_{10}(\mathrm{peakval}^2/\mathrm{MSE}(\mathbf{x},\mathbf{y})) \tag{7}$$

where

$$\mathrm{MSE}(\mathbf{x},\mathbf{y}) = \frac{1}{MN}\sum_{i=1}^{M}\sum_{j=1}^{N}(x_{ij} - y_{ij})^2 \tag{8}$$

and peakval represents the maximum possible value among all pixels of both images. In this case, considering that the $\mathbf{x}$ and $\mathbf{y}$ are velocity images, then the highest value allowed is peakval = VENC. It should be noted that in this context the images $\mathbf{x}$ and $\mathbf{y}$ correspond, respectively, to the reference velocity image and the velocity image calculated as the results of the optimizations in Eqs. (5)–(6).

Finally, it should be noted that PSNR values approach infinity as the MSE approaches zero, indicating a direct relationship between the calculated value and the similarity between the compared images.

On the other hand, SSIM is a subjective quality metric that is calculated between two images $\mathbf{x}$ and $\mathbf{y}$ as [37]:

$$\mathrm{SSIM}(\mathbf{x},\mathbf{y}) = \frac{(2\mu_x\mu_y + C_1)(2\sigma_{xy} + C_2)}{(\mu_x^2 + \mu_y^2 + C_1)(\sigma_x^2 + \sigma_y^2 + C_2)} \tag{9}$$

where μ_x and μ_y are the local averages, σ_x and σ_y the standard deviations, σ_{xy} the cross covariance and C_1 and C_2 are regularization constants. Furthermore, as with PSNR, the value associated with SSIM is directly related to the similarity between the images in question.

3 Results and Discussion

Figure 3 show the velocity images encoded on the z and x axes, respectively, resulting from the reconstructions via DCT and $p = 1$ for different values of the parameter λ. Note how the progression of the λ parameter makes the images closer to the original, shown in Fig. 1.

On the other hand, Table 1 show the quantitative results of the quality metrics applied to the phase images obtained with the proposed reconstruction algorithm. As expected, from the visual result in Fig. 3, both PSNR and SSIM increases as the parameter λ increases.

The qualitative and quantitative behavior shown can be explained based on the constraint added on the CS optimization problem. The new term $\|(\mathbf{x_1} \odot \bar{\mathbf{x}}_\mathbf{2} - \mathbf{C}) \odot \mathbf{m}\|_2^2$ helps on forcing the reconstructed bipolar images $\mathbf{x_1}$ and $\mathbf{x_2}$ to have velocity closer to the one supplied to the solver, through this term. Note that, the solution given by Eqs. (5)–(6) shows that the parameter λ acts like a weight on a weighted mean. Thus, as λ increases is expected that $\mathbf{x_1}$ and $\mathbf{x_2}$ will be those that guarantee the reconstructed velocity image is close as it can get to the provided *a priori* velocity map.

The results presented here show that the optimizer can integrate signals of different natures. As expected, the method is able to generate velocity images that are closer to the reference velocity image. Clearly, there are limitations to the approach presented, since several parameters are not being tested here. However, such details will be dealt with in an extension of this work.

Table 1. Results from the objective quality metrics (PSNR and SSIM) calculated for the phase constrast reconstructed images as a function of the λ parameter.

Phase contrast image	PSNR (dB)				
	$\lambda = 10^0$	$\lambda = 10^1$	$\lambda = 10^2$	$\lambda = 10^3$	$\lambda = 10^4$
echo x	11.7	11.9	13.0	16.1	22.0
echo z	7.9	8.1	8.4	15.8	25.7
			SSIM		
echo x	0.761	0.782	0.875	0.959	0.990
echo z	0.817	0.838	0.871	0.970	0.997

4 Conclusion

This paper aimed into the development of a method of reconstruction of flow MR images by phase contrast based both on Compressive Sensing and on velocity images calculated by a different technique, defined here as *a priori* information.

Here was possible to confirm the validity of the proposed optimization. Although the tests were carried out under the adoption of the original velocity image as the *a priori* information, it was possible to notice how the reconstructed image approached the model image as the weight referring to the velocity term was increased. Also, important to notice that both PSNR and SSIM indicated an improvement in the quality of the reconstructed phase contrast images, when compared to the original ones, when the value of the velocity term weight was increased.

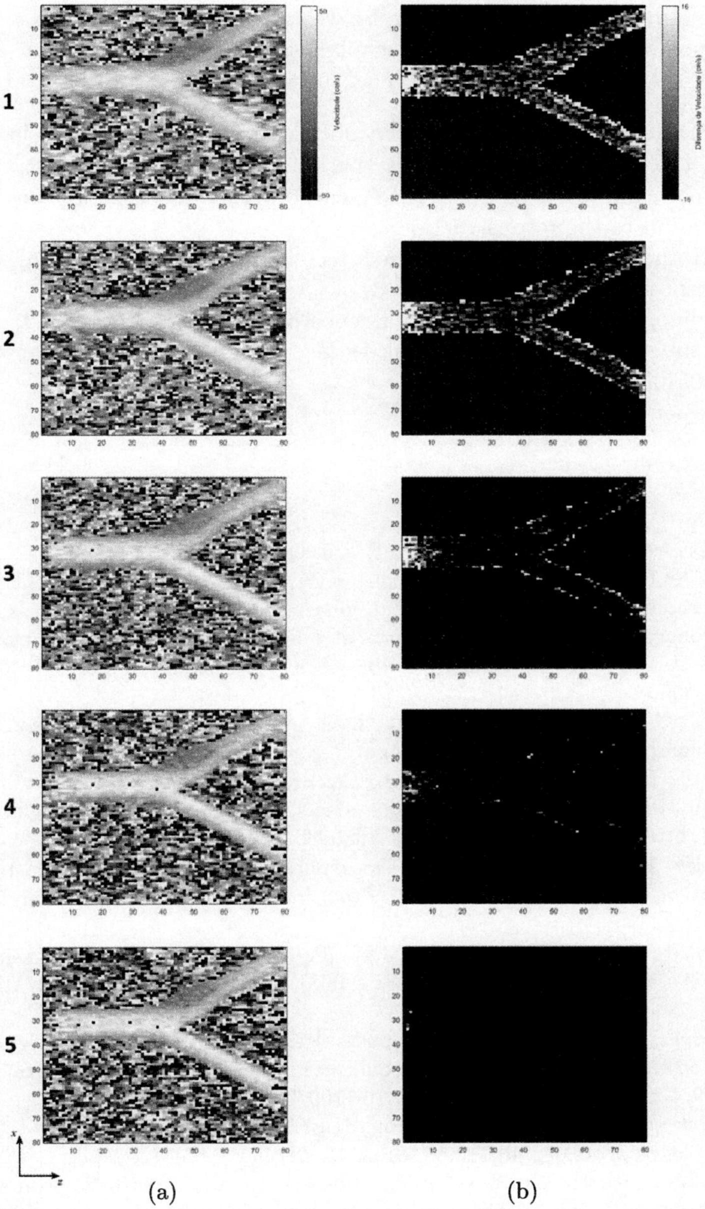

Fig. 3. Phase contrast image from the z axis as a function of weight λ. Obtained from the reconstruction of phase images using DCT and $p = 1$. (a) Reconstructed PC images. (b) Absolute error between the corresponding reconstructed image and the original image 1a. Lines 1 to 5 represents, respectively, λ equal to: 10^0, 10^1, 10^2, 10^3 and 10^4

The optimizer proposed in this work expands the universe of possibilities for research in the area of integration between magnetic resonance flow imaging and alternative flow quantification techniques. Among the topics that were not addressed in this work and that serve as suggestions for future work are:

1. Carry out experiments and possible modifications in the methods in order to reduce the number of samples in k-space necessary for the reconstruction;
2. Implement and test a CFD solver to incorporate model based *a priori* velocity information to the reconstruction;
3. Extend the usage of the proposed technique to the context of artificial intelligence-based solvers (e.g. [30,31]);
4. Determine, in more detail, the influence of different sparsifying transforms better suited for phase contrast images (e.g. total variation [9] or phase images complex difference [34,35]).

References

1. Pelc, N.J., Herfkens, R.J., Shimakawa, A., Enzmann, D.R.: Phase contrast cine magnetic resonance imaging. Magn. Reson. Q. **7**, 229–254 (1991)
2. Clarke, G.D., Hundley, W.G., McColl, R.W., Eckels, R., Chaney, D.S.C., Li, H.-F., Peshock, R.M.: Velocity-encoded, phase-difference cine MRI measurements of coronary artery flow: Dependence of flow accuracy on the number of cine frames. J. Magn. Reson. Imaging **6**, 733–742 (1996). https://doi.org/10.1002/jmri.1880060506
3. Man, L.-C., Pauly, J.M., Nishimura, D.G., Macovski, A.: Nonsubtractive spiral phase contrast velocity imaging. Magn. Reson. Med. **42**, 704–713 (1999)
4. Tang, C., Blatter, D.D., Parker, D.L.: Accuracy of phase-contrast flow measurements in the presence of partial-volume effects. J. Magn. Reson. Imaging **3**, 377–385 (1993). https://doi.org/10.1002/jmri.1880030213
5. Gonzalez, E.G., Carvalho, J.L.A.: Does phase contrast MRI provide the mean velocity of the spins within a voxel? Proc. Intl. Soc. Mag. Reson. Med. **22**, 2480 (2014)
6. Lustig, M., Donoho, D.L., Santos, J.M., Pauly, J.M.: Compressed sensing MRI. IEEE Signal Process. Mag. **25**, 72–82 (2008). https://doi.org/10.1109/MSP.2007.914728
7. Candés, E., Romberg, J., Tao, T.: Robust uncertainty principles: exact signal reconstruction from highly incomplete frequency information. IEEE Trans. Inf. Theory **52**, 489–509 (2006). https://doi.org/10.1109/TIT.2005.862083
8. Baraniuk, R.G.: Compressive sensing. IEEE Signal Process. Mag. **24**, 118–121 (2007). https://doi.org/10.1109/MSP.2007.4286571
9. Kowalik, G., Knight, D., Steeden, J., Muthurangu, V.: Perturbed spiral real-time phase-contrast MR with compressive sensing reconstruction for assessment of flow in children. Magn. Reson. Med. **83**, 2077–2091 (2020). https://doi.org/10.1002/mrm.28065
10. Zhang, J., Brindise, M.C., Rothenberger, S.M., Markl, M., Rayz, V.L., Vlachos, P.P.: A multi-modality approach for enhancing 4D flow magnetic resonance imaging via sparse representation. 19, 20210751 (2022). doi: 10.1098/rsif.2021.0751

11. Garreau, M., Puiseux, T., Toupin, S., Giese, D., Mendez, S., Nicoud, F., Moreno, R.: Accelerated sequences of 4D flow MRI using GRAPPA and compressed sensing: A comparison against conventional MRI and computational fluid dynamics. Magn. Reson. Med. **88**, 2432–2446 (2022). https://doi.org/10.1002/mrm.29404
12. Lauterbur, P.C., Liang, Z-P.: Magnetic resonance imaging with a priori constraints: possibilities and limitations. Conf. Proc. IEEE Eng. Med. Biol. Soc. 5, 2170–2171 (1996). https://doi.org/10.1109/IEMBS.1996.646481
13. Miosso, C.J., von Borries, R., Pierluissi, J.H.: Compressive Sensing With Prior Information: Requirements and Probabilities of Reconstruction in ℓ_1-Minimization. IEEE Trans. Signal Process. **61**, 2150–2164 (2013). https://doi.org/10.1109/TSP.2012.2231076
14. Zhang, X. Cui, W. Liu, Y.: Compressed sensing with prior information via maximizing correlation. IEEE Int. Symp. Inf. Theory, 221–225 (2017). https://doi.org/10.1109/ISIT.2017.8006522
15. Lei, M., Kleinstreuer, C., Truskey, G.A.: Numerical investigation and prediction of atherogenic sites in branching arteries. J. Biomech. Eng. **3**, 350–357 (1995). https://doi.org/10.1115/1.2794191
16. Steinman, D.A.: Image-based computational fluid dynamics: a new paradigm for monitoring hemodynamics and atherosclerosis. Curr. Drug Targets Cardiovasc. Haematol. Disord. **4**, 183–197 (2004). https://doi.org/10.2174/1568006043336302
17. Steinman, D.A., Taylor, C.A.: Flow imaging and computing: large artery hemodynamics. Ann. Biomed. Eng. **33**, 1704–1709 (2005). https://doi.org/10.1007/s10439-005-8772-2
18. Boussel, L., Rayz, V., Martin, A., Acevedo-Bolton, G., et al.: Phase-contrast magnetic resonance imaging measurements in intracranial aneurysms in vivo of flow patterns, velocity fields, and wall shear stress: comparison with computational fluid dynamics. Mag. Reson. Med. **61**, 409–417 (2009). https://doi.org/10.1002/mrm.21861
19. Canstein, C., Cachot, P., Faust, A., Stalder, A.F., et al.: 3D MR flow analysis in realistic rapid-prototyping model systems of the thoracic aorta: comparison with in vivo data and computational fluid dynamics in identical vessel geometries. Mag. Reson. Med. **59**, 535–546 (2008). https://doi.org/10.1002/mrm.21331
20. Marshall, I., Zhao, S., Papathanasopoulou, P., Hoskins, P., Xu, X.Y.: MRI and CFD studies of pulsatile flow in healthy and stenosed carotid bifurcation models. J. Biomech. **37**, 679–687 (2004). https://doi.org/10.1016/j.jbiomech.2003.09.032
21. Long, Q., Xu, X.Y., Ariff, B., Thom, S.A., Hughes, A.D., Stanton, A.V.: Reconstruction of blood flow patterns in a human carotid bifurcation: a combined CFD and MRI study. Mag. Reson. Med. **11**, 299–311 (2000). https://doi.org/10.1002/(sici)1522-2586(200003)11:3⟨299::aid-jmri9⟩3.0.co;2-m
22. de Hoon, N., van Pelt, R., Jalba, A., Vilanova, A.: 4D MRI Flow Coupled to Physics-Based Fluid Simulation for Blood-Flow Visualization. Comput. Graph. **33**, 121–130 (2014). https://doi.org/10.1111/cgf.12368
23. Rispoli, V., Nielsen, J., Nayak, K., Carvalho, J.: Computational fluid dynamics simulations of blood flow regularized by 3D phase contrast MRI. Biomed. Eng. 110 (2015). https://doi.org/10.1186/s12938-015-0104-7
24. Bakhshinejad, A., Baghaie, A., Vali, A., Saloner, D., Rayz, V.L., D'Souza, R.M.: Merging computational fluid dynamics and 4D Flow MRI using proper orthogonal decomposition and ridge regression. J. Biomech. 62–173 (2017). https://doi.org/10.1016/j.jbiomech.2017.05.004

25. Töger, J., Zahr, M., Aristokleous, N., Bloch, K., Carlsson, M., Persson, P-O.: Blood flow imaging by optimal matching of computational fluid dynamics to 4D-flow data. Magn. Reson. Med. 2231–2245 (2020). https://doi.org/10.1002/mrm.28269
26. Wüstenhagen, C., John, K., Langner, S., Brede, M., Grundmann, S., Bruschewski, M.: CFD validation using in-vitro MRI velocity data - Methods for data matching and CFD error quantification. Comput. Biol. Med. **131**, 104230 (2021). https://doi.org/10.1016/j.compbiomed.2021.104230
27. Lu, Y., Wang, Y.: Physics-Based Compressive Sensing Approach to Monitor Turbulent Flow. AIAA J. **58**, 1–9 (2020). https://doi.org/10.2514/1.J058990
28. Jayaraman, B., Mamun, A.: On Data-Driven Sparse Sensing and Linear Estimation of Fluid Flows. Sensors. **20**, 3752 (2020). https://doi.org/10.3390/s20133752
29. Mohammadi, A., Shimoyama, K., Karimi, M.S., Raisee, M.: Efficient uncertainty quantification of CFD problems by combination of proper orthogonal decomposition and compressed sensing. Appl. Math. Model. **94**, 187–225 (2021). https://doi.org/10.1016/j.apm.2021.01.012
30. Ferdian, E., Suinesiaputra, A., Dubowitz, D., Zhao, D., Wang, A., Cowan, B., Young, A.: 4DFlowNet: Super-Resolution 4D Flow MRI Using Deep Learning and Computational Fluid Dynamics. Front. Phys. **8**, 138 (2020). https://doi.org/10.3389/fphy.2020.00138
31. Peper, E. S., Kozerke, S., van Ooij, P.: Magnetic Resonance Imaging-Based 4D Flow: The Role of Artificial Intelligence. In: Artificial Intelligence in Cardiothoracic Imaging (2022). https://doi.org/10.1007/978-3-030-92087-6_34
32. Miosso, C.J., Von Borries, R., Pierluissi, J.H.: Compressive Sensing Method for Improved Reconstruction of Gradient-sparse Magnetic Resonance Images. ACSSC, 799–806 (2009). https://doi.org/10.1109/ACSSC.2009.5469970
33. Miosso, C.J., von Borries, R., Argaez, M., Velàzquez, L., Quintero, C., Potes, C.M.: Compressive sensing reconstruction with prior information by iteratively reweighted least-squares. IEEE Trans. Signal Process. **57**, 2424–2431 (2009). https://doi.org/10.1109/TSP.2009.2016889
34. Kwak, Y., Nam, S., Kissinger, K.V., et al.: Accelerated phase contrast imaging using compressed sensing with complex difference sparsity. J. Cardiovasc. Magn. Reson. 14 (2012). https://doi.org/10.1186/1532-429X-14-S1-W24
35. Kwak, Y., Nam, S., Akçakaya, M., et al.: Accelerated aortic flow assessment with compressed sensing with and without use of the sparsity of the complex difference image. Magn. Reson. Med. **70**, 851–858 (2013). https://doi.org/10.1002/mrm.24514
36. Wang, Y., Yang, J., Yin, W., Zhang, Y.: A new alternating minimization algorithm for total variation image reconstruction. SIAM J. Imaging Sci. **1**, 248–272 (2008). https://doi.org/10.1137/080724265
37. Hore, A., Ziou, D.: Image quality metrics: PSNR vs. SSIM. ICPR 2366–2369 (2010). https://doi.org/10.1109/ICPR.2010.579

Comparison of Cerebral Autoregulation in Patients with Mild and Severe Arterial Hypertension

Renata Romanelli[1(✉)], Matheus Zamai[1], M. F. Machado[2], V. A. Costa-Hong[3], L. A Bortolotto[2], R. C. Nogueira[2], J. Salinet[1], and A. S. M. Salinet[1,2]

[1] Federal University of ABC, São Bernardo do Campo, Brazil
romanelli.c@aluno.ufabc.edu.br

[2] Neurology Department, School of Medicine, University of São Paulo, Hospital das Clínicas, São Paulo, Brazil

[3] Heart Institute (Incor), University of São Paulo Medical School-Hypertension Unit, São Paulo, Brazil

Abstract. Autoregulation of cerebral circulation (CA) is a mechanism at arteriolar level that ensures a constant cerebral blood flow (CBF) during changes in arterial blood pressure (ABP). Systemic arterial hypertension (SAH) is the leading cause of morbidity and mortality in the population worldwide, and may lead to structural and functional changes in the arteriolar wall resulting in an impairment of CBF regulation. While some studies have demonstrated that CA is retained in hypertensive patients, further associations with respect to SAH severity are limited. Therefore, this study aimed to evaluate CA response in SAH patients of differing severity. The data were collected from 15 participants with mild and 15 with severe hypertension. CA was assessed using four methods: autoregulation index (ARI), auto regressive moving average autoregulation index (ARMA-ARI), transfer function analysis (TFA) and mean flow index (Mx method). The results between SAH groups were compared through a statistical t test, and Spearman's correlation was used to analyze the CA methods results. Results: No statistical difference was observed between mild and severe SAH groups. In the inter-method analysis, ARMA-ARI and Mx presented moderate correlation with ARI (p-values < 0.001, $\text{rho}_{\text{ARMA-ARI}} = 0.44$ and $\text{rho}_{\text{Mx}} = -0.4$). TFA presented weak correlation with ARI in the measures of LF range (p-values < 0.03, $\text{rho}_{\text{Coherence LF}} = 0.28$, $\text{rho}_{\text{Gain LF}} = 0.28$ and $\text{rho}_{\text{Phase LF}} = 0.33$). Conclusion: Sustained arterial hypertension does not alter CA within the ABP limits studied and the reproducibility of CA methods is poor to moderate when all SAH cases were used for the analysis.

Keywords: Cerebral autoregulation · Hypertension · Hemodynamic circulation

J. L. B. Marques et al. (Eds.): CLAIB 2022/CBEB 2022, IFMBE Proceedings 98, pp. 245–255, 2024.
https://doi.org/10.1007/978-3-031-49401-7_24

1 Introduction

Autoregulation of the cerebral circulation is a mechanism at arteriolar level that ensures a constant cerebral blood flow (CBF) during changes in arterial blood pressure (ABP) [1].

Due to the importance of cerebral blood circulation, many studies have focused on the effects of chronic modulation of ABP on cerebral hemodynamics [2, 3]. Systemic arterial hypertension (SAH) may lead to structural and functional changes in the arteriolar wall resulting in an impairment of CBF regulation [4]. It is associated with an increased cerebrovascular resistance achieved through reduced luminal diameter and increased cerebrovascular resistance [5].

SAH is the major risk factor for stroke and is one the leading causes of morbidity and mortality worldwide [4–6]. Therefore, there is an urgent need to understand the impact on systemic and cerebral hemodynamics of this devastating disease. This study aims to evaluate cerebral autoregulation (CA) response in subjects with controlled and uncontrolled SAH (mild and severe SAH, respectively), in order to assess whether CA mechanism is able to adapt and produce a positive CA response even in subjects with uncontrolled CA.

The evaluation of CA is commonly done by mathematical approaches that use the mean values of ABP and CBF collected in the same period of time. Though, currently no gold standard test exits, many different methods are now available for CA assessment, such as autoregulation index (ARI), auto regressive moving average autoregulation index (ARMA-ARI), transfer function analysis (TFA) and mean flow index (Mx method). ARI and ARMA-ARI methods are based in the same mathematical principle. They compare changes in cerebrovascular resistance per second with the physiological changes in ABP, stipulating a value between 0 to 9, where 0 is the lack of CA and 9 is an extreme CA—in normal conditions, ARI is approximately 5 [7]. TFA calculates parameters of gain, phase and coherence that indicate the dynamic of CA by assuming the relation of spontaneous ABP variations and CBF velocity (CBFv) as a linear system [6]. Lastly, the Mx method uses Pearson's correlation among mean values of ABP and CBFv to report an index where positive values suggest a disturbed CA and correlations close to zero or negative, a preserved CA [7]. However, the understanding of relationships among CA indices remains limited.

Therefore, more information is needed about the metric convergence between these methods. To address this, the present work has also the objective to perform a comparative analysis between the results of these different techniques to assess the CA in patients with mild and severe SAH. The evaluation will be made through a program provided by the Medical Physics Department at the University of Leicester (UK) [8] which is a software used as a gold-standard for calculus of these parameters.

2 Methods

2.1 Data Collection

Figure 1 summarizes the methodological process for CA analysis. Data were collected from 30 participants diagnosed with SAH, being 15 with mild hypertension (medicated-normotensive SAH patients) and 15 with severe hypertension (no-responsive to medication, medicated-SAH patients). Participants with mild hypertension have mean age of 47 ± 12 years and participants with severe hypertension have mean age of 51 ± 12 years. Continuous beat-to-beat ABP (Ohmeda 2300; Finapres, Louisville, CO, USA) and bilateral CBFv (DWL Doppler Box, Germany) were registered during 5 min at rest (sampling frequency: 100 Hz). The procedures were performed in the Neurosonology and Brain Hemodynamics laboratory, Neurology Department at the University of São Paulo.

2.2 Pre-processing Signals

ABP signals were calibrated based on the brachial ABP signal collected prior to the signal acquisition. ABP and CBFv signals were digitally filtered by a 3rd order median filter [9]. The detection of each systole peak and end-diastole was carried out in both ABP and CBFv signals [9]. The mean value of ABP and CBFv of each cardiac cycle was computed and associated to the time instant of the central sample of the cardiac cycle [6]. It is followed, by a upsamplig at 5 Hz through a cubic spline interpolation [9].

2.3 Evaluating the Cerebral Autoregulation Parameters

CA assessment was performed by four methods, which are ARI, ARMA-ARI, Mx and TFA, as follows.

Autoregulaiton Index (ARI). For ARI estimation, the inverse Fourier transform of the Transfer Function (Eq. 1), was applied, converting data back to the time domain. The Transfer Function for ARI estimation H1(f) was obtained by Welch's method, used to calculate the auto spectrum of ABP and the cross spectrum of ABP and CBFv, represented by Eqs. 2 and 3—respectively, resulting in the Eq. 4. In the time domain, the impulse response function (IRF) was computed and it was used to predict the CBFv step response to ABP changes. ARI was determined to the CBFv recording by using the minimum root mean square error (RMSE) between the IRF and each one of the 10 template curves proposed by Tiecks [10] to provide an ARI estimate [7]. In summary, this model uses a second-order differential equation to predict CBFv response, V(t), corresponding to a relative ABP modulation, dP(t), by the Eq. 5, where K represents a gain parameter, an × 2(t) is a state variable obtained based in three parameters gain, time constant and damping factor. The template curve corresponding to the minimum RMSE determines the corresponding value of ARI of the subject.

$$Transfer\,function = \frac{inputs}{outputs} \tag{1}$$

$$S_{xx}(f) = E\{E_w(f)X_w^*(f)\} \tag{2}$$

$$S_{xy} = E\{X_W(f)Y_W^*(f)\} \tag{3}$$

$$H_1(f) = \left(\frac{S_{xy}(f)}{S_{xx}(f)}\right) \tag{4}$$

$$V(t) = 1 + dP(t) - kx_2(t) \tag{5}$$

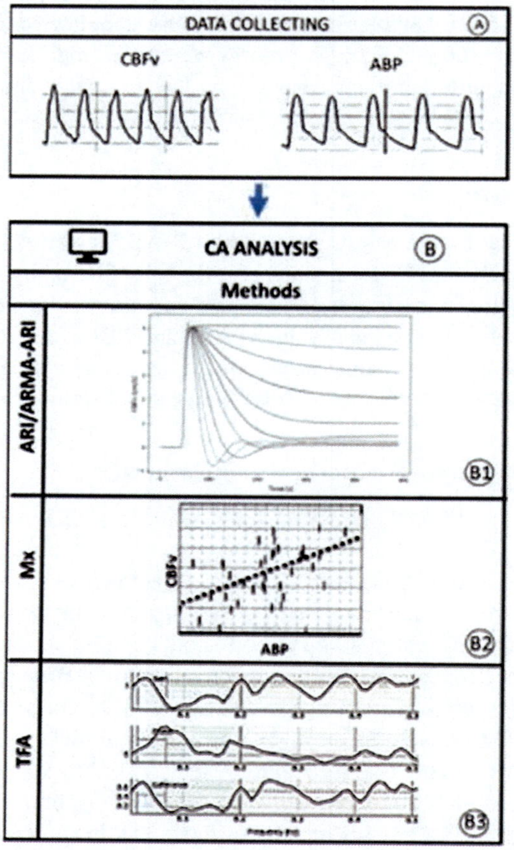

Fig. 1. Summary of the methodological process for CA analysis. A) Bilateral ABP and CBFv signal collection from patients at rest. B) CA assessment methods. B1) ARI and ARMA-ARI graphic which result in curves between 0 to 9. B2) Pearson's correlation graph of Mx method that compares how similar the CBFv and ABP signals are. B3) Illustration of TFA in which we obtain the values of gain, phase and coherence

Auto Regressive Moving Average ARI (ARMA-ARI). ARMA-ARI can be divided in autoregressive (AR) part, which estimates the value of the variable as a function of its past in the series, and the the moving average (MA) part, which is basically the weighted average of the terms of the time series (Eq. 6). Then, the method compares step response of CBFv with the 10 proposed step response curves by Tiecks [10] through the correlation coefficient. Then the transfer function estimation $H_2(f)$ is obtained by the ARMA coefficients, resulting in Eq. 7. Furthermore, ARMA-ARI value is obtained in the same way as ARI method, comparing the subjects IRF with the 10 template curves proposed by Tiecks [10].

$$V(t) = \sum_{i=1}^{p} a_i V(t-i) + \sum_{j=0}^{q-1} b_j P(t-j) \tag{6}$$

$$H_2(f) = \frac{\sum_{i=1}^{p} a_i}{\sum_{j=1}^{q} b_i} \tag{7}$$

Mean Flow Index (Mx). Mx index consists in a correlation coefficient analysis. This method takes a sequence of consecutive samples of two time series and calculates Pearson's correlation coefficient over this period. To calculate that, the correlation coefficient is done between the mean of CBFv and ABP signals by Eq. 8 [12]. Positive association among them indicates a passive dependence of CBFv on ABP, presenting the CA as damaged. On the other hand, a negative or zero value of Mx implies active cerebrovascular responses to the variation of ABP, indicating that CA is preserved [12].

$$r_{Mx} = \frac{1}{(N-1)\sigma_{ABP}\sigma_{CBFv}} \sum_{n=1}^{N} \left(ABP_n - \overline{ABP}\right)\left(CBFv_n - \overline{CBFv}\right) \tag{8}$$

where σ is the variance of each signal and $\overline{ABP}$ and $\overline{CBFv}$ are the mean of ABP and CBFv [13].

Transfer Function Analysis (TFA). TFA, is based on the Fourier decomposition of stationary input and output signals into sums of sines and cosines of multiple frequencies and the TFA is obtained by the division between auto and cross spectrum (Eqs. 2 and 3 respectively) [14]. With the switch of domain from time to frequency by TFA (Eq. 9), we obtain three parameters 1) Coherence (Eq. 10) that shows how reliable the values of gain and phase are [7], for this parameter preserved CA reflects in values nearby to zero. 2) Gain (Eq. 11) which presents the difference of amplitude, for this parameter preserved CA presents low values of gain and 3) Phase (Eq. 12) indicating the changes in the signals' domain, for this parameter we expect large phase interval in preserved CA. In the frequency domain, CA system works as a high-pass filter and it is mostly assessed in the frequency bands of 0.02–0.07 Hz (Very Low Frequency, VLF), 0.07–0.2 Hz (Low Frequency, LF) and 0.2–0.5 Hz (High Frequency, HF) [16].

$$H(f) = u(f) + iv(f) \tag{9}$$

$$MQC(f) = \frac{\left|S_{xy}(f)\right|^2}{S_{xx}(f)S_{yy}(f)} \tag{10}$$

$$G(f) = |H(f)| = \sqrt{u(f)^2 + v(f)^2} \tag{11}$$

$$\Phi(f) = tan^{-1}\left(\frac{v(f)}{u(f)}\right) \quad (12)$$

where u(f) and v(f) are the real and imaginary parts of the transfer function.

2.4 Statistical Analysis

Data was divided in two groups according with its hypertension degree, as described previously. Shapiro–Wilk and Levene's methods were performed to test the normality and homogeneity of the data, respectively. Statistic t test unpaired was performed between the groups: T student and Mann–Whitney U were performed in the methods, accordingly to its adherence to the normal curve. Furthermore, interclass analysis was performed comparing the ARI with others through Spearman's correlation. Being considered the ranges of absolute numbers [0.1–0.29], [0.3–0.49] and [0.5–1] as weak, moderate and strong correlations, consecutively. For all statistical methods, values of $p \leq 0.05$ were considered statistically significant.

3 Results

Table 1 shows the results stratified by SAH group. It can observe that the mean values between mild and severe groups are very similar for all CA methods. Despite no method presented p-value in the confidence interval (CI = 95%), ARI-ARMA and Mx method presented mean values slightly higher for the mild hypertension group.

Figure 2 presents de difference between the mean values of mild and severe SAH groups, for all CA methods, for better visualization.

Table 2 summarizes the correlation analysis between ARI and the remaining three CA methods. ARI was chosen as a basis for comparing due to its clear threshold to classify CA. There was a statistically significant correlation between ARI and the methods ARMA-ARI, Mx and TFA (at LF range). Of these correlations, ARMA-ARI, Mx, Coherence VLF and Phase LF, presented a moderate correlation with ARI. TFA parameters in general presented a weak correlation with ARI.

The results showed that ARMA-ARI and Mx has a moderate correlation with ARI. Moreover, TFA has a weak correlation with ARI in the measures of LF range. Subject presented preserved CA, on average, for both groups and both cerebral hemispheres by the results of the methods ARI and ARMA-ARI.

Figure 3 presents the correlations best fit between ARI versus the other CA methods with moderate correlation with ARI. In the figure, the points and correlation lines are stratified by mild and severe hypertension for better comparison between the groups. It can be seen that the slope of the correlation line is different in all cases, with the severe group correlation line showing a lower slope in the four comparisons. Moreover, the slope of the correlation line presents a great difference between the mild and severe groups.

Table 1. Descriptives stratified by degree of hypertension

			Mild hypertension n = 15		Severe hypertension n = 15	
Method		Cerebral hemisp.	Mean	SD	Mean	SD
ARI		Left	5.33	2.78	5.47	2.32[a]
		Right	5.99	1.87	6.00	2.07[a]
ARMA-ARI		Left	5.56	2.15	5.77	2.41[a,b]
		Right	5.86	2.28	6.03	1.68[a,b]
Mx		Left	0.25	0.38	0.29	0.22
		Right	0.15	0.35	0.26	0.32
Coherence	VLF	Left	0.59	0.19	0.95	0.50
		Right	0.61	0.21	0.55	0.23
	LF	Left	0.74	0.16	0.67	0.16
		Right	0.76	0.12	0.70	0.16
	HF	Left	0.65	0.16	0.66	0.11
		Right	0.61	0.15	0.61	0.14
Gain (cm/s mmHg)	VLF	Left	1.51	0.55	1.59	0.93[a,b]
		Right	1.58	0.52	1.47	0.84[a,b]
	LF	Left	1.95	0.50	2.27	1.16[a,b]
		Right	2.07	0.60	2.41	0.83[a,b]
	HF	Left	2.69	1.19	2.90	1.82[a,b]
		Right	2.40	0.67	2.51	1.04[a,b]
Phase (∘)	VLF	Left	0.73	0.68	0.71	0.96
		Right	0.73	0.51	1.01	0.62
	LF	Left	0.63	0.28	0.73	0.33[a,b]
		Right	0.71	0.28	0.71	0.51
	HF	Left	0.08	0.28	−0.03	0.27
		Right	0.07	0.24	−0.04	0.23

Note H_a μ mild>μ severe. [a]Nonparametric test (Mann-Whitney U). [b]H_a μ mild<μ severe

4 Discussion

In this study, it was evaluated the impact of the arterial hypertension severity in CA assessment through four analysis methods (ARI, ARMA-ARI, Mx and TFA). Individual results of each CA method showed a preserved CA for the two groups (mild and severe hypertension) for both cerebral hemispheres. CA represents the ability of cerebral arteriolar vessels to regulate their on-blood supply during changes in ABP (between 50 and

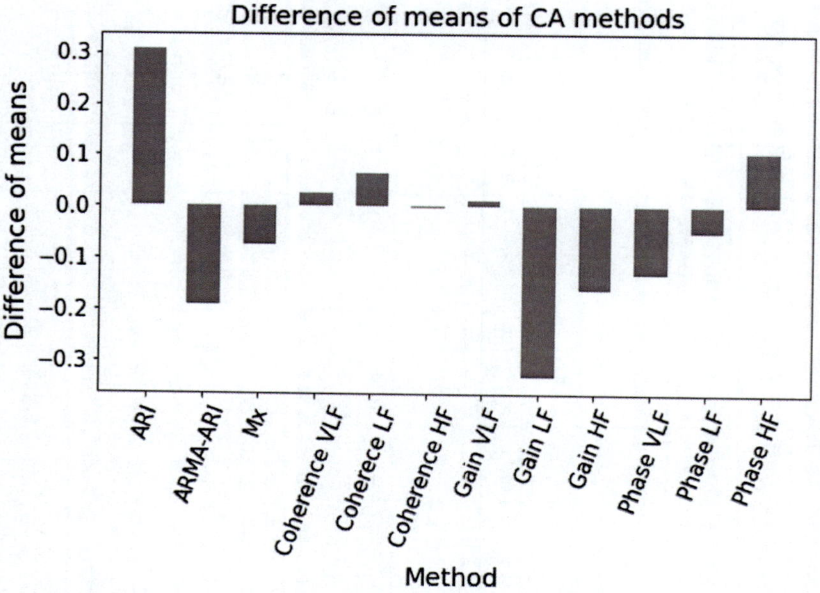

Fig. 2. Difference between the mean values of mild and severe SAH groups, for CA method

Table 2. Correlation between CA methods

Method compared with ARI		Correlation	p-value
ARMA-ARI		0.44	< 0.001[a]
Mx		−0.40	0.001[a]
Coherence	VLF	−0.34	0.008[a]
	LF	0.28	0.031[a]
	HF	0.20	0.123
Gain	VLF	−0.13	0.321
	LF	0.28	0.031
	HF	0.06	0.665
Phase	VLF	0.18	0.173
	LF	0.33	0.011[a]
	HF	0.08	0.527

[a]p-value < 0.05

150 mmHg). In SAH, the lower limit of CA is raised leading to a shift of the pressure-flow relationship to the right [1]. This may be the result of the adaptive changes on the structure of the cerebral vessels, including hypertrophic remodeling [16]. As consequence of this limit shifting, SAH patients may be more susceptible to cerebral ischemia during

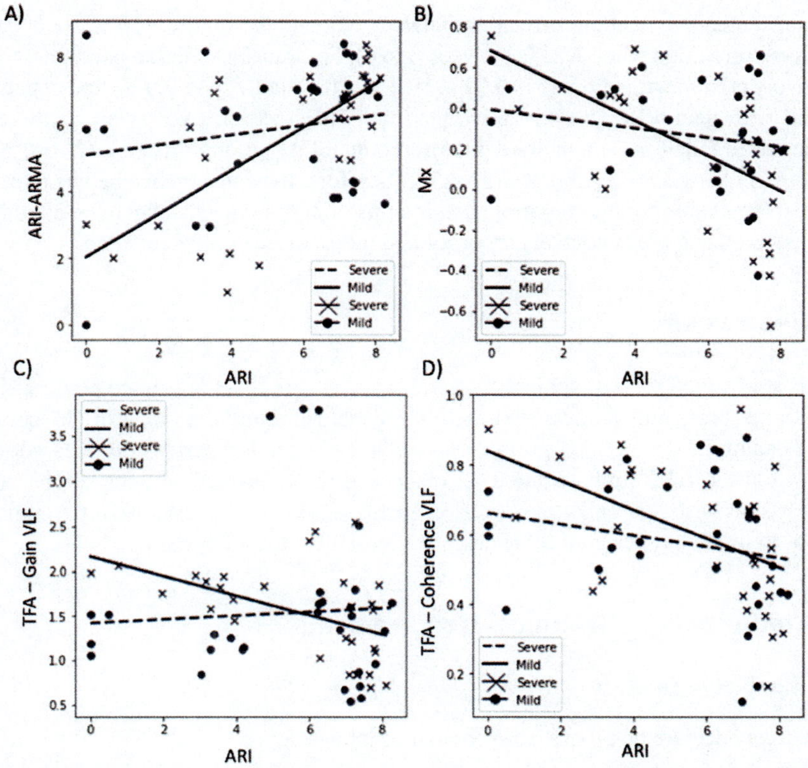

Fig. 3. Correlations between CA methods stratified by SAH degree. A) ARI versus ARMA-ARI; B) ARI versus Mx; C) ARI versus TFA—Gain VLF; D) ARI versus TFA—Coherence VLF

sustained decreased in ABP [17]. Long-term treatment of hypertension may partially reverse this rightward shift [18].

In line with our results, previous studies have demonstrated no CA impairment in SAH patients [17, 19]. Moreover, our findings also suggest that CA is well preserved irrespective of the severity. Despite of these results, we also observed in Fig. 2, a difference in the dispersion of the data according to each group. When the correlations between CA methods are compared, the severe group presents a lower slope of the correlation line when compared with the mild group. Though, even with this difference in the slope among the correlation lines, the CA results for severe groups remains in a range considered to be preserved CA. Data should be interpreted with caution considering the relatively small sample of each group. A study with a larger sample may bring a clearer relationship between the degrees of hypertension and CA.

Furthermore, a standardized method for dynamically assessing autoregulation has not been established, limiting direct interpretation of existing studies [20]. ARI method was arbitrarily chosen for comparisons with other calculation methods in this study, based in a previous study [21]. In line with our results, Czosnyka et al. have showed moderate correlation (r = 0.62) between ARI and Mx methods [7]. Whereas Liu et al.

[21] presented even better correlation between Mx and ARI ($r = -0.95$, $p < 0.001$, stronger correlation when ARI is between 3 to 6). For transfer function parameters, ARI positively related with phase ($r = 0.99$ at VLF and 0.93 at LF, $p < 0.001$) and negatively related with gain VLF ($r = -0.98$, $p < 0.001$). However, two recent multi-center studies have failed to demonstrate good reproducibility among different CA methods, particularly between ARI and Mx [22, 23]. Therefore, the convergence between the CA methods needs further investigation, helping this important mechanism to be effectively implemented and are adequately incorporated into routine clinical practice.

5 Conclusion

In this study, the group of subjects with mild and severe SAH presented preserved CA in both cerebral hemispheres, leading to the conclusion that the range of CA curve is able to adapt in severe SAH group, i.e., sustained arterial hypertension does not alter CA within the ABP limits studied. In respect of the inter-method analysis, based on comparisons with Spearman's correlation estimates, the reproducibility of CA methods is poor to moderate when all SAH cases were used for the analysis.

6 Compliance with Ethical Requirements

6.1 Conflict of Interest

The authors declare that there is no conflict of interest.

6.2 Statement of Informed Consent

All enrolled subjects gave informed consent for aquisitionof CBFv and ABP signals.

6.3 Statement of Human and Animal Rights

The study had local ethical approval (ethics comitee committee research protocol n FMUSP—CAPPesq #4.386.526).

References

1. Strandgaard, S., et al.: Autoregulation of brain circulation in severe arterial hypertension. BMJ **1**(5852), 507–510 (1973). https://doi.org/10.1136/bmj.1.5852.507
2. Pires, P.W et al.: The effects of hypertension on the cerebral circulation. Amer. J. Physiol. Heart and Circulatory Physiol. **304**(12) (2013). https://doi.org/10.1152/ajpheart.00490.2012
3. Oliveira, M. et al.: Intracranial hypertension and cerebral autoregulation: a systematic review and meta-analysis. World Neuro Sugery **113** (2018). https://doi.org/10.1016/j.wneu.2018.01.194
4. Shashank, S et al.: Cerebral autoregulation in hypertension and ischemic stroke: a mini review. J. Pharmaceut. Sci. **1**, 21–27 (2017)

5. Michel Ferreira, M. et al.: Evaluation of cerebral autoregulation performance in patients with arterial hypertension on drug treatment. J. Clinical Hypertension (Greenwich, Conn.) **22**(11), 2114–2120 (2020). https://doi.org/10.1111/jch.14052
6. Duarte, A.M. et al.: Evaluation and systematization of the transfer function method for cerebral autoregulation assessment. In: Bastos-Filho. IFMBE Proceedings, vol. 83 (2014)
7. Czosnyka, M., et al.: An assessment of dynamic autoregulation from spontaneous fluctuations of cerebral blood flow velocity: a comparison of two models, index of autoregulation and mean flow index. Anesth. Analg. **106**, 234–239 (2008)
8. Panerai, R.B., et al.: Analysis of cerebral blood flow autoregulation in neonates. IEEE Trans. Biomed. Eng. **43**, 779–788 (1996)
9. Caldas, J., et al.: Dynamic cerebral autoregulation is impaired during submaximal isometric handgrip in patients with heart failure. Am. J. Physiol. Heart Circ. Physiol. **315**, H254–H261 (2018)
10. Tiecks, F.M., et al.: Comparison of static and dynamic cerebral autoregulation measurements. Stroke **26**, 1014–1019 (1995)
11. Panerai, R.B., et al.: Variability of time-domain indices of dynamic cerebral autoregulation. Physiol. Meas. **24**(2), 367–381 (2003). https://doi.org/10.1088/0967-3334/24/2/312
12. Payne, S.: Cerebral autoregulation: control of blood flow in the brain (2016)
13. Semmlow, J., Griffel, B.: In: Biosignal ans Medical Images. 3rd edition. CRC Press
14. Claassen, et al.: Transfer function analysis of dynamic cerebral autoregulation: a white paper from the international CARNet . Blood Flow Metab **36**, 665–680 (2016)
15. Zhang, R. et al.: Transfer function analysis of dynamic cerebral autoregulation in humans. Am J. Physiol. **274**, H233–H241 (1998)
16. Paulo, W.P., et al.: The effects of hypertension on the cerebral circulation. Dorrance Amer. J. Physiol.-Heart and Circulatory Physiol. **304**(12), H1598–H1614 (2013)
17. Traon, A.P., et al.: Dynamics of cerebral blood flow autoregulation in hypertensive patients. J. Neurol. Sci. **195**(2), 139–144 (2002)
18. Tryambake, D. et al.: Intensive blood pressure lowering increases cerebral blood flow in older subjects with hypertension. **61**(6), 1309-15 (2013). https://doi.org/10.1161/HYPERTENSIONAHA.112.200972
19. Eames, P.J., et al.: Cerebral autoregulation indices are unimpaired by hypertension in middle aged and older people. Am. J. Hypertens. **16**, 746–753 (2003)
20. Ruland, S., et al.: Cerebral autoregulation and blood pressure lowering. Hypertension **49**(5), 977–978 (2007). https://doi.org/10.1161/HYPERTENSIONAHA.107.087502
21. Liu, X., et al.: Assessment of cerebral autoregulation indices—a modelling perspective. Sci. Rep. **10**, 9600 (2020)
22. Sanders, M.L., et al.: Dynamic cerebral autoregulation reproducibility is affected by physiological variability. Front. Physiol. **10**, 865 (2019). https://doi.org/10.3389/fphys.2019.00865
23. Sanders, M.L. et al.: Reproducibility of dynamic cerebral autoregulation parameters: a multicentre, multi-method study. Physiol Meas **39**,125002 (2018). https://doi.org/10.1088/1361-6579/aae9fd

Prediction of Stem Cell Differentiation for Cardiac Tissue Engineering Using the Finite Element Method (FEM)

Thaís Souza[1], Ronny Carbonari[2](✉), and Márcia Simbara[1]

[1] Faculty of Electrical Engineering, Federal University of Uberlândia, Uberlândia, Brazil

[2] Center of Engineering, Modelling and Applied Social Sciences, Federal University of ABC, São Bernardo do Campo, Brazil

ronny.carbonari@ufabc.edu.br

Abstract. Tissue engineering aims to repair damaged tissues, and to do so it needs three basic elements: cells, scaffolds and signals. Understanding the relationship between these elements, i.e. how the scaffold material and different types of stimuli affect cell behavior, is crucial to form a functional tissue. Mechanotransduction is the transformation of mechanical stimuli into biochemical cues, and many researchers have conducted experiments and created models to better understand how this mechanism works. The Mechanical Bidomain Model (MBM) predicts that mechanotransduction and, therefore, cellular differentiation occur in regions of differences between intra and extracellular displacements. In this work, we applied the MBM to a Matrigel® scaffold seeded with human induced pluripotent stem cells (hiPSCs), and compared the results to a Finite Element (FE) model we developed, considering integrins as rigid bonds between the cell and the scaffold, with an effect similar to a truss link between them. Results showed that both methods suggest a larger difference between displacements (and, therefore, differentiation) on the edge of the cell colonies compared to points closer to the center, which was in accordance with experimental data from the literature. However, in the FE analysis, differentiation should also occur in the vicinity of the center of the colonies, a phenomenon described by Myers et al. (2013), who showed that distinct differentiation strategies can lead to that type of differentiation pattern. In conclusion, the proposed FE model was able to replicate the MBM results, but also predict another type of behavior that was not foreseen before.

Keywords: Cell differentiation · Tissue engineering · Finite element method · Mechanical bidomain model

1 Introduction

Acute myocardial infarction is a disease that causes damage to heart tissue, which is not reversible with currently available treatments. Thus, tissue engi-

J. L. B. Marques et al. (Eds.): CLAIB 2022/CBEB 2022, IFMBE Proceedings 98, pp. 256–265, 2024.
https://doi.org/10.1007/978-3-031-49401-7_25

neering (TE), a relevant alternative to repair damaged tissues and organs, is a future hope of treatment [1].

TE is formed by a triad of elements: the biomaterial, fundamental constituent of the scaffold (important for initial cellular support) [1]; the cells, that populate the scaffolds and form the engineered tissue; and signals, such as growth factors, to induce the desired responses for the application [2,3]. In this way, it is necessary to: 1) select the appropriate cells; 2) select the scaffold material, considering its biological and mechanical characteristics; 3) fabricate the scaffold; and 4) cultivate the cells in the scaffold [1].

In this context, the mechanical properties of the scaffold (such as the longitudinal and transverse elasticity modulus, E and G, respectively) can influence cell conformation on the substrate and differentiation. It has been shown that mesenchymal stem cells (MSCs) can undergo adipogenesis or osteogenesis depending on the E of the scaffold [4].

This transformation of mechanical stimuli into biochemical cues is called mechanotransduction, a process mediated by transmembrane proteins, with emphasis on integrins, which make the connection between the extracellular matrix (ECM) and the cell [4]. Integrins bind to specific ligands, such as RGD peptide sequence (Arginine-Glycine-Aspartate), present in ECM proteins, and to intracellular adaptor proteins, which mediate the connection with the cytoskeleton [5]. In this way, they act in the propagation of a force, stress or strain between the intra and extracellular spaces.

Based on the importance of integrins, the Mechanical Bidomain Model (MBM) predicts that mechanotransduction and, therefore, cellular differentiation, occur in regions of differences between intra and extracellular displacements [6]. The consequence of these differences are forces acting on the membrane, inducing a cascade of biochemical events related to differentiation. So, it is possible to estimate the most likely place of cellular differentiation and its extension, as well as their variations in response to different substrates.

The mechanotransduction effects could also be simulated using a Finite Element Method (FEM) structural analysis, determining the displacements, stresses and strains, especially in the region between the intra and extracellular media. This approach has been used in numerous works, however, generally seeking the direct relationship of the strains [7] and stress magnitudes [8] with cell differentiation, without considering the difference between the displacements of the two media and the influence of the transmembrane proteins in this process.

The present work aimed to: 1) propose a new model that better represents the integrin coupling to the intra and extracellular domains in the FE mesh, by modifying the stiffness matrix of this region (instead of simply adjusting its Young Modulus); 2) simulate, using MBM and FE analysis, the differentiation of stem cells (with potential application in myocardial TE) subjected to mechanical stimulation and cultivated *in vitro*, and compare the results of these two methods. The application of the proposed model would allow the observation of the effects of simplifications in the MBM equations validated by Auddya and Roth [6] and allow greater knowledge of the interaction between the cell and its medium.

2 Methodology

The implementations of both models were made in Python and are available at https://github.com/thais-souza/Dif.-Celular.git. The results obtained here were also compared with simulations made in Abaqus software (Dassault Systèmes, France).

Besides that, the MBM and FEM inputs were included in such a way that the results could be compared with experimental standards present in the literature. For this, a Matrigel® scaffold and human induced pluripotent stem cells (hiPSCs) were applied in the models [9].

As for the symbology used to describe the input data of the models, the properties are specified according to the accompanying subscript: s, referring to the scaffold; c to the seeded cells; i, to the integrin.

2.1 Mechanical Bidomain Model

In the MBM, the intra and extracellular domains are represented as spring sequences, and so are the connection between these two spaces (integrins). In this context, displacements are differentiated into: 1) related to the monodomain, characterized by the equality of these quantities in the two spaces; 2) related to the bidomain, in which the difference between the displacements occurs, deforming the integrins and producing the mentioned forces, therefore causing cellular differentiation [10].

Here, the equations analytically obtained by Auddya and Roth [6] were used, and the tissue was considered an incompressible fluid and with isotropic characteristics. Other considerations for the model were: the thickness of the cell layer was much smaller than the radius, and colony growth would occur in a radial and symmetrical manner (regardless of the θ coordinate).

The following parameters were established in the MBM: colony radius, R; the uniform stress term, T, relative to the growth and crowding of cells; the scaffold transverse elasticity modulus, G_s; the cell transverse elasticity modulus G_c. Also, a length constant, σ (assumed equal to $\frac{R}{10}$ by Auddya) was used [6]. These parameters are related according to the following equations:

$$u_c = -\frac{T}{6\left(G_c + G_s\right)}\left[r + \frac{3}{2}\left(\frac{G_s}{G_c}\right)\sigma e^{\frac{r-R}{\sigma}}\right], \tag{1}$$

$$u_s = -\frac{T}{6\left(G_c + G_s\right)}\left[r - \frac{3}{2}\sigma e^{\frac{r-R}{\sigma}}\right], \tag{2}$$

in which u is the displacement and r is the radial coordinate of the point of interest. By the considerations made, there is not a angular displacement, and therefore, the above equations give just the radial displacement.

Following that, the stresses showed in the results were obtained by $\epsilon_c^{rr} = \frac{\partial u_c}{\partial r}$ and $\epsilon_s^{rr} = \frac{\partial u_s}{\partial r}$, while the strains were calculated by $\tau_c^{rr} = G_c\left(\frac{2u_c}{r} + 4\varepsilon_c^{rr}\right) + T$ and $\tau_s^{rr} = G_s\left(\frac{2u_s}{r} + 4\varepsilon_s^{rr}\right)$.

2.2 Finite Element Model

Meanwhile, in the FEM formulation, it was assumed that the engineered tissue formed was made up of 4-noded axisymmetric solids elements (isoparametric formulation) of isotropic materials, characteristics that were used in previous studies [7]. Thus, it was considered that each constituent part of the engineered tissue (cell, integrin and scaffold) was formed by a single hookean material with the input data obtained from the literature.

In this model, a linear and static analysis was applied, following, therefore, Eq. 3:

$$K \cdot u = F, \tag{3}$$

where K, u and F are the global stiffness matrix, the nodal displacement vector for r and z directions, and the external force vector, respectively.

The axisymmetric formulation was taken from Õnate [11]. However, for the integrin elements, the axisymmetric formulation was modified to seek greater proximity to the behavior showed by these proteins in reality after the application of a displacement, stress or strain in the tissue. Here, we proposed the formation of a rigid bond between the cell and the scaffold, with an effect similar to a truss link between the upper and lower nodes within the elements that define integrin. To that end, the impacts of node crosslinking on the local stiffness matrix, k, were minimized.

In general, the local stiffness matrix is calculated by:

$$k = 2\pi \int_{-1}^{1} \int_{-1}^{1} B(\xi, \eta)^T \cdot D \cdot B(\xi, \eta) \cdot |J| \cdot r(\xi, \eta)\, d\xi d\eta, \tag{4}$$

in which B is the strain-displacement matrix obtained from the shape functions; D is the constitutive matrix, variable according to the material; and J is the Jacobian matrix.

For the application of the algorithm, a local coordinate system (ξ,η) is considered in each element. This system is related to the global coordinates through the standard displacement shape functions (N):

$$r(\xi, \eta) = \sum_{j=1}^{4} N_j(\xi, \eta) \cdot r_j,\ z(\xi, \eta) = \sum_{j=1}^{4} N_j(\xi, \eta) \cdot z_j,$$

in which (r_j, z_j) are the global coordinates for the j node in the element and $N_j = \frac{1}{4}(1 + \xi\xi_j)(1 + \eta_j\eta)$. In the local coordinates (ξ,η), for a normalized geometry, the nodes are placed in: $(-1, -1)$, $(1, -1)$, $(1, 1)$, $(-1, 1)$.

The strain-displacement matrix, on the other hand, consists of a sequence of 4×2 items determined by each node according with Eq. 5,

$$B_j = \begin{bmatrix} \frac{\partial N_j}{\partial r} & 0 \\ 0 & \frac{\partial N_j}{\partial z} \\ \frac{N_j}{r(\xi,\eta)} & 0 \\ \frac{\partial N_j}{\partial z} & \frac{\partial N_j}{\partial r} \end{bmatrix}, \quad (5)$$

meanwhile the Jacobian matrix is given by Eq. 6,

$$J = \sum_{j=1}^{4} \begin{bmatrix} \frac{\partial N_j}{\partial \xi} r_j & \frac{\partial N_j}{\partial \xi} z_j \\ \frac{\partial N_j}{\partial \eta} r_j & \frac{\partial N_j}{\partial \eta} z_j \end{bmatrix}. \quad (6)$$

As for the proposed modification in the integrin elements to include the effects of this protein in the FEM model, the described changes results in the following local stiffness matrix:

$$k = \begin{bmatrix} a & a & a & a & a & a & k_{1,7} & k_{1,8} \\ a & a & a & a & a & a & k_{2,7} & k_{2,8} \\ a & a & a & a & k_{3,5} & k_{3,6} & a & a \\ a & a & a & a & k_{4,5} & k_{4,6} & a & a \\ a & a & k_{5,3} & k_{5,4} & a & a & a & a \\ a & a & k_{6,3} & k_{6,4} & a & a & a & a \\ k_{7,1} & k_{7,2} & a & a & a & a & a & a \\ k_{8,1} & k_{8,2} & a & a & a & a & a & a \end{bmatrix},$$

with $a = 0.0001 \cdot max(k)$.

In this method, the model input parameters were: the structure radius, R; the thicknesses t_s, t_i and t_c; the number of elements in the direction of the radius, n_r, and in the direction of the thickness (z), n_z; the E_s, E_i and E_c, as well as Poisson's ratios ν_s, ν_i and ν_c. In addition, the application features of the prestress were defined: the limits, in z, of the action region (l_i is the lower limit and l_s the upper), the prestress module and its direction.

2.3 Model Geometry and Properties

It is possible to see that both models have some similar input properties. Thus, the same structure of $R = 2\ mm$ with the scaffold and the cell, with the same mechanical properties, was used in both MBM and FEM, remembering that, because of the Hookean material in question, the relation $E = 2 \cdot G \cdot (1 + \nu)$ is valid.

Scaffold Thereby, the values of E_s and ν_s were identified in the literature, allowing the calculation of G_s. Soofi et al. measured the aforementioned mechanical properties of Matrigel® and obtained an average value of $E_s = 443\ Pa$ by adopting $\nu_s = 0.5$ [12].

However, $\nu_s = 0.49$ was used to avoid mathematical indeterminacy in the equations solved with the FEM. This same strategy was used for cells. From these data, $G_s = 149\ Pa$ was calculated. Regarding the thickness, the scaffolds can have their dimensions delimited according to design and production technique. In this case, $t_s = 0.3\ mm$ was adopted (based on scaffolds which were previously produced by the authors).

Cells A similar process was used to determine the cell properties. Hammerrick et al. [13] measured some mechanical properties of induced pluripotent stem cells derived from human adipose-derived stromal cells (hASC-iPSCs), and obtained $E_c = 0.9\ kPa$, by adopting $\nu_c = 0.5$. Therefore, considering $\nu_c = 0.49$ (for the same reason described in section C1), $G_c = 0.3\ kPa$. In addition, these cells have the diameter of, approximately, 15 μm, thus, for a monolayer of cells, $t_c = 15\ \mu m$.

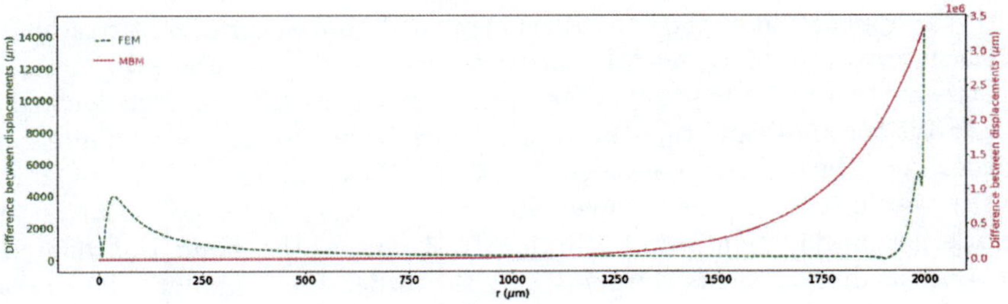

Fig. 1. Difference between intra and extracellular displacements by MBM and FEM

Integrins After the scaffold and cell definitions, it was necessary to model the connection between these two components. With the generation of stresses over the cell-matrix junctions, also part of the natural mechanism of stiffness perception of the ECM by the cells, there is a reinforcement in the connection between the ECM and the cytoskeleton, through the recruitment of adjacent proteins that enhance the binding of integrin to actin [5], and the increase in the density of integrins in the activation region [5].

In this context, a high stiffness value for integrins is important, so a stiffness of $E_i = 1000 \cdot E_c$ was considered. As for the thickness of this protein, it is known that the maximum size of the integrin binding between a collagen fiber and the cell membrane is approximately 56 nm [14]. However, due to restrictions on RAM memory, limiting of the number of finite elements, $t_i = 56\ nm$ was not feasible.

Thus, according to the maximum number of divisions in z, the minimum possible size for E_i (from a single line of integrins) was 1 μm. In this context, $n_r = n_z = 500$, a value greater than the minimum expressed for n_r and n_z in the mesh convergence analysis.

2.4 Prestress and Boundary Conditions

Finally, a prestress was applied to the structure simulating the stiffness perception of the ECM by the cells. In practice, this is directly related to myosin retraction, with the generation of a stress on the substrate. In a collagen substrate, for fibroblasts, this generation reaches 10 Pa [15].

Therefore, a value of $T = -10\ Pa$ was assumed and, in the FEM, a prestress in r, directed to the center of the structure with the same magnitude and between $l_i = 310\ \mu m$ and $l_s = 316\ \mu m$, was applied. Regarding the boundary conditions, a displacement equal to zero on the inferior surface of the scaffold was considered.

3 Results and Discussion

With the application of the conditions expressed above, intra and extracellular displacements, as well as the difference between them, were obtained (Fig. 1). In the case of the FEM, the results were taken from the rows of elements located at three positions above and three positions below the region composed of integrins. Besides that, the difference between the intra and extracellular displacements in FEM was calculated by the euclidean distance between the displacement vectors considering the directions r and z. In the MBM, on the other hand, this difference between the displacements considers only the radial direction, once there is not a angular displacement in the model.

In both models, there is a peak of the difference of displacements in the edges of the cell colony, behavior also presented in both numerical and experimental works [8,16]. However, in the FE results, a maximum value also occurs at $(36.0, 3934.68)$. The presence of more than one probable region of differentiation is in accordance with Myers et al. [9].

Myers et al. observed that hiPSCs cultivated on a patterned Matrigel® substrate could have different differentiation patterns depending on the growth factors and/or biomolecules added to culture medium. Some showed prominent differentiation between the edge and the middle of the substrate, while others showed a pattern of two concentric rings, with the interior ring at roughly the center of the original colony [9].

This behavior was not foreseen in the model by Auddya and Roth [6], especially because it was built around the theory that differentiation occurs first at the edge of cell colonies. The proposed FE modeling, on the other hand, assumed nothing regarding cell behavior (except that the pattern would have some kind of axial symmetry). Despite that, more experimental data are needed to determine which model is more suitable to represent the differentiation pattern of the cell colony.

Besides the analysis of displacements, it is also necessary to discuss the behavior of stresses and strains, expressed in Fig. 2. Considering that we do not have experimental data to validate the model (except for what could be found in the literature), the viable option would be to check whether the stress and strain curves obtained with the FE analysis would be similar to the ones obtained with the MBM. Disregarding more pronounced numerical instabilities at 0 and 2000 μm, all curves present a similar behavior.

The stress and strain magnitudes, on the other hand, are considerably different, which could be caused by the simplifications in the MBM. It was the limitations regarding the numerical resolution of the problem and the imposition of characteristics about the integrins and about the cell colony itself, such as the consideration of the plane stress, that have led to the suggestion of a model based on the FEM in the first place.

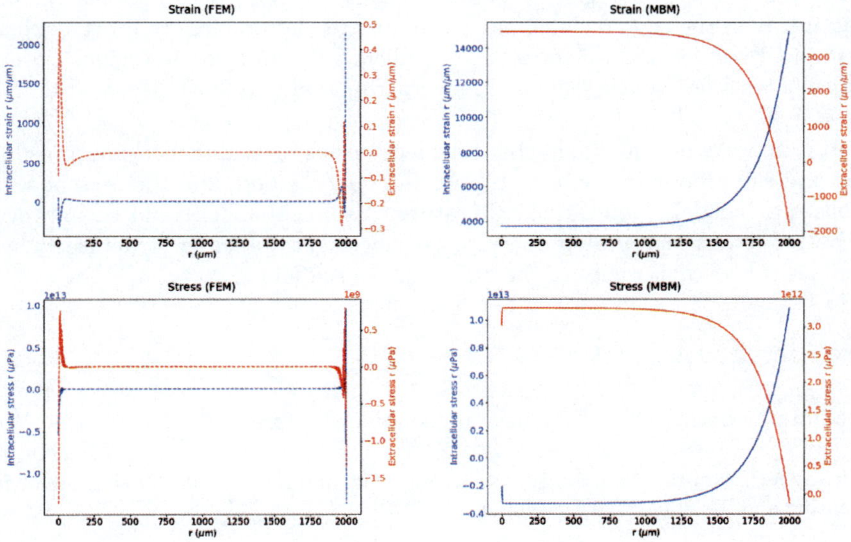

Fig. 2. Intra and extracellular stresses and strains by MBM and FEM

It is important to highlight that there are other hypotheses that consider other mechanical properties as having the most representative influence on the spatial distribution of cell differentiation. Li et al., for example, propose that the stress magnitude would be this property. However, their experimental results also show that differentiation naturally occurs first at the edge of the colony, and if the scaffold is stretched, differentiation occurs in the central area as well. However, they indicate that there would be a "window stress threshold" for a cell colony to respond to mechanical stresses and form cell differentiation patterns [8]. Despite this, none of the models were able to predict this behavior.

4 Conclusion

Cell differentiation is influenced not only by the insertion of biomolecules and growth factors in culture media, but also by the characteristics of the scaffold, including its stiffness. In this way, mechanotransduction is a mechanism that induces the differentiation. Therefore, the connections between the cell and the scaffold, mediated mainly by integrins, are vital.

The proposed FE model maintained the MBM's essence, but was able to overcome certain limitations and disregard possibly faulty assumptions. The MBM was built around the theory that cell differentiation first occurs at the edge of cell colonies, yet experimental data has shown that this is not always the case: adding different biomolecules to culture media, or submitting the engineered structure to certain strain values may impact the cell differentiation pattern.

Simulating the MBM and the FE model, it has been shown that the culture of hiPSCs cells on Matrigel® substrates results in differences between the displacements in the edge regions, pattern of differentiation already proven experimentally. However, the FE model also predicts differentiation in regions close to the colony center, which was observed experimentally by both Myers et al. and Li et al. [8,9].

It is important to highlight that the mechanisms regarding mechanotransduction and cell differentiation are still not fully understood, and the present work is based on models found in the literature. Also, the model could be improved to better reflect material properties, considering a dynamic structure, which includes the viscoelasticity of the cells and the scaffold.

Conflict of Interest

The authors declare that they have no conflict of interest.

Acknowledgments. This study was financed in part by the National Council for Scientific and Technological Development (CNPq).

References

1. Rodrigues, I., Kaasi, A., Filho, R., Jardini, A., Gabriel, L.: Cardiac tissue engineering: current state-of-the-art materials, cells and tissue formation. einstein (São Paulo) **16**, 1–9 (2018). https://doi.org/10.1590/S1679-45082018RB4538
2. O'Brien, F.: Biomaterials and scaffolds for tissue engineering. Mater. Today **14**, 88–95 (2011). https://doi.org/10.1016/S1369-7021(11)70058-X
3. Fleischer, S., Dvir, T.: Tissue engineering on the nanoscale: lessons from the heart. Curr. Opin. Biotechnol. **24**, 664–671 (2013). https://doi.org/10.1016/j.copbio.2012.10.016
4. Wheeldon, I., Farhadi, A., Bick, A., Jabbari, E., Khademhosseini, A.: Nanoscale tissue engineering: spatial control over cell-materials interactions. Nanotechnology **22**, 212001 (2011). https://doi.org/10.1088/0957-4484/22/21/212001

5. Alberts, B., Johnson, A., Lewis, J., Morgan, D., Raff, M., Roberts, K., Walter, P., Wilson, J., Hunt, T.: In: Molecular Biology of the Cell. Garland Science, New York (2015)
6. Auddya, D., Roth, B.: A mathematical description of a growing cell colony based on the mechanical bidomain model. J. Phys. D Appl. Phys. **50**, 105401 (2017). https://doi.org/10.1088/1361-6463/aa59b5
7. Sen, S., Engler, A., Discher, D.: Matrix strains induced by cells: Computing how far cells can feel. Cell. Mol. Bioeng. **2**, 39–48 (2009). https://doi.org/10.1007/s12195-009-0052-z
8. Li, B., Li, F., Puskar, K., Wang, J.: Spatial patterning of cell proliferation and differentiation depends on mechanical stress magnitude. J. Biomech. **42**, 1622–1627 (2009). https://doi.org/10.1016/j.jbiomech.2009.04.033
9. Myers, F., Silver, J., Zhuge, Y., Beygui, R., Zarins, C., Lee, L., Abilez, O.: Robust pluripotent stem cell expansion and cardiomyocyte differentiation via geometric patterning. Integr. Biol. **5**, 1495–1506 (2013). https://doi.org/10.1039/c2ib20191g
10. Roth, B.: A Mathematical Model of Mechanotransduction. arXiv (2016). 10.48550/ARXIV.1611.08287
11. Õnate, E.: In: Structural Analysis with the Finite Difference Method: Linear Statics. Springer, Barcelona (2009)
12. Soofi, S., Last, J., Liliensiek, S., Nealey, P., Murphy, C.: The elastic modulus of Matrigel^{TM} as determined by atomic force microscopy. J. Struct. Biol. **167**, 216–219 (2009). https://doi.org/10.1016/j.jsb.2009.05.005
13. Hammerick, K., Huang, Z., Lam, M., Longaker, M.: Elastic properties of induced pluripotent stem cells. Tissue Eng. Part A **17**, 495–502 (2011). https://doi.org/10.1089/ten.TEA.2010.0211
14. Lepzelter, D., Bates, O., Zaman, M.: Integrin clustering in two and three dimensions. Amer. Chem. Soc. Public. **28**, 5379–5386 (2012). https://doi.org/10.1021/la203725a
15. Moore, S., Roca-Cusachs, P., Sheetz, M.: Stretchy proteins on stretchy substrates: the important elements of integrin-mediated rigidity sensing. Developm. Cell Rev. **19**, 194–202 (2010). https://doi.org/10.1016/j.devcel.2010.07.018
16. Sharma, K., Al-Asuoad, N., Shillor, M., Roth, B.: Intracellular, extracellular, and membrane forces in remodeling and mechanotransduction: the mechanical bidomain model. J. Coupled Syst. Multiscale Dynam. **3**, 200–207 (2015). https://doi.org/10.1166/jcsmd.2015.1079

Simplified Computational Model of the Cervical Region for Transcutaneous Spinal Direct Current Stimulation

L. O. Fernandes[1,2](✉), C. M. Germer[3], and P. X. de Oliveira[1]

[1] FEEC/DEEB, University of Campinas (UNICAMP), Campinas, Brazil
lofernandes@uesc.br

[2] DCET, University of Santa Cruz (UESC), Ilhéus, Brazil

[3] DEBM, University of Pernambuco (UFPE), Recife, Brazil

Abstract. Transcutaneous spinal direct current stimulation (tsDCS) is a neuromodulation technique used for the rehabilitation of spinal cord disorders and injuries. Despite its potential effect, the stimulation parameters are not well established, and the underlying effects of tsDCS on the neural mechanisms in the spinal cord remain unknown. A common approach to these problems is the use of computer models to simulate both the electric field in the targeted region and the neural response to the stimulus. Nevertheless, these models are limited, especially for tissues in the cervical region. Thus, this paper presents a model of the volume conductor for the cervical region and optimized simulations using the finite element method in the solution of the current flow problem induced by the application of tsDCS. It was possible to obtain a volume conductor model representing different tissues: skin, subcutaneous fat, muscle, trachea, esophagus, cartilage, vertebral ligaments, vertebrae, vertebral arteries, intervertebral discs, duramater, cerebrospinal fluid, white and gray matter of the spinal cord. Three meshes with different element densities were produced (low, intermediate, and high resolution). The lower density mesh showed an electric field with a mean square error between 6.21% and 2.50% in the spinal cord tissues when compared to the higher resolution mesh. Therefore, the optimization proposed reached acceptable errors and allows multiple simulations, using different setups (conductivities, current intensity, geometric and positions of electrode) with less computational cost (88 min to 17 min for runtime and 4 GB to 800 MB for local storage).

Keywords: Electric field · Spinal cord · Finite element method · Optimization

1 Introduction

Electrical stimulation has been widely employed for therapeutic purposes throughout history. Late reports (43–48 A.D.) indicate the use of electrical discharges from torpedo fish in the treatment of headache [1]. Despite the diversity of techniques for stimulation (e.g., invasive, non-invasive, electrical, magnetic, alternating, and continuous), all of them are similar in their purpose of altering the functioning of the nervous system.

J. L. B. Marques et al. (Eds.): CLAIB 2022/CBEB 2022, IFMBE Proceedings 98, pp. 266–276, 2024.
https://doi.org/10.1007/978-3-031-49401-7_26

In general, techniques using DC stimulation are safe and inexpensive, which allows for different experimental treatment options in patients with disorder of central nervous system [2–4]. While stimulation at high intensities can directly evoke neural activation via action potential (AP), low-intensity stimulation influences neural functioning without the direct generation of APs. Recently, direct current stimulation (DCS) has been employed in a noninvasive way to promote modulatory effects on spinal motor circuitries [2]. However, the effects elicited by transcutaneous spinal direct current stimulation (tsDCS) still being poorly understood [5].

A low-intensity direct current (DC) applied to the skin at a targeted region of the spinal cord induces an electric field which causes short or long-term functional changes in the neural pathways [6]. Thus, the main goal of tsDCS in the cervical region is to modulate the behavior of the local neural circuits and the spinal pathways' excitability [6–8].

Due to the nature of the electric field interaction on tissues, different results are observed depending on the dose, the region of interest, and the positioning and shape of the electrodes. Also, the conditions of inhomogeneity, anisotropy, and the geometric complexity of the tissues difficult the analytical solution to the volume conductor problem [9, 10]. Therefore, computational modeling of the electrical stimulation of the central nervous system (computational neurostimulation) is an important tool for understanding how and which spinal mechanisms are affected by the application of tsDCS, and how it can be employed to promote the desired changes.

The finite element method (FEM) is one of the main computational modeling techniques employed to predict the current flow and the electric field in different segments of the spinal cord. It allows considering the different characteristics, properties, and locations of the tissues and electrodes [6]. Thus, an important advantage of FEM is its application in problems with high geometric complexity and/or a diversity of electrical properties.

The partial differential equations (EDP) problems, like presented above, are solved by FEM, where the solution domain is discretized into several finite elements, uniform or not, connected by nodes [10]. From this breakdown, usually into tetrahedral elements for the three-dimensional model, it is possible to compute the electric field in each element. The solution can be approximated by numerical methods, considering the material property in a system of simplified equations. However, in addition to the availability of the geometric model of the volume conductor, the number of mesh elements is directly linked to the accuracy of the solution, and this requires more computational capacity (runtime, memory, and storage) to FEM solver.

Computational models of volume conductors are essential for predicting the setups needed so that the clinical target of tsDCS can be safely achieved. However, there are still few studies that use computational modeling of tsDCS application, and those are focused on the thoracic and lumbar regions of the spinal cord [11, 12], whereas few works have focused on the cervical region [13, 14]. Perhaps one of the main reasons is the complexity involved in the construction process or the lack of availability of volume conductor models of the cervical region. Despite the models provided by the ViP (Virtual Population) library being accessible and based on real segmentation of human body tissues [15, 16], they present simplifications in their composition, when provided in their open format version, that may affect the desired simulation result in the region of interest.

The objective of this study is to present a simplified volume conductor model that would represent the human cervical region. Furthermore, we aim to evaluate how the model optimization interferes with the solution of the problem using FEM.

2 Materials and Methods

2.1 Volume Conductor Model of Cervical Region

The simplified model proposed in this study was built from anatomical data of the cervical region [16], using the geometric tool from COMSOL (v5.3, Stockholm, Sweden) and performed in a notebook computer with a 2.21 GHz Intel® Core™ i7-8750H processor, 64 GB DDR4 RAM and 512 GB NVMe storage. For this purpose, the following tissues were considered: the cervical vertebrae C2 to C7 [16–18]; the white (WM) and gray (GM) matter of the spinal cord [19]; the cerebrospinal fluid (CSF); the intervertebral discs [20]; the vertebral arteries [21]; the duramater [22]; the longitudinal, capsular, supraspinal, interspinal, and flavum ligaments [23, 24]; cartilage of the articular facet [25]; walls of the esophagus [26] and trachea [27]; musculature [16]; subcutaneous fat [16]; and skin [16].

Figure 1 shows the modeled vertebra overlayed by the anatomical C5 vertebra of the Duke V2.0.1 model from the ViP library [16] for comparison. Based on the average value of the anatomical parameters of the C2 to C7 vertebrae, a single model shows: body height of 12 mm; spinal canal width of 25 mm and depth of 17 mm; total length and width of 55 mm and 52 mm, respectively. In addition, each vertebra had both cortical and trabecular bones modeled [18], where the cortical shell thickness has an average value of 0.527 mm.

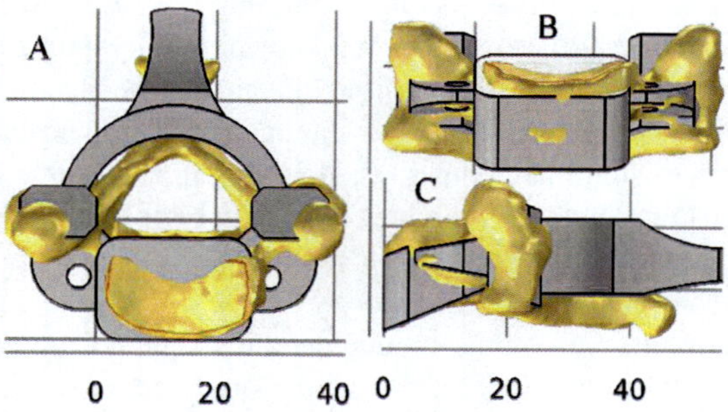

Fig. 1. Model used to represent all vertebrae in the cervical region (gray) and reference model built from real segment of C5 vertebra (yellow). Views with dimension in mm: transverse (A), anterior (B), and lateral (C).

Figure 2.A shows eight bidimensional segments of the cervical spinal cord segmented from diagrams based on Nissl images [19] and used to compose the tridimensional model of the cervical region (Fig. 2.B). The lordosis (curvature) of the cervical region was ignored. The initial (C1) and final (C8) segments of the spinal cord have been extended beyond their intercervical limits, allowing the GM geometry to end in these regions, avoiding direct contact with the CSF.

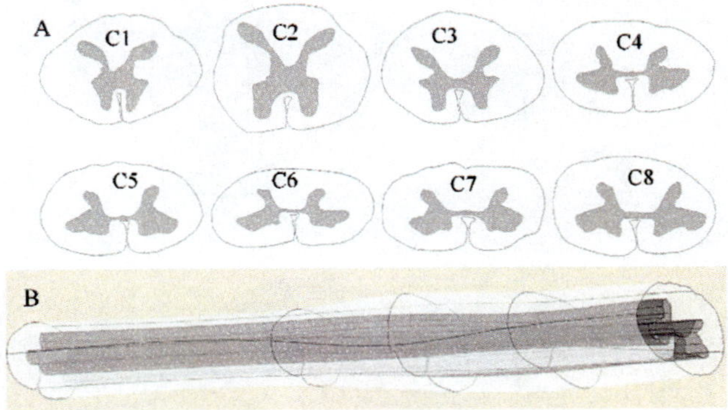

Fig. 2. A) Bidimensional segments for eight cervical sections of the spinal cord. B) Tridimensional model of the cervical region up to segment C5.

The geometric building of the remaining tissues was based on anatomical sizes (e.g., width, thickness, length, or height) or by approximation using the Duke V2.0.1 model [16]. The average value of the tissue dimensions in each segment of the cervical region was used in the individual geometry that forms the vertebra assemblies. The tissues that exhibit a large extension along the longitudinal axis, hold constant anatomical dimensions. Figure 3 shows the volume conductor model, it is composed of 13 vertebra assemblies (vertebra, disc, articular facet cartilage, flavum, and capsular ligaments) and other tissues that composes the cervical region, as presented above. In total, the geometric model had a volume of 2,953,800 mm^3 (2.96 L), with a diameter of 127 mm and height of 232 mm, presenting a total of 173 domains (geometric entities).

The standard position of the electrodes is given in the anterior (anode) and posterior (cathode) central region of the volume conductor. The electrode set (anode and cathode), with parameterizable dimension and positioning, is formed by the electrode ($49 \times 49 \times 1$ mm^3), the conductive gel/sponge ($50 \times 50 \times 2.5$ mm^3), and a representation of wire conductors (bottom edge of electrode) with a length of 1 mm and a cross-sectional area of 1 mm^2.

2.2 Mesh Parameterization

COMSOL (v5.3, Stockholm, Sweden) has five parameters for generating the FEM mesh of the geometric model: maximum element size; minimum element size; maximum element growth rate; curvature factor and resolution of narrow regions. These parameters can be filled from predefined sets (e.g., Extremely Fine, Fine, Normal, and Extremely Coarse) or individual values, uniformly or individually applied on the geometric entities and furthermore, control the mesh resolution (number of elements in the mesh).

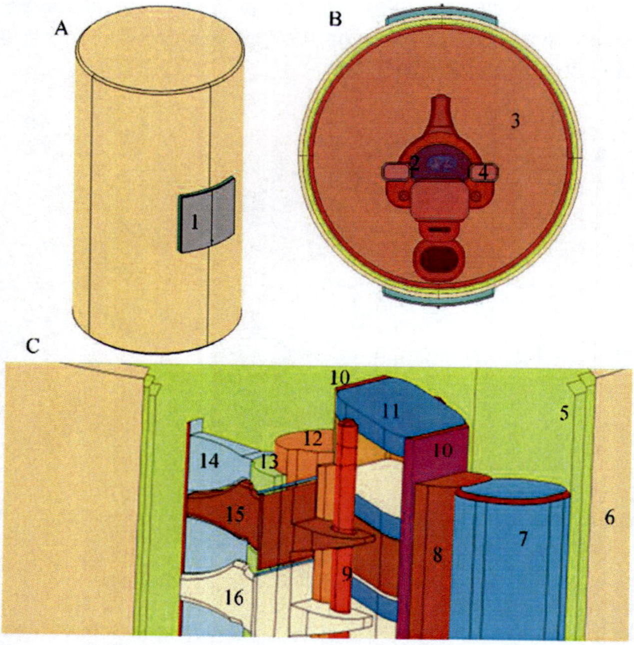

Fig. 3. Volume conductor proposed and its views with some tissues: 3D isometric (A), Transverse with tissues transparence (B), and 3D isometric with tissues details (C). The electrode (1) is placed on the anterior region. The view B shows: the spinal cord, composed by gray and white matters, and evolved by the cerebrospinal fluid (2); the articular cartilage facet and the capsular ligament (4). The vertebrae (16) are surrounded by muscle (3) and ligaments (4, 10, 13 and 14). More externally is the subcutaneous fat (5), rounded by skin (6). There are the: trachea (7), the esophagus (8), vertebral artery (9), longitudinal ligaments (10), disk (11), duramater (12), flavum ligament (13), interspinal ligament (14), trabecular bone of vertebrae (15), and the cortical shell vertebrae (16).

Additionally, COMSOL (v5.3, Stockholm, Sweden) provides different quality measures, in which the "Skewness" measure is the default and most suitable for most types of meshes. The "Skewness" is a measure of the equiangular skew and has values between 0 and 1, where 1 represents a perfectly regular element, and 0 represents an element with high or small angles (between edges). While for values under 0.01 is considered very low, the values above 0.1 are suitable for simulations. Since the accuracy of results is also associated with mesh quality, COMSOL (v5.3, Stockholm, Sweden) allows us automatically to measure, in the whole volume conductor or individual tissues, the minimum and average quality during the mesh building process.

From the geometric models, three mesh types were produced with different resolutions (Fig. 4). Mesh 1 was considered the reference for its high resolution and standard predefinition "Extremely Fine". The tissues were parameterized individually and manually for meshes 2 and 3 using the five parameters described above, attending for the minimum (> 0.01) and average (> 0.1) quality values during the mesh generation process.

Fig. 4. Cross-section in the central region of the volume conductor showing the distribution of tetrahedral elements for high, intermediate, and low-resolution mesh (left to right). The distortion quality of each element is indicated by the tone level, green for higher and red for lower quality.

For each mesh, it is possible to observe the differences in the elements density of some tissues, indicated by the delineation pattern formed between the interfaces of the elements. In mesh 1, there is a higher density of elements per tissue when compared to meshes 2 and 3. While mesh 2 shows a density reduction in some tissues (e.g., vertebra, trachea, and esophagus), mesh 3 shows a considerable reduction in the elements density throughout the tissues when compared to mesh 1.

As shown in Table 1, mesh 3 presented the smallest processing time, the lowest number of elements, and a small distortion in the total volume. The meshes 2 and 3 maintained the Minimum Quality but slightly reduced the Average Quality in comparison with mesh 1. Even though the number of elements in mesh 2 was intermediate, the processing time was the longest.

Table 1. Processing time (single runtime) and mesh data

Mesh	Time (min)	Volume (mm3)	Elements	Minimum Quality	Average Quality
1	37	2.96×10^6	5.46×10^7	0.02	0.68
2	99	2.96×10^6	4.47×10^7	0.02	0.65
3	11	2.95×10^6	5.85×10^6	0.02	0.65

2.3 Tissue Dielectric Properties

The tsDCS electric field distribution across the neck region was described by Maxwell's equations and, considering the static regime of the phenomenon, follows the formulations of electrostatics. The stationary electric potential in the inhomogeneous conductive volume model was computed using the AC/DC Electric Currents module of the COMSOL (v5.3, Stockholm, Sweden) for solving the EDP of the current flow in the volume conductor, neglecting the neurophysiology of the nervous tissue [29]:

$$\nabla \cdot (-\sigma V) = 0 \tag{1}$$

where, V is the electric potential, σ is the electrical conductivity of volume conductor. The boundary conditions have been configured in COMSOL (v5.3, Stockholm, Sweden) to treat all outside boundaries as electrically isolated (except by the surface of electrodes) and all internal boundaries as continuous between the interfaces. For the electrodes in bipolar configuration, the applied current was defined as cathode Ic = −2.5 mA and anode Ia = 2.5 mA. The stationary solver in the conjugate gradient configuration was used during the simulations and adjusted with a relative tolerance of 10^{-6}.

2.4 Electric Field Analysis

A tsDCS simulation was performed for each mesh, assuming the properties and configurations presented in the Sects. 2.1 and 2.2. Data analysis was performed in MATLAB (R2021a, Massachusetts, USA). The electric field was obtained at the spatial nodes of the whole volume of WM and GM tissues, and the average magnitude of the electric field E was estimated in volumetric sections of 0.5 mm. In order, to evaluate the effect of the mesh resolution on the electric field at the spinal cord, mesh 1 was taken as reference, and the electric field error of meshes 2 and 3 was estimated with the mean absolute error (MAE), the mean absolute percentage error (MAPE), and the normalized root mean square error (NRMSE).

$$MAE = (1/n)\,\Sigma\left|y_i - \hat{y}_i\right| \tag{2}$$

$$MAPE = (100/n)\,\Sigma\left|y_i - \hat{y}_i\right|/y_i\% \tag{3}$$

$$NRMSE = (100/\overline{y})\sqrt{\left((1/n)\Sigma\left(y_i - \hat{y}_i\right)^2\right)}\% \tag{4}$$

where, n is the number of observations, y_i is the observed value, $\hat{y}_i$ is the reference value, and $\overline{y}$ is the mean of observations.

3 Results

As seen in Table 2, the FEM solution for mesh 3 required the shortest runtime, with an electric field MAE lower than 2% relative to the model using the mesh 1. Even though there is an increase in MAE with the reduction in mesh size, the NRMSE for the mesh with the lowest number of elements is lower than 7%. No changes were seen in E for GM, while WM presented a 0.001 and 0.002 V/m reduction in the E magnitude for meshes 2 and 3, respectively.

Figure 5 shows the distribution (on the surfaces) and the average magnitude of electric field for the three meshes of the WM and GM. There was a small variation of E with the mesh resolution, and the average value is higher for WM than GM with the same modulation pattern (oscillatory amplitude).

Figure 6 shows the MAPE and NRMSE along the longitudinal axis of the WM and GM. For the MAPE, it is possible to verify changes of less than 2.5% in WM and 1.5% in GM. For the NRMSE of mesh 2, there is a variation between 2 and 8% in WM but under 1% for GM, while the model with mesh 3 shows a variation between 3 and 14% for WM but below 3% in GM. All these values occur except for the extremity regions (rostral and caudal), which show higher errors due to the greater geometric distortion relative to the mesh 1 caused by local lower resolution.

Table 2. Runtime (single run) to solve the model and electric field data.

Mesh	Time (min)	Tissues	E (V/m)	MAE (V/m)	MAPE (%)	NRMSE (%)
1	51	WM	0.356	–	–	–
		GM	0.254	–	–	–
2	53	WM	0.355	0.002	0.59	2.50
		GM	0.254	0.001	0.29	0.68
3	6	WM	0.354	0.007	1.77	6.21
		GM	0.254	0.003	1.06	2.05

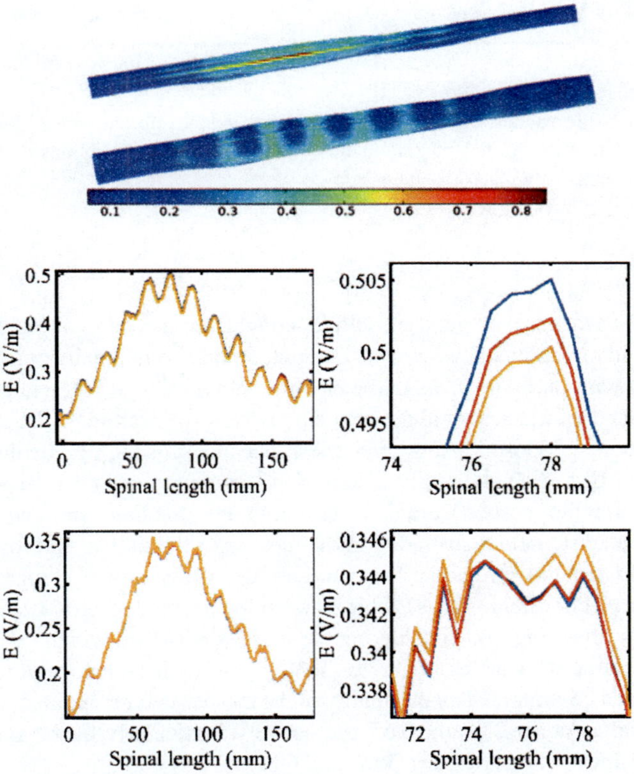

Fig. 5. The magnitude of the electric field distribution on the surfaces of gray matter (first row, top) and white matter. Average magnitude of the electric field across volumetric Sects. (0.5 mm) of the cervical region for the white matter (WM, second row) and gray matter (GM, third row) for three meshes (blue: mesh 1, red: mesh 2, and orange: mesh 3). Electric field zoomed at the C5 vertebra (second column).

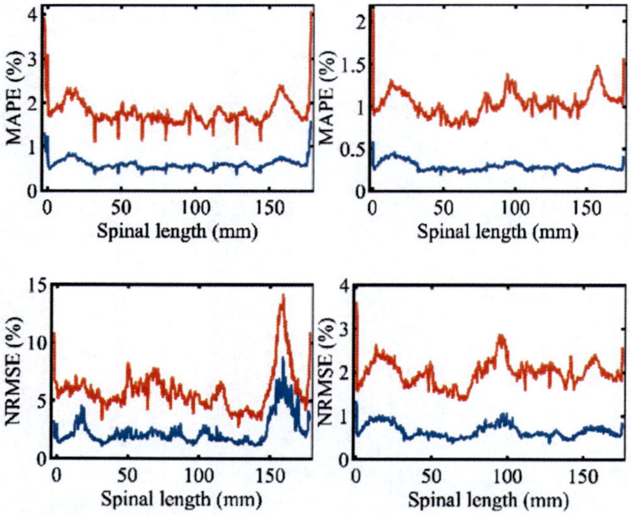

Fig. 6. Electric field error between the simplified meshes and the reference mesh along the cervical region for the white matter (left) and gray matter (right). The blue and red lines show the error for mesh 2 and 3, respectively.

4 Discussion

The goal of our work was to get a simplified model from the cervical region for tsDCS simulations and to implement a low-cost computational simulation by optimizing of the FEM solution with minimal errors in the electric field measured in the target tissues.

We simulated the electric field generated by tsDCS application on the surface of the volume conductor proposed, using three mesh densities for the FEM solution. As the density of the FEM mesh is directly linked to the computational cost necessary for the problem solution, the meshes were built with high, intermediate, and low density. The magnitude of electric field distribution in the meshes of the tissues that form the spinal cord was evaluated and compared. We could not determine any discrepant differences in the average electric field values (Table 2) or in the average electric field distribution between the meshes (Fig. 5). Considering the highest NRMSE, which penalizes larger errors, the overall error was less than 7% (Table 2), while local errors were under 15% (Fig. 6). This can be explained by the quality of the meshes that maintained the Minimum and Average values above 0.02 and 0.65, respectively. Particularly, the Minimum Quality was above 0.1 for the target tissues (WM and GM).

The optimization done in mesh 3 allowed a reduction from 88 to 17 min in processing time (meshing and FEM solution), and from 4 GB to 800 MB (geometry, mesh, and FEM solution) in storage. Additionally, it facilitates variation in parameters such as conductivity, current intensity, and electrode characteristics (e.g., quantity, geometric form, and dimensions) of the tsDCS setup, since any geometric/parametric change requires a new mesh building and solution reprocessing. However, the reduction in the number of elements obtained in mesh 2 did not show the same pattern of reduction in the processing time of the mesh and the FEM solution. This can happen due to the order of tissue processed and the combination of parameter values used during the mesh 2 setup.

Overall, the electric field distribution shown in Fig. 5 has values in the same range reported by Kuck (max 0.82 V/m) [11] and Fernandes (max 0.5 V/m) [14]. Despite the limitations that the proposed model may present quantitatively in the results, it is believed to provide the necessary information for the intended purpose.

5 Conclusions

It was possible to estimate the electric field in the spinal cord induced by tsDCS using a simplified tissue model of the cervical region. With the parameterization performed during the mesh construction, it was possible to obtain an optimized mesh with a reduced number of elements. The optimized mesh (mesh 3) allowed the runtime for mesh generation to be reduced by three times, and ten times for processing the solution through FEM, while keeping the electric field comparable to the highest resolution one (mesh 1) throughout the entire volume of spinal cord.

Conflict of Interest. The authors declare that they have no conflict of interest.

References

1. Gebodh, N., Esmaeilpour, Z., Adair, D., et al.: Practical Guide to Transcranial Direct Current Stimulation. Springer, Cham (2019)
2. Cogiamanian, F., Ardolino, G., Vergari, M., et al.: Transcutaneous spinal direct current stimulation. Front. Psy. **3**, 63 (2012)
3. Schweizer, L., Meyer-Frießem, C.H., Zahn, P.K., et al.: Transcutaneous spinal direct current stimulation alters resting-state functional connectivity. Brain Connectivity **7**(6), 357–365 (2017)
4. Awosika, O.O., Sandrini, M., Volochayev, R., et al.: Transcutaneous spinal direct current stimulation improves locomotor learning in healthy humans. Brain Stimul. **12**(3), 628–634 (2019)
5. Kaczmarek, D., Ristikankare, J., Jankowska, E.: Does trans-spinal and local DC polarization affect presynaptic inhibition and post-activation depression? J. Physi. **595**(5), 1743–1761 (2017)
6. Grecco, L.H., Li, S., Michel, S., et al.: Transcutaneous spinal stimulation as a therapeutic strategy for spinal cord injury: state of the art. J Neurorestoratology **3**(1), 73–82 (2015)
7. Ahmed, Z.: Trans-spinal direct current stimulation modulates motor cortex-induced muscle contraction in mice. J. Appl. Physiol. **110**(5), 1414–1424 (2011)
8. Bączyk, M., Krutki, P., Zytnicki, D.: Is there hope that transpinal direct current stimulation corrects motoneuron excitability and provides neuroprotection in amyotrophic lateral sclerosis? Physio. Rep. **9**(2), e14706 (2021)
9. Cardoso, J.R.: Electromagnetics through the finite element method: a simplified approach using Maxwell's equations. Crc Press (2016)
10. Dokos, S.: Modelling organs, tissues, cells and devices: using Matlab and Comsol multiphysics. Springer (2017)

11. Kuck, A., Stegeman, D.F., Van Asseldonk, E.H.F.: Modeling trans-spinal direct current stimulation for the modulation of the lumbar spinal motor pathways. J. Neural Eng. **14**(5), 056014 (2017)
12. Fernandes, S.R., Salvador, R., Wenger, C., et al.: Transcutaneous spinal direct current stimulation of the lumbar and sacral spinal cord: a modelling study. J. Neural Eng. **15**(3), 036008 (2018)
13. Dongés, S.C., Bai, S., Taylor, J.L.: Concurrent electrical cervicomedullary stimulation and cervical transcutaneous spinal direct current stimulation result in a stimulus interaction. Exp. Physiol. **102**(10), 1309–1320 (2017)
14. Fernandes, S.R., Pereira, M., Salvador, R., et al.: Cervical trans-spinal direct current stimulation: a modelling-experimental approach. J Neuroeng. Rehabilit. **16**(1), 1–14 (2019)
15. Christ, A., Kainz, W., Hahn, E.G., et al.: The virtual family—development of surface-based anatomical models of two adults and two children for dosimetric simulations. Phys. Med. Biol. **55**(2), N23 (2010)
16. Gosselin, M.C., Neufeld, E., Moser, H. et al.: Development of a new generation of high-resolution anatomical models for medical device evaluation: the virtual population 3.0. Phys. Med. Biol. **59**(18), 5287 (2014)
17. Panjabi, M.M., Duranceau, J., Goel, V. et al.: Cervical human vertebrae. Quantitative three-dimensional anatomy of the middle and lower regions. Spine **16**(8), 861–869 (1991)
18. Panjabi, M.M., Chen, N.C., Shin, E.K., Wang, J.L.: The cortical shell architecture of human cervical vertebral bodies. Spine **26**(22), 2478–2484 (2001)
19. Sengul, G., Watson, C., Tanaka, I., Paxinos, G.: Atlas of the spinal cord: mouse, rat, rhesus, marmoset and human (2013)
20. Frost, B.A., Camarero-Espinosa, S., Foster, E.J.: Materials for the spine: anatomy, problems, and solutions. Materials **12**(2), 253 (2019)
21. Sureka, B., Mittal, M.K., Mittal, A., et al.: Morphometric analysis of diameter and relationship of vertebral artery with respect to transverse foramen in Indian population. Indian J. Rad. Ima. **25**(02), 167–172 (2015)
22. Kwon, S., Suh, S.W., Kim, D., et al.: Analysis of dural sac thickness in the human cervical spine. Anat. Sci. Int. **93**(2), 284–290 (2018)
23. Panjabi, M.M., Oxland, T.R., Parks, E.H.: Quantitative anatomy of cervical spine ligaments. Part II. Middle and lower cervical spine. J. Spinal Dis. **4**(3), 277–285 (1991)
24. Sayit, E., Daubs, M.D., Aghdasi, B., et al.: Dynamic changes of the ligamentum flavum in the cervical spine assessed with kinetic magnetic resonance imaging. Glob. Spi. J. **3**(2), 69–73 (2013)
25. Womack, W., Woldtvedt, D., Puttlitz, C.M.: Lower cervical spine facet cartilage thickness mapping. Osteoarthritis Cartilage **16**(9), 1018–1023 (2008)
26. Schmalfuss, I.M., Mancuso, A.A., Tart, R.P.: Postcricoid region and cervical esophagus: normal appearance at CT and MR imaging. Radiology **214**(1), 237–246 (2000)
27. Furlow, P.W., Mathisen, D.J.: Surgical anatomy of the trachea. Annals of cardiothoracic surgery **7**(2), 255 (2018)
28. Hasgall, P.A. et al.: IT'IS Database for thermal and electromagnetic parameters of biological tissues (2022).https://doi.org/10.13099/VIP21000-04-1
29. Plonsey, R., Heppner, D.B.: Considerations of quasi-stationarity in electrophysiological systems. Bull. Math. Biophys.Biophys. **29**(4), 657–664 (1967)

Thermographic Analysis in a Multilayer Medium Cortical Bone/Cancellous Bone/Bone Cement Heated by Therapeutic Ultrasound

B. T. M. T. dos Santos[1], L. V. C. de Oliveira[1], A. M. C. Lopes[1], L. F. Pereira[1], A. J. F. Pereira[1,3], T. P. Omena[2(✉)], and W. C. A. Pereira[1]

[1] UFRJ/COPPE/ Biomedical Engineering Program, Av. Horácio Macedo 2030, Centro de Tecnologia, COPPE/UFRJ, Bloco H, S. 327, Rio de Janeiro, Brazil

[2] UFRJ/Physiotherapy Department, Rio de Janeiro, Brazil
thaisomena@gmail.com

[3] University Centre Serra Dos Órgãos, Rio de Janeiro, Brazil

Abstract. The thermal effects of therapeutic ultrasound are used to produce analgesia or reduce inflammation in biological tissues. However, it is contraindicated on regions with cemented prostheses, due to the supposed heating of it. The present study aims to evaluate the heating pattern in a multilayer medium, using cortical bone phantom, cancellous bone phantoms and, bone cement specimen. It was observed that the cancellous bone phantom was the one that warmed the most, reaching an average temperature of 33.89 ± 1.89 °C, while the bone cement specimen barely changed. In the multilayer medium, the heating pattern was more superficial than expected. More studies are needed to elucidate the influence of application time and intensity in ultrasonic irradiation in places with the presence of bone cement in treatment protocols.

Keywords: Phantoms · Bone tissue · Bone cement · Heating · Therapeutic ultrasound

1 Introduction

One of the main objectives of Therapeutic Ultrasound (UST) is to heat biological tissues. To promote physiological effects such as analgesia, reduction of muscle spasm, and reduction of inflammation, it is necessary to raise the temperature to a range from 40 to 45 °C, and this range should be maintained for at least for 5 min [1–4].

Literature in Physiotherapy contraindicates the use of UST over regions of cemented prostheses because, according to some authors, [5, 6] there is a high risk of overheating of bone cement and may damage the material and its surroundings.

The use of phantoms in research, with similar acoustic properties to biological tissues, makes it possible to improve the detection of bone changes [7]. Possibly, the investigation of the UST safety can also be performed.

Thus, the present study aims to evaluate the heating pattern in a multilayer medium, using materials that mimic ultrasonic properties of biological tissues (ultrasonic phantoms) and bone cement.

J. L. B. Marques et al. (Eds.): CLAIB 2022/CBEB 2022, IFMBE Proceedings 98, pp. 277–283, 2024.
https://doi.org/10.1007/978-3-031-49401-7_27

2 Material and Methods

The experiment was carried out at the Ultrasound Laboratory - LUS/COPPE/UFRJ and, the following devices were used: 1) UST Sonopulse (IBRAMED, Amparo, SP, Brazil), Nominal Effective Radiation Area (ERA) 7 cm2, at 1 MHz, intensity (ISATA) 0.8 W.cm-2, irradiating in continuous mode during 1 min; 2) Circular cortical bone phantom (Sawbone, Vashon Island, WA, USA) of dimensions 74 mm Ø and 1 mm thick; 3) Phantom of cancellous bone (Sawbone, Vashon Island, WA, USA) of a parallelepiped shape of dimensions 74-mm wide, 95-mm long and, 13-mm thick; 4) cylindrical bone cement, with a thickness of 5.24 mm and a diameter of 51.0 mm; 5) FLIR-E64501 infrared camera (Teledyne FLIR LLC, Wilsonville, OR, USA), perpendicular and lens far from the absorber plate at 24.0 cm, temperature range from 20 to 35°C; 6) APTFLEX absorber plate (Precision Acoustics, Dorchester, West Dorset, UK) to avoid reirradiation. The phantoms and the bone cement specimen are presented in Fig. 1.

In the experimental assembly (Fig. 2), the following materials were positioned in contact with each other by a layer of ultrasonic conductive gel: cortical bone phantom/ cancellous bone phantom/ bone cement specimen and absorber plate.

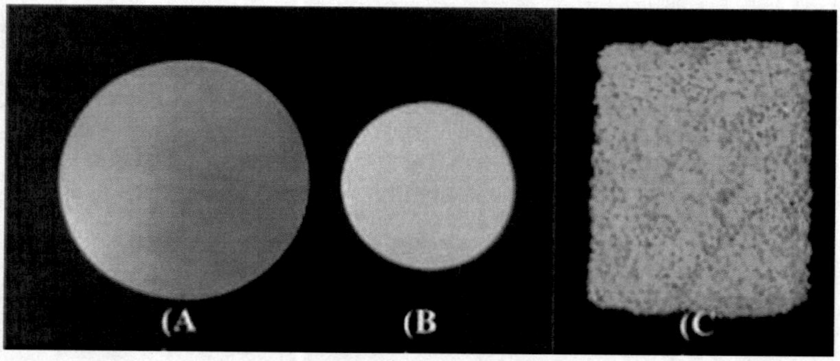

Fig. 1. Phantoms and bone cement specimen. (A): Cortical bone phantom; (B): Bone cement specimen; (C): Cancellous bone phantom.

The ultrasonic transducer was fixed in direct contact with the cortical bone phantom with a thin gel layer. The power curve was previously raised in a Radiation Force balance (UPM-DT-1, Ohmic Instruments, St. Charles, MO, USA).

In addition, to maximize the propagation of the ultrasonic beam, the cancellous bone phantom was impregnated with conductive gel and placed in the vacuum chamber, to remove the air bubbles contained in the gel, for 5 min.

The experiment was conducted in air and ambient temperature measured with digital thermocouple (Thermo-hygrometer MTH-1362W) was 24 ± 2 °C.

The thermal images were recorded before and shortly after irradiation, at interfaces: 1) transducer/surface of the cortical bone phantom; 2) phantom of cortical bone/surface of cancellous bone phantom and, 3) cancellous bone phantom/bone cement specimen. After ultrasonic irradiation, the transducer was removed to carry out the acquisition of

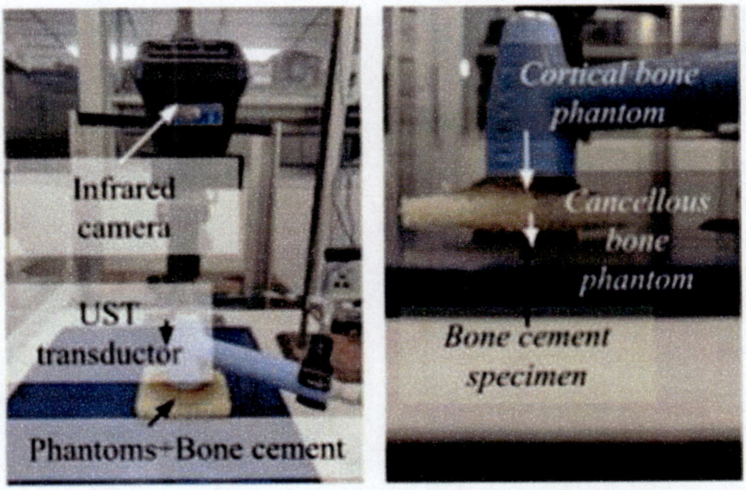

Fig. 2. Experimental assembly: Infrared camera, transducer, cortical bone phantom, cancellous bone phantom, bone cement and, absorber plate.

thermal images at interfaces 1, 2 and, 3. A 10-min interval between ultrasonic irradiations was performed for cooling of the parts.

This procedure was performed 10 times and data were analyzed quantitatively through a spreadsheet in Excel (Microsoft Corporation, Redmond, WA, USA).

The acoustic and thermal properties of the cortical bone and the cement phantoms are respectively: attenuation coefficient at 1 MHz: 6.15 $dB.cm^{-1}$ and, 5.79 $\pm$ 0.22 $dB.cm^{-1}$; wave propagation speed: 2924.31 $m.s^{-1}$ and, 1984.83 $\pm$ 3.49 $m.s^{-1}$; thermal conductivity: 0.47 $W.m^{-1}.°C^{-1}$ and, 0,229 $\pm$ 0,005 $W.m^{-1}.°C^{-1}$; specific heat: 1,26.103 $J.kg^{-1}.K^{-1}$ and, $1{,}46.10^{-3}$ $J.kg^{-1}.K^{-1}$.

3 Results

The results of the thermal readings from the surface of the cortical and cancellous bone phantoms, and from the bone cement specimen are shown in Tables 1, 2 and 3, respectively, as well as their means and standard deviations.

Taking into account Tables 1, 2 and 3, it can be seen that the cancellous bone phantom reached the highest temperatures after ultrasonic irradiation, of 33.89 $\pm$ 1.8 6°C, when compared with the cortical bone phantom, 31.93 $\pm$ 3.27 °C, and the bone cement specimen, 24.21 $\pm$ 1.06 °C.

The cancellous bone phantom also presented the highest mean difference between the previous (Ti) and posterior (Tf) temperatures at irradiation. However, it is the cortical bone phantom that presents the greatest difference, and the greatest dispersion between temperature differences.

One can observe that the interval between the irradiations was enough for the temperature before the irradiation return.

Table 1. Comparison of the surface temperature values of the cortical bone phantom before (Ti) and after (Tf) the ultrasonic irradiation.

Cortical bone phantom			
	Ti (°C)	Tf (°C)	Difference (°C)
1	22.60	32.00	9.40
2	22.40	29.70	7.30
3	22.50	33.50	11.00
4	22.40	29.60	7.20
5	21.80	30.40	8.60
6	21.70	33.10	11.40
7	22.30	39.00	16.70
8	23.10	26.60	3.50
9	22.90	33.10	10.20
10	22.60	32.30	9.70
Mean ± Std	22.43 ± 0.43	31.93 ± 3.27	9.50 ± 3.42

Table 2. Comparison of surface temperature values of the cancellous bone phantom before (Ti) and after (Tf) ultrasonic irradiation.

Cancellous bone phantom			
	Ti (°C)	Tf (°C)	Difference (°C)
1	22.20	35.30	13.10
2	22.70	32.30	9.60
3	23.00	34.80	11.80
4	21.50	31.00	9.50
5	22.20	32.70	10.50
6	22.60	35.00	12.40
7	22.90	37.20	14.30
8	22.70	32.10	9.40
9	23.20	34.60	11.40
10	23.30	33.90	10.60
Mean ± Std	22.63 ± 0.54	33.89 ± 1.86	11.26 ± 1.65

Figure 3 illustrates the cortical and cancellous bone phantoms, and the bone cement specimen, after UST irradiation.

The bone cement specimen presented the smallest changes in temperature after ultrasonic irradiation.

Table 3. Comparison of surface temperature values of the cancellous bone phantom before (Ti) and after (Tf) ultrasonic irradiation.

Bone cement specimen			
	Ti (°C)	Tf (°C)	Difference (°C)
1	22.00	23.70	1.70
2	22.50	24.00	1.50
3	21.70	25.10	3.40
4	22.50	23.20	0.70
5	21.00	23.40	2.40
6	21.50	23.80	2.30
7	22.00	24.70	2.70
8	21.70	24.90	3.20
9	21.90	26.40	4.50
10	22.20	22.90	0.70
Mean ± Std	21.90 ± 0.46	24.21 ± 1.06	2.31 ± 1.21

4 Discussion

It is widely accepted that ultrasonic irradiation at 1 MHz produces deep heating in tissues (from 30 to 50 mm) [5, 6].

The higher temperatures in the cancellous bone phantom are supposed to be due to backscattering. A previous study carried out by this laboratory [8] confirmed the correlation between the parameters Apparent Integrated Backscattering (AIB) and the inclination of the Frequency of Apparent Backscattering (FSAB). Meanwhile, the bone cement specimen did not exhibit a noteworthy variation in relation to phantoms. This may be due to the characteristics of bone attenuation, resulting in little acoustic energy reaching the bone cement layer.

It is necessary to pay attention to the fact that physiotherapy treatments last more than a minute, and the application time can lead to a considerable increase in tissue temperature [2–4]. Therefore, further studies should be performed considering this parameter to ascertain the safety of the UST application.

The cancellous bone phantom impregnation with conductive gel, as well as the vacuum process, were performed to extract the air from the empty spaces. Although the coupling gel is not composed of the same elements of the bone matrix, it was assumed that this procedure would allow the irradiated UST to be diffused through the intertrabecular space. Since air has a great difference in acoustic impedance with bone tissue, it causes acoustic irradiation to be transmitted only by the trabeculae. With the filling of the spaces, it was conjectured that the energy would be distributed in a more similar way to the biological environment, and not only along the trabeculae, which presumably happened. However, further studies should be conducted to investigate this assumption.

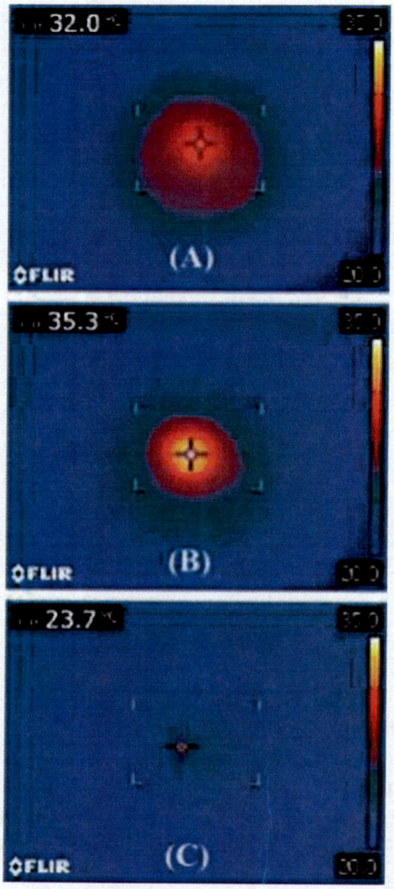

Fig. 3. Thermal images of the interfaces. (A): Surface of the cortical bone phantom; (B): Surface of cancellous bone phantom; (C): Surface of the bone cement specimen.

Studies of this type are important to ensure the physical integrity of patients during treatment with UST, since many protocols neglect this interaction with bone tissue. In addition, it broadens horizons for the use of this technique in analgesia, increase of elasticity, and accelerate healing in places that there is bone cement.

5 Conclusions

It was observed that, in a multilayer medium composed of bone phantoms and bone cement specimen, the heating pattern was more superficial than expected, with a small temperature increment indicating that, for the ultrasonic dose used, cement did not cause any important difference that could raise any safety concerns.

However, more studies are needed to clarify the influence of application time and ultrasonic intensity in places with the presence of bone cement in treatment protocols.

Acknowledgments. The authors would like to thank the Brazilian agencies CNPq, CAPES and FAPERJ, for their financial support.

Conflict of Interest The authors declare no conflict of interests.

References

1. Speed, C.A.: Therapeutic ultrasound in soft tissue lesions. Rheumatology **40**(12), 1331–1336 (2001)
2. Itakura, D.A.: Alteração da temperatura nos tecidos biológicos com a aplicação do ultrassom terapêutico: uma revisão. Fisioterapia em Movimento **25**(4), 857–868 (2012)
3. Merrick, M.A.: Identical 3-MHz ultrasound treatments with different devices produce different intramuscular temperatures. J. Orthop. Sports Phys. Ther. **33**(7), 379–385 (2003)
4. Miller, M.G.: Intramuscular temperature differences between the mid-point and peripheral effective radiating area with ultrasound. J. Sports Sci. Med. **7**(2), 286–291 (2008)
5. Bélanger, A.: Therapeutic electrophysical agents, 2nd edn. Lippincott Williams & Wilkins, Philadelphia, EUA (2010)
6. Cameron, M.: Agentes físicos na reabilitação, 3rd edn. Elsevier, São Paulo, Brasil (2009)
7. Falardeau, T.: Ultrasound tomography in bone mimicking phantoms: simulations and experiments. J. Acoust. Soc. America **144**(5), 2937–2946 (2018)
8. Fontes-Pereira, A.: Monitoring bone changes due to calcium, magnesium, and phosphorus loss in rat femurs using quantitative ultrasound. Sci. Rep. **8**(1), 11963 (2018)

Proteomic Profile of Colorectal Cancer Cell Treated with Coffee

Laura Zuluaica(✉), David Osorio-Mera, Johanna Pedroza-Díaz, and Sarah Röthlisberger

Departamento de Ciencias Aplicadas, Instituto Tecnológico Metropolitano, Medellín, Colombia
laura.zuluaica@hotmail.com

Abstract. The risk of developing colorectal cancer is greatly reduced by dietary polyphenols and bioactive compounds. Coffee is an important source of these compounds, which have antioxidant, anti-inflammatory and anticancer properties. The aim of this study was to visualize the proteomic profile of SW480 human colorectal adenocarcinoma cells treated with green and roasted coffee extracts, compared to untreated cells. Image analysis showed that when cells were treated with coffee extracts there were noticeable changes in protein expression, especially in those treated with green coffee. These preliminary results could help shed some light on the biological mechanism that coffee induces in colorectal cancer cells.

Keywords: Coffee · Colorectal cancer · Proteome · Two-dimensional gel electrophoresis

1 Introduction

According to the Global Cancer Observatory, colorectal cancer (CRC) is the second cause of death by cancer worldwide, and as such, represents a global public health problem. Dietary patterns such as a high consumption of red and processed meats, low intake of fruits, vegetables and fibre, contribute significantly to the risk of developing this disease. Coffee is one of the most widely consumed beverages worldwide and is an important source of polyphenols and bioactive compounds. A great number of epidemiological studies have been conducted over the last decade revealing a significant protective effect against CRC [2], as well as *in vitro* and *in vivo* studies that show that coffee reduces cell proliferation and induces apoptosis [1]. However, there are very few studies that report the effect of coffee on protein expression in CRC cells, so there is still no clear vision of the biological mechanisms involved in the chemopreventive effect of coffee. In this study we visualized the proteomic profile of CRC cells treated with green (unroasted) coffee and roasted coffee, compared to untreated cells.

J. L. B. Marques et al. (Eds.): CLAIB 2022/CBEB 2022, IFMBE Proceedings 98, pp. 284–287, 2024.
https://doi.org/10.1007/978-3-031-49401-7_28

2 Materials and Methods

2.1 Cell Culture

The human colorectal adenocarcinoma cell line SW480 (ATCC® CCL-228) was used and cultured according to ATCC guidelines. Cells were treated for 48 h as follows: untreated cells (control), cells treated with green coffee extract (750 μg/μl), cells treated with roasted coffee extract (500 μg/μl). These subtoxic concentrations were selected in order to observe the effect of the compounds without actually causing death in the entire sample. The extracts of Coffea Arabica were previously prepared by our research group [3].

2.2 Protein Extraction

Approximately 10×10^6 cells were collected for each cell culture condition and incubated with lysis buffer (Urea 7M, CHAPS 4%, protease inhibitor cocktail) for protein extraction. Proteins were precipitated from the lysed samples using cold acetone (1:4 v/v) overnight, centrifuged for 1 h at 13.200 rpm, and the pellet was resuspended in the lysis buffer and stored at -70 °C.

2.3 Two-Dimensional Gel Electrophoresis

Proteins (50μg) were loaded by passive rehydration onto 7cm ZOOM™ IPG (Immobilized pH gradient) strips pH 3–10 NL. Isoelectric focusing was carried out using the following voltage ramp: 100 V for 1 h, 150 V for 1 h, 200 V for 5 min, 450 V for 5 min, 600 V for 5 min, 750 V for 5 min, 950 V for 5 min, 1200 V for 10 min, 1400 V for 10 min, 1600 V for 10 min, and 2000 V for 45 min. The IPG strips were then reduced with 100 mM DTT and alkylated with 2.5% iodoacetamide according to the manufacturer's recommended protocol. After this, the IPG strips were loaded onto SDS-PAGE NuPAGE™ Novex™ 4–12% Bis-Tris Protein Gels, 1,5 mm, and run at 200V for 45 min. After electrophoresis these were stained with SYPRO® Ruby (Invitrogen™) and the gel images were acquired using the ChemiDocTM MP System (Biorad).

3 Results and Discussion

Figure 1 shows the proteomic profile of SW480 CRC cells treated with green and roasted coffee extracts, compared to untreated cells.

The number of protein spots in each gel was counted using the PdQuest 2D analysis software, as shown in Table 1. These preliminary results show that protein expression is altered in cells treated with both green and roasted coffee as compared to untreated cells. The changes in protein expression are more evident in cells treated with green coffee extract, which could be due to the higher concentration of polyphenols, such as chlorogenic acid, in green (not roasted) versus roasted coffee beans. *In vitro* studies of Caco-2 cells derived from colon carcinoma have shown that treatment with coffee extracts causes a reduction in proliferation, in intracellular ROS and in KRAS expression

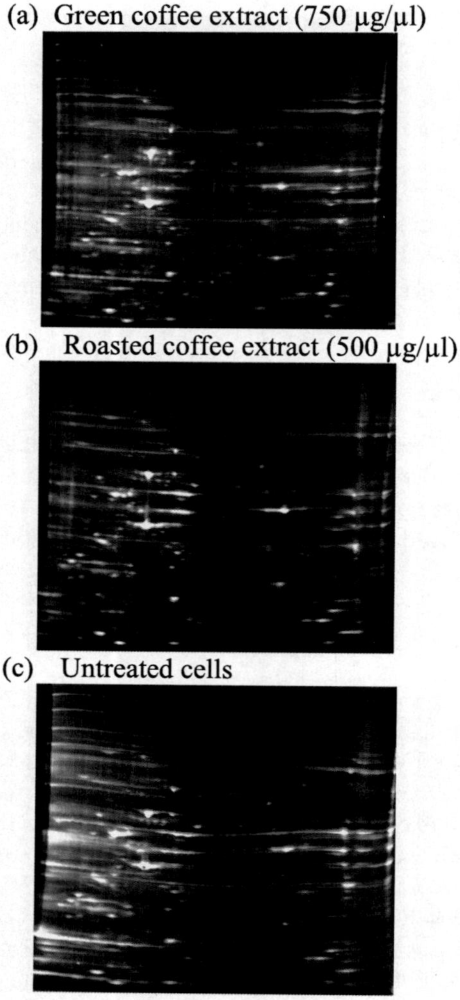

Fig. 1. Proteomic profiles of SW480 cells treated with (a) green coffee extract, (b) roasted coffee extract, and (c) untreated cells.

in these cells [1]. While in other cell lines such as HT-29 and LoVo, treatment with coffee derivatives such as Kahweol, caffeic acid, or chlorogenic acid, caused cell cycle arrest, reduced cell proliferation, and increased apoptosis [1]. However, whole proteome analysis has not been carried out to determine which proteins are responsible for these effects, and as such, these preliminary results could help shed some light on the biological mechanism that coffee induces in these cells.

Table 1. Number of protein spots in each gel

Gel	# spots
Green coffee extract (750 μg/μl)	282
Roasted coffee extract (500 μg/μl)	217
Untreated cells	165

4 Conclusions

These preliminary results show that when human colorectal adenocarcinoma SW480 cells are treated with coffee extracts protein expression changes. In the near future these results will be validated and changes in protein expression identified by mass spectrometry in order to correlate these results with a potential biological mechanism induced by coffee in colorectal cancer cells.

Acknowledgements. The authors would like to acknowledge support received from Minciencias grant (project code: 115080763215 CT 811–2018) and the Instituto Tecnológico Metropolitano.

References

1. Moreno-Ceballos, Arroyave, J.C., Cortes-Mancera, F.M., Röthlisberger, S.: Chemopreventive effect of coffee against colorectal cancer and hepatocellular carcinoma. Int. J. Food Properties **22**(1), 536–555 (2019)
2. Poole, R., Kennedy, O.J., Roderick, P., et al.: Coffee consumption and health: umbrella review of meta-analyses of multiple health outcomes. Br. Med. J. **359**, 1–17 (2017)
3. Villota, H., Moreno-Ceballos, M., Santa-González, G.A., Uribe, D., Castañeda, et al.: Biological impact of phenolic compounds from coffee on colorectal cancer. Pharmaceuticals **14,** 761 (2021)

Check for updates

Computational Simulation of the Human Eye with Glaucoma to Identify Early Changes in Anterior Segment

Luis Mario Rosales[1], Ivonne Bazan[1(✉)], and Luis Fernando Barba-Gallardo[2]

[1] Biomedical Engineering Department, Universidad Autónoma de Aguascalientes, Aguascalientes Ags 20392, México
ivonne.bazan@edu.uaa.mx

[2] Optometry Department, Universidad Autónoma de Aguascalientes, Aguascalientes Ags 20392, México

Abstract. Maintaining the greatest amount of visual field of a glaucoma patient is still a challenge today. It is difficult to diagnose the disease in its early stages, due to the lack of symptoms in many cases. Therefore, in this work, an eyeball model was developed to evaluate the behavior of some anterior segment structures of the eye against slight changes in intraocular pressure. An axisymmetric model of the eyeball was used to contemplates the behavior of the cornea, part of the sclera, iris and ciliary bodies against small changes over intraocular pressure values. The results of our model showed total displacements of 0.43, 0.451 and 0.459 mm respectively in the cornea, lens, and iris related to the increase of intraocular pressure for values above 25 mmHg. Therefore, the small increase of interocular pressure can cause changes in certain ocular measurements such as the depth of the anterior chamber that, until now, are not commonly utilized and that could be, which can be measured and characterized in the future in real cases to support glaucoma diagnosis.

Keywords: Finite element method · Glaucoma · Intraocular pressure · Simulation

1 Introduction

Glaucoma is today one of the main causes of blindness in the world [1]. This disease is usually associated with elevated intraocular pressure (IOP) caused by an abnormal increase in the volume of aqueous humor in the anterior segment of the eye [2].

Primary open-angle glaucoma (POAG) is the most frequent variant of this disease, in which damage to the cells axons in the retina occurs progressively. The person usually becomes aware of their illness when they have already lost an important part of their visual field [3].

To see the effects of IOP on different structures of the eye, several authors have proposed the use of computational models to simulate the behavior of different tissues under pressure loads. These models often use the finite element method (FEM) which

J. L. B. Marques et al. (Eds.): CLAIB 2022/CBEB 2022, IFMBE Proceedings 98, pp. 288–298, 2024.
https://doi.org/10.1007/978-3-031-49401-7_29

is based on dividing a continuous object into a mesh of multiple elements with their material properties, to be simulated under specific boundary conditions to which they are subjected in the real cases are assigned [4, 5].

Peishan Dai et al. developed a 3D model of the human eye with many structures of both segments of the eye. The simulations were based on observing Von Mises stress, displacements, and strains in a general and specific way in the area around the head of the optic nerve. The results showed that the cornea, the lens and the optic nerve head (ONH) were the structures with the greatest strains and displacements [6].

Vivek Suganthan et al. also developed a 3D model that considered both se ments of the eye. In this case, the results mainly considered the anterior segment, showing that the lens and the zonules were the structures that were subjected to the greatest stress. The lens and the cornea were the structures with greater displacement [7].

Elsheikh built different models of the cornea based on slightly changes of its shape, thickness, Young's modulus, and layered conformation to see the influence in stress, strain, and displacement. Aspects such as to consider the cornea as elastic, the shape of the structure, and the differentiation of the different layers were some of the parameters that presented the greatest changes in stress and strain [5].

Therefore, the purpose of this article is to develop a simulation of the entire eyeball to address the behavior of anterior segment tissues with slight increases in IOP.

2 Method

The FEM was used to carry out the simulations. To carry out the eyeball model and the simulations, the program COMSOL Multiphysics® was used. The computer used to carry out the entire process was a 64 bits Lenovo Ideapad 330s laptop with 8 GB of RAM, 1.80 GHz CPU and Windows 11 operating system.

2.1 Geometry Model

The ocular structures considered for the simulations were: cornea, sclera, choroid, retina, optic nerve, lamina cribrosa, iris, lens, ciliary bodies, zonules and vitreous humor of the eye. Figure 1 shows the result of the geometry made.

The eye corresponds to a structure of the human body which is not symmetrical [8]. The optic nerve, which mostly is composed by the set of axons of the retina nerve cells, has a slight angle away from the central axis of the eye, displaced to the nasal one. The elaborated model simplified this characteristic considering the optic nerve parallel to the optic axis [9]. This feature is commonly used in models' construction, and it allows an axisymmetric geometry, which obtain results like those with a 3D model, but with less computational requirements [6].

The ocular globe geometry was based on an ellipse with a semi-major axis of 12 mm and a semi-minor axis of 11.5 mm, [10, 11]. The sclera layer was constructed with variable thickness having 1mm at its posterior zone, near to the optic nerve, a thickness of 0.5 mm at its equator, and 0.65 mm near to the cornea [10]. The outer surface of the cornea was constructed using an ellipse with a semi-major axis of 9.5 mm and a semi-minor axis of 7.84 mm that had a prolate shape (with more curvature in its center than in

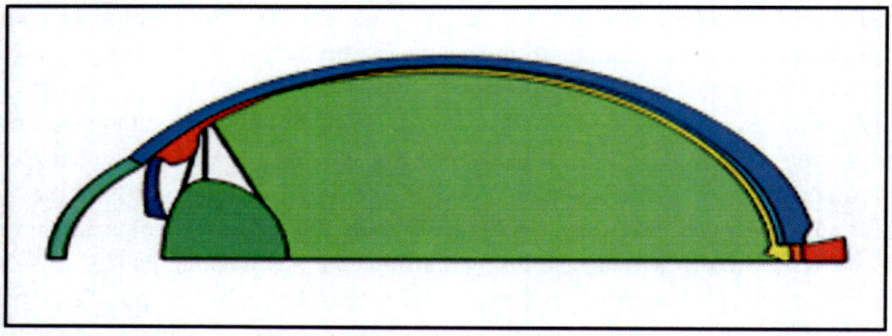

Fig. 1. The geometry of the model. The different parts have different colors to differentiate them from each other.

its edges) with a sphericity factor of -0.27, which is the most common in the population [12]. Its thickness was 0.55 mm in its center, increasing progressively to 0.625 mm near to the sclera [13]. The length from the anterior surface of the cornea to the posterior surface of the sclera was 25 mm [14].

The thickness of the choroid at the region near the optic nerve was defined of 0.22 mm, 0.15 mm at the height of the equator and 0.22 mm at the height of the connection with the ciliary body [15]. The ciliary bodies were included as a structure of 6mm of length, divided in 4 sections with the following descending thickness each one: 1, 0.55, 0.4, and 0.27 mm respectively, from the limit of the cornea to its junction with the choroid [18, 19].

A retina with a thickness of 0.25 mm in its posterior section and 0.2 mm at the level of the equator was included. It subsequently becomes smaller as it approaches the ciliary bodies [15]. At the posterior segment of retina, a protuberance followed by a small excavation was modeled to give form to the optic nerve cupping with a diameter of 1.8 mm and a cup of 0.54 mm with a depth of 0.475 mm [16].

Lamina cribosa was modeled as a constant thickness region of 0.38 mm thickness which closed the model at its back. The distance between lamina cribosa and Brunch membrane was 0.4 mm. The external portion of the optic nerve was patterned to increase in thickness to 3 mm at 1.5 mm from the posterior surface of the lamina cribosa [17, 18].

The lens was generated from two different ellipses which have a common semiaxis of 4.5 mm simulating the width of 9 mm of the structure [8]. The other semi axes of both ellipses were 1.36 and 2.73 mm respectively to form the anterior and posterior face of the lens. Therefore, the thickness lens was 4.1 mm and its anterior face was placed 3.15 mm from the posterior face of the cornea [13, 19, 20]. The vitreous chamber had a length from the posterior face of the lens to the retina of 15.8 mm [19]. The zonules were generated as 3 fibers with thicknesses of 50 μm inserted at the equator and on the anterior and posterior face of the lens.

The iris consisted in a structure that comes out from the ciliary bodies having a maximum thickness at the level of the pupil of 0.55 mm, 0.34 mm at its base and with a curvature of 0.21 mm. 0.05 mm from the posterior face of the iris to the anterior face of the lens was considered [21, 22]. The anterior chamber angle was modeled to be 28° [23].

2.2 Material Proprieties

All tissues were considered as linear elastic and isotropic. The Young's modulus, the Poisson's ratio and density for the different tissues were used. The retina, the head of the optic nerve, and the post-laminar tissue were considered as the same tissues type. Material tissues propriety values are presented in Table 1.

2.3 Mesh

For this case, the COMSOL Multiphysics® software offers different types of meshing in two dimensions. A triangular mesh type was considered with a density level of triangular elements of the "Extremely fine" type, this type of mesh generated for the eyeball geometry a total of 115195 triangular elements, 40 vertex type elements and 2818 contour elements. With a maximum size of 71 μm and a minimum size of 0.23 μm. The mean quality of elements was of 0.9323 and the minimum quality was 0.0751.

2.4 Boundary Conditions and Loads

The normal range of values of IOP is between 10 to 21 mmHg and it depends of a great number of factor like the position of the person, the hour of day, race and age. Values over 21 mmHg and until 35 mmHg were considered suspect to glaucoma [2].

Table 1. Material proprieties

Structure	Young modulus (MPa)	Poisson's ratio	Density (g/cm 3)
Cornea	0.3 [24]	0.42 [24]	1.024 [25]
Sclera	3 [10]	0.47 [24]	1.049 [25]
Lens	0.004 [6]	0.49 [25]	1.056 [25]
Iris	0.027 [26]	0.49 [25]	1.1 [25]
Choroid	0.6 [24]	0.49 [25]	1.002 [25]
Retina	0.02 [25]	0.47 [25]	1.008 [25]
Ciliar body	0.582 [27]	0.49 [6]	1.09 [28]
Zonules	0.34 [29]	0.49 [6]	1.09 [37]
Lamina Cribosa	0.3 [10]	0.49 [24]	1.049 [35]
Vitreous Body	0.042 [6]	0.49 [6]	1.002 [25]

For this study a small range of pressure values (10 to 35 mmHg) was considerate trying to simulate early stages of the disease when the elevation of IOP is not so high

[6, 9]. These values were applied as a distributed pressure load to the surfaces which the aqueous humor is in contact. For the vitreous body, the walls of the retina and the ONH were affected by a hydrostatic pressure and considering the eye of a person in a recumbent state (with the optic nerve being the lowest point) [11, 30].

The hydrostatic pressure is given by the following expression:

$$Ph = pgz \tag{1}$$

where Ph corresponds to the total pressure on the surfaces in contact with the vitreous humor, p corresponds to the density of the vitreous humor, g to the constant of gravity and z to the height at which pressure is being applied (the pressure generated at the anterior limit of the vitreous body due to the uniformly distributed pressure load was also considered as a initial pressure).

It was also considered allowing only radial movement of the sclera at 6.5 mm from the corneal limbus to simulate the presence of the extraocular rectus muscles. [6, 9].

2.5 Solution

A parametric study with steps of 5mmHg within the previously described range (10 to 35 mmHg) was considerate. The simulation took a total time of 16 min and 14 s.

The obtained results consider the fields for Von Mises equivalent stress, displacements (general and specific in some axis) and for 1st and 3rd principal strains. The structures of main interest were the cornea, lens, iris, sclera (anterior segment) and ciliary bodies.

3 Results

In Fig. 2, it is possible to see the equivalent Von Mises stress increases mainly in the sclera and in the cornea as the IOP increases. Within the same image it can be appreciated that the cornea moves away from the posterior segment while the lens moves toward it. The maximum appreciable tension at the value of 35 mmHg occurs in the sclera with a value of 140.3 kPa.

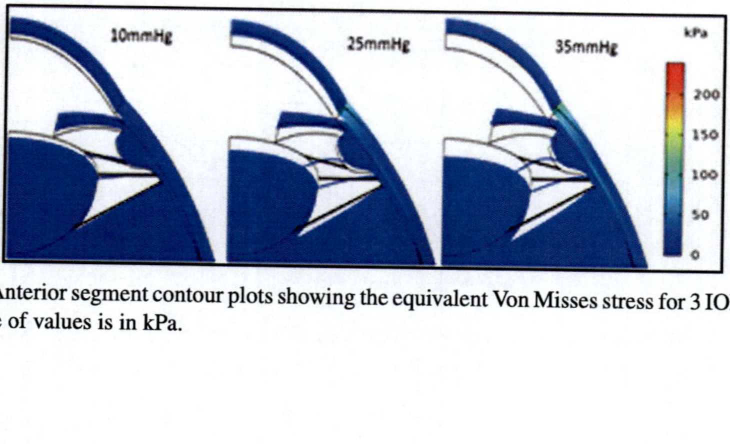

Fig. 2. Anterior segment contour plots showing the equivalent Von Misses stress for 3 IOP values. The scale of values is in kPa.

It was also possible to see movement in the ciliary bodies and in the zonules (although the latter are not of interest in our study). The largest displacement when the simulation reaches 35 mmHg was in the iris with a total displacement of 0.64 mm.

The total displacements in Fig. 3. Showed that the cornea, the anterior part of lens and a part of the iris, were the structures with the biggest change. The maximum value of total displacements was at the tip of the iris with 0.643 mm at 35 mmHg.

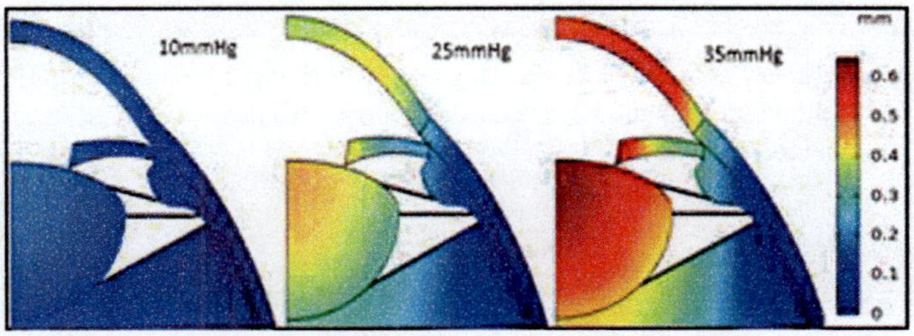

Fig. 3. Anterior segment contour plots showing the total displacement for 3 IOP values. The scale values is in mm.

For the case of the 1st principal strain shown in Fig. 4. The union zones between the lens and zonules, ciliary bodies and zonules have been identified as the most affected, as well as for a large part of the cornea and the tip of the iris. The maximum percentage of 1st principal strain occurs in the lens with 194%.

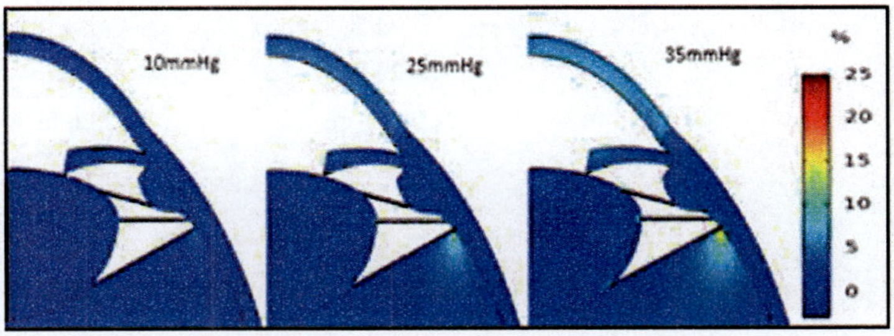

Fig. 4. Anterior segment contour plots showing the 1st principal strain for 3 IOP values. The scale values is in percentage (%) and it was configured to show changes in most structures leaving out the maximum value.

Finally, for the 3rd principal strain shown in Fig. 5. Again, the lens was the structure that presents the lowest value of - 190% in its union with the zonules. Some other structures with considerable strains were the cornea and the base of the iris near the iridocorneal angle.

The maximum displacement, stress and strain data for all the structures at the indicated IOP values (10, 25 and 35 mmHg) are found in Table 2.

Additionally, the length of the anterior chamber (derived from the displacements of the cornea and lens) was measured by each value of IOP. Figure 6 shows the length

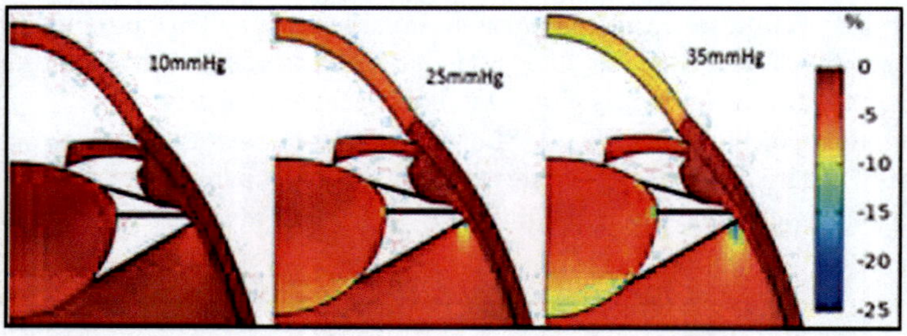

Fig. 5. Anterior segment contour plots showing the 3rd principal strain for 3 IOP values. The scale values is in percentage (%) and it was configured to show changes in most structures leaving out the minimum value.

Table 2. Main results

Structure	10 mmHg	25mmHg	35mmHg
Cornea	17.8 kPa	44.4 kPa	62.1 kPa
	0.171 mm	0.43 mm	0.602 mm
	5.4%	13.4%	18.8%
	−4.3%	−10.7%	−15%
Sclera	108.9 kPa	100.2 kPa	140.3 kPa
	0.111 mm	0.28 mm	0.393 mm
	3.3%	2.7%	3.8%
	−2.7%	−2.9%	−4%
Lens	2.5 kPa	6.2 kPa	8.7 kPa
	0.185 mm	0.451 mm	0.628 mm
	55.5%	138.7%	194.3%
	−52.4%	−131.1%	−183.6%
Iris	1.8 kPa	4.5 kPa	6.3 kPa
	0.183 mm	0.459 mm	0.643 mm
	5.6%	14%	19.6%
	−5.9%	−14.8%	−20.7%
Ciliary body	59.6 kPa	145.6 kPa	203 kPa
	0.092 mm	0.233 mm	0.327 mm
	9.1%	22.2%	31%
	−8.4%	−20.7%	−28.9%

The first value for each structure corresponds to the maximum Von Mises stress. The second value corresponds to the maximum total displacement. The third value is the maximum 1st principal strain. The last value is 3rd principal strain increase related to the IOP rise. The length of the anterior chamber was 3.48, 3.98 and 4.31 mm for the values of 10, 25 and 35 mmHg, respectively.

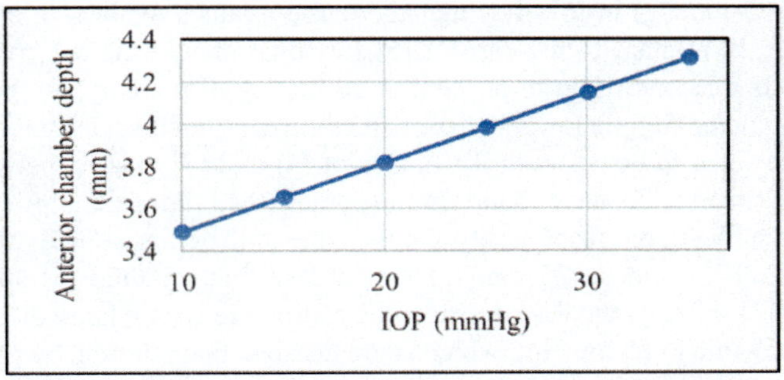

Fig. 6. Trend line showing the relationship between the increase in anterior chamber depth and IOP.

4 Discussion

The contour graphs shown in Figs. 2 and 3 have a behavior like that exhibited in other works for the same parameters, that is, equivalent stress and displacements [6]. An increase in the length of the anterior chamber was measured. It is generated through the displacement of the cornea and lens in opposite directions which this is also already mentioned by other authors of simulations [7]. The explanation for this may be because the increase in aqueous humor causes an increase in IOP, which causes the displacement and strains of the structures that are in contact with the liquid [31].

The elongation and compression given through the 1st and 3rd principal strains respectively, are commonly analyzed in works focused on the optic nerve head (the portion of axon within the eye) behavior [10, 32], however, in the present study the analysis of both parameters was focused on structures belonging to anterior segment. The structures with greater elongation and visible compression as the IOP value increases are the cornea, the lens at its junction with the zonules, and part of the iris.

In all the structures of interest and according to the results of Table 2, there are si ilar levels of both strains. This behavior could be reflected, for example, in the cornea (shortening its thickness and increasing its length). However, as a counterargument, today there is still a debate about whether central corneal thickness is related to glaucoma. More studies in the future could corroborate the changes in thickness and length in the structures of interest.

Strains in the lens at its junction with the zonule is also presented in [6]. Although in that case with a more accentuated behavior throughout the structure. The zonules are structures that allow the lens to be attached and therefore adhere to it with a certain tension, so increasing the IOP value could cause greater stress in the union areas.

There are research works that perform simulations focused on iris. Most of the consulted works take an approach to closed-angle glaucoma, where the shape of the iris and its position can be altered to simulate conditions of this disease.

In research works such as [26], it is exhibited a greater total strain at the tip of the iris like that shown in Fig. 4.

Tension and displacements reported in Table 2 are in the same value range reported (44.4 kPa at 25 mmHg) in [6], they manage tension values for the cornea between 0 to 56 kPa for 25 mmHg. In the case of [5] for some of its different cornea models, it also exhibits tensions of up to 50 kPa, at 30 mmHg of mercury, so their values at 25 mmHg should be slightly lower. On the part of the lens, our displacements (0.451 mm at 25 mmHg) seem to be between the ranges of [6] and [7], that is, between 0.3 and 0.6 mm. The same results were found for the cornea (0.43 mm at 25 mmHg) which are between 0.3 and 0.48 mm respectively. On the zone of the ciliary bodies, the measured displacement (0.233 mm at 25 mmHg) was double that reported (as approximately 0.1 mm) per [7]. Finally, the iris appears to be a structure with a large displacement in our case (0.459 mm at 25 mmHg), which is greater than that reported by [26], where at 27 mmHg it has values of up to 0.33 mm.

The increase in length of the anterior chamber corresponds to a parameter that has already been studied in people and has been mentioned in simulations. Studies such as [33, 34] have shown that for people with POAG there is an increase in the anterior chamber length. For a mean anterior chamber length of 3 mm at a IOP between 10 to 15 mmHg (values obtained from other sources to their study), an increase to 3.47 ± 0.37 mm for a mean IOP of 25 mmHg was estimated [34]. In our case, for the same value of 25 mmHg, a 3.98 mm chamber length was obtained, which represents an increase of 0.5 mm respect to 3.48 mm to 10 mmHg (Fig. 6), which is within the range of increase given for measurements on people. This aspect is also mentioned by other authors, and it is assumed that it is a consequence of the increase in aqueous humor in the anterior segment [6, 7].

It is important to note that all the variations in measurements between our models can be due to many factors. The properties of the different tissues, the geometry of each structure, their meshing considerations and boundary conditions influence the results of the simulations, however, the results show general behaviors like those shown by other authors, especially from of the contour plots.

5 Conclusions

The increase in IOP in small intervals can cause marked variations in the structures of the anterior segment that are in contact with the aqueous humor. The displacements and strains in structures such as the cornea, the lens, and the iris, stand out as the most notable among structures studied in this work. These changes can lead to variations in measurements such as the length of the anterior chamber, which may be a factor that could have a greater relationship for glaucoma variants such as open-angle glaucoma. Due to the above, it is necessary a greater number of simulations, varying the properties of the tissues and geometric aspects to corroborate if the results presented here are confirmed, although the evidence described by other authors seems to affirm the presence of these changes in the anterior segment.

Acknowledgement. We thank the support given by Consejo Nacional de Ciencia y Tecnología (CONACYT) and by Universidad Autónoma de Aguascalientes to develop the research work presented in this article.

References

1. Tham, Y.C., Li, X., Wong, T.Y., Quigley, H.A., Aung, T., Cheng, C.Y.: Global prevalence of glaucoma and projections of glaucoma burden through 2040: a systematic review and meta-analysis. Ophthalmology **121**(11), 2081–2090 (2014)
2. Shcacknow, P.N., Samples, J.R.: The Glaucoma Book, 1st edn. Springer, New York (2010)
3. Potop, V., et al.: The importance of corneal assessment in a glaucoma suspect—a review. Rom J Ophthalmol. **63**(4), 321–326 (2019)
4. Chandrupatla, T.R., Belegundu, A.D., Ramesh, T., Ray, Chaitali.: Introduction To Finite Elements in Engineering. 4th edn. Pearson Education, Harlow (2012)
5. Elsheikh, A.: Finite element modeling of corneal biomechanical behavior. J. Refract. Surg. **26**(4), 289–300 (2010)
6. Dai, P., Zhao, Y., Sheng, H., Li, L., Wu, J., Han, H.: Simulating the effects of elevated intraocular pressure on ocular structures using a global finite element model of the human eye. J Mech Med Biol. **17**(2), 1–18 (2017). https://doi.org/10.1142/S0219519417500385
7. Vivek Suganthan, R., Meenatchi Sundaram, S., Ramesh, S.V., Rinu, T., Pai, R., Khader, S.M.A., Hazarika, M., Girish, H.: Finite element analysis of the human eye for a range of intraocular pressure. Lect. Notes Electr. Eng. **660**, 173–181 (2020)
8. Remington, L.A.: Clinical Anatomy and Physiology of the Visual System, 3rd edn. Butterworth-Heinemann, Saint Louis (2012)
9. Sigal, I.A., Flanagan, J.G., Tertinegg, I., Ethier, C.R.: Finite element modeling of optic nerve head biomechanics. Investig. Ophthalmol. Vis. Sci. **45**(12), 4378–4387 (2004)
10. Norman, R.E., Flanagan, J.G., Sigal, I.A., Rausch, S.M.K., Tertinegg, I., Ethier, C.R.: Finite element modeling of the human sclera: Influence on optic nerve head biomechanics and connections with glaucoma. Exp. Eye Res. **93**(1), 4–12 (2011)
11. Uchio, E., Ohno, S., Kudoh, J., Aoki, K., Kisielewicz, L.T.: Simulation model of an eyeball based on finite element analysis on a supercomputer. Br. J. Ophthalmol. **83**(10), 1106–1111 (1999)
12. Xiong, Y., et al.: The analysis of corneal asphericity (Q value) and its related factors of 1,683 Chinese eyes older than 30 years. PLoS ONE **12**(5), 1–9 (2017)
13. Feizi, S., Jafarinasab, M.R., Karimian, F., Hasanpour, H., Masudi, A.: Central and Peripheral corneal thickness measurement in normal and keratoconic eyes using three corneal pachymeters. J. Ophthalmic Vis. Res. **9**(3), 296–304 (2014)
14. Bhardwaj, V., Rajeshbhai, G.P.: Axial length, anterior chamber depth-a study in different age groups and refractive errors. J Clin Diagnostic Res. **7**(10), 2211–2212 (2013)
15. Dai, P., Han, H., Zhao, Y., Fan, M.: Finite element analysis of the mechanical characteristics of glaucoma. J Mech Med Biol. **16**(4), 1–16 (2016)
16. Kolb, H., Nelson, R.F., Ahnelt, P.K., Ortuño-Lizarán, I., Cuenca, N.: The Architecture of the Human Fovea, 1st edn. University of Utah Health Sciences Center, Salt Lake (2020)
17. Seo, J.H., Kim, T.W., Weinreb, R.N.: Lamina cribrosa depth in healthy eyes. Investig. Ophthalmol. Vis. Sci. **55**(3), 1241–1250 (2013)
18. Kim, Y.W., et al.: Clinical assessment of lamina cribrosa curvature in eyes with primary open-angle glaucoma. PLoS ONE **11**(3), 1–15 (2016)
19. Roy, A.: Variation of Axial ocular dimensions with age, sex, height, bmi -and their relation to refractive status. J Clin Diagnostic Res. **9**(1), 1–4 (2015)

20. Barlatey, M.S., Koga-Nakamura, W., Moreno-Londoño, M.V., Takane-Imay, M., González-González, M.C.: Distribución de la longitud axial y factores relacionados en una población adulta de la Ciudad de México. Rev Mex de Oftalmol. **93**(5), 233–237 (2019)
21. Kapnisis, K., van Doormaal, M., Ross Ethier, C.: Modeling aqueous humor collection from the human eye. J. Biomech. **42**(15), 2454–2457 (2009). https://doi.org/10.1016/j.jbiomech.2009.07.025
22. Lee, R.Y., Huang, G., Porco, T.C., Chen, Y.C., He, M., Lin, S.C.: Differences in iris thickness among African Americans, caucasian Americans, hispanic Americans, Chinese Americans, and filipino-Americans. J. Glaucoma **22**(9), 673–678 (2013)
23. Feng, M.T., Belin, M.W., Ambrósio, R., Grewal, S.P.S., Yan, W., Shaheen, M.S., McGhee, C., Maeda, N., Neuhann, T.H., Burkhard Dick, H., Alageel, S.A., Steinmueller, A.: Anterior chamber depth in normal subjects by rotating scheimpflug imaging. Saudi J. Ophthalmol. **25**(3), 255–259 (2011)
24. Ayyalasomayajula, A., Park, R.I., Simon, B.R., Vande, J.: A porohyperelastic finite element model of the eye: the influence of stiffness and permeability on intraocular pressure and optic nerve head biomechanics. Physiol. Behav. **176**(12), 139–148 (2017)
25. Tahavvor, A.R., Gholami, M.: Numerical simulation of the early stages of glaucoma in human eye. J. Rehabil. Res. Dev. **7**(3), 134–140 (2020)
26. Wang, W., Qian, X., Song, H., Zhang, M., Liu, Z.: Fluid and structure coupling analysis of the interaction between aqueous humor and iris. Biomed. Eng. Online. **15**, (2016)
27. Michael, R., Mikielewicz, M., Gordillo, C., Montenegro, G.A., Pinilla Cortés, L., Barraquer, R.I.: Elastic properties of human lens zonules as a function of age in presbyopes. Investig. Ophthalmol. Vis. Sci. **53**(10), 6109–6114 (2012)
28. Hasgall, P.A., Di Gennaro, F., Baumgartner, C., Neufeld, E., Lloyd, B., Gosselin, M.C., Payne D., Kligenbock, A., Kuster, N.: IT'IS database for thermal and electromagnetical parameters of biological tissues. https://itis.swiss/virtual-population/tissue-properties/overview/. Last Accessed 18 Oct 2021
29. Wang, K., Venetsanos, D.T., Hoshino, M., Uesugi, K., Yagi, N., Pierscionek, B.K.: A modeling approach for investigating opto-mechanical relationships in the human eye lens. IEEE Trans. Biomed. Circuits Syst. **67**(4), 999–1006 (2020)
30. Young, H.D., Freedman, R.A.: Física Universitaria. 12th edn. Pearson Education, Naucalpan de Juárez (2013)
31. Simonato Alonso, R., Ambrósio Junior, R., Junior, A.P., Sakata, L.M., Ventura, M.P.: Glaucoma anterior chamber morphometry based on optical Scheimpflug images. Arq. Bras. Oftalmol. **73**(6), 497–500 (2010)
32. Eilaghi, A., Flanagan, J.G., Simmons, C.A., Ethier, C.R.: Effects of scleral stiffness properties on optic nerve head biomechanics. Ann. Biomed. Eng. **38**(4), 1586–1592 (2010)
33. Marchini, G., Pagliarusco, A., Toscano, A., Tosi, R., Brunelli, C., Bonomi, L.: Ultrasound biomicroscopic and conventional ultrasonographic study of ocular dimensions in primary angle-closure glaucoma. Ophthalmology **105**(11), 2091–2098 (1998)
34. Sihota, R., et al.: Is combined mechanism glaucoma a distinct entity? Graefe's Archive for Clinical and Experim. Ophthalmol. **256**(10), 1961–1969 (2018)

Development of a Flexible Algorithm for Simulations with Human-Exoskeleton Models: An Approach to the Robot-Assisted Therapy

Denis Mosconi[1,2](✉) and Adriano A. G. Siqueira[2]

[1] Industry Department, Federal Institute of São Paulo, Catanduva, SP, Brazil
[2] Center of Engineering Applied to Healthy, São Carlos School of Engineering, University of São Paulo, São Carlos, Brazil
denis.mosconi@ifsp.edu.br

Abstract. Robot-assisted therapy is a promising approach to overcome the challenge that is to recover the motor capabilities of cerebrovascular accident victims. Simulation algorithms are widely used for the development of robots and controls used in this type of therapy. However, such algorithms are of specific use, needing for ones that are both flexible and able to simulate not only the robot performance, but also human behavior and its interaction with the machine. In this work we proposed a flexible simulation algorithm useful to perform analysis, assessment and development of human-interaction controls applied to robotic neurorehabilitation. Simulations using experimental data were performed, proving that the algorithm is effective and also a promising approach for the development of features related to the recovery of motor skills lost due to stroke.

Keywords: OpenSim · robotic neurorehabilitation · stroke · forward dynamics

1 Introduction

Approximately 15 million people suffer a cerebrovascular accident (CVA) worldwide, per year. Of these, 5 million die and 5 million remains with some impairment. These numbers place the stroke in second position among the top 10 causes of death, and as the leading cause of motor disability [15]. For the next ten years, the number of impaired people due stroke is expected to exceed 200 million [14]. Bearing that in mind, it can be observed an increasing demand for rehabilitation treatments, with the robotic neurorehabilitation being one of them.

With the use of robotics in rehabilitation treatments, it is possible to perform more intensive and repetitive movements during long duration routines, demanding less effort from the therapist as well as reducing the number of therapists per patient, mainly in gait rehabilitation. In addition, it is also possible

J. L. B. Marques et al. (Eds.): CLAIB 2022/CBEB 2022, IFMBE Proceedings 98, pp. 299–309, 2024.
https://doi.org/10.1007/978-3-031-49401-7_30

to recording data which allow a more precise biomechanical analysis and a more accurate prognosis on the part of the physiotherapist [1,6].

However, although robotic rehabilitation reduces the therapists effort in rehabilitation centers, to develop robots and interaction controls is not a simple task, since it is necessary to ensure not only the effectiveness of the treatment but also the safety of the patients. In addition to that, the human-robot interaction is a nonlinear mechanical system with high complexity.

Therefore, the use of computational resources is a suitable approach, either with regards to the mechanical design of the robot, with respect to the development of interaction controls or for movement analysis, before they are actually implemented in the real setup. Several approaches have been taken, focusing on these objectives in isolation [3,7,10–12], however there is still a lack of a system that is flexible and can meet the largest number of possible demands, such as developing interaction controls, simulating human behavior, developing and analyzing interaction models between humans and exoskeletons.

This work proposes a simulation algorithm based on forward dynamics, which is, at the same time, able to support development and tuning of interaction controls, biomechanical data analysis as well as the development and evaluation of human-exoskeleton interaction models. With that, it is expected to make the development of such resources within the context of robotic rehabilitation more agile and flexible.

The paper is organized as follows: Sect. 2 introduces the simulation algorithm, the human-exoskeleton interaction model used and describes the experimental procedures. The results and discussion are contained in Sect. 3 and the conclusion is in Sect. 4.

The study of the interaction between humans and exoskeleton was approved by the Ethics Committee of the Federal University of São Carlos (Number 26054813.1.0000.5504). All subjects gave their informed consent before the experimental session.

2 Methodology

In this section we present the development of the simulation algorithm proposed and the experimental procedure conducted to test it.

2.1 Simulation Algorithm

The proposed algorithm was developed in MATLAB®, using the libraries and models from OpenSim. Its operation mode is based on the Forward Dynamics.

OpenSim[1] is an open-source platform developed by [3] for modeling, simulating and analyzing movements related to the human neuromusculoskeletal system. Through its Application Programming Interface (API), it is possible to make MATLAB® access OpenSim codes, enabling the use of libraries and models in addition to making possible the development of controls (e.g. interaction

[1] http://opensim.stanford.edu.

controls between human and exoskeleton). In this work the OpenSim was used to provide the neuromusculoskeletal model that was used to develop the computational model of a human wearing an exoskeleton and to fit its anthropometry to that of a specific subject.

As the human-exoskeleton computational model is a biomechanical system, if the forces, torques or control signal inputs (e.g. muscular activations, electrical current of the robot motors) are available, given the initial conditions we can also determine its kinematics by integrating its dynamics equations, a procedure that is called *Forward Dynamics*. In this work we used the Forward Dynamics to determine the knee angular position of the human-exoskeleton computational model when subjected to certain torque inputs.

A flowchart of the proposed simulation algorithm can be seen in the Fig. 1. The **Input Files** can be .mat, .sto, .mot or .txt files containing the initial conditions, reference position, torques or forces involved or another variables, according to the requirements of the simulation to be performed (.mat are files native to the MATLAB® while .sto and .mot are native to the OpenSim). In this work we used a .mat file containing the position reference, the robot torque and the feedforward human torque (more about these variables will be explained in the following sections).

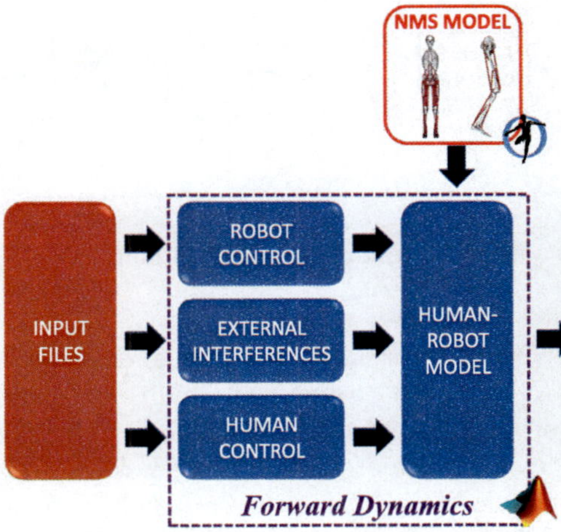

Fig. 1. Flowchart of the proposed algorithm

The **Robot Control** contains the definition of how the robot interacts with the human. This definition can be an explicit control law (e.g. impedance control law) or a vector containing the time-based torque values. In this work we used a vector that is introduced through the input files, containing the robot torques measured experimentally.

External Interferences can be external forces or torques, like ground reaction forces (GRF) or contact forces, which can influence the movement realization. In this work no external interference were considered.

The **Human Control** block contains the definition of the human behavior and can be either an explicit control law or a vector containing the human torque or muscular activations or even a combination of both. In this work a feedforward-feedback human control was used [4,13], where the feedforward loop applies the premeditated torque to perform the movement, determined through the *Inverse Dynamics Tool* from OpenSim, whereas the feedback loop consists of a conventional PID control that seeks to eliminate the tracking errors that can occur during the execution of the movement. This type of human control is based on the human internal model concept (Fig. 2), which states that the subject tries to predict the torque necessary to perform a certain movement, based on an internal model of his body contained in the cerebellum, and when tracking errors occur during the execution of such movement, a correction torque is applied in order to to eliminate such errors [4,13].

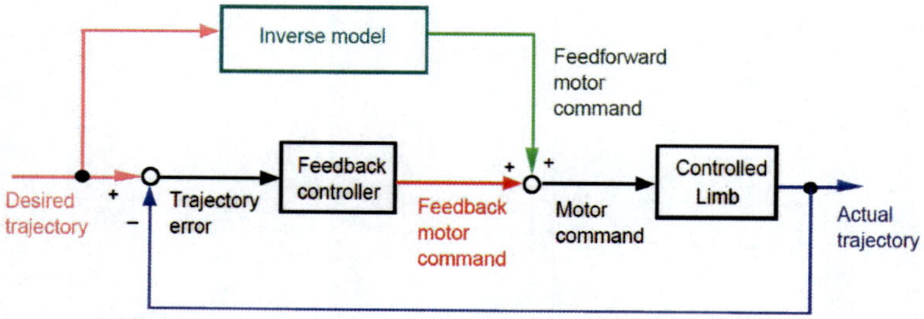

Fig. 2. Feedforward-feedback human control. Adapted from [13]

The **Human-robot Model** is an interaction model developed using a neuromusculoskeletal model (NMS model) from OpenSim in whose joints are placed coordinate actuators to produce a computational model of an exoskeleton (the actuators seek to emulate the DC motors from an exoskeleton). In this work the human-robot model was developed using the *gait2392* (a three-dimensional, 23 degree-of-freedom model of the human lower limbs, [2]) in whose right knee joint an active exoskeleton joint computational model was coupled, based on the ExoTao (a modular exoskeleton for lower limbs, composed of a tubular structure with six free and independent joints [9]). In order to reduce computational cost as well as modeling complexities, we considered the actuators as ideal: no response delay and no losses. The axes of the robot joint and the user ones are considered collinear. The ExoTao and the NMS gait2392 are depicted in Fig. 3.

The box drawn with a dashed line and surrounding the three blue blocks in Fig. 1 represents the **Forward Dynamics**, which integrates the dynamic equations of the human-robot model that receives the torques, forces or another input variables from the previous blocks. In this work we used the ODE45 function

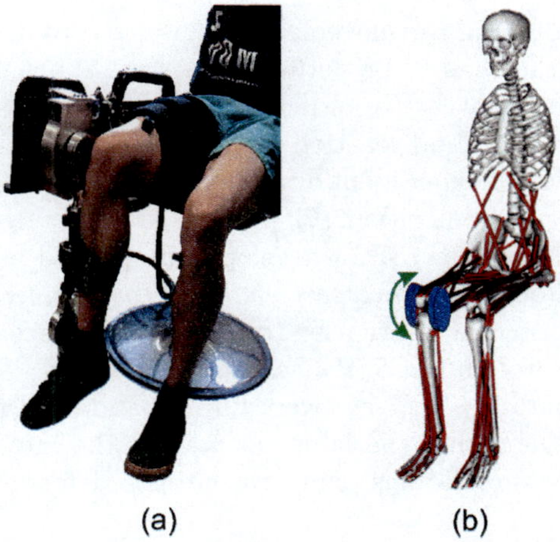

Fig. 3. (a) A user wearing only the knee joint of the ExoTao, (b) the correspondent human-robot model (red lines denote the musculotendon actuators, the blue disc represents the robot actuator and the green arrow indicates the sagittal direction of the movements)

from MATLAB® to produce the outputs that are saved in the **Output Files**. Such files can be .mat, .sto, .mot or .txt and can both contain the kinematic data as well as other pertinent variables such as muscle fiber length, penance angle and others defined according to the need of the researcher.

2.2 Experimental Procedure

A physical experiment was carried out, generating data that were later used to test the simulation algorithm. To perform this task, a healthy subject, 1.75 m tall, 85 kg, wearing the active right knee joint from ExoTao, in a seated position, performed movement of extension and flexion of the knee. The exoskeleton was governed by an impedance control, producing a force in response to an imposed movement [5]. The control law is expressed by Eq. (1), where τ_R is the robot torque, K_R is the virtual stiffness, B_R is the virtual damping, θ^d is the desired position, θ_R is the real position and $\dot{\theta}_R$ is the angular velocity.

$$\tau_R = K_R(\theta^d - \theta_R) - B_R\dot{\theta}_R \tag{1}$$

This physical experiment was divide in two steps:

- **Step I:** The exoskeleton behaved as active-assistive, helping the user to perform the desired movement. For this, both the desired trajectory of the user and the robot are the same.
- **Step II:** The exoskeleton behaved as active-resistive, opposing to the movement performed by the user. For this, the desired trajectory of the robot is shifted of 180°C in relation to the desired trajectory of the user.

During the execution of the movement, the user received a visual reference of the desired motion that had to be performed. As the robot joint was supported by a chair (as depicted in Fig. 3) and the user did not contact the ground, the actuator mass and the ground reaction forces were neglected.

Through the physical experiment described above, the torque of the exoskeleton (τ_R) and the angular position (θ_R) of the knee were measured using the rotary series elastic actuator (SEA) developed by [8] and used to estimate the user torque (τ_H) through the *Inverse Dynamics Tool* from OpenSim (Fig. 4).

The anthropometric data were used to fit the human-exoskeleton interaction model to the user, through the *Scale Tool* from OpenSim.

The measured data (θ_R and τ_R) were filtered and used to the estimate the user torque (τ_H). Then, these variables as well as the interaction model were used to perform the simulations using the proposed forward dynamics-based algorithm.

Two simulations batches were made: the first one to tune the PID control of the human control and the second, using another data set, to verify the algorithm convergence. The resulting PID gains were obtained by trial and error method, being: $Kp = 6.3$, $Ki = 5.8$ and $Kd = 5.5$.

In the simulations, the interaction model is expected to perform the same movement that the user performed during the physical experiment, that is: $\hat{\theta} = \theta_R$, being $\hat{\theta}$ the knee position obtained with the simulation.

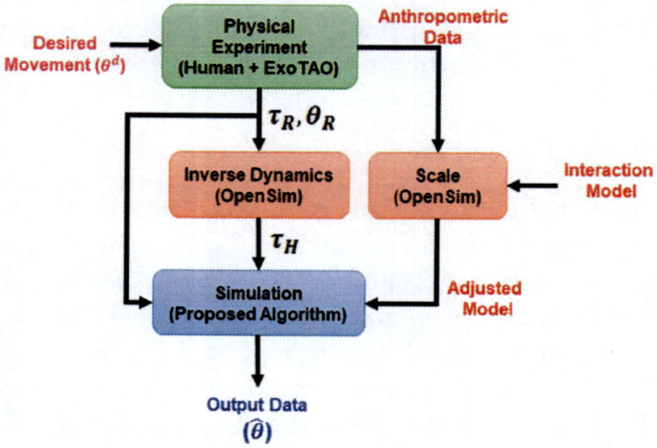

Fig. 4. A flowchart that depicts the experimental procedure.θ_R is the angular position of the knee (that is the same both for the robot and the user), τ_R is the robot torque, τ_H is the user estimated torque and $\hat{\theta}$ is the knee angular position obtained with the simulation

The flexibility of the algorithm in lead with different types of data, its complexity in to be prepared and used for the simulations, as well as its applicability were evaluated. To this, we analyzed the movements, errors torques obtained with each simulation, together with the time spent to prepare and execute the simulations.

All simulations were carried out on a computer with Intel® Core™i7-5500 2.40 GHz processor, 8.00 GB of RAM, 2.00 GB dedicated video card, Windows 10 Home Single Language 64 bits. The OpenSim version 3.3 and the MATLAB R2017b were the platforms where the simulations took place.

3 Results

In this section we present the results from simulations of the second batch, when the PID is already correctly tuned to the specific subject. The shaded regions in the figures represent the maximum and minimum interval value of the variables and the bold lines are de average value. In Figs. 6 and 8, **Exo** is the torque of the exoskeleton, **User FF** is the feedforward torque of the user (estimated through Inverse Dynamic Tool) and **User FB** is the feedback torque of the user determined by the PID loop of the human control.

The Fig. 5(a) shows the knee position, comparing the results between the physical experiment Step I (Reference in the Fig. 5(a)) and the simulation (Simulated in the Fig. 5(a)). It is possible to notice the extension and flexion phases. Comparing the two results, it can be seen that the human-exoskeleton interaction model executed accurately the movement performed by the subject, and that the simulation algorithm was able to realize the forward dynamics.

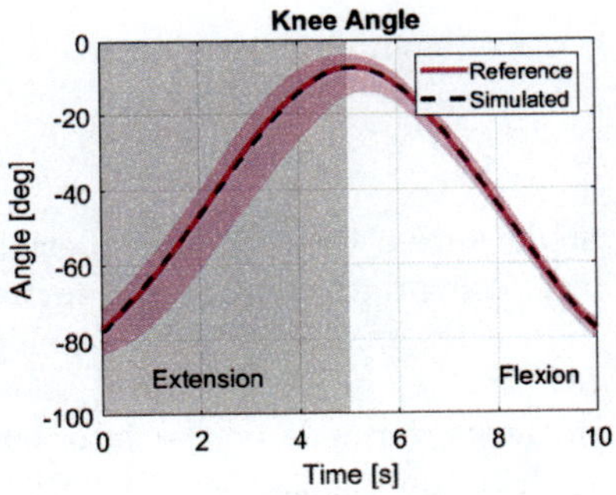

(a) Knee angle

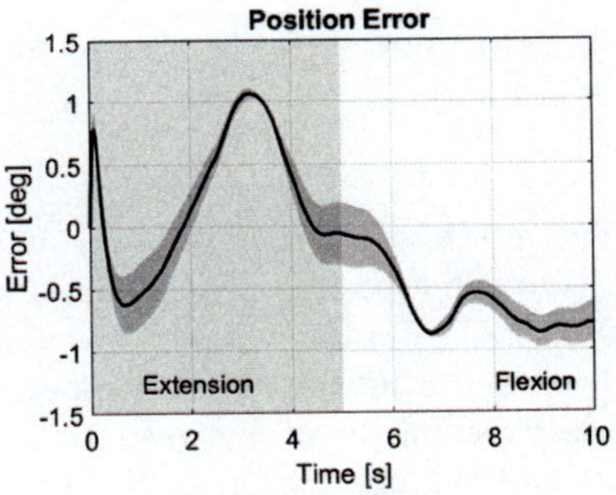

(b) Position error

Fig. 5. (a) Knee angular position obtained with the simulations using data from physical experiment Step I and (b) error between the movement performed by the interaction model and the one performed by the user during the physical experiment Step I

The position error between the movement performed by the interaction model and the one performed by the user during the physical experiment Step I is depicted in Fig. 5(b). In this case the root mean square (RMS) and maximum errors are 0.61 and 1.06 degrees, respectively.

In Fig. 6 is possible to see the torques exerted by the user and the robot. The robot torque was measured from the physical experiment Step I, the User FF is the feedforward torque developed by the user and estimated from the Inverse

Dynamics Tool from OpenSim, the User FB is the feedback user torque obtained from de PID of the human control loop. As the user is a health subject and the robot is the active-assistive mode, the torque applied by the exoskeleton is near to zero, since the user did not need additional help to perform the movement during the physical tests step I (the position error is small). As the subject developed a feedforward torque able to perform well the movement, the feedback torque was also small.

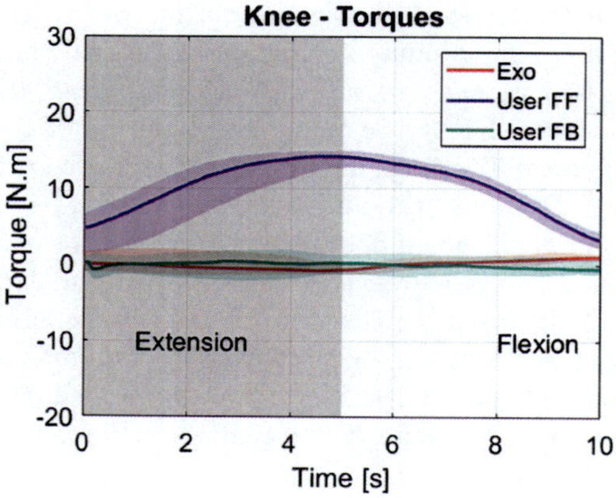

Fig. 6. Torques performed by the user and the robot (input data from physical experiment Step I)

The results of position obtained with the data from the physical experiment Step II are showed in Fig. 7(a). As in the tests using data from Step I, here the human-robot interaction model was able to repeat the same movement developed by the user during the physical experiment.

The tracking error between the movement performed by the interaction model and the one performed by the user during physical experiment Step II can be seen in the Fig. 7(b). In this case the RMS and maximum errors are 0.43 and 1.16 degrees, respectively.

The torques involved in Step II are depicted in Fig. 8. It is possible to notice that in this case the exoskeleton performed a higher torque than in the Step I, besides that the robot torque is shifted in relation to the one performed by the subject (and also in relation to the movement performed by the user) since the robot is configured to behave as active-resistive. Due to this operation mode of the exoskeleton, the user had to perform a higher torque than in Step I. As no additional interference was imposed on the movement, the user feedback torque was also low in this case.

The obtained results makes it possible to state that the model was well adjusted to the subject and the proposed algorithm is able to perform the simulations using inputs of torque and position, relating the interaction of the human

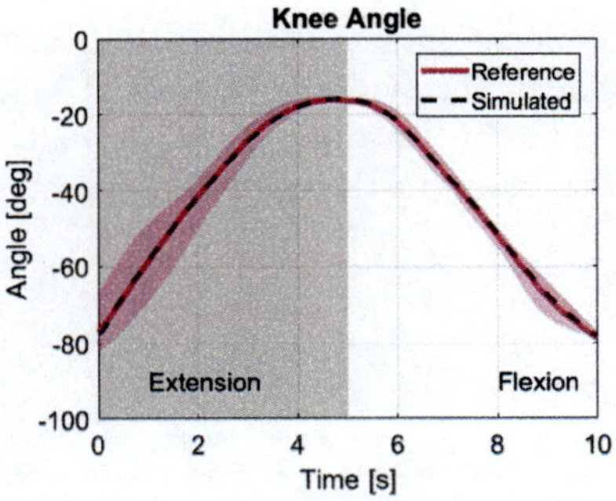

(a) Knee angle

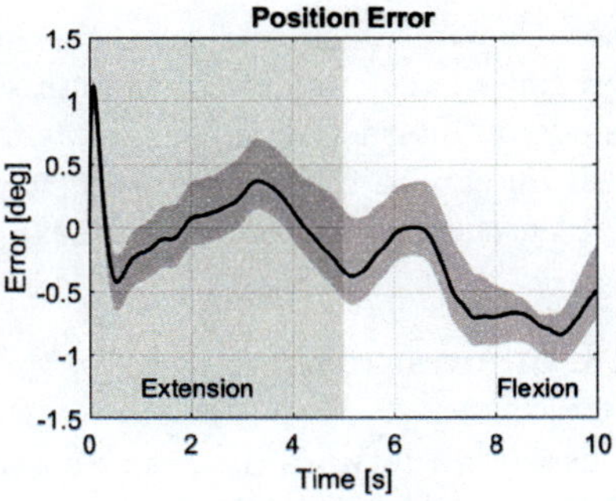

(b) Position error

Fig. 7. (a) Knee angular position obtained with the simulations using data from physical experiment Step II and (b) error between the movement performed by the interaction model and the one performed by the user during the physical experiment Step II

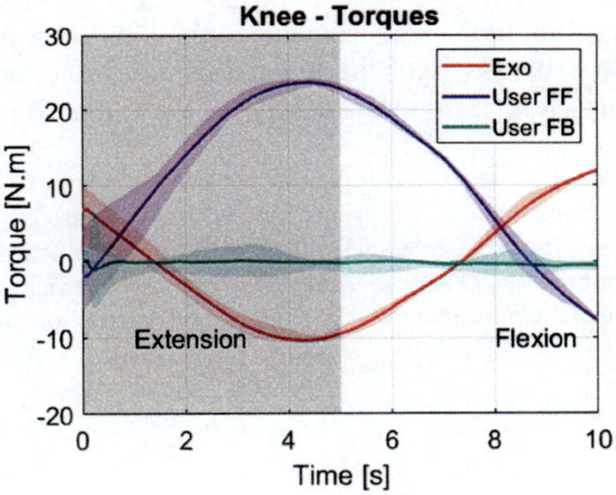

Fig. 8. Torques performed by the user and the robot (input data from physical experiment Step II)

and exoskeleton as well as reproducing computationally what occurred in the physical experiment. The tracking errors are less the 1.5 degree, what can be considered negligible for the case in question.

Each simulation took about 15 min to be accomplished, what can be considered a good time when compared to the time spent during physical tests that can be more than one hour. Based on this is possible to affirm that the proposed algorithm is a resource that allow the development of solutions in the context of neurorehabilitation relatively quickly.

The proposed algorithm can be used in different ways than the one presented in this work. For example, a researcher may be interested in testing different control laws for a given human-robot set, to do so just vary the contents of the

Robot Control block and keep the other blocks intact. One can also use the algorithm to test models of human control, other types of models of human-exoskeleton interaction (for example for studies of the upper limbs). Finally, the algorithm can be used in different ways, according to the need and objective of the researcher, which makes it a flexible tool.

4 Conclusions

In this work a forward dynamics-based simulation algorithm was proposed and its purpose is to be an environment able to promote test, tuning and analysis of interaction control, human control and human-exoskeleton interaction models. The tests using experimental data proved that the algorithm is able to reproduce computationally the data from physical experiments with accuracy, which leads to the conclusion that it presents itself as a flexible and feasible approach to aid the development of resources related to robotic rehabilitation.

For future work we expect to use the proposed algorithm to run simulations with another interaction models, movements and rehabilitation controls, furthermore, we intend to apply physical characteristics to interaction model actuators (i.e. mass, damping, delay), in order to bring them closer to real actuators.

Acknowledgments. This work is supported by Pro-Rectory of Research of University of São Paulo, Coordenação de Aperfeiçoamento de Pessoal de Nível Superior - Brasil (CAPES) - Finance Code 001, PGPTA, under grant 3457/2014, and São Paulo Research Foundation (FAPESP) under grant 2013/14756-0. This study was approved by the Ethics Committee of the Federal University of São Carlos (Number 26054813.1.0000.5504).

References

1. Androwis, G.J., Pikar, R., Ramanujam, A., Nolan, K.J.: Electromyography assessment during gait in a robotic exoskeleton for acute stroke. Front. Neurol. **9**, 630 (2018). https://doi.org/10.3389/fneur.2018.00630
2. Delp, S., Loan, J., Hoy, M., Zajac, F., Topp, E., Rosen, J.: An interactive graphics-based model of the lower extremity to study orthopaedic surgical procedures. IEEE Trans. Biomed. Eng. **37**(8), 757–767 (1990). https://doi.org/10.1109/10.102791
3. Delp, S.L., Anderson, F.C., Arnold, A.S., Loan, P., Habib, A., John, C.T., Guendelman, E., Thelen, D.G.: Opensim: Open-source software to create and analyze dynamic simulations of movement. IEEE Trans. Biomed. Eng. **54**(11), 1940–1950 (2007). https://doi.org/10.1109/TBME.2007.901024
4. Gillespie, R.B., Ghasemi, A.H., Freudenberg, J.: Human motor control and the internal model principle. IFAC-PapersOnLine **49**(19), 114–119 (2016). https://doi.org/10.1016/j.ifacol.2016.10.471
5. Ibarra, J.C.P., Siqueira, A.A.G.: Impedance control of rehabilitation robots for lower limbs, review. In: 2014 Joint Conference on Robotics: SBR-LARS Robotics Symposium and Robocontrol, pp. 235–240 (2014). https://doi.org/10.1109/SBR.LARS.Robocontrol.2014.53

6. Jutinico, A.L., Jaimes, J.C., Escalante, F.M., Perez-Ibarra, J.C., Terra, M.H., Siqueira, A.A.G.: Impedance control for robotic rehabilitation: a robust Markovian approach. Front. Neurorobot. **11**, 43 (2017)
7. Peña, G.G., Consoni, L.J., dos Santos, W.M., Siqueira, A.A.G.: Feasibility of an optimal EMG-driven adaptive impedance control applied to an active knee orthosis. Robot. Auton. Syst. **112**, 98–108 (2019). https://doi.org/10.1016/j.robot.2018.11.011
8. dos Santos, W.M., Caurin, G.A.P., Siqueira, A.A.G.: Design and control of an active knee orthosis driven by a rotary series elastic actuator. Control. Eng. Pract. **58**, 307–318 (2017)
9. dos Santos, W.M., Nogueira, S.L., de Oliveira, G.C., Pena, G.G., Siqueira, A.A.G.: Design and evaluation of a modular lower limb exoskeleton for rehabilitation. In: 2017 International Conference on Rehabilitation Robotics (ICORR). IEEE (2017). https://doi.org/10.1109/icorr.2017.8009288
10. de Sousa, A.C.C.: Control of Functional Electrical Stimulation and Active Orthoses in Lower Limbs Using Electromyography and Muscles Co-activation. Ph.D. thesis, Universidade de Brasília, Brasília (2018)
11. de Sousa, A.C.C., Freire, J.P.C.D., Bo, A.P.L.: Integrating hip exosuit and FES for lower limb rehabilitation in a simulation environment. IFAC-PapersOnLine **51**(34), 302–307 (2019). https://doi.org/10.1016/j.ifacol.2019.01.030
12. de Sousa, A.C.C., Ramos, F.M., Dorado, M.C.N., Fonseca, L.O., Bó, A.P.L.: A comparative study on control strategies for FES cycling using a detailed musculoskeletal model. IFAC-PapersOnline, pp. 204–209 (2016). https://doi.org/10.1016/j.ifacol.2016.12.215
13. Wolpert, D.M., Miall, R.C., Kawato, M.: Internal models in the cerebellum. Trends Cogn. Sci. **2**(9), 338–347 (1998)
14. World Health Organization: World Report on Disability (2011). https://www.who.int/disabilities/world_report/2011/report.pdf
15. World Health Organization: The Top 10 Causes of Death (2018). https://www.who.int/news-room/fact-sheets/detail/the-top-10-causes-of-death

Characterization and Adjustment Model of the Cricopharyngeal Muscle Potential from Surface Electromyographic Signals in the Swallowing Process

Maria Alejandra Zapata, Valeria Cadavid, and Fabian Andres Castaño(✉)

Bioinstrumentation and Clinical Engineering Research Group—GIBIC, Bioengineering Department, Engineering Faculty, Universidad de Antioquia UdeA, Calle 70 No. 52-21, Medellín 050010, Colombia
fabian.castano@udea.edu.co

Abstract. Parkinson's disease is a neurodegenerative disorder that can cause denervation in the pharyngeal muscles such as the cricopharyngeus and alterations in the neuronal signal transmission, which causes motor alteration during the swallowing process known as dysphagia. This article seeks to characterize the behavior of this muscle through physical simulation and surface electromyography during the swallowing process. For this, the acquisition of electromyographic signals was carried out in healthy subjects in the neck at the level of C5. Analog and digital conditioning were implemented to reduce artifacts and place the signal in the bandwidth of interest. Additionally, the characterization of the signals was carried out to implement a multiphysics simulation to obtain the potential distribution in the volume conductor. The fundamental frequency of the electromyographic signal was found to be 50 Hz, which allowed establishing the frequency value of the electrical properties of the tissues involved in the model. The cricopharyngeus muscle was modeled as a bioelectric source using the cable equation and the Laplace equation to find the potential distribution in the volume conductor. In the model, a voltage reduction from the source to the surface was evidenced, thus, an equivalence between the potential amplitude of the cricopharyngeal muscle of the implemented model and the potentials recorded by invasive electromyography reported in the literature was found. This result shows the feasibility of implementing similar models in studies that seek to evaluate by surface electromyographic and the state of specific muscle areas as new approaches to support the diagnosis and treatment of pathologies.

Keywords: Electromyography · Dysphagia · Biomedical signal processing · Medical simulation

J. L. B. Marques et al. (Eds.): CLAIB 2022/CBEB 2022, IFMBE Proceedings 98, pp. 310–321, 2024.
https://doi.org/10.1007/978-3-031-49401-7_31

1 Introduction

Neurodegenerative diseases constitute a group of conditions associated with the progressive degeneration of the nervous system. These diseases produce the loss of voluntary control of movement, tremors, and balance disorders, significantly reducing people's quality of life [1]. Parkinson's disease is a neurodegenerative disorder with an incidence rate between 9.1 and 13.8 per 100000 inhabitants per year, and its symptoms become evident between 55 and 65 years affecting the nigral dopaminergic neurons [1,2].

Swallowing dysfunction occurs as a consequence of alterations of the anatomical structures (loss of neural connections) or physiological mechanisms (alterations in neuronal signal transmission) involved in the swallowing phases, where diseases of the nervous, muscular, and metabolic systems may be present, amount others [3,4]. Cricopharyngeal muscle dysfunction is a motor disorder evidenced by a permanent contraction of this muscular structure during swallowing. The upper esophageal sphincter closes when it should open to let food through, this desynchronization causes intraesophageal stasis (retention of the food bolus) that anatomically coincides with the intervertebral space C5–C6 [4].

The Electromyographic (EMG) signal is the electrical manifestation of neuromuscular activation associated with a contraction, which is generated by the Motor Unit Action Potential (MUAP) that reaches the muscle through a nerve [5]. The information contained in this signal is associated with the strength of the muscles that obey two mechanisms: 1) the recruitment of Motor Units (MUs), and 2) the increase in the firing rate of the MUs. Thus, it is possible to find the relationship between the characteristics of the EMG signals and the strength of the muscle [5]. The EMG signal presents an amplitude that varies from μ V to a low range of m V (10m V); depending on the associated muscle size and the level of activation during the measurement [5].

To register action potentials from a specific muscle invasive Electromyographic (iEMG) is used. However, the insertion of the needle electrodes is a painful process, which also requires medical supervision and surgical intervention [6]. This procedure involves high hospital costs and a tedious recovery process for the patient.

On the other hand, surface Electromyographic (sEMG) makes use of surface electrodes that are placed non-invasively on the skin. This technique registers action potentials from muscle fibers that are in the sensitive volume of the lead field, acquiring information from a muscle region and not from a specific muscle [6]. sEMG detects potentials that are influenced by neighboring sources, which are considered artifacts in the signal. Additionally, the information must pass through different tissues, which present different conductivity and electrical permittivity, leading to drops in electric potential [7]. Despite the sEMG is not considered a definitive diagnostic technique, multiple studies agree that it is a simple and safe method to detect dysphagia, because if the nerves or muscles involved in swallowing are damaged, the signal differs significantly from a healthy patient's recording [6].

This article presents an adjustment model for non-invasively obtaining the potential generated in the cricopharyngeal muscle from the measurement of the potential on the surface of the neck during the swallowing process. This model involves a study through multiphysics simulation on the section of the neck at the C5 level where the muscle of interest participates, it considers as input some characteristics of signals obtained through sEMG, such as: the amplitude of the signal (m V) and the fundamental frequency (Hz) [8], in addition to the electrical properties of the tissues involved in the neck. This model allows obtaining the potential on the surface of the cricopharyngeal muscle from the measurement of the potential obtained by sEMG. This study proposes a new approach to determine the behavior of the cricopharyngeal muscle in a healthy person and presents its potential to be used in future research where muscle function is affected by Parkinson's disease in the swallowing process.

2 Materials and Methods

This section describes the acquisition of sEMG signals from the muscles involved in swallowing. The conditioning, acquisition, processing, characterization of the sEMG signals, and the multiphysics simulation to obtain the potentials in the neck are presented.

2.1 Analog Conditioning

The sEMG signal conditioning involves the amplification, the filtering in the bandwidth of interest between 10Hz and 250Hz [5], and the DC level adjustment of the interest signal before digitization. This process was performed to significantly reduce external disturbances [9]. Figure 1 presents the block diagram of the conditioning implemented for the sEMG signal. The designed circuit has a total gain of 8800V/V.

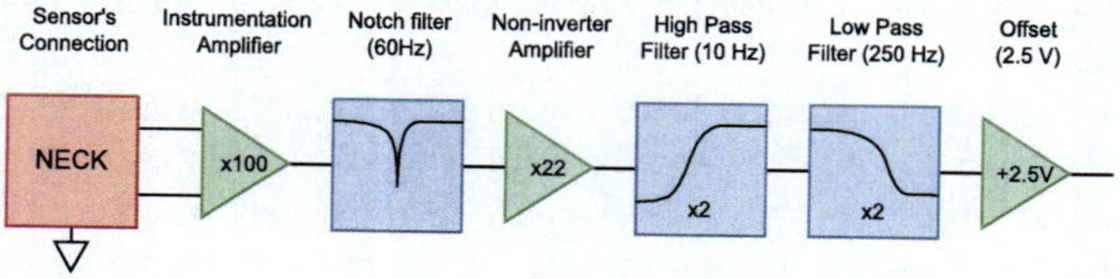

Fig. 1. Block diagram of the analogic conditioning of the sEMG signal acquired in the neck of the subjects

Sensing and preamplification: Three Ag/ClAg surface electrodes were used to detect potentials on the neck's surface. Two electrodes were placed on the left and right side of the thyroid cartilage to register the effector muscle differential potential [10]. The third electrode was used as a reference and was placed on the mandible. This stage was performed using an AD620 instrumentation amplifier with a gain of 100V/V, in which the inputs were connected to the neck electrodes in differential mode.

The reference electrode was in contact with the reference terminal of the amplifier and was connected to GND. This amplifier was used due to its high common-mode rejection, low offset voltage, minimal input current, linear gain, high input impedance, and low output impedance [9].

Signal filtering in the bandwidth of interest: A Notch filter was implemented to eliminate the influence of the 60Hz powerline interference on the signal of interest [9], as was selected in previous studies [5]. Also, a high-pass filter with 10Hz low-pass filter with 250Hz cutoff frequencies were implemented. The filters' had second-order Sallen Key topology with a gain of 2V/V. The frequency response was chosen for a Butterworth approximation due to its flat response in the passband and its smooth slope in the transition band [11].

Additionally, a 22V/V gain amplification stage was added between the Notch filter stage and the high-pass and low-pass filters, as shown in Fig. 1. This amplification allowed to improve the signal acquisition interval in the digital system [9].

2.2 Digitization and Digital Conditioning

sEMG acquisition and processing: Arduino Mega 2560 microcontroller was used to digitize the sEMG signal. This microcontroller has a 10-bit Analog-to-Digital Converter (ADC) with reference voltages of 0V and 5V. The acquisition of the sEMG signal was performed with a sampling frequency of 1k Hz, according to the Nyquist theorem that requires a sampling frequency of at least twice the maximum frequency of interest of the signal to reduce aliasing. Acquisition of sEMG signals was performed in three healthy subjects. For each, 10 signal recordings with a duration of 15s were performed, and for each recording, the subjects carried out 3 water swallows. The digital processing was performed with the corresponding voltage value of the EMG signal considering the amplification given by the analog conditioning stage. Likewise, the offset was removed to obtain the voltage value of the signal.

Digital filtering was performed using FIR high-pass and low-pass filters with cut frequencies of 10Hz and 250Hz respectively. **firwin** and **filtfilt** functions of **scipy** were used to synthesize the filters. Additionally, the signal was denoised using the Wavelet function db6 with 5 decomposition levels. In this process the Wavelet coefficients were calculated, the thresholding was performed and the signal was reconstructed from the thresholded coefficients [12].

2.3 Characterization of sEMG signals

Frequency characterization: To find the fundamental frequency of the signals, the frequency domain analysis using the Welch periodogram was performed, which by applying the Fast Fourier Transform (FFT) calculates the Power Spectral Density (PSD) [13]. Welch's method is carried out by dividing the time signal into successive windows, composing the periodogram for each window, and averaging [14]. The **welch** function of **scipy** was used, which considered the sampling frequency and a Hamming window of 2000 samples.

Once the Welch periodogram was obtained, the fundamental frequency of the signal was determined as the frequency where the PSD is highest. The highest signal energy is concentrated at this frequency, which may be related to the depolarization frequency of the muscles involved in swallowing.

Amplitude characterization: To find the amplitude interval in which the potentials of the swallowing process occur, the Hilbert transform was applied using the **hilbert** function of **scipy**. This function eliminates the fast oscillations of the signal to obtain the amplitude of the slow oscillations [15]. This result is used to know the maximum and minimum amplitudes of the registered signals providing the upper and lower envelope for each signal.

Also, the Moving Average (MA) was calculated using the pandas **rolling** function. The amplitude of the sEMG signal was obtained from the difference between the upper and lower envelope. It should be noted that the scale of the signals to find the amplitude considered the gain given at the fundamental frequency, due to the gain lost for the analog filters' defects.

2.4 Modeling of the Cricopharyngeal Muscle as a Bioelectrical Source

The CAD model design of the cross-section of the neck at the C5 level was performed following the anatomical structure shown in Fig. 2.

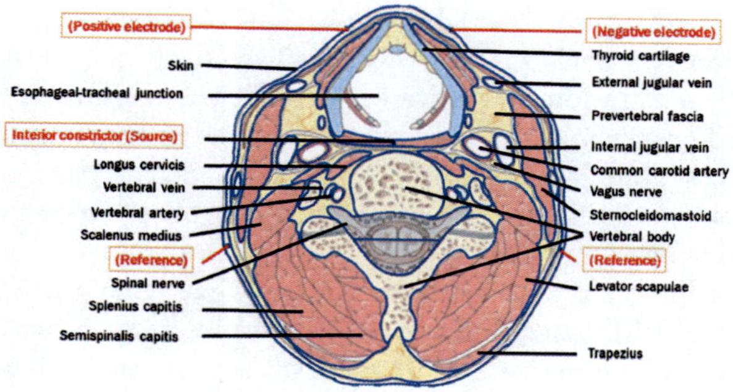

Fig. 2. Transversal section of the neck at C5 level [16]. The tissue sections are delimited by blue lines and the electrodes' location is shown in red

The CAD model is composed of different sections corresponding to the anatomical structures of the neck. The interest muscle (inner constrictor) and the other muscles in the neck section were considered to have identical electrical properties [1,17]. The model was scaled to precise dimensions by taking the transverse diameter of the vertebral foramen equal to 23.3mm. A straight extrusion of 12.4mm corresponding to the height of the C5 vertebra was performed [18].

Electrical parameters of the tissues such as conductivity and relative permittivity depend on frequency, thus the configuration of these properties depends on the result of the characterization of sEMG signals obtained from healthy subjects [3].

Once the fundamental frequency of the sEMG signals was determined, the electrical properties were assigned to the different sections of the CAD model of the volume conductor in COMSOL. The multiphysics model considers two physical interaction domains; (1) the volume conductor was modeled by the Laplace equation to study the electrical behavior of the tissues around the muscle of interest (Eq. 1), and 2) the cricopharyngeal muscle which was modeled as a bioelectric source using the cable equation (Eq. 2) [7].

$$\nabla \cdot (-\sigma_0 \nabla V) = 0 \tag{1}$$

where σ_0 represents the electrical conductivity of the biological tissues involved in the model.

$$-\lambda^2 \left(\nabla^2 V\right) + \tau \frac{\partial V}{\partial t} + V = 0 \tag{2}$$

To obtain a resembled behavior of a bioelectric source, squid axon parameters were taken as a reference, where the characteristic length λ is equal to 0.50cm and the time constant τ is equal to 0.50ms [7].

3 Results

To observe the potential behavior in the different tissues of the neck cross-section at C5 level, and in the areas where the sEMG signals were taken, point electrodes were used as were presented in Fig. 2.

3.1 Characterization of the sEMG signals

Frequency characterization: It was found that the fundamental frequency of the signal corresponds to 49.56 ± 16.86 Hz on average from the PSD analysis. Figure 3 shows the average PSD of all sEMG signals. It is observed that the highest power amplitudes are between 40Hz and 75Hz.

Amplitude characterization: Fig. 4 shows: A) the envelopes obtained for a signal after applying the Hilbert transform and B) the envelopes after performing the smoothing by Moving Average.

To obtain the sEMG signals amplitude, they were scaled with the gain of the analog circuit at 50Hz, which was 2800. Then, envelopes were calculated and the mean maximum amplitude of a swallow in healthy subjects was found to be 0.119 ± 0.043 m V.

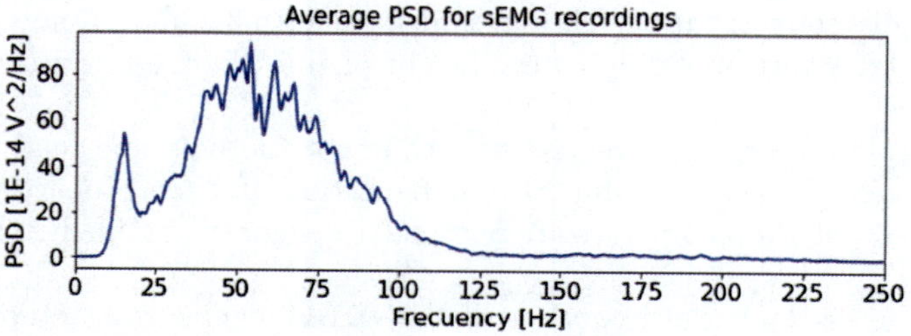

Fig. 3. Average PSD for sEMG signals acquired from healthy subjects

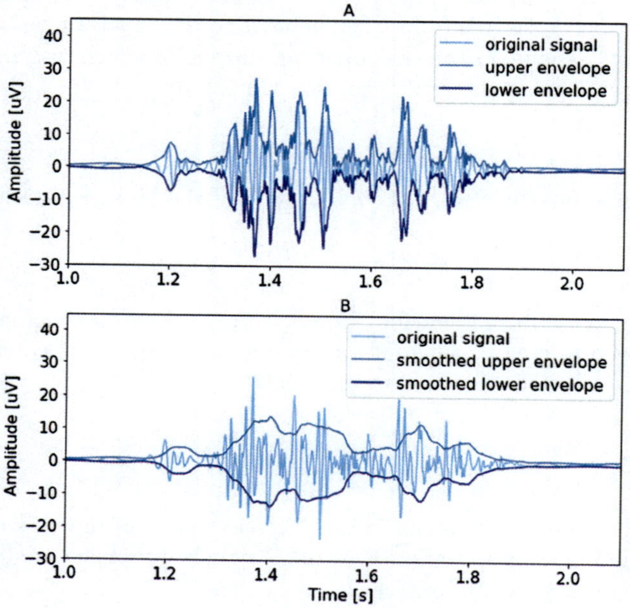

Fig. 4. Example of a sEMG signal envelope acquired from a healthy volunteer

3.2 Model of the Cricopharyngeal Muscle as a Bioelectrical Source

After performing the characterization of the sEMG signal and since the electrical parameters of the tissues depend on the frequency of the muscle depolarization, the values of the electrical parameters of the tissues were found at a frequency of 50Hz corresponding to the fundamental frequency of the signal. Table 1 presents the electrical permittivity and conductivity of the tissues included in the multi-physics model, as shown in Fig. 2.

Additionally, Fig. 5 shows the distribution of conductivity in the different tissues of the model as specified in Table 1. The tissues with dark colors correspond to areas with lower conductivity, while the tissues with light colors correspond to areas with higher conductivity.

Table 1. Relative permittivity and conductivity of the tissues in the neck section at a frequency of 50Hz [19]

Tissue	Rel. Permittivity	Conductivity [S /m]
Fatty tissue	$4.58E+5$	$4.04E-2$
Cervical vertebra	$8.87E+3$	$2.01E-2$
Thyroid gland	$5.66E+4$	$5.24E-1$
Muscle	$1.77E+7$	$2.33E-1$
Trachea	$8.87E+4$	$3.01E-1$
Nerves	$1.61E+6$	$2.74E-2$
Blood	$5.26E+3$	$7.00E-1$
Spinal cord	$1.61E+6$	$2.74E-2$
Connective Tissue	$1.71E+7$	$2.70E-1$

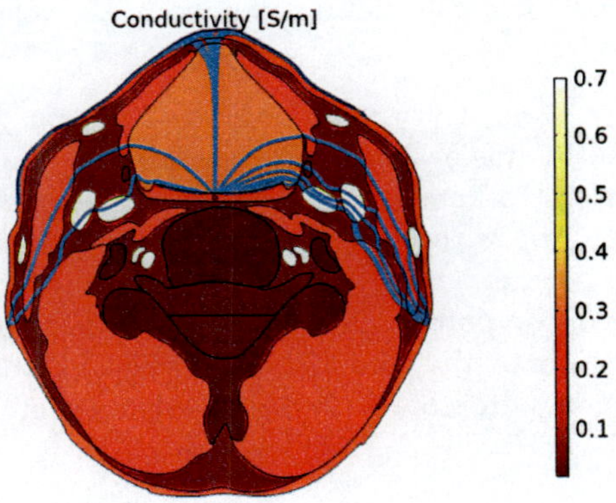

Fig. 5. Conductivities associated with tissues of the neck cross section and current flow lines distribution in the neck section (Blue lines)

3.3 Potential Calculation in Point Electrodes

Figure 5 presents the distribution of flow lines from the source in the cricopharyngeal muscle. The flow lines describe the trajectories of the current from the source to the rest of the internal tissues. Likewise, their distribution is not uniform, and they tend to be directed towards the anterior part of the neck.

In Fig. 6, the electric potential distribution in the neck section was evaluated. The surface plot shows the area where it is possible to find the highest amount of information coming from the source.

Figure 6 shows a lower potential drop in the anterior zone of the neck section. According to this, the best location to place the electrodes for the potentials' detection coming from the cricopharyngeal muscle is the anterior area. This result is in accordance with the behavior obtained from the current flow lines.

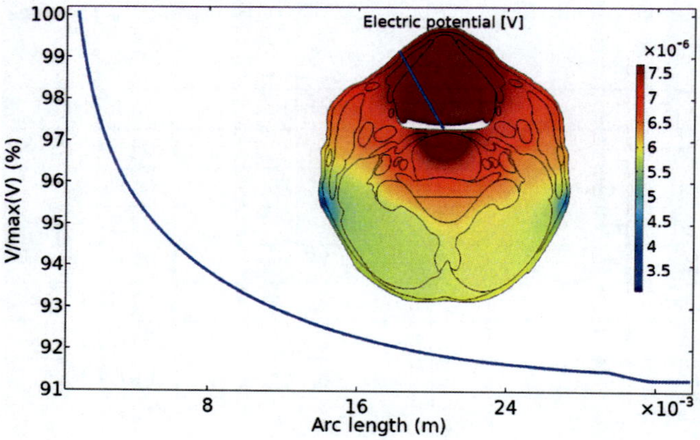

Fig. 6. Percentage of electric potential with respect to the maximum electric potential across arc length and electric potential distribution in the volume conductor of the cross-section of the neck

Finally, the potential was evaluated in the trajectory from the source to the electrode on the surface to know the ratio of voltage drop. Figure 6 shows the ratio of the potential drop with respect to the potential found at the surface of the cricopharyngeus muscle.

As shown in Fig. 6, the potential decreases from the source to the electrode and this behavior is given by the potential drop in the different tissues. Likewise, it was observed a voltage drop of 9% from the volume source to the surface.

4 Discussion

In this study, the frequency with the highest depolarization energy was found at 50Hz, matching the frequency interval in which the EMG signals occur and is related to the amplitude of the potentials in time when swallowing water [5]. It is well known that the electrical parameters of the tissues depend on the frequency. For this reason, the results obtained by the simulation depend on the electrical properties of the tissues [20].

In the current density graph (Fig. 5), the non-uniform behavior in flow lines is mainly due to a phenomenon called bioimpedance that represents the opposite of biological tissue to the current flow. Bioimpedance has two components: 1) The resistance that is conditioned by the opposition of the different tissues to the electric current flow. 2) The reactance is due to the insulating effect of the cell membranes, which behave as capacitors that charge and discharge with the flow of the current [21]. For this reason, the amount of flow lines that pass through blood, connective, and muscle tissue is greater than the amount passing through bone and fat tissue.

The electric current behavior is related to the distribution of the electric potential (Fig. 6). The higher potential values were found in the anterior area, this is because the vertebral body and fatty tissue present lower conductivity compared to the trachea and muscles, as was shown in Table 1 and Fig. 5. A higher potential was distributed in the anterior area, it is because there is less opposition to depolarization in these tissues and consequently less voltage drop. It should be noted that the anatomical structures associated with swallowing are in the anterior area, consequently a drastic loss of potential would not generate the muscular synergy necessary for swallowing.

The amplitude characterization showed that the maximum potential present in the sEMG signals during swallowing was 0.119 ± 0.043 m V. Furthermore, from the results obtained it was found that from the cricopharyngeal muscle to the positive electrode on the skin, there is a potential drop of 9%, which indicates that potentials in the muscle are around 0.130 ± 0.047 m V for healthy subjects.

Studies regarding the potential generated in the cricopharyngeal muscle in healthy subjects using iEMG report average amplitudes of 0.41 ± 0.13 m V [22] and 0.24 ± 0.18m V [23]. These values are close to the results obtained from the simulation, since they are in the same order of magnitude. It should be noted that there is an intrinsic error in the amplitude of the acquired signals, due to errors generated in the analog and digital conditioning.

Based on the results obtained in the present study, it is possible to propose new models that allow knowing the remote state of different muscular areas using sEMG that can help in the diagnosis of muscular dysfunctions due to Parkinson's disease, as well as the design of therapeutic strategies that promote the recovery of these motor alterations.

5 Conclusions

The simulation of a muscle as a bioelectric source using the cable equation allowed to know the distribution of the potential generated in the volume conductor. Depending on the electrical properties of the tissues, a specific behavior was presented by the potential drop in the different tissues. Likewise, this model fulfilled the expectations of the study, since it allowed to obtain potentials of the cricopharyngeal muscle comparable to those presented in iEMG [22,23].

Although the model proposed in this research was designed for the study of the cricopharyngeal muscle during the swallowing process, it can be implemented in other sections of the body, to develop new tools to help in the diagnosis or therapy of pathologies.

As future work, it is proposed the development of models that allow for the establishment of new support strategies for the recovery of motor alterations in people with Parkinson's disease, such as the study of electrical stimulation in therapeutic processes.

Acknowledgments. This work has been supported in part by "Convocatoria 727 – Programa de Becas de Doctorado Nacionales 2015 de Colciencias" and the project "Fortalecimiento de la plataforma tecnológica para la formación especializada en el área de la salud y el desarrollo de tecnología biomédica", code Ruta N 139C of Fondo Nacional de Regalías, Republic of Colombia.

Conflict of interest. The authors declare that there is no conflict of interest regarding the publication of this paper.

References

1. Bellows, S., Jankovic, J.: Parkinsonism and tremor syndromes. J. Neurolog. Sci. **433** (2022). www.pubmed.ncbi.nlm.nih.gov/34686357/
2. Organization, W.H. (ed.): Neurological Disorders: public Health Challenges (2006). www.who.int/publications/i/item/9789241563369
3. Warnecke, T., Dziewas, R., Langmore, S.: Neurogenic Dysphagia. Springer International Publishing, Cham (2021). www.ink.springer.com/10.1007/978-3-030-42140-3
4. Venegas, M., Navia, R., Fuentealba, I., de Medina, M.D., Kunstmann, P.: Clinical management of elderly with dysphagia. Revista Médica Clínica Las Condes **31**(1), 50–64 (2020)
5. Haberman, M.: Procesamiento de señales aplicado a dispositivos de ayuda para personas con discapacidades motoras. Tesis de doctorado, Universidad Nacional de La Plata (2016). www.hdl.handle.net/10915/53022www.sedici.unlp.edu.ar/handle/10915/53022
6. Lourelio Casadelrrey, M.B.: Estudio de la sincronización de la actividad bioeléctrica de los músculos del proceso deglutorio. Ph.D. thesis, Universitat Politécnica de Valéncia, Valencia (2019). www.riunet.upv.es/handle/10251/127319
7. Malmivuo, J., Plonsey, R.: Bioelectromagnetism: Principles and Applications of Bioelectric and Biomagnetic Fields. Oxford University Press (2012)
8. Rodríguez Gil, F.J., Martínez Prieto, C., De Prado Serrano, R.: Oropharyngeal dysphagia and esophageal motor disorders. Medicine - Programa de Formación Médica Continuada Acreditado **11**(1), 26–34 (2012)
9. Alva, C.: Procesamiento de señales de electromiografía superficial para la detección de movimiento de dos dedos de la mano. Grado, Universidad Ricardo Palma (2012). www.repositorio.urp.edu.pe/bitstream/handle/urp/416/Alva_ca.pdf?sequence=1&isAllowed=y
10. Vaiman, M.: Standardization of surface electromyography utilized to evaluate patients with dysphagia. Head & Face Medicine 3(1), 26 (2007), www.pmc/articles/PMC1904196/pmc/articles/PMC1904196/?report=abstractwww.ncbi.nlm.nih.gov/pmc/articles/PMC1904196/
11. Orts Blázquez, J.: Diseño de un sistema de adquisición de señales por electromiografía superficial para la interpretación del movimiento de la mano. Universitat Politécnica de Valéncia, Tesis de fin de grado (2019)
12. Ballesteros Larrota, D.M.: Aplicación de la transformada wavelet discreta en el filtrado de señales bioeléctricas. Umbral Científico **5**, 92–98 (2004)
13. Meigen, T., Bach, M.: On the statistical significance of electrophysiological steady-state responses. Documenta ophthalmologica. Adv. Ophthalmol. **98**(3), 207–232 (1999). www.pubmed.ncbi.nlm.nih.gov/10945442/
14. Smith, J.O.: Spectral audio signal processing. In: Center for Computer Research in Music and Acoustics (CCRMA), pp. 1–674 (2011). www.ccrma.stanford.edu/~jos/sasp/www.books.w3k.org/
15. Kjaer, B., Thrane, N., Wismer, J., Konstantin-Hansen, H., Gade, S.: Application note practical use of the "Hilbert transform". Brüel & Kjær (2000). www.bk.dk
16. Netter, F.H.: Atlas of human anatomy. Elsevier Health Sciences (2014)

17. Ryu, S., Shuer, L., Brown, A., Larson, P.: Spinal neurosurgery (2016). www.aneskey.com/spinal-neurosurgery/
18. Beltrán, E.: Catálogo de osteología de la columna vertebral: Evaluación osteométrica y morfológica de las muestras esqueléticas del Departamento de Morfología Facultad de Medicina. Universidad Nacional de Colombia. Magíster en morfología humana, Universidad Nacional de Colombia Facultad (2020)
19. Foundation, I.: Dielectric Properties of the physiological tissues. https://itis.swiss/virtual-population/tissue-properties/, pp. Visited on 2022-02-02 (1996). www.itis.swiss/virtual-population/tissue-properties/database/dielectric-properties/
20. Valencia, B.S.: Diseño e implementación de un sistema de medición in-vitro que permita sensar el tránsito de sodio en epitelio de piel utilizando la tecnica de bioimpedancia. Proyecto de grado para optar al título de ingeniero biomédico, Universidad Autónoma de occidente (2017). www.linkinghub.elsevier.com/retrieve/pii/S0167273817305726
21. Sánchez Iglesias, A., Fernández Lucas, M., Teruel, J.: Fundamentos eléctricos de la bioimpedancia. Nefrologia **32**(2), 133–135 (2012)
22. Jaradeh, S.S., Shaker, R., Toohill, R.B.: Electromyographic recording of the cricopharyngeus muscle in humans. Am. J. Med. **108** Suppl 4a(4 SUPPL. 1), 40–42 (2000). www.pubmed.ncbi.nlm.nih.gov/10718450/
23. Alkan, Z., Demir, A., Yigit, O., Adatepe, T., Kesici, B., Kocak, I., Gor, A.P., Taskin, U., Uzun, N.: Cricopharyngeal muscle electromyography findings in patients with gastroesophageal reflux disease. In: Otolaryngology-Head and Neck Surgery : Official Journal of American Academy of Otolaryngology-Head and Neck Surgery, pp. 147, no. 2, pp. 295–301 (2012). www.pubmed.ncbi.nlm.nih.gov/22434101/

Modeling the Viscoelastic Load Share Between Knee Porcine Ligaments

P. P. Kenedi[1,2(✉)], B. M. Silveira[2], S. A. S. de Barros[2], L. de Souza Pinto[2], J. E. Silva[1], and R. R. P. Rodarte[1,3]

[1] Mechanical Engineering and Materials Technology (PPEMM), CEFET/RJ, Av. Maracanã, 229, Rio de Janeiro, Brazil
paulo.kenedi@cefet-rj.br

[2] Mechanical Engineering Department (CCGMEC), CEFET/RJ, Av. Maracanã, 229, Rio de Janeiro, Brazil

[3] National Institute of Traumatology and Orthopedics, INTO, Av. Brazil, 500, Rio de Janeiro, Brazil

Abstract. A former analytical elastic model, of the load share between porcine knee ligaments, is enhanced with the inclusion of the viscoelastic approach. Experimental tests were performed to obtain the necessary Fung's model constants. The following knee porcine ligaments were used: lateral collateral ligament (LCL), anterior cruciate ligament (ACL), posterior cruciate ligament (PCL) and medial collateral ligament (MCL). The final objective of the proposed analytical model is to estimate the load share between porcine knee ligaments, in both elastic and viscoelastic phases. Preliminary conclusions show that the load share rearrangement after viscoelastic phase is quite discreet, but the reduction of the ligaments force resistance capacity becomes very significative. In fact, the force values that the ligaments can resist drop around half of its initial value. Probably, this fact explains why newly operated knee ligaments, which are apparently tight just after the operations, can develop an unstretched state, in a couple of weeks.

Keywords: Knee ligaments · Analytic model · Viscoelasticity

1 Introduction

Zheng et al. [1] proposed an analytical model of the knee joint to estimate the forces at the knee during exercise. Muscle forces were estimated based upon electromyographic activities, during exercise and during maximum voluntary isometric contraction (MVIC). Shelburne and Pandy [2] proposed analytical models, where a quadriceps force was applied to extend an intact cadaveric knee quasi-statically, for a given angle of knee flexion, which defined a system of five equations with five unknowns. The solution of this system of equations was obtained through the iteratively use of a nonlinear programming algorithm. Other authors have concentrated in the description of the ligaments behavior. Peña et al. [3] present both analytical and Finite Element Models (FEM), with complete results, using principal stresses developed in the knee ligaments.

J. L. B. Marques et al. (Eds.): CLAIB 2022/CBEB 2022, IFMBE Proceedings 98, pp. 322–331, 2024.
https://doi.org/10.1007/978-3-031-49401-7_32

In the description of the ligaments behavior, the recognition of the importance of viscoelasticity response, to different loadings conditions, have been investigated by many authors. Provenzano et al. [4] state that a non-linear study of ligaments viscoelasticity requires a more general description, than the most used in quasi-linear models. Also, evidenced by experimental studies, in a rat MCL, that stress relaxation and creep are nonlinear in time. Ristaniemi et al. [5] studied the viscoelasticity of knee ligaments and hypothesized that viscous properties are similar between ligaments. In their study of biomechanics of knee ligaments, Woo et al. [6] presented a good review of the viscoelasticity theories and properties. Considered the quasi-linear theory developed by Fung [7]. This theory combines, in a convolution integral, the non-linear elastic response and the time-dependent relaxation function, using a continuum mechanics theory approach.

In this paper a mechanical description of statically loaded porcine knee ligaments, with an imposed vertical force, that generates a vertical displacement, is presented. The relationship between angles, displacements, strains, forces and stresses, involved in the sharing load between knee ligaments, are shown in detail.

2 Analytical Model

The proposed analytical model deals with the sharing load between the porcine ligaments assuming an imposed initial vertical force of 100 N. Also, it is considered, that this force generates a vertical displacement in the knee, which increases as fast as a step, and then, the displacement is maintained constant, as a relaxation test. Moreover, it was done based on Fung's approach. This analytical model improves the former one, Silva et al. [8], which had assumed that the ligaments increase its lengths (but not changes angles), just in elastic phase. This proposed analytic model considers that the ligaments lengths and angles vary continuously and encompasses both, elastic and viscoelastic phases.

The proposed analytic model uses mechanics of solids to estimate the load share between four ligaments, as shown schematically in Fig. 1.a, namely: LCL, ACL, PCL and MCL. Figure 1.b shows a schematic representation of the displacement accomplished, because of the imposed vertical force, on the four principal porcine knee ligaments.

Note that in Fig. 1.b the ligaments are represented in a proportional way, where the cosine directors were estimated.

The ligament geometrical description is shown in Fig. 2. The geometric relations for each ligament are shown:

$$L = L_0 + \delta_L \quad z = z_0 + \delta_z \quad L^2 = z^2 + d^2 \tag{1}$$

$$\sin(\theta_0) = \frac{d}{L_0} \quad \cos(\theta_0) = \frac{z_0}{L_0} \quad \tan(\theta_0) = \frac{d}{z_0} \tag{2}$$

$$\sin(\theta) = \frac{d}{L} \quad \cos(\theta) = \frac{z}{L} \quad \tan(\theta) = \frac{d}{z} \tag{3}$$

where, L_0, L, z, d, θ_0 and θ represents, respectively, the ligament's initial and final length, the height, the projection in x-y plane axis, the initial and final angle of a ligament with the vertical axis z. The variables with zero subscript represent the initial value. δ_L and δ_z

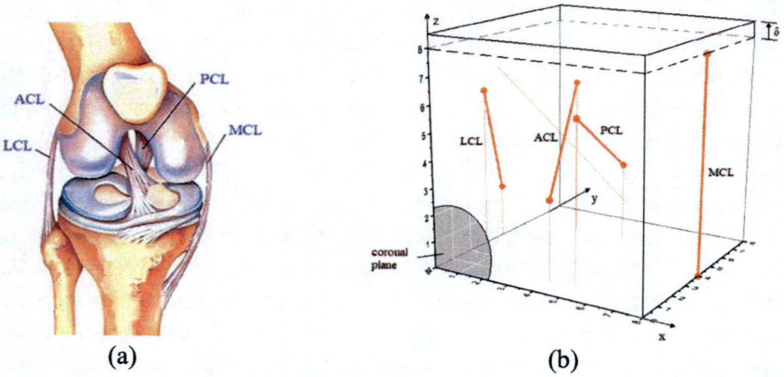

Fig. 1. Left knee: (a) view from behind figure of Knee's ligaments (adapted from vitalitychiropractor.com.au) and (b) schematic representation of the geometry of the 1D ligament model submitted to a vertical displacement, (adapted from [8]).

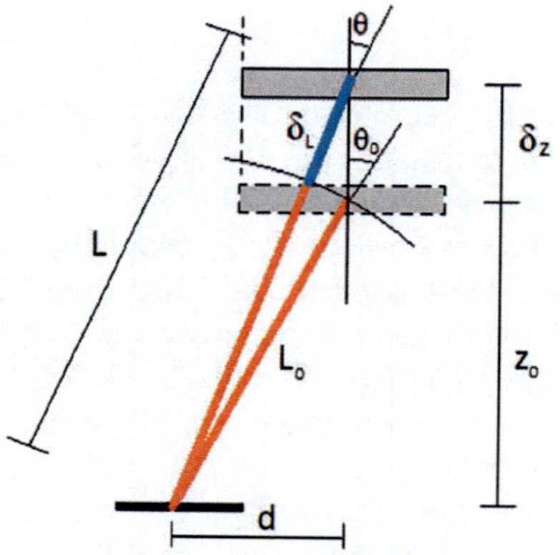

Fig. 2. Ligament geometrical characterization.

represents, displacement in ligament direction and displacement projection in the vertical direction. The equations for a final ligament angle θ and the displacement projection in the vertical direction δ_z are shown next.

$$\theta = \arctan\left[\frac{L_0 \cdot \sin(\theta_0)}{L_0 \cdot \cos(\theta_0) + \delta_z}\right] \tag{4}$$

$$\delta_z = \sqrt{(L_0 + \delta_L)^2 - [L_0 \cdot \sin(\theta_0)]^2} - L_0 \cdot \cos(\theta_0) \tag{5}$$

Moreover, the model uses the following vertical equilibrium condition, Eq. (6), and vertical compatibility, Eq. (7).

$$\sum_1^4 F_n \cdot \cos(\theta_n) = P_z \tag{6}$$

$$\delta_z = \delta_{1z} = \delta_{2z} = \delta_{3z} = \delta_{4z} \quad (7)$$

where P_z is imposed vertical force and F_n is the force in ligament *n*. Note that the ligaments are renamed as follows: n = 1 for LCL, n = 2 for ACL, n = 3 for PCL and n = 4 for MCL. In this paper a viscoelastic model is used to describe the force-displacement and the stress-strain relationships. Moreover, it was considered that the cross-section area maintains constant and it is applied the small strains theory. Also, the equations of Fung's quasi-linear model are used, even though it presents some limitations [9].

Fung [7] stated the Eq. (8), in function of time, to describe the stress-strain relation for the general case of relaxation.

$$\sigma(t) = \int_0^t G(t-\tau)\frac{d\sigma^e(\tau)}{d\tau}d\tau \quad (8)$$

where *G* and σ^e represent, respectively, the reduced relaxation function (of viscous part of equation) and elastic response and written according with Eqs. (9) and (10).

$$G(t) = G_0 + \sum_{i=1}^{3} G_i \cdot e^{\frac{-t}{\tau_i}} \quad (9)$$

where G_0 and G_i are material dimensionless constants, called relaxation modulus and represents the amplitude of the stress curve in relaxation, and τ_i is the relaxation time in seconds, also a material constant. This equation is based on Prony Series [10, 11]. The initial elastic part of the loading is:

$$\sigma^e(t) = A \cdot \left(e^{B\cdot\varepsilon(t)} - 1\right) \quad (10)$$

where constants *A*, in MPa, and *B*, dimensionless, are material constants. Represents, respectively, an elastic stress constant and an elastic power constant. For this paper the Eq. (8) was rewritten, since the ramp time is not considered and assuming a constant strain during the relaxation phase.

$$\sigma(t) = \sigma_0^e \cdot G(t), \sigma_0^e = A \cdot \left(e^{B\cdot\varepsilon_0} - 1\right) \quad (11)$$

So, the force-displacement relation could be stated.

$$F(t) = F_0^e \cdot G(t), F_0^e = S \cdot A \cdot \left(e^{B\cdot\frac{\delta_0}{L_0}} - 1\right) \quad (12)$$

where *S* is the ligament cross section area. Moreover, the initial and final displacements can be, respectively, obtained:

$$\delta_0 = \frac{L_0}{B} \cdot \ln\left[\frac{F(t)}{S \cdot A \cdot G(t)} + 1\right] \quad (13)$$

$$\delta_L = \sqrt{L_0^2 + 2 \cdot \delta_z \cdot L_0 \cdot \cos(\theta_{0k}) + \delta_z^2} - L_0 \quad (14)$$

To perform the necessary calculations, for the proposed model, it was assembled a system of equations:

$$\begin{cases} \sum_{1}^{4} F_n \cdot \cos(\theta_n) = P_z \\ F_n(t) = S_n \cdot A_n \cdot \left(e^{B_n \cdot \frac{\delta_{nL}}{L_{n0}}} - 1 \right) \cdot G(t) \\ \delta_z = \delta_{nz} \\ \delta_{nz} = \sqrt{(L_{n0} + \delta_{nL})^2 - [L_{n0} \cdot \sin(\theta_{n0})]^2} - L_{n0} \cdot \cos(\theta_{n0}) \\ \theta_n = \arctan\left[\frac{L_{n0} \cdot \sin(\theta_{n0})}{L_{n0} \cdot \cos(\theta_{n0}) + \delta_{nz}} \right] \end{cases} \tag{15}$$

To solve this system of equations, it was implemented a numerical routine that performs the necessary steps. Initially, the knee displacement range was estimated. The minimum value is equal to zero. The maximum value was calculated according to the following steps:

First, calculates the displacement for each ligament through Eq. (13), using the knee force and the initial value for reduced relaxation function, G(0) = 1. Second, calculates the knee vertical displacement through Eq. (5), using the ligament displacement calculated previously. Finally, gets the biggest knee displacement calculated previously. The problem step was arbitrary chosen dividing this maximum value by 10^6.

The knee displacement was iterated using the range and step calculated previously. For each value, the following parameters were calculated: final angle, Eq. (4); ligament displacement, Eq. (14); ligament force, Eq. (12); and knee force, Eq. (6). After calculating the knee force, this is compared with the imposed initial force and, if the percentage difference is smaller than 1%, this is considered an acceptable result.

The result calculated contains the angle, strain, displacement, force and stress of each ligament, and also time, knee displacement and knee force.

3 Results

It must be observed that geometrical and elastic properties and Fung's model constants of porcine ligaments, presented in Tables 1 and 2, have a wide variability, as can be seen, for instance, in [4]. The porcine age, diseases, the ligament preparation (including freezing and campling methods) and test strain rates, can lead these characteristics to be quite scattered.

The ligaments' lengths, cross section area and angles were obtained in [12]. The ligaments' Young Modules and Fung's model constants were obtained through the experimental data [13][14].

Figures 3, 4, 5, 6, 7 and 8 show the results for, respectively, final angle, final displacement, final strain, initial force, final force and final stress, for elastic [8] and proposed models.

In Table 2 the experimentally obtained Fung's model porcine knee ligaments constants are presented.

The knee vertical displacement for elastic [8] and proposed model are, respectively, 1.64 mm and 1.38 mm.

Table 1. Estimated knee porcine ligaments geometrical and elastic properties

Ligament	LCL	ACL	PCL	MCL
S_n (mm^2)	63.62	63.62	63.62	20
L_n (mm)	38	26	30	37
θ_n (rad)	0.357	0.812	0.644	0.233
E_n (MPa)	4.4	5.8	25.1	3.2

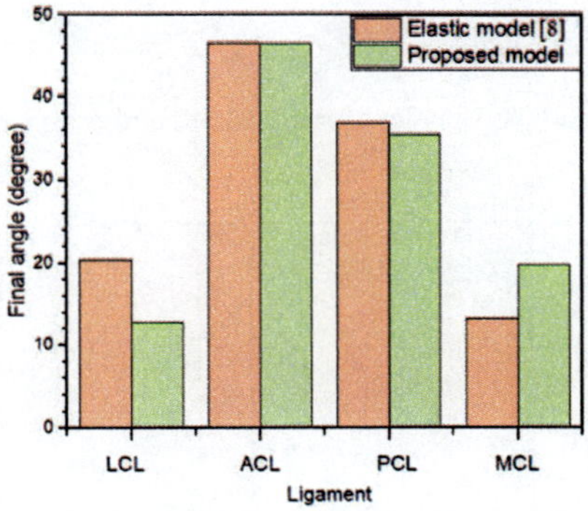

Fig. 3. Final angle for elastic [8] and proposed models.

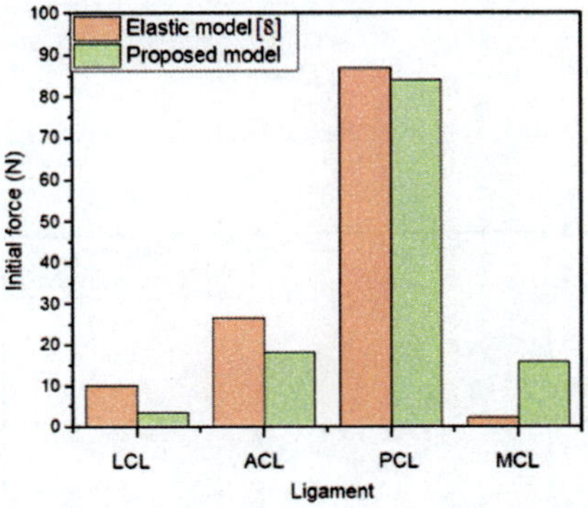

Fig. 4. Initial force for elastic [8] and proposed models.

The final angle of ligaments, of Fig. 3 are quite similar for both models. The final displacement of ligaments, shown in Fig. 5, the final strains, shown in Fig. 7 and the initial force, of Fig. 4, show a modest performance of the elastic model [8] if compared with the proposed model. For ligaments final force (after of the viscous phenomenon has been developed), shown in Fig. 6, the errors were even larger, as in Fig. 8, for final stresses. The overall conclusion is that the elastic model [8], and similar, although of

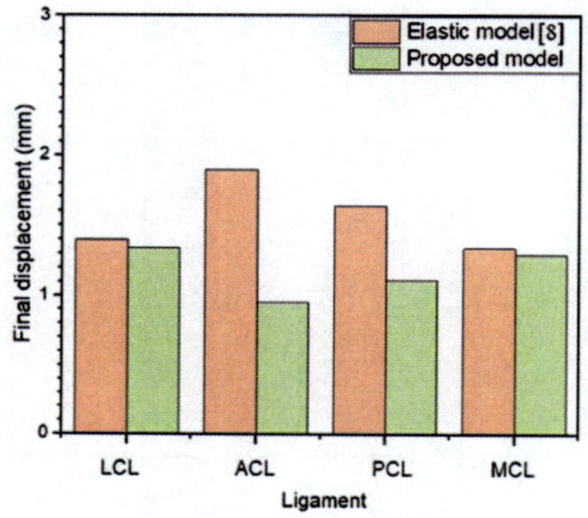

Fig. 5. Final displacement for elastic [8] and proposed models.

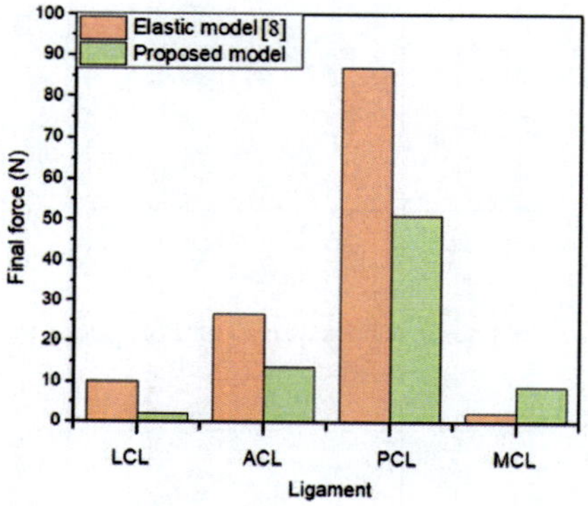

Fig. 6. Final force for elastic [8] and proposed models.

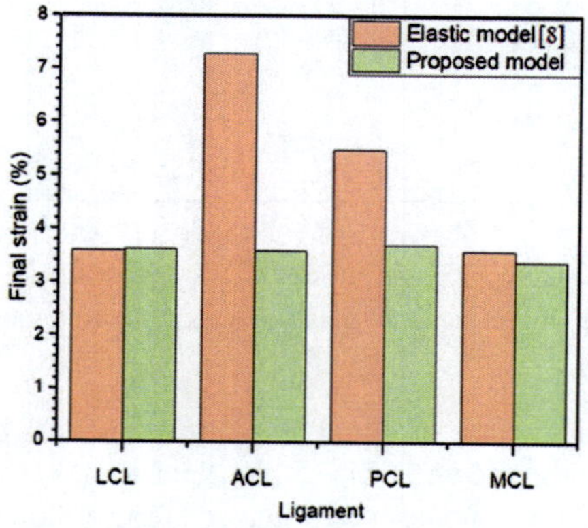

Fig. 7. Final strain for elastic [8] and proposed models.

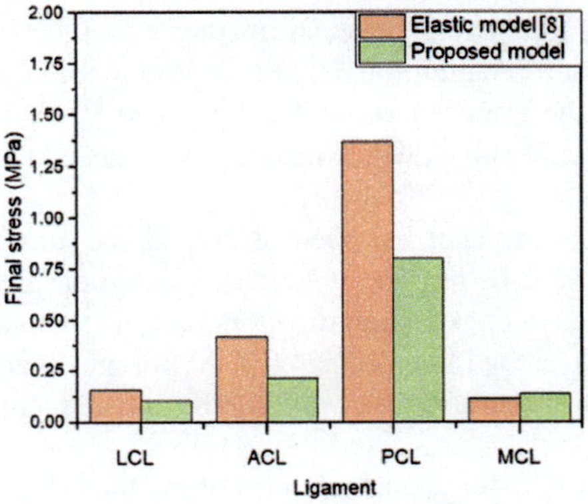

Fig. 8. Final stress for elastic [8] and proposed models.

Table 2. Fung's model porcine knee ligaments constants

Ligament	LCL	ACL	PCL	MCL
A_n (MPa)	0.297	0.548	0.616	0.366
B_n	17.93	11.71	30.90	11.34
G_{0n}	0.552	0.747	0.604	0.578
G_{1n}	0.144	0.063	0.103	0.138
G_{2n}	0.119	0.062	0.095	0.100
G_{3n}	0.150	0.111	0.165	0.148
τ_{1n} (s)	1.486	2.448	2.176	1.750
τ_{2n} (s)	23.663	33.400	30.883	26.130
τ_{3n} (s)	277.932	364.830	342.936	323.415

easier application, are not suitable to model soft tissues, as porcine knee ligaments, mainly because not recognizes the viscous effects.

Table 3 contains the percentage of force decrease, for the proposed model, from the initial force to the final force.

Table 3. Force decreases in the proposed model along the porcine knee ligaments axis.

Variable	Ligament			
	LCL	ACL	PCL	MCL
Initial force (N)	15.80	18.90	83.43	3.70
Final force (N)	8.76	14.20	50.81	2.15
Force decrease (%)	44.56	24.86	39.10	41.82

The summation of initial force of each ligament, of Table 3, results in 121.83 N. Note that each ligament force multiplied by its respective initial cosines directors result in the application of the imposed vertical initial force $P_z = 100$ N (the summation actually makes 98.2 N, not 100 N, which inaccuracy is related to limits of the employed interactive method).

After the viscous phenomenon has been developed, the summation of the final force of each ligament, of Table 3, results in 71.15 N. Also, note that each ligament force multiplied by its respective final cosines directors result in vertical final force of 55.7 N.

Analyzing the Table 3, the force decreases in the porcine ligaments, after the viscous phenomenon has been developed, are very significant, reducing the ligaments force resistance by half.

Table 4 show the results for the calculations of the load share at initial time (before the ligaments are affected by the viscous effects) and load share at final time (after the ligaments developed the viscous effects).

Table 4. Load share between porcine knee ligaments at the beginning and at the end of relaxation in vertical direction

Variable	Ligament			
	LCL	ACL	PCL	MCL
Initial shared force (%)	14.87	13.50	68.02	3.61
Final shared force (%)	13.31	16.39	66.91	3.39

Although the ligament load sharing percentages do not change much, as shown in Table 4, the vertical sum of the ligaments forces were reduced from 100 N to 55.7 N.

4 Results

An 1D load share model, in tridimensional space, of porcine knee ligaments was presented. The viscous phenomenon was considered in the proposed model through the utilization of Fung's equations and experimental constants.

The results for the proposed model presented quite important differences when comparing with former elastic model [8], except for the final angles. For final displacements, final strains, initial forces, final forces and final stresses, the results diverged considerably, when comparing both models.

The sharing loading of the porcine ligaments before and after the viscous phenomenon has been developed, presented a great reduction of the ligaments capacity of force resistance. In fact, the resisting forces were reduced to about half the initial values. In ACL and PCL, the force resistance reduction were slightly less pronounced.

In short, it is fundamental to consider the viscoelastic behaviour for a soft tissue, as knee porcine ligaments, to estimate, for instance, the sharing loading between porcine knee ligaments.

Conflict of Interest. The authors declare that they have no conflict of interest.

References

1. Zheng, N., et al.: An analytical model of the knee for estimation of internal forces during exercise. J. Biomech. **31**, 963–967 (1998). https://doi.org/10.1016/S0021-9290(98)00056-6
2. Shelburne, K.B., Pandy, M.G.: A musculoskeletal model of the knee for evaluating ligament forces during isometric contractions. J. Biomech. **30**(2), 163–176 (1997). https://doi.org/10.1016/S0021-9290(96)00119-4
3. Penã, E., et al.: A three-dimensional finite element analysis of the combined behaviour of ligaments and menisci in the healthy human knee joint. J. Biomech. **39**, 1686–1701 (2006). https://doi.org/10.1016/j.jbiomech.2005.04.030
4. Provenzano et al.: Nonlinear ligament viscoelasticity. Ann. Biomed. Eng. **29**(10), 908–914 (2001). https://doi.org/10.1114/1.1408926
5. Ristaniemi, A., et al.: Comparison of elastic, viscoelastic and failure tensile material properties of knee ligaments and patellar tendon. J. Biomech. **79**, 31–38 (2018). https://doi.org/10.1016/j.jbio-mech.2018.07.031
6. Woo, S.L.Y., et al.: Biomechanics of knee ligaments: injury, healing, and repair. J. Biomech. **39**, 1–20 (2006). https://doi.org/10.1016/j.jbio-mech.2004.10.025
7. Fung, Y.: Biomechanics: mechanical properties of living tissues. Springer, New York, University of Michigan (1993)
8. Silva, J.E., Rodarte, R.R.P., Segmiller, L.S.P.P., Barros, S.A.S., Kenedi, P.P.: An analytical model for knee ligaments. In: IFMBE Proceedings of XXVII Brazilian Congress on Biomedical Engineering (2020). https://doi.org/10.1007/978-3-030-70601-2_82
9. Silveira, B., Barros, S.A.S., Rodarte, R.R.P., Kenedi, P.P.: Viscoelastic modeling of porcine ligaments. In: Proceedings of 26th International Congress of Mechanical Engineering, Florianopolis—SC (2021). https://doi.org/10.26678/ABCM.COBEM2021.COB2021-1122
10. Babaei, Behzad, et al.: Efficient and optimized identification of generalized maxwell viscoelastic relaxation spectra. J. Mech. Behav. Biomed. Mater. **55**, 32–41 (2015). https://doi.org/10.1016/j.jmbbm.2015.10.008
11. Funk, J.R., Hall, G.W., Crandall, J.R., Pilkey, W.D.: Linear and quasi-linear viscoelastic characterization of ankle ligaments. J. Biomech. Eng. **122**(1), 15–22 (2000). https://doi.org/10.1115/1.429623
12. Guedes, R.M., Silva, J.A., Viriato, N., Pinto, V.: Ligaments' clamping: a novel solution to prevent soft tissue slippage. J. Mech. Eng. Biomech. **2**(5), 36–42 (2018). https://doi.org/10.24243/JMEB/2.5.160
13. Rodarte, R.R.P., Barros, S.A.S., Silveira, B., Duarte, B.T., Kenedi, P.P.: Experimental characterization of porcine ligaments. In: Proceedings of 26th International Congress of Mechanical Engineering, Florianopolis—SC (2021). https://doi.org/10.26678/ABCM.COBEM2021.COB2021-1114
14. Rodarte, R.R.P.: Characterization of the mechanical behavior of the knee ligaments: an experimental study in a porcine animal model. Doctoral thesis. Rio de Janeiro—RJ (2022)

Classification of Spatiotemporal Features of Time Series in Post-fatigue Gait

R. S. Gomide[1,2(✉)], E. M. Mesquita[1], G. A. G. Villa[1], and M. F. Vieira[1]

[1] Federal University of Goiás, Goiânia, Brazil
sousagomide@gmail.com
[2] Federal Institute Goiano—UnU Trindade, Trindade, Brazil

Abstract. Muscle fatigue affects gait to the point of causing changes in your patterns. People with muscle training have more endurance and better recovery time from muscle fatigue than people without training. The study compared three classification algorithms in the analysis of gait data under normal conditions and different levels of muscle fatigue. Spatio-temporal data from a group of people who do and do not do weight training were analyzed. The result showed that the classification accuracy of the k-nearest neighbor algorithm had the best result with 86.78% accuracy. The results indicated by the classification algorithms show a difference in the muscle fatigue recovery process between the groups, similar to the clinical results discussed in the literature.

Keywords: Muscle fatigue · Classification · kNN · Random forest · SVM

1 Introduction

Muscle fatigue can be explained by the loss of contractile capacity of the muscle as a result of muscle activity and the decrease in the ability to produce force as a result of continuous or repeated activation or intense work [1]. Severe muscle fatigue can contribute to the increase in the occurrence of falls accidents [2], interruption of the afferent feedback system, alteration of joint awareness [3], alteration of reflexes and changes in the pattern of muscle activation [4].

According to [5], gait adjustment as a consequence of muscle fatigue seems to be influenced: by the muscles involved in fatigue; by the type and duration of the exercise; by the pre-existing physical condition and by physiological differences between the sexes. The calf muscles, in particular, play an important role in postural control and locomotion, and muscle fatigue in this region can significantly influence gait behavior.

Other studies also seek to observe the impacts that muscle fatigue causes on postural control and gait [6, 7].

Also, it is known that subjects without physical conditioning, referring to activities related to muscle condition, can reach the degree of muscle fatigue more quickly than active subjects. Weight training has been widely adopted by people with the aim of increasing muscular endurance and also in rehabilitation protocols.

J. L. B. Marques et al. (Eds.): CLAIB 2022/CBEB 2022, IFMBE Proceedings 98, pp. 332–340, 2024.
https://doi.org/10.1007/978-3-031-49401-7_33

The objective of this study is to assess the classification performance of different machine learning algorithms by verifying gait data from groups of people who practice weight training and non-practitioners, under normal conditions and at different levels of muscle fatigue. Although statistical methods can also be used to establish a classification of groups, the use of machine learning allows a computational gain after training the model. This ad-vantage is most evident in large databases. Furthermore, a set of characteristics can be analyzed with the trained model, while in statistics there would be a need for a re-analysis.

Our hypothesis is that the effects of fatigue and the level of muscle recovery can affect the spatio-temporal characteristics to the point of allowing the classification between a trained and untrained person. This study may contribute to the construction of intelligent applications that seek to help health and physical education professionals, making it possible to identify the levels of fatigue between people with and without experience during muscle physical training, allowing strategic decision-making during training.

2 Methods

2.1 Subjects

Forty-one young women, 20 weight training practitioners (TG) (22.00 $\pm$ 3.27 years; 1.60 $\pm$ 0.04 m; 57.10 $\pm$ 6.35 kg) and 21 non-practitioners (UTG) (21.76 $\pm$ 3.01 years; 1.62 $\pm$ 0.05 m; 62.35 $\pm$ 8.50 kg) participated in this study. The TG group was composed of women who had performed weight training for less than 12 months. The inclusion criteria for the participants were: (i) age between 18 and 27 years; (ii) not present functional impairment, pain or orthopedic pathology in the last six months; (iii) not present any cardio-vascular, pulmonary or neurological disease. Participants were instructed not to perform strenuous physical activity within 48 h of data collection and not to consume stimulants.

The study was conducted in accordance with the Declaration of Helsinki after local approval by the Ethics Committee. All subjects signed a consent form before the experiment.

2.2 Procedures

Each participant walked 4 min on the treadmill for ambiance. Their preferred walking speed (PWS) on the tread-mill was assessed according to the protocol [8] followed by 4 min of rest. The participant walked for 4 min at PWS speed before the fatigue protocol (PreF). Afterwards, a isometric maximum voluntary contraction (MVC) test was performed, followed by the fatigue protocol and a second MVC test. Then, the participant performed three walks (0-PostF; 6-PostF; and 12-PostF) of 4 min at PWS speed with 2 min of rest between them. Finally, the participant per-formed another MVC test.

Gait kinematic data were collected using a motion capture system, with 10 infrared cameras operating at 100 Hz (Vicon Nexus, Oxford Metrics, Oxford, UK). Thirteen reflexive markers were placed on the lateral malleoli, heels, heads of the second and fifth metatarsals (bilaterally), the right and left anterior superior iliac spine, the right and left posterior superior iliac spine, and the spinous processes of the first thoracic vertebrae (T1).

The MVC test and the fatigue protocol were conducted according to [5]. The fatigue protocol adopted ensured that all participants had the same level of muscle fatigue during the *N*-PostF collections.

2.3 Data Analysis

Kinematic data were filtered using a fourth order low pass band, zero lag Butterworth filter with 10 Hz cutoff frequency. All collections were calculated for intermediate 150 strides, discarding the initial and final steps. The steps were detected from the zero-cross velocity of the heel marker [9]. Data analysis was performed using MatLab (R2020A, MathWorks, Natick, MA).

2.4 Classification

The spatio-temporal variables of the kinematic data were used as a set of input characteristics for training and testing the classification algorithms. The output variable consists of determining two types of classes: TG or UTG. Characteristics related to spatio-temporal variables were chosen due to the diversity of equipment that can collect these types of data. Nowadays it is possible to obtain spatio-temporal data through cameras, accelerometer, smartphone, among others. This facility can contribute to the development of future commercial applications.

In this study, three learning machine methods were used to classify the proposed dataset: k-nearest neighbors (kNN); random forest (RF); and support vectors machine (SVM). kNN is an instance-based learning algorithm that classifies unlabeled observations assigning a class based on the characteristics of the nearest neighbors. The nearest neighbors are determined calculating the distance between them and the term *k* indicates the number of neighbors used to deter-mine the class of the observed object [10]. Based on ensemble learning, the RF algorithm aims to build several decision trees where each one is considered a classifier, the majority of votes determine the final classification prediction [11]. SVM is a supervised learning method that seeks to minimize classification errors and maximize the geometric margin between classes expanding the dimensional space by building a hyperplane that can separate the dataset [12].

2.5 Steps for Data Processing

Before applying the classification algorithms, some procedures were performed to prepare the data and define the best fit parameters for the construction of the learning models. First, the spatio-temporal data were submitted to feature selection methods, considering: the correlation coefficient; the degree of importance of the feature; and the low importance of the feature. After choosing the most relevant characteristics for the classification task, the data passed a characteristic standardization through a scaling of the mean and standard deviation. Tests were performed to define the best parameters for the three algorithms through GridSearch.

The training and prediction of the algorithms were executed $n = 30$ times. In each round, a cross-validation with 10-Fold was used [13]. The accuracy of the algorithms is the result of the averages of cross-validation and executions.

Dataset accuracy was analyzed in all gait conditions (PreF; 0-PostF; 6-PostF; and 12-PostF) and separately in each group.

3 Results and Discussion

3.1 Feature Selection

The initial dataset of this study consisted of eight characteristics: Step Time (STEP_TIME); Step Frequency (STEP_FREQ); Cadence (CADENCE); Step Length (STEP_LENG); Stride Width (STRIDE_WIDTH); Support Phase Duration (SUPPORT_PD); Swing Phase Duration (SWING_PD); and Double Support Phase Duration (DOUBLES_PD). There are also two columns that identify the gait condition (0—PreF; 1—0-PostF; 2—6-PostF; 3—12-PostF) and the group (0—UTG; and 1—TG).

The Fig. 1 presents the graph of the autocorrelation be-tween the spatio-temporal characteristics. It is possible to notice a very strong correlation of the variable STEP_TIME with SUPPORT_PD, SWING_PD since the duration of the gait phases are essential components to determine the step time. The variables STEP_LENG and STRIDE_WDTH are related to step frequency, while STEP_FREQ and CADENCE are strongly correlated.

The most relevant characteristics for the classification task were also analyzed, Fig. 2. The three best features evaluated were: STRIDE_WDTH; STEP_LENG; and DOUBLE_PD.

It is also possible to notice that the CADENCE variable does not appear on the chart since it had zero importance. The degree of importance was calculated based on the missing values, on the characteristics that have only unique values, on the correlation between the characteristics, on the zero and low importance methods.

The feature selection showed that only the variables: STEP_TIME; STEP_LENG; and STRIDE_WIDTH are required to perform the classification task. Therefore the other features were removed from the dataset.

Figure 3 presents the pair plot of the characteristics used in the classification algorithms. The graph presents the data separated by group. It is possible to notice a greater separation of the data by group when the variable STRIDE_WDTH is compared with the others.

3.2 Classification

The hyper-parameters were defined through the grid search. For the kNN algorithm, the number of neighbors and different types of metric were analyzed to calculate the distance. In the RF algorithm, the tree separation criteria, the number of trees generated, the minimum quantity of necessary samples generated and the minimum quantity of samples necessary to form a node were analyzed. In the SVM algorithm, the tolerance criterion, the regularization parameter and different types of kernel were analyzed.

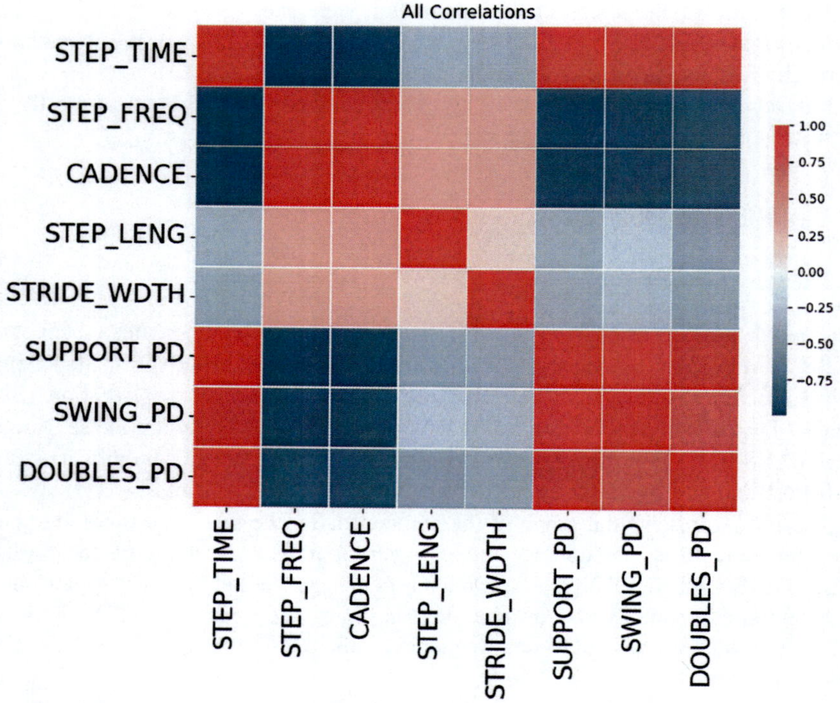

Fig. 1. Plot of autocorrelation between characteristic variables

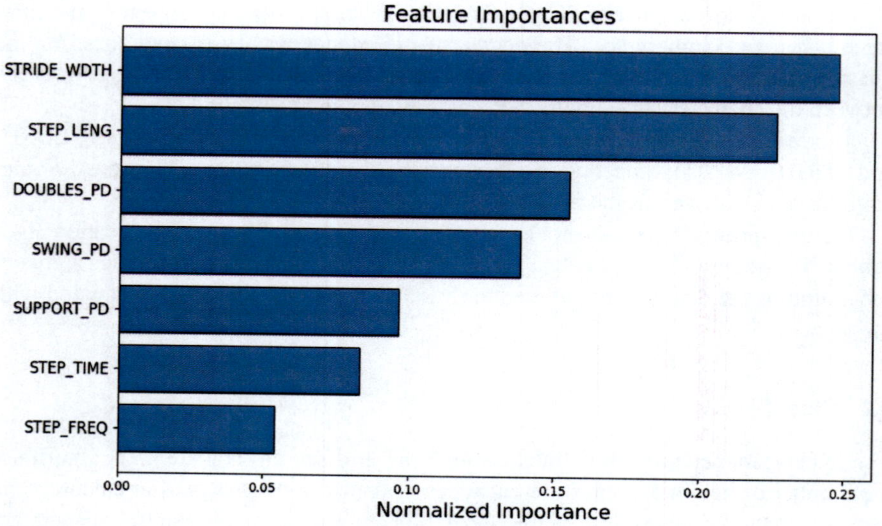

Fig. 2. Rank chart of the most important characteristics for classification task

In this first test, classification algorithms were applied to estimate the accuracy in predicting which group a given set of spatio-temporal data belongs to. In this dataset all conditions were considered. Figure 4 presents the results of the three algorithms. Considering the mean and the highest value of the thirty executions, for this dataset, the kNN was the one that obtained the highest accuracy in the classification prediction, with 86.78% and 90.88%, respectively.

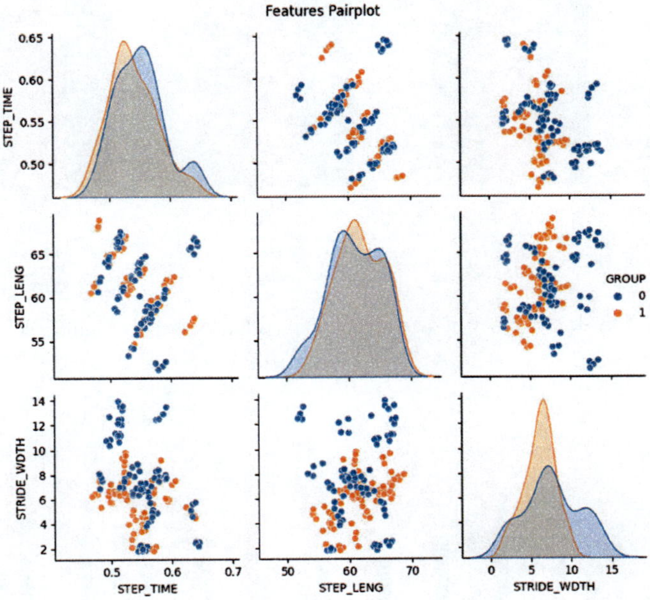

Fig. 3. Pair plot of selected features

From these results, it is possible to state that the strategy of counting the nearest neighbor classes for this data distribution, Fig. 3, was more advantageous than seeking the separation of groups with the creation of a new hyper plane or creating a set of decision trees that seek the best path regarding classification prediction. We believe that the STRIDE_WIDTH variable contributed to the best result of the kNN strategy to stand out in relation to the other algorithms.

3.3 Classification by Condition

The classification algorithms were also submitted to the dataset under the conditions separately. The accuracy of each algorithm in classifying the four conditions (PreF; 0-PostF; 6-PostF; and 12-PostF) can be observed in Fig. 5. The best results of accuracy in the prediction can be observed in the 12-PostF condition. In this condition, the SVM was the algorithm that presented the best accuracy (65.36%).

Overall, all algorithms reduced their ability to predict this dataset. To verify if the dimensionality reduction in this case could be compromising the classification process, we ran the algorithms again with the complete dataset. However, there was no progress.

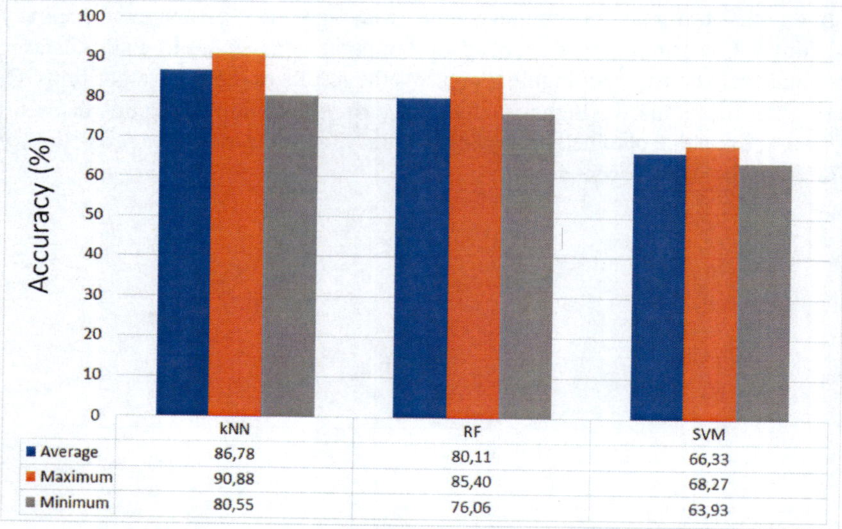

Fig. 4. Accuracy of classification algorithms under all conditions

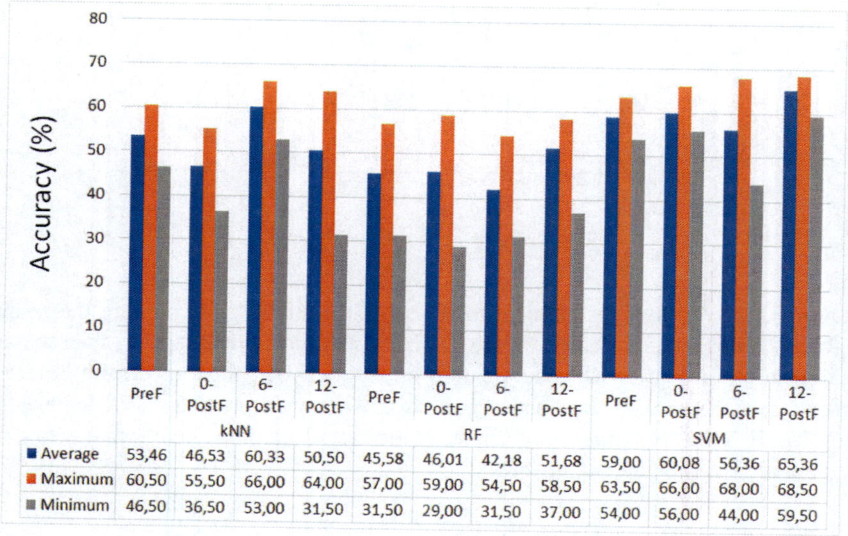

Fig. 5. Accuracy of condition classification algorithms

The fact that the RF and SVM algorithms obtain greater accuracy in the 12-PostF condition means a greater spacing of the analyzed characteristics (STRIDE_WDTH, STEP_LENG, STEP_TIME) between the groups.

From a biomechanical point of view, this difference may mean that a given group has a greater capacity to recover from muscle fatigue than the other, causing the distance in spatio-temporal features between the groups, facilitating the classification task. Through a clinical analysis, [5] it was found that the TG group presented better conditions for the execution of the fatigue protocol and better recovery of the MVC. Also in Lehnen, [5], it is possible to notice a lower stability in the TG group in relation to the UTG and in the 12-PostF condition for both groups.

3.4 Limitations

Dataset size may have been a limiting factor in the analysis under separate conditions. The lack of more strongly correlated attributes and the sparse distribution shown in Fig. 3 may also have been an aggravating factor. Another cross-validation method indicated for small databases, Leave-One-Out Cross-Validation (LOOCV) was tested, there was an improvement in the results, but we believe that this improvement was due to over-fitting. Also, the researchers in [5] report that the analyzed data did not show statistical differences between the groups.

4 Conclusion

In this study, groups of women practitioners and non-practitioners of weight training walked on a treadmill before and after the application of a fatigue protocol. After the fatigue protocol, data were collected at different rest times to verify the muscle recovery process. The kNN algorithm was able to produce the best classification result in relation to RF and SVM for all gait conditions.

The results show that the use of classification algorithms in these cases can help in the development of an autonomous diagnostic system for muscle fatigue discrimination based on spatio-temporal characteristics of the gait. This study has potential applicability in physical therapy and weight training, classifying and understanding the level of muscle fatigue can contribute to more efficient and concentrated training. Also, the classification of these spatio-temporal variables during muscle fatigue can help in the diagnosis of diseases related to muscle weakness and neuromuscular disorders. Clinical studies are of fundamental importance to understand the object of study and the results generated by the classification algorithms. Although the results are promising, the research can be improved by parameterizing these algorithms in larger databases.

For future work, it is suggested to parameterize the models with larger databases. Test other combinations and cross-validation methods. Use other sets of features such as fractal properties, maximum Lyapunov exponent, linear gait variability and stability margin.

Conflict of Interest. The authors declare that they have no conflict of interest.

References

1. Bigland-Ritchie, B., Woods, J.J.: Changes in muscle contractile properties and neural control during human muscular fatigue. Muscle Nerve **7**(9), 691–699 (1984). https://doi.org/10.1002/mus.880070902

2. Swaen, G.M.H., Van Amelsvoort, L.G.P.M., Bült-mann, U., Kant, I.J.: Fatigue as a risk factor for being injured in an occupational accident: results from the Maastricht Cohort Study. Occup. Environ. Med. **60**(SUPPL. 1), 88–93 (2003). https://doi.org/10.1136/oem.60.suppl_1.i88
3. Forestier, N., Teasdale, N., Nougier, V.: Alteration of the position sense at the ankle induced by muscular fatigue in humans. Med. Sci. Sports Exerc. **34**(1), 117–122 (2002). https://doi.org/10.1097/00005768-200201000-00018
4. Gates, D.H., Dingwell, J.B.: The effects of muscle fatigue and movement height on movement stability and variability. Exp. Brain Res. **209**(4), 525–536 (2011). https://doi.org/10.1007/s00221-011-2580-8
5. Lehnen, G.C.S., Rodrigues, F.B., Galvão, J.M.A., Andrade, A.O., Vieira, M.F.: Effects of triceps surae fatigue and weight training level on gait variability and local stability in young adults. Med. Biol. Eng. Comput. **58**(8), 1791–1802 (2020). https://doi.org/10.1007/s11517-020-02196-8
6. Hamacher, D., Törpel, A., Hamacher, D., Schega, L.: The effect of physical exhaustion on gait stability in young and older individuals. Gait Posture **48**, 137–139 (2016). https://doi.org/10.1016/j.gaitpost.2016.05.007
7. Steib, S., Hentschke, C., Welsch, G., Pfeifer, K., Zech, A.: Effects of fatiguing treadmill running on sensorimotor control in athletes with and without functional ankle instability. Clin. Biomech. **28**(7), 790–795 (2013). https://doi.org/10.1016/j.clinbiomech.2013.07.009
8. Dingwell, J.B., Marin, L.C.: Kinematic variability and local dynamic stability of upper body motions when walking at different speeds. J. Biomech. **39**(3), 444–452 (2006). https://doi.org/10.1016/j.jbiomech.2004.12.014
9. Vieira, M.F., et al.: Gait stability, variability and complexity on inclined surfaces. J. Biomech. **54**, 73–79 (2017). https://doi.org/10.1016/j.jbiomech.2017.01.045
10. Zhang, Z.: Introduction to machine learning: K-nearest neighbors. Ann. Transl. Med. **4**(11), 1–7 (2016). https://doi.org/10.21037/atm.2016.03.37
11. Ali, J., Khan, R., Ahmad, N., Maqsood, I.: Random forests and decision trees. IJCS Int. J. Comput. Sci. Issues **9**(September), 272–278 (2012)
12. Srivastava, D., Bhambhu, L.: Data classification using support vector machine. J. Theor. Appl. Inf. Technol. 6 (2009)
13. Bates, S, Hastie, T., Tibshirani, R.: Cross-validation: what does it estimate and how well does it do it? 1–38 (2021)

Numerical Simulation of a Non-linear Control of Pulsed Electrical Fields for Non-thermal Pasteurization

Pedro Paulo Santos(✉), Guilherme Brasil Pintarelli, and Daniela Ota Hisayasu Suzuki

Department of Electrical Engineering, Federal University of Santa Catarina, Florianópolis, Brazil
pedro.santos.silva@posgrad.ufsc.br

Abstract. Several liquids and semi-liquid foods, such as eggs derivatives, milk derivatives, juices, beers, wines, and ice creams are thermally pasteurized. Pasteurization purposes to eliminate pathogenic bacteria that could cause spoilage of the product. Nevertheless, it is known that thermal pasteurization has effects on food's nutrient composition, flavor, texture, and color. Non-thermal Pulsed Electrical Fields pasteurization is a promising process that can improve the pasteurization process and offer consumers micro-biologically safe, minimally processed, and fresh-like products. However, there are difficulties in imposing electric fields in non-homogeneous solutions, and treatment failure could cause food poisoning and even be lethal. This work achieves a sufficient hysteresis control system for the application of different high voltage waveforms in non-homogeneous loads through numerical simulation of a pasteurization system of orange juice using COMSOL and electronic simulation using PSIM. Due to the undefined characteristic of the load the nonlinear control is the most indicated. This concept is a part of a non-thermal pasteurization system that can be used for non-thermal pasteurization processes in industries and research.

Keywords: Hysteresis control · Food preservation · Pulsed high voltage · Orange juice

1 Introduction

Perishable food deteriorates due to the influence of bacteria, enzymes, mold, yeast, moisture, temperature, and chemical reactions. Commonly, living microorganisms do a metabolic breakdown of food units turning them into simpler components, which depreciate the flavor, aroma, color, sensory, and textural properties of foods [1]. This process is called food spoilage, in other words, food deteriorates to the stage in which it is not edible to humans or its quality of edibility becomes reduced. Also, there are harmful microbes that could be a public health concern. Those cause food-borne illnesses and can even be

J. L. B. Marques et al. (Eds.): CLAIB 2022/CBEB 2022, IFMBE Proceedings 98, pp. 341–348, 2024.
https://doi.org/10.1007/978-3-031-49401-7_34

lethal. For instance, *Listeria monocytogenes, Escherichia coli O157, Salmonella, Staphylococcus aureus, Bacillus cereus, Campylobacter, Clostridium perfringens, Aspergillus niger,* and *Saccharomyces cerevisiae* [1].

Killing living microorganisms, mainly bacteria increases food safety and quality for public consumption. Nevertheless, the request for low processed, long-term storable, and fresh food, leads to challenges in the food industry, regarding food safety and quality. Since the consumption of natural and fresh food brings health benefits [2].

Heat processing of foods is one of the earliest and still one of the most important preservation methods [3]. Thermal pasteurization is the food heating processing that aims to kill pathogenic bacteria. Thus, make food safe to eat and store. In Brazil, the Ministry of Agriculture, Livestock and Food Supply (MAPA) standardizes the use of pasteurization for low-processed milk products, and the National Sanitary Surveillance Agency (ANVISA) standardizes pasteurization for milk and egg derivatives processed products. Thermal pasteurization is a binomial between process time and temperature that depends on the type of food and the objective. The treatment can be done in Low-Temperature Long-Time Treatment (LTLT, 63° / 30 min), High-Temperature Short Time (HTST, 75–120° / 15 s), and Ultra High Temperature (UHT, 135° / 4 s).

It is known that thermal pasteurization has effects on food's nutrient composition and sensory characteristics, such as aroma, flavor, texture and color [2,4,5]. Formerly, final customers are continuously looking forward to those characteristics, as they are often related to freshness and authenticity [5]. Non-thermal Pulsed Electrical Fields (PEF) pasteurization is a promising tool that can provide consumers with micro-biologically safe, minimally processed, and fresh-like quality foods, preserving original aroma, flavor, texture, and color. PEF uses the non-thermal electroporation phenomenon to inactivate cells, and usually consists of electric fields with a magnitude of 10 to 50 kV/cm with duration of 1–100 μs [2–4,6,7].

The complete single-pass PEF processes is shown in Fig. 1 which was proposed by Qin et al. [4] and Vega-Mercado et al. [3]. Recently, the technique was combined with UHT by Reineke [8] for food sterilization. Single-pass PEF processes can be applied for eggs derivatives, milk derivatives, juices, beers, wines, and ice creams. The electrical field amplitude is dependent on the electrical characteristics of the substance and the desired effect (desired decontamination). This work aims to present a control system sufficient to impose and control electric fields on a not defined liquid food. The proposed system is easy to install, and its feedback is from the final electric field applied, or an electrical current in the load. We use finite element modeling to calculate the chamber impedance with orange juice (*i.e.*, load). Electronic circuit simulation software was used to test the proposed non-linear control with the calculated load.

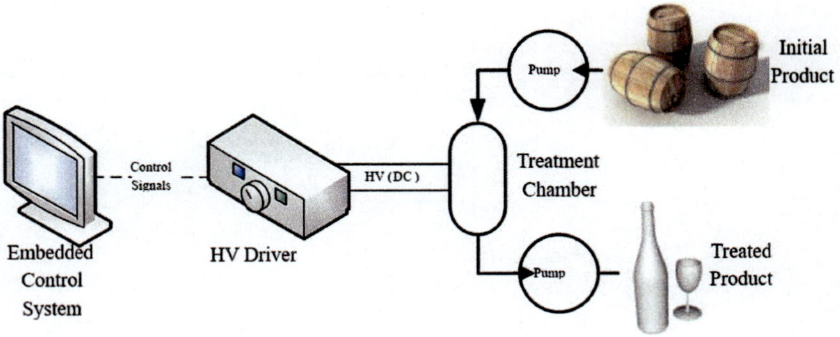

Fig. 1. Single-pass PEF operation.

2 PEF System Design

2.1 Hysteresis Non-linear Control

Food conductivity can change with temperature, gasification, or due to variations in the hydraulic flow system, and this change is not immediate. Due to the undefined characteristic of the load and, possibly, its related plant parameters variation, the hysteresis control was chosen in a preliminary study to control the output voltage. This control consists of turning on and off specific sets of interrupters in such a way that the output voltage follows a defined voltage reference slope within a range error [9,10].

Figure 2 shows the implementation of the hysteresis control. Where Vin is the reference signal, Vhys is the hysteresis voltage set, *i.e.* the maximum value that the output can exceed the reference signal, and Vfeed is the differential voltage measured at the load (block C).

2.2 Treatment Chamber Model

The electric field amplitude is dependent on the size of the treatment chamber. The treatment chamber geometry was simulated by three dimensional models using COMSOL Multiphysics 4.4 (COMSOL AB, Sweden). Two food-industry-compatible stainless steel [6], parallel circles were built with 245 cm radius and 3 mm thickness, inter spaced 2.5 cm by nylon insulation [11]. The chamber is shown in Fig. 3. The food (yellow) enters and exists the system electrode by pressure.

A steady-state regime was considered, and the applied voltage in the electrode was 2 kV which is 0.8 kV/cm. The Dirichlet boundary condition (contact between electrodes and beverage), and Neumann (insulating external surfaces) were applied. Therefore, the Laplace equation (1) was solved by the COMSOL. The impedance data were calculated as equivalent resistor and capacitor parallel.

$$\mathbf{J} = \sigma\mathbf{E} + j\omega\epsilon_0\epsilon_R\mathbf{E} \qquad (1)$$

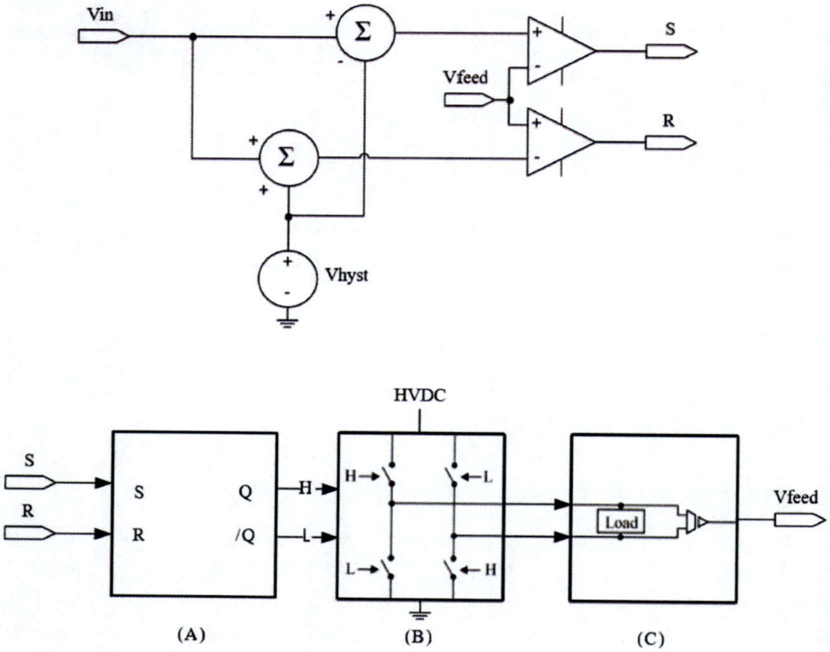

Fig. 2. Preliminary hysteresis control for PEF. (A) is a flip flop type D, (B) is a high voltage H-bridge and (C) is a differential voltage measure at the load.

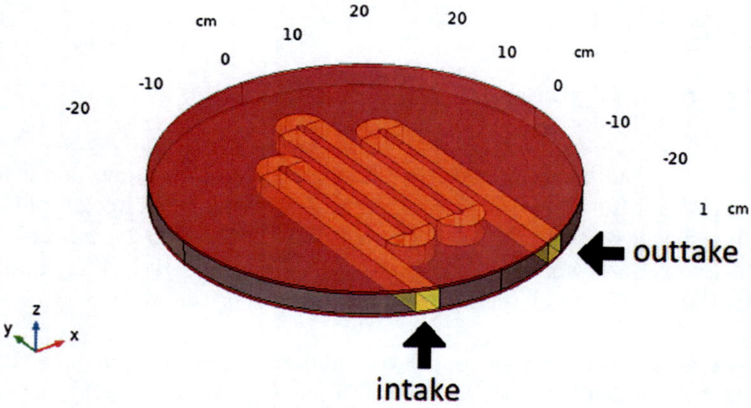

Fig. 3. PEF treatment chamber. The electrodes are represented in red, the nylon insulation delimiter in gray, and the orange juice path in yellow. The arrows represent the food intake and outtake.

where σ is the beverage conductivity [S/m], $\mathbf{E}$ is the electric field [V/m] produced by the applied voltage V [V], j is the imaginary unit ($j = \sqrt{-1}$), $\mathbf{J}$ is the current density [A/m^2] and ω is the simulation frequency [rad/s]. The beverage was considered homogeneous. The conductivity σ used was 0.36 S/m [12] and the relative permittivity ϵ_R used was 80 [13].

3 Results

In Fig. 4 is shown the electric field distribution on beverage for 2000 V between electrodes (40 kV/m). As the plates are parallel it is expected a linear voltage drop, and a homogeneous electrical field. The chamber is described by resistor (R = 0.14 Ω) and capacitor (C = 5.01 nF) parallel. Those values were used to simulate the non-linear control of the chamber's voltage.

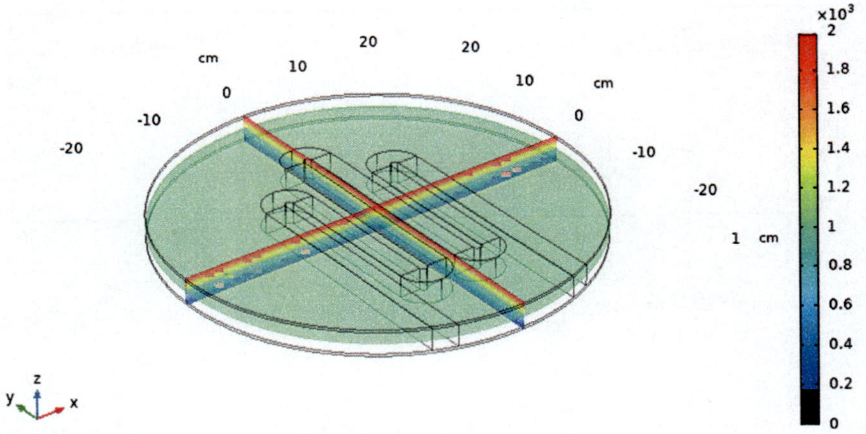

Fig. 4. Voltage distribution across the PEF treatment chamber. This 3D image shows two 2D plane cuts. Red color are regions which voltage is near 2000 V and blue color are regions which voltage is near zero (ground electrode).

Figure 5 shows the characteristic output voltage waveforms of sine, monopolar, and by-polar square waves. These are the desired wave-forms for industrial PEF. The hysteresis non-linear control can output other arbitrary waveforms.

4 Discussion

Pasteurizing is an important process in the food processing. Since, the process should make food safe to be eaten and stored. The pasteurization should have the smallest possible impact on food's nutrients composition and sensory characteristics. Thermal pasteurization is known to affect on nutrients and sensory

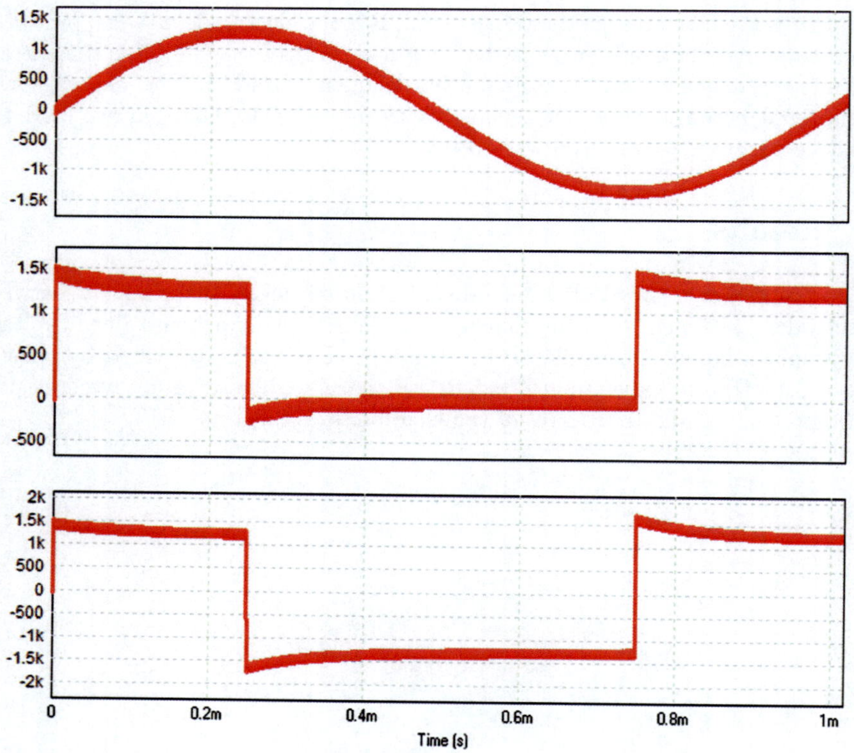

Fig. 5. Output voltage wave forms of sine(on top) and mono-polar(middle) and by-polar square wave(bottom).

characteristics of food [2,4,5]. The PEF process may be an alternative for some food processing, especially delicate food as juice, wine and beer. The pasteurizing process cannot show faults. If ineffectiveness occurs in the treatment process or bottling of beverages, it can cause harm to the health of the consumer, or even be lethal [1,14].

The PEF is mostly studied with high voltage pulse generators without any feedback from the load. For instance, exponential decay pulses [3], rectangular pulse [15], square pulse generator with filter LC, and over-damped oscillator pulse [16]. Better control of the voltage applied to the load can probably increase the quality of PEF treatment. It is known that the PEF-pulses have several frequency components that may possibly generate different responses in the food, and, also, the load is nonlinear, as it affects temperature, gasification and hydraulic flow pressure.

Due to the nonlinear characteristic of the load, the nonlinear control is the most indicated to use [17]. Non-linear control systems offer high-performance designs for non-linear dynamics in order to detect plant behavior. Moreover, with relatively simple implementation, the methodological proposal has a robust realization without stability problems. We demonstrate using computer simulations a solution which can impose arbitrary high-voltage wave forms on the food. Although hysteresis non-linear control can lead to intense converter switching loss and poor voltage tracking, it provides the necessary accuracy to test this paper premise. As this works develops, a more suited control method can be implemented.

This work aimed at a preliminary study of a control system to operate with an unknown PEF plant. Future work is expected to better understand the PEF plant through experiments and modeling. Moreover, to finish the embedded device of non-thermal pasteurization and examine its effectiveness. Yet, the power electronics and the controller in a digital signal processing (DSP), which are required to operate the PEF, must be developed and implemented. Note that switching and propagation delays must be observed in the development of the PEF apparatus. Therefore, effectiveness in the microbial elimination of different waveforms could be examined. After that, a study should be carried out to see whether other waveforms do produce less intervention, or not, in the sensory and nutritional characteristics of the food.

5 Conclusion

The work was able to evaluate that the non-linear control system may be a promising tool to satisfy the problem of controlling high voltage application in non-homogeneous and not well-defined plants. This concept is a part of a system that can be used for non-thermal pasteurization processes in industries and research.

References

1. Lucera, A., et al.: Food applications of natural antimicrobial compounds. Front. Microbiol. **3**, 287 (2012)
2. Branch, D.: Pulsed electric Fields for food preservation: an update on technological progress. In: Progress in Food Preservation, pp. 277–295. Wiley, Oxford (2012)
3. Vega-Mercado, H., et al.: Pulsed electric fields in food preservation. In: Handbook of Food Preservation, Second, pp. 783–813 (2007)
4. Qin, B.L., et al.: Inactivating microorganisms using a pulsed electric field continuous treatment system. IEEE Trans. Ind. Appl. **34**(1), 43–50 (1998)
5. Charles-Rodriguez, A.V., et al.: Comparison of thermal processing and pulsed electric fields treatment in pasteurization of apple juice. Food Bioprod. Process. **85**(2), 93–97 (2007)
6. Putri, R.I., Syamsiana, I.N., Hawa, L.C.: Design of high voltage pulse generator for pasteurization by pulse electric field (PEF). Int. J. Comput. Electr. Eng. **2**(5), 1793–8163 (2010)
7. Golberg, A., Fischer, J., Rubinsky, B.: The Use of Irreversible Electroporation in Food Preservation, pp. 273–312. Springer, Berlin Heidelberg (2010)
8. Reineke, K., et al.: Sterilization of liquid foods by pulsed electric fields-an innovative ultra-high temperature process. Front. Microbiol. **6**, 1–11 (2015)

9. Shamsi-Nejad, M.A., et al.: Modelling and design of an hybrid modulated hysteresis current controller Application to a single phase voltage source inverter. In: 37th IEEE Power Electronics Specialists Conference, pp. 1–6 (2006)
10. Prasad, V.: Hysteresis current controller for a micro—grid application. In: 2016 International Conference on Energy Efficient Technologies for Sustainability, pp. 73–77 (2016)
11. Martin, O., et al.: Inactivation of *Escherichia Coli* in skim milk by high intensity pulsed electric fields. J. Food Process Eng. 317–336 (1997)
12. Amiali, M., et al.: Electrical conductivities of liquid egg products and fruit juices exposed to high pulsed electric fields. Int. J. Food Prop. 533–540 (2006)
13. Archer, D.G., Wang, P.: The dielectric constant of water and debye- Hückel limiting law slopes. J. Phys. Chem. Ref. Data 371–411 (1990)
14. Food and Drug Administration (FDA).: Science and our food supply: food safety A to Z reference guide. eprint: SilverSprin (2017)
15. Buckow, R., Ng, S., Toepfl, S.: Pulsed electric field processing of orange juice: a review on microbial, enzymatic, nutritional, and Sen-sory Quality and Stability. Compr Rev Food Sci Food Saf **12**(5), 455–467 (2013)
16. Syamsiana, I.N., Putri, R.I.: High voltage pulse generator design with voltage control for pulse electric field (PEF) pasteurization. In: Proceedings of the 2011 International Conference on Electrical Engineering and Informatics, pp. 1–5 (2011)
17. Slotine, J.-J.E., Li, W.: Applied Nonlinear Control. Prentice-Hall, Englewood Cliffs, New Jersey (1991)

Evaluation of Electroporation Memory in Electrochemotherapy Protocols: *Ex Vivo* Liver Preliminary Study

Isabela B. Paro[1], Lucas B. Lopes[1], Daniella L. L. S. Andrade[1], Marco A. Lorenzo[2], and Daniela Ota Hisayasu Suzuki[1](✉)

[1] Biomedical Engineering Institute, Federal University of Santa Catarina, Florianópolis, Brazil
daniela@ppgeel.ufsc.br

[2] Central Animal Facility, Federal University of Santa Catarina, Florianópolis, Brazil

Abstract. Electrochemotherapy (ECT) is a cancer treatment that uses electroporation (EP) to boost chemotherapeutic drug effectiveness due to increased cell membrane permeability. The membrane effect is a phenomenon of incremented conductivity increasing beneath consecutive pulses. This preliminary study tested whether the membrane effect would occur between three consecutive ESOPE protocol applications. Further, we analyzed if it would affect the safety of ECT treatments by unsafely increasing electrical currents in the tissue. Our findings showed that electroporated membranes did not fully recover in twenty seconds after an ECT protocol. The phenomenon was well observed at the ECT levels. Still, it had a minor influence on conductivity rising, not representing a risk to ECT safety.

Keywords: ESOPE · Electrical conductivity · Rat liver · Tissue dielectric dispersion

1 Introduction

Electroporation (EP) is the phenomenon of transitory increased cell membrane permeability due to a sufficiently high applied electric field. Under the effect, EP is perceived as an enhancement of the electrical conductivity of the cells [1]. Below the reversible electroporation threshold (E_{rev}), the phenomenon does not occur, and the conductivity follows a linear model. EP is considered reversible until a very high electric field (E_{irev}) is reached and a plateau of conductivity is observed. At this level, the disorder leads to membrane lysis and cell apoptosis.

In combination with chemotherapeutic drugs, EP is responsible for the selectivity of electrochemotherapy (ECT) [2]. The treatment improves the effectiveness of the medication, favoring its entrance into the cells not possible in homeostatic status. ECT is a well-established technique that has been optimizing

J. L. B. Marques et al. (Eds.): CLAIB 2022/CBEB 2022, IFMBE Proceedings 98, pp. 349–356, 2024.
https://doi.org/10.1007/978-3-031-49401-7_35

efficiency and safety parameters over the years [3,4]. During ECT treatments, it is a common practice to apply multiple protocols in target tissues (i.e., sequenced EP protocols application without electrode position changing), which causes an overlap of areas subjected to the electric field.

The time for membrane recovery from pulse application is not instantaneous. Between consecutive pulses within the same protocol, one can observe a membrane relaxation period, when the membranes work to regain the homeostatic level (i.e., pores recovery). However, basal membrane permeability is not restored after the next pulse, as the application frequency appears to disagree with the required time for pore closure. The result is an increased electrical conductivity at the beginning of the next pulse, leading to conductivity increment of each pulse until saturation (with applied potential correspondence) [5]. The duration of pore closure is yet unknown. This influence, known as the memory effect, is an essential subject of study in the electroporation field, especially amidst single EP protocols [6–8], proving to be harmless to the effectiveness of the treatment. However, evaluating the memory effect between subsequent applications of the ECT protocol is necessary to assess electrical effects in tissues.

In this work, we evaluate the EP memory effect among ECT standardized protocols. Furthermore, we intend to analyze whether the increased conductivity caused by the memory effect compromises the efficiency of ECT treatment.

2 Materials and Methods

2.1 Experimental Setup

Female Wistar rats weighing 200–250 g were euthanized prior to tests in accordance with the guidelines of Resolution 1000 of the Federal Council of Veterinary Medicine of 11 May, 2012. The animal experiments are in accordance with the local Ethical Committee for Animal Use. Livers were removed, cut into 10 mm x 10 mm samples and stored in petri dishes. Due to different liver sizes, three sample thicknesses were obtained: 2 mm, 3 mm and 4 mm. Table 1 shows study groups classification by sample thickness and electric field. For each thickness, five samples were used as control (G1, G2 and G3) and electroporated (G4, G5, and G6) groups. A total of 30 samples were used. We adjusted the applied voltage to keep constant electric fields. The EF of 30 kV/m was applied to control samples (i.e., under the reversible electroporation threshold, no effect expected), while the electroporated samples received 150 kV/m.

We followed ESOPE standard protocol for ECT treatment, which consists of 8 square-wave pulses of 100 μs with repetition rate of 1 Hz [9], using a custom electroporation system [10]. The EP protocol was applied three times in each sample, and an interval of 20 s was given between each EP protocol. We used ESOPE standard parallel plates electrode (10 mm $\times$ 10 mm). Figure 1 shows the experimental setup. A conductive gel (Carbogel ULT, produced by Carbogel) [11] was used to improve contact between electrodes and liver tissue [3,12]. After the complete application, the samples were discarded and the electrodes were cleaned. All measurements were performed within 2 h of animal death, at

Table 1. Sample thickness, applied voltage and electric field of each group. G1, G2 and G3 are the control groups. G4, G5 and G6 are the electroporation groups.

	Groups					
	G1	G2	G3	G4	G5	G6
Sample thickness [mm]	2	3	4	2	3	4
Applied voltage [V]	60	90	120	300	450	600
Electric field [kV/m]	30	30	30	150	150	150

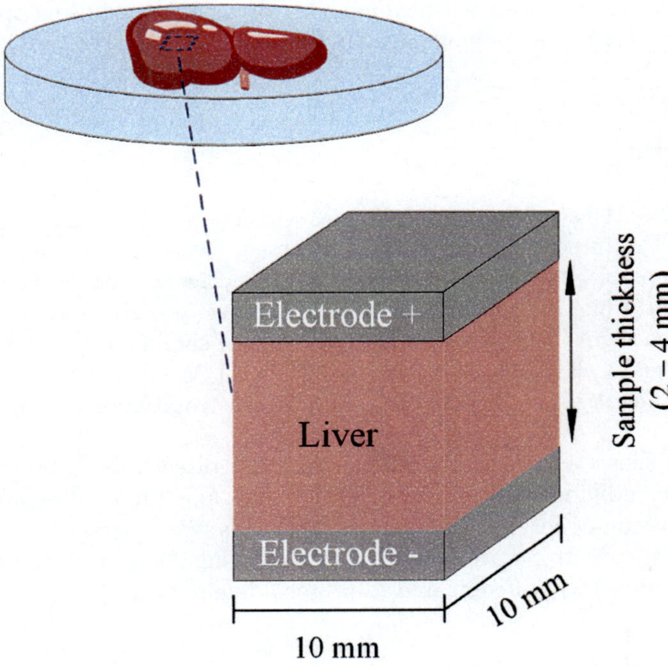

Fig. 1. Ilustration of the experimental setup for ECT pulse applications in liver samples.

25°C. Changes in the liver dielectric parameters are insignificant within 10 h of death [13].

In the last delivered pulse of each EP protocol, current was acquired by a Tektronix TDS 2004C digital oscilloscope and a Tektronix A622 current probe (Tektronix Inc., Beaverton, OR, USA). The data was then analyzed on a personal computer and each current was extracted by calculating the mean amplitude of the current pulse. Following, electrical conductivity was calculated for each experimental sample as shown in simplified Eq. 1.

$$\sigma = \frac{l}{A} \times \frac{I}{V} \tag{1}$$

where l represents sample length (2 to 4 mm), A the electrode width (10 mm), V the applied electric potential, and I the acquired electric current.

2.2 Statistical Analysis

We submitted the experimental electric currents of all groups to the Shapiro-Wilk normality test. Since the groups had normal distribution, they could be represented as means and 95% confidence intervals (CI). Electrical conductivity conversion occurred after data normality testing (i.e., no data was transformed to adjust for normality). All statistical analyses were performed using open source software $R^{©}$ v.4.1.0, Rcmdr package (R Foundation for Statistical Computing, USA).

3 Results

Figure 2 shows the evolution of the electrical conductivity (means and CI = 95%) for each of the three EP protocols. As expected, electrical conductivity in the control groups (G1–G3) stayed at initial values as the applied electric field was not sufficiently high to provoke tissue permeabilization. However, groups G4–G6 expressed higher electrical conductivity values, similar to those reported in the literature for electroporated liver tissue [14,15]. We illustrated in Fig. 2 the conductivity electric-field dependency (i.e., the electroporation curve) described by [14].

Figure 3 shows a dot chart of electric currents of each EP protocol in liver samples (i.e., individual values of the eighth pulse), and Table 2 presents the percentage variation of currents between the protocols. We observed the increasing behavior in the electroporated groups, with more significant variations between protocol 2 and 1 (Δ_{21}) if compared to protocol 3 and 2 (Δ_{32}).

4 Discussion

Electroporation studies aim to understand and reproduce the rising electrical conductivity effect between consecutive pulses [5,16]. Still, there is no conclusion on how long the effects of EP last in biological tissues. Applying 24 sequential pulses (3 ESOPE sequences 20 s apart) had a minor influence on conductivity changes due to electroporation. We found no statistical difference between the electric currents within the same groups (control G1-G3; electroporated G4-G6). However, the increasing behavior of electric current within applied EP protocols shown in our results suggests an accessory influence on electroporation which we believe to be the memory effect in the membrane cells.

Although literature explains the result of closing pores from the interruption of EP pulses, there is still no consensus on the exact time of full membrane recovery. As shown in Fig. 3, cell membranes could not fully recover in 20 s, as we observed an increasing tendency among the electric currents. In control

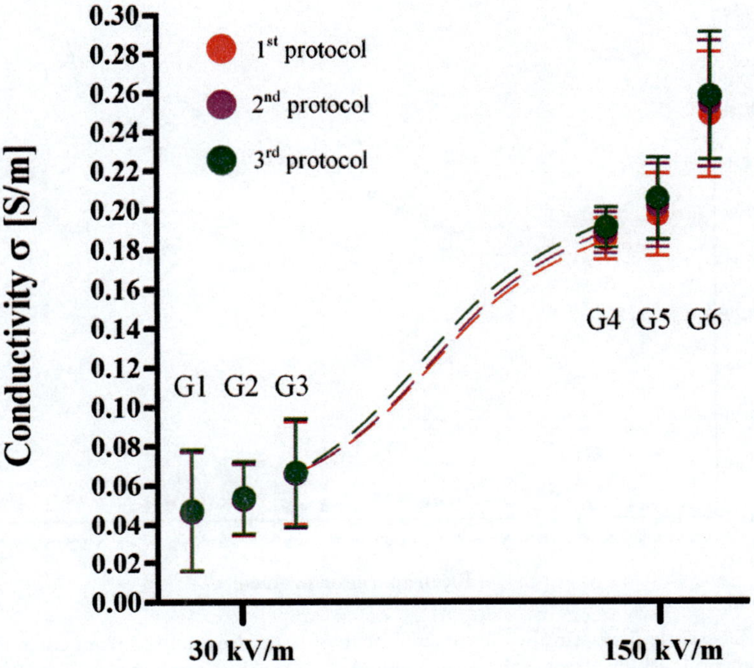

Fig. 2. Electrical conductivity (means and 95% CI) for each EP protocol separated by the experimental groups. Dashed lines indicate the tendency of the conductivity-electric field dependency in electroporation described by [14] for liver tissue. Red = 1st protocol; magenta = 2nd protocol; green = 3rd protocol. G1 to G3 EF = 30 kV/m; G4 to G6 EF = 150 kV/m.

groups (G1, G2, and G3), we observed no increased tendency in currents. As presented in Fig. 2, the applied electric field is too low to trigger EP, and a memory effect is not expected. In EP groups (G4, G5, and G6), we presume that the noticeable electric current increase is due to a memory effect - cell membranes are still permeabilized when the following protocol is applied (i.e., there are still open pathways for electric current within the cell membrane). The maximum time interval between applications for the occurrence of the memory effect is still unknown, and new comparative studies are necessary for the precise determination of the limits of this effect.

We observed a reduction in the current increase (%) in Δ_{32} compared to Δ_{21} in electroporated groups (see Table 2), suggesting the beginning of conductivity saturation on the liver around the sixteenth pulse. Similar results in a previous study of sequenced EP protocols in the liver showed a conductivity plateau after a dozen pulses [8]. Although we showed that the memory effect does not significantly affect the conductivity rising, these results shed light on whether multiple protocol applications are necessary for ECT treatments. For instance, Suaréz and colleagues [6] showed that multiple pulse applications did

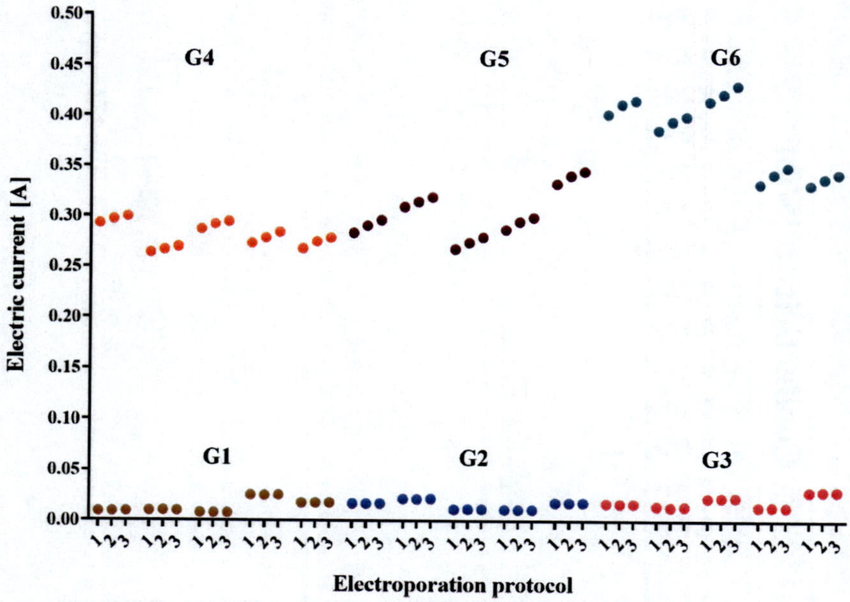

Fig. 3. Electric current amplitudes of the last (eighth) pulse resulted from three ESOPE protocol applications. Protocols were applied 20 s apart. Electric currents of electroporated groups (G4 - G6) present a membrane memory effect (an increasing behavior over the protocols). G1 = brown; G2 = dark blue; G3 = magenta; G4 = orange; G5 = purple; G6 = light blue.

not significantly increase electroporated volumes. However, only fifteen pulses were sufficient to increase more than 10° the initial temperature, which could be an issue when surrounding healthy tissues are affected by overlapping electric pulses. Therefore, we aim to study conductivity changes with additional ECT protocols, including temperature effects in liver tissue.

We found an unexpectedly increased conductivity in the G6 group. The EP model [17] describes a dependency of conductivity on the electric field and current density given by $J = \sigma(E) \cdot E$. Although we preserved the electric field by increasing the applied potential in thicker samples, the conductivity results in G6 (see Fig. 2) indicate that the distance between the plates and/or the applied voltage affects the flow of electric current. We are interested in investigating such an effect on tissue conductivity EP-dependent.

5 Conclusion

This work investigated the memory effect on liver tissue after sequenced electrochemotherapy pulses and its outcomes. Our results show that the electrical conductivity is enhanced for at least twenty seconds after standard EP protocol. The enhanced conductivity, however, has no substantial impact on ECT

Table 2. Percentage differences between the three EP protocols applied to the groups. A positive delta means an increase in current, while a negative delta means a decrease in current

G1	Δ_{21}	Δ_{32}	Δ_{31}	**G4**	Δ_{21}	Δ_{32}	Δ_{31}
	−1,0%	1,9%	0,9%		1,4%	0,9%	2,3%
	2,2%	−1,9%	0,2%		1,1%	1,1%	2,2%
	−0,5%	−1,0%	−1,5%		1,7%	0,9%	2,6%
	−1,2%	1,6%	0,4%		1,9%	2,0%	3,9%
	0,8%	0,7%	1,5%		2,6%	1,3%	3,9%
G2	Δ_{21}	Δ_{32}	Δ_{31}	G5	Δ_{21}	Δ_{32}	Δ_{31}
	−0,6%	0,3%	−0,3%		1,7%	1,4%	3,1%
	−1,5%	2,0%	0,5%		2,4%	1,9%	4,4%
	3,5%	−0,2%	3,2%		2,8%	1,4%	4,2%
	−0,9%	0,6%	−0,3%		2,4%	1,3%	3,7%
	−0,3%	−0,1%	−0,4%		2,5%	1,9%	4,5%
G3	Δ_{21}	Δ_{32}	Δ_{31}	**G6**	Δ_{21}	Δ_{32}	Δ_{31}
	−1,2%	−1,4%	−2,6%		2,4%	0,9%	3,4%
	−6,2%	4,8%	−1,7%		2,2%	1,2%	3,5%
	1,7%	0,4%	2,1%		1,7%	2,0%	3,7%
	0,2%	−2,0%	−1,9%		3,0%	1,9%	4,9%
	1,6%	−0,4%	1,2%		1,9%	1,3%	3,3%

success, ensuring the procedure's safety and effectiveness even in cases of three ESOPE applications in the same area. We intend to continue the investigation of membrane resealing after electroporation imposition.

Acknowledgements. The authors would like to thank the Central Animal Facility-UFSC for technical support of animals care management used on this work. This study was financed in part by the *Coordenačão de Aperfeičoamento de Pessoal de Nível Superior - Brasil* (CAPES). The authors would like to thank the Brazilian research funding agencies CAPES and CNPq for the scholarships granted to the post-graduate students participating in the study and for the *Programa Institucional de Bolsas de Iniciačão Científica* (PIBIC) scholarship granted to the undergraduate student.

Conflict of Interest. The authors declare that they have no conflict of interest.

References

1. Suzuki, D.O.H., et al.: Theoretical and experimental analysis of electropo-rated membrane conductance in cell suspension. IEEE Trans. Biomed. Eng. **58**(12), 3310–3318 (2011)

2. Mir, L.: Bases and rationale of the electrochemotherapy. Eur. J. Cancer, Supplement **4**(11), 38–44 (2006)
3. Bertinetti Lopes, L., et al.: Computer optimization of conductive gels for electrochemotherapy. Med. Eng. Phys. **98**, 133–139 (2021)
4. Andrade, D.L.L.S., et al.: Electrochemotherapy treatment safety under parallel needle deflection. Sci. Rep. **12**(1), 1–10 (2022)
5. Weinert, R.L., Ramos, A.: Electroporation threshold, conductivity and memory effect in rat liver. Biomed. Sig. Proc. Cont. **64**, 102275 (2021)
6. Suárez, C., et al.: The role of additional pulses in electro permeabilization protocols. In: PloS One **9**(12), e113413 (2014)
7. Lv, Y., et al.: Analysis of the electric field-dependent current during electroporation pulses. IEEE Acess **8**, 93850–93856 (2020)
8. Zhao, Y., et al.: Development of a multi-pulse conductivity model for liver tissue treated with pulsed electric fields. In: Frontiers in Bioengineering and Biotechnology, vol. 8, p. 396 (2020)
9. Mir, L.M., et al.: Standard operating procedures of the electrochemotherapy: instructions for the use of bleomycin or cisplatin administered either systemically or locally and electric pulses delivered by the CliniporatorTM by means of invasive or non-invasive electrodes. Eur. J. Cancer Suppl. **4**(11), 14–25 (2006)
10. Berkenbrock, J., et al.: In vitro simulation of electroporation using potato model. CMBES Proc. **40** (2017)
11. Pintarelli, G.B., et al.: Computer simulation of commercial conductive gels and their application to increase the safety of electrochemotherapy treatment. Med. Eng. Phys. **74**, 99–105 (2019)
12. Suzuki, D.O.H., Marques, C.M.G., Rangel, M.M.M.: Conductive gel increases the small tumor treatment with electrochemotherapy using needle electrodes. Artif. Organs **40**(7), 705–711 (2016)
13. Surowiec, A., Stuchly, S.S., Swarup, A.: Radiofrequency dielectric properties of animal tissues as a function of time following death. Phys. Med. Biol. **30**(10), 1131 (1985)
14. Sel, D., et al.: Sequential finite element model of tissue electropermeabilization. IEEE Trans. Biomed. Eng. **52**(5), 816–827 (2005)
15. Garcia-Sanchez, T., et al.: Physiological changes may dominate the electrical properties of liver during reversible electroporation: Measure-ments and modelling. In: Bioelectrochemistry **136**, 107627 (2020)
16. Weinert, R., Pereira, E., Ramos, A.: Inclusion of memory effects in a dynamic model of electroporation in biological tissues. Artif. Organs **43**, 688–693 (2019)
17. Corovic, S., et al.: Modeling of electric field distribution in tissues during electroporation. In: BioMedical Eng. Online **12**(16) (2013)

Prediction Models of Sudden Death from Chronic Chagas Cardiomyopathy

Dayana Vieira(✉), Johana Gomez, and Jurandir Nadal

Federal University of Rio the Janeiro, Pedro Calmon 550, Rio de Janeiro, Brazil
dayana.vieira@peb.ufrj.br

Abstract. Chagas disease is an endemic disease in Latin America that cause cardiomyopathy in its chronic stage. The symptoms include heart rhythm disturbance that can evolve to sudden cardiac arrest. This paper presents methods for stratifying and predicting the risk of sudden death in chronic Chagas' cardiomyopathy, retrospectively analyzing the 24 h Holter records of patients. The database includes 60 cases of alive patients and 22 cases that evolved to sudden death. For this study, 13 predictor variables were found from an ECG signal. They were extracted from the heart rate turbulence and heart rate variability parameters in the time domain. These variables were used as input to seven classifier models, these being Support Vector Machine, Multilayer Perceptron, Python's Keras Library, K-nearest Neighbors, Logistic Regression, Randon Forest, and Decision Trees. The best model (Randon Forest) provided 95% accuracy, 100% sensitivity, and an area under the ROC curve of 95.8%, fitting the assessment of an excellent classifier.

Keywords: Chagas disease · Machine learning · Risk stratification · Sudden cardiac death

1 Introduction

According to the World Health Organization, Chagas disease is a life-threatening infection caused by the parasite Trypanosoma cruzi, which is a progressive parasitic disease that does not regress and may have a latency period between the two phases of evolution, acute and chronic, thus being one of the major public health problems in Latin America. It is estimated that there are 7 to 8 million people worldwide with Chagas disease, most of them in Mexico, Central and South America, with an average of more than 10.000 deaths per year [1].

Trypanosoma cruzi tends to lodge in the muscle fibers of the myocardium, causing Chagas' disease. The parasites in their "amastigote" form inside the cells, multiply by successive binary divisions, forming "pseudocysts" and starting a chronic fibrosing myocarditis picture, which causes alterations in the myocardium, damaging the structures that conduct the cardiac impulses and also the innervation of the autonomous system, unbalancing the cardiac output and frequency. This progression leads to the occurrence of severe arrhythmias, common in the chronic phase of the disease [2].

J. L. B. Marques et al. (Eds.): CLAIB 2022/CBEB 2022, IFMBE Proceedings 98, pp. 357–366, 2024.
https://doi.org/10.1007/978-3-031-49401-7_36

One of the methods to diagnose Chagas disease is through microscopic inspection of blood smears, a very common and convenient technique, but sometimes it requires a lot of time and effort because it involves the analysis of many blood samples; for this type of analysis, a machine learning method can help automate microscopic observation activities, because it encompasses techniques and technological tools that allow computers to perform complex tasks with high accuracy [3–5].

In the area of machine learning, a limited number of works have been reported for the detection and segmentation of the Trypanosoma cruzi parasite [6].

From the presented references, Soberanis-Mukul proposes an automated algorithm to detect the infectious agent using various segmentation and classification techniques in images of blood samples. His database has 120 images, of which 60 show the presence of parasites and as a result of training and testing, the authors reported 98% sensitivity and 85% specificity [6].

On the other hand, the Support Vector Machine (SVM) classifier [7], was trained with several features. The various experiments performed were with the proposed classifiers: (1) AdaBoost, (2) AdaBoost + post-processing, (3) SVM with linear kernel, (4) SVM with polynomial kernel, and (5) SVM with radial basis function (RBF) kernel. The classification using AdaBoost plus post-processing obtained the best results and reported 100% sensitivity and 93.25% specificity [7].

Moreover, the FIOCRUZ risk score has been proposed for sudden death in Chagas' cardiomyopathy, according to clinical profiles and non-invasive exams [8–10]. This score uses four factors (QT dispersion, syncope, presence of ventricular extrasystoles, and left ventricular systolic dysfunction) to predict risk: low risk, intermediate risk, and high risk. Of these factors, the "QT Variance" index requires a special multi-lead electrocardiogram (ECG) analysis program, its reproducibility is low, and it varies from observer to observer.

In a previous study, different techniques for assessing heart rate variability (HRV) on 24 h Holter have revealed that decreased HRV during the night correlates with a higher risk of Sudden Cardiac Death (SCD) [11], so it is important to develop effective means to identify a subset of patients with Chagas' cardiopathy at high risk for sudden death for patients with Chagas' disease.

This study presents a method for risk stratification and prediction of sudden death in chronic Chagas cardiopathy by retrospectively analyzing 24 h of Holter records of patients with Chagas disease applying SVM, Multilayer Perceptron (MLP), Python's Keras library, K-nearest Neighbors (KNN), Randon Forest, Logistic Regression and Decision Tree methods. A database consisting of classical heart rate turbulence (HRT) and heart rate variability (HRV) parameters in the time domain was studied. The respective parameters were obtained from 12 h periods corresponding to day and night.

2 Material and Methods

2.1 Dataset

Eighty-two Holter ECG records from patients with acute and chronic Chagas' disease heart disease were selected for the study, from a set of 236 patients followed up at the Chagas' Disease Outpatient Service of the Federal University of Rio de Janeiro, Brazil

(1992–2012) [10]. The dataset was sampled with a frequency of 128 Hz, 8-bit resolution, and a 20 Hz anti-aliasing filter. The data are for male and female patients between 39 and 53 years old. The database contains a set of 14 variables of which we have 82 values each referring to one of 82 patients, of which 60 patients are alive and 22 have died from sudden cardiac death.

The dependent variable is the output and indicates whether the individual is dead (= 1) or alive (= 0). The predictor variables available in the database are gender (Male = 0 and Female = 1), Chagas disease class (classes A = 1, B1 = 2, B2 = 3, C = 4, and D = 5), age, and ten parameters of HRV and HRT: turbulence onset (TO24h); turbulence slope (TS24h); standard deviation of all NN intervals (SDNN); the square root of the mean of the successive differences in NN intervals (rmsSDNN); the ratio of the count of NN intervals greater than 50 ms divided by the total number of NN intervals (pNN50); Heart Rate (HR); the cardiac acceleration rate (CAR); the cardiac deceleration rate (CDR); the CAR/CDR ratio, and the left ventricular ejection fraction (LVEF), as described below. To obtain the parameters, the heart rate variability technique was used, which consists of the coherent averaging of the tachogram segments using as anchors the short RR intervals due to each isolated premature ventricular contraction (PVC) detected throughout the 24 h Holter ECG recording. From this average tachogram, two indices are obtained: turbulence onset (TO), which measures the immediate increase in heart rate; and turbulence slope (TS), which measures the compensatory and progressive decrease in heart rate [11]. TO was measured as the ratio between the first two RR intervals after PVC and the two immediately before PVC, while TS is recovered from the maximum slope of the regression line obtained over any five consecutive RR intervals within the 15 intervals after PVC at the coherent mean. The following HRV parameters were also calculated [11]: SDNN, rmsSDNN, the number of NN intervals greater than 50 ms (NN50) and pNN50. In addition to these classical parameters, it was also measured the CAR, CDR, CAR/CDR ratio, and LVEF.

2.2 Methology

This study was organized into three stages: exploratory analysis, data resampling, and classification. The first stage aimed at identifying variables that have high correlation among themselves, missing data, and unbalanced data. According to the results obtained in this analysis, the second stage was found necessary, where techniques for balancing the data distribution to improve the predictive power of the final classifier. For oversampling, the synthetic minority oversampling (SMOTE) technique was adopted, which generates synthetic samples for the minority class of the existing samples. After that, in the normalization step, the variables were resampled between 0 and 1 to ensure that there are no discrepancies in magnitudes.

Then, the variables with high cross-correlation to others were removed to test if this lack of independence may harm the interpretation and the results of the models. Finally, the remaining variables were used as input to the classification algorithm.

2.3 Data Balancing

The original dataset is small, counting only 82 samples, and presents a considerable unbalance (being 60 samples for the living class and 22 for the dead class). A dataset is said to be unbalanced in classification when there are many more cases of one class than another [12]. Traditional classifiers are sensitive to this unbalance and tend to value the majority classes and ignore the minority classes. This is because classifiers in this situation have a high false-negative rate for the minority classes. Therefore, there are pre-processing techniques that resample the data. These methods consist of eliminating cases from the majority class (undersampling) and replicating (or synthetically generating) cases from the minority class (oversampling) of the training data to obtain better classifiers than those obtained from the original distribution. In the present case, the methods of downsampling by selecting only an equal or approximately equal number of samples for each class are not the preferable choice, since they reduce the already small sample size.

A method commonly cited in the literature and used in this work is SMOTE. Python's imblearn SMOTEENN library was used for balancing the data.

2.4 Classifiers Creation

The proposed work consists of the prediction and classification of patients as alive (= 0) or dead (= 1), and seven classification models were chosen, namely SVM, MLP, Python's Keras library, KNN, Logistic Regression, Randon Florest, and Decision Tree. The methodology will be implemented using the SkLearn library for Python.

Separation of Dataset into Training and Testing. The database was divided into a training and test group, as the libraries and code were used in Python even though the nets require a validation group, the functions in the program only require one training and one test group. The data were separated randomly, 85% of the samples were selected for the training group and 15% of the data set was reserved for checking the accuracy of the trained model.

SVM. SVM is generally known for its application as a classifier, however, it can be used for both classification and regression. This classifier constructs a hyperplane in multidimensional space seeking to separate the different classes. The central goal of SVM is to find a maximum marginal hyperplane that best bounds the dataset into classes to minimize errors [12]. In practice, the SVM algorithm is implemented using a kernel. The kernel is responsible for transforming an input data space into the desired format, i.e., the kernel takes a small dimensional input space and transforms it into a higher-dimensional space. In this work, an SVM network with a linear kernel was developed.

Multilayer Perceptron (MLP). A multilayer perceptron (MLP) is a feed-forward artificial neural network model that maps input data sets into a set of appropriate outputs. This is also called a fully connected feed-forward artificial neural network [13]. An MLP consists of several layers of nodes in a directed graph, with each layer fully connected to the next. Except for the input nodes, each node is a neuron (or processing element) with a nonlinear activation function. MLP uses a supervised learning technique called backpropagation to train the network. MLP is a modification of the standard linear perceptron and can distinguish data that are not linearly separable. In this work, an MLP network with a hyperbolic tangent activation function and three hidden layers with two neurons in each of the first two layers and one neuron in the last hidden layer was used.

Keras Library. Keras is a deep learning library for Python. This is a highly powerful library capable of building abstract building blocks for deep learning network construction. The building blocks that Keras provides are built using TensorFlow. Keras supports CPU and GPU computing and is a great tool for quickly prototyping ideas [14]. To build a neural network proposed in this work with Keras, the following steps were followed:

1. A model has been defined using the sequential build, which allows the user to add/configure layers.
2. The first layer was added, using Dense, and the input dimensionality is specified. In this case, the model will have an input of dimensionality 11 and produce an output of dimensionality 1.
3. The selected activation function was the sigmoid.
4. Next, the second layer was defined using Dense. In this case, the dimensionality of the output was defined to be 1. The dimensionality of the input is the same as the dimensionality of the output of the previous layer.
5. After this, the optimizer and the loss function were defined, and the model was compiled, trained, and evaluated.

An appropriate loss function was chosen according to the task at hand; in this case, with a binary classification problem, binary cross-entropy was selected. The optimization algorithm chosen was 'adam'.

K-Nearest Neighbors (KNN). The K-nearest neighbor (KNN) method was one of the first and most fundamental applied to classification that has emerged. Although it is a classic method, it should be one of the first when not much is known about the dataset to be studied. The KNN classifier initially identifies the K neighbors in the training bank that is closest to the new input to be classified. The distance between these neighbors and the new input is the classical Euclidean distance [12]. Then, to classify the new input between 0 or 1 for example, where the largest number of neighbors belong. One way to optimize the classification and decrease the error is to search for the number of K (number of close neighbors). In this work, the ideal value of K was found by searching in a range from 1 to 100 (this value being chosen as a number smaller than the total number of samples), and the value of K that resulted in the best accuracy was 18, as can be seen later in Fig. 3.

Binary Decision Tree. To create the decision tree model in Python, the supported criteria "Gini" was chosen as the impurity index and a maximum depth was set to three (max_depth=3).

Random Forest. For the Randon forest model, the "Gini" criterion was also used, with a maximum depth equal to 10 (max_depth=10).

Logistic Regression. For logistic regression, there was no definition of hyperparameters.

3 Results

Initially, the data analysis step was performed to verify the need for cleaning, normalization, and balancing. There were no missing data, however, the dataset was unbalanced. Applying the SMOTE technique, 43 more samples were synthetically generated, 9 of them with class 0 and 34 with class 1. These generated data were added to the original dataset and the new dataset had a total of 125 samples, 69 for output class 0 and 56 for output class 1.

To evaluate the performance of the proposed models using the methodology already exposed, four scenarios were developed using algorithms based on state-of-the-art machine learning algorithms. These scenarios allowed the evaluation of the effects of some factors on classifier performance. In scenario 1, there was neither variable selection nor resampling; in scenario 2, only resampling was performed to balance the classes; in scenario 3, variables were selected and no resampling was performed; and finally, in scenario 4, resampling and variable selection were applied. The data were divided as described in the methodology and the hyperparameters for each model were selected based on previous studies, selection by testing different parameters, and specific selection techniques, depending on the model analyzed.

After running the experiments, the results were pooled and tabulated. Accuracy, precision, recall, AUC (area under the ROC curve), and Kappa were the model evaluation metrics used in this work to analyze the performance of the models. For the first scenario (Table 1), the performance indicators are very affected by the lack of data treatment. The accuracy of the models was above 76%, except for MLP and decision tree. However, the models are biased to classify the data as the majority class (in this case the output class 0) and for this reason, the other performance values were null and it was not possible to calculate the AUC because the models were classifying the data in only one class.

Data balancing generated a significant improvement in results (Table 2), not only regarding the accuracy, since in some cases accuracy decreased, but regarding performance metrics in general. All classifiers had a significant increase in their sensitivities (recall) and accuracy. KNN and Randon Forest performed very well, as both the accuracy and AUC of both classifiers were above 0.85. The Randon Forest had a much better performance, with all indicators above 0.88, and the accuracy and AUC were in the range of 0.95. Figure 1 shows the ROC curve for Randon Forest in scenario 2. Following the tables of metric value ranges as a basis for evaluating a model, the model obtained with Randon Forest has a kappa index (≥ 0.8) [15], an area on the ROC curve (≥ 0.9) [16], and excellent accuracy.

Adding the feature selection step without any data balancing technique did not provide any improvement in the results (Table 3). It can be seen that the accuracy values do not vary in most of the classifiers, except for MLP and Keras.

The application of variable selection and data balancing techniques resulted in good indicators for most classifiers (Table 4), except for the MLP which had an accuracy of 0.578 and a sensitivity of 0.5. KNN was the best performing classifier in scenario 4, with an accuracy of 0.894 and an area under the ROC curve of 0.909, fitting the assessment of an excellent classifier (ROC curve ≥ 0.9) [16]. Figure 2 shows the area under the curve for KNN and based on this we can conclude that the classifier has excellent classification power. Figure 3 shows the graph used in the hyper-parameter selection step of the KNN

Table 1. Scenario 1: no variable selection and no resampling

	Accuracy	ROC curve	Precision	Recall	Kappa
SVM	0.769	NaN	0.0	0.0	0.0
Keras	0.846	NaN	0.0	0.0	0.0
MLP	0.615	NaN	0.0	0.0	0.0
KNN	1	NaN	0.0	0.0	0.0
Logistic regression	0.846	NaN	0.0	0.0	0.0
Decision tree	0.692	NaN	0.0	0.0	0.0
Randon forest	0.769	NaN	0.0	0.0	0.0

Table 2. Scenario 2. Without variable selection and with resampling

	Accuracy	ROC curve	Precision	Recall	Kappa
SVM	0.6	0.604	0.5	0.625	0.199
Keras	0.8	0.833	0.66	1	0.615
MLP	0.55	0.520	0.42	0.375	0.042
KNN	0.85	0.875	0.727	1	0.705
Logistic regression	0.65	0.645	0.555	0.625	0.285
Decision tree	0.75	0.708	0.8	0.5	0.444
Randon forest	0.95	0.958	0.888	1	0.897

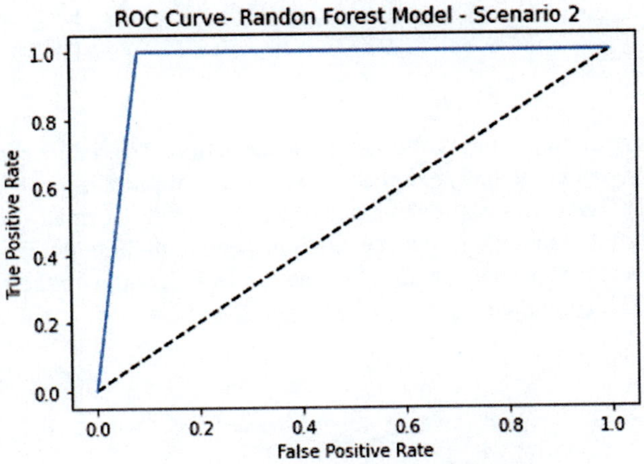

Fig. 1. ROC curve of the randon forest model—scenario 2

Table 3. Scenario 3: with selected variables and without resampling

	Accuracy	ROC curve	Precision	Recall	Kappa
SVM	0.769	NaN	0.0	0.0	0.0
Keras	0.692	NaN	0.0	0.0	0.0
MLP	1	NaN	0.0	0.0	0.0
KNN	1	NaN	0.0	0.0	0.0
Logistic regression	0.846	NaN	0.0	0.0	0.0
Decision tree	0.692	NaN	0.0	0.0	0.0
Randon forest	0.769	NaN	0.0	0.0	0.0

classifier in scenario 4, used to seek the number of neighbors that provided the highest accuracy. The value used was 18 neighbors.

Table 4. Scenario 4: with the selected variables and with resampling

	Accuracy	ROC curve	Precision	Recall	Kappa
SVM	0.789	0.784	0.75	0.75	0.568
Keras	0.736	0.755	0.636	0.875	0.486
MLP	0.578	0.568	0.5	0.5	0.136
KNN	0.894	0.909	0.8	1	0.791
Logistic regression	0.842	0.829	0.857	0.75	0.670
Decision tree	0.631	0.664	0.538	0.875	0.303
Randon forest	0.894	0.892	0.875	0.875	0.784

The comparison of scenarios shows that the selection of variables did not promote significant changes in the analyzed classifiers, since it did not improve the prediction power of the analyzed models for this database. Conversely, a small decrease in performance occurred. Among the four scenarios presented, the Randon Forest algorithm presented the best performance, without variable selection and with the balancing step using the SMOTE technique.

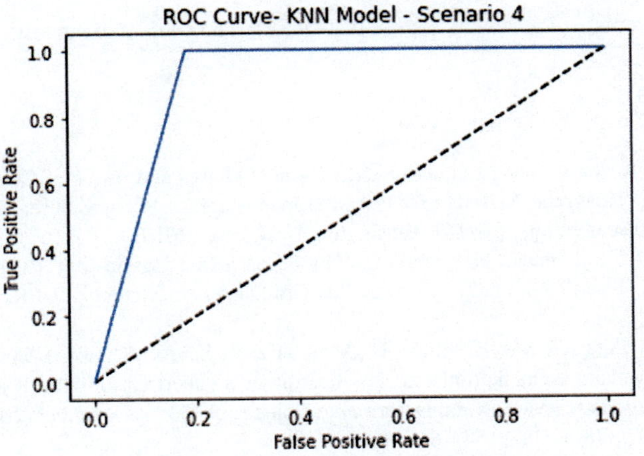

Fig. 2. ROC curve of the KNN model—scenario 4

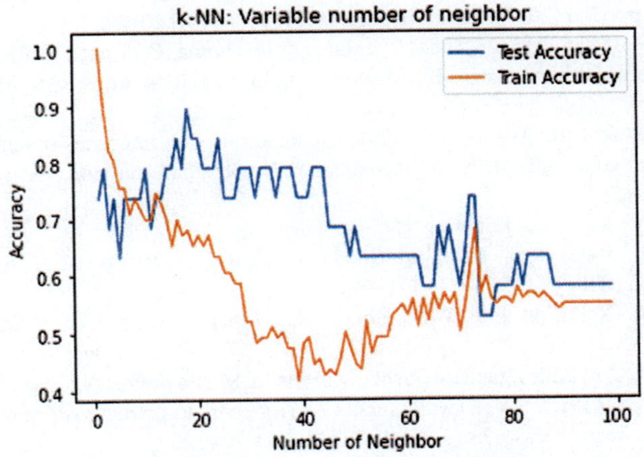

Fig. 3. Selection of the number of neighbors for KNN for scenario 4

4 Conclusion

In this study, predictors of sudden death were analyzed based on 13 variables that can be easily found from an ECG signal. The models developed using Randon Forest in scenario 2 and KNN in scenario 4 perform excellently and can be useful for assessing the risk of sudden death or be an aid to the healthcare professional. From the four studied scenarios, the best results were obtained by Randon Forest, with 0.950 accuracy and 0.958 area under the ROC curve. It was also possible to conclude how important the balancing step is, the classifier models created with the unbalanced database presented a large number of false positives that resulted in a biased and unreliable classifier.

Acknowledgment. The authors thank the support of the Brazilian Agencies FINEP, CAPES, and CNPq.

Conflict of Interest. The authors declare that they have no conflict of interest.

References

1. Rassi, A., Rezende, J.M., Luquetti, A.O.: Clinical phases and forms of Chagas disease. In: Telleria, J., Tibayrenc, M. (eds.) American Trypanosomiasis (Chagas Disease). One Hundred Years of Research, pp. 709–741. Burlington, New York (2010)
2. Pedrosa, R.C.: Evolução eletrocardiográfica desde a fase Aguda (6 a 90 anos) da Doença de Chagas. Tese de Mestrado, Curso de Pós-Graduação em Medicina, UFRJ, Rio de Janeiro (1987)
3. Górriz, M., Aparicio, A., Raventós, B., Vilaplana, V., Sayrol, E., López-Codina, D.: Leishmaniasis parasite segmentation and classification using deep learning. In: Perales, F., Kittler, J. (eds.) Articulated Motion and Deformable Objects, pp. 53–62. Springer (2018). https://doi.org/10.1007/978-3-319-94544-66
4. Peñas, K.E.D., Rivera, P.T., Naval, P.C.: Malaria parasite detection and species identification on thin blood smears using a convolutional neural network. In: IEEE/ACM International Conference on Connected Health: Applications, Systems and Engineering Technologies (CHASE), pp. 1–6 (2017). https://doi.org/10.1109/CHASE.2017.51
5. Poostchi, M., Silamut, K., Maude, R.J., Jaeger, S., Thoma, G.: Image analysis and machine learning for detecting malaria. Transl. Res. **194**, 36–55 (2018). https://doi.org/10.1016/j.trsl.2017.12.004
6. Soberanis-Mukul, R.: Detección de Trypanosoma cruzi en imágenes obtenidas a partir de muestras sanguíneas (Tesis de pregrado inédita). Universidad Autónoma de Yucatán, Mérida, Yucatán (2012)
7. Vapnik, V.N.: Statistical Learning Theory. Wiley, New York (1998)
8. Rassi, A., Jr., Rassi, S.G., Rassi, A.: Sudden death in Chagas' disease. Arq. Bras. Cardiol. **76**(1), 75–85 (2001)
9. Rassi, A., Jr., Rassi, A., Lite, W.C.: Chagas' heart disease. Clin. Cardiol. **23**(12), 883–889 (2000)
10. De Souza, A.C.J., Salles, G., Hasslocher-Moreno, A.M., de Sousa, A.S., do Brasil, P.E.A.A., Saraiva, R.M., Xavier, S.S.: Development of a risk score to predict sudden death in patients with Chagas heart disease. Int. J. Cardiol. **187**, 700–704 (2015)
11. Alberto, A.C., Pedrosa, R.C., Zarzoso, V., Nadal, J.: Association between circadian holder ECG changes and sudden cardiac death in patients with Chagas disease. Physiol. Meas. **41**(2), 025006 (2020). https://doi.org/10.1088/1361-6579/ab6ebc
12. Dal Pozzolo, A., Caelen, O., Johnson, R.A., Bontempi, G.: Calibrating probability with undersampling for unbalanced classification. In: Symposium on Computational Intelligence and Data Mining (CIDM). IEEE (2015)
13. Andre, A.B., Beltrame, E., Wainer, J.: A combination of support vector machine and k-nearest neighbors for machine fault detection. Appl. Artif. Intell. **27**(1), 36–49 (2013)
14. Souza, F.: Análise de desempenho da rede neural artificial do tipo multilayer perceptron na era multicore. Natal, RN (2012)
15. Araújo, F.H.D., Carneiro, A.C., Silva, R.R.V., Medeiros, F.N.S., Ushizima, D.M.: Redes Neurais Convolucionais com Tensorflow: Teoria e Prática. Sociedade Brasileira de Computação. III Escola Regional de Informática do Piauí. Livro Anais-Artigos e Minicursos, v. 1, pp. 382–406 (2017)
16. Matos, P.F.E.A.: Relatório técnico "Metricas de Avaliação. Tech. Rep., Universidade Federal de São Carlos (UFScar) (2009)
17. Metz, C.E.: ROC methodology in radiologic imaging. Investig. Radiol. **21**, 720–733 (1986)

Non-destructive Analysis of Spheroids Using Artificial Intelligence: Preliminary Study of Convolutional Neural Networks Performance

Danilo França[1], Andrea Dorion Rodas[1], Juliana Daguano[1,2], and Anderson Gabriel Santiago[1(✉)]

[1] Center for Engineering, Modelling and Applied Social Sciences, Federal University of ABC, São Bernardo do Campo, Brazil
gabriel.santiago@ufabc.edu.br

[2] Renato Archer Information Technology Center–CTI, Campinas, Brazil
juliana.daguano@cti.gov.br

Abstract. With an increasing research of new 3D cell cultures, specially spheroids, Convolutional Neural Networks (CNNs) have proven their potential for different classification tasks in the analysis of these biological images, and choosing the most suitable CNN for these tasks is a challenge due to a variety of algorithms specially concerning features extraction. This paper presents a non-destructive spheroid cell classification using CNN's and three different image datasets of fibroblast NIH-3T3 spheroids acquired in different culture conditions, each consisting of approximately 300 cell samples. The first two sets possess four possible cellular structures: living cells inside spheroids; living cells outside spheroids; dead cells; and background, while the third consists only of living cells inside the spheroid and background. Each dataset was sampled with a proportion of 70–20–10% for training-testing-validation. In order to improve the classification task and avoid overfitting, i.e., the lack of generalization of the network, several strategies were applied, such adding blurred images to the training dataset. The CNN architectures chosen were AlexNet, GoogleNet, MobileNetV3 Large, ResNeXt, SqueezeNet, Wide ResNet-50–2 and VGG16 with Batch Normalization, each presenting a different technology for the feature extraction layer and the classification layer as well. ResNeXt presented the highest F1-score with mean value of 0.9915 and was chosen as the best CNN for this application.

Keywords: Cellular spheroids · Nih-3t3 fibroblast · Non-destructive · Cnn architectures · Neural network

J. L. B. Marques et al. (Eds.): CLAIB 2022/CBEB 2022, IFMBE Proceedings 98, pp. 367–377, 2024.
https://doi.org/10.1007/978-3-031-49401-7_37

1 Introduction

The study of drugs in 3D tumors cultivated using spheroids technology have increased in Biomaterials and Regenerative Medicine research fields [20], but in order to rely on those results, spheroids must have their characteristics standardized according to their size and shape, since the variation in such parameters are an evidence of drug effect [1].

Aiming Regenerative Medicine, cellular spheroid structures must have their efficiency analyzed. For example, an adequate osteogenic profile must be presented for bone replacements [8], and the caracterization of spheroid cells is performed by employing destructive techniques. For this reason, new non-destructive methodologies such as impedance cytometry [5], gradient light interference microscopy (GLIM) [4] and micro-scaled cell sheet [12] have been developed. All these techniques are based on microscopy techniques, thus, resulting in high-quality images that can be used to access information regarding the spheroid quality, such as its size, shape, relation between living cells and dead cells.

With the advance in the Artificial Intelligence algorithms and in the context of Machine Learning, Artificial Neural Networks (ANN's) are described as computational processing systems which encompass a significant amount of interconnected computational nodes, also called neurons. Each of these nodes presents its own input and output, performs an operation over the input data [16] and establishes a communication with other nodes in the environment [6]. Convolutional Neural Network (CNN) appears as a specialized class of ANN applied to image pattern recognition [16], in a way that it has surpassed human capacity not only in terms of speed of the analysis but also of efficiency, finding remarkable applications in the analysis of biological images [15].

This paper presents an application of CNN to the classification of different structures that may be present in spheroid cultures to characterize them in a non-destructive way. Seven different CNN architectures were tested, each one chosen due to its unique convolutional layer algorithm, which is responsible for the image feature extraction: AlexNet [14], GoogLeNet [19], MobileNetV3 Large [9], ResNeXt [21], SqueezeNet [10], VGG 16 with Batch Normalization [18] and Wide ResNet-50–2 [22]. Since the problem consists of a multiclass classification, F1-score was the classification metric chosen and ResNeXt architecture was able to achieve a F1-score value of 0.991, followed by SqueezeNet and VGG 16 both with 0.981. The paper also presents the hyperparameters used for training-testing procedures, the computational time as well as the Confusion Matrix for ResNeXt, which depicts the relation between correct and incorrect classification for each considered class during the test procedure.

2 Methodology

2.1 Spheroid Images

To prepare the spheroids to be imaged, the ideal form and characterization had to be achieved. Standard cell culture method was used to cultivate and maintain NIH-3T3 cells, a fibroblast continuous cell line obtained from mice, and were kept in specific conditions, such as a DMEM medium supplemented with antibiotic, antimycotic and fetal bovine serum and were put in a humidified incubator at 37°C with 5% of CO_2 atmosphere. The cells were then harvested and subcultured at suspension of cells, which was prepared to be seeded in the micromolds to build the spheroids cell culture plate and placed back in the incubator under the same conditions of temperature and humidity. Subsequently, micromolds structures were prepared based by non-adhesive hydrogel technique using microwell arrays, developed at Renato Archer Information Technology Center (CTI), to accommodate cells and keep the shape of the spheroid.

Basically, the plates were covered with agarose and a specific mold stamped the wells to form a spheroid cradle. The NIH-3T3 cells were then poured in the microwells, producing the spheroids over time. The fibroblasts NIH-3T3 spheroids were cultivated in the micromolds previously prepared and captured spheroids were characterized by the nondestructive methodology which is the image analysis using Convolutional Neural Networks.

2.2 Dataset Preparation

Three different image datasets of fibroblast NIH-3T3 spheroids acquired in different culture conditions, each consisting of 300 cell samples. Figure 1a, b and c present three different images used in the training/testing procedures.

The spheroids in dataset A, presented in Fig. 1a, four possible cellular structures were defined to encompass as much as possible from the image:

- Background: well plate area, no cell structure is identified.
- Dead cells: single-cell, characterized by high brightness and very light central formation, identifies as the cell nucleus.
- Living cell inside spheroids: the largest cell aggregate is the spheroid; it comprises cells with lower brightness values and a very well defined membrane with a dot placed at its center, characterizing its nucleus.
- Living cell outside spheroids: cells characterized by their lower brightness values, very well-defined boundaries, the cell membrane, and easy identification of the nucleus, well-defined boundaries, which looks like a dot in the center of the cell; those cells are found in the aggregates.

Dataset B presented in Fig. 1b, consists only of living cells inside the spheroid and background, while dataset C, Fig. 1c, however, consists of blurred images and was used only during the training procedure in order to improve the classification task and avoid overfitting, i.e., the lack of generalization of the network. Considering dataset A and B, the number of images for training, testing and validation followed the proportion 70–20–10%. The amount of samples from dataset C was considered as an hyperparameter and discussed in the following section. Data augmentation was used in order to artificially broad the dataset and several transformations were considered, being the color channel normalization and image rotation those with greater impact over the training procedure [3].

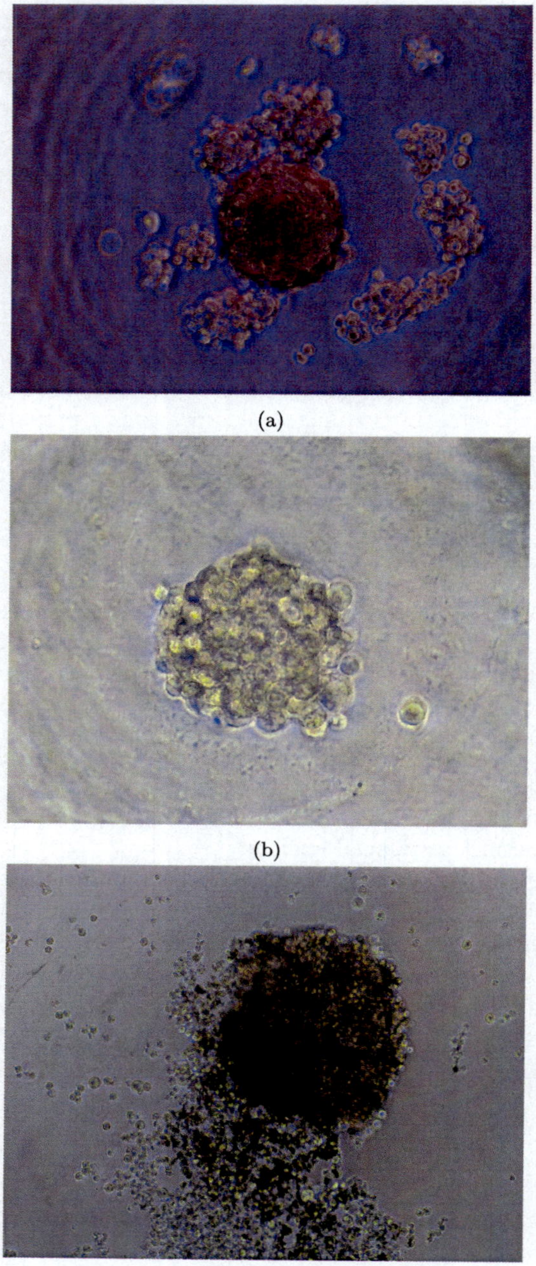

Fig. 1. Spheroid image samples used in the training procedure. Each image presents a spheroid sample of a different dataset: a) dataset composed by all four classes; b) dataset composed only by background and living cells inside the spheroid; c) blurred cell images used for training procedure to reduce overfitting.

2.3 CNN Implementation

For the assignment of spheroid classification, seven CNN architectures were evaluated and the number of trainable parameters for each one is given below:

- AlexNet: 61,100,840
- GoogLeNet: 13,004,888
- MobileNet V3 Large: 3,504,872
- ResNeXt: 25,028,904
- SqueezeNet: 1,248,424
- Wide ResNet-50–2: 68,883,240
- VGG 16 BN: 138,365,992

No transfer learning was applied and only the output of last classification layer was set to match the number of classes considered in this paper. All CNN architectures were implemented using Python 3.7 Anaconda distribution[1] and the PyTorch[2] Application Programming Interface (API). Concerning the hardware, an Intel i7 Core, 16Gb of RAM and a Tesla K80 Nvidia Graphics Processing Unit (GPU) with 24Gb of RAM running Linux Ubuntu 20.04 was used.

2.4 Training and Testing Process

Each class described in the Dataset Preparation was labeled from 0 to 3.To follow up the training process, Adam [13] optimizer was used by all the CNN's, in addition to the Cross Entropy cost function, defined by Eq. 1, where y_i is the predicted value for the i-eth class and $\bar{y}_i$ its true value.

$$CE = \sum_{i=0}^{N_{class}} [y_i \log(\bar{y}_i) + (y_i - 1) \log(\bar{y}_i - 1)] \quad (1)$$

In order to achieve a broad variety of results for the image classification, each image was resized to 256 × 256 and four different hyperparameters were chosen to have their values modified within a previously established range:

- Epochs: 55–70;
- Learning Rate: 1×1.0^{-6}–1×1.0^{-3};
- Rotation Angle: 10°–20°;
- Batch Size: 10—25;
- Blurred images percentage: 5–20%;

The combination of these hyperparameters were applied randomly until and an average F1-score above 0.9 was obtained for each architecture.

1 https://www.anaconda.com.
2 https://pytorch.org/.

2.5 Results Evaluation

Due to the fact that the training dataset comprises an unbalanced number of classes and that these classes form a multi-class classification system, F1 score (Eq. 2) was the measure chosen as a performance metric. It consists of the harmonic mean of *precision* (Eq. 3), which expresses the model's ability of correctly classifying a sample as positive and *recall* (Eq. 4), which expresses the model's ability of detecting positive samples among all the ones that were correctly classified (true positives and false negatives). Thus, the reliability of the result on the performance metric chosen is directly to the F1 score: the CNN architecture with a score closest to one is most likely to be the most adequate for the task of spheroid-cell classification.

$$F1 = 2 \times \frac{Precision \times Recall}{Precision + Recall} \tag{2}$$

$$Precision = \frac{TP}{TP + FP} \tag{3}$$

$$Recall = \frac{TP}{TP + FN} \tag{4}$$

at which TP stand for true positive, FP for false positive and FN for false negative.

3 Results

The highest F1-score achieved for each of the CNN architectures considered in this paper are displayed on Table 1, which encompasses both the individual F1 values for each class and their average, as well as the computational time (CT) evaluated for training procedure for each CNN considered in this work, while Table 2 displays the hyperparameters used during the training phase in order to achieve such results, in which LR, θ and BS represent the learning rate, rotation angle and batch size, respectively.

Table 1. F1-score achieved for each class and CNN architecture

Architectures	0	1	2	3	Average	CT (min)
AlexNet	1.000	0.965	0.930	0.941	0.951	3.890
GoogleNet	0.895	0.870	0.919	0.955	0.910	3.590
MobileNet V3 Large	1.000	0.965	0.965	0.985	0.979	5.640
ResNeXt	1.000	0.983	0.983	1.000	0.991	27.630
SqueezeNet	1.000	0.983	0.977	0.985	0.986	2.300
Wide ResNet-50–2	1.000	0.965	0.954	0.970	0.972	9.080
VGG 16	1.000	0.983	0.970	0.985	0.986	19.140

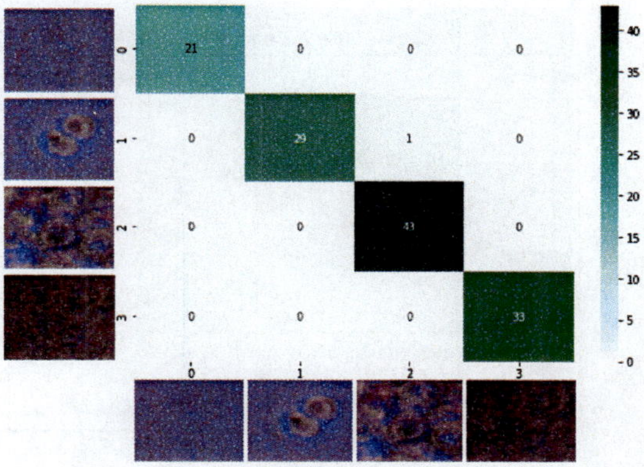

(a)

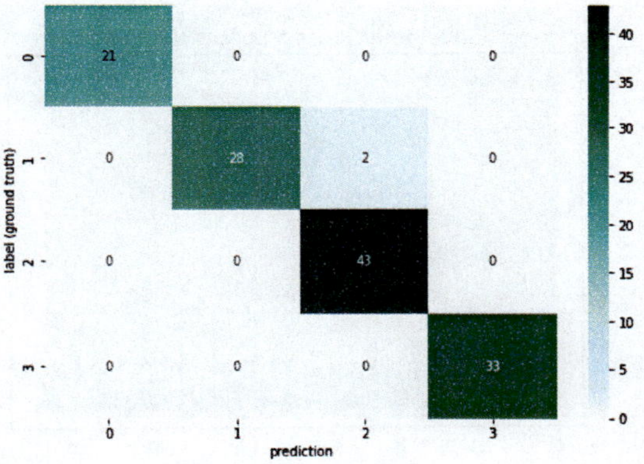

(b)

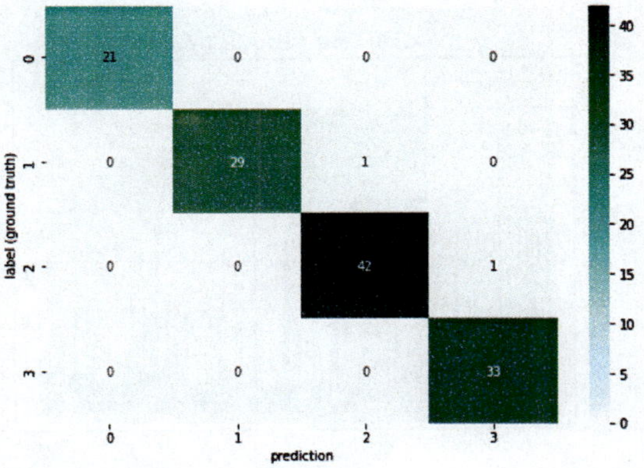

(c)

Fig. 2. Confusion matrix for the top three highest F1 average values results for test dataset: 0-Background; 1-Dead Cells; 2-Living Cells Outside Spheroid; 3-Living Cells Inside Spheroid. a) ResNeXt; b) SqueezeNet c)VGG 16

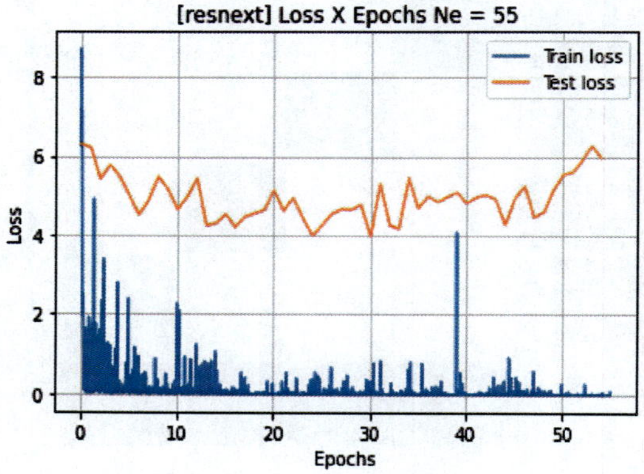

(a)

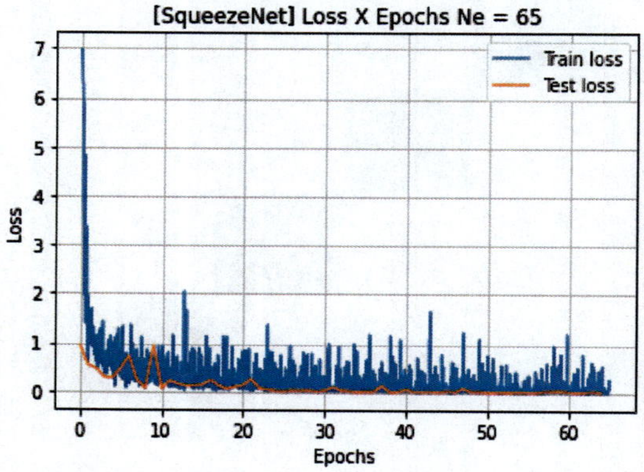

(b)

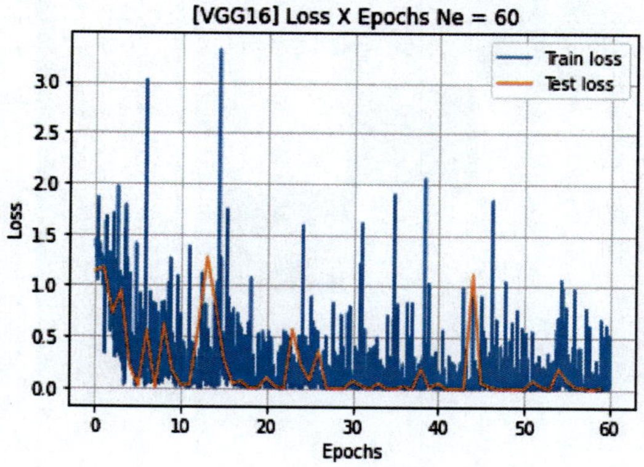

(c)

Fig. 3. Loss function for the top three highest F1 average values considering the test dataset: a) ResNeXt; b) SqueezeNet c)VGG 16

Since ResNeXt, SqueezeNet and VGG 16 present the highest F1 average value out of the 7 CNN architectures, Fig. 2 displays their confusion matrices, which summarizes the correct and incorrect predictions of the results in the model during the testing procedure. The vertical axis represents the samples classified by the expert, while the horizontal axis the ResNeXt classification.

Table 2. Optimal hyperparameters set for each CNN architecture

Architectures	Epochs	LR	θ	BS	% of blurred images
AlexNet	55	10^{-4}	20°	10	20
GoogLeNet	60	10^{-3}	20°	10	20
MobileNet V3 Large	55	10^{-4}	20°	10	20
ResNeXt	55	10^{-4}	20°	10	20
SqueezeNet	65	10^{-4}	20°	10	20
Wide ResNet-50-2	70	10^{-4}	20°	15	20
VGG 16	60	10^{-4}	15°	10	20

4 Discussion

CNNs have been used in medicine to help in image diagnostics as X-rays for COVID-19 infected patients [2], skin cancer [7], early gastric cancer [11] among other specialities. The first investigation of a more suitable CNN for labeled cell images was published by Santiago et al. [17] but with using four different architectures and more suitable hyperparameters optimization was needed.

Table 1 provides an overview of the average F1 values, measure chosen as a performance metric to evaluate the models. ResNeXt presented to be the most adequate CNN to identify the structures of the spheroid considering the range of hyperparameters chosen and the dataset used, with the highest F1 average value of 0.991, followed closely by SqueezeNet and VGG 16, both with an average of 0.986.

Ordinarily, the identification and classification of single-cells in such images are performed by a qualified professional, a task which demands a significant amount of time and expertise. For instance, the classification of some of the images in the dataset used in this paper took about 40 to 120 min. Besides, due to the manual nature of this task, it is common to observe a high variability regarding the classification results, which stems from the bias of the professional performing the task. Therefore, automating the process of spheroid classification or single-cells in general would not only increase the precision of the results but also reduce the time spent on this task, providing as an important tool to assist professionals of the field.

Regarding the computational time for the training-testing procedure, it is important to mention that although ResNeXt presented the highest F1 average value among the considered CNN's, it also had the highest computational time, 27.63 min, while SqueezeNet, which presented a F1 average value of 0.986, was able to perform the training-testing procedures with a considerably smaller computational time of 2.30 min. Considering this, if the computational time is an aggravating factor due to any computational resource restriction, SqueezeNet is more suitable CNN for image classification.

According to Fig. 2a, ResNeXt successfully identified 100% of the background and 100% of the living cells outside and inside spheroid, besides being able to identify 96.6% of the dead cells. SqueezeNet and VGG16 in Fig. 2b and c also presented similar performance and according Fig. 3b and 3c, smaller values for the loss function in the test dataset indicating that those architectures produces greater generalization than ResNeXt. Therefore, the CNN functions as a reliable tool to operate the task of identifying different structures in cells, in a process that is both practical and non-destructive.

The use of CNN to identify live and dead cells in an image is important in in vitro experiments where the same spheroid is followed during a defined time. Those spheroids can be used in drug test where they can die or stimulated to grow in special treatment, because of this, it is important the CNN identifies 99% the cells.

5 Conclusion

This paper delves into the performance of seven different CNN's in cell identification of structures in fibroblast NIH-3T3 spheroids. Among all the CNN's, ResNeXt presented the highest classification metrics. Using CNN's for such tasks allows a faster and more precise classification of cells.

Acknowledgements. The authors would like to acknowledge the Federal University of ABC, researcher Daniel Perez from IPEN and Renato Archer Information Technology Center.

References

1. Astrauskas, R. et al.: Modeling the uptake of fluorescent moleculesinto 3D cellular spheroids. In: Nonlinear Analysis: Modelling and Control, vol. 24.5, pp. 838–852. (2019)
2. Bukhari, S.U.K. et al.: The diagnostic evaluation of convolutional neural network (CNN) for the assessment of chest X-ray of patients infected with COVID-19. In: MedRxiv (2020)
3. Csink, L. et al.: Color normalization and object localization (1998)
4. Fanous, M.J. et al.: Effects of substrate patterning on cellular spheroid growth and dynamics measured by gradient light interference microscopy (GLIM). J. Biophotonics **12** 12 (2019) e201900178
5. Gong, L. et al.: Direct and label-free cell status monitoring of spheroids and microcarriers using microfluidic impedance cytometry. Small **17.21**, 2007500 (2021)

6. Grossi, E., Buscema, M.: Introduction to artificial neural networks. Europ. J. Gastroenterol. Hepatol. **19**, 1046–54 (2008). https://doi.org/10.1097/MEG.0b013e3282f198a0
7. Haenssle, H.A. et al.: Man against machine: diagnostic performance of a deep learning convolutional neural network for dermoscopic melanoma recognition in comparison to 58 dermatologists. Annals of Oncol. **29.8**, 1836–1842 (2018)
8. Heo, D.N., Hospodiuk, M., Ozbolat, I.T.: Synergistic interplay between human MSCs and HUVECs in 3D spheroids laden in collagen/fibrin hydrogels for bone tissue engineering. Acta Biomater. **95**, 348–356 (2019)
9. Howard, A. et al.: Searching for MobileNetV3. In: Proceedings of the IEEE/CVF International Conference on Computer Vision (ICCV) (2019)
10. Iandola, F.N. et al. S:queezeNet: AlexNet-level accuracy with 50x fewer parameters and lt;0.5MB model size (2016). https://doi.org/10.48550/ARXIV.1602.07360. https://arxiv.org/abs/1602.07360
11. Ikenoyama, Y. et al.: Detecting early gastric cancer: comparison between the diagnostic ability of convolutional neural networks and endoscopists. Digest. Endoscopy **33.1**, 141–150 (2021)
12. Kim, E.M. et al.: Fabrication of spheroids with uniform size by self-assembly of a micro-scaled cell sheet (μCS): the effect of cell contraction on spheroid formation. ACS Appl Mater Interfaces **11.3**, 2802–2813 (2018)
13. Kingma, D., Ba, J.: Adam: a method for stochastic optimization. In: International Conference on Learning Representations (2014)
14. Krizhevsky, A., Sutskever, I., Hinton, G.E.: ImageNet classification with deep convolutional neural networks. In: Pereira, F. et al. (ed.) Advances in Neural Information Processing Systems, vol. 25, Curran Associates, Inc. (2012). https://proceedings.neurips.cc/paper/2012/file/c399862d3b9d6b76c8436e924a68c45b-Paper.pdf
15. Kusumoto, D., Yuasa, S.: The application of convolutional neural network to stem cell biology. In: Inflammation and Regeneration, vol. 39 (2019)
16. O'Shea, K., Nash, R.: An introduction to convolutional neural networks. In: CoRR abs/1511.08458 (2015). http://dblp.uni-trier.de/db/journals/corr/corr1511.html#OSheaN15
17. Santiago, A.G. et al.: Non-destructive fibroblast NIH-3T3 spheroid classification using convolutional neural network. In: Medical Imaging 2021: Biomedical Applications in Molecular, Structural, and Functional Imaging, vol. 11600, pp. 116000N. International Society for Optics and Photonics (2021)
18. Simonyan, K., Zisserman, A.: Very deep convolutional networks for large-scale image recognition. In: International Conference on Learning Representations (2015)
19. Szegedy, C. et al.: Going deeper with convolutions. In: 2015 IEEE Conference on Computer Vision and Pattern Recognition (CVPR). 2015, pp. 1–9. https://doi.org/10.1109/CVPR.2015.7298594
20. Wong, C.-W. et al.: Biomaterial substrate-derived compact cellular spheroids mimicking the behavior of pancreatic cancer and microenvironment. Biomaterials **213**, 119202 (2019)
21. Xie, S. et al.: Aggregated residual transformations for deep neural networks. In: Proceedings of the IEEE Conference on Computer Vision and Pattern Recognition (CVPR) (2017)
22. Zagoruyko, S., Komodakis, N.: Wide residual networks. ArXiv abs/1605.07146 (2016)

Biomaterials, Tissue Engineering and Artificial Organs

Biological Evaluation of in vitro Cytotoxicity of Extracts of *Lafoensia glyptocarpa* Koehne (Lythraceae)

J. T. Faria[1(✉)], F. N. Ambrosio[2], R. A. Lombello[2], and C. B. Lombello[1]

[1] UFABC/CECS, São Bernardo Do Campo, Brazil
juliana.toledo@outlook.com.br
[2] UFABC/CCNH, São Bernardo Do Campo, Brazil

Abstract. Secondary metabolites, organic compounds produced by plants, are currently studied in relation to their medicinal potential, such as antioxidant, anti-inflammatory, antibiotic and antiviral effects. Among the plants of clinical interest is *Lafoensia glyptocarpa* Koehne, an arboreal species of the Lythraceae family widely used in Brazilian urban landscaping. This plant has scientific data indicative of medicinal potential and its genus presents studies related to the therapeutic potential. Considering the importance of new studies that aim to expand the scientific knowledge of species of native flora, this work has the objective to test the safety of the use of different *L. glyptocarpa* extracts through an in vitro cytotoxicity assay in Vero cells. They were exposed to concentrations (25, 50, 100 and 1000 μg/mL) of aqueous and ethanolic extracts of *L. glyptocarpa* leaves for 24h and then a solution of MTT (3-(4,5-dimethylthiazol-2-yl)-2,5-diphenyltetrazolium bromid) was added for 2 h. The method is indicative of mitochondrial activity related to cellular viability and the absorbance is read at 570 nm. In addition, photomicrographs were observed in phase contrast microscopy for cell culture evaluation of cytotoxicity at different concentrations. Cytotoxicity assays were performed using positive (cells with DMSO 50%) and negative controls (only growth medium with cells). The results showed that the extracts of *Lafoensia glyptocarpa* (aqueous and ethanolic) do not present cytotoxicity in the range of 25 to 100 μg/mL. This work reinforces the importance of the study with medicinal plants and demonstrates a high sensitivity and reliability in in vitro cytotoxicity studies.

Keywords: Cytotoxicity · in vitro · *Lafoensia glyptocarpa* · Plant extracts

1 Introduction

Plants naturally produce components that are essential for their development, called primary metabolites, such as fatty acids, proteins and polysaccharides. Some species also produce secondary metabolites, which are important for their adaptation to the environment, defense and competition, such as phenolic compounds and terpenoids [1]. These secondary substances are widely studied and demonstrate medicinal potential

J. L. B. Marques et al. (Eds.): CLAIB 2022/CBEB 2022, IFMBE Proceedings 98, pp. 381–387, 2024.
https://doi.org/10.1007/978-3-031-49401-7_38

when used in humans, such as antioxidant, anti-inflammatory, antibiotic, antiviral effects, among other actions on the nervous system. In addition, the commercialization of isolated compounds or crude extracts of medicinal plants can be a low-cost and high-quality alternative, as research on herbal medicines provides quantitative results and easy test reproducibility [2].

However, before using herbal medicine, it is necessary to know its properties and test the safety of these substances. One of the tests that should be performed initially is the in vitro cytotoxicity assay [3]. This method consists of incubating cultured cells in contact with a device and/or extracts from a device, directly or by diffusion, to determine the biological response of mammalian cells in vitro using the appropriate biological parameters.

The plant of interest in this work, *Lafoensia glyptocarpa* Koehne (Fig. 1), popularly known in Brazil as mirindiba rosa, is a native tree species of Atlantic Forest, widely used in urban landscaping, that belongs to the Lythraceae J. St.-Hil. family [4]. There are still no publications related to its medicinal potential and toxicity, there are only records of flavonoids and saponins isolated from its leaves [5, 6]. Despite this, there are several data indicating the anti-inflammatory, antioxidant, antimicrobial and antiviral properties of species of the same genus, such as *Lafoensia pacari* A. St.-Hil. [7–10] and *Lafoensia replicata* Pohl [11].

The antiviral, antimicrobial and regenerative potential of species of the genus *Lafoensia* direct the objectives of this project, which aims to apply in vitro cytotoxicity tests, evaluating the effects of different *L. glyptocarpa* extracts (aqueous and ethanolic) [12] on Vero cells, seeking to demonstrate the biological safety of these extracts.

Fig. 1. *Lafoensia glyptocarpa* Koehne tree (R.A. Lombello, 2020).

In this work it was expected that the aqueous and ethanolic extracts would not be cytotoxic. The comparison of different extraction methods on in vitro cytotoxicity test results allows to test a greater variety of compounds and increases knowledge of native flora species of medicinal interest.

2 Materials and Methods

2.1 Plant Extracts

The extracts were prepared from plant material of *L. glyptocarpa* collected in Campinas - SP (RAL85, 22°52′32.8"S/47°03′36.4"W, SisGen A8A3C26). This material was dried in an oven at 40 °C. The leaves were crushed, and the crude extracts were prepared as previously described in the literature and briefly mentioned below [13–15].

The aqueous extract was obtained through a decoction, adapting the methodology of Fonsêca [13], where the crushed leaves were kept in a beaker with distilled water, heated at 90 °C, for 30 min, and then the beaker was transferred to a bowl of ice water for another 10 min. After this procedure, simple filtration of the extract was performed on filter paper, and the solvent was removed by lyophilization [14].

For the ethanol extract, the crushed leaves were kept in capped flasks, with the solvent, for one week, with periodic manual agitation. Then, the solution was filtered through filter paper and kept in uncapped flasks in a hood, at room temperature, for evaporation [15].

Finally, the powdered extracts were weighed to prepare use concentrations of 1000, 100, 50 and 25 μg/mL (diluted in distilled water). These extracts were filtered through a 0.22 μm syringe filter and stored with appropriate identification in a −20 °C freezer.

2.2 Cell Culture

For this assay, Vero cells, fibroblastic type, originating from the kidney epithelium of the African green monkey (*Cercopithecus aethiops*) [16] were used, respectively cultured in HAM F10 medium, and supplemented with 10% fetal bovine serum (FBS). Cells were maintained in culture bottles at 37 °C and 5% CO_2.

The culture medium was changed every two to three days, subcultures were performed when necessary, i.e. part of the cells from the bottles were kept in new ones, so that the number of cells in each bottle reached a confluence of approximately 80% (there were no more cells than the medium could supplement).

For each assay performed, cells from the culture bottles were suspended by trypsinization (trypsin/EDTA solution – 0.25%, Cultilab) in order to be able to count the cell density of each assay.

2.3 Cytotoxicity Assays

Vero cell cultures were used in direct contact with the extracts (aqueous and ethanolic) and were evaluated by light microscopy (qualitative) and quantitative method using the 3-(4,5-dimethylthiazol-2-yl)-2,5-diphenyltetrazolium bromide (MTT) assay [17].

For the cytotoxicity analysis, the cells were seeded in 96-well plates (approximately 1500 cells/well), and cultivated under ideal conditions for 24 h, being then exposed to the different concentrations of the extracts, 1000 μg/mL, 100 μg/mL, 50 μg/mL and 25 μg/mL, for another 24 h.

The cell culture light microscopy evaluation was performed with inverted light microscopy and phase contrast. Then, the assessment of cell viability by MTT, with the removal of the culture medium, and replacement with 0.5 mg/mL of MTT solution. The cells were incubated for a period of 2 h, followed by removal of the MTT solution and inoculation of DMSO (dimethylsulfoxide). The determination of cell viability was obtained with absorbance at 570 nm and presented in percent normalized with negative control (considered 100% viability). As a negative, non-cytotoxic control, cells were cultivated without treatment. As a positive, cytotoxic control, a 50% DMSO solution (diluted in medium) was used.

The results were analyzed by description of cell culture light microscopy observation, with the presentation of the respective micrographs, and in the form of graphs with quantitative data of cell viability, using statistical analysis by One-way ANOVA (Fisher) and Tukey's Test ($p < 0.05$), for comparison between the groups studied, through the Jamovi program. To be considered non-cytotoxic, extracts must maintain a cell viability greater than 70% [3].

3 Results and Discussion

The aqueous and ethanolic extracts were not considered cytotoxic, as they showed cell viability greater than 70%, except at a concentration of 1000 μg/mL (Fig. 2). It was possible to observe that there was no significant difference in cell viability between the concentrations of 100, 50, 25 μg/mL compared to the negative control (group b), while there was a significant difference when compared to the positive control (group c) and the concentration of 1000 μg/mL (Group a). These comparisons were analyzed using Fisher's test (One-Way ANOVA) and Tukey's post hoc test, with $p < 0.001$.

In the qualitative assay it was also possible to observe the cell culture characteristics after 24 h of exposure (Figs. 3 and 4). For the aqueous and ethanolic extracts, the negative control cell culture observation is similar to those at concentrations of 100, 50 and 25 μg/mL, while in the concentration of 1000 μg/mL and in the positive control it is possible to observe mild damage to the monolayer and the appearance of cell agglomerations, which makes it considered slightly cytotoxic. Although cells present at 1000 μg/mL and positive control shows a slight modification in culture pattern these experimental groups resulted in a significant reduction in cell viability, indicating that the observed cells are not viable.

There are no published articles about cytotoxicity tests of *Lafoensia glyptocarpa* extracts, but there are several studies about *Lafoensia pacari.* Assays performed on mouse bone marrow micronucleus showed a slight cytotoxic effect at all doses of *L. pacari* stem bark ethanolic extract at 24 h and the authors suggested that this could be observed by the presence of saponins in the extract [18]. Podolak reports that saponins have cytotoxic activity, stimulating the apoptotic process and causing cell cycle arrest and disintegration of the cytoskeleton in tumor cells [18].

The work by Da Silva Marcondes et al. [19], with leukemic cell lines (U-937), showed results similar to ours, where solutions of the methanolic extract of stem bark of

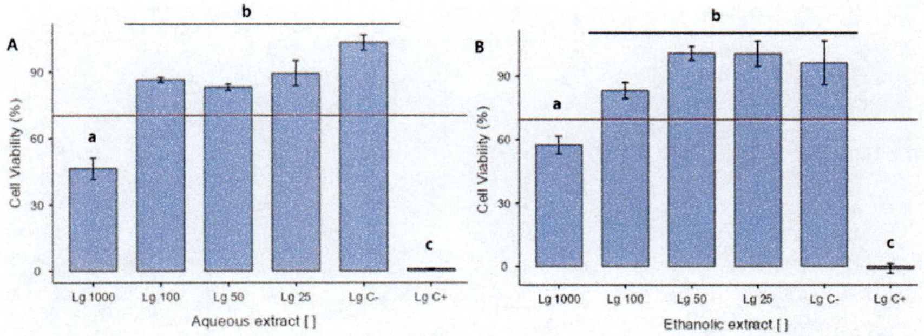

Fig. 2. Cytotoxicity test of: A - aqueous extract of *L. glyptocarpa* and B - ethanolic extract of *L. glyptocarpa*; the statistically different groups. Lg 1000: 1000 μg/mL, Lg 100: 100 μg/mL, Lg 50: 50 μg/mL, Lg 25: 25 μg/mL, Lg C-: negative control, Lg C+: positive control. Red line represents 70% viability.

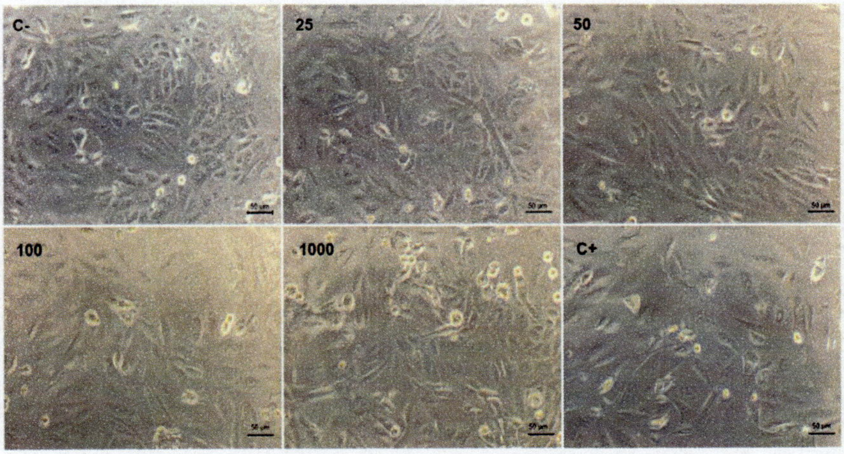

Fig. 3. Photomicrographs of Vero cell monolayers after 24 h exposure to aqueous extract of *L. glyptocarpa* at 1000 μg/mL (1000), 100 μg/mL (100), 50 μg/mL (50) and 25 μg/mL (25), in addition to negative (C-) and positive (C+) controls. Bar 50 μm.

L. pacari with concentrations lower than 100 μg/mL were not cytotoxic, but in cultures where concentration was 1000 μg/mL mild cytotoxicity was observed. Otherwise, a study using stem barks of *L. pacari* in macrophage lineages showed different toxicities in fractions of the ethanolic extract and maximum cytotoxicity of 100 μg/mL in the crude ethanolic extract [20]. These publications with species of the same genus corroborate the results obtained in this work and the non-cytotoxicity of *L. glyptocarpa* extracts lower than 100 μg/mL demonstrates the importance of further studies on this plant for medicinal purposes.

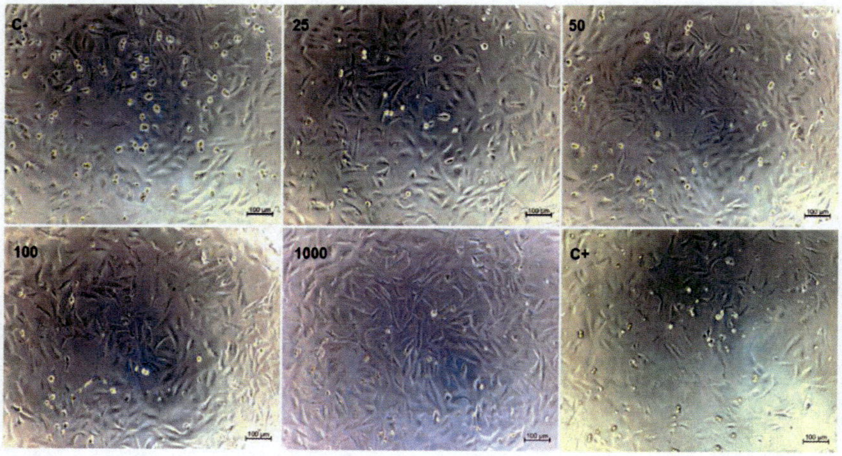

Fig. 4. Photomicrographs of Vero cell monolayers after 24 h exposure to ethanolic extract of *L. glyptocarpa* at 1000 μg/mL (1000), 100 μg/mL (100), 50 μg/mL (50) and 25 μg/mL (25), in addition to negative (C-) and positive (C+) controls. Bar 100 μm.

4 Conclusions

In this study, it was possible to conclude that the extracts of *Lafoensia glyptocarpa* (aqueous and ethanolic) do not present cytotoxicity in the range of 25 to 100 μg/mL. The results presented in this work reinforce the importance of basic research related to the study of medicinal plants, which are taxonomic or biological assessment. Furthermore, it was possible to compare them with other studies that use in vivo and in vitro evaluations and demonstrate a high sensitivity and reliability of in vitro cytotoxicity studies.

Acknowledgments. We thank the Universidade Federal do ABC and CAPES for the structure and funding; to the SBC Multiuser Center and to the students and professors of the L603 laboratory (SA) for the equipment used and for the enriching exchanges of knowledge.

Conflict of Interest The authors declare that they have no conflict of interest.

References

1. Vizzoto, M., et al.: Metabólitos secundários encontrados em plantas e sua importância. Documento: Embrapa Clima Temperado **316**, 7–15 (2010)
2. Kaziyama, V.M., et al.: Atividade antiviral de extratos de plantas medicinais disponíveis comercialmente frente aos herpesvírus suíno e bovino. Rev Bras Pl Med **14**, 522–528 (2012)
3. International Organization for Standardization ISO 10993-5: Biological evaluation of medical devices—Part 5: Tests for in vitro cytotoxicity, p. 24 (2009)
4. Carvalho, P.E.R.: Espécies Arbóreas Brasileiras: Mirindiba-Rosa: *Lafoensia glyptocarpa.* Embrapa **5** (2014)
5. De Carvalho, G.J.A., De Carvalho, M.G., Braz-Filho, R.: A triterpenoid saponin isolated from *Lafoensia glyptocarpa.* Phytochem **52**, 1617–1619 (1999)

6. Santos, D.Y.A.C., Salatino, L.F.M.: Salatino A Foliar flavonoids of *Lafoensia* (Lythraceae). Biochem. Syst. Ecol. **28**(5), 487–488 (2000)
7. Ávila, R.I.D., Ferreira, C.C., Alvarenga, C.B.M., et al.: Toxicological evaluation of *Lafoensia pacari* A. St.-Hil. (Lythraceae) stem bark extract: acute and subchronic studies in mice. Braz. J. Pharm. Sci. **55**, e17289 (2019)
8. Campos, J.S., Frasson, A.P.Z.: Avaliação da atividade antioxidante do extrato aquoso de *Lafoensia pacari* A. ST-HIL. em emulsão não-iônica. Rev. Ciênc. Farm. Básica Apl. **32**, 363–368 (2012)
9. Firmo, W.C.A., Miranda, M.V., Olea, R.S.G.: Caracterização do "estado da arte" de *Lafoensia pacari* A. St.-Hil. (Lythraceae). Natureza Online **14**, 12–22 (2016)
10. Silva Junior, I.F., Raimondi, M., Zacchino, et al.: Evaluation of the antifungal activity and mode of action of *Lafoensia pacari* A. St.-Hil., Lythraceae, stem-bark extracts, fractions and ellagic acid. Rev. Bras. Farmacog. **20**(3), 422–428 (2010)
11. Sobrinho, F.C.B., Almeida, A.L.S., Monteiro, J.M.: Estudo etnofarmacológico sobre *Lafoensia replicata* Pohl. no leste do Maranhão, Brasil: uma promissora espécie para bioprospecção. Desenvolv Meio Amb **39**, 207–216 (2016)
12. Stanojević L P, Stanković M Z, Cvetković D J et al. The effect of extraction techniques on yield, extraction kinetics, and antioxidant activity of aqueous-methanolic extracts from nettle (*Urtica dioica* L.) leaves. Separat Sci Technol 51(11):1817–1829 (2016)
13. Da C. Fonsêca, S.G.: Farmacotécnica de Fitoterápicos at https://farmacotecnica.ufc.br/wp-content/uploads/2019/11/farmacot-fitoterapicos.pdf
14. Suárez, A.I., Compagnone, R.S., Salazar-Bookaman, M.M., et al.: Antinociceptive and anti-inflammatory effects of *Croton malambo* bark aqueous extract. J. Ethnopharmacol. **88**(1), 11–14 (2003)
15. Milhem, M.M., Al-Hiyasat, A.S., Darmani, H.: Toxicity testing of restorative dental materials using brine shrimp larvae (*Artemia salina*). J. Appl. Oral Sci. **16**(4), 297–301 (2008)
16. Ammerman, N.C., Beier-Sexton, M., Azad, A.F.: Growth and maintenance of Vero cell lines. Curr. Protoc. Microbiol. **11**, 4EA (2008)
17. Masson, A.O.: Estudo comparativo de ensaios de citotoxicidade aplicados à biomateriais: metodologias e condições de ensaio. Dissertação de Mestrado. Curso de Engenharia Biomédica. Universidade Federal do ABC (2016)
18. Lima, D.C.S., Silva, C.R., Sampaio, B.L., et al.: Antigenotoxic, and anticytotoxic activities of an ethanolic extract of *Lafoensia pacari* (Lythraceae) stem bark in bacteria and mice. Genet. Mol. Res. **12**(3), 3887–3896 (2013)
19. Da Silva, M.D.B., et al.: Cytotoxicity and apoptogenic effects of *Lafoensia pacari*. J. Ethnopharm. **157**, 243–250 (2014)
20. Fonseca, S.A., et al.: Molecular docking studies and evaluation of the antiretroviral activity and cytotoxicity of the species *Lafoensia pacari* Saint-Hilaire. Braz. J. Biol.Biol. **82**, e256261 (2022)

Towards the Development of 3D-Printed Model of the Human Aorta: Comparison of Mechanical Properties of Biological and Polymeric Materials

M. J. S. Maizato[1(✉)], P. S. Gutierrez[2], S. Bacht[1], I. N. Cestari[1], E. Sales[1], M. Mazzetto[1], V. D. Aiello[2], and I. A. Cestari[1]

[1] Divisao de Bioengenharia, Instituto do Coracao (InCor), Hospital das Clinicas HCFMUSP, Faculdade de Medicina, Universidade de Sao Paulo, Sao Paulo, Brasil
marina.maizato@incor.usp.br

[2] Servico de Anatomia Patologica, Instituto do Coracao (InCor), Hospital das Clinicas HCFMUSP, Faculdade de Medicina, Universidade de Sao Paulo, Sao Paulo, Brasil

Abstract. The determination of mechanical properties of biological materials is a relevant and interdisciplinary topic of research. In this work we compared the stress–strain relationship of human thoracic aortic wall samples with polymeric 3D printing materials with vessel wall properties. Mechanical tests were performed on a standard test equipment (Instron 3365) in dogbone shaped samples. The physiological modulus of elasticity, the maximum modulus of elasticity, maximum stress and the deformation at the point of maximum stress were determined. The tests showed that mechanical failure consistently occurred in the middle region of the sample. Results of aortic samples obtained from two aortas (ao1 and ao2) showed a physiological elastic modulus of 0.68 ± 0.2 and 0.58 ± 0.4 MPa and maximum elastic modulus of 2.44 ± 0.3 and 2.18 ± 1.1 MPa, for ao1 and ao2, respectively. 3D printed samples of different material composition did not show a tensile stress in the physiological range with elastic modulus and maximum elastic modulus of 0.85 ± 0.04 and 0.95 ± 0.07 MPa, respectively. The results suggest the need of using a combination of materials or microstructures for modeling blood vessel wall mechanical properties.

Keywords: Biomechanical properties · Aortic wall · 3D printed materials · Physiological elastic modulus · Maximum elastic modulus

1 Introduction

The use of 3D printing for constructing anatomic models is being increasingly used in medicine with applications in pre-surgical planning, surgical training, and the design of medical devices [1–3]. In line with this, studies of the mechanical characteristics of biological materials such as trachea, cartilage, skin, tendons and vessels show a renewed interest.

Adding to that, recent progress in tissue engineering techniques have expanded the promise of obtaining new materials for organ replacement contributing to the increased need to better understand the mechanical behavior of biological tissues.

J. L. B. Marques et al. (Eds.): CLAIB 2022/CBEB 2022, IFMBE Proceedings 98, pp. 388–395, 2024.
https://doi.org/10.1007/978-3-031-49401-7_39

Aging and the appearance of cardiovascular diseases are related to changes in the biomechanical properties of blood vessels and the study of the elastic behavior of these vessels may contribute to understand the physiological basis of many diseases.

As a simplification, blood vessel walls can be considered as containing three main components, collagen, elastin and smooth muscle cells. Morphologically, the wall has 3 layers or tunics which are the intima, media and adventitia. The tunica intima is the innermost layer made up of endothelial cells. The media is thicker, composed of non-striated muscle cells, between these cells there is an extracellular matrix composed of elastic and reticular fibers. The tunica adventitia is the outer layer and is formed by elastic fibers and collagen [4]. The overall mechanical characteristics of a vessel depend on these layers that make up the blood wall. The stress–strain behavior of blood vessels shows nonlinearities which makes it impossible to define a single elastic modulus representing the stiffness of the vessel [4–7].

Theoretically, a 3D printed vase can reproduce anatomical details with high precision and if designed with materials with elasticity and flexibility similar to that of the human vessel, it can be used to simulate the properties of the vessel.

The objective of this work is to obtain mechanical properties of 3D printing materials considered adequate to produce vessel models for surgical planning and device design. Dogbone 3D printed samples were characterized and compared to the human aorta wall samples obtained from autopsies.

2 Materials and Methods

2.1 Mechanical Testing

Mechanical tensile tests were performed using a standard test equipment (Instron 3365, Norwood, MA, USA) and its data analysis tool (Bluehill 3). The stress–strain test was performed according to ASTM D638 [8] using V size dogbone shaped specimens. Two test protocols were performed considering: (a) human thoracic aorta in dogbone shape cut in the circumferential direction and (b) 3D printed samples with different hardness and elasticity.

Before testing, the biological specimens were preconditioned with 3 cycles from zero to 0.1N to allow the alignment of the vascular tissue fibers, and then stretched until rupture. To follow the same pattern as biological tissues, the 3D printed specimens also underwent similar preconditioning.

2.2 Ethics Committee Approval

The use of autopsy material from human subjects was approved by the Ethics Committee of Hospital das Clinicas da FMUSP, under protocol number 14642.

2.3 Testing Samples

For the present study, samples from aorta were harvested from two autopsies (ao1, ao2) within 24 h after death and stored in physiological solution at 4 °C [9]. The samples were collected from normal aortas, with no signs of vascular diseases. The segments were cleaned and after removal of fat adhered to their surface, they were cut using a dogbone shape knife size V. Three samples were obtained from circumferential rings of the thoracic aortic segments (ao1.1, ao1.2, ao1.3 and ao2.1, ao2.2, ao2.3). All tests, carried out with the aim of obtain information on the biomechanical behavior of the arterial vessel, were performed before 72 h of removal from the body.

The 3D printed dogbone shape samples were obtained by printing 1–3 different materials with various hardness and elasticity in a layer-by-layer fashion using a 3D Printer (J750 Digital Anatomy Equipment, Stratasys, Israel). The materials were selected using the General Digital Material and Anatomy Blood Vessel for Thoracic Artery option in the printer configuration.

Figure 1 shows a 3D printed dogbone samples and the testing of a biological sample positioned for the stress–strain test.

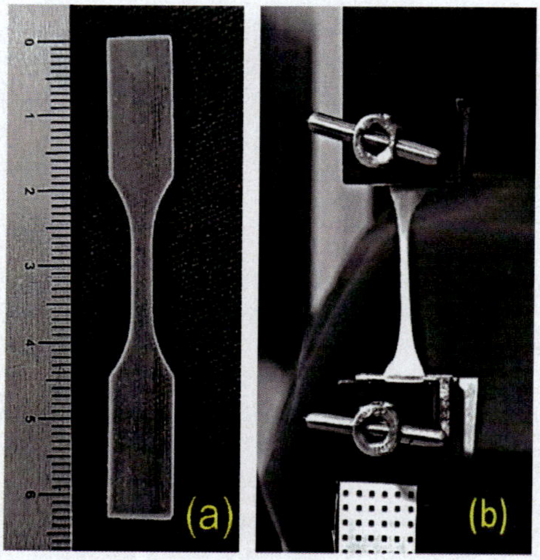

Fig. 1. Dogbone samples for stress–strain test using 3D printed materials (a) and thoracic aorta wall (b).

2.4 Data Analysis

Using the results of the mechanical assay, the physiological modulus and the maximum elastic modulus were determined, based on the study by Duprey et al. [5] illustrated in Fig. 2.

2.5 Statistical Analysis

Statistical analysis was performed by T test and ANOVA with statistical differences considered significant for $p < 0.05$ (Sigma Plot, Systat Software Inc., USA).

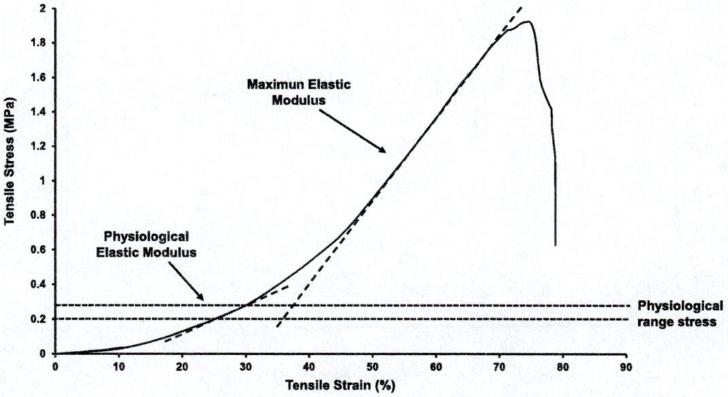

Fig. 2. Regions for calculation of the physiological and maximum elastic modulus, adapted from Duprey et al. [5].

3 Results and Discussion

3.1 Biological Samples

The results of testing biological samples are presented in Figs. 3 and 4. The data consists of measurements obtained from the thoracic aorta segment of two subjects considering three samples of each subject. All dogbone shaped samples failed in the middle region, showing that the test was consistent.

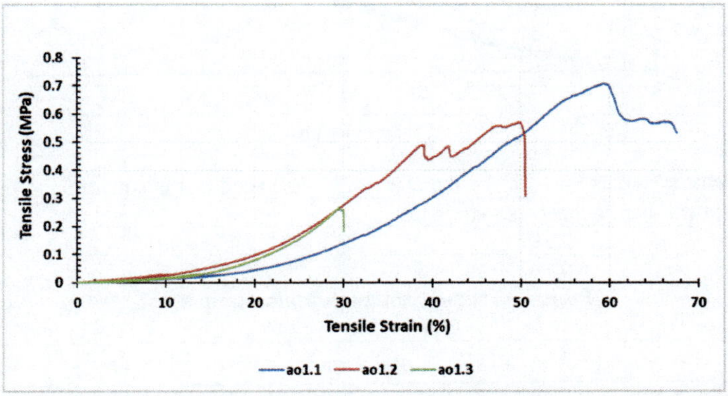

Fig. 3. Stress–strain test of thoracic wall dogbone shaped samples (ao1.1, ao1.2, ao1.3) cut in the circumferential direction of the vessel.

Figure 5 illustrates extension versus the load curves obtained during conditioning of the aortic wall samples. Mechanical results are presented in Table 1.

The T test showed no significant difference ($p > 0.05$) between aorta1 and aorta2 regarding the mechanical properties.

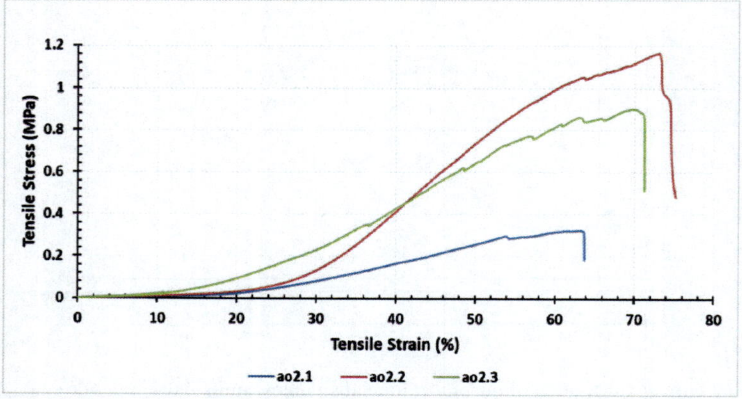

Fig. 4. Stress–strain test of thoracic aorta wall dogbone shaped samples (ao2.1, ao2.2, ao2.3) cut in the circumferential direction of the vessel.

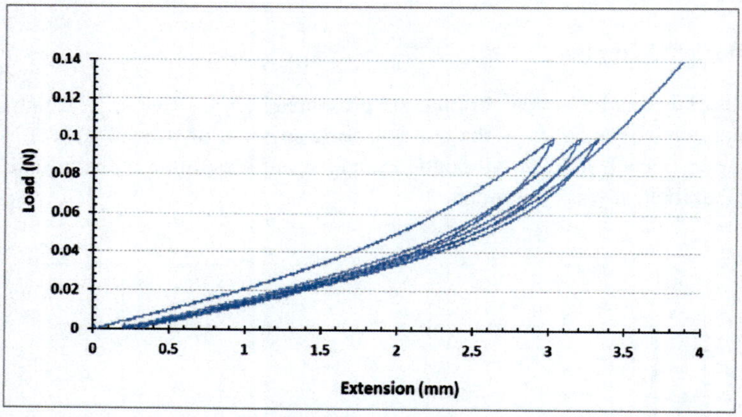

Fig. 5. Extension versus load for the thoracic aortic wall used as the conditioning protocol (3 cycles) performed before the stress–strain test.

Table 1. Thoracic aorta mechanical properties

	$aorta_1$	$aorta_2$
Physiological elastic modulus (MPa)	0.68 ± 0.20	0.58 ± 0.40
Maximum elastic modulus (MPa)	2.44 ± 0.30	2.18 ± 1.10
Maximum tensile stress (MPa)	0.73 ± 0.18	0.80 ± 0.43
Tensile strain to maximum stress (%)	45.87 ± 15.14	68.50 ± 4.92

In Figs. 3 and 4 it is possible to identify the physiological elastic modulus and the maximum elastic modulus in the tensile stress range from 0 to 0.2 MPa as presented in the study by Duprey et al. [5].

3.2 3D Printed Material

Results of mechanical characterization of samples made using different 3D materials with 1, 3 and 5 layers, with different hardness and elasticity are shown in Figs. 6, 7 and 8. The samples of 3D materials broke in the median region during the mechanical tests.

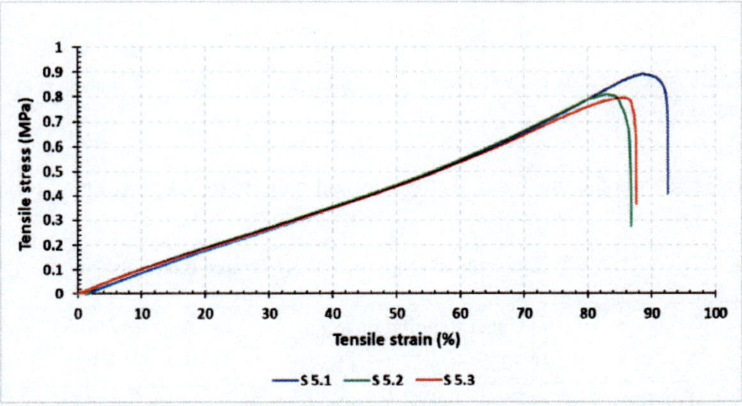

Fig. 6. Stress–strain curves obtained with 3D printed samples made of 5 layers of 30/40/30/70/30 shore A materials.

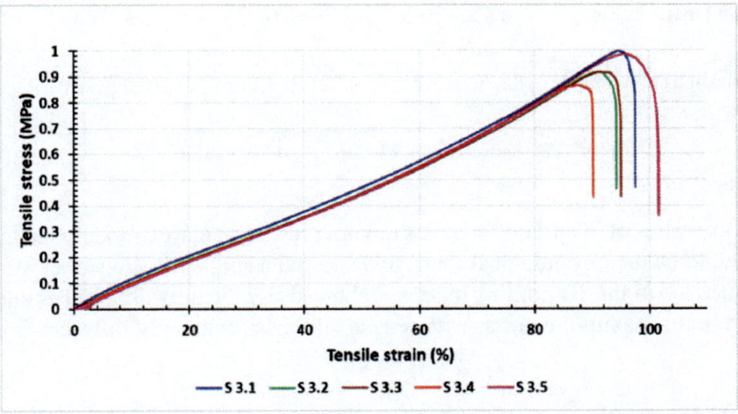

Fig. 7. Stress–strain curves obtained with 3D printed samples made of 3 layers of 30/70/30 shore A materials.

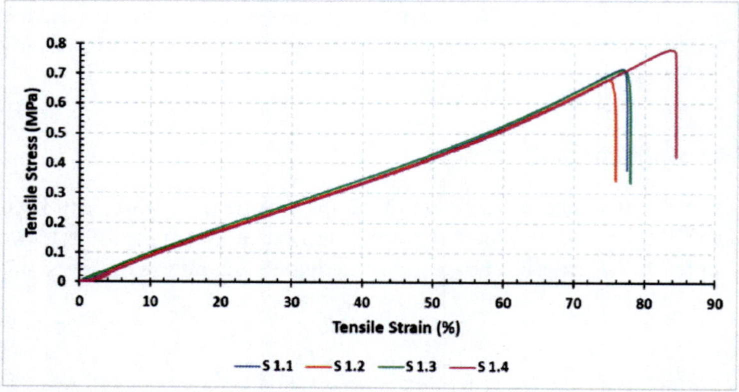

Fig. 8. Stress–strain curves obtained with 3D printed samples made of materials specific for thoracic arterial vessel.

Table 2 presents the mechanical characterization data to 3D printing material assay.

Table 2. Mechanical properties of 3D printed materials

	General digital material		Anatomy blood vessel for thoracic artery
	5 layers	3 layers	
Physiological elastic modulus (MPa)	0.88 ± 0.06	0.92 ± 0.04	0.85 ± 0.04
Maximum elastic modulus (MPa)	1.06 ± 0.03	1.05 ± 0.06	0.95 ± 0.07
Maximum tensile stress (MPa)	0.83 ± 0.05	0.94 ± 0.07	0.72 ± 0.04
Tensile strain to maximum stress (%)	85.76 ± 2.74	92.21 ± 3.19	78.07 ± 3.77

Note 5 layers = 30/40/30/70/30 shore A and 3 layers = 30/70/30 shore A

The properties of 3D printed samples did not present an elastic modulus in the range considered to be the physiological range and showed a single modulus of elasticity.

Comparison of the 3D printing materials showed significantly differences only in the tensile strain to maximum stress, with tensile strains significantly different from each other.

4 Conclusions

Our results suggest that the 3D printing materials showed that absolute maximal values of strain and stress were comparable to data obtained for human arterial wall. However, the materials did not show the viscoelastic behavior seen in the aortic wall samples.

Further investigation is necessary to tune more realistic materials for the modeling of blood vessels to be used for surgical planning or the design of medical devices.

Acknowledgments. The authors thank FINEP (01/170054–02), Fundacao Zerbini and Fundacao Faculdade de Medicina da Universidade de Sao Paulo (LIM 65) for the Financial Support.

References

1. Wú, C.A., Squelch, A., Sun, Z.: Investigation of three-dimensional printing materials for printing aorta model replicating type B aortic dissection. Current Med. Imaging **17**, 834–849 (2021)
2. Kim, T., Hong, D., Ock, J., et al.: Utilizing patient specific 3D printed guides for graft reconstruction in thoracoabdominal aortic repair. Nat. Portfolio Sci. Rep. **11**, 18027 (2021). https://doi.org/10.1038/s41598-021-97541-8
3. Coles-Black, J., Bolton, D., Chuen, J.: Accessing 3D printed vascular phantoms for procedural simulation. Front. Surg. **7**, 626212 (2021). https://doi.org/10.3389/fsurg2020.62612
4. Holzapfel, G.A., Gasser, T.C.: A new constitutive framework for arterial wall mechanics and a comparative study of material models. J. Elast. **61**, 1–48 (2000)
5. Duprey, A., Khanafer, K., Schilicht, M., et al.: In vivo characterization of physiological and maximum elastic modulus of ascending thoracic aortic aneurysms using uniaxial tensile testing. Eur. J. Vasc. Endovasc. Surg. **39**, 700–707 (2010)
6. Pejcic, S., Hassan, S.M.A., Rival, D., et al.: Characterizing the mechanical properties of the aortic wall. https://doi.org/10.20517/2574-1209.2019.18 (2019)
7. Wuyts, F.L., Vanhuyse, V.J., Langewouters, G.J., et al.: Elastic properties of human aortas in relation to age and atherosclerosis; a structural model. Phys. Med. Biol. **40**, 1577–1597 (1995)
8. ASTM D638: Standard Test Method for Tensile Properties of Plastics (2014)
9. Weisbecker, H., Pierce, D., Regitning, P., et al.: Layer-specific damage experiments and modeling of human thoracic and abdominal aortas with non-atherosclerotic intimal thickening. J. Mech. Behav. Biomed. Mater. **12**, 93–106 (2012)

Modification of Polymeric Scaffolds Made of PEG and Nanohydroxyapatite Embedded with Carbon Dots for Applications in Bone Tissue Regeneration

S. Degiovanni Morales[1], J. S. Ballestas Pineda[1], M. A. Molina Daza[1], P. Mancipe Celis[1], A. Ondo-Mendez[2], D. C. Rodríguez Burbano[3(✉)], and M. Múnera[4]

[1] Research Incubator Synergia, Colombian School of Engineering Julio Garavito/Universidad del Rosario, Bogota, Colombia
[2] Clinical Research Group, School of Medicine and Health Sciences, Universidad del Rosario, Bogota, Colombia
[3] Gibiome, School of Medicine and Health Sciences, Universidad del Rosario, Bogota, Colombia
Dianaco.rodriguez@urosario.edu.co
[4] Department of Biomedical Engineering, Colombian School of Engineering Julio Garavito, Bogota, Colombia

Abstract. Composite scaffolds have been used to treat bone defects from severe trauma, pathologies, or congenital malformations. Polymers such as polyethylene glycol (PEG) and ceramics such as nanohydroxyapatite (nHA) have been used for scaffold fabrication due to their mechanical and biological properties, which favor osteoconduction as well as mimic physical and structural characteristics of natural bone. Nanomaterials, such as carbon dots, have been implemented for this application due to their biocompatible nature and ability to improve mechanical properties and enhance biological interactions. Hydroxyapatite nanoparticles embedded with carbon dots (nHA-CD) were obtained by a rapid continuous precipitation method. Carbon dots (CDs) were synthesized by the microwave-assisted method. Characterization performed by UV-Vis spectrophotometry, Scanning electron microscopy (SEM), and Attenuated total reflectance Fourier transform infrared spectroscopy (FTIR-ATR) confirmed the incorporation of CDs in nHA. A scaffold based on PEG and nHA embedded with carbon dots (PEG-nHA-CD) was synthesized using a direct mixing method. The scaffold presented a density of 0,48 g/cm^3, similar to that of natural bone, indicating similar physical and structural characteristics. The cytotoxicity of the synthesized samples was evaluated by MTT and Trypan Blue assays. The results showed the potential of the PEG-nHA-CD scaffold fabricated for applications in tissue engineering aimed at future bone tissue applications.

Keywords: Carbon dots · Nanohydroxyapatite · Polyethylene glycol · Scaffolds · Tissue engineering

J. L. B. Marques et al. (Eds.): CLAIB 2022/CBEB 2022, IFMBE Proceedings 98, pp. 396–404, 2024.
https://doi.org/10.1007/978-3-031-49401-7_40

1 Introduction

Bones are tissues of organic-inorganic composition that perform essential functions such as protecting internal organs and soft tissues and offering support to the human body [1]. Even though this tissue has a natural capacity to regenerate, age or pathologies such as bone cancer, trauma, or congenital malformations can alter this ability. Consequently, the lack of tissue in a given area may occur, known as a bone defect [2, 3].

Bone tissue engineering (BTE) seeks to improve the health of the musculoskeletal system by providing solutions for the repair of bone defects by materials in the form of scaffolds that act as cell carriers to induce regeneration and formation of bone tissue [4]. For applications in BTE, the implemented scaffolds must meet ideal characteristics such as surface properties that favor cell migration, adhesion, proliferation, and growth. Additionally, these three-dimensional structures must have a distribution of interconnected and open pores that allow nutrient transport and vascularization to facilitate cell migration [1, 5].

The design of composite scaffolds allows the development of three-dimensional, porous, and interconnected structures, commonly composed of a dispersed ceramic phase and a polymeric matrix [6]. Thus, the scaffold may be tailored to possess desirable physicochemical and mechanical properties by altering compositions used for their synthesis [7]. PEG is a biocompatible and non-toxic polymer, which has been shown to promote cell adhesion, proliferation and growth by fostering the creation of interconnected pores while exhibiting improved mechanical and morphological properties compared to different polymers [8, 9]. Despite these properties, they exhibit poor osteoconductive properties and low compressive resistance [10]. Due to this, the fabrication of scaffolds composed of polymers and ceramics, such as hydroxyapatite (HA), a biomaterial that comprises about 50–70% of the ceramic phase of bone tissue, has been explored. HA has been implemented to improve the mechanical and biological properties of the resulting composite, establishing a robust interfacial bond strength between the ceramic phase and the polymeric matrix [4, 11]. This ceramic at a nanometric scale has been found to exhibit biocompatibility properties, bioactivity, and a high osteoconduction and osteoinduction capacity, making it ideal for applications in the treatment of bone defects [12]. However, HA presents limited mechanical properties such as brittleness and low load-bearing capacity, limiting its practical applicability as a bone implant material [13].

In recent years, nanotechnology-based approaches have been implemented to improve biomaterial's mechanical and biological properties, such as HA, by designing and fabricating nanohybrid systems in conjunction with different nanomaterials [13]. The addition of carbon-based nanomaterials, such as CDs, has been found to improve the properties of HA [14]. CDs are nanoparticles with an average size of 10 nm, obtained from a wide range of carbon precursors, exhibiting attractive optical properties such as fluorescence [15]. The nanohybrid composed of HA and CDs has proven to be effective for fabricating composite structures due to the presence of functional groups that allow interaction with the polymeric matrix [13]. Consequently, this work is focused on the synthesis, characterization, and biological and mechanical evaluation of a PEG-nHA-CD scaffold to establish its potential for applications in bone tissue regeneration.

2 Materials and Methods

2.1 Chemicals and Reagents

Citric acid monohydrate, ethanol, methanol, and N,N-dimethylformamide were purchased from Sharlau. Acetonitrile, meta-phosphoric acid and polyethylene glycol (Mw 3350) were purchased from Sigma Aldrich. Calcium hydroxide was obtained from Emsure. Silica gel (70–230 mesh) was obtained from Fluka Analytical. Distilled water was produced in-house. All reagents were of analytical grade and used as is, without requiring further purification.

2.2 Synthesis of Carbon Dots

CDs were synthesized by the microwave-assisted method. Briefly, 0.5 g of citric acid was dispersed in a 1:1 (vol: vol) solution of ethanol and N, N-dimethylformamide. The reaction of the resulting solution was carried out using an Anton Paar Monowave 50 reactor at 175°C for 5 min in as fast as possible mode. The solution was centrifuged at 10000 rpm for 10 min and filtrated (syringe filter 0.2μm, Minisart NY 25, Polytetrafluoroethylene). A modified chromatography column method reported by Hinterberger et al. for further purification was implemented [16]. The column's stationary phase consisted of silica gel with a high degree of purity, a pore size of 4 nm, and a particle size of 0.063–0.200 mm. A 1:1 (vol: vol) solution of ethanol and distilled water was used as an eluent. Finally, the solution obtained was dried in a vacuum concentrator oven (CentriVap micro IR Vacuum Concentrator, Labconco) at 65°C in the IR/HEAT configuration.

2.3 Synthesis of Nanohydroxyapatite Embedded with Carbon Dots

The synthesis of nHA-CD nanoparticles was carried out by a modified rapid continuous precipitation method [17]. Typically, a suspension of calcium hydroxide (2.5g in 20mL of distilled water) was stirred at 85°C for 5 min. 2.22g of meta-phosphoric acid was added to a 30mL aqueous solution with a concentration of 2.5 mg/mL of CDs. The solution of meta-phosphoric acid and CDs was stirred for 2 h and added dropwise to the calcium hydroxide suspension at 85 °C, followed by constant stirring for 24 h at room temperature. The obtained precipitate was separated by centrifugation (4000 rpm for 10 min) and washed with ethanol and distilled water. Finally, the paste was dried in an oven (UM100, Memmert, The Netherlands) at 110°C for 12 h.

2.4 Fabrication of the PEG-NHA-CD Scaffold

The PEG-nHA-CD scaffold was fabricated using the direct blending method [1]. First, 1.5 g of PEG was dispersed in 50 mL of distilled water, and 1 g of nHA-CD was dispersed in 50 mL of methanol. The PEG solution was slowly added to the nHA-CD methanolic suspension and stirred continuously for 4 h. Subsequently, the mixture was stirred at 40°C for 2 h. The final mix was centrifuged at 5000 rpm for 1 min to remove the slurry excess. The obtained precipitate was dried in an oven (UM100, Memmert, The Netherlands) at 80°C for 24 h. Finally, the saturated mixture was sintered for six hours at 200°C. The final powder was compacted using the Shimadzu AGS-X Universal Machine. The force used for the compaction of the samples was varied (2000 N, 3000 N, and 4000 N) to analyze its impact on the density of the resultant scaffold. The density was measured through the "Standard Test Methods for Density and Specific Gravity (Relative Density) of Plastics by Displacement" according to standard D792–13 [18].

2.5 Cytotoxicity Studies

Cytotoxicity was evaluated using MTT and Trypan Blue assays on 3T3 (fibroblasts) cell lines. Cells were cultured in Dulbecco's Modified Eagle's medium (DMEM, Gibco), supplemented with 10% Fetal Bovine Serum (FBS, Biowest) and 0.01 mg/ml penicillin/streptomycin (Gibco) at 37 °C with 5% CO_2 in a humid atmosphere. For MTT, 3T3 cells were seeded in 96-well plates at a density of 10000 cells per well and incubated at 37°C in a 5% CO_2 environment. Solutions of nHA, nHA-CD and nHA-CD-PEG at 50 and 500 μL/mL were added to the seeded cells and incubated for 24 h. Cells incubated without any samples were used as the negative control, and cells exposed to hydrogen peroxide were used as a positive control. Subsequently, 20 μL of MTT solution was added to each well used and cultured for 4 h at 37°C. The MTT solution was removed, and 200 μL of dimethyl sulfoxide was added to dissolve the formed formazan salts. Finally, the optical density at 560 nm was measured. For the Trypan Blue assay, 3T3 cell lines were seeded and exposed to nHA, nHA-CD, and nHA-CD-PEG, following the same procedure carried out for the MTT assay. After this, 10 μL of Trypan Blue solution was added to each well, and subsequently, a viable cell count was performed by an automatic cell counter (Cyto SMART^{TM}, , The Netherlands). Data was represented as the percentage of cell survival (mean ± standard deviation, n = 3) compared to negative control wells.

2.6 Apparatus

UV-Vis absorption and fluorescence spectroscopy. The optical properties of CDs and nHA-CD were characterized using a FLAMES-S-XR1-ES spectrophotometer using a 1cm quartz cuvette. The absorbance measurements were collected over a 200–800nm scan range. Fluorescence spectra were recorded using an excitation wavelength ranging from 250 to 700nm. All the data processing and contour plots were generated using Origin Lab® and Prism®.

Fourier transform infrared spectroscopy – attenuated total reflectance (FTIR-ATR). FTIR-ATR spectra were collected on the dried solid powder of CD, nHA, nHA-CD, and PEG-nHA-CD samples, using a Thermo Scientific Nicolet iS5 equipped with an iD5 ATR accessory. Analysis was performed on a laminate-diamond crystal window using 64 scans at 0.4 cm^{-1} resolution. Data processing was carried out using the Thermo Scientific Nicolet Omnic 9® software.

Scanning electron microscopy (SEM). Scanning electron microscopy was collected using a TESCAN Lyra 3 microscope. Analysis was performed on a non-conductive coated aluminum foil holder. Two readout methods were developed to analyze the samples, with the dried samples placed directly on the holder and with the ethanolic solutions of CD, nHA, nHA-CD, and PEG-nHA-CD. The results were processed with OriginLab® and ImageJ® to analyze grain size.

3 Results and Discussion

3.1 Synthesis and Characterization of CDs and NHA-CD and the Fabrication of PEG-NHA-CD Scaffold

CDs were synthesized *via* a *bottom-up* method. For the synthesis process, citric acid was implemented as a carbon precursor. The obtained CD suspension was clear under ambient light, but it exhibited a blue fluorescence upon UV irradiation (365 nm) (Fig. 1a). nHA-CD nanoparticles were obtained through a rapid continuous precipitation method (Fig. 1b). The presence of CDs in the nHA matrix was confirmed when the samples were excited at a wavelength of 365 nm and exhibited the fluorescence characteristic of CDs (Fig. 1. C). Furthermore, the PEG-nHA-CD mixture was obtained following the direct blending procedure reported by Kumar and coll [1]. Forces of 2000N, 3000N, and 4000N were applied for the compaction of the nHA-CD-PEG samples. These forces were selected considering the values reported in the literature based on the nature of polymeric-ceramic composites [19].

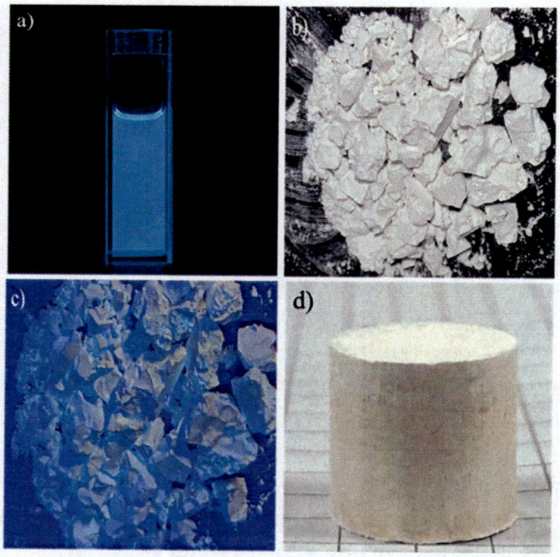

Fig. 1. Images of a) aqueous solution of CDs (40 mg/mL) upon 365 nm irradiation; b) and c) nHA-CDs in the absence and exposed to UV light (365 nm), respectively; and d) PEG-nHA-CD scaffold obtained after compaction (4000 N).

The optical properties of CDs and nHA-CD were characterized by UV-Vis absorption and fluorescence spectroscopy. UV-Vis spectroscopy revealed an absorbance peak for CDs and nHA-CD within the 200–300 nm range attributed to the $\pi \longrightarrow \pi^*$ transition involving aromatic carbons sp^2 with C = C bonds (Fig. 2a). As this transition is shown in both spectra, the incorporation of the CDs in the nHA matrix is confirmed. Additionally, a second peak is observed at 280 nm corresponding to n⟶π* transitions, attributed to C = O in the CDs. In the case of nHA-CD, this peak is also attributed to the contribution of the oxygen atomic species in the nHA. The emission spectrum of CDs samples showed an emission peak with a maximum at 500 nm (Fig. 2b). It is essential to mention that CDs exhibit fluorescence properties spanning the 400 to 600 nm range when excited with 365 nm irradiation. Similar emission bands are shown with a lower intensity when comparing this emission spectrum with that obtained for nHA-CDs. Since nHA does not possess fluorescent properties, the only source of emission corresponds to the CDs embedded within the ceramic matrix.

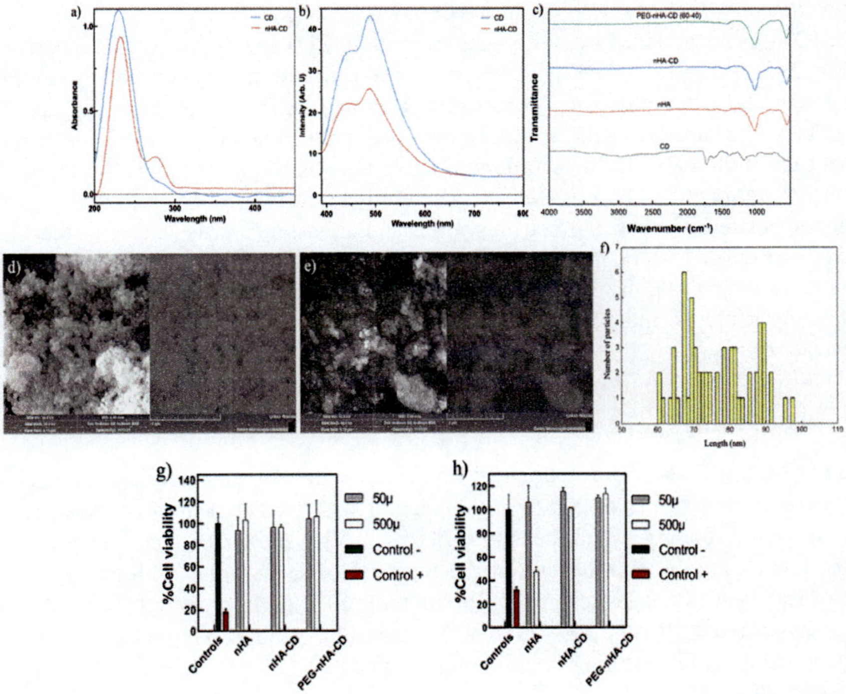

Fig. 2. a) Absorption and b) emission spectrum (λ exc = 365 nm) of aqueous solutions of CDs and nHA-CD samples, c) FTIR spectra of CD, nHA-CD, and nHA-CD-PEG samples, SEM images of d) nHA-CD and e) PEG-nHA-CD samples (The morphology is shown on the left and the density distribution on the right), f) Particle size distribution of nHA-CD, g) MTT and h) Trypan Blue assays on 3T3 cell line exposed to (grey) 50 and (white) 500 μg/mL of nHA, nHA-CD, and PEG-nHA-CD (Negative and positive controls correspond to cells without any of the materials and in H_2O_2 10%, respectively).

To confirm the results obtained from the UV-Vis characterization and perform a chemical description of the surface of the synthesized samples, an analysis of the different functional groups was carried out by FTIR-ATR (Fig. 2c). CDs spectrum revealed vibrational features ascribed to the -OH stretch at 3300 cm^{-1}, along with carboxylic acid C = O and acyl C-O stretches at 1720 and 1200 cm^{-1}, respectively. These oxygen-containing functional groups play an essential role in incorporating CDs in the nHA since it is a process favored by maximizing polar interactions. Moreover, the sp^2 C-H stretching was observed between 2900 and 3100 cm^{-1}, attributed to the sp^2-sp^3 character of CDs prepared using citric acid as a precursor [20].

On the other hand, the nHA spectrum exhibit stretches at 1010 cm^{-1} at and at 560 cm^{-1} attributed to $PO_4{}^{3-}$ and O-P-O, characteristic of HA [21]. nHA-CD and PEG-nHA-CD reveal similar vibration spectra to those of nHA. The bands of CDs are expected to decrease in intensity as it is a surface characterization technique, and the CDS are embedded in the nHA matrix. Furthermore, the execution of Raman Spectroscopy would be beneficial as a complimentary characterization method to corroborate the data obtained using FTIR and UV-Vis spectroscopy.

SEM images obtained from the samples of nHA-CDs and PEG-nHA-CD are shown in Fig. 2d) and 2e), respectively. The particle size distribution is determined by counting the longitudinal dimension of 60 particles. The distribution is shown in Fig. 2f. The particle size obtained from the nHA-CDs could favor the improvement of the mechanical properties of the nHA itself, according to the reports of Kumar et al. [1]. Furthermore, surface roughness can be observed, a characteristic that favors the bioactivity of the nanocomposite by benefiting cell migration and adhesion. Furthermore, in the SEM image of PEG-nHA-CD, a layer surrounding the nHA-CD nanoparticles is observed. This layer is attributed to PEG, which acts as a binding material, packing the nHA-CD nanoparticles forming an agglomerate and supporting its use in tissue regeneration applications.

3.2 Bulk Density

The PEG-nHA-CD scaffolds fabricated via the direct blending method were compacted using compaction forces of 2000 N, 3000 N, and 4000 N. The density of the scaffolds was measured, finding a bulk density of 0.553, 0.598, and 0.619 g/cm^3 respectively. When comparing the bulk density of the PEG-nHA-CD scaffolds with the bulk density of trabecular bone corresponding to an adult male (0.48 g/cm^3 [22]), it is possible to hypothesize that the PEG-nHA-CD scaffold has similar physical and structural characteristics to natural bone, reinforcing its potential for applications in bone tissue regeneration.

3.3 Cytotoxicity

The *in-vitro* cytotoxicity of nHA, nHA-CD, and PEG-nHA-CD toward 3T3 cell lines was examined using MTT and Trypan Blue assay (Fig. 2g and 2h). The cytotoxicity of the materials used for synthesizing polymeric scaffolds was rated based on fibroblast cell viability. The ratio PEG:nHA-CD is 60:40.

The results of the MTT assay showed cell viability higher than 96% when exposed to nHA, nHA-CD, and PEG-nHA-CD at concentrations of 50 and 500 mg/mL (Fig. 2g). On the other hand, the results of the Trypan Blue assay showed cell viability higher than 94% when the materials are at a concentration of 50 mg/mL (Fig. 2h). However, reduced cell viability when exposed to nHA at a concentration of 500 mg/mL is observed. This reduction may indicate that the cell membrane is compromised when exposed to high concentrations of nHA due to high calcium concentration. Despite this, when the ceramic matrix begins to dilute, as in the case of nHA-CD and PEG-nHA-CD, where its proportion is lower, viabilities above 90% are obtained. Fibroblast cell viability is an indication of the non-toxic nature of the composites synthesized, acting as a preliminary result of the biocompatibility of the samples.

4 Conclusions

The synthesis of a polymeric scaffold made from PEG and nHA-CD is carried out to mimic natural bone's structural and chemical characteristics. The fabricated scaffolds exhibited a bulk density similar to bone tissue, indicating that the scaffolds obtained showed similar physical properties to those of natural bone. Additionally, the cytotoxicity evaluation showed the non-toxic nature of nHA, nHA-CD, and nHA-CD-PEG against fibroblasts, showing the preliminary biocompatibility of the samples and their potential for applications in BTE. Furthermore, the optical and physicochemical characterizations confirmed the incorporation of CDs in the nHA matrix. Adding PEG helped improve various characteristics, including morphology and stability of the synthesized composites. The results indicate that the composite-based PEG-nHA-CD has the potential for bone tissue engineering applications.

Acknowledgment. M.M, D.C.R.B, and A.O.M. thank the Colombian School of Engineering Julio Garavito for financial support.

Conflict of Interest. There are no conflicts to declare.

References

1. Kumar, P., Dehiya, B.S., Sindhu, A.: Synthesis and characterization of nHA-PEG and nBG-PEG scaffolds for hard tissue engineering applications. Ceram. Int. **45**(7), 8370–8379 (2019). https://doi.org/10.1016/j.ceramint.2019.01.145
2. Dilogo, I.H., et al.: Allogeneic umbilical cord-derived mesenchymal stem cells for treating critical-sized bone defects: a translational study. Eur. J. Orthop. Surg. Traumatol. **31**(2), 265–273 (2021). https://doi.org/10.1007/s00590-020-02765-5
3. Smrke, D., Roman, P., Veselko, M., Gubi, B.: Treatment of bone defects—allogenic platelet gel and autologous bone technique. In: Regenerative Medicine and Tissue Engineering, InTech (2013).https://doi.org/10.5772/55987
4. Tetteh, G., Khan, A.S., Delaine-Smith, R.M., Reilly, G.C., Rehman, I.U.: Electrospun polyurethane/hydroxyapatite bioactive Scaffolds for bone tissue engineering: the role of solvent and hydroxyapatite particles. J. Mech. Behav. Biomed. Mater. **39**, 95–110 (2014). https://doi.org/10.1016/j.jmbbm.2014.06.019
5. Meskinfam, M. et al.:Polyurethane foam/nano hydroxyapatite composite as a suitable scaffold for bone tissue regeneration. Mater. Sci. Eng. C **82**, 130–140 (2018). https://doi.org/10.1016/j.msec.2017.08.064

6. Ghassemi, T.E. et al.: Current concepts in scaffolding for bone tissue engineering. Arch. Bone Jt Surg. **6**(2), 90–99 (2018)
7. Pereira, H., Cengiz, I., Silva, F., Reis, R., Oliveira, J.: Scaffolds and coatings for bone regeneration. J. Mater. Sci. Mater. Med. **31**(27) (2020)
8. Yan, D., et al.: Study on the properties of PLA/PBAT composite modified by nanohydroxyapatite. J. Market. Res. **9**(5), 11895–11904 (2020). https://doi.org/10.1016/j.jmrt.2020.08.062
9. Li, H., et al.: Bioengineered three-dimensional scaffolds to elucidate the effects of material biodegradability on cell behavior using POSS-PEG hybrid hydrogels. Polym. Degrad. Stab. **164**, 118–126 (2019). https://doi.org/10.1016/j.polymdegradstab.2019.04.008
10. Wang, J.Z., You, M.L., Ding, Z.Q., Bin Ye, W.: A review of emerging bone tissue engineering via PEG conjugated biodegradable amphiphilic copolymers. Materials Science and Engineering C, vol. 97, pp. 1021–1035. Elsevier Ltd, Apr. 01, 2019. https://doi.org/10.1016/j.msec.2019.01.057
11. Sadat-Shojai, M., Khorasani, M.T., Dinpanah-Khoshdargi, E., Jamshidi, A.: Synthesis methods for nanosized hydroxyapatite with diverse structures. Acta Biomaterialia **9**(8), 7591–7621 (2013) Elsevier Ltd. https://doi.org/10.1016/j.actbio.2013.04.012
12. Maia, F., Correlo, V., Oliveira, J., Reis, R.: Natural origin materials for bone tissue engineering: Properties, processing, and performance (2019). https://doi.org/10.1016/c2015-0-02433-9
13. Gogoi, S., Kumar, M., Mandal, B.B., Karak, N.: A renewable resource based carbon dot decorated hydroxyapatite nanohybrid and its fabrication with waterborne hyperbranched polyurethane for bone tissue engineering. RSC Adv. **6**(31), 26066–26076 (2016). https://doi.org/10.1039/c6ra02341j
14. Eivazzadeh-Keihan, R., et al.: Carbon based nanomaterials for tissue engineering of bone: Building new bone on small black scaffolds: a review. J. Adv. Res. **18**(March), 185–201 (2019). https://doi.org/10.1016/j.jare.2019.03.011
15. Pramanik, S., Kar, K.K.: Nanohydroxyapatite synthesized from calcium oxide and its characterization. Int. J. Adv. Manuf. Technol. **66**(5–8), 1181–1189 (2013). https://doi.org/10.1007/s00170-012-4401-z
16. Hinterberger, V., Damm, C., Haines, P., Guldi, D.M., Peukert, W.: Purification and structural elucidation of carbon dots by column chromatography. Nanoscale **11**(17), 8464–8474 (2019). https://doi.org/10.1039/c9nr01029g
17. Mercado, D.F., et al.: Paramagnetic iron-doped hydroxyapatite nanoparticles with improved metal sorption properties. A bioorganic substrates-mediated synthesis. ACS Appl. Mater. Interfaces **6**(6), 3937–3946 (2014). https://doi.org/10.1021/am405217j
18. ASTM Standard F2150.: Standard guide for characterization and testing of biomaterial scaffolds used in regenerative medicine and tissue-engineered medical products (2019)
19. Cui, Y., Liu, Y., Cui, Y., Jing, X., Zhang, P., Chen, X.:The nanocomposite scaffold of poly(lactide-co-glycolide) and hydroxyapatite surface-grafted with L-lactic acid oligomer for bone repair. Acta Biomaterialia **5**(7), 2680–2692 (2009)
20. Manioudakis, J., et al.: Effects of nitrogen-doping on the photophysical properties of carbon dots. J. Mater. Chem. C **7**(4), 853–862 (2019). https://doi.org/10.1039/c8tc04821e
21. Bose, S., Roy, M., Bandyopadhyay, A.: Recent advances in bone tissue engineering scaffolds. Trends Biotechnol. **30**(10), 546–554 (2012). https://doi.org/10.1016/j.tibtech.2012.07.005
22. Caeiro, J., González, P., Guede, D.: Biomecánica y hueso (y II)- Ensayos en los distintos niveles jerárquicos del hueso y técnicas alternativas para la determinación de la resistencia ósea. Revista de Osteoporosis y Metabolismo Mineral **5**(2), 99–108 (2013)

Synthesis and Evaluation of the Potential Use of an Electro-Responsive Hydrogel for the Encapsulation of Carbon Dot-Curcumin Nanoplatforms

Nicolas J. Bonilla Fajardo[1], M. Natalia Ochoa Paipilla[1], Michelle D. Soto Florido[1], Sandra Ramirez-Clavijo[2], D. Fabio Mercado[3], Marcela Munera[4], and Diana C. Rodriguez Burbano[5](✉)

[1] Research Incubator Synergia, Colombian School of Engineering Julio Garavito/Universidad del Rosario, Bogota, Colombia
[2] Director of Basic Medical Sciences Group, Department of Biology, Faculty of Natural Sciences, Universidad del Rosario, Bogota, Colombia
[3] University Grenoble Alpes, CNRS, Grenoble-INP2, LMGP, 38000 Grenoble, France
[4] Department of Biomedical Engineering, Colombian School of Engineering Julio Garavito, Bogota, Colombia
[5] GIBIOME, School of Medicine and Health Sciences, Universidad del Rosario, Bogota, Colombia
dianaco.rodriguez@urosario.edu.co

Abstract. Curcumin is a polyphenolic compound recently studied as a potential therapeutic agent for cancer treatment. However, this molecule possesses low solubility in water which complicates its direct use in the organism, making it essential to find a drug delivery system to carry this molecule through hydrophilic environments. To achieve this, carbon dots (CDs) with a polar surface were synthesized from citric acid and urea by solvothermal method, followed by the surface functionalization of the CDs via the adsorption of curcumin molecules (CD-Cur). This nanoplatform was encapsulated inside an electro-responsive hydrogel based on dextran and aniline (Dex/HDI/AT), which acts as a vehicle for CD-Cur. To evaluate the improvement of solubility of curcumin, a drug encapsulation test was carried out to compare its encapsulation rate to a molecule of great hydrophilicity, fluorescein. The results of FTIR, UV-Vis spectroscopy and zeta potential tests enabled to identify the composition of CDs and the functionalization of CD-Cur. The composition and structure of Dex/HDI/AT was confirmed by FTIR and SEM. Finally, the drug encapsulation test demonstrated the capacity of CDs to improve the interaction of hydrophobic curcumin molecules with aqueous-based environments such as the one offered by the DEX/HDI/AT hydrogel.

Keywords: Carbon dots · Curcumin · Drug delivery · Hydrogel · Encapsulation of hydrophobic molecules

J. L. B. Marques et al. (Eds.): CLAIB 2022/CBEB 2022, IFMBE Proceedings 98, pp. 405–415, 2024.
https://doi.org/10.1007/978-3-031-49401-7_41

1 Introduction

Curcumin is a molecule extracted from turmeric (*Curcuma longa*) whose potential biomedical applications have been known for centuries. In recent years, this molecule has gained interest within the scientific community to explore its use in cancer treatment due to its chemotherapeutic and chemopreventive properties [7]. This compound belongs to the family of polyphenols and is characterized by its low solubility in water, which causes difficulties in its administration in the organism. Thus, it is necessary to find solutions through controlled drug release systems to improve curcumin's therapeutic performance by modifying the drug's polarity and, consequently, its interaction with aqueous-based environments [1,6].

A novel alternative is the use of carbon dots. CDs are carbon-based nanoparticles that, depending on the carbon or other heteroatom precursors used in the synthesis procedure, possess chemical properties on the surface that may allow the functionalization of drugs using different intermolecular forces [11]. Moreover, by taking advantage of their large surface area, the adsorption of a hydrophobic drug on the surface of the CDs can not only improve its solubility but also a more significant amount of a drug can be incorporated into a single nanoparticle, which may imply a positive impact on the dose reduction [1]. This strategy has been used before to potentially deliver curcumin and other drugs [14]. Although, the improvement of its solubility has not been reported yet.

As it is based on types of interactions that favor the adsorption of curcumin on the surface of CDs, it is necessary to encapsulate these nanoplatforms in a vehicle that allows maintaining a good interaction with the external medium without disrupting the functionalization [12]. The vehicles reported in the literature have been synthesized in different forms and from various materials, such as liposomes, micelles, polymeric nanoparticles and, recently, hydrogels. The latter ones are based on hydrophilic polymeric networks that can encapsulate high amounts of water and have attracted attention due to the high biocompatibility and especially for the capacity to add functionalities during the crosslinking of the precursors, which would give the hydrogel the ability to respond to external stimuli such as light, temperature, pH, enzymes, magnetic, or electric fields [2,8]. Special attention in developing controlled drug delivery systems has been given to those that could respond to electrical current as a form of remote control [5].

In this work, a controlled drug release system is proposed that allows the encapsulation of low solubility drugs such as curcumin using CDs to improve the interaction of this molecule with hydrophilic environments by the adsorption of this molecule on the surface of CDs. In addition, an electro-responsive hydrogel based on dextran and aniline will be used as a vehicle for the encapsulation of the nanoplatforms for its potential application as a controlled drug release system.

2 Materials and Methods

2.1 Chemicals and Reagents

Citric acid monohydrate was purchased from Scharlau (Scharlab, Spain). Urea, formamide, N-phenyl-p-phenylenediamine, stannous octoate ($Sn(Oct)_2$), hexamethylene diisocyanate (HDI), ammonium persulfate ($(NH_4)_2S_2O_8$), dimethylsulfoxide (DMSO) and Dextran were obtained from Sigma Aldrich (Merck KGaA, USA). Distilled water was produced in-house. All reagents were of analytical grade and were used as is, without the need for further purification.

2.2 Apparatus

Fourier transform infrared spectroscopy—attenuated total reflectance (FTIR-ATR). FTIR-ATR spectra were collected on the dried solid powder of CDs, CD-Cur and AT and, on the DEX/HDI/AT hydrogel swelled in distilled water using a Thermo Scientific Nicolet iS5 equipped with iD5 ATR accessory (Thermo scientific, USA). Analysis were performed on a laminate-diamond crystal window using 64 scans at 0.4 cm^{-1} resolution from 400 cm^{-1} to 4000 cm cm^{-1} for 2 min. All data were processed using the Thermo Scientific Nicolet Omnic 9 software.

UV-Vis and fluorescence spectroscopy. CDs, CD-Cur, and AT optical properties were characterized using a FLAMES-S-XR1-ES spectrophotometer (Ocean Optics, UK), using a 1cm quartz cuvette. The absorbance measurements were collected over a 200–800nm scan range. Fluorescence spectra were recorded using an excitation wavelength ranging from 250 to 700 nm. Data processing and all contour plots were generated using MathWorks MATLAB R2020b.

Z potential measurement—Z potential was measured on CDs dispersed in distilled water using an Antoon Paar Litesizer 500 equipped with cmPALS technology that allows a susceptible Z potential measurement by electrophoretic light scattering (Antoon Paar, Austria).

Scanning electron microscopy (SEM). Scanning electron microscopy was collected using a TESCAN Lyra 3 microscope (Oxford Instruments, UK). The readout was taken with the Dex/HDI/AT samples swelled with distilled water at low vacuum and magnified 1500x and 5000x.

2.3 Synthesis and Characterization of CD-Cur Nanoplatforms

Carbon dots (CDs) were synthesized by the solvothermal method. Briefly, 1 g of citric acid and 1 g of urea were dispersed in 15 mL of formamide. The mixture was transferred to a 20 mL Teflon-lined stainless-steel autoclave and then heated at 160°C for 12 h, resulting in a dark-red solution. The obtained solution was centrifuged at 4000 rpm for 10 min, followed by two centrifugation processes at 10,000 rpm for 10 min. The collected supernatant was dialyzed in distilled water for 48 h, changing the water every 12 h. The CDs were dried in an oven (UM100, Memmert, The Netherlands) at 100°C for 12 h.

Synthesized CDs were functionalized with curcumin via adsorption. Typically, 3.75 mg CDs were dispersed in 7.5 mL of distilled water. Then, 0.5 mg of curcumin was dispersed in 5 mL of ethanol. Both solutions were mixed and placed under magnetic stirring for 48 h at room temperature. The obtained solution was centrifuged at 12500 rpm for 15 min, followed by the dialysis of the supernatant in distilled water for 72 h, changing the water every 12 h. The dialysate was filtered (0.22 μ m) and dried in an oven (UM100, Memmert, The Netherlands) at 100°C for 12 h.

2.4 Synthesis and Characterization of the DEX/HDI/AT Electro-Responsive Hydrogel

The electro-responsive hydrogel was synthesized following the procedures reported by Qu et al. and Guo et al. [3,10].

First, aniline tetramer (AT) was synthesized as the building block that provides the hydrogel electro-responsive characteristic. This is formed from the reaction of amine groups between N-phenyl-p-phenylenediamine. Briefly, 460.9 mg of N-phenyl-p-phenylenediamine were dissolved in a mixture of HCl 1M, H_2O and acetone (1:1:1.5). Then, it was placed in an ice bath at 0°C under magnetic stirring while 5 ml of $(NH_4)_2S_2O_8$ 1M were added by dripping to the mixture. The mixture reacted for 4 h. Later, the dark-blue colored product was vacuum filtered and washed with 5 ml of HCl 1M. Then, NH_4OH 1M was added until the final product was neutralized. Finally, the product was vacuum dried for 24 h at 50°C.

For the DEX/HDI/AT hydrogel synthesis, a reaction between dextran and AT using HDI as crosslinking agent was carried out. Briefly, 202.5 mg of dextran were dissolved in a mixture of DMSO, HDI and $Sn(Oct)_2$ (200:2.4:1). 20,2 mg of AT were added to the mix and placed in a 50 ml round bottom flask at 80°CC for 8 h.

2.5 Drug Encapsulation Test

The drug encapsulation test used fluorescein as a hydrophilic molecule model, CDs, and CD-cur. The calibration curves for each agent were established to find a relation between the concentration and absorbance values. The pre-weighted hydrogel was immersed in a solution of known concentration containing the compound of interest to be encapsulated. Every hour the absorbance values of the solution surrounding the hydrogel were measured. For fluorescein, absorbance values were measured at 430 nm and 340 nm for CDs and CD-cur. The amount of compound encapsulated inside the hydrogel was found using Eq. 1, where D.E is the percentage of compound encapsulated in the hydrogel, Conc0 is the initial compound concentration in the solution and Concx is the concentration of compound a the time of the measure.

$$D.E(\%) = \frac{(Conc0 - Concx)}{Conc0} * 100 \tag{1}$$

3 Results and Discussion

3.1 Synthesis and Characterization of CDs-Cur Nanoplatforms

Since the nanoplatform is composed of CDs with curcumin adsorbed on the surface, the characterization of CDs should demonstrate the presence of functional groups that could form a negative electrostatic potential that could attract this molecule and a highly hydrophilic surface to carry hydrophobic curcumin molecules through aqueous environments.

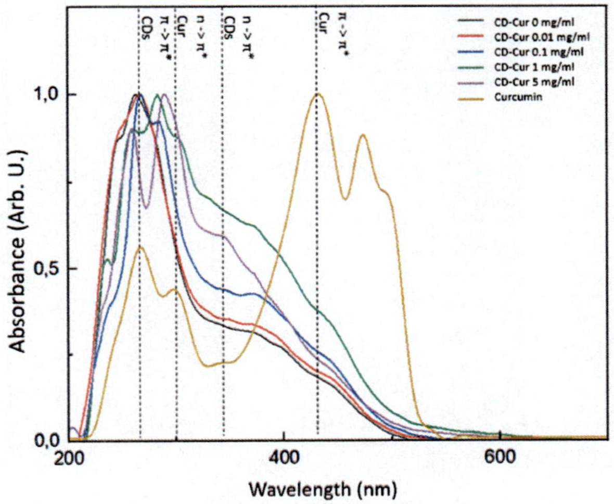

Fig. 1. Absorption spectra of Curcumin dispersed in ethanol and, CDs and CD-Cur dispersed in distilled water.

In this way, the absorption spectrum of CDs exhibits two absorption peaks (Fig. 1). The first peak at 250 nm is attributed to the π -> π^* transition of aromatic rings in the core of the CDs. The second peak at 340 nm is attributed to the n -> π^* transition from the C-N and the C-O bonds present in the core of the carbon dots and the functional groups on the surface of the CDs. Regarding the curcumin absorption spectrum, two absorption peaks at 300 and 400 nm are found. Although these peaks are also attributed to the π -> π^* transition, a red shift is observed due to the interaction of the curcumin with -OH groups of the ethanolic medium [4].

The functionalization reaction mixed CDs and curcumin to form a homogeneous solution that could allow both agents to interact and, due to the differences in their electrostatic potential, adsorb the curcumin molecules onto the CDs surface. Consequently, the CD-Cur UV-Vis absorption spectra should exhibit some of the relevant absorption peaks for both CDs and curcumin to determine the result of this reaction.

Thus, the absorption spectra for all CD-Cur showed that the process was satisfactory in assays with initial concentrations of curcumin higher than 0.1 mg/ml as all exhibit peaks at 250 nm and 340 nm from CDs and the 300 nm absorption peak of curcumin. Nevertheless, the assays with concentrations of 1 mg/ml and 5 mg/ml showed low efficiency in the reaction by precipitating a high amount of this molecule at the end of the functionalization reaction. Thus, subsequent characterization of CD-Cur and the drug encapsulation test were performed using the nanoplatforms functionalized using an initial concentration of 0.1 mg/ml of curcumin.

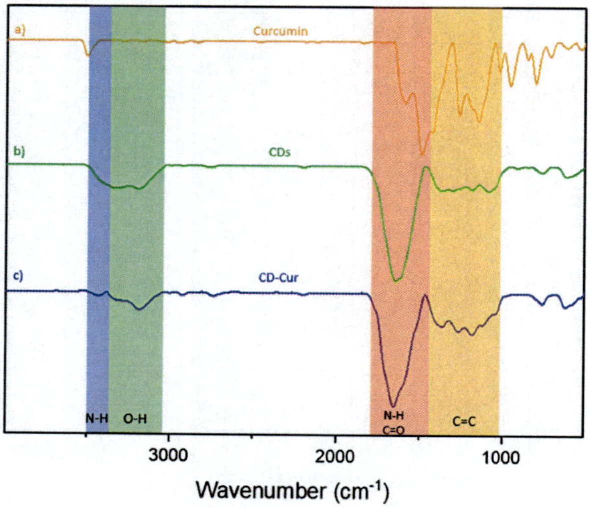

Fig. 2. FTIR of curcumin, CDs and CD-Cur

Moreover, curcumin and the functional groups of the surfaces of the CD and CD-cur were characterized by FT-IR (Fig. 2). The vibrational spectrum for curcumin exhibits the C=C stretch in the range of 1000 to 1400 cm cm^{-1} from the aromatic rings of the molecule and, the C=O stretch in the field of 1400 to 1700 cm cm^{-1}. The CDs spectrum exhibits the C=C stretch of the aromatic rings from the core of CDs in the range of 1000 and 1400 cm cm^{-1}, a prominent band attributed to the overlapping of N-H and C=O stretching is exhibited from 1500 cm cm^{-1} to 1800 cm cm^{-1} and, the bands from the N-H and O-H bending are shown in the range of 3000 cm cm^{-1} to 3500 cm cm^{-1}. The presence of amine and hydroxyl groups is relevant for forming this drug delivery system because they will establish the negative electrostatic environment that allows the functionalization with curcumin and the formation of a highly polar surface of the CDs. The functionalization of CDs and curcumin is confirmed by the presence of the peaks of the C=C stretching and the C=O stretchings from curcumin appearing in the C=C and the C=O/N-H stretching bands of CDs, respectively, in the CD-Cur spectrum.

Furthermore, Z potential measurements of CDs revealed a surface charge of -22.75 $\pm$ 0.79 mV. This proves that CDs obtained possess many amines and hydroxyl groups on the surface making them negatively charged nanoparticles. An increased distribution in the magnitude of the electrostatic charge of CDs demonstrated that these nanoparticles obtained have a non-uniform amount of polar functional groups on the surface.

3.2 Synthesis and Characterization of Dex/HDI/AT

Dex/HDI/AT hydrogel was synthesized in two phases: the synthesis and characterization of AT and, the crosslinking of the hydrogel. From the characterization of AT, it is necessary to determine the composition of the compound and its potentially conductive properties. Then, from Dex/HDI/AT hydrogel, it is essential to confirm its composition and the favorability of its structure for the encapsulation of CD-Cur nanoplatforms.

In Fig. 3, the UV-Vis absorption spectrum of AT shows two absorption peaks. The first one, with a maximum at 320 nm, is attributed to the $\pi- > \pi^*$ transition of the C=C bonds found in the aromatic ring of the compound. The second one, with a maximum in 600 nm, relates to the benzene-quinone transition of the aromatic rings [13]. This transition is provoked by the electron transfer from alkenes in the benzene ring to the C-N bond. This electron transfer in the AT, is the phenomenon that will allow the AT to conduct electricity and, thus, provide an electro-responsive functionality to the Dex/HDI/AT hydrogel.

The FTIR spectrum of AT shown in Fig. 4b exhibits a vibrational band at 800 cm^{-1} attributed to the C-H bond bending from the aromatic rings, the 1100 cm cm^{-1} and 1300 cm cm^{-1} bands can be related to the C-N bonds from the amines found in aniline and, the peaks at 1500 and 1600 cm cm^{-1} refer to the C=C bonds found in the aromatic rings from the same compound. The band at 3300 cm cm^{-1} confirms the presence of the secondary amines that are formed in the polymerization of aniline [3]. Thus, it is confirmed that the presence of the aromatic rings that are relevant to the conductivity of the compound, and the formation of secondary amines from the polymerization of aniline that will allow the crosslinking of this tetramer with dextran.

In Fig 4c, it is shown the FT-IR spectrum of the synthesized DEX/HDI/AT hydrogel. The first band at 1000 cm cm^{-1} is attributed to the C-O bonds from the ethers found in natural polymers like dextran, as shown in Fig. 4a. The peaks located at 1300 cm cm^{-1} and 1600 cm cm^{-1} are attributed to the C-N and C=C bonds found in AT which are exhibited in Fig. 4b. Lastly, the band at 3300 cm cm^{-1} is attributed to the O-H and N-H bonds from dextran and AT, respectively [10].

SEM images shown in Fig. 5 confirm the porous structure of the hydrogel that favors the encapsulation of the different agents inside the hydrogel. It is possible to observe that most pores sizes range from 5 μ m to 10 μ m. This is convenient for the proposed application herein since the synthesized CD-cur are expected to have size values on the nanometric scale.

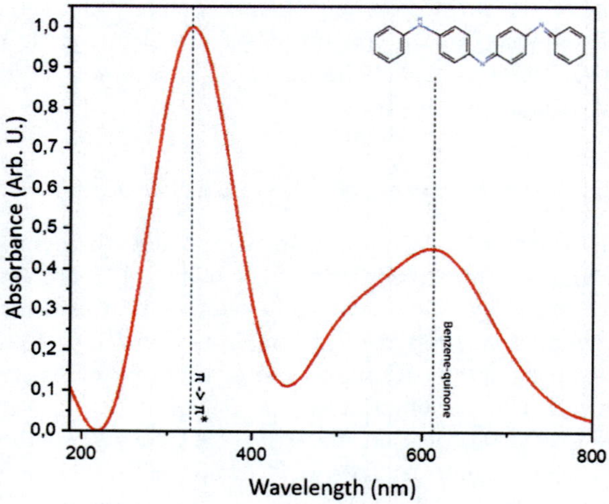

Fig. 3. Absorption spectrum of aniline tetramer dispersed in DMSO. The figure also displays the AT structure on the upper right part of the graph.

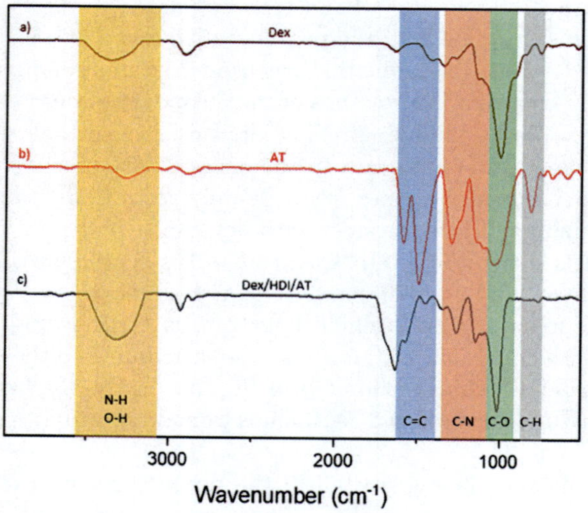

Fig. 4. FTIR spectrum of Dex, AT and Dex/HDI/AT.

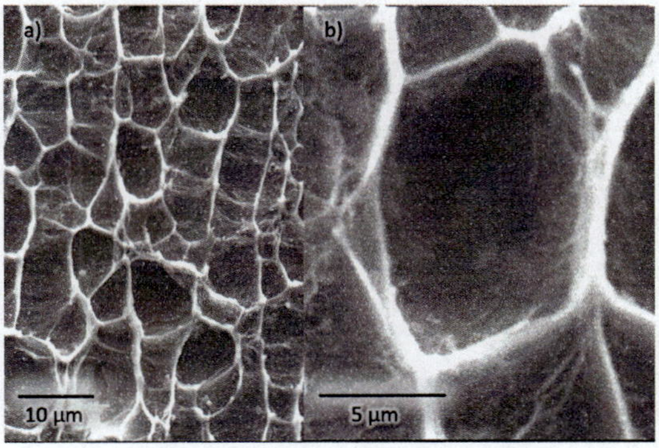

Fig. 5. SEM images of the DEX/HDI/AT hydrogel surface at a) 1500x and b) 5000x.

3.3 Drug Encapsulation Test

For the drug encapsulation test, it was necessary to establish the calibration curves for all the agents used (fluorescein, CDs, and CD-Cur). These curves showed the correlation between concentration and absorbance for each agent (Table 1). The calibration curves are used to calculate the amount of agent encapsulated inside the hydrogel (Eq. 1).

Table 1. Linear regression parameters of the calibration curves

Agent	Slope	Intercept	R^2
Fluorescein	35.8311	0.0091	0.9978
CDs	4.1878	0.0234	0.9894
CD-Cur	1.4799	0.0792	0.9961

From Fig. 6 is possible to observe that fluorescein was encapsulated faster inside the hydrogel than CDs and CD-Cur. As the fluorescein is highly hydrophilic and since the polymeric network of the hydrogel creates a highly polar environment on the inside, this molecule is very likely to migrate to the interior of the matrix. CDs are also encapsulated inside the hydrogel, initially, at a similar rate as the fluorescein. However, after 150 min, a lower encapsulation percentage was observed. This occurs probably due to the formation of aggregates during the dispersion of CDs or the lack of uniformity in the distribution of functional groups on the surface of CDs. Finally, the CDs-Cur nanoplatform exhibited similar behavior as the CDs.This aspect validated the hypothesis that using CDs as a vehicle for curcumin delivery can improve this molecule's interaction with hydrophilic environments. These results are similar to the study performed by Pooresmaeil and Coll. The authors showed how the co-loading of curcumin and doxorubicin inside a hydrogel could be achieved by adding carbon dots, that through electrostatic interactions facilitate the retention of the curcumin molecule [9].

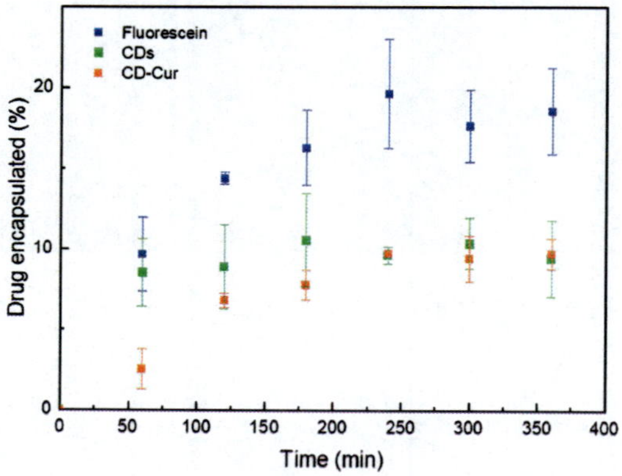

Fig. 6. Drug encapsulation test. The graph displays the encapsulation profile of each agent in the hydrogel. The blue, green and orange points represent fluorescein, CDs and CD-Cur, respectively. (Color figure online)

4 Conclusions

After successfully synthesizing the electro-responsive hydrogel, CD and CD-Cur, the encapsulation test was performed. The results established the hydrogel's capacity to encapsulate fluorescein efficiently due to the favorable interactions with the polymeric matrix and aqueous environment. Lower values are observed in the incorporation profiles and accumulation percentage of CD and CD-Cur in the hydrogel, which is expected due to the physicochemical properties of each of these agents. However, the polar interactions of the CDs with the polymeric matrix allow the incorporation into the hydrogel despite the presence of curcumin molecules, validating the use of CDs as drug delivery vehicle for hydrophobic drugs like curcumin. Further studies will be carried out to characterize the hydrogel's electro-responsive drug-releasing mechanism.

Conflict of Interest The authors declare that they have no conflict of interest.

Acknowledgements. D.C.R.B., D.F.M. and S.R.C., are grateful to the Universidad del Rosario for financial support. M.D.S.F acknowledges financial support from MinCiencias under Convocatoria 874 - Jovenes Talento.

References

1. Chen, Y., Lu, Y., Lee, R.J., Xiang, G.: Nano encapsulated curcumin: and its potential for biomedical applications. Int. J. Nanomed. **15**, 3099–3120 (2020)

2. Dreiss, C.A.: Hydrogel design strategies for drug delivery. Current Opinion in Colloid Interface Sci. **48**, 1–17 (2020)
3. Guo, B., Finne-Wistrand, A., Albertsson, A.C.: Facile synthesis of degradable and electrically conductive polysaccharide hydrogels. Biomacromol **12**(7), 2601–2609 (2011)
4. Kim, H.J., Kim, D.J., Karthick, S.N., Hemalatha, K.V., Raj, C.J.: Curcumin dye extracted from Curcuma longa L. used as sensitizers for efficient dye-sensitized solar cells. Int. J. Electrochem. Sci. **8**(9) (2013)
5. Kolosnjaj-Tabi, J., Gibot, L., Fourquaux, I., Golzio, M., Rols, M.P.: Electric field-responsive nanoparticles and electric fields: physical, chemical, biological mechanisms and therapeutic prospects. Adv. Drug Deliv. Rev. **138**, 56–67 (2019)
6. Nagaraju, G.P., Benton, L., Bethi, S.R., Shoji, M., El-Rayes, B.F.: Curcumin analogs: Their roles in pancreatic cancer growth and metastasis. Int. J. Cancer **145**(1), 10–19 (2019)
7. Nessa, M.U., Huq, F.: 8–Curcumin in cancer prevention and therapy. In: Kabir, Y. (ed.) Functional Foods in Cancer Prevention and Therapy, pp. 137–158. Academic Press (2020). https://doi.org/10.1016/B978-0-12-816151-7.00008-9, https://www.sciencedirect.com/science/article/pii/B9780128161517000089
8. Onaciu, A., Munteanu, R.A., Moldovan, A.I., Moldovan, C.S., Berindan-Neagoe, I.: Hydrogels based drug delivery synthesis. Characterization and Admin. Pharmaceutics **11**(9), 432 (2019)
9. Pooresmaeil, M., Namazi, H.: Folic acid-modified photoluminescent dialdehyde carboxymethyl cellulose crosslinked bionanogels for pH-controlled and tumor-targeted co-drug delivery. Int. J. Biol. Macromol. **200**, 247–262 (2022)
10. Qu, J., Liang, Y., Shi, M., Guo, B., Gao, Y., Yin, Z.: Biocompatible conductive hydrogels based on dextran and aniline trimer as electro-responsive drug delivery system for localized drug release. Int. J. Biol. Macromol. **140**, 255–264 (2019)
11. Sagbas, S., Sahiner, N.: Carbon dots: preparation, properties, and application. In: Khan, A., Jawaid, M., Inamuddin, Asiri, A.M. (eds) Nanocarbon and its Composites, Woodhead Publishing Series in Composites Science and Engineering, pp 651–676. Woodhead Publishing (2019). https://doi.org/10.1016/B978-0-08-102509-3.00022-5, https://www.sciencedirect.com/science/article/pii/B9780081025093000225
12. Vigata, M., Meinert, C., Hutmacher, D.W., Bock, N.: Hydrogels as drug delivery systems: a review of current characterization and evaluation techniques. Pharmaceutics **12**(12), 1188 (2020)
13. Wang, H., Guo, P., Han, Y.: Synthesis and surface morphology of tetraaniline-block-poly(L-lactate) diblock oligomers. Macromol. Rapid Commun. **27**(1), 63–68 (2006)
14. Yan, H., Zhang, B., Zhang, Y., Su, R., Li, P., Su, W.: Fluorescent carbon dot-curcumin nanocomposites for remarkable antibacterial activity with synergistic photodynamic and photothermal abilities. ACS Appl. Bio Mater. **4**(9), 6703–6718 (2021)

Characterization of Mesenchymal Stem Cells from Tooth Pulp

J. S. Katayose[1,2](✉), D. C. Ferraraz[2], and C. B. Lombello[3]

[1] Federal University of ABC (UFABC), São Bernardo Do Campo, SP, Brazil
jackeline.s@ufabc.edu.br
[2] Criogênesis, São Paulo, Brazil
[3] Center for Engineering, Modeling and Applied Social Sciences (CECS), Federal University of ABC (UFABC), São Bernardo Do Campo, SP, Brazil

Abstract. Advances in regenerative medicine have shown a promising perspective in the use of mesenchymal stem cells in the treatment of various diseases and other conditions, such as in the cellular healing process. Mesenchymal stem cells are undifferentiated cells, with high proliferative capacity and differentiation in other cell lines, and when manipulated *in vitro*, can undergo several passages without losing their characteristic. Among the various sources of obtaining this cell type the tooth pulp has shown to be more promising, due to its easy obtainment (it is a non invasive procedure) and isolation, than other cell sources. The mesenchymal stem cells present some characteristics, such as adhesion to the plastic, the possibility of multiplying and specific surface markers expression pattern by flow cytometry labeling that allows its characterization. The aim of the present study is to characterize the mesenchymal stem cells obtained from the third molar tooth pulp, according to the parameters described above. The results showed adherence of the cultured cells to the plastic, 70% of the cells adhered 30 min after seeding. The quantitative data was confirmed with light microscopy observations. The immunophenotypic labeling was consistent with MSCs flow cytometry pattern. In addition the karyotype analysis demonstrated cell chromosome stability. These results indicate a promissing source of cells to be used in tissue engineering.

Keywords: Mesenchymal stem cells · Tooth pulp stem cells · Flow cytometry · Tissue engineering

1 Introduction

In 2006, the Mesenchymal and Tissue Stem Cells Committee of the International Society for Cellular Therapy (ISCT) proposed a minimum criterion for the characterization of mesenchymal stem cells (MSC). These are considered undifferentiated cells, and should have a great potential for proliferation and a high capacity for differentiation into specialized cells, such as, for example, osteocytes, chondrocytes and adipocytes. And, under ideal conditions of temperature (37 °C) and CO_2 (5%), they have the ability to adhere to plastic and proliferate [1–3].

J. L. B. Marques et al. (Eds.): CLAIB 2022/CBEB 2022, IFMBE Proceedings 98, pp. 416–424, 2024.
https://doi.org/10.1007/978-3-031-49401-7_42

Due to this ability to differentiate into distinct mesenchymal phenotypes, mesenchymal stem cells (MSCs) have been extensively studied in tissue repair in regenerative medicine and cell-based therapy [4]. Especially tooth pulp cells are promising tools in tissue engineering as a promising alternative treatment for pulpitis and apical periodontitis [5].

Adult MSCs are widely studied because of their main advantages, that is to have low immunogenicity, ease of obtaining and isolation, as well as *in vitro* expansion, in addition to the possibility of being stored cryopreserved. One possible origin of MSCs is tooth pulp stem cells (DPSCs), obtained in children from deciduous teeth, or in adult humans from the third molar [6, 7].

Tooth pulp stem cells (DPSCs) are used because teeth are often extracted during dental treatment, being an excellent source of stem cells when compared to the collection of bone marrow mesenchymal stem cells (BMMSCs). DPSCs are located in the dental crown, more specifically in the pulp chamber. Studies have shown that these cells can be cultured for long periods of time without affecting their characteristics, such as specific stem cell labeling, plasticity and morphology [8, 9].

The morphological characteristics of MSCs are different at each passage, according to their source. Mesenchymal cells in menstrual blood are flattened; a more polygonal shape and with cytoplasmic granulations can be seen in cells coming from the umbilical cord, whereas those coming from the tooth pulp have a fibroblast-like morphology along the passages [10].

The profile of cell proliferation between the MSC from the pulp of primary teeth and from the third molar presents some variations, some studies have shown that PDSC of permanent teeth have a greater proliferative potential when compared to those of deciduous teeth, while other studies show that the latter have greater ability to proliferate [11, 12].

Another characteristic of MSCs is their characterization by flow cytometry, which consists of the analysis and separation of cells, obtaining a qualitative and quantitative result on each cell in particular. In this process, antibodies conjugated to different fluorochromes are used that react with specific epitopes, on the surface or inside the cells, binding or adhering to them, resulting in specific labeling [13]. Mesenchymal stem cells labeling pattern show expression of CD105, CD90 and CD73 markers, being negative for CD34 and CD45 markers of hematopoietic cells, CD14 (monocyte marker), or CD11b, CD79, or CD19 and HLA-DR in more than 95% of cells in culture [14].

The control of genetic stability after cell manipulation is performed by the karyotype, which aims to detect possible alterations in the chromosomes obtained after cell culture passage. For karyotype evaluation the metaphases are observed chromosomes are arranged and analyzed, based on the "International System for Human Cytogenetic Nomenclature" (ISCN) [15, 16].

For the development of this study, post-thawed mesenchymal stem cells were used, obtained from third molar tooth pulp, from healthy patients aged between 20 and 37 years. The project was submitted to the Plataforma Brasil system, in order to obtain approval from the Research Ethics Committee, following the normative recommendations CAAE 80572417.0.0000.5594. The aim of the study is to characterize the mesenchymal stem cells obtained from the tooth pulp, taking into account the adhesion to the plastic, the possibility of multiplying and labeling by flow cytometry, to evaluate these cells as a possible source to be used in tissue engineering.

2 Materials and Methods

2.1 Cell Culture

The mesenchymal stem cells (MSCs) used in the experiments were isolated from tooth pulp (CAAE 80572417.0.0000.5594) digested with collagenase type IV and cultivated in culture medium containing 89% DMEM low glucose with phenol red (Gibco, USA), 10% heat inactivated fetal bovine serum (FBS) (Gibco, USA) and 1% antibiotic/antimycotic (Gibco, USA), and kept in an incubator at 37 °C with 5% CO_2 until growth. The culture medium was changed every 3 days.

2.2 Adherence Curve

After their growth, the mesenchymal stem cells were washed with 1× PBS pH 7,4 (Gibco, USA) and released from the T25 culture bottles, using 0.25% Trypsin-EDTA (Gibco, Canadá).

These were then washed with culture medium and centrifuged for 5 min and 1200 rpm and plated in a 96-well cell culture plate (Baby I Centrifuge Mod 206—FANEM). It was inoculated 1000 cells per well, in a 96 well culture plate.

Cells in suspension were counted in periods of 05, 10, 20 and 30 min, using a Neubauer chamber with trypan blue. The adhesion was observed indirectly, once non-adherent cells were counted.

2.3 Cell Morphology

The morphological analysis of the cell culture was evaluated under an inverted microscope, without phase contrast, after 24, 48, and 72 h. An overall observation of the monolayer was made for a qualitative analysis.

2.4 Immunophenotyping

The immunophenotypic characterization of DPSCs was analyzed by Flow Cytometry in a BD FACS-calibur device using commercially available monoclonal antibodies—BD's Human MSC Analysis Kit.

Cells were suspended and washed and subsequently labeled. The expression profiles used in this characterization were CD73-APC, CD90-FITC, CD44-PE, in addition to unlabeled tubes (negative control) and labeled positive and CD44 (positive control).

2.5 Karyotype

After two days of plating in a T25 culture bottle, another 05 mL of culture medium and 100 uL of colcemid were added, obtaining a final concentration of 0.1 ug/mL and the bottles were then incubated at 37 °C for 6 h. Cell trypsinization was performed and the cells were washed with culture medium in a centrifuge, with rotation of 1600 rpm for 05 min.

After washing, the supernatant was discarded and 04 mL of hypotonic KCl solution [0.075 M] was added and incubated for 12 min. 04 mL of the fixing solution of methanol and acetic acid was added in a proportion of 3:1 and centrifuged again for 05 min (1600 rpm). The supernatant was discarded and 02 mL of fixative solution was added to the pellet to drip the slides.

For use in this experiment, the slides were washed with extran neutral detergent and kept refrigerated in distilled water, the chromosomes were stained with Giemsa and later visualized and photographed under a microscope (Global Optics).

3 Results and discussions

The MSCs count in suspension showed its capacity for cell adhesion to plastic, by a decrease in the cell count in the supernatant, analyzed with trypan blue die in a Neubauer chamber (Fig. 1). The mean and standard deviation were also analyzed, as shown in Table 1. After 5 min 82.5% of the cells remained in suspension. The percentage of suspension cells decreased during the experiment, being 76.75, 46.25, 30% respectively at 10, 20 and 30 min after cell seeding. After 30 min 70% of the cells adhered to plastic substrate, indicating a good substrate affinity.

Table 1. Adherence data (cels/well).

Time (mins)	05 min	10 min	20 min	30 min
Number of samples (n)	4	4	4	4
Mean	8250	6375	4625	3000
Variance	416667	229167	562500	500000
Standard deviation	645	479	750	707
Coefficient of variation (%)	8	8	16	24

The decrease in cell counts in the supernatant can also be visualized under an inverted microscope (Fig. 2). The fewer cells in suspension, either in quantitative or qualitative approaches, indicates adhesion to the plastic substrate, as confirmed by light microscopy monolayer MSCs culture observation (Fig. 3). The adhesion pattern characterization is important to further studies of substrate-cell interactions with different substrates viewing, for example, tissue engineering applications [5, 17].

Cell adhesion and proliferation were observed at 24 h, 48 h and 72 h, showing the cells with their most elongated shape and monolayer formation (Fig. 3), following cell pattern culture previously described.

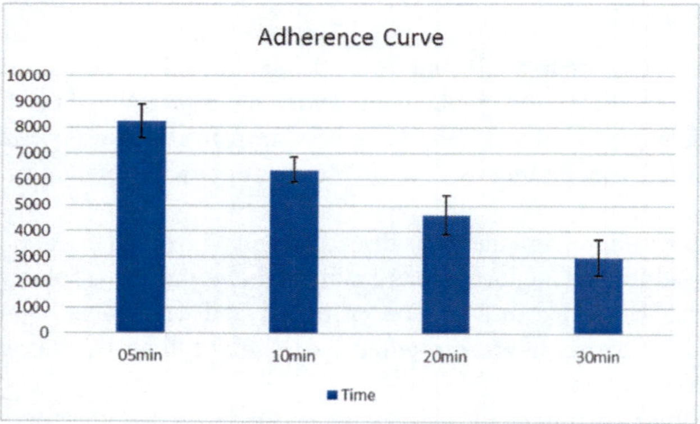

Fig. 1. Adherence curve showing the decrease in the average number of cells present in the supernatant

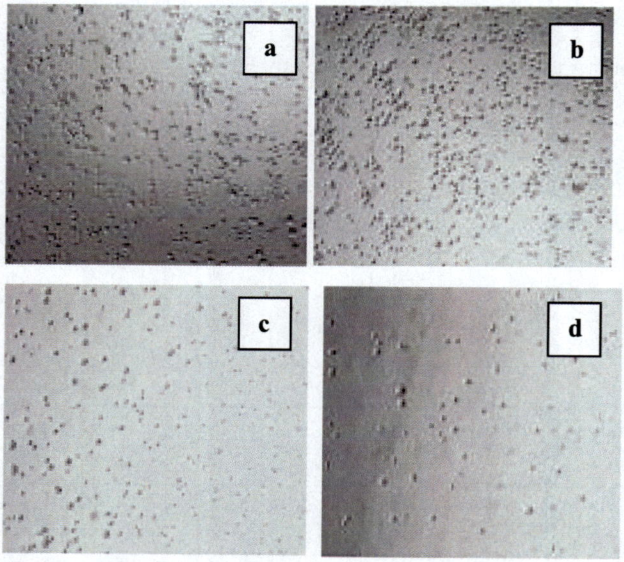

Fig. 2. Cells visualized under an inverted microscope at times of 05 min **(a)**, 10 min **(b)**, 20 min **(c)** and 30 min **(d)**.

Flow cytometry immunophenotyping (Fig. 4) showed positive staining for CD90, CD44 and CD73 antibodies, compared with the results of positive and negative controls (unlabeled). It can be seen that the marking of the samples followed the positive control, as shown in Table 2.

Cell chromosome stability control performed by means of cytogenetics showed that there was no numerical alteration of the chromosomes. Conventional staining was performed, therefore, no type of structural alteration could be observed, as it would be possible in staining with G band. Chromosomes were arranjed according to their group

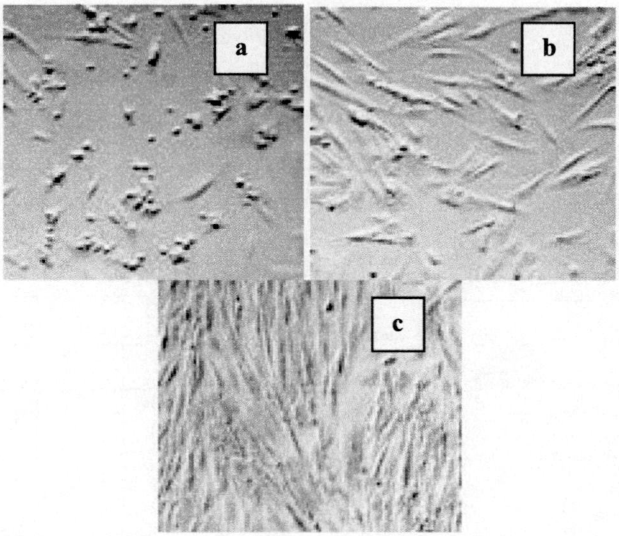

Fig. 3. Proliferation of mesenchymal stem cells from tooth pulp at 24 h **(a)**, 48 h **(b)** and 72 h **(c)**. The morphological analyzes allow us to identify that the cells showed adhesion to the plastic, complying with one of the criteria of the International Society for Cell Therapy (ISCT) and their morphology presents an elongated shape, similar to a fibroblast.

(A–G), observing the 46 chromosomes, being the 22 pairs of homologous (somatic) chromossomes and the pair of sex chromosomes (Fig. 5). The technique used to isolate and proliferate MSCs did not result in chromosome instability or alterations, so the protocol can be used to safely expand these cells viewing tissue engineering future applications. MSC chromosome alteration during proliferation,or culture expansion, may result in oncogenic potential or altered cellular physiology, so its characterization, such as the adhesion pattern, cell viability and substrate adhesion, are fundamental to assure a correct and safe clinical cell application.

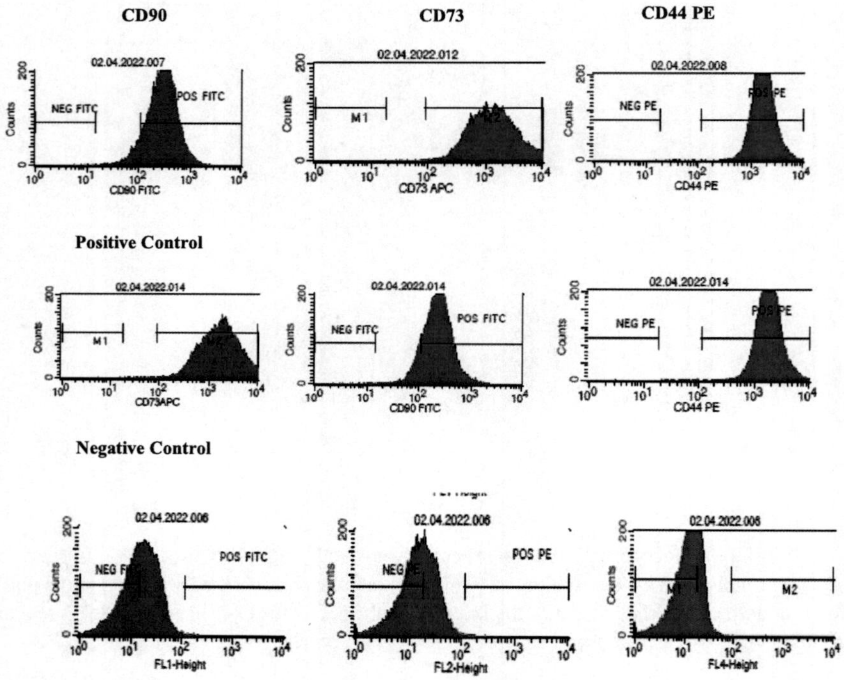

Fig. 4. Flow cytometry of DPSCs, using FITC Mouse Anti-Human CD90 markers; PE Mouse Anti-Human CD44; APC Mouse Anti-Human CD73; Positive Isotypic Control and CD44 (Positive Control); Unmarked (Negative Control).

Table 2. Percentage of flow cytometry labeling of mesenchymal stem cells, compared to positive and negative controls

Antibody	Positive marker (% gated)	Negative control (% gated)	Positive control (% gated)
$CD44^+$ PE	99.93	0.08	99.8
$CD73^+$ APC	99.86	0.01	99.85
$CD90^+$ FITC	91.33	0.08	86.65

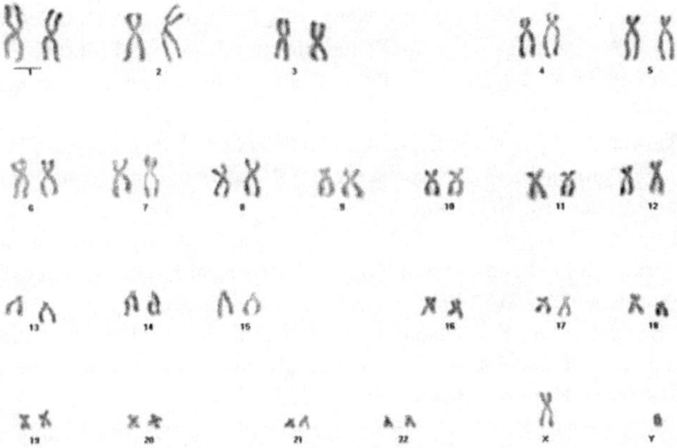

Fig. 5. Cytogenetics of MCS obtained from third molar tooth pulp, with conventional Giemsa staining.

4 Conclusion

Mesenchymal cells from tooth pulp showed their proliferative potential, showing their cell growth after 48 h of plating. The cells used presented cellular stability observed in chromosomal analysis. The immunophenotypic profile seen by flow cytometry resulted in cell labeling above 90% in the analyzed profiles. The permanent teeth stem cell growth and morphology pattern indicates that these are immature cell populations.

Acknowledgment. We thank the Universidade Federal of ABC (UFABC) and Criogenesis for all the support in its infrastructure, in carrying out this work under CAAE 80572417.0.0000.5594.

Conflict of Interest The authors declare that there is no conflict of interest.

References

1. Park, J.W., Fu, S., Huang, B., et al.: Alternative splicing in mesenchymal stem cell differentiation. Stem Cells **38**, 1229–1240 (2020)
2. Suchánek, J., Visek, B., Soukup, T., et al.: Stem cells from a human exfoliated deciduous teeth—isolation, long term cultivation and phenotypical analysis. Acta Medica (Hradec Kralove) **53**, 93–99 (2010)
3. Dominici, M., Le Blanc, K., Mueller, I., et al.: Minimal criteria for defining multipotent mesenchymal stromal cells the international society for cellular therapy position statement. Cytotherapy **8**, 315–317 (2006)
4. Kobayashi, T., Torii, D., Iwata, T., et al.: Characterization of proliferation, differentiation potential, and gene expression among clonal cultures of human dental pulp cells. Hum. Cell **33**, 490–501 (2020)
5. Li, W., Mao, M., Hu, N., et al.: In vitro evaluation of periapical lesion-derived stem cells for dental pulp tissue engineering. FEBS Open Bio. **12**, 270–284 (2022)
6. Prosper, F., Peres, A., Merino, J., et al.: Adult stem cells for myocardial repair. Basic Appl. Myol. **13**, 15–22 (2003)

7. Arora, V., Arora, P., Ak, M.: Banking stem cells from human exfoliated deciduous teeth (SHED): saving for the future. J. Clin. Pediatr. Dent. **33**, 289–294 (2009)
8. Mortada, I., Mortada, R., Al, B.M.: Dental pulp stem cells and neurogenesis. Adv. Exp. Med. Biol. **1083**, 63–75 (2018)
9. Kerkis, I., Kerkis, A., Dozortsev, D., et al.: Isolation and characterization of a population of immature dental pulp stem cells expressing oct-4 and other embryonic stem cell markers. Cells Tissues Organs **184**, 105–116 (2006)
10. Ren, H., Sang, Y., Zhang, F., et al.: (2015) Comparative Analysis of human mesenchymal stem cells from umbilical cord, dental pulp, and menstrual blood as sources for cell therapy. Stem Cells Intern. **1–13**, 2015 (2016)
11. Kunimatsu, R., Nakajima, K., Awada, T., et al.: Comparative characterization of stem cells from human exfoliated deciduous teeth, dental pulp, and bone marrow derived mesenchymal stem cells. Biochem. Biophys. Res. Commun. **501**, 193–198 (2018)
12. Shekar, R., Ranganathan, K.: Phenotypic and growth characterization of human mesenchymal stem cells cultured from permanent and deciduous teeth. Indian J. Dent. Res. **23**, 838–839 (2012)
13. Bacal, N. S., Faulhaber, M. H. W.: Aplicação prática em citometria de fluxo. São Paulo: Editora Atheneu (2003)
14. Bydlowski, S.P., Debes, A.A., Mselli, L.M.F., et al.: Características biológicas das células-tronco mesenquimais. Rev. Bras. Hematol. Hemoter. **21**, 25–35 (2009)
15. Payão, S.L.M., Segato, R., Santos, R.: Controle genético das células-tronco humanas cultivadas. Rev. Bras. Hematol. Hemoter. **31**, 15–18 (2009)
16. Shaffer, L. G., Mggowan-Jordan, J., Schmid, M.: ISCN: An internacional system for human cytogenetic nomenclature (2013)
17. Xie, Z., Shen, Z., Zhan, P., et al.: Functional Dental pulp regeneration: basic research and clinical translation. Int. J. Mol. Sci. **22**, 8991 (2021)

Biological Evaluation of *Himatanthus Drasticus*: Citotoxicity

V. Cauzzo[1(✉)], J. T. Faria[1], F. N. Ambrosio[2], and C. B. Lombello[1]

[1] UFABC/CECS, São Bernardo Do Campo, Brazil
vitoriacauzzo@gmail.com
[2] UFABC/CCNH, São Bernardo Do Campo, Brazil

Abstract. Phytotherapy is the oldest technique for diseases treatment and prevention, and its practice has persisted through generations and maintained principally by popular knowledge. A phytotherapic is a product obtained exclusively from active vegetable raw material. In Brazil, the National Program of Medicinal Plants and Herbal Medicines and the Live Pharmacy are means of promoting traditional culture and making sustainable use of the high diversity of Brazilian flora. Among the families known for their medicinal potential is the Apocynaceae, in which there are species with chemical compounds that are useful in medicine, such as indolic alkaloids, tannins, triterpenes, coumarins, and saponins. Among these species, *Himatanthus drasticus* (janaúba) is popularly used to treat various efermities and, through the substances it presents, points to be a promising plant for medicinal studies. This study aims to evaluate the cytotoxicity of *Himatanthus drasticus*, to ensure the safe use of this plant by the population. The janaúba extract was prepared with 8 drops of latex for 125 mL of ultrafiltered water, and it was stored in a refrigerator until use. The culture of Vero cells was use to the citotoxicity assays, qualitative and quantitative, using extracts concentrations of 25, 50, 100 μg/mL. Standard cell culture (negative control) and dimethylsulfoxide (DMSO, positive control) were used as controls. Qualitative evaluation was made with the analysis of cell morphology, and quantitative evaluation was performed by MTT assay. Through qualitative and quantitative cytotoxicity analysis, the extracts with concentrations of 25, 50 and 100 μg/mL were non-cytotoxic, and safe for use.

Keywords: Cytotoxicity · *Himatanthus sp* · In vitro techniques · Tissue engineering

1 Introduction

The term phytotherapy derives from the Greek therapeia, treatment, and phyton, vegetable. Phytotherapy is the oldest of the techniques for the treatment and prevention of diseases, and its practice has persisted since 8500 B.C. [1], crossing generations and being maintained mainly through popular knowledge [2].

The term phytotherapic is defined by ANVISA [3] as "the product obtained exclusively from active vegetable raw material (…) with prophylactic, curative or palliative

J. L. B. Marques et al. (Eds.): CLAIB 2022/CBEB 2022, IFMBE Proceedings 98, pp. 425–432, 2024.
https://doi.org/10.1007/978-3-031-49401-7_43

purposes". The persistence in the use of phytotherapy is a result of its efficiency in the cure and treatment of several diseases reported in the literature, results not only speculated but also scientifically proven. Its use throughout history has been gradually replaced by synthetic drugs, however, such replacement is slowly receding, mainly due to the support given by the World Health Organization (WHO), which encourages the use of herbal medicines, especially in underdeveloped countries [4].

In Brazil, some policies such as the creation of the National Program of Medicinal Plants and Herbal Medicines are ways to keep alive the traditional culture and to take advantage of the high diversity of Brazilian flora in a sustainable way, besides guaranteeing the rational and safe use of medicinal plants [5]. In addition to this program, the Ministry of Health invested more than R$7.9 million in Live Pharmacy in 2020, which promote the stages of cultivation, collection, processing, storage of medicinal plants, preparation and dispensation of magistral and officinal products of medicinal plants and phytotherapics, with the objective of increasing and facilitating access to phytotherapics [6].

The Pharmacopoeia Herbal Form [3, 7–9] created and updated since 2011 has more than 80 species, including popular species such as chamomile (*Matricaria chamomilla L.*), babosa (*Aloe vera L. Burm.f.*), arnica (*Arnica montana L.*), artichoke (*Cynara scolymus L.*), boldo (*Peumus boldus Molina*). Besides the presentation, the form also exposes: guidelines for preparation, packaging, and storage, warnings, indications, and how to use each species.

Among the families popularly known for their therapeutic properties is the Apocynaceae, of the order Gentiales and subclass Asteridae [10] which consists of about 300 genera and more than 2000 species [11], which manifest themselves mainly as shrubs and latescent trees, with a greater presence in the tropics and subtropics. In Brazil, there are about 41 genera and 376 species of this Family [12]. Some compounds are common among the Apocynaceae species that give them medicinal potential, among them: indolic alkaloids, tannins, triterpenes, coumarins, and saponins.

Indole alkaloids are derived from the amino acid L-tryptophan [11] and can be used as an arterial hypotensive, diuretic, peripheral vasoconstrictor, respiratory stimulant, anesthetic, adrenergic blocking agent, intestinal spasmogen, sedative, and skeletal muscle relaxant [10]. Besides the reported gastroprotective, antitumor, antiplasmodium, antimicrobial and antimalarial actions [13, 14]. It is found, for example, in the bark of the *Aspidosperma quebracho-blanco*, which is rich in this compound, and even has an analgesic action [10].

Tannins are phenolic compounds responsible for the astringency of several plant products and are crucial in the plant-ecosystem relationship [15]. Among its functions, the following are reported in the literature: photoprotective [16], bactericidal, fungicidal, antiviral and antitumor [14].

Triterpenes are secondary metabolites of the terpene class and are identified chemically by the presence of six isoprene units. They are known to have anti-inflammatory, anti-adipogenic, antinociceptive, antimicrobial, and antitumor activities [17].

Coumarins, in turn, are secondary metabolites derived from phenylalanine metabolism [18]. They are noted for their antimicrobial, anti-inflammatory, antiviral, and antioxidant potential [14].

Saponins are glycosides of steroids or polycyclic terpenes and are indicated as phytoprotectants because they participate in the plant's defense system. In addition, they are traditionally used as diuretics [15].

A member of the Apocynaceae family, *Himatanthus drasticus* (Fig. 1) popularly known as janaúba, has both alkaloids, tannins and coumarins [14], as well as triperpenes [19] and saponins [20]. Such composition makes it a target of interest in view of the numerous therapeutic potentials already mentioned and those discussed below.

Fig. 1. Himatanthus drasticus

Among the main medicinal potentials of *Himatanthus drasticus*, the latex has an antimicrobial action [21]; the delay in progression of the ascitic form of Sarcoma 180 [22] and antitumor activity [23]. While research on bark has shown sedative and neuroleptic effects [24]. There is also a related potential to have anti-inflammatory, analgesic, ant nociceptive, and healing effects, besides applications in the treatment of diabetes, ulcers, and cancer [25].

Moura [19] states that through histopathological analysis for acute toxicity with a single dose of *Himatanthus drasticus* latex at the concetration of 2000 mg/kg there was no damage to the liver, kidneys or heart of male Swiss albino mice.

Mattos [26] concludes that the whole *Himatanthus drasticus* (LH) latex and its isolated protein fraction show no toxic effects, even at concentrations higher than popularly recommended and can be considered completely safe for consumption. In addition, LH showed anti-inflammatory activity orally. However, Santos [27] concludes that commercially obtained latex induces toxicity and shows no topical anti-inflammatory activity.

In view of the scarcity of studies that address *the Himatanthus drasticus*, of which there is almost exclusively the focus on the latex extracted from this species, and considering the presence of promising chemical compositions for phytotherapeutic application, this research aims to deepen the data and learning about the biological activities of the janaúba focusing mainly on cytotoxicity to evaluate its viability in other studies of clinical proposition of tissue engineering.

2 Materials and Methods

2.1 Extracts

The janaúba samples were collected in São Paulo, in January 2022, at the coordinates: latitude −23.4841411 and longitude −46.7107164 (SisGen A273ECE) and taken to the Tissue Engineering and Biomolecules Laboratory at the Federal University of ABC. In a sterile environment, the latex was removed from small cuts made with a scalpel. Crude extracts of *Himatanthus drasticus* were prepared from latex with collection at room temperature. In view of the popular use of latex in the preparation using 32 drops per liter of water [28], the extract was prepared at a higher concentration concentration with 8 drops of latex for 125 mL of ultrafiltered water, mixing them manually until a homogeneous solution was formed, in order to analyze cytotoxicity at different concentrations. The extract was stored in a refrigerator until use. The phytochemical profile of the samples was performed, following the methodology described in the literature [29].

2.2 Cell Culture

The culture of Vero cells (CCIAL 057) was performed in HAM F10 medium (Sigma Aldrich), containing 10% fetal bovine serum (Cultilab), and the culture was maintained at 37 °C with 5% CO_2 [30]. Culture medium changes were performed every three days, and subcultured when there was confluence of the cell monolayer. Cultures were monitored by phase microscopy, under an inverted microscope (Axiovert A1/Zeiss). This cell line is considered the standard for testing cellular response for cytotoxicity [31].

Standard cell culture was used as negative control and dimethylsulfoxide (DMSO, Sigma Aldrich) was used as positive control.

2.3 Citotoxicity: Qualitative Assay

A direct cytotoxicity test was performed using a protocol adapted from the ISO 10993–5 [31] standard, for the crude extracts. Test concentrations were 25, 50, 100 μg/mL [19, 32].

For cytotoxicity, Vero cells were inoculated into 96-well plates at a concentration of 10^4 cells/well. Cell cultures were incubated for 24 h, and then exposed to the action of the extract at the concentrations described above.

The qualitative evaluation was made through the analysis of cell morphology after 24 h of contact with the extract, using light microscopy, with phase contrast (Axiovert, Zeiss).

2.4 Citotoxicity: Quantitative Assay

Quantitative evaluation was performed by MTT assay (3-[4,5-dimethyl thiaxolone-2-yl]-2,5-diphenyltetrazolium bromide) (Sigma Aldrich) [33]. The results were obtained with spectrophotometric reading at 570 nm in an Elisa reader (Thermo Scientific, Multiskan Go). Data on percentage of cell viability relative to the negative control were presented. Statistical analysis of the data was performed, with one-way ANOVA test followed by Tuckey's test for $p < 5\%$.

3 Results and Discussion

3.1 Qualitative Analysis

The direct cytotoxicity test for the different concentrations were observed individually by phase-contrast light microscopy, in comparison to the positive control (Fig. 2a) and negative control (Fig. 2b).

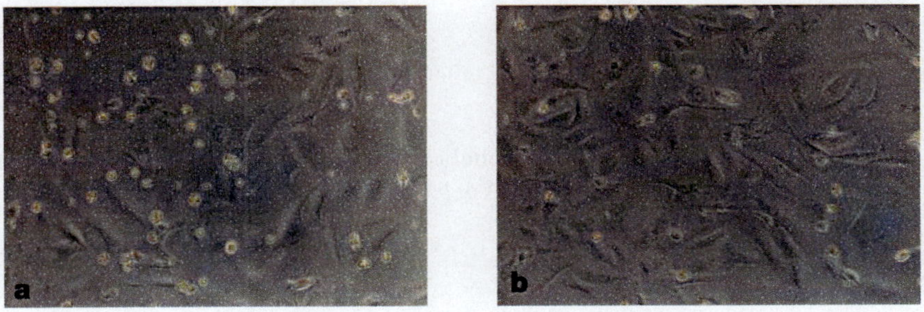

Fig. 2. Positive (**a**) and negative control (**b**)

As can be seen in Fig. 3, the crude extract concentration of 25 μg/mL maintained the morphological integrity of the cells, with a polygonal shape, nucleus and nucleolus evident, in addition to the confluent monolayer remaining adherent. Thus, it is concluded that, for this concentration, there is no cytotoxicity [34].

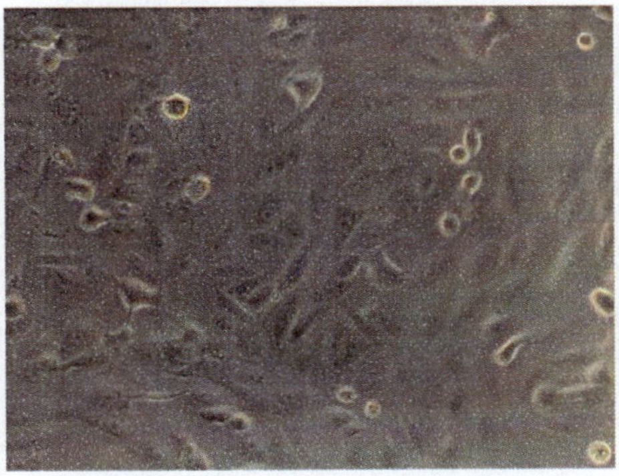

Fig. 3. *Himatanthus drasticus* extract (25 μg/mL)

As for the concentrations of 50 and 100 μg/mL, as can be seen in Fig. 4, there were also no indications of cytotoxicity, in view of the same parameters observed in the concentration of 25 μg/mL.

In the literature review Moura [19] states that cytotoxicity tests performed on Sarcoma 180 cells and Ehrlich cells using the cell culture method, showed cytotoxicity of 46.21 and 41.79% respectively, at the concentration of 200 μg/mL. Coutinho [20]

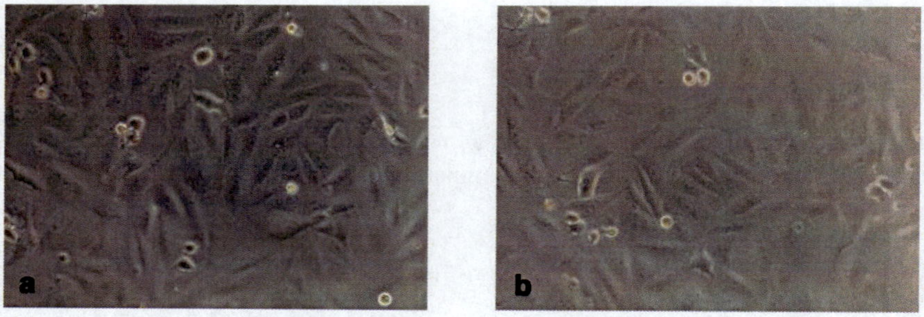

Fig. 4. *Himatanthus drasticus extract* (**a** 50 µg/mL, **b** 100 µg/mL)

conducted the study from the hydrooalcoholic extract of the leaves (996.48 g of the dry material) in sheep hemocyte analysis and determined that there is no cytotoxicity.

3.2 Quantitative Analysis

The quantitative analysis considered concentrations in which the cell viability score was higher than 70% as non-cytotoxic [31]. Therefore, the results (Fig. 5) show that the concentrations of 25, 50 and 100 µG/ml are not cytotoxic, highlighting the same results obtained in the qualitative analysis.

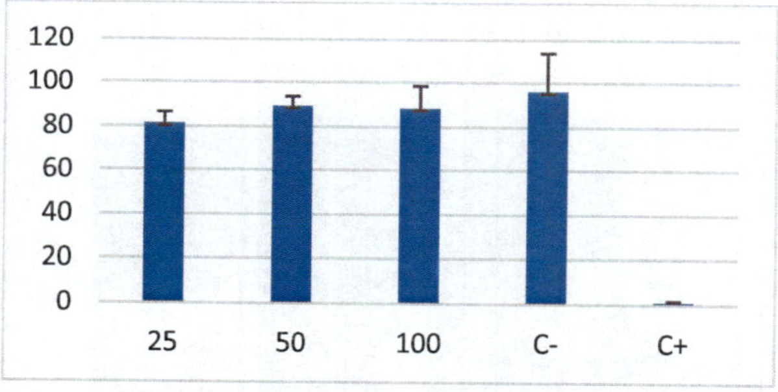

Fig. 5. Quantitative evaluation of the cytotoxicity of *Himatanthus drasticus* extract at a concentration of 25, 50 and 100 µg/mL

4 Conclusions

The latex of *Himatathus drasticus* was considered non-citotoxic at concentrations of 25, 50 and 100 µg/mL in Vero cell culture assay. Thus, it becomes feasible to maintain the studies for future clinical proposition in addition to further studies to observe the other effects and potentials of the species viewing tissue engineering.

Acknowledgment. The authors would like to thank the UFABC-CNPq for Institutional Scientific Initiation Scholarship Program (PIBIC).

Conflict of Interest The authors declare that they have no conflict of interest.

References

1. UFJF: Universidade Federal de Juiz de Fora Programa de Plantas Medicinais e Terapias Não-convencionais. Atividades: Fitoterapia. Disponível em: https://www.ufjf.br/proplamed/atividades/fitoterapia (2021). Acesso em: 21 de maio de 2021
2. Brasil: Ministério da Saúde Programa Nacional de Plantas Medicinais e Fitoterápicos (2009)
3. Brasil: Agência Nacional de Vigilância Sanitária. Formulário de Fitoterápicos da Farmacopeia Brasileira. 2ª ed (2021a)
4. Mattos, G., et al.: Plantas medicinais e fitoterápicos na atenção primária em saúde: percepção dos profissionais. Ciênc. Saúde Coletiva **23**, 3735–3744 (2018)
5. Brasil: Ministério da Saúde Política e Programa Nacional de Plantas Medicinais e Fitoterápicos (2016)
6. Brasil: Governo investe R$ 7,9 milhões em ampliação de Farmácias Vivas. Disponível em: https://www.gov.br/pt-br/noticias/saude-e-vigilanciasanitaria/2021/02/governo-investe-r-7-9-milhoes-em-ampliacao-de-farmacias-vivas (2021b). Acesso em: 21 de maio de 2021
7. Brasil: Agência Nacional de Vigilância Sanitária (2011) Formulário de Fitoterápicos da Farmacopéia Brasileira.1ª ed (2011)
8. Brasil: Agência Nacional de Vigilância Sanitária. Formulário de Fitoterápicos da Farmacopéia Brasileira. Primeiro Suplemento (2018a)
9. Brasil: Agência Nacional de Vigilância Sanitária. Consolidado de normas de registro e notificação de fitoterápico (2018b)
10. Pereira, M.M., Jácome, R.L.R.P., Alcântara, A.F.C., et al.: Alcalóides indólicos isolados de espécies do gênero Aspidosperma (Apocynaceae). Quím Nova **30**, 970–983 (2007)
11. Baratto, L. C.: Estudo químico-analítico e morfoanatômico de espécies medicinais brasileiras da família Apocynaceae: *Himathantus lancifolius* (MULL. ARG.) Woodson e *Rauvolfia sellowii* MULL. ARG.. Dissertação de Mestrado. Pós-graduação em Ciências Farmacêuticas. Universidade Federal do Paraná (2010)
12. Kinoshita, L.S.: Parte integrante da Flora Fanerogâmica do Estado de São Paulo. Flora Fanerogâmica do Estado de São Paulo. Instituto de Botânica, São Paulo **4**, 35–92 (2005)
13. Henrique, M.C., Nunomura, S.M., Pohlit, A.M.: Alcaloides indólicos de cascas de Aspidosperma vargasii e A. desmanthum. Quím Nova **33**, 284–287 (2010)
14. Luz, H.S., Santos, A.C.G., Lima, F.C., et al.: Prospecção fitoquímica de Himatanthus drasticus Plumel (Apocynaceae), da mesorregião leste maranhense. Rev Brasil Plantas Medic **6**, 657–662 (2014)
15. Castejon, F. V.: Taninos e Saponinas. Seminário de mestrado apresentado ao Programa de Pós Graduação em Ciência Animal. Universidade Federal de Goiás (2011)
16. Violante, I.M.P., Souza, I.M., Venturini, C.L., et al.: Avaliação in vitro da atividade fotoprotetora de extratos vegetais do cerrado de Mato Grosso. Rev. Bras. Farmacogn. **19**, 452–457 (2009)
17. Ferreira, S.F.C.O., Silva, M.K.A., et al.: Bioatividades de Triterpenos isolados de plantas: Uma breve revisão. Rev Virtual Quím **12**, 234–247 (2020)
18. Montagner, C.: Atividades antifúngica, citotóxica (células tumorais humanas) e hemolítica de cumarinas naturais e semi-sintéticas. Universidade Federal de Santa Catarina, Programa de Pós Graduação em Biotecnologia (2007)

19. Moura, D. F.: Avaliação da toxicidade e efeitos biológicos do látex extraído de Himatanthus drasticus (Mart.) PLUMEL, Dissertação de Mestrado. PósGraduação em Saúde Humana e Meio Ambiente, Universidade Federal de Pernambuco (2016)
20. Coutinho, G.S.L.: Bioprespecção das folhas, casca e látex da espécie vegetal Himatanthus drasticus (janaúba). Universidade Federal do Maranhão, Programa de Pós Graduação de Saúde e Ambiente (2013)
21. Figueiredo, C.S.S.: Folhas de Himatanthus drasticus: Caracterização química e avaliação de suas atividades antimicrobianas, antibiofilmes e antiproliferativas. Molecules (Basel, Suíça) **22**, 910 (2017)
22. Santos, G.J.L.: O látex de Himatanthus drasticus (Apocynaceae) reduz o estresse oxidativo e modula o CD4+, CD8+, FoxP3+ e HSP-60+ expressões em camundongos portadores de Sarcoma 180. J. Ethnopharmac. **220**, 159–168 (2018)
23. Mousinho, K.C., Oliveira, C.C., Ferreira, J.R.O., et al.: Efeito antitumoral de proteínas laticíferas de Himatanthus drasticus (Mart.) Plumel—Apocynaceae. J. Ethnopharmacol. **137**, 421–426 (2011)
24. Pinto, B.A.S.: Estudo pisicofarmacologico do extrato bruto das cascas de Himatanthus drasticus MART. Universidade Federal do Maranhão, Dissertação de Mestrado. Pós-graduação em saúde e Ambiente (2011)
25. Cauzzo, V.: Capítulo 23: Propriedades biológicas de Himatanthus drasticus: Revisão de literatura. In: Medicina: A ciência e a tecnologia em busca da cura. Ponta Grossa, PR. Editora Atena (2021)
26. Matttos, M. P. V.: Aspectos bioquímicos e etnofarmacológicos do látex de Himatanthus drasticus Mart. (Plumel). Programa de Pós Graduação em Bioquímica. Universidade Federal do Ceará (2013)
27. Santos, G.J.L.: Modulação da resposta cicatricial e antitumoral do látex de Himatanthus drasticus (Apocynaceae) em modelos experimentais in vivo. Universidade Estadual do Ceará, Programa de Pós Graduação em Ciências Veterinárias (2016)
28. Sousa, K. O.: Látex do Avelós (Euphorbia tirucalli) e da Janaúba (Himathanthus drasticus) modificam a expressão de proteínas intracelulares nas células de Carcinoma Pulmonar de Lewis in vitro. Relatório Final. Fatec Sorocaba, Centro Paula Souza (2018)
29. Nascimento, E.M., et al.: Estudo fitoquímico e potencial antibacteriano do látex de Himatanthus drasticus (Mart.) Plumel. Biota. Amazônia. **8**, 28–32 (2018)
30. Ammerman, N. C. et al.: Growth and maintenance of Vero cell lines. Curr. Protoc. Microb. 11:4E:A.4E.1–A.4E.7 (2008)
31. International Standards Organization: ISO 10993–5. Biological evaluation of medical devices—Part 5: Tests for in vitro cytotoxicity. p. 34 (2009)
32. Morais, F. S.: Chemical profiling of secondary metabolites from Himatanthus drasticus (Mart.) Plumel latex with inhibitory action against the enzymes αamylase and α-glucosidase: In vitro and in silico assays. J. Ethnopharmacol. **253**, 112644 (2020)
33. Mosmann, T.: Rapid colorimetric assay for cellular growth and survival: application of proliferation and cytotoxicity assays. J. Immunol. Meth. **65**, 55–65 (1983)
34. Masson, A. O., Lombello, C. B.: Metodologias de avaliação citotóxica: estudo comparativo segundo tempo de exposição. 9° COLOAB (2016)

Check for updates

Nanostructured Ti Grade 2 for Biomedical Applications: Evaluation of Cellular Interaction

A. F. Martins(✉), L. F. Costa, A. Mendes Filho, and C. B. Lombello

Center for Engineering, Modeling and Applied Social Sciences (CECS), Universidade Federal Do ABC (UFABC), São Bernardo Do Campo, Brazil
andressa.francine@aluno.ufabc.edu.br

Abstract. Tissue engineering has as one of its objectives the development of biocompatible materials, capable of interacting with biological tissues, promoting regeneration or function repair. Titanium is widely used in dental and medical implants, due to its mechanical properties and biocompatibility. Alternatives of relatively low-cost titanium, with low elastic modulus, non-toxic and that maintain the mechanical properties are currently intended. Pure titanium grade 2 presents desired characteristics such as low density, excellent biocompatibility, and good corrosion resistance, although mechanical resistance may be improved. This work is focused on the biological evaluation of commercialized pure titanium grade 2 (Ti CP2), and its characteristics as a biocompatible material, viewing medical applications, such as osseointegration.

Keywords: Biocompatibility · Cellular adhesion · Cell culture techniques · Titanium · Nanostructure

1 Introduction

Tissue engineering has as one of its objectives the development of biocompatible materials, capable of interacting with biological tissues, and promoting regeneration or function repair. Simultaneously, the biomaterials mimetize the mechanical and biological properties of the specific tissue to be engineered [1].

Biomaterials also aim to be bioinert and bioactive, since it is necessary to minimize adverse effects on patients, generate an effective tissue interaction, be functional, and promote cell growth [2, 3]. In addition, metallic biomaterials have been idealized to be efficiently used in the development of medical devices, used as an alternative in lesions caused in bone tissues that need implants, in order to withstand high loads for fracture fixation, and also as substitutes for some joints. [3, 4]. Therefore, among the most used metallic biomaterials, stainless steels, cobalt-chromium alloys, and titanium alloys can be mentioned [4, 5].

Titanium is widely used in dental and medical implants, due to its mechanical properties and biocompatibility. However, some titanium related toxicity is known, especially in titanium alloys containing aluminum and vanadium, as Ti-6Al-4V [6, 7]. Alternatives of relatively low cost titanium, with low elastic modulus, non-toxic and that maintain

J. L. B. Marques et al. (Eds.): CLAIB 2022/CBEB 2022, IFMBE Proceedings 98, pp. 433–438, 2024.
https://doi.org/10.1007/978-3-031-49401-7_44

the mechanical properties are currently intended. Pure titanium grade 2 presents desired characteristics such as low density, excellent biocompatibility, and good corrosion resistance, although mechanical resistance may be improved. Nanostructured pure titanium can result in a more resistant material, without chemical changes that could result in associated toxicity.

This work is focused on the biological evaluation of commercialized pure titanium grade 2 (Ti CP2), and its characteristics as a biocompatible material. It is based on morphological analysis of initial cellular interaction with nanostructured pure titanium grade 2 samples, viewing medical applications, such as osseointegration.

2 Methodology

2.1 Titanium

To perform the cell interaction analysis, commercially pure titanium grade 2 (Ti CP2) was used, which has its chemical composition according to ASTM F67 requirements. The nanostructured Ti was obtained through severe plastic deformation routes involving ECAP (Equal Channel Angular Pressing), a combination of ECAP with lamination at different temperatures (cryogenic deformations) and HPT (High Pressure Torsion) [8–10].

The samples used were: 0X, control, without treatment; 2X, two ECAP process, at 300oC; 2X + CRLT, two ECAP process, at 300oC and lamination at −100 °C; 2X + HPT, two ECAP process, at 300oC and HPT at room temperature. For more details on the ECAP and HPT processing please see reference 8. The preparation of samples was carried out from initial sterilization. Initially, the samples were cut into four pieces in a low-speed cutter (LabCut 1010) using a diamond blade. At sequence, using a sander and an ARAPOL-E universal polisher, the samples were sanded in water sandpaper with granulometry between 120 and 2000. The cleaning protocol of the samples pre-biological evaluation started with an ultrasonic bath with the presence of detergent lasting 20 min for each sample, from the ultrasound vat UltraCleaner 750 A. After these steps, the quadruplicates of each type were placed in beakers of 50 mL with the presence of distilled water and remain at rest for 24 h. At the end of this period, the samples were covered in packaging for autoclave sterilization.

2.2 Cell Culture

The in vitro cell interaction test, cells of the Vero lineage were used (Instituto Adolfo Lutz/SP, CCIAL 057) [11]. These cells originally are obtained from the renal epithelial tissue of an African green monkey (Cercopithecus aethiops), present a fibroblast-like type and they are anchorage-dependent, being widely used in biological evaluation, since it has a high rate of cell proliferation, do not have oncogenic factors and does not present risks to human health [12]. The cell cultures were maintained with HAM-F10 culture medium, supplemented with 10% Fetal Bovine Serum (FBS) and antibiotic, and were incubated at 37 °C and 5% CO_2.

The Ti samples used were sterilized in an autoclave. The samples were positioned in 24-well culture plates. For the evaluation of the initial cellular interaction, the cells were directly inoculated onto the Ti samples, and were kept in culture for two hours. After this period, it was possible to observe initial cellular interaction on Ti surface with Scanning Electron Microscope (SEM) (FEI, Quanta 250). For this, the samples were fixed in 2.5% glutaraldehyde and were washed with phosphate buffer following distilled water, dehydrated in critical point equipment (Leica EM CPD 300) and coated with gold (Leica ACE 200), so that they could be visualized with SEM.

Furthermore, for the morphological control of the experiment, light microscopy with phase contrast (AxioVert A1, Zeiss) was used.

3 Results and Discussion

Vero cells from the kidney of the African Green Monkey (Cercopithecus aethiops) present fibroblast-like morphology, and are widely used in biological evaluation assays [13]. In the ideal cell culture condition the characteristic morphology is polygonal, or elongated, with cytoplasmic spreading, forming a cell monolayer (Fig. 1) [11, 14].

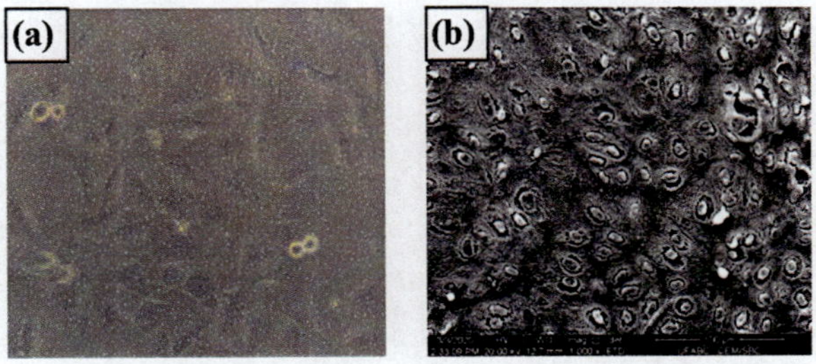

Fig. 1. Morphology of Vero cells after 7 days of culture, standard characteristic morphology for the lineage. **a** Phase-contrast light microscopy (200x); **b** Scanning electron microscopy (SEM).

The morphological analysis of the cells in the presence of the different Ti samples evidences morphological changes of cells in initial cellular interaction, characterizing adhesion assays. The cell attachment and spreading can occur within a 24 h period, depending on factors such as the solid substrate and cell line [15, 16]. Although it can characterize the initial interaction, the cell adhesion with the substrate can be observed in the first hours of culture for Vero cells [16–18].

In all samples, rounded cells were observed, with the beginning of adhesion onto the Ti substrate, and regions with less or less great cell density (Fig. 2).

In the Ti samples in which ECAP passes were performed (Fig. 2a, b) and HPT (Fig. 2c, d) the cells were preferentially deposited in regions of deformities or cutoffs of Ti, or close to the edges of the samples, with rounded morphology, and the little surface of contact between the cells and the metal. Few cells were observed in the process of spreading, mainly in samples with HPT (Fig. 2d). This morphology pattern was not observed in the other samples, being representative of a more stable interaction with the material. In samples with ECAP and cryogenic treatment (Fig. 2e, f) and in the control condition (Fig. 2g, h) the areas with the highest number of cells were observed in interaction with

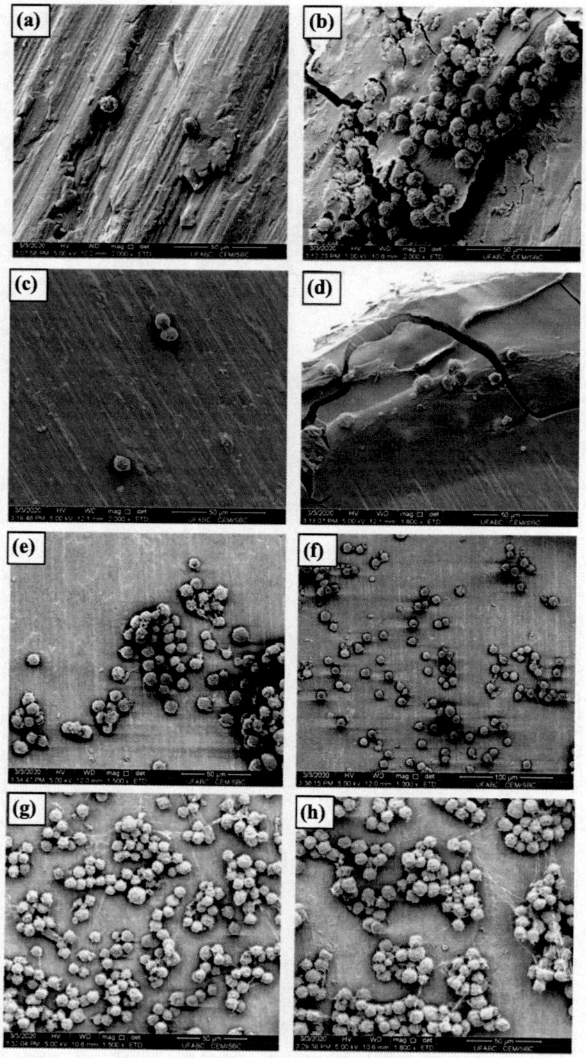

Fig. 2. Scanning electron micrographs of Vero cell morphology after 2 h of interaction with Ti samples. (**a, b**) 2X; (**c, d**) 2X + HPT; (**e, f**) 2X ECAP + CRLT; (**g, h**) Control (0X).

Ti, in addition to the specific treatment (Fig. 2e, f), the geometry of these samples may also have favored the preferential interaction of cells, as they are surfaces more regular and flat. The samples submitted to ECAP and cryogenic lamination, represented by Fig. 2e, f, were the ones that showed the best adhesion to Vero cells, if not considering the control condition. Thus, these samples with better mechanical properties maintain the characteristics of biocompatibility.

As evidenced in the literature, nanostructured samples usually show greater adhesion than samples with natural microstructure and are related to surface roughness [15, 16].

However, in this mentioned situation, the titanium had been processed by four passes ECAP, and an even higher cell volume result was reported in the sample with eight passes. This may be the reason for this work to present a smaller number of cells in the sample with just two ECAP passes. In this way, more passes of both ECAP and HPT, could result in greater cellular adhesion. In a similar way, the cells were present in the regions that had grooves, according to what was evidenced by other authors [17, 18].

4 Conclusions

This work aimed to understand the interaction of nanostructured Ti CP2 with simulated biological conditions, simulated with in vitro experimentation using Vero cells, to consider the viability of cell growth for applications in Ti implants. With the biological tests performed, greater initial cell adhesion was observed in samples that underwent greater improvement in mechanical properties, indicating promising future osseointegration. Characterization of long term cellular interaction may be developed in the future.

Acknowledgements. The authors would like to thank the Multiuser Central Facilities (UFABC) for the experimental support.

Conflict of Interests The authors declare that they have no conflict of interest.

References

1. Nascimento, M.H.M., et al.: Hyaluronic acid and chitosan based hydrogels for cartilage tissue engineering. Polymers [online] **26**, 360–370 (2016)
2. de Menezes, A.W.A.: Estudo da deposição de hidroxiapatita por PEO em amostras de TI CP2 e aço inoxidável AISI 316L TI CP2 e AISI 316LVM, p. 57p. Universidade Federal do Rio Grande do Norte, Pós-graduação em Engenharia Mecânica (2021)
3. Cruz, M.N.C.F.: Inovação em Dispositivos Médicos, p. 47p. Universidade de Lisboa, Mestrado Integrado em Ciências Farmacêuticas. Faculdade de Farmácia (2017)
4. Moreno, M.S.M.S.: Engenharia de Tecidos na substituição de tecido ósseo, p. 57p. Universidade Fernando Pessoa Faculdade de Ciências da Saúde, Porto (2014)
5. Moreira, A.C.: Análise da influência da morfologia porosa de implantes de titânio no processo de crescimento ósseo, p. 157p. Universidade Federal de Santa Catarina, Pós-Graduação em Ciência e Engenharia de Materiais (2013)
6. Sidambe, A.T.: Biocompatibility of advanced manufactured titanium implants-a review. Materials (Basel) **7**, 8168–8188 (2014)
7. Kim, K.T., Eo, M.Y., Nguyen, T.T.H., Kim, S.M.: General review of titanium toxicity. Int. J. Implant Dentistr. **5**(1), 1–12 (2019). https://doi.org/10.1186/s40729-019-0162-x
8. Mendes Filho, A.A., et al.: The effects of severe plastic deformation on some properties relevant to Ti implants. Mater. Res. **15**, 27–31 (2012)
9. Antonialli, A.I.S., et al.: The machinability of ultrafine-grained grade 2 Ti processed by equal channel angular pressing. J. Mater. Res. Tech. **1**, 148–153 (2010)
10. Sordi, V.L., et al.: Microstructure and tensile strength of grade 2 titanium processed by equalchannel angular pressing and by rolling. J. Mater. Sci. **47**, 7870–7876 (2012)
11. Ammerman, N. C. et al.: Growth and maintenance of Vero cell lines. Curr. Prot. Microb. 11:4E:A.4E.1–A.4E.7 (2008)

12. Vero (ATCC® CCL-81TM). In: https://www.atcc.org/products/all/CCL-81.aspx
13. International Standards Oganization.: ISO 10993–5. Biological evaluation of medical devices—Part 5: Tests for in vitro cytotoxicity, pp. 34 (2009b)
14. Marlina, S., et al.: Development of a Real-Time Cell Analysing (RTCA) method as a fast and accurate screen for the selection of chikungunya virus replication inhibitors. Parasit. Vect. **8**, 579 (2015)
15. Chen, N., et al.: Cell adhesion on an artificial extracellular matrix using aptamer-functionalized PEG hydrogels. Biomater. **33**, 1353–1362 (2012)
16. Moussa, H.I.: Pattern-dependent mammalian cell (Vero) morphology on Tantalum/Silicon Oxide 3D nanocomposites. Materi. **11**, 1306 (2018)
17. Esposito, A.R., et al.: Benefits of oxygen and nitrogen plasma treatment in Vero cell affinity to poly(lactide-co-glycolide acid). Mater. Res. **16**, 695–702 (2013)
18. Nascimento, M.H.M., et al.: Evaluation of cell interaction with polymeric biomaterials based on hyaluronic acid and chitosan. J. Mater. Sci. Mater. Med. **28**, 68 (2017)
19. Estrin, Y., et al.: Accelerated stem cell attachment to ultrafine grained titanium. Acta Biomater. **7**, 900–906 (2011)
20. Wennerberg, A., et al.: Effects of titanium surface topography on bone integration: a systematic review. Clin. Oral Impl. Re. **20**, 172–184 (2009)
21. Qu, J., et al.: The use of micromachined surfaces to investigate the cell behavioral factors essential to osseointegration. Oral Dis. **2**, 102–115 (1996)
22. Anselme, K., et al.: Effect of grooved titanium substratum on human osteoblastic cell growth. J. Biomed. Mater. Res. **60**, 529–540 (2002)

Surface Treatment of Polyurethane Membranes to Increase the Hemocompatibility

H. T. T. Oyama[1(✉)], A. P. Godoy[2], L. R. X. Cortella[1], G. J. M. Fechine[2], H. Ribeiro[2], and I. A. Cestari[1]

[1] Divisao de Bioengenharia, Instituto do Coracao (InCor), Hospital das Clinicas, HCFMUSP, Faculdade de Medicina, Universidade de Sao Paulo, Sao Paulo, Brazil
helena.oyama@incor.usp.br

[2] Mackenzie Institute of Research in Graphene and Nanotechnologies – MackGraphe, Mackenzie Presbyterian University, Sao Paulo, SP, Brazil

Abstract. Over the past decades, the hemocompatibility of polymeric materials that are used in contact with blood remains one of the main challenges in the development of medical devices. As the interactions of material with blood are strongly dependent on the chemical and physical properties of the material's surface, surface modification has been an interesting strategy used to improve their hemocompatibility. In this work, a method to modify the surface of polyurethanes (PU) membranes was investigated. Therefore, the benzalkonium heparinate complex grafting over PU surface using UV light approach was employed. The PU membranes produced were characterized by a set of techniques as infrared spectroscopy, colorimetry, contact angle and surface micrographs. The results showed that the presence of heparin grafted onto the surface produced the modification of the structure as presented by FTIR results, as well as the increasing of the hydrophilicity was verified. These changes suggested that modifications reduce the chances of protein adsorption on the PU surface.

Keywords: Hemocompatibility · Surface modification · Polyurethane · Grafting · UV light

1 Introduction

A good interaction between blood and synthetic materials is desirable in cardiovascular devices for short term use and demanded in long term devices. Despite the current progress in material sciences, the use of synthetic materials may generate significant risk of thrombosis and other complications [1].

Material-blood contact may cause platelet consumption, activation of the coagulation cascade complement, depletion and denaturation of plasma proteins. Material calcification that leads to loss of mechanical properties is also a problem. Also, clot and thrombus formation and thromboembolization in the case of implants, remain the most serious problem in the development of blood contacting devices. When blood comes into contact with any surface that is not covered by endothelium, a set of changes

J. L. B. Marques et al. (Eds.): CLAIB 2022/CBEB 2022, IFMBE Proceedings 98, pp. 439–447, 2024.
https://doi.org/10.1007/978-3-031-49401-7_45

occur, resulting in the conversion of proteins into a complex of insoluble proteins, fibrin and platelet aggregates. The latter, together with trapped red blood cells, constitute the thrombus, the final stage of the coagulation phenomena [2].

As blood material interactions are strongly dependent on the chemical and physical properties of the materials´ surface, surface modification has been a strategy used to improve its hemocompatibility [3, 4]. Among the already used strategies, surface passivation stands out by the insertion of hydrophilic and zwitterionic polymers [5] that proved to be strongly "protein resistant", incorporation of bioactive molecules, (anticoagulants such as platelet inhibitors) and the mimicry of vascular endothelium [6].

This work aims to investigate a method for surface modification of polyurethanes by photografting process. The strategy is based on the use of benzalkonium chloride as an intermediate component, associated with heparin. Surface modification characterization was assessed by Scanning Microscopy, Fourier Transform Infrared Spectroscopy by Attenuated Total Reflectance, Colorimetry and Contact Angle.

2 Materials and Methods

2.1 Materials

Polyurethane Copolymer (BioSpan®, Royal DSM N.V., Netherlands); Sodium heparin (Adeste Animal Products Ltd., Pr, Brazil); Benzalkonium heparinate solution (Celsus Lab, Inc. USA); Toluidine blue P.A (LabSynth, SP, Brazil); Benzalkonium chloride (LabSynth); Omnirad 2959 (1- [4- (2- hydroxyethoxyl)- phenyl]- 2- hydroxy.

-2- methyl propanone (IGM Resins, Netherlands); Sodium chloride, Hexane (Sigma-Aldrich, Germany). DMAc (N, N-Dimetylacetamide) (Sigma-Aldrich, Germany).

2.2 Methods

PU Membranes (BioSpan®) Preparation. Membranes were prepared using the solution mixing and evaporation techniques according to the manufacturer. Briefly, the polyurethane (PU) copolymer in gel form was dissolved in DMAc in a ratio of 1:1 by weight. The mixture was deposited on a steel mold to obtain a regular flat polymeric membrane. The solution was kept in laminar flow with exhaustion for 4 hours at 55 °C. After that, membranes were washed in ultrapure water and heated at 60 °C. This process was repeated 3 times and then membranes were left in a vacuum desiccator for 24 h to ensure complete elimination of the solvent.

Treatment with Benzalkonium Heparinate (HepB) by Photographing Process. The PU membranes were placed in a mixture of a solution of 15 mL of HepB in isopropanol, used as purchased, and 0.1% of Omnirad 2959 curing agent for 30 min. Subsequently, the samples were dried and placed in the dark in a vacuum desiccator.

The PU membranes with HepB complex were submitted to UV radiation for 10 min at a distance of 30 mm from the UV lamp, with an intensity of 200 mW/cm^2, at a wavelength of 400 nm. Thereafter, they were washed repeatedly with ultrapure water.

After drying, the membranes were immersed during 3 s in saline solution and allowed to dry again for 60 min, for removal of free heparin.

Quantification of Heparin by the Colorimetric Method. Toluidine blue metachromatic dye was used to colorimetrically determine the amount of heparin bound to the PU membrane. As toluidine blue is adsorbed onto the heparin polymer, the method involves monitoring the depletion of the dye in the supernatant at the maximum wavelength of dye [7, 8].

Preparation of Solutions.

- Toluidine Blue—25 mg of toluidine blue and 1 g of NaCl were added to 500 mL of 0.01 N HCl solution.
- 0.2% sodium chloride solution—0.2 g of NaCl was diluted to 100 mL with ultrapure water.

Heparin Calibration Curve. To obtain the calibration curve of absorbance of heparin solution, seven measurements were performed, adding 2.0 ml of toluidine blue solution and 10, 20, 30, 40, 50, 70, 90, 100, 150 μg of heparin in each tube. Afterwards, their contents were diluted to a total volume of 3 mL with the 0.2% NaCl solution and shaken, so that the dye-heparin complex was formed. To remove the complex formed, 3 mL of hexane was added to each tube and they were shaken vigorously for 30 s on a Vortex shaker. The complex phase formed was removed and the remaining solutions were diluted (1:10) with absolute ethanol. The absorbances of alcoholic solutions were measured by a spectrophotometer at the maximum wavelength of absorption of the toluidine blue dye obtained. The calibration curve was obtained by absorbance and heparin quantities were determined using the absorbance values measured [9].

Quantification of Heparin on PU Surface. 18 samples measuring 1cm^2 approximately were separated and placed in 18 test tubes containing 2.5 mL of the toluidine blue standard solution diluted to a total volume of 5 mL of 0.2% NaCl solution. After 1, 8, 24, 48 and 96 h, the complex toluidine blue and heparin was removed adding 3 mL of hexane to each tube. After stirring on a Vortex shaker for 30 s, the complex phase was removed and the remaining solutions were diluted (1:10) with absolute ethanol and read at a wavelength of 626 nm, considering as "blank" a solution of 0.2% NaCl in absolute ethanol. Thus, the areas of all samples were calculated.

To analyze the stability of the immobilized heparin, a sample of 3 by 6 cm^2 area was placed in a glass vial containing phosphate-buffered saline (PBS, 12 mL, pH 7.4) at 37°C for 4 days. Periodically, the supernatant (2 mL) of the PBS solution was removed and mixed with a 0.005% toluidine blue solution (3 mL). After precipitation of the dye with heparin, the absorbance of the solution was measured with a spectrophotometer, and the concentration of released heparin was calculated from the heparin calibration curve and the toluidine blue concentration.

The stability of the immobilized heparin was analyzed by placing the sample was placed in a glass vial containing phosphate-buffered saline (PBS, 12 mL pH 7.4) at 37 °C for 4 days. Periodically, the supernatant (2 mL) of the PBS solution was removed and mixed with a 0.005% toluidine blue solution (3 mL). After precipitation of the dye with heparin, the absorbance of the solution was measured.

In all evaluations, solutions samples and PU were prepared in triplicate for each incubation time.

Wettability and Surface Energy. The contact angles related to the surface free energy were measured to determine the adhesion between materials. This method involves the contact angle that a drop of a given liquid makes with the surface, with the increase or decrease in hydrophilicity, as showed by Young's Eq. (1).

$$\gamma_{LV}\cos\theta_y = \gamma_{SV} - \gamma_{SL} \tag{1}$$

where θ_y is the contact angle [10].

According to Fowkes theory, the total surface free energy is the sum of the dispersive and polar components, presented in Equation [2]:

$$(\gamma_S) = \left(\gamma^d\right) + (\gamma^P) \tag{2}$$

Owens-Wendt's [11, 12] approach included the non-dispersive surface liquid's interaction to calculate the surface free energy substituting the value of the contact angle of each sample measured with water, obtaining the Eq. (3).

$$\gamma_{SL} = \gamma_S + \gamma_L - 2(\gamma_S^d \gamma_L^d)^{1/2} - 2(\gamma_S^P \gamma_L^P)^{1/2} \tag{3}$$

where γ_{SL} is the surface free energy at the solid-liquid interface.

The contact angle measurements of the treated PU and PU membrane were performed by Drop Shape Analyzer Goniometer- DSA100 (KRÜSS Germany) at room temperature and evaluated by the software Advance. The values of surface free energy, as well as the dispersive and polar components, were obtained according to the Owens-Wendt method [12] by measuring the static contact angle with two liquids: water and ethylene glycol. Ten measurements were performed at different sites on the air-facing surfaces of the films of each sample with the two liquids using the sessile drop method and the mean static contact angle and surface energies were determined.

Qualitative Surface Analysis by Fourier Transform Infrared Spectroscopy with attenuated total reflectance (FTIR-ATR). The attenuated total reflectance (ATR) mode is not highly sensitive to the surface, but offer a detailed molecular analysis of the outermost 1-5 microns of a sample. This technique was carried out to obtain information about chemical structures of the of PU surface before and after treatment to identify the involved bonds [13]. For this analysis, a Shimadzu IR Affinity-1 FTIR Spectrometer with ATR mode and resolution 4 in a range from 400 to 4000 cm^{-1} was employed.

Scanning Electron Microscopy (SEM). This assay was performed to characterize the surface morphology of PU membranes before and after treatment. The samples were cut into 5.5 mm diameter discs. The surfaces of the samples were examined using a scanning electron microscope (Hitachi TM 3000, Japan) at a voltage of 5 kV.

Statistical Analysis. The ANOVA statistical analysis was used to verify if there is a difference between the distribution of a measure between the groups (($P < 0.05$) and NSD ($P > 0.05$)).

3 Results and Discussion

The alcoholic solutions with toluidine blue absorbances were read with the aid of a spectrophotometer at the visible wavelength, obtaining the toluidine blue dye maximum wavelength of absorption at 626 nm (see in Fig. 1).

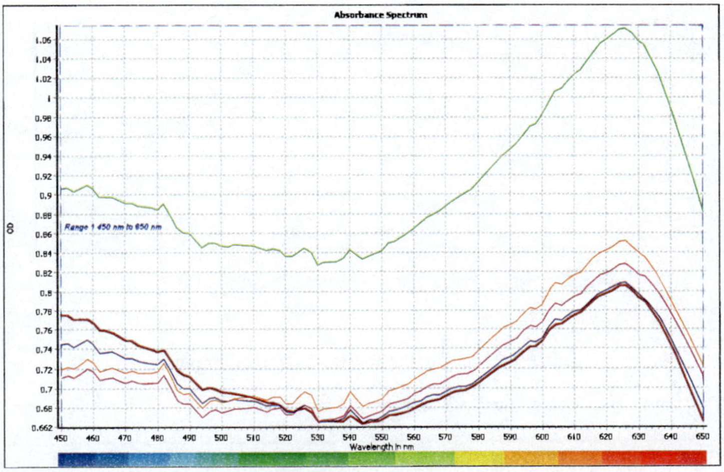

Fig. 1. Absorption spectrum of 5 samples of alcoholic solutions with toluidine blue.

The heparin calibration curve showed a linear relationship between the absorbance at 626 nm caused by residual toluidine blue and the concentration of heparin in the solution (see in Fig. 2). The Table 1 shows the stability of immobilized heparin as a function of time and the Table 2 presents the amount of heparin on the treated surface obtained using the colorimetric method.

Surface wettability, often referred as hydrophobicity and hydrophilicity, affects the biological response to an implanted material. It influences protein adsorption, platelet adhesion and activation, blood clotting and cell and bacterial adhesion [14].

In general, hydrophobic surfaces are considered to be more protein adsorbents than hydrophilic surfaces and in direct contrast to the repulsive solvation forces arising from the strongly bound water on the hydrophilic surface and that inert surfaces [15–17]. The contact angles were obtained with water and ethylene glycol droplets on the surface of the PU before and after treatment with HepB by UV (see Fig. 3a). The respective surface free energy results, as well as the polar and dispersive energy components, are presented in Fig. 3b.

The results showed a decrease in the contact angle after the treatment, that is, the surface after the treatment became more hydrophilic due to the presence of heparin, which is an anionic mucopolysaccharide with a large amount of H^+ and SO_3^- groups.

The surface energy is an important parameter that affects polymer adhesion, material wettability and even biocompatibility [18]. According to Owens, Wendt, Rabel and Kaelble the interfacial tension can be divided into two components: dispersive and polar

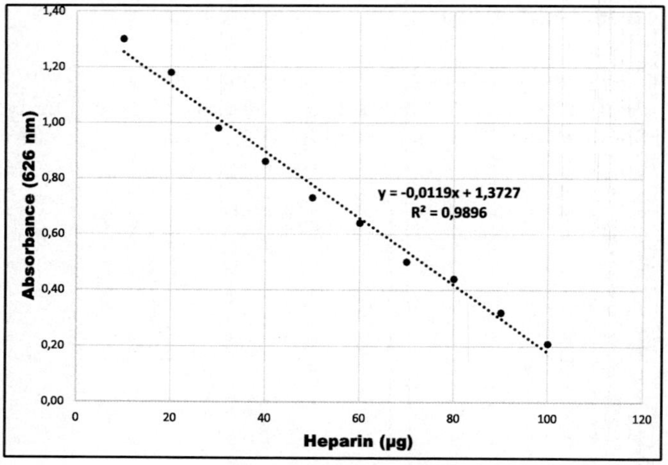

Fig. 2. Heparin calibration curve obtained at 626 nm.

Table 1. Stability of the immobilized heparin.

Time (h)	ABS (626 nm)
1	0.806 ± 0.298
3	0.809 ± 0.044
9	0.851 ± 0.077
24	0.829 ± 0.299
48	0.703 ± 0.007

Table 2. Quantification of heparin obtained by the colorimetric method.

PU treated	Average area (cm^2)	HEPARIN (µg)	HEPARIN/area ($\mu g/cm^2$)
1	0.99	1.358	1.344
8	1.10	1.358	1.494
24	1.00	1.358	1.358
48	1.09	1.357	1.371
72	1.10	1.3558	1.491
96	1.10	1.3556	1.371

interactions [12]. Polar interactions comprise the Coulomb interactions between permanent and induced dipoles. Interactions caused by charge distribution time fluctuations within molecules are called dispersive interactions.

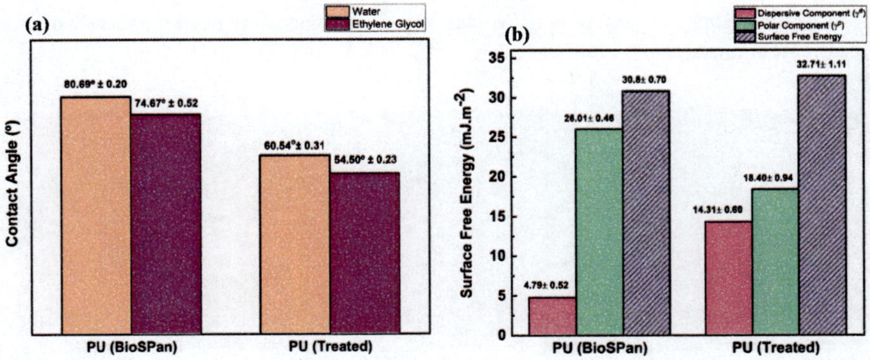

Fig. 3. **a** Contact angles obtained with water droplet and ethylene glycol, on the surface of the PU (BioSpan®) before and after treatment. **b** Surface free energy and polar and dispersive components of the surface tensions of liquids.

The FTIR spectra of PU membranes and PU membranes treated by UV and heparin are depicted in Fig. 4.

The FTIR of the PU-HepB membrane showed the presence of bands related to both PU and heparin functionalities. Regarding the PU, the band at 802.6 cm^{-1} refers to the C = C bending mode; the region comprising between the 1099.2 and 1217.8 cm^{-1} presents the stretching mode of the C-O bond and refers to the alkyl, aryl and vinyl ether radicals; and at 2853.5 cm^{-1}, there is the stretching band C = C. A weak intensity, in the region of 1300 to 1700 cm^{-1}, was observed with bands resulting from the stretching modes of the S = O bonds of heparin.

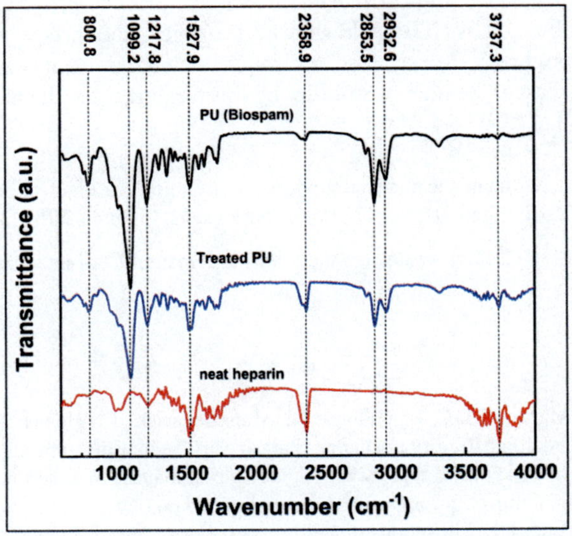

Fig. 4. FTIR spectra of PU Biospan membrane surface, membrane treated and neat heparin.

The micrographs obtained by SEM (see Fig. 5) showed more homogeneous surfaces after the treatment.

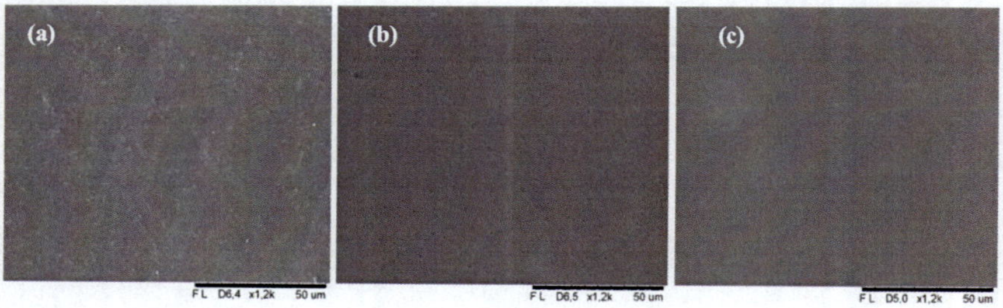

Fig. 5. SEM: **a** PU, **b** PU after photografting and **c** PU after 72 h in PBS.

4 Conclusions

The hemocompatibility of materials used in contact with blood remains one of the main challenges in the development of medical devices. A decrease in the contact angle can be observed after grafting with HepB. The contact angle decreases from 80.69° for the unmodified surface to 60.58° after UV grafting of the heparinate. In addition, the results obtained show an increase in dispersive energy and a decrease in polar energy in the samples after treatment.

The observed decrease in contact angles may represent an indirect measure of the extent of structural change. The increment in hydrophilicity suggests that there will be less adsorption of proteins, thus increasing the surface hemocompatibility of the PU.

The spectra obtained by FTIR-ATR showed a change in the surface of the treated PU. Thus, however, only with the physicochemical characterizations, it was not possible to conclude the increase in hemocompatibility by this treatment, but it showed a tendency to improve it.

Acknowledgment. We thank the financial support of São Paulo State Foundation (FAPESP, Grant 2012/50283–6), FINEP (Grant 01.14.0177.00), FINEP (1253/13) and CNPQ (313846/2021–9).

Conflict of Interest "The authors declare that they have no conflict of interest".

References

1. Maitz, M. F., Martins, M. C. L., Grabow, N., Matschegewski, C., Huang, N., Chaikof, E. L. et al.: The blood compatibility challenge. Part 4: Surface modification for hemocompatible materials: Passive and active approaches to guide blood-material interactions. Acta Biomater. **94**, 33–43 (2019). https://doi.org/10.1016/j.actbio.2019.06.019
2. Jaffer, I. H., Weitz, J. I.: The blood compatibility challenge. Part 1: blood-contacting medical devices: the scope of the problem. Acta Biomater. [Internet] **94**, 2–10 (2019). https://doi.org/10.1016/j.actbio.2019.06.021

3. Ishihara, K.: Blood-compatible surfaces with phosphorylcholine-based polymers for cardiovascular medical devices. Langmuir **35**(5), 1778–1787 (2019). https://doi.org/10.1021/acs.langmuir.8b01565
4. Amiji, M., Park, K.: Surface modification of polymeric biomaterials with poly (ethylene oxide), albumin, and heparin for reduced thrombogenicity. J. Biomater. Sci. Polym. Ed. **4**(3), 217–34 (1993). https://doi.org/10.1163/156856293x00537
5. Ito, Y.: Antithrombogenic heparin-bound polyurethanes. J. Biomater. Appl. **2**(2), 235–265 (1987). https://doi.org/10.1177/088532828700200204.PMID:3333067
6. Qiu, H., Qi, P., Liu, J., Yang, Y., Tan, X., Xiao, Y., et al.: Biomimetic engineering endothelium-like coating on cardiovascular stent through heparin and nitric oxide-generating compound synergistic modification strategy. Biomater. 10–22 (2019). https://doi.org/10.1016/j.biomaterials
7. Smith, P.K., Mallia, A.K., Hermanson, G.T.: Colorimetric method for the assay of heparin content in immobilized heparin preparations. Anal. Biochem. **109**(2), 466–473 (1980). https://doi.org/10.1016/0003-2697(80)90679-x
8. Bae, J.S., Seo, E.J., Kang, I.K.: Synthesis and characterization of heparinized polyurethanes using plasma glow discharge. Biomaterials **20**(6), 529–537 (1999). https://doi.org/10.1016/S0142-9612(98)00204-X
9. Harada, N.S., Oyama, H.T.T., Bártoli, J.R., Gouvêa, D., Cestari, I.A., Wang, S.H.: Quantifying adsorption of heparin on a PVCsubstrate using ATR-FTIR. Polym. Int. **54**, 209–214 (2005). https://doi.org/10.1002/pi.1685
10. Agrawal, G., Negi, Y. S., Pradhan, S., Dash, M., Samal, S. K.: Wettability and contact angle of polymeric biomaterials. In: Tanzi, M. C., Farè, S. (eds.) Characterization of Polymeric Biomaterials, Woodhead Publishing. Charact Polym Biomater, pp. 57–81 (2017)
11. Ren, Z., Chen, G., Wei, Z., Sang, L., Qi, M.: Hemocompatibility evaluation of polyurethane film with surface-grafted poly (ethylene glycol) and carboxymethyl-chitosan. J. Appl. Polym. Sci. **127**, 308–315 (2013)
12. Kozbial, A., Li, Z., Conaway, C., McGinley, R., Dhingra, S., Vahdat, V., et al.: Study on the surface energy of graphene by contact angle measurements. Langmuir **30**(28), 8598–8606 (2014)
13. Rudawska, A., Jacniacka, E.: Analysis for determining surface free energy uncertainty by the Owen-Wendt method. Int. J. Adhes. Adhes. **29**(4), 451–457 (2009)
14. Junkar, I.: Interaction of cells and platelets with biomaterial surfaces treated with gaseous plasma. Adv. Biomembr. Lipid Self-Assemb. **23**, 25–59 (2016)
15. Noh, H., Vogler, E. A.: Volumetric interpretation of protein adsorption : Mass and energy balance for albumin adsorption to particulate adsorbents with incrementally increasing hydrophilicity. **27**, 5801–5812 (2006)
16. Ostuni, E., Chapman, R. G., Holmlin, R. E., Takayama, S., Whitesides, G. M.: A survey of structure—property relationships of surfaces that resist the adsorption of protein. **17**(9), 5605–20 (2001)
17. Luan, Y., Li, D., Wang, Y., Liu, X., Brash, J.L., Chen, H. I.: Radiolabeling, surface plasmon resonance, and quartz crystal microbalance with dissipation: three tools to compare protein adsorption on surfaces of different wettability. **30**(4), 1029–1035 (2014)
18. Kwok, S.C.H., Wang, J., Chu, P.K.: Surface energy, wettability, and blood compatibility phosphorus doped diamond-like carbon films. Diam. Relat. Mater. **14**(1), 78–85 (2005)

Cytotoxic Evaluation of Aqueous Extracts Obtained from Two Brazilian Native Plant Species with Medicinal Potential, *Casearia Sylvestris* and *Hyptis Lacustris*

F. N. Ambrosio[1](✉), F. D. da Silva[1], R. A. Lombello[1], and C. B. Lombello[2]

[1] Center for Engineering, Modeling and Applied Social Sciences (CECS), Universidade Federal Do ABC (UFABC), São Bernardo Do Campo, Brazil
felipe.nogueira@ufabc.edu.br

[2] Center for Natural and Human Sciences (CCNH), Universidade Federal Do ABC (UFABC), São Bernardo Do Campo, Brazil

Abstract. Brazil is home to a wide variety of plants species and, as such, represent a considerable source of new vegetal secondary metabolites with medicinal potential. Two examples of such species are *Casearia sylvestris* and *Hyptis lacustris*, which are native to the country and with the first already being used by some people in folk medicine and the later belonging to a genus containing other species with some health care utilities. However, it is important to assess the safety of the usage of products derived from them and thus a study about the cytotoxicity of aqueous extracts obtained from leaves of both species was performed. Extracts were obtained through the method of dried leaves decoction in water. The work was executed *in vitro* using Vero cells and the MTT viability assay method. It was observed that extracts from both plants did not present cytotoxicity potential in concentrations $\leq$ 1000 μg/ml. Therefore, further investigations may advance using this range of non-cytotoxic concentrations.

Keywords: Aqueous plant extract · *Casearia sylvestris* · Cytotoxicity · *Hyptis lacustris* · MTT

1 Introduction

Historically, the human kind has often relied on plants as sources of treatment for numerous ailments [1] and still does, but as with synthetic drugs, it is important to guarantee the safety and right dosage for natural products [2, 3].

The Brazilian flora has a great potential for discovery of new active compounds with medicinal uses, such as antioxidant [4], anti-inflammatory [5], antimicrobial [6] and antiviral [7].

This work focus on two Brazilian native plant species that have potential medicinal uses, *Casearia sylvestris* Sw. (*Salicaceae*) and *Hyptis lacustris* A.St.-Hil. (*Lamiaceae*).

J. L. B. Marques et al. (Eds.): CLAIB 2022/CBEB 2022, IFMBE Proceedings 98, pp. 448–453, 2024.
https://doi.org/10.1007/978-3-031-49401-7_46

C. sylvestris is used in folk medicine, mainly in the form of infusions, as treatment for snake bite, burns, gastrointestinal problems, among others [8].

Meanwhile *H. lacustris* is still little studied in regards of its medicinal uses, but it belongs to a genus containing other species with antiviral activity [9].

Due to the interest in finding ways to apply extracts from both plants in health related services, it is imperative to first determine a safe dosage by *in vitro* assays. Such information is still scarce in the scientific literature.

Altough there are studies showing the cytotoxic potential of alcoholic [10] and hidroalcoholic [11] extracts of *C. sylvestris*, the present paper aims to explore the cytotoxicity of a aqueous extract, which contains predominantly hydrophilic molecules and could provide useful information about the safety of the extracts in forms such as infusions and other preparations that involves the use of water as solvent.

In parallel, there is no work published about the cytotoxic potential of *H. lacustris*, which highlights the importance of further studies to determine safe dosages of extracts obtained from such species.

Casearia sylvestris leaves and *H. lacustris* infusions may contain compounds that may be toxic to those who consume them. Therefore, this work investigated the safe dosage of aqueous extracts obtained from *C. sylvestris* and *H. lacustris* by *in vitro* assays. Our results showed that both extracts are not cytotoxic against Vero cells in concentrations $\leq 1000\ \mu g/ml$.

2 Methodology

2.1 Plant Extract Preparation

Specimens of *C. sylvestris* and *H. lacustris* were gathered at Campinas, SP (S 22678585 W 47313141) and Santo André, SP (S 23773830 W 36313098), respectively and both with recording at National System for the Management of Genetic Heritage and Associated Traditional Knowledge (Sistema Nacional de Gestão do Patrimônio Genético e do Conhecimento Tradicional Associado—SisGen), A5BDEFF and AED0BE8, respectively.

The plants were dryed at 40 °C for 48 h and then the leaves were separated and crushed.

The plant extracts were obtained by decoction method [12], by adding the crushed dryed leaves at a beaker with distilled water at concentration of 0.1 g/ml. The soaked leaves were heated until 95 °C for 30 min, followed by an ice bath for another 30 min and filtered.

Then, the extracts were dried in a freeze dryer (LIOTOP, K105) and stored at −20 °C.

At the moment of the assays, the extracts were resuspended in distilled water at concentrations of 10 mg/ml.

2.2 Cytotoxicity Assay with MTT Viability Test

Epithelial Vero cell line, derived from the kidney tissue of an African green monkey (*Cercopithecus aethiops L.*) [13], were maintained in cell culture flasks with HAM-F10 medium supplemented with 10% bovine fetal serum (BFS), inside an incubator set to preserve the temperature at 37 °C and atmospheric CO_2 at 5%. When necessary the cells were put through the process of subculturing, following ATCC guidelines [13].

The cytotoxicity assays were performed in accordance with the international standards stipulated by ISO 10993 set of guidelines and, as such, the results were presented in the form of a percentage of cell viability determined through absorbance reading at 570 nm and normalized by the mean absorbance of the non-treated cells (negative control C−). It was considered as cytotoxic any condition that would result in a decrease in cell viability to lower than 70% of the non-cytotoxic negative control group [14, 15].

For the tests, the cells in culture flasks were resuspended with trypsin and seeded in 96-wells plates, with approximately 1.5 x 10^4 cells per well with an available area of 0.32 cm^2. After 24 h of incubation, the culture medium was replaced by a solution of medium containing plant extract, or a solution of 50% dimethyl sulfoxide (DMSO) for positive control (C+), or fresh medium for the negative control (C−).

The extracts were tested at the concentrations of 1000, 100, 50 and 25 μg/ml.

After 24 h of cell culture with the plant extracts, the culture medium was removed and replaced by a solution of 0.5 mg/ml of 3-(4.5-dimethylthiazol-2-yl)-2.5-diphenyltetrazolium bromide (MTT) diluted in HAM-F10 medium without BFS and incubated for 2 h.

Since the MTT is metabolized into formazan crystals inside the mitochondria of viable cells, its concentration in each well, indirectly assessed through light absorbance measurements, can be used as a way to estimate cell viability after the experimental conditions. For this, the plates are taken from the incubator and the MTT solution was removed from the wells, which were then washed with a phosphate buffer solution (PBS) and, finally, it was added 100% DMSO to solubilize the formazan crystals, that acquire a purple coloration, with a darker color for higher concentrations.

Using a plate reader (Thermo Scientific, Multiskan Go) it was possible to determine the absorbance values at 570 nm as an indirect measure of the formazan concentration and, thus, the cell viability in each well.

For the analysis of the results it was used one-way ANOVA followed by Tukey test for multiple comparisons with a P value of 5%, with the aid of the software GraphPad Prism.

3 Results and Discussion

Plant extracts tend to be rich in a wide variety of molecules, some of them toxic for humans and other animals. The scientific literature says that it is possible to find sesquiterpenes in extracts from *C. sylvestris* leaves, along with flavonoids, phenolic compounds [16] and diterpenoids [11] and some species of the genus *Hyptis* contain phenolic acids, flavonoids, chlorogenic acid derivatives, cinnamic acid derivatives, rosmarinic acid and other phenolic compounds [17]. Although some of these compounds may initiate cytotoxic responses [18–23], the cytotoxicity assays showed that both plant aqueous extracts were not cytotoxic in the range of concentrations tested, observed in Figs. 1 and 2, for *C. sylvestris* and *H. lacustris*, respectively.

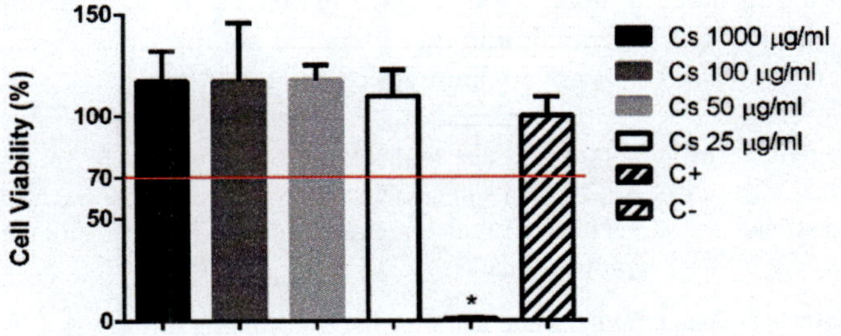

Fig. 1. Vero cells viability 24 h after treatment with aqueous extract from *Casearia sylvestris* (Cs) in comparison with 50% DMSO (C +) and HAM-F10 culture medium (C−), with the horizontal red line indicating the 70% cell viability value, which represents the threshold below which a sample is considered cytotoxic. The * indicates the group with a significant statistical difference, with $p < 5\%$ in an one-way ANOVA test followed by Tukey test.

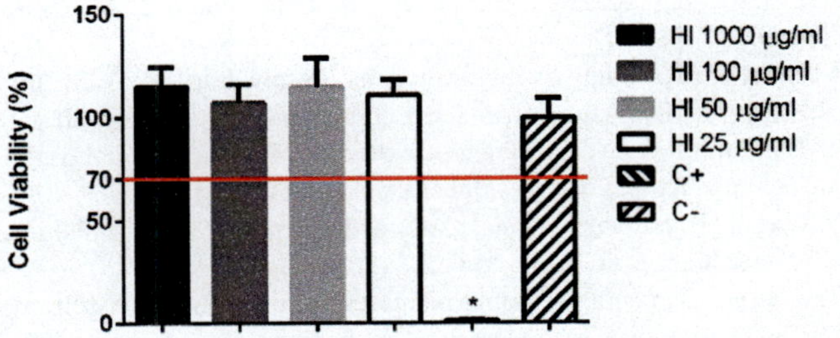

Fig. 2. Vero cells viability 24 h after treatment with aqueous extract from *Hyptis lacustris* (Hl) in comparison with 50% DMSO (C +) and HAM-F10 culture medium (C−), with the horizontal red line indicating the 70% cell viability value, which represents the threshold below which a sample is considered cytotoxic. The * indicates the group with a significant statistical difference, with $p < 5\%$ in an one-way ANOVA test followed by Tukey test.

Comparing the results with the available material from the scientific literature, it is clear that aqueous extracts from *C. sylvestris* shows lower cytotoxic potential than alcoholic ones [10, 11].

Since there is little information about *H. lacustris* available, this results contribute to the determination of a safe concentration range for future applications.

4 Conclusions

According to the results, both the extracts obtained by the decoction of dried leaves from *C. sylvestris* and *H. lacustris* are non-cytotoxic at concentrations ≤ 1000 μg/ml and this is the first step to develop a safe and effective application with them. For *C. sylvestris* this is an important result in comparison with alcoholic extracts, while for *H. lacustris* this allows for an expansion of the knowledge about extracts obtained from this particular species, still little studied. Future works in investigations of biological activities of this extracts should take this information in consideration and limit the studies conditions inside concentrations already proven non-cytotoxic.

Acknowledgments. The authors would like to thank the *Coordenação de Aperfeiçoamento de Pessoal de Nível Superior* (CAPES) for its financial support in the form of a scholarship.

Also, the authors are grateful to the Multiuser Central Facilities (UFABC) for the experimental support, especially for the support with the freeze dryer.

Conflict of Interest The authors declare that they have no conflict of interest.

References

1. Petrovska, B.B.: Historical review of medicinal plants' usage. Pharmacogn. Rev. **6**(11), 1–5 (2012)
2. Akelma, H., Tarikci Kiliç, E., Karahan, Z.A.: Rare burn cases treated traditionally: folk medicine: review of eight cases. J. Burn Care Res. Offic. Publicat. Am. Burn Assoc. **40**(4), 520–526 (2019)
3. Farzaei, M.H., et al.: Poisoning by medical plants. Archiv. Iran. Med. **23**(2), 117–127 (2020)
4. Andrade, J.K.S., et al.: Evaluation of bioactive compounds, phytochemicals profile and antioxidant potential of the aqueous and ethanolic extracts of some traditional fruit tree leaves used in Brazilian folk medicine. Food Res. Int. **143**, 110282 (2021)
5. Ribeiro, V.R., et al.: Brazilian medicinal plants with corroborated anti-inflammatory activities: a review. Pharmaceutic. Boil. **56**(1), 253–268 (2018)
6. Holetz, F.B., et al.: Screening of some plants used in the Brazilian folk medicine for the treatment of infectious diseases. Memorias do Instituto Oswaldo Cruz **97**(7), 1027–1031 (2002)
7. Salles, T.S., et al.: Chemical composition and anti-Mayaro virus activity of Schinus terebinthifolius fruits. Virusdisease **32**(3), 526–534 (2021)
8. Ferreira, P.M.P., et al.: Folk uses and pharmacological properties of Casearia sylvestris: a medicinal review Anais da Academia Brasileira de Ciências scielo (2011)
9. Idrees, M., et al.: Effect of the phytochemical agents against the SARS-CoV and some of them selected for application to COVID-19: a mini-review. Curr. Pharm. Biotechnol. **22**(4), 444–450 (2021)
10. Dos Santos, A.G., et al.: Casearin X, its degradation product and other clerodane diterpenes from leaves of Casearia sylvestris: evaluation of cytotoxicity against normal and tumor human cells. Chemistr. Biodiv. **7**(1), 205–215 (2010)
11. Bueno, P.C.P., et al.: Infraspecific chemical variability and biological activity of Casearia Sylvestris from different Brazilian biomes. Planta medica **87**(1–02), 148–159 (2021)
12. Farmacotécnica de Fitoterápicos. https://farmacotecnica.ufc.br/wp-content/uploads/2019/11/farmacot-fitoterapicos.pdf
13. Vero (ATCC® CCL-81TM). https://www.atcc.org/products/all/CCL-81.aspx
14. ISO 10993–1:2009: Biological evaluation of medical devices—part 1: evaluation and testing within a risk management process. https://www.iso.org/standard/68936.html
15. ISO 10993–5:2009—Biological evaluation of medical devices—part 5: Tests for in vitro cytotoxicity. https://www.iso.org/standard/36406.html
16. Carvalho, F.A., et al.: Casearia sylvestris essential oil degradation products generated by leaf processing. Chemistr. Biodiv. **18**(6), e2000880 (2021)

17. Dos Santos, K.P., et al.: Biological activity of Hyptis Jacq. (Lamiaceae) is determined by the environment. Industr. Crops Product. **112**, 705–715 (2018)
18. Amorim, M.H.R., et al.: Sesquiterpene lactones: adverse health effects and toxicity mechanisms. Critic. Rev. Toxicol. **43**(7), 559–579 (2013)
19. Efferth, T., Kaina, B.: Toxicity of the antimalarial artemisinin and its dervatives. Crit. Rev. Toxicol. **40**(5), 405–421 (2010)
20. Erdenetsogt, U., et al.: Bioactive flavonoids from plant extract of Pyrethrum pulchrum and its acute toxicity. Nat. Prod. Res. **35**(24), 5960–5963 (2021)
21. Chen, Y., et al.: Phytochemical profiles and antioxidant activities in six species of ramie leaves. PloS one **9**(9), e108140 (2014)
22. Meng, J., et al.: Cytotoxic Diterpenoids from Euphorbia fischeriana. Chemistr. Biodiv. **18**(2), e2000919 (2021)
23. Abotaleb, M., et al.: Therapeutic potential of plant phenolic acids in the treatment of cancer. Biomolecules **10**(2) (2020)

Check for updates

Development of a Computational Simulator of the Physiological Control of Ventricular Assist Devices (VADs)

B. J. Santos[1,2(✉)] and I. A. Cestari[1,3]

[1] Biomedical Engineering Graduate Program, Escola Politécnica de Engenharia, Universidade de Sao Paulo, Sao Paulo, Brazil
brunojsantos@usp.br
[2] Federal Institute of Sao Paulo, Sao Paulo, Brazil
[3] Bioengineering Division—Heart Institute, Hospital das Clinicas HCFMUSP, Faculdade de Medicina, Universidade de Sao Paulo, Sao Paulo, Brazil

Abstract. Congestive Heart Failure (CHF) is a significant global health issue and considered the most debilitating form of Heart Failure with the heart unable to efficiently pump enough blood to the body, resulting in a high risk of mortality. Left Ventricular Assist Devices (LVADs) are utilized for providing mechanical circulatory assistance for patients with CHF. LVADs can generate pulsatile or continuous flows. Continuous flow LVADs may induce a more physiological flow pattern with some degree of pulsatility when the device is operated using a physiological type of controller. We present a computational simulator that can be used to compare hemodynamic responses of a simulated patient implanted with a LVAD, operated with or without a physiological controller. The computational simulator described here includes: (i) simulations of the patient's condition; (ii) mathematical models of the estimators and sensors; and (iii) LVAD physiological control algorithms. The application of the simulator allowed us to investigate the performance of physiological control strategies considering various scenarios for LVAD applications.

Keywords: Ventricular assist devices (VADs) · Physiological control · Mock circulation loop · Cardiovascular computer simulation

1 Introduction

Heart failure (HF) is a significant global health issue, affecting approximately 26 million people worldwide [1]. Congestive Heart Failure (CHF) is considered the most debilitating form of HF with the heart unable to efficiently pump enough blood to the body, resulting in a high risk of mortality [2]. Heart transplantation is considered the standard therapy, but the scarcity of available donors and possible contraindications limit heart transplantation [3]. Implantable rotary blood pumps are utilized to assist the left (Left Ventricular Assist Device, LVAD) and/or right (Right VAD, RVAD) side of the heart in CHF, leading to good clinical outcomes. However, in the long-term support, these

J. L. B. Marques et al. (Eds.): CLAIB 2022/CBEB 2022, IFMBE Proceedings 98, pp. 454–461, 2024.
https://doi.org/10.1007/978-3-031-49401-7_47

pumps are associated with adverse events and subsequent hospital readmissions [4]. Currently, axial and centrifugal continuous-flow (cfLVADs) rotary pumps with preset speeds used clinically prohibits automatic adaptation of pump flow to meet the body's demands, which can lead to congestion or suction [5]. Improvement of LVAD performance may necessitate physiological control of the active flow in response to the demands, resembling the Frank-Starling mechanism, and utilizing physiological control of LVADs (PC-LVAD) [6].

Despite the reported promising outcomes, significant challenges persist in identifying, validating, and experimentally testing the flow adjustment approach for cfLVADs in an environment that accurately reflects physiological responses "in vivo" [7]. Computer simulations were widely used to evaluate the proposed PC-LVADs, in which, the cardiovascular system is numerically modeled and used as a virtual patient for the design and preliminary evaluation of PC-LVAD, because circulatory parameters can be extensively modified for validation under different patient conditions and scenarios before practical and clinical evaluations. This paper presents a computer simulator to examine the application of a PC-LVAD system and the resulting hemodynamic response of a simulated patient. The paper is organized as follows: Introduction; Sect. 2 outlines the fundamental principles associated with modeling patient hemodynamics. Section 3 outlines the experiment; Sect. 4 presents the results; and the conclusion is presented in Sect. 5.

2 Modeling Patient Hemodynamics

Lumped parameter models have been utilized to simulate the significant hemodynamic relationship between a ventricular assist pump and the native circulation [8, 9].

Contemporary cardiovascular modeling is grounded on the work of Arthur Guyton, who laid the groundwork for comprehending ventricular-vascular coupling by providing a relatively straightforward means of explaining the essentials of cardiovascular hemodynamics [10, 11]. This model suggests that hydraulic networks, such as the circulatory system, can be represented by electronic circuits that draw analogies between fluid flow through pipes with electric currents, pressure with voltage, hydraulic and electrical resistance, and hydraulic and electrical capacitance [12].

Models that describe vascular properties by a series of resistances and capacitances were based on the work of Otto Frank [13, 14]. Frank likened the vasculature (Fig. 1a) to a Windkessel (air chamber) system employed to transform the pulsating water flow generated by a hand pump into a constant flow from a fire hose (Fig. 1b). This model, which was later refined by Westerhof et al. [15], could be expanded and represented by an electrical circuit analog (Fig. 1c). The pressure-flow relationships at different locations within this model can be easily expressed in analytical and graphical terms.

3 Computational Simulator of PC-LVAD (CSPC-LVAD) Development

This section describes the computational simulator developed to investigate the PC-LVADs (CSPC-LVAD). The software used to develop the CSPC-LVAD is LabVIEW (V.2018, National Instruments, Austin, USA).

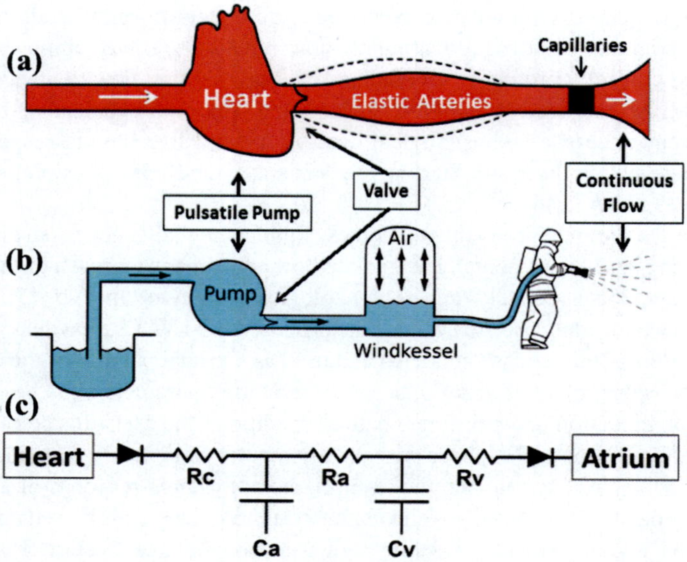

Fig. 1. Correlations in dynamic modeling of the circulation system: **a** vasculature, **b** Windkessel chamber, and **c** electrical circuit analog. Adapted from [16].

The physiological parameters of the patients are simulated in CSPC-LVAD using the Harvi platform (PVLoops LLC, Baltimore, USA), which integrates a set of complex interactive models of the cardiovascular system. The cardiovascular model employed by Harvi is based on the electrical system (as shown in Fig. 2), which enables the derivation of time-varying simultaneous differential equations to represent the circulatory system [8]. A cardiovascular model based on the electrical system employs analogies between electrical components, such as diodes, capacitors, and resistors, and the physiological properties of the cardiovascular system, such as blood flow, pressure, and resistance. In this model, the heart acts as a pump that generates blood flow through a network of blood vessels, which is modeled as an electrical circuit. The arteries and arterioles are represented as electrical resistors that generate a pressure drop, while the veins and venules are represented as capacitors that store blood volume. The heart chambers are modeled as electrical capacitors that store blood volume and generate pressure. Blood flow and pressure in this model are considered analogous to current and voltage in electrical circuits, respectively. The mathematical equations that describe the electrical properties of the cardiovascular system are derived from Kirchhoff's laws, which govern the flow and distribution of electrical current in circuits. By applying these laws to the cardiovascular system, it is possible to simulate the hemodynamic behavior of the cardiovascular system and to predict how changes in physiological parameters or the introduction of medical devices such as VADs affect cardiovascular function. These equations are solved in real-time, producing real-time pressure, volume, and flow values that form the basis of the simulations performed by Harvi. The model can simulate various cardiovascular features such as cardiac contractility, coronary vasculature, myocardial oxygen consumption, the effects of cardiovascular drugs, and advanced features such as VAD implantation, regurgitant and stenotic lesions in either heart valve [9].

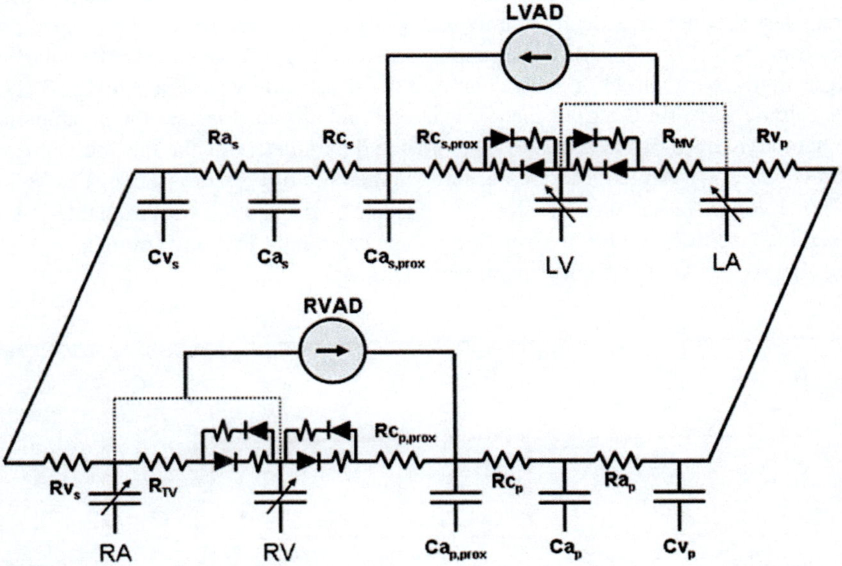

Fig. 2. Electrical circuit representation of the entire circulatory system on which the Harvi modeling is based. Adapted from [16].

The mathematical model of the centrifugal pump in the CSPC-LVAD system is based on the Look-Up Table (LUT) technique that uses 2-dimensional data of the pump's speed and flow or pressure from the empirical hydraulic performance curve of a commercial centrifugal pump cfLVADs available in the Harvi simulator. The estimator's mathematical model of CSPC-LVAD uses the LUT of the healthy patient as a reference, which was established by correlating the pump parameters with the hydrodynamic performance curve test of a cfLVAD placed using a Mock Circulation Loop (MCL). In this process, appropriate LVAD speeds are selected, and LVAD flow outlet is closed in steps to obtain the hydraulic resistance equivalent to a LVAD implanted in a healthy patient. The sensor's mathematical model in CSPC-LVAD uses the LUT of the chronic CHF patient as a reference, as it can accurately read variations in heart disease in a patient implanted with an artificial heart with no drifts over time, based on in situ monitoring of a calibrated sensor. The integration of the CSPC-LVAD operational blocks was performed using the LUT technique. This technique allowed the description of the patient's physiological behavior and the LVAD's flow and pressure in non-linear regions of operation in tables to be interpolated in real-time based on simulation reference variables (for example, to infer the LVAD's flow, the LVAD's pressure and speed are used as reference variables to interpolate the LVAD's LUT values stored within the relevant region).

4 Results

The Harvi simulator was used to model different conditions, in which, chronic CHF without and with LVAD implantation were chosen. Figures 3 and 4 presents physiological parameters that are graphically subdivided according to right atrial loop (Figs. 3, 4a), left atrial loop (Figs. 3, 4b), right ventricular loop (Figs. 3, 4c), left ventricular loop (Figs. 3, 4d), aortic and left ventricular and atrial pressure wave (in mmHg) (Figs. 3, 4e), volume wave (in l/min) of the left ventricle and atrium (Figs. 3, 4f), parameters of the hemodynamics of the heart (Figs. 3, 4h), and parameters of the metabolism of the heart (Figs. 3, 4i). All loops are represented in the ratio of Pressure (mmHg) vs Volume (ml) in 1 complete heartbeat (systole and diastole). For patients with implantation, the model also includes a loop of Flow (l/min) vs Differential Pressure (mmHg) of LVAD (Fig. 4g) and LVAD functioning parameters (Fig. 4j).

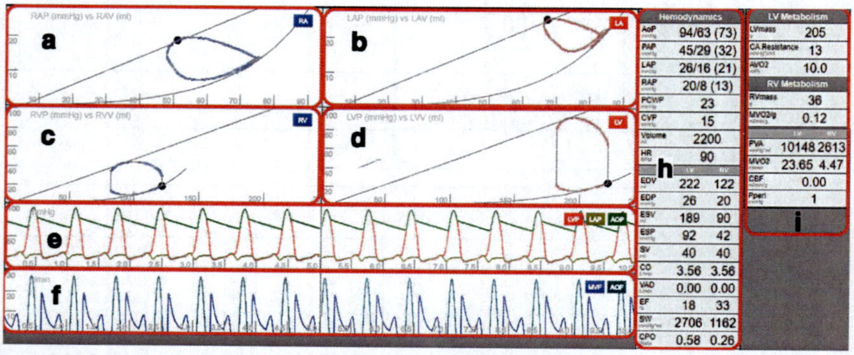

Fig. 3. Results of simulation of hemodynamic conditions of a patient extracted from the Harvi platform: chronic CHF without LVAD implantation.

The Harvi simulator is a web-based platform that allows for automatic data collection and adjustment through an algorithm, enabling integration with other computational and hybrid simulators [17]. The physiological parameters of the patients' computational models were transferred from the Harvi simulator to the CSPC-LVAD using an algorithm developed through the python language for collecting data from the web browser where the Harvi platform runs (refer to Fig. 5). The algorithm facilitates real-time transmission and retrieval of data in the computer simulator, enabling the interaction of the LVAD physiologic control for adjusting the hemodynamic response of the patient.

The simulations allow for a detailed analysis of the physiological response to the control strategy, enabling the optimization of the controller parameters to achieve the desired hemodynamic outcomes [18]. In the CSPC-LVAD (Fig. 6), the parameters for the simulation can be adjusted, such as: (i) physiological control technique and its setting values; clinical prescription; (ii) characteristics of the patient implanted by the LVAD; (iii) effect of the circadian cycle; (iv) patient activity level; and (v) dataloggers.

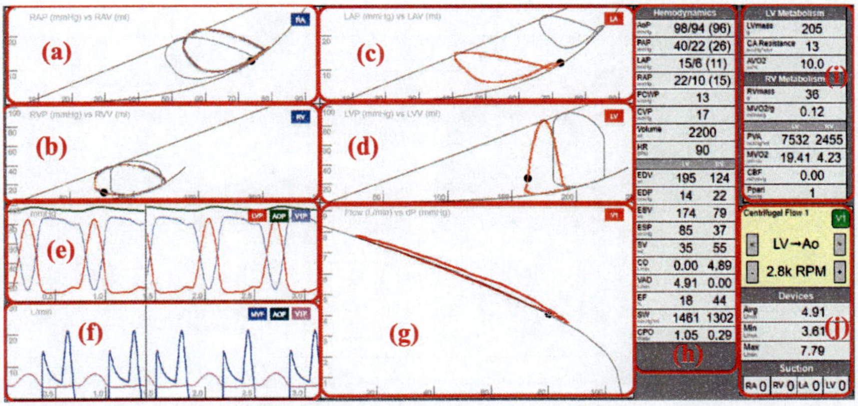

Fig. 4. Results of simulation of hemodynamic conditions of a patient extracted from the Harvi platform: chronic CHF with LVAD implantation.

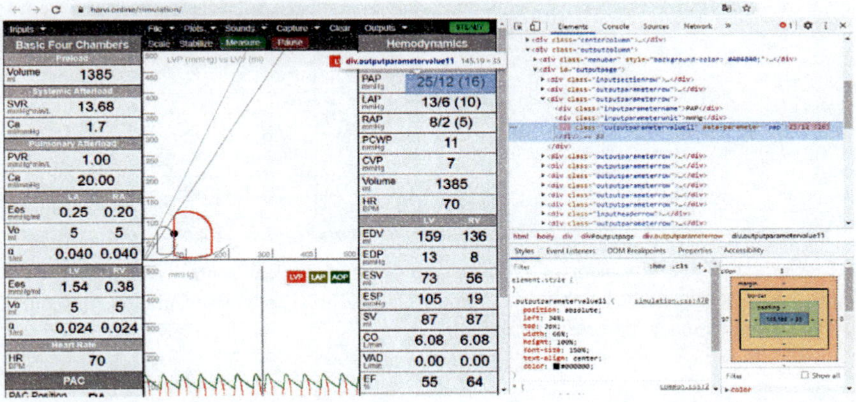

Fig. 5. Web browser development environment running the Harvi simulator with the data management algorithm.

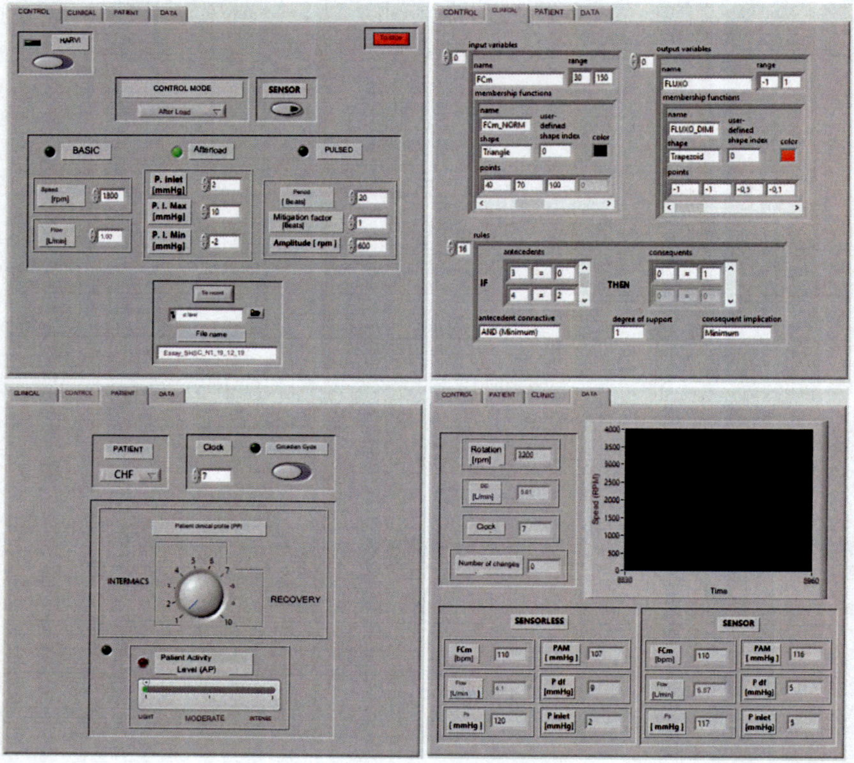

Fig. 6. Computational simulator of performance of physiological control techniques (CSPC-LVAD). In this simulator, the labels were originally in Brazilian Portuguese and were subsequently translated using an image processing program.

5 Conclusion

The simulator developed enables the evaluation of the various control strategies for VADs. The integration of this simulator with the physiological data provided by Harvi allows the assessment of physiological control in a range of scenarios relevant to both adult and pediatric patients, as well as variations in the clinical progression of heart disease. These scenarios include pulsatile and continuous flow VAD, axial and centrifugal pumps, LVAD and RVAD, and estimators and sensors. Future work will investigate the correlation between the data obtained from the computational simulator and results presented from an "in vitro" circulation bench.

References

1. Savarese, G., Becher, P.M., Lund, L.H., Seferovic, P., Rosano, G.M., Coats, A.J.: Global burden of heart failure: a comprehensive and updated review of epidemiology. Cardiovasc. Res. **118**(17), 3272–3287 (2022)

2. Zipes, D.P., Libby, P., Bonow, R.O., Mann, D.L., Tomaselli, G.F.: Braunwald's heart disease E-Book: a textbook of cardiovascular medicine. Elsevier Health Sciences (2018)
3. Mantha, A., Lee, R.O., Wolfson, A.M.: Patient selection for heart transplant: balancing risk. Curr. Opin. Organ Transplant. **27**(1), 36–44 (2022)
4. Yuzefpolskaya, M., Schroeder, S.E., Houston, B.A., Robinson, M.R., Gosev, I., Reyentovich, A., D'Alessandro, D.A.: The society of thoracic surgeons intermacs 2022 annual report: focus on the 2018 heart transplant allocation system. Ann. Thoracic Surg. **115**(2), 311–327 (2023)
5. Stephens, A.F., Gregory, S.D., Burrell, A.J., Marasco, S., Stub, D., Salamonsen, R.F.: Physiological principles of starling-like control of rotary ventricular assist devices. Expert Rev. Med. Dev. **17**(11), 1169–1182 (2020)
6. Stevens, M.C., Stephens, A., AlOmari, A.H.H., Moscato, F.: Physiological control. In: Mechanical Circulatory and Respiratory Support, pp. 627–657. Academic Press (2018)
7. Pauls, J.P., Stevens, M.C., Bartnikowski, N., Fraser, J.F., Gregory, S.D., Tansley, G.: Evaluation of physiological control systems for rotary left ventricular assist devices: an in-vitro study. Ann. Biomed. Eng. **44**, 2377–2387 (2016)
8. Doshi, D., Burkhoff, D.: Cardiovascular simulation of heart failure pathophysiology and therapeutics. J. Cardiac Fail. **22**(4), 303–311 (2016)
9. Bronicki, R.A., et al.: The cardiovascular system in severe sepsis: insight from a cardiovascular simulator. Pediatr. Crit. Care Med. **23**(6), 464–472 (2022)
10. Guyton, A.C.: Determination of cardiac output by equating venous return curves with cardiac response curves. Physiol. Rev. **35**(1), 123–129 (1955)
11. Guyton, A.C., Pl, C.: Pressure-volume curves of the arterial and venous systems in live dogs. Am. J. Physiol. **184**(2), 253–258 (1956)
12. Owen, B., Bojdo, N., Jivkov, A., Keavney, B., Revell, A.: Structural modelling of the cardiovascular system. Biomech. Model. Mechanobiol. **17**(5), 1217–1242 (2018)
13. Sagawa, K., Lie, R.K., Schaefer, J.: Translation of Otto frank's paper" Die Grundform des arteriellen Pulses" zeitschrift für biologie 37: 483–526 (1899). J. Mol. Cell. Cardiol. **22**(3), 253–254 (1990)
14. Frank, O.: Grundform des Pulses. Z. Biol. **37**, 483–526 (1899)
15. Stergiopulos, N., Westerhof, B.E., Westerhof, N.: Total arterial inertance as the fourth element of the windkessel model. Am. J. Physiol. Heart Circulat. Physiol. **276**(1), H81–H88 (1999)
16. Burkhoff, D.: Brief history of cardiovascular modeling: insights into pathophysiology and therapeutics. cardiovascular simulation in clinical research and practice (2015)
17. Torres, D.S., Mazzetto, M., Cestari, I.A.: A novel automated simulator of pediatric systemic circulation: design and applications. Biomed. Signal Process. Control **70**, 102926 (2021). https://doi.org/10.1016/j.bspc.2021.102926
18. Melo, T.R., Neto, J.S.R., Cestari, I.A., Lima, A.M.N.: Feedback controller for restoring the basal hemodynamic condition with a rotary blood pump used as left ventricular assist device. Biomed. Sign. Process. Control **62**, 102–136 (2020). https://doi.org/10.1016/j.bspc.2020.102136

Freeze-Drying Effect on the Hydrodynamic Performance of Bovine Pericardium Biological Valves

Eraldo de Sales(✉), Marina Junko Shiotsu Maizato, Marcelo Mazzetto, and Idágene Aparecida Cestari

Divisao de Bioengenharia, Faculdade de Medicina, Instituto Do Coracao, Hospital das Clinicas HCFMUSP, Universidade de Sao Paulo, Sao Paulo, SP, Brazil
eraldo.sales@hc.fm.usp.br

Abstract. Freeze-drying is a process that allows biological tissues and tissue based devices such as biological valves to be stored dry, making their storage more convenient. The purpose of the present work is to evaluate the alterations in hydrodynamic performance of biological prosthetic valves after freeze-drying. Bovine pericardium biological heart valves with 25 mm obtained from two manufacturers were tested on a pulse duplicator (5.0 L/min) before and after freeze-drying. The instantaneous signals of pressures and flows were recorded simultaneously with images of opening and closing of the valves. The results show that freeze-drying did not caused significant modifications in transvalvular pressure gradients (12 $\pm$ 3 and 11 $\pm$ 2 mmHg), effective orifice areas (1.4 $\pm$ 0.2 and 1.5 $\pm$ 0.1 cm^2) and regurgitation fractions (5 $\pm$ 2 and 6 $\pm$ 3%), respectively, considering devices produced by the same manufacturer. Our findings show that the lyophilization process did not significantly modify the performance of the prostheses tested, suggesting the process can be applied to bovine pericardium. Further investigation is necessary to evaluate the influence of lyophilization in fatigue and the long term performance of prosthetic heart valves.

Keywords: Prosthetic heart valves · Hydrodynamic performance · Freeze-drying · Bovine pericardium

1 Introduction

Bovine pericardium biological heart valves (BPV) are widely used as heart valve replacements with the advantage of not requiring anticoagulation therapy and therefore do not contribute to bleeding problems. Despite these advantages, BPV are not free from complications: leaflet calcification and leaflet tearing can happen [1].

BPV can be manufactured with bovine pericardium tissue treated with glutaraldehyde and preserved in formaldehyde. Before implantation BPV are washed to remove residues from these materials. An alternative to preservation in formaldehyde is the freeze-drying process, also called lyophilization [2].

J. L. B. Marques et al. (Eds.): CLAIB 2022/CBEB 2022, IFMBE Proceedings 98, pp. 462–471, 2024.
https://doi.org/10.1007/978-3-031-49401-7_48

Freeze-drying is a process of drying a previously frozen material or device with most of the water or solvent removed by sublimation. It can be considered as part of a third generation in drying technology, as it was developed to overcome structural damage and loss of volatile compounds, observed in conventional processes [2, 3].

When BPV undergo freeze-drying, their fundamental properties are preserved, making future handling and storage more convenient. Freeze dried BPV can be stored at room temperature and easily transported [2–4]. When immersed in water, the lyophilized valve can be readily reconstituted, regaining a size and shape close to the original [2].

Preliminary research has revealed that lyophilization results in less cytotoxicity, less chemical residues and a potential for less calcification [5]. It should be noted that calcification is the most important factor in limiting the useful life of bioprostheses.

In this work we assessed the effect of freeze-drying on the hydrodynamic performance of BPV by measuring their pressure gradient, effective orifice area (EOA) and regurgitation fraction on a pulse duplicator.

2 Materials and Methods

2.1 Bioprostheses

BPV with the same size (25 mm) provided by two manufacturers were compared: manufacturer #A (Biocardio, Brazil) and #B (Braille Biomedica, Brazil), Fig. 1. A total of 4 valves of each manufacturer (N = 4) were evaluated, 8 prostheses in total. Valve prostheses were removed from the original packaging (preserved in formaldehyde) and hydrated in NaCl saline solution at 0.9% for 30 min.

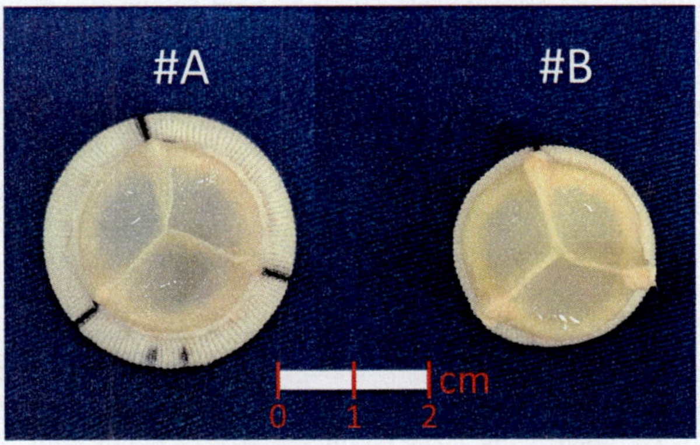

Fig. 1. Bioprostheses from manufacturer #A and #B before freeze-drying process.

2.2 Hydrodynamic Performance Tests

Hydrodynamic performance of each valve was studied on a pulse duplicator ("The Shelhigh Pulse Duplicator®", Shelhigh Inc, New Jersey, USA) after hydration. Figure 2 shows a prosthesis placed in the aortic position of the pulse duplicator for measurements of pressure gradients and flows. The pulse duplicator was adjusted to simulate physiological conditions, with aortic systolic pressure set at 120 mmHg and diastolic pressure at 80 mmHg. Other parameters adjusted were ventricular pressure, stroke volume and flow.

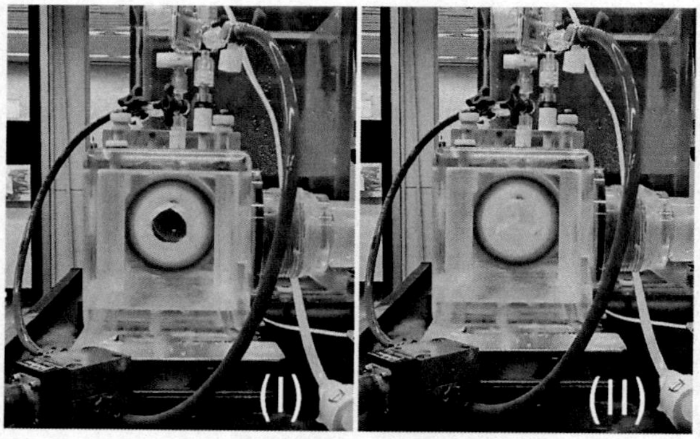

Fig. 2. Biological valve being tested in the pulse duplicator. In this image it is possible to observe the moment when the leaflets are open (I) and closed (II).

After placement in the aortic position of the pulse duplicator, flow was adjusted to 5.0 L/min and a stroke volume of 70 ml at a beating rate of approximately 90 cycles per minute. All experimental settings followed ISO 5840 [6].

A 25 mm bileaflet mechanical heart valve (St. Jude Medical® Regent®, USA) was placed in the mitral position of the pulse duplicator. The pulse duplicator was filled with a blood analog fluid (33.33% of glycerin and 66.67% of NaCl saline solution at 0.9%).

A differential transducer Validyne DP15-TL (Validyne Engineering Corporation, USA) was used to measure the transvalvular pressure gradient (difference between ventricular and aortic pressure). The pressure gradient was determined by the average value in the positive differential pressure period, that is, when ventricular pressure (before the aortic valve, following the flow direction) is greater than the aortic pressure (after the valve).

The ventricular and aortic pressures were measured with absolute pressure transducers (Edwards PX12N, Edwads LifeScience, USA) and the instantaneous flow was measured by an ultrasonic flow meter (Transonic System Inc., USA).

Signals were recorded using DataQ WinDaq DI-220 acquisition system (DataQ Instruments, USA) at a sampling rate of 500 samples per second per channel.

2.3 Determination of BPV Parameters

The effective orifice area (EOA) was calculated using the equation proposed by Gorlin and Gorlin [7]:

$$EOA = \frac{q_{V_{RMS}}}{51.6\sqrt{\frac{\Delta p}{\rho}}} \tag{1}$$

where Δp is the mean positive differential pressure (mmHg), ρ is the fluid density (g/cm^3) and q_{V_RMS} is the root mean square (RMS) of the anterograde flow (ml/s).

RMS of the anterograde flow along a pulse is given by:

$$q_{V_{RMS}} = \sqrt{\frac{\int_{t_i}^{t_f} q_V(t)^2 dt}{t_f - t_i}} \tag{2}$$

where $q_V(t)$ is the pulsatile flow (ml/s) and, t_i and t_f are the initial and final instants of the anterograde flow, respectively.

The regurgitation fraction ($R_{Fraction}$, in percentage) is determined by the following equation:

$$R_{Fraction} = \frac{C_{Vol} + L_{Vol}}{F_{Vol}} \cdot 100 \tag{3}$$

where C_{Vol} is the closing volume, L_{Vol} is the leakage volume, and F_{Vol} is the total forward volume.

Pressure gradient, regurgitation fraction and effective orifice area for each valve have been calculated from 10 pulses randomly chosen from the set of measurements performed.

2.4 Freeze-Drying Process

The biological valve prosthesis preserved in formaldehyde were washed for 30 min in saline solution, with 3 changes of solution. The excess water was slightly dried and weighed on a precision balance (Shimadzu UX 420H, Japan). The valves were previously frozen in liquid nitrogen and lyophilized (Fig. 3) in a Thermo Modulyod 115 (Thermo Fisher Scientific, USA) at −50 °C and a vacuum of about 800 μBar for 6 h (#A) and 7 h (#B). A shorter freeze-drying time was applied to prostheses from #A to compensate for difference in thickness of bovine pericardium tissue.

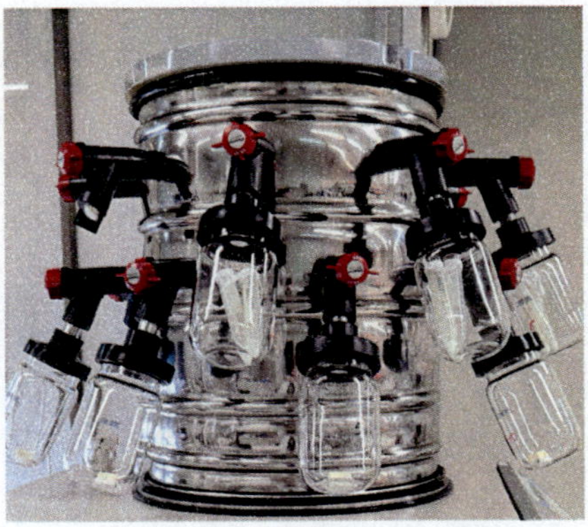

Fig. 3. BPV during freeze-drying process.

After lyophilization, each prosthesis was weighed again and stored in vacuum packs overnight to prevent them from accumulating humidity. Then, 24 h after lyophilization, each valve was hydrated in saline solution for 30 min and weighed in order to verify the amount of liquid that the valve could absorb during hydration. After that the hydrodynamic performance was evaluated under the same experimental conditions used in control valves.

Figure 4 shows the same valves as in Fig. 1 (before lyophilization), but now after freeze-drying (I) and then rehydrated (II).

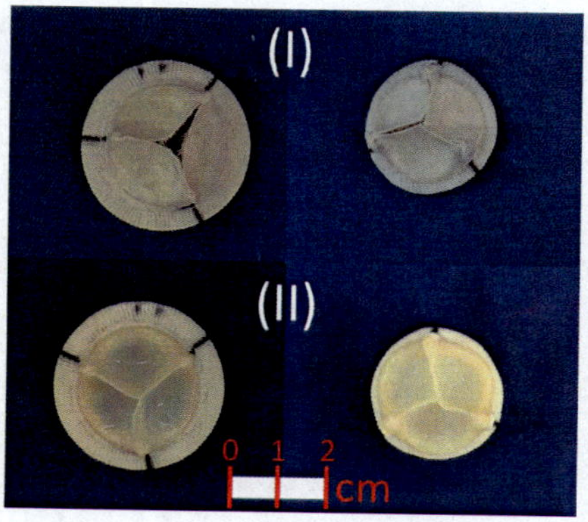

Fig. 4. Bioprostheses freeze-dried (I) and hydrated after freeze-drying (II).

2.5 Statistical Analysis

Statistical analysis was performed using SigmaPlot (Systat Software Inc.), where a statistical significance ($p < 0.05$) was determined by paired t-test for each valve before and after lyophilization, with a confidence limit of 95%.

3 Results

3.1 Hydrodynamic Performance

Representative signals of valve operation recorded on the pulse duplicator are depicted in Fig. 5.

Hydrodynamic performance was determined for all valves before and after the lyophilization process. The results for all BPV for each hydrodynamic parameter for both situations are shown in Tables 1 and 2.

The characteristics of BPV of the same manufacturer were pooled and compared before and after freeze-drying, in order to determine the effect of this process on the hydrodynamic performance of the prostheses. The results of the paired statistical analysis with the p values obtained for all BPV by group is shown in the Table 3.

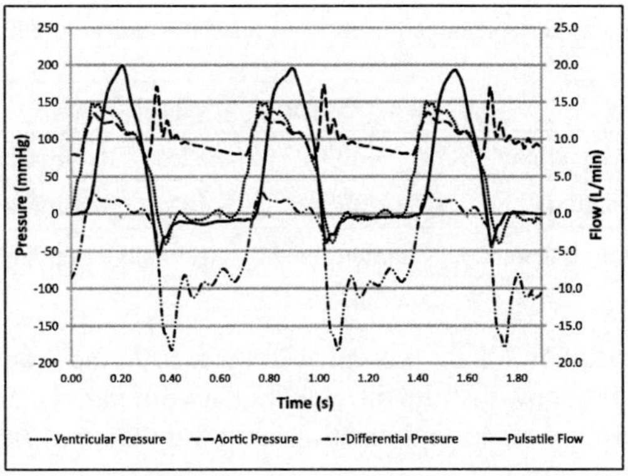

Fig. 5. Representative hydrodynamic recordings of pressure and flows of a biological valve.

Table 1. BPV hydrodynamic measurements (mean ± standard deviation) before lyophilization.

Valve	Pressure gradient (mmHg)	EOA (cm^2)	Regurgitant fraction (%)
#A1	9.4 ± 0.2	1.57 ± 0.05	5 ± 3
#A2	13.9 ± 0.2	1.30 ± 0.04	6 ± 2
#A3	8.6 ± 0.3	1.66 ± 0.04	6 ± 2
#A4	15.4 ± 0.3	1.27 ± 0.03	5 ± 2
#B1	9.5 ± 0.2	1.56 ± 0.03	6 ± 3
#B2	11.9 ± 0.3	1.40 ± 0.04	8 ± 3
#B3	9.4 ± 0.2	1.50 ± 0.04	5 ± 3
#B4	10.5 ± 0.2	1.47 ± 0.04	7 ± 4

Table 2. BPV hydrodynamic measurements (mean ± standard deviation) after lyophilization.

Valve	Pressure gradient (mmHg)	EOA (cm^2)	Regurgitant fraction (%)
#A1	8.9 ± 0.1	1.63 ± 0.05	4 ± 2
#A2	12.5 ± 0.2	1.35 ± 0.04	7 ± 2
#A3	9.6 ± 0.2	1.60 ± 0.04	7 ± 3
#A4	15.1 ± 0.2	1.32 ± 0.03	6 ± 3
#B1	11.9 ± 0.2	1.48 ± 0.04	13 ± 2
#B2	11.2 ± 0.2	1.62 ± 0.05	9 ± 4
#B3	9.1 ± 0.1	1.64 ± 0.04	7 ± 2
#B4	8.7 ± 0.2	1.64 ± 0.04	7 ± 3

Table 3. p values obtained in each comparison carried out by group.

BPV group	Property	p value	Property	p value	Property	p value
#A	Pressure gradient	0.587	EOA	0.625	Regurgitant fraction	0.375
#B	Pressure gradient	0.918	EOA	0.188	Regurgitant fraction	0.206

A statistically significant difference is characterized by a p-value less than 0.05.

The average results for #A are presented in Figs. 6, 7, and 8 and for #B in Figs. 9, 10, and 11. The results show that the differences between parameters measured in both groups of BPV before and after lyophilization are lower than the dispersion given by the variability in the data.

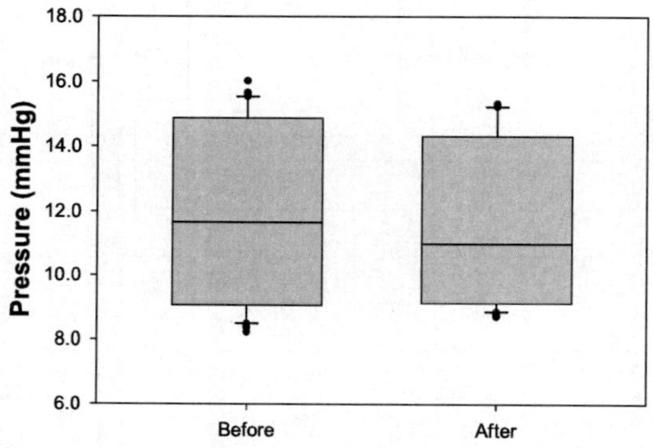

Fig. 6. Pressure gradient for BPV from manufacturer #A before and after lyophilization.

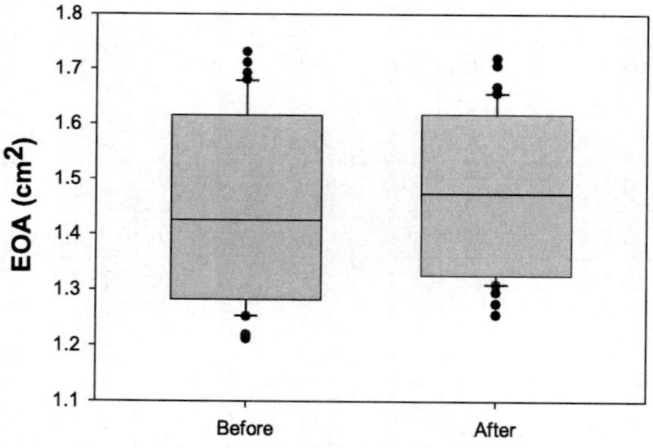

Fig. 7. Effective orifice area for BPV from manufacturer #A before and after lyophilization.

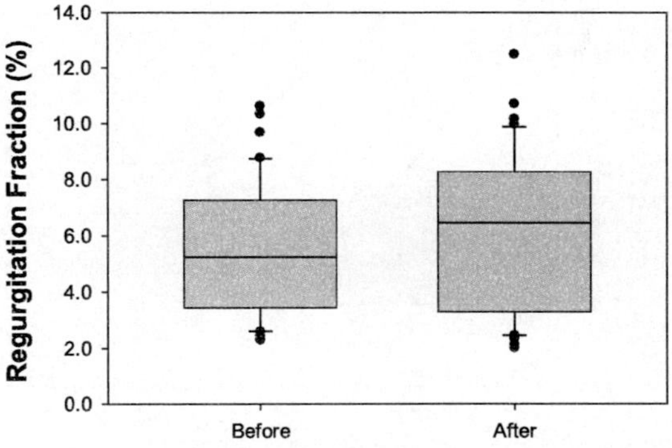

Fig. 8. Regurgitation fraction for BPV from manufacturer #A before and after lyophilization.

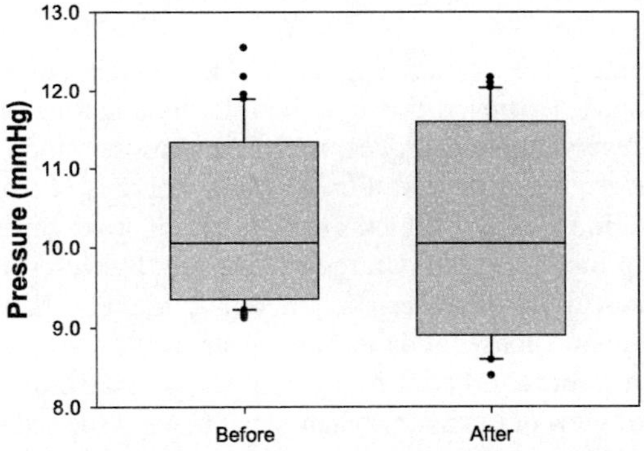

Fig. 9. Pressure gradient for BPV from manufacturer #B before and after lyophilization.

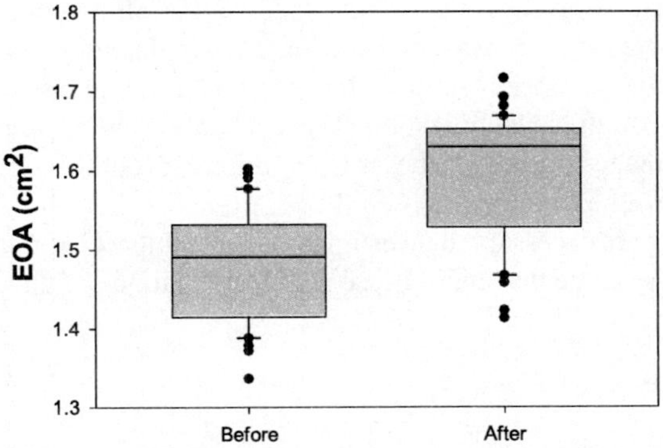

Fig. 10. Effective orifice area for BPV from manufacturer #B before and after lyophilization.

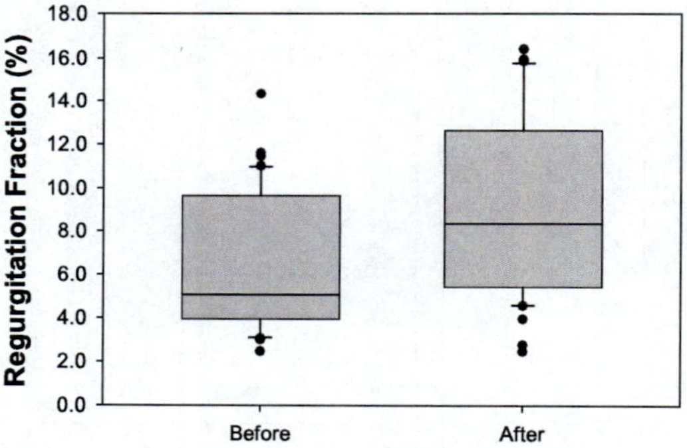

Fig. 11. Regurgitation fraction for BPV from manufacturer #B before and after lyophilization.

3.2 Discussion

BPV are usually treated with glutaraldehyde in order to stabilize the tissue and reducing its thrombogenicity. However, this treatment is toxic and can induce calcification in vivo [8]. Some preservation techniques, such as cryopreservation or vitrification, can be applied, but both require storage at ultra-low temperatures and the need of be transported frozen in addition to other problems such as risk of cross-contamination, constant maintenance, among others [9, 10]. On the other hand, freeze-drying can cause some undesirable processes in the biomaterials, potentially making the tissue vulnerable to denaturation, osmotic imbalance, and mechanical stress [9]. Despite that, freeze-drying has been studied as a viable alternative for the biological tissue preservation [2–5, 9, 10].

From the point of view of the hydrodynamic parameters studied in the present work, the values found are in agreement with those reported in the literature [11–13] and in compliance with the minimum performance requirements established by ISO 5840 [6].

Due to data variability within the same group, no significant variations in hydrodynamic performance of BPV were found before and after lyophilization and also considering possible differences given by different manufacturers.

Two of the eight prostheses studied showed an increase in pressure gradient and a decrease in the EOA after lyophilization. These alterations were anticipated by visualization of video images of the prostheses taken prior to lyophilization with leaflets not opening completely.

Further studies are necessary to investigate possible impacts of freeze-drying in long term changes in the tissue that could cause anticipated fatigue of the material.

4 Conclusions

Lyophilization did not cause significant changes in hydrodynamic performance of the BPV studied. The results, however, are based on a small number of prostheses tested.

Future work includes investigation of the lyophilization process applied to a larger number of BPV to better define the parameters which could contribute to using it for storage and preservation.

Acknowledgment. We thank the financial support of São Paulo State Foundation (FAPESP, Grant 2012/50283–6), FINEP (Grant 01.14.0177.00) and CNPQ (313846/2021–9) for financial support.

Conflict of Interest The authors declare that they have no conflict of interest.

References

1. Dasi, L.P., Simon, H.A., et al.: Fluid mechanics of artificial heart valves. Clin. Exp. Pharmacol. Physiol. **36**(2), 225–237 (2009)
2. Borgognoni, C.F., Maizato, M.J.S., et al.: Effect of freeze-drying on the mechanical, physical and morphological properties of glutaraldehyde-treated bovine pericardium: evaluation of freeze-dried treated bovine pericardium properties. J. Appl. Biomater. Biomech. **8**(3), 186–190 (2010)
3. Hafeez, Y.M., Zuki, A.B.Z., et al.: Effect of freeze-drying and gamma irradiation on biomechanical properties of bovine pericardium. Cell Tissue Bank. **6**(2), 85–89 (2005)
4. Borgognoni, C.F., Bevilacqua, J.S., et al.: Freeze-drying microscopy in mathematical modeling of a biomaterial freeze-drying. Braz. J. Pharm. Sci. **48**(2), 203–209 (2012)
5. Maizato, M.J.S., Higa, O.Z., et al.: Glutaraldehyde-treated bovine pericardium: Effects of liophilization on cytotoxicity and residual aldehydes. Artif. Organs **27**(8), 692–694 (2003)
6. American National Standard: Cardiovascular implants—Cardiac valve prostheses. ISO/FDIS 5840–1:2020(E) (2020)
7. Gorlin, R., Gorlin, S.G.: Hydraulic formula for calculation of area of the stenotic mitral valve, other cardiac valves, and central circulatory shunts. Am. Heart J. **41**(1), 1–29 (1951)
8. Polak, R., Pitombo, R.N.M.: Care during freeze-drying of bovine pericardium tissue to be used as a biomaterial: a comparative study. Cryobiology **63**(2), 61–66 (2011)
9. Merivaara, A., Zini, J., Koivunotko, E., et al.: Preservation of biomaterials and cells by freeze-drying: change of paradigm. J. Control. Release **336**, 480–498 (2021)
10. Zouhair, S., Aguiari, P., et al.: Preservation strategies for decellularized pericardial scaffolds for off-the-shelf availability. Acta Biomater. **84**, 208–221 (2019)
11. Kuetting, M., Sedaghat, A., et al.: In vitro assessment of the influence of aortic annulus ovality on the hydrodynamic performance of self-expanding transcatheter heart valve prostheses. J. Biomech. **47**(5), 957–965 (2014)
12. Gerosa, G., Tarzia, V., et al.: Small aortic annulus: the hydrodynamic performances of 5 commercially available tissue valves. J. Thorac. Cardiovasc. Surg. **131**(5), 1058–1064 (2006)
13. Bottio, T., Tarzia, V., et al.: The changing spectrum of bioprostheses hydrodynamic performance: considerations on in-vitro tests. Interact Cardio Vascul Thoracic Surg **7**(5), 750–754 (2008)

Conjunction of Bioelectrical Impedance Spectroscopy with Archie's Law to Characterize Scaffolds Porosity

Carlos Vinicius Nascimento Barbosa(✉), Marcio Nogueira de Souza, and Alexandre Visintainer Pino

COPPE/UFRJ/Biomedical Engineering Program, Rio de Janeiro, Brazil
eng.carlosvinicius@outlook.com

Abstract. Tissue Engineering is a multidisciplinary field that uses two and three dimensional structures called scaffolds to create a physical support for growth, gas change and nutrition of the cell. Those scaffolds can be synthetic, biological or a combination of both types of material, but the porosity, the interconnectivity of pores and pore size are among the most important physical parameters of the scaffolds. The estimation of these characteristics is still a challenge in Tissue Engineering, as it involves techniques and methods that need to be non-toxic and non-destructive, and that must operate in structures with dimensions of the order of micrometers (μm). The bioimpedance spectroscopy (bis) technique has been reported as a potential non-invasive monitoring of scaffold porosity and biological tissue growth. This work aims to study how bioimpedance spectroscopy in conjunction with Archie´s law data can be used to characterize the porosity and pore connectivity of 3D cellular structures.

Keywords: Tissue engineering · Archie Law · Porous measurement · Bioimpedance

1 Introduction

The Tissue Engineering is a new field of Biomedical Engineering with a lot of challenges that need to be investigated. The possibility of cell growth *in vitro* and the transposition for the human body with feasibility and without rejection are the mains goals for Tissue Engineering.

Tridimensional structures used for cell growth are denominated scaffolds (Fig. 1) and the deposit of cells in these structures mimics the function of a biological extracellular matrix. Despite having a 3D structure to support the cells, many aspects of cell growth are difficult to control because parameters such as cell nutrition and gas exchange need to be monitored for an appropriate growth.

The scaffolds can be of three basic types: decellularized, polymeric or synthetic. Independent of types, they need to be biocompatible and bioresorbable, and their microstructures must be constituted by a porous network. The relationship between porous network

J. L. B. Marques et al. (Eds.): CLAIB 2022/CBEB 2022, IFMBE Proceedings 98, pp. 472–479, 2024.
https://doi.org/10.1007/978-3-031-49401-7_49

and porous size, characterize others two parameters: the permeability and the interconnectivity. These parameters define the capacity of cell in gas exchange and nutrition [1, 2].

Bioimpedance technique has been reported in the literature as a promising tool for evaluating the porosity (ratio of the volume of pores to the volume of the scaffold) of scaffolds in their building phase [2] as well as for monitoring the change during the 3D growing of the biological tissue [2, 3].

The present study investigates the conjunction of a BIS (Bioelectrical Impedance Spectroscopy) technique with the Archie´s law to evaluate the porosity characteristics of 3D porous structures that tried to mimic the porosity of the scaffolds. We used this type of scaffold phantoms to achieve a porosity evaluation by a classic volume method that could be compared with the results achieved by the BIS method. After validated, the developed technique aims to be applied in real scaffolds used in Tissue Engineering.

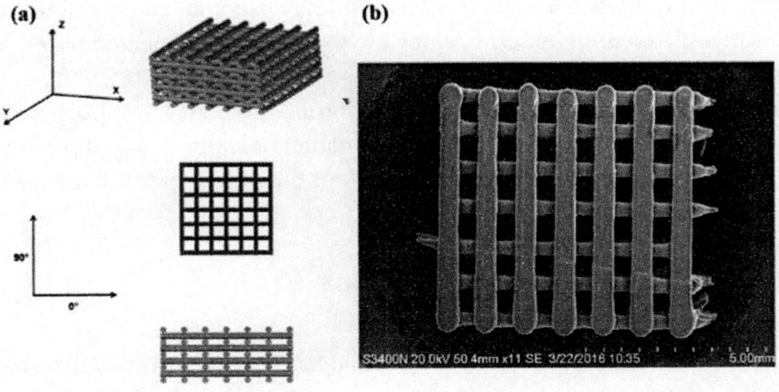

Fig. 1. (a) A 3D scaffold perspective view, top view and side view; (b) PLA (Poly lact acid) scaffold [1].

1.1 The Archie's Law

Proposed in 1942 by Gus E. Archie [2], this law has been applied for decades to relate the conductivity of a medium sutured by a conductive fluid with the conductivity of the fluid and the porosity of the medium (Eq. 1).

$$\sigma = \sigma_f \emptyset^m \quad (1)$$

where:

σ – effective DC conductivity of porous medium;

σ_f – DC conductivity of the fluid that fills the pore;

ϕ – porosity of the medium (a dimensionless parameter that can be expressed as a percentual value);

m – cementation exponent, which is related to level of compaction of the material that composes the porous medium.

Another important parameter is formation factor of Archie (F), which indicates the ratio between the electrolyte conductivity solution and the medium effective conductivity (Eq. 2).

$$F = \frac{\sigma_f}{\sigma} = \frac{\rho}{\rho_f} \tag{2}$$

The formation factor can also be written using the cementation exponent (m), according to Eq. 3.

$$F = \frac{\sigma_f}{\sigma} = \emptyset^{-m} \tag{3}$$

The cementation exponent is affected by many factors, with highlight on sizes, shapes and profiles of the pores, and also size/numbers of unliked porous. The m exponent increases when an interconnected porous network is observed, and it can be even lower than unity when the medium exhibits many closed and non-connected pores. Typically, media with connected pores produce m values between 1.0 and 3.0.

The characteristic called connection of a porous medium (G) [3, 4] is defined by physics disponible of the transports routes and mathematically it is the ratio between the conductivity of the medium and the conductivity of the fluid present in the pores, i.e. the inverse of the Archie formation factor (Eq. 4).

$$G = \frac{1}{F} = \frac{\sigma}{\sigma_f} = \emptyset^{m} \tag{4}$$

An additional parameter of interest is the connectivity of a porous medium (χ) [5], which is related to how the space of the pores is organized in the medium (Eq. 5).

$$\chi = \emptyset^{m-1} \tag{5}$$

It is possible to see the dependence between connection, porosity, and connectivity (Eq. 6).

$$G = \emptyset\chi \tag{6}$$

2 Materials and Methods

Because Archie's law relates the electrical conductivity of the medium to its porosity, it could be applied to estimate the porosity of a scaffold by using BIS to determine the conductive medium if one assumes to know the conductivity of the fluid that fills the medium pores (complete saturation of the pores).

Original Archies's law uses DC conductive values but since we have an electrode-electrolyte interface the DC parameters cannot be easily obtained. Moreover, the BIS technique can be used to study the behavior of conductivity along the frequency, which can provide additional information about porosity and connectivity profiles.

Considering a measuring chamber composed of two metal electrode (Fig. 2) where the porous medium is inserted and the pores are filled with a conductive fluid, the BIS data can be used to estimate the values of conductivity. In this study we used a impedance analyzer Hioki IM3536 to measure the impedance data in 50 frequencies geometrical spaced in the range from 10 Hz to 1 MHz.

After investigating the impedance data of all the 50 frequencies (data not shown here) the lowest frequency for which the model conductivity reaches a nearly constant value was chosen to apply Archie's law.

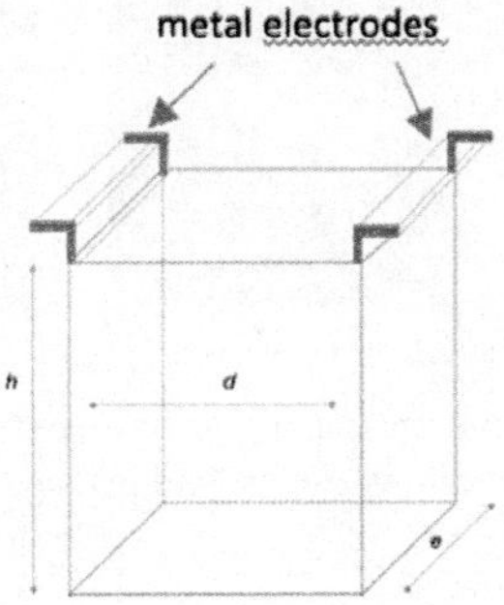

Fig. 2. Measuring chamber with the metal electrodes in each side.

Considering the dimensions of the measuring chamber (space between electrodes $= d$; deep $= e$, and high $= h$), the geometrical factor of the chamber can be defined as $K = \frac{h.e}{d}$.

Using the BIS equipment to get the impedance data **Z** at an adequate low frequency, the admittance ***Y*** of the system (the inverse of the impedance **Z** measured by equipment) can be write as (Eq. 7):

$$Y = G + jB = G + j\omega C = K(\sigma + j\omega\varepsilon) = K(\sigma + j\omega\varepsilon_r\varepsilon_0) \tag{7}$$

where:

- G is the real part of admittance, i.e., conductance, in Siemens (S);
- B is the imaginary part of admittance, i.e., susceptance, in Siemens (S);
- C is the capacitance in Farads (F);
- K is geometrical factor of measuring chamber (cm);
- σ is conductivity of material between the electrodes (S/cm);
- ε is permittivity of material between the electrodes (F/cm);
- ε_r is relative permittivity of material between the electrodes with respect of the vacuum permittivity ($\varepsilon_0 = 8.8 \times 10^{-14}$ F/cm).

If geometrical factor (K) is known, the conductivity ($\sigma = G/K$) and the relative permittivity ($\varepsilon_r = \varepsilon/\varepsilon_0$) ($\varepsilon = B/(\omega K) = B/(2\pi f.K)$) of porous medium can be obtained by the conductance (G) and the susceptance (B). In our assays we had $K = 9.3$ cm.

To mimic porous media with different porosity characteristics we used three materials: construction sand (Fig. 3 (a)), small size aquariums stones (Fig. 3 (b)) and medium size aquariums stones (Fig. 3 (c)). The measuring chamber was filled with one of them and after a saline solution (0.9%) was used to complete the chamber volume and sature the material, filling the pores space. Such fluid has a conductivity $\sigma_f = 14$ mS/cm [6, 7].

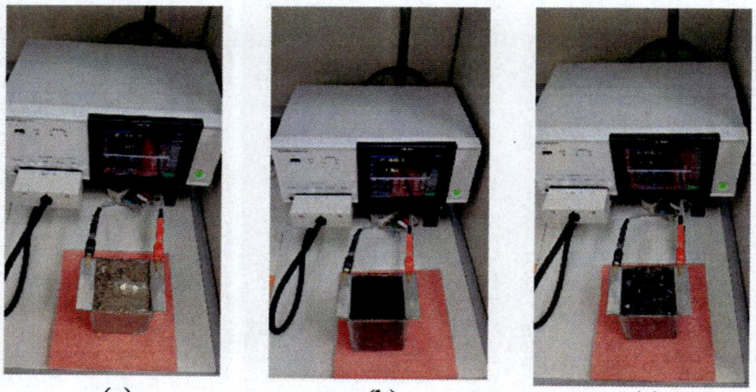

(a) (b) (c)

Fig. 3. The three types of porous materials using to have different porosity characteristics. (a) construction sand, (b) small-size aquariums stones, and (c) medium-size aquariums stones.

As porosity can also be understood as the ratio between the volume of the empty spaces (spaces of the pores) and the total volume of the medium, one can estimate the porosity of each phantom as the volume of the saline solution used to complete the measuring chamber and the total volume of the chamber.

For each phantom after determining the conductivity of the porous medium (σ) by the rational above-mentioned, we calculated the formation factor (F) (Eq. 2), porosity (Ø) (Eq. 3, using a heurisitic value of $m = 1.8$), and connectivity (Eq. 5). Finally, we compared the porosity estimates obtained the volumes ratio (ϕ_V) with the ones achieved by conjunction of BIS and Archie's law.

3 Results

Corroborating some previous work [8] our conductivity and permittivity data exhibited the first two out of the three regions: (i) low frequency dependent region, (ii) medium frequency plateau, and (iii) high frequency dependent region (Fig. 4). These first two regions are associated to electrode–electrolyte interface and the bulk conductivity, respectively. According to this behavior we developed a program to detect the frequency at which the permittivity became stable (the one indicated by a red dot in the graphics of Fig. 4) and used the conductivity value for this frequency to procedure the calculations of porosity and connectivity.

Table 1 shows the results for the three studied porous media.

4 Discussion

The results of the present study corroborate some previous works [9] that point out that the bioimpedance technique seems to be a good tool to estimate porosity characteristics in non-invasive and non-destructive assays. Unlike previous studies, we investigated

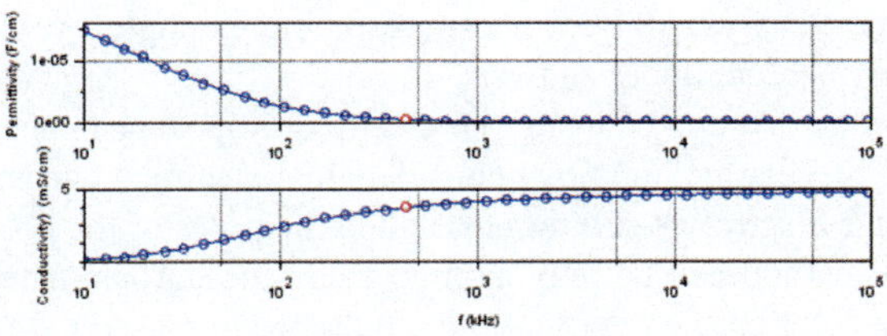

Fig. 4. Example of and permittivity (a) and conductivity (b) data for the medium-size aquariums stones.

Table 1. Results for construction sand, small size aquariums stones and medium size aquariums stones, assuming $\sigma_f = 14$ mS/cm.

Parameters from BIS technique			
Parameters	Construction sand	Small-size stones	Medium-size stones
Z	(37.002 – 5.548i)Ω	(24.998 – 6.743i)Ω	(26.191 – 6.594 i)Ω
Y	(0.0264314 + 0.0039631 i) S	(0.03729 + 0.0100587 i) S	(0.0359052 + 0.0090397 i) S
G	0.026	0.037	0.036
Conductivity	2.842 mS/cm	4.0096 mS/cm	3.861 mS/cm
Formation factor	4.926	3.492	3.626
Porosity	0.412 0.4(vol.estimate)	0.499 0.5(vol.estimate)	0.489 0.5(vol.estimate)
Connectivity	0.492	0.574	0.564

the behavior of conductivity at different frequencies. As expected at low frequencies, a dispersion was found due to the electrode interface and a wide range of frequencies showed a plateau in the conductivity and permittivity values. In this study we only investigated the behavior of Archie´s law applied to only one frequency out of the 50 frequencies acquired by the impedance analyzer, but we showed that several others could be used, which can facilitate the analysis with more common AC systems.

Although the cementation exponent was estimated heuristically, the porosity estimates by two methods (BIS and volume) were close. As the agreement between the estimates was considered satisfactory, we believe that the estimates for connectivity are also consistent, but further analysis is necessary. In a practical application of the method, the cementation exponent can be estimated for all types of materials used in scaffolds. The same approach has been used successfully in geology and does not represent a difficulty for the use of the technique. Higher frequencies can be explored to look for other dispersions that can add information to the analysis allowing the determination of other important characteristics. This frequency increase may be important only for samples with small pore size, as in real scaffolds, and not with large pore size, as in our experiments. In any case, we believe that this increase in frequency can lead to other dispersions and, as in the analysis of body composition, for example, provide additional information about the phantom under analysis.

Despite the encourage results of using the electrical impedance spectroscopy and the Archie's Law, the present methodology considers that electrical current in the measurement cell is due the electrolyte solution that filled the pores of the porous medium. This implies that the method can be only applied when the porous material does not absorb the electrolyte solution, i.e., the scaffold would be composed by a hydrophobic and isolate material. When the scaffold absorbs the electrolyte solutions its electrical characteristics, as e.g., conductivity, can change and a significant amount of the electrical current would also flow though the porous material, adding to the current that flows though the pores.

Future studies are needed to investigate the relationship of the porosity profile, pore size, interconnectivity and others feature as a function of frequency. The application of the method in real scaffolds used in Tissue Engineering is also necessary.

5 Conclusions

Based on ours results, we can say the bioimpedance technique can be used as a potential tool to evaluate porosity of scaffolds during building phase.

Acknowledgment. The authors would like to thank CNPq, CAPES and FINEP by the financial support of the research.

References

1. Chong, C., Loh, Q.L.: Three-dimensional scaffolds for tissue engineering applications: role of porosity and pore size. Tissue. Eng. Part B Rev. **2013**, 485–502 (2013)
2. Ahmed, S., Chauhan, V.M., Ghaemmaghami, A.M., Aylott, J.W.: New generation of bioreactors that advance extracellular. Biotechnol. Lett. **41**, 1–25 (2019)
3. Canali, C., Mohanty, S., Heiskanen, A., Muhammad, H.B., Martinsen, O.G., Dufva, M., Emnus, A. W. a. J.: Impedance spectroscopic characterisation of porosity in 3D cell culture scaffolds with different channel networks. Electroanalysis **27**, 193–199 (2015)
4. Teixeira, B.N., et al.: Evaluation of bone marrow stem cell response to PLA scaffolds. J. Biomed. Mater. Res. **2018**, 1–13 (2018)
5. Arcchie, G.E.: The electrical resistivity log as an aid in determining some reservoir characteristics. Trans. AIME **146**(01), 54–62 (1942)
6. Canali, C. et al.: Impedance-Based monitoring for tissue engineering applications. In: II Latin American Conference on Bioimpedance, pp. 36–39 (2016)
7. Glover, P.W.: A generalized Archie's law for n phases. Geophysics **75**(6), E247–E265 (2010)
8. Martinsen, O.G., et al.: (2008), Bioimpedance and bioelectricity basics. Department of Physics, University of Oslo and Department of Biomedical, Oslo (2008)
9. Canali, C., Heiskanen, A., Muhammad, H.B., Hoyum, P., Pettersen, F.-J., Hemmingsen, M., Wolf, A., Dufva, M., Emnéus, O. G. M. a. J.: Bioimpedance monitoring of 3D cell culturing-complementary electrode configurations for enhanced spatial sensitivity. Biosens. Bioelectron. **63**, 72–79 (2014)

10. Fuzlin, A.F.A., Misnon, I.I., Samsudin, A.S.: Conduction properties study on alginate incorporated with glycolic acid based solid biopolymer electrolytes. In: Materials Science Forum, vol. 981. Trans Tech Publications Ltd, (2009)
11. Ivorra, Bioimpedance monitoring for physicians: an overview, Centre Nacional de Microeletrònica (2002)
12. Liu, X., et al.: Novel magnetic silk fibroin scaffolds with delayed degradation for potential long distance vascular repair. Bioact. Mater. 126–143 (2022)
13. LNCS Homepage, http://www.springer.com/lncs. Last accessed 21 Nov 2016

Study of the Collagen Study of the Collagen Mineralization in Simulated Body Fluid (SBF) Medium

Christiane Ribeiro(✉) and Luiz Felipe Ribal

Center for Engineering, Modeling and Applied Social Sciences (CECS), Federal University of ABC (UFABC), São Bernardo do Campo, SP 09606-045, Brazil
christiane.ribeiro@ufabc.edu.br

Abstract. Mineralized collagen is an organic–inorganic composite material formed by a biopolymer and a calcium phosphate-based ceramic phase. The formation of hard connective tissues, such as bone, involves the deposition of calcium phosphate in a collagenous matrix, which forms a common building block for these tissues. The obtention of biomaterials by *in vitro* mineralization, has been investigated as an alternative for the development of methods that mimic biological systems. Furthermore, the understanding of the mechanism of collagen mineralization is of great significance for the design of materials for hard tissue regeneration and may also have important implications for the treatment of mi-neralization related diseases. In this work, was explored the viability of *in vitro* mineralization of the collagen by simulated body fluid (SBF) medium. The star-ting material was previously characterized in terms of its morphology by scanning electron microscopy (SEM), its structural features by Fourier Transformed Infrared Spectroscopy (FTIR) and its surface charge by zeta potential measurements. The mineralization ability of collagen fibers with SBF was investigated by SEM and FTIR as a function of immersion time in SBF, as well as pH monitoring during the time of the test. The results showed evidence of mineralization on surface of the collagen and presence of the precursor phase of the biological apatite, proving the existence of a suitable microenvironment for the nucleation and maturation of calcium phosphate.

Keywords: Collagen · Mineralization · SBF · Calcium phosphate

1 Introduction

Biomimetic materials, composed of hydroxyapatite on an organic matrix, such as collagen are promising candidates for developing and constructing smart organic/inorganic hybrid biomaterials through a calcium-based mineralization [1, 2].

In recent years, the biomineralization process has attracted much interest from academics and industries for potential technological application, mainly to produce highly biocompatible composites resembling natural hard tissues, inspiring bio-interactive materials with different microstructures and enhanced physical, chemical and biological

J. L. B. Marques et al. (Eds.): CLAIB 2022/CBEB 2022, IFMBE Proceedings 98, pp. 480–490, 2024.
https://doi.org/10.1007/978-3-031-49401-7_50

functions for implantable biomaterials [3, 4]. The mineralized collagen, for example, has shown a significant improving in its mechanical strength and makes it a suitable matrix for bone repair [5]. In addition, the mineralized matrices are a valuable option for regulating diverse phenotypic activities of stem and progenitor cells in the repair and regeneration of bone tissues, as well as useful in the study of bone diseases, such as osteoporosis or cancer metastasis [1, 5].

Several *in vitro* approaches have been used to investigate the mechanisms underlying the specificity of apatite formation in various collagen substrates, ranging from reconstituted collagen to natural tissues [5–7]. A promising approach to mineralization surface is using simulated body fluid (SBF), a supersaturated solution that comprised ions at similar concentrations to those found in blood plasma. In this method, the dri-ving force for mineralization is associated with a variety of ions existing in the mine-ralizing medium not only as bystander, but an active player that directly regulates the precipitation process and nature of biogenic apatite's [6–8].

These *in vitro* systems cannot replicate the dynamic biochemical and cellular processes that occur in the living tissue. However, the insights obtained from the *in vitro* models constitute the first step in unraveling the mechanisms of collagen mineralization *in vivo* [5, 7].

The SBF technique has advantages over other apatite mineralization methods in that it is simple and results in the formation of a bone-like apatite mineral, which is chemically and structurally equivalent to the mineral in bone, is similar in crystallites size and number of defects to that of bone and provides an interfacial bonding between tissue and material. In addition, this apatite layer can be formed successfully on flat surfaces as well as on 3D scaffolds of various synthetic biomaterials, with favorable cell viability and normal spreading morphology of the osteoblasts [1, 7].

The factors that direct nucleation of the collagen remain briefly describable. Therefore, the main objective of this study was to produce *in vitro* mineralized collagen using simulated body fluid (SBF) and physicochemical characterize the material developed, aiming at a new alternative of mineralized biomaterials to enable its applicability to bone repair, promoting cell support and accelerating the osseointegration.

2 Materials and Methods

The fibrous type I collagen (NovaProm®, Brazil) was used as starting material. The collagen fibers were observed by scanning electron microscopy (SEM, FEI Quanta 250, USA) and analyzed the zeta potential to determine their surface electrical charge in microelectrophoresis equipment (Zetasizer Nano ZS- Malvern Instruments, UK), and measurement cell, DTS 1060C – Clear Disposable Zeta Cell. For this, the samples of collagen were diluted to 0.2% (volume) in NaCl (Synth-Brazil) solution 0.001 M, as a background electrolyte, using MilliQ water. The suspension was centrifuged, and the pH was adjusted with 0.01M HCl (Dinâmica química contemporânea-Brazil) and 0.01M NaOH (Synth-Brazil), obtaining a pH range between 3 and 11 for measurement.

This system adopts laser Doppler electrophoresis to measure the electrophoretic mobility of the material. The zeta potential (ζ) is given by the Smoluchowski model:

$$\zeta = 4\pi\eta U/\varepsilon \quad (1)$$

where U is the electrophoretic mobility of the material, η is the viscosity of the solution and ε is the dielectric constant of the solution.

For the study of mineralization, a solution simulated of body fluid (SBF) was used for *in vitro* modelling, which has ion concentrations nearly equal to those of human blood plasma (Na^+ 142.0, K^+ 5.0, Mg^{2+} 1.5, Ca^{2+} 2.5, Cl^- 148.8, HCO^{3-} 4.2, and $HPO_4{}^{2-}$ 1.0 mM). The SBF was prepared by dissolving reagent-grade chemicals (NaCl, $NaHCO_3$, KCl, $K_2HPO_4{\cdot}3H_2O$, $MgCl_2{\cdot}6H_2O$ $CaCl_2$, and Na_2SO_4 into deionized water, and the SBF solution was adjusted to physiological pH 7.4 with tris-HCl buffer) according to the protocol described in Standard ISO 23317:2014 [9]. The mineralization was evaluated in the period of 7, 15 and 21 days.

The collagen fibers were incubated and soaked in SBF, maintained in polyethylene bottles, previously cleaned and dried. For the test was adopted the ratio of 1mg (collagen)/mL(SBF) and the bottles containing the samples were incubated at 37 °C ± 0.5 in a shaker (Solab Científica - SL-223, Brazil), with constant orbital agitation of 42 rpm. The analyzes were performed in triplicates for each time period and the medium was replenished every week. At each replenished, the supernatant solution was removed, and the material was rinsed 2 times in deionized water, and then another 2 times with fresh SBF solution, respectively. Subsequently, was added the appropriate volume of the SBF medium. It is worth to point out that the pH was monitored in every SBF medium change. For further analysis, the samples were centrifuged and dried at 42 °C/48h in an oven (Sterilifer indústria e comércio Ltda - Brazil).

The surface of the collagen fibers was analyzed by SEM (FEI Quanta 250, USA) to assess the mineralization ability. For this, the samples were sputter-coated with a layer of 20 nm gold using an ion sputtering (AC200, Leica, Germany).

Furthermore, structural features of collagen samples, before and after mineralization test, were assessed with Fourier Transformed Infrared Spectroscopy (FTIR). The FTIR spectra were obtained in Attenuated Total Reflection mode (ATR) using a SPOTLIGHT 400 FTIR – USA, spectrometer. The spectra were recorded in the range between 510 and 4000 (cm^{-1}), and spectra plots represented 60 scans.

3 Results

A preliminary characterization was conducted to assess the structural and morpholo-gical properties of native collagen, being analyzed by SEM, as shown in Fig. 1.

SEM images indicate the presence of a microfibrillar structure with different diameters for type I collagen, commercially available (NovaProm®, Brazil). The varied appearance of its fibers is a common and inherent characteristic of the extraction process of the insoluble native collagen, according to the manufacturer's information.

Additionally, was obtained FTIR spectrum of the starting collagen, Fig. 2. The FTIR spectrum revealed the presence of typical bands of native collagen. The band at 1238 cm^{-1} is due to vibrations of the amine III, sensible to the presence of tropocollagen secondary structure (α-helix), while the band around 1457 cm^{-1} corresponds to the C-H vibrations of the pyrrolidine rings of proline and hydroxyproline [10, 11].

The ratio between these two bands (1238–1457 cm^{-1}), when near to 1, represents the conservation of the triple helix structure, in this material the ratio was 0.97. The

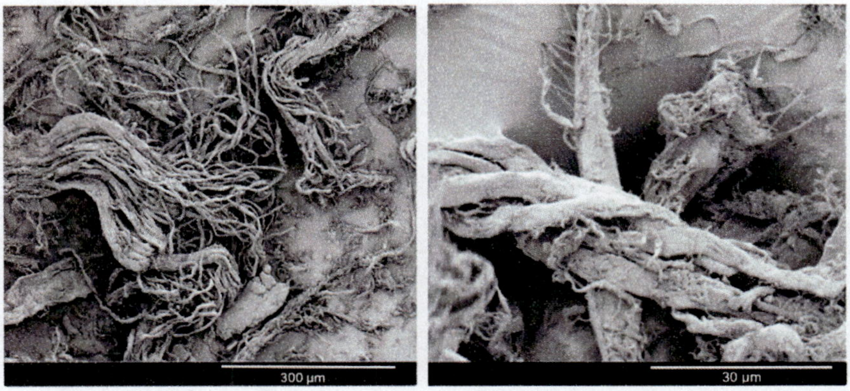

Fig. 1. Micrographs obtained by SEM of type I collagen (NovaProm®) at different magnifications.

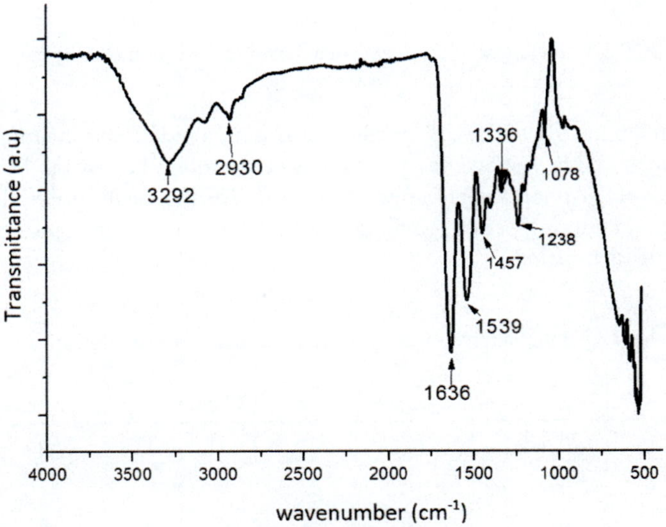

Fig. 2. FTIR spectrum of the starting collagen (NovaProm®).

bands at 3292 and 2930, 1636, 1539, 1336, 1078 (cm^{-1}) corresponds to O-H stretch and to specific vibrations in the C–N plane of the amine; vibration of the C = O of the amide I; the vibration of the C = O of the amide II; the C–N amine strain; and C-N axial deformation, respectively. In the region between 600–750 (cm^{-1}), was observed a high intensity band, typical of non-hydrolyzed collagen [10, 12].

Prior to *in vitro* mineralization test by soaking to SBF, the collagen was characterized in terms of its surface charge. According to the zeta potential curve (Fig. 3), a significant negative charge density was observed at higher pH values.

The isoelectric point (pI) of the starting collagen was around 5.5 and only in the a-cid range below of the pI, the surface showed a positive potential.

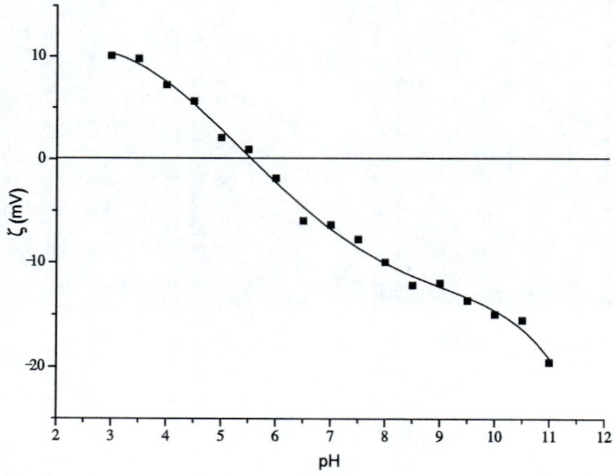

Fig. 3. Zeta potential curve as a function of the pH for the collagen.

In this study, the pH of the SBF medium was monitored during the mineralization test every week. Table 1 shows the pH data at each replenished of the SBF medium (initial $pH_{SBF} = 7.4$) and the pH after 24h of test. For this analysis, triplicates were used for each monitoring period, being observed a small deviation measured between the samples of each period.

Table 1. pH measurements as a function time during mineralization test.

Time period	24 h	7 days	14 days	21 days
pH	6.25 ± 0.1	7.43 ± 0.08	7.41 ± 0.05	7.45 ± 0.02

According to the pH data, was possible to prove that collagen exerts an influence on the pH of the medium, changing to a slightly acid range (pH = 6.25) during an initial period of 24h, showing an approximate variation of 1.15. However, for longer periods, such as 7, 14 and 21 days, no relevant pH variations were detected between the different periods in which the medium was replenished and compared to the original pH of the SBF (pH = 7.4).

The micrographs obtained during the mineralization test are presented in the Fig. 4.

There was a gradual increase of the mineral deposited as a function of the time defined for the test. A similarity in the morphological characteristics was observed between the precipitated material with the apatite phase, that presents a typical globular aspect.

The spectra obtained by FTIR (Fig. 5) of the collagen samples during mineralization test in SBF, reveal the typical bands of collagen, in the region of 3294, 3082, 2926, 1634 and 1420 (cm^{-1}) [10, 12]. In addition to the typical collagen bands, with lower intensity,

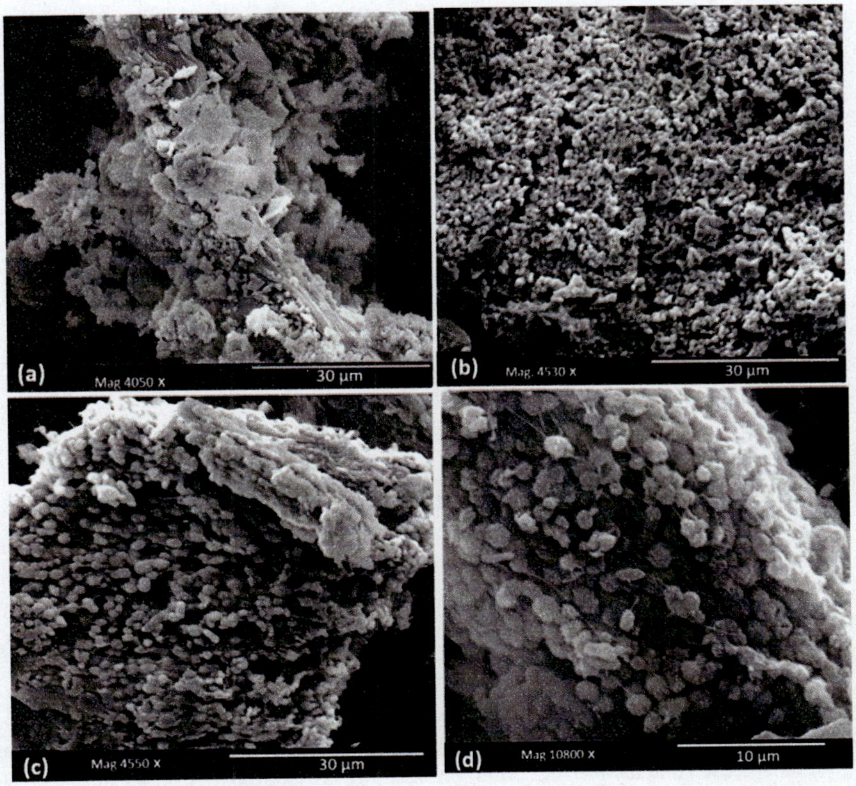

Fig. 4. Surface of the collagen fibers observed by SEM after soaking in SBF at different periods; (a): 7 days; (b): 14 days; (c) and (d): 21days, in different magnification.

the typical phosphate bands were observed, including the band around 1078 and 1033 (cm^{-1}), revealing an evidence of mineralization associated by presence of a phase with typical morphology of the apatite [10, 13, 14], as shown in Fig. 4.

4 Discussion

It has been shown that collagen can act as an active scaffold which provides an appropriate environment for mineral infiltration through charge interactions with the amorphous calcium phosphate (ACP) and mediates its crystallization into apatite crystals [3, 15]. This behavior allows a better performance of mineralization techniques that use solutions rich in calcium and phosphorus, as proposed in this work, than those techniques based on coating, such as thermal spraying, sputter coating, sol–gel deposition, dip coating and others [1, 7]. The biomimetic mineralization technique using SBF can be applied to various types of materials, including organic/ inorganic composites and at this point it resembles the other method that uses a sequential mineralization by calcium and phosphate solutions [1].

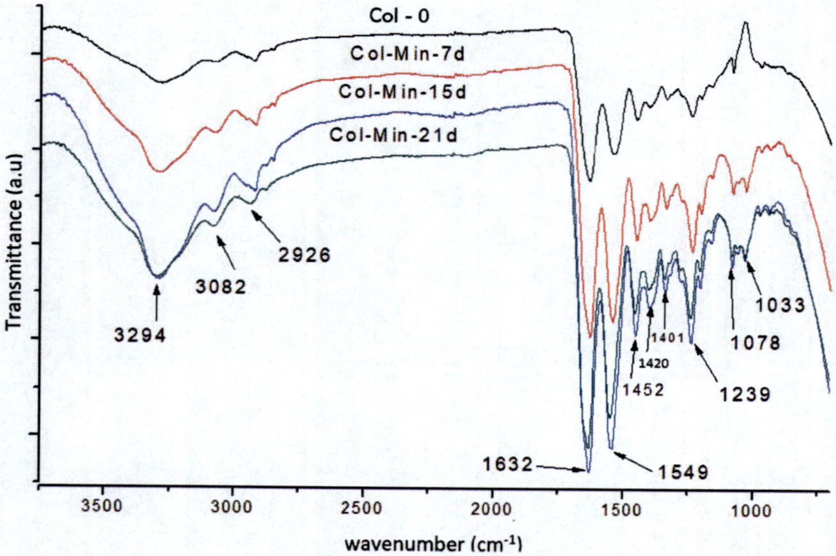

Fig. 5. FTIR spectra of the collagen, before (Col – 0) and after mineralization test (Col-Min) for 7, 15 and 21 days in SBF.

Comparatively, it was observed by microscopy images that there was a change in the fibrous morphology of the collagen structure before (Fig. 1) and after the SBF test (Fig. 4). It is known that the fibrillar collagen structure is essential to produce an organized array of crystals with the same crystallographic orientation [6, 11].

In aqueous saline solutions, such as SBF medium, collagen can absorb a considerable amount of water. As its isoelectric point is around of 5.5, any deviation from this range (i.e., change in the isoelectric point) can cause non-specific swelling, which can promote significant volume changes, inducing morphological changes in its structure [7, 10]. In this study was verified that the pH variations during the mineralization test (Table 1), may have promoted an alteration of the original morphology of the collagen, being observed, an arrangement in the form of plates (Fig. 4) different of the fibrous original form, in the Fig. 1. Also, this change may have been favored by the movement attributed to the shaker during the test.

According to the pH monitoring, it was possible to observe that collagen hinders pH changes to alkaline ranges, an unusual tendency for tests involving SBF [7, 14] which, due to the high concentration of alkaline ions in its medium, easily induce a pH increase.

Based on the literature, neutral pH value or low ionic strength medium can suppress the interaction between collagen molecules and another phase, because some ionizable functional groups of the collagen molecule are inaccessible due to changes in their conformation [5, 6].

During the mineralization test in SBF, a typical medium that induces an expressive ionic strength, the pH oscillated between a low acidity range in the initial stage, behavior associated with groups presents in the structure of collagen and a range close to neutrality, in the intermediate and final phase of the test (Table 1), but distant from its pI of 5.5 (Fig. 3). Even with the morphological changes observed in the collagen structure, there may be specific sites for nucleation, associated with both positive and negative charged clusters, but with a predominance of a surface potential negatively charged during the mineralization tests in SBF, due to the pH being around of 7.4. Thus, the collagen surface may have a high affinity with calcium ions present in the SBF solution, favoring the interfacial interaction and formation of the apatite [14].

Also is important to emphasize that the changes in pH observed in this study are not sufficient to cause the denaturation of this collagen. In addition, it is highlighted that as the mineralization test occurs at a pH like the physiological and different from its isoelectric point, there is the possibility of ionization of the $-COO^-$ (carboxylate) groups present in collagen which remain accessible on its surface, thus enabling HAp adsorption through the formation of the HAp-$Ca^{2+} \ldots {}^-$ OOC–.

The literature reports that the side chains of these amino acids could form an environment capable of coordinating calcium and phosphate ions and thus providing nucleation sites for apatite [3, 6], as the groups detected in the starting collagen by FTIR analysis (Fig. 2).

These conditions of the test in SBF promote calcium ion accumulation on the carboxyl group of collagen molecules, leading to calcium phosphate nucleation, as shown in the sequence for surface after mineralization (Fig. 5).

The characteristics of the collagen used in this study associated with the *in vitro* mi-neralization approach with SBF allowed the deposition of fine calcium phosphate particles with different morphologies, as shown in the images in Fig. 4 (a, b) in the initial stage. The flexible structure of collagen can be modified during the test, exposing domains with higher or lower concentration of charged species and random organization of structures, making organized mineralization difficult, being a disadvantage of this mineralization technique [6, 14].

These morphological changes observed on the collagen surface during the test may culminate in differences in the process of uniform growth of the newly formed surface layer by mineral phase due to the presence of both plate-like crystals more evidence in the early period, in addition to globular particles, similar at apatite [7, 14]. However, at longer test periods (21d), the image at a higher magnification, seen in Fig. 4-d, showed a predominant morphology of calcium phosphate, grouped in spherical clusters. This morphology is similar to crystal deposit, reported to be characteristic of carbonated hydroxyapatite (CHA), an advantage of this technique [13, 14].

Furthermore, *in vitro* studies have reported that the size of the calcium phosphate aggregates is important for the mineralization, in that smaller aggregates result in faster infiltration into the fibril and apatite formation [6]. It has also been proposed that the rate of collagen mineralization is related to the stability of the ACP and when the overall rate of mineral deposition was slow, only ACP was formed, while with a higher rate of mineral deposit, oriented apatite crystals formed. Recent studies suggest that the more rapidly formed ACP is less stable, and therefore more rapidly transforms into apatite [3, 4, 7–10].

The formation of a continuous layer of mineral on material is a process that takes longer than other conventional CaP coating methods. The nonuniform coating can be particularly problematic when the material is large, and the pore size is small [7].

According to the FTIR spectra, collagen denaturation was not observed (Fig. 5), confirmed by the presence of bands at 1239 cm^{-1} and 1452 cm^{-1} (referring to amine III and the C–H bond of the pyrrolidine, respectively), and the bands near to the region 3294 cm^{-1}, 2926 cm^{-1}, referring to the vibrations of the stretching of the O–H, C–N and C = O grouping of the primary and secondary amines [10, 12], so the collagen integrity was maintained.

Few changes in the FTIR spectra could be noticed between the samples soaking in SBF. The occurrence of typical bands of the calcium phosphate phase, probably apatite, at 1033 and 1078 (cm^{-1}) were observed during the mineralization assay in SBF (Fig. 5), but not in the starting collagen (Fig. 5 - Col-0). This probably means that a precipitation process occurred in equilibrium with the dissolution mechanism of the amorphous calcium phosphate precursor phase, thus, increasing the concentration of calcium and carbonate ions in the medium during the mineralization assay [16]. Additionally, the intensity of the band around to 3294 cm^{-1} (stretching of the O-H by presence of water) increased after immersion in SBF, compared to the starting material (col-0). This behavior reinforces the hydrophilic characteristic of collagen, furthering the incorporation of water in its structure, during *in vitro* mineralization test, agreeing with swelling of its fibers, followed by morphological changes, a common characteristic for mineralization techniques that use solutions [1, 7].

The integrity of the collagen structure can be determined by the transmittance ratio measured around 1238 cm^{-1} and 1451 cm^{-1}, and values for this ratio close to 1, prove that the integrity of the secondary structure of the tropocollagen and when the value is around 0.5 it is indicative of denaturation of collagen structure [10, 17]. According to spectra obtained in the Figs. 2 and 5 - Col-Min 21d, the relationship between the bands was calculated, showing a result equal to 0.9 and 0.85, respectively, which means that the secondary structure of tropocollagen was maintained, since its extraction (star-ting material), and also indicating that the drying temperature (42 °C) and the pH during the mineralization test, did not promote degradation of the molecule.

However, further research is needed in order to contribute to characterize this mineralization process, including a higher number of the samples to prove the stability and maturation of calcium phosphate phases throughout the test in SBF, with more accurate data.

5 Conclusion

According to the results obtained, can be concluded that the commercial collagen (Novaprom) presented an isoelectric point of 5.5, and an expressive negative charge is obtained at alkaline pH, allowing a great affinity with calcium ion, inducing the formation of the mineral layer on surface.

The conditions adopted for mineralization in SBF, as the constant shaking, promoted a change in the original morphology of the collagen, but its denaturation did not occur.

In vitro mineralization approach using SBF was possible at the proposed times by this study. The nucleation was observed on the surface and peripheral regions of the collagen, being noticed the formation of nanoparticles agglomerates with predominant globular morphology, typical of hydroxyapatite phase. However, the presence of the amorphous calcium phosphate phase cannot be ruled out due to the accelerated mineral deposition mineral. Furthermore, the distinction between the different phases of calcium phosphates cannot be accurately made by analysis of the FTIR.

The results obtained with this study were consistent with data reported in the literature. However, further studies with a higher number of samples are needed to prove the identification of the mineral phase of biological apatite, as well as the presence, stability, and maturation of others precursor phases on the collagen surface.

Acknowledgment. The authors would like to thank the Novaprom® for the supply of collagen and the Federal University of ABC and Multiuser Experimental Centers (CEM-UAFBC) and the technical team of the Biomaterials Laboratory, by the infrastructure and technical support.

Conflict of Interest. The authors declare that they have no conflict of interest.

References

1. Wu, X., Walsh, K., Hoff, B.L., Camci-Unal, G.: Mineralization of biomaterials for bone tissue engineering. Bioeng (Basel) **7**(4), 132 (2020). https://doi.org/10.3390/bioengineering7040132
2. Campodoni, E., Montanari, M., Artusi, C., et al.: Calcium-Based biomineralization: a smart approach for the design of novel multifunctional hybrid materials. J. Compos. Sci. **5**, 278–303 (2021)
3. Du, T., Niu, X., Hou, S., Li, Z., Li, P., Fan, Y.: Apatite minerals derived from collagen phosphorylation modification induce the hierarchical intrafibrillar mineralization of collagen fibers. J. Biomed. Mater. Res. A **107**(11), 2403–2413 (2019)
4. Zou, L., Zhang, Y., Liu, X., Chen, J., Zhang, Q.: Biomimetic mineralization on natural and synthetic polymers to prepare hybrid scaffolds for bone tissue engineering. Colloids Surf. B Biointerfaces **178**, 222–229 (2019)
5. Tomoaia, G., Pasca, R.D.: On the collagen mineralization. A Rev. Clujul Med. **88**(1), 15–22 (2015). https://doi.org/10.15386/cjmed-359
6. Nudelman, F., Lausch, A.J., Sommerdijk, N.A., Sone, E.D.: In vitro models of collagen biomineralization. J. Struct. Biol. **183**(2), 258–269 (2013)
7. Shin, K., Acri, T., Geary, S., Salem, A.K.: Biomimetic mineralization of biomaterials using simulated body fluids for bone tissue engineering and regenerative medicine. Tissue Eng. Part A **23**(19), 1169–1180 (2017)
8. Zhou, Z., Zhang, L., Li, J., Shi, Y., Wu, Z., Zheng, H., et al.: Polyelectrolyte–calcium complexes as a pre-precursor induce biomimetic mineralization of collagen. Nanoscale **13**, 953–967 (2021). https://doi.org/10.1039/D0NR05640E
9. International Organization for Standardization—ISO 23317:2014 (E).: Implants for Surgery: In Vitro Evaluation for Apatite-Forming Ability of Implant Materials, 3rd edn, pp. 1–13 (2020)

10. Du, T., Niu, Y., Liu, Y., Yang, H., Qiao, A., Niu, X.: Physical and chemical characterization of biomineralized collagen with different microstructure. J. Funct Biomater **13**, 2–57 (2022). https://doi.org/10.3390/jfb13020057
11. Dimasi, E., Gower, L.B.: Biomineralization sourcebook: characterization of biominerals and biomimetic materials. CRC Press, Boca Raton (2017)
12. Kuzmany, H.: Solid-State Spectroscopy: An Introduction. Springer, Berlin Heidelberg. (2009). https://doi.org/10.1007/978-3-642-01479-6
13. Elliot, JC.: Structure crystal chemistry and density of enamel, apatites. Wiley, Chichester (1997).https://doi.org/10.1002/9780470515303.ch5
14. Kim, H.M., Himeno, T., Kawashita, M., Kokubo, T., Nakamura, T.: The mechanism of biomineralization of bone-like apatite on synthetic hydroxyapatite: an in vitro assessment. J. R. Soc. Interface. **1**(1), 17–22 (2004). https://doi.org/10.1098/rsif.2004.0003
15. Parenteau, RB., Gauvin, R., Berthod, F.: Collagen-Based biomaterials for tissue engineering applications. Materials (Basel) **3**(3), 1863–1887 (2010)
16. Weiner, S.: Biomineralization: a structural perspective. J. Struct. Biol. **163**(3), 229–234 (2008). https://doi.org/10.1016/j.jsb.2008.02.001
17. Silverstein, R.M., Webster, F.X.: Identificação espectrométrica de compostos orgânicos, 6th edn. LTC, Rio de Janeiro (2007)

Special Topics

Application of an Automotive Relay as Optical Beam Shutter

F. T. C. S. Balbina[1,2(✉)], F. J. H. Moraes[1,2], and E. Munin[1,2]

[1] Biomedical Engineering Center, Anhembi Morumbi University (UAM), Estrada Doutor Altino Bondensan, 500, Eugênio de Mello, São José dos Campos, São Paulo, Brazil
ftcoimbra@gmail.com

[2] Center for Innovation, Technology and Education (CITE), São José dos Campos, São Paulo, Brazil

Abstract. Laser beam modulation is often required in many applications. Continuous wave (cw) gas lasers can be modulated only with devices external to the resonant laser cavity. Beam modulation can be achieved with several types of instruments like rotating wheel choppers, rotating blade shutters and electromagnetically actuated iris. In this work, an automotive relay model DNI0102 was modified to operate as an optical beam shutter. The electrical and optical characterization of the modified device are presented. The intrinsic resistance and inductance of the relay coil was measured as 80 Ω and 117.1 mH, respectively. The switching time obtained with the use of a 0.65 mm helium-neon gas laser beam was approx. 250 μs. The DNI0102 relay electrical characterization must substantiate and enforce future discussions about the relation between coil inductance and relay's switching time performance as shutters devices. In conclusion, the built device showed a fast switching speed that fits the demands of mostly experimental applications.

Keywords: Optical beam shutter · Automotive relay · Laser optical modulation · Biomedical engineering

1 Introduction

Photodynamic Therapy (PDT) is an effective modality for cancer treatment and diseases of bacterial, fungal or even viral origins. Photosensitization occurs through the interaction between a nontoxic photosensitizer and visible light at the wavelength of absorption of the photosensitizing agent. In the presence of oxygen, the interaction generates cytotoxic substances that cause oxidative damage to targeted cells. Nowadays, laser is a major source of light applied to activate photosensitizers in PDT treatments and experiments [1–4], as well as to investigate the photophysical and photochemical molecular properties of photosensitizers.

J. L. B. Marques et al. (Eds.): CLAIB 2022/CBEB 2022, IFMBE Proceedings 98, pp. 493–500, 2024.
https://doi.org/10.1007/978-3-031-49401-7_51

Laser stands for light amplification by stimulated emission of radiation. It means intensified light generated by the stimulation of a material inside a chamber, emitting energy in a continuous or pulsed way. There are different types of lasers (solid state lasers, liquid lasers, gas lasers and, semi-conductor lasers) [5,6]. Some of these types of lasers can be pulsed by intracavity devices, while others can be pulsed only by using extracavity devices. Semiconductor lasers have a major advantage of allowing modulation by electronically controlling the external current.

Gas lasers, in which the pumping mechanism is generally an electrical discharge in a gas filled tube, have a simpler design if compared to other types, finding a lot of use in medicine. Usually can be modulated only with the use of devices external to the cavity, which in simple words, modulate the gas laser beam periodically blocking or letting it pass [7].

Commercial electromechanical devices, known as optical beam shutters and choppers, fulfill this mission, allowing their users to interrupt or release a beam by inserting or removing an absorbing/reflecting actuator, frequently a blade or a rotating disc. These apparatuses are able to grant switching times of the order of 1–50 ms for shutters and under 1 ms for choppers. Many laser experiments requires fast switching, but commercial devices are expensive [8,9].

A relay is an electrically operated switch, commonly actuated by an electromagnet, ensuring complete electrical isolation between a primary driver circuit and a secondary (controlled circuit) [10]. The primary circuit provides the control signal to operate the relay, typically connected to a low voltage DC supplier. It consists in an electromagnetic coil which generates a magnetic field due to current passing through the wire. At the end of the coil, there is a pivoted arm containing a switch contact, which will be attracted or repelled (returning to the original position) by the magnet field action. The secondary circuit is connected to the load which needs to be switched [11]. The relay is very useful when there is a need to control a high current circuit using a low power signal (low current) or when the electrical potential from the secondary to the primary is relatively large. In addition, the relay permits the control of several secondary circuits with a single primary circuit.

Electromagnetic relays have a large array of applications, from basic mass-produced products to specialized and critical system components [12]. In particular, automotive relays find use in the control of the ignitions wires, headlights, horns and ought to handle currents up to 30–40 A [13].

The present work investigates the application of the DNI0102 automotive relay as a low-cost alternative optical beam shutter for gas laser-based photosensitizers molecular characterization experiments. The shutter constructive method is disclosed and the electrical and optical characterization data of the designed device are shown.

2 Methodology

The DNI0102 is a $12V/40A$ normally open single pole relay, meaning the load on the secondary side is normally off and there is only one secondary circuit to be energized by the actuator. Figure 1 shows the employed automotive relay as purchased and with its case totally removed.

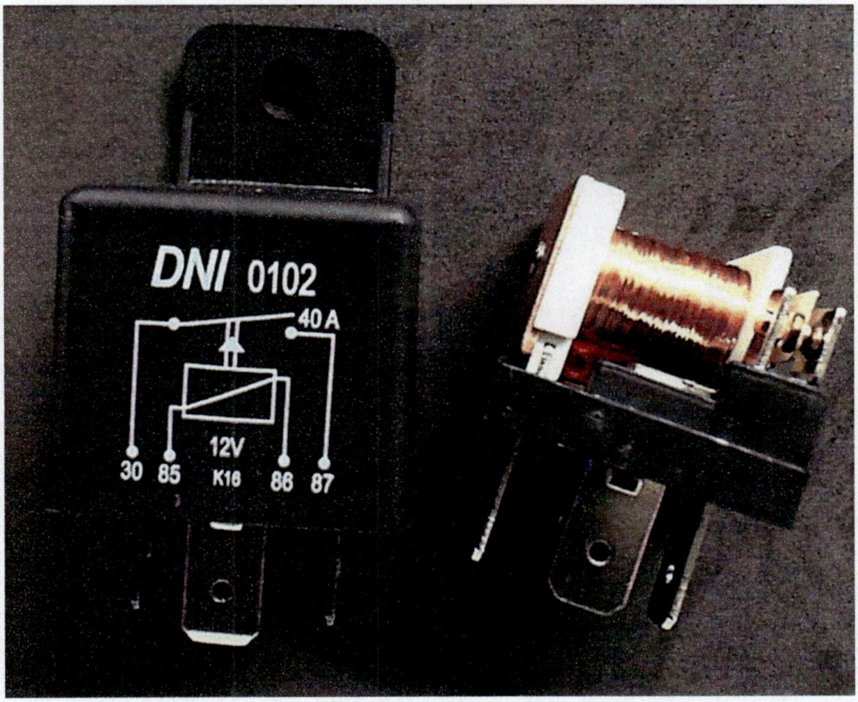

Fig. 1. DNI0102 automotive relay as purchased (left) and the device removed from its case (right).

To adapt the automotive relay to the intended optical shutter application, the top electric contact was removed, allowing fully arm movement amplitude. Also, an aluminum square flag (approx. 1 cm × 1 cm) was glued to pivoted arm, making it longer (4 cm) and able to block/transmit the beam, as shown in Fig. 2.

The coil intrinsic resistance was measured using ET-1002 Digital Multimeter (Minipa). A 632.6 nm Helium-Neon laser source with 10 mW output power and 0.65 mm beam diameter (model: 05-LHP-991 Melles Griot, USA) was utilized for optical characterization. The optical signal was captured by a SGD-444 fast photodetector with 100 mm^2 active area (EG & G Optoelectronics, USA) and processed by TBS1102B Digital Oscilloscope 100 MHz (Tektronix, USA). All data were analyzed and plotted using OriginPro graphical software. A dual channel, 12 MHz/12 bits, function/arbitrary waveform signal generator (model: FY2300 Feeltech, P.R.C.) was used for the electrical characterization of the relay's coil.

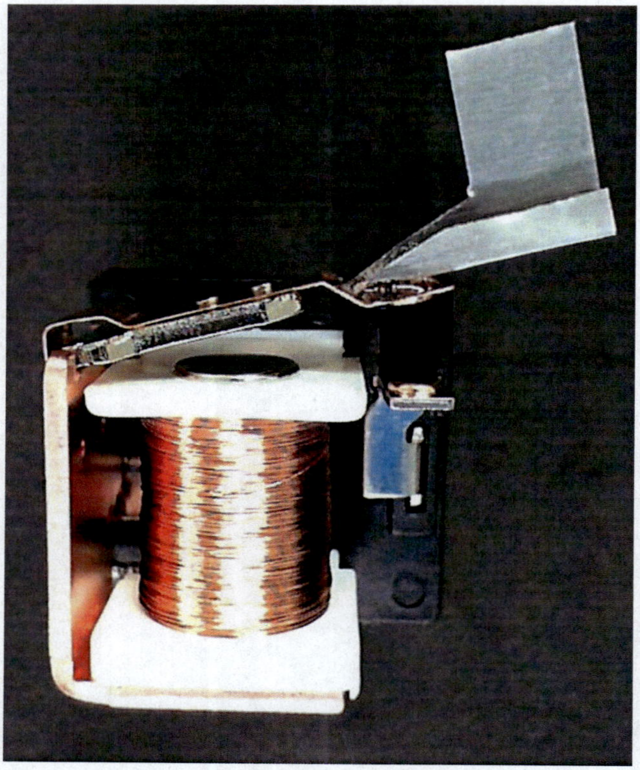

Fig. 2. Modified DNI0102 automotive relay with glued flag and top electrical contact removed.

2.1 Electrical Characterization

The relay's coil inductance L was measured by using the test circuit shown in Fig. 3.

While keeping source voltage at $V_S = 5V$, the signal generator frequency was varied from low frequency up to 3 kHz, and the Voltage V_{ab} was plotted as a function of the frequency. The frequency f for which V_{ab} equals half the source voltage amplitude value was identified. For such a frequency, the following relation holds:

$$R_i + X_L = R \tag{1}$$

where R_i stands for intrinsic resistance. R is an external resistor with nominal value of 320 Ω (322 Ω measured). X_L is the inductive reactance given by:

$$X_L = 2\pi f L \tag{2}$$

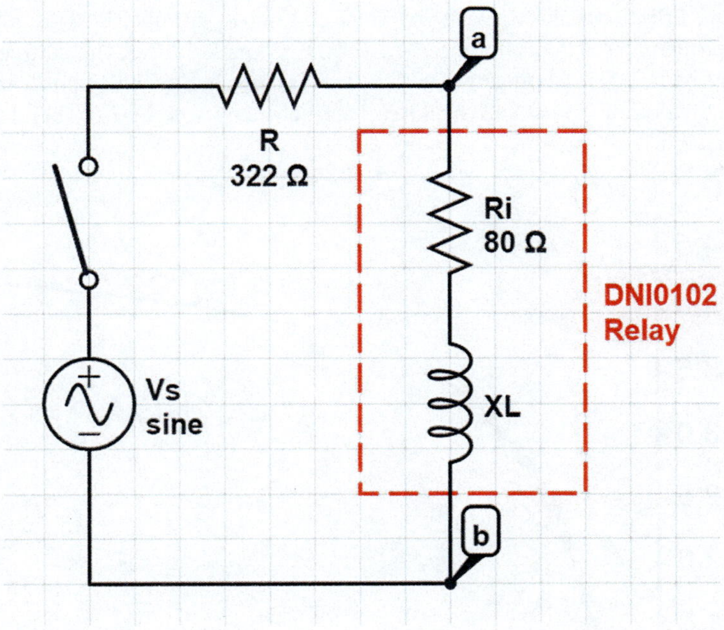

Fig. 3. Test circuit for inductance investigation

where L stands for the coil inductance, that can be given by combining Eqs. 1 and 2 as:

$$L = \frac{R - R_i}{2\pi f} \tag{3}$$

2.2 Optical Characterization

A shutter fundamental purpose is to switch between full laser beam blockage to full beam transmission state as fast as possible. Switching time was examined by positioning the modified DNI0102 relay at the near field of the laser source (30 cm distant from the source output), with a totally eclipsed Helium-Neon 0.65 mm beam positioned as far as possible from to the eclipsing edge of the flag, allowing the flag to gain speed. When the relays was actuated by a 12 V single pulse, the produced optical transient signal was captured by a photodetector. Signal processing with Fast Fourier Transform (FFT) applied to the raw signal filtered undesirable noise.

3 Results

The resistance of the coil wire (intrinsic resistance) measured with an ohm-meter was $R_i = 80\,\Omega$.

Figure 4 discloses the V_{ab} voltage (refer to the test circuit Fig. 3) as a function of the frequency f. It can be observed that V_{ab} equals half of the source voltage amplitude when the frequency $f = 329$ Hz. By using Eqs. 1, 2 and 3, inductive reactance value of $X_L = 242\,\Omega$ and coil inductance value of $L = 117.1$ mH can be calculated.

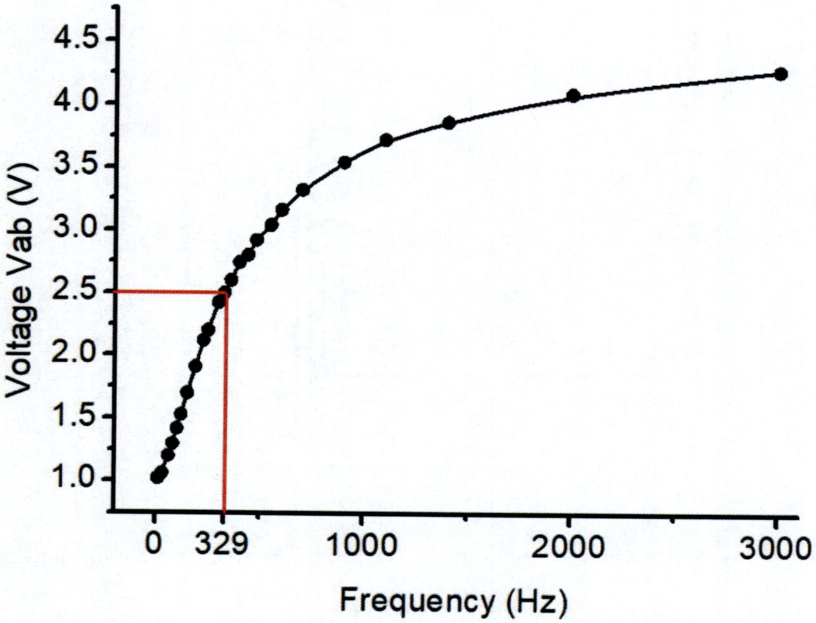

Fig. 4. Frequency X V_{ab} using Test Circuit at $V_S = 5V$.

Figure 5 shows the transmitted optical signal as a function of time. The achieved switching time of the fabricated shutter device was $\sim 250\,\mu$s.

4 Discussion and Conclusion

Optical beam shutters are essential for laser-based experiments and reliable solutions can be designed from cheaper elements [14–16].

The results obtained in this work suggest that DNI0102 automotive relay is an inexpensive and compact alternative for commercial beam shutters.

In the case of rotating wheel choppers and shutters, the adjustment of duty cycle requires replacement of to wheel and switching time is not constant relying upon disc/blade velocity, reaching 1–50 ms for commercial shutters and under $1ms$ for commercial choppers. Electromagnetic shutters based on electromagnetic actuators such as the modified automotive relay used in this work, allow

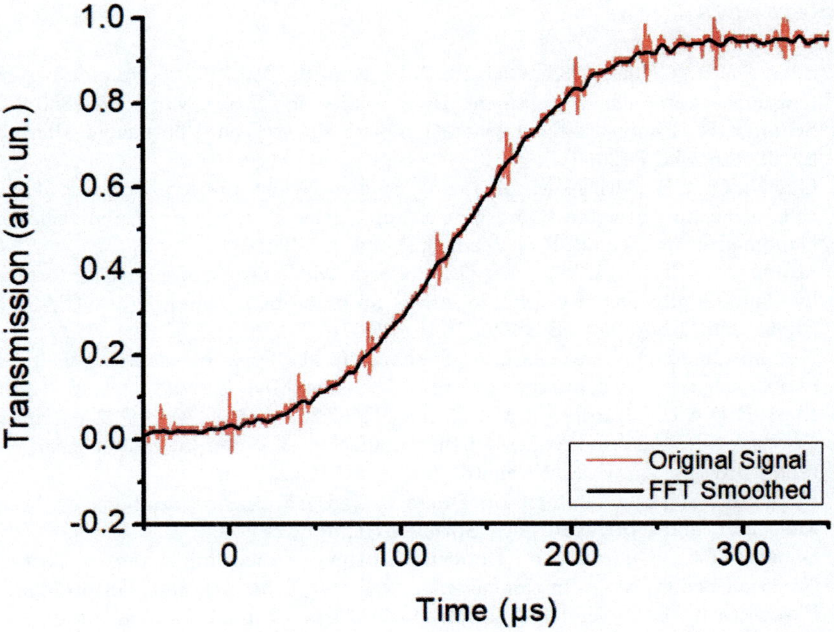

Fig. 5. Temporal evolution of the optical signal captured by the SGD-444 photodetector as the shutter flag moves away from the laser beam. The signal transit time correspond to the laser beam switching time of the electromagnetic relay actuator.

full control of operation frequency and duty cycle with a simple signal generator adjustment, while keeping constant the laser beam switching time in the order of $\sim 250\,\mu$s.

DNI0102 relay electrical characterization must substantiate and enforce future discussions about the relation between coil inductance and relay's switching time performance as shutters.

Published works have been discussed ways to analyze, protect and enhance electromagnetic relays, addressing all types of problems, *e.g.* maximum life span, reliability, and more direct subjects like frequency of bouncing, number of bounces, contact degradation, and arm velocity impact [10,12], but these issues were not the concern of this work.

Future works shall explore the development of an dedicated electronic drive to operate the modified DNI0102 automotive relay in a more versatile way.

Acknowledgements. This study was financed in part by the Coordenačão de Aperfeičoamento de Pessoal de Nível Superior - Brasil (CAPES) - Finance Code 001.

Conflict of Interest. The authors declare that they have no conflict of interest.

References

1. Sibata, C.H., Colussi, V.C., Oleinick, N.L., Kinsella, T.J.: Photodynamictherapy: a new concept in medical treatment. Braz. J. Med. Biol. Res. **33**(8), 869–880 (2000)
2. Schmidt, R.: Photosensitized generation of singlet oxygen. Photochem. Photobiol. **82**(5), 1161–1177 (2006)
3. Guo H., Qian H., Idris N. M., Zhang Y.: Singlet oxygen-induced apoptosis of cancer cells using upconversion fluorescent nanoparticles as a carrier of photosensitizer. Nanomed. Nanotechnol. Biol. Med. **6**(3):486–495 (2010)
4. Munin, E., Giroldo, L.M., Alves, L.P., Costa, M.S.: Study of germtube formation by Candida albicans after photodynamic antimicrobial chemotherapy (PACT). J. Photochem. Photobiol., B **88**(1), 16–20 (2007)
5. Nazemisalman, B., Farsadeghi, M., Sokhansanj, M.: Types of lasers and their applications in pediatric dentistry: a review. J. Lasers. Med. Sci. **6**(3), 96–101 (2015)
6. Lins, R.D.A.U., Dantas, E.M., Lucena, K.C.R., Catao, M.H.C.V., Granville-Garcia, A.F., Carvalho, N.L., G.: Biostimulation effects of low-power laser in the repair process. An. Bras. Dermatol. **85**(6), 849–855 (2010)
7. Powell R.. Types of Lasers. In: Drake G. (eds) Springer Handbook of Atomic, Molecular, and Optical Physics. Springer Handbooks (2006)
8. Bonetti, Y.C., Gobrecht, J.: Rotating shutters: a mechanical way of flattening Gaussian beam profiles in time average. Appl. Opt. **39**(31), 5806–5810 (2000)
9. Paschotta R.: Beam shutters. Encycl. Laser Phys. Technol. 1 (2008)
10. Witter, G., Polevoy, I.: Contact Erosion and Malerial Transfer For Contacts in Automotive Relays. IEEE. **9**, 223–228 (1996)
11. Evans P.: How Relays Work at https://theengineeringmindset.com/how-relays-work/ (2021)
12. Kirschbaum, L., Robu, V., Swingler, J., Flynn, D.: Prognostics for electromagnetic relays using deep learning. IEEE Acess **10**, 4861–4895 (2022)
13. Bhatt, M., Kamdar, J.: Automation in automotive electric system. Int. J. Sci. Eng. Sci. **3**(6), 1–4 (2019)
14. Martínez, S., Hernandez, L., Reyes, D., et al.: Note: fast, small, and low vibration mechanical laser shutters. Rev. Sci. Instrum. **82**, 046102 (2011)
15. Maguire, L.P., Szilagyi, S., Scholten, R.E.: High performance laser shutter using a hard disk drive voice-coil actuator. Rev. Sci. Instrum. **75**(9), 3077–3079 (2004)
16. Zhang G.H., Braverman B., Kawasaki A., Vuletic V.. Note: fast compact laser shutter using a direct current motor and three-dimensional printing. Rev. Sci. Instrum. **86**, 126105 (2015)

Effect of Ultrasound on Sunflower Seeds *Helianthus Annuus* L. Planted on Organic Compound and Clay Substrate

I. O. Santos[1](✉), R. A. N. Souza[1], L. E. Maggi[1], R. F. Lacerda[2], W. C. A. Pereira[1,3], and M. C. Silva[1]

[1] CCBN/Biophysics Laboratory, Federal University of Acre, Rio Branco, AC, Brazil
isaacoliveira.tk@gmail.com

[2] CCBN/Cell Biology and Biochemistry Laboratory, Federal University of Acre, Rio Branco, AC, Brazil

[3] COPPE/PEB/Ultrasound Laboratory, Federal University of Rio de Janeiro, Rio de Janeiro, RJ, Brazil

Abstract. The species *Helianthus annuus* L., (sunflower) have a great impact on the world economy, moving billions. They can be adapted to different conditions in relation to other botanical species cultivated in the country. The objective of the work was to evaluate the effect of ultrasound as a treatment for germination and increase in the emergence performance of sunflower seeds. Ultrasound was applied in seven groups with 75 seeds each except CG for 1 min in groups T1, T2 and T3 with a frequency of 1 MHz and intensity of 0.5; 1.0 and 1.5 $W{\cdot}cm^{-2}$ respectively. Groups T4, T5 and T6 received a 3-MHz frequency ultrasound at the same intensities and time. After application, the seeds were washed in solution for 10 min for disinfection and then washed in distilled water for the same period. They were immersed in distilled water with 2.5 times its weight for 24 h. They were sown one day after ultrasound application, conditioned in plastic bags with organic compost and clay, in a 1:1 ratio. Regarding germination, all treatments presented lower indexes than CG, except T5 in AGT. In the emergency, T1 was the best treatment for %E and ESI, while in AET, T2 presented the best value. The T4 treatment was the worst among all for 99% of the analyzed variables, except for AET. New experiments with ultrasound at different doses and with a greater number of seeds per treatment, to minimize the weight of the physiological effects on the uniform development of the seedlings.

Keywords: Ultrasound in seeds · Germination · Emergency · Agricultural production

1 Introduction

The sunflower (*Helianthus annuus* L.) belongs to the botanical family Asteraceae and was one of the most commercialized seeds in the world in 2016, moving around 1.0 billion dollars [1]. These species have great ability to adapt to various soil and climatic

J. L. B. Marques et al. (Eds.): CLAIB 2022/CBEB 2022, IFMBE Proceedings 98, pp. 501–510, 2024.
https://doi.org/10.1007/978-3-031-49401-7_52

conditions, in addition to relative resistance to water stress, cold and heat, compared to other botanical specimens cultivated in Brazil, linked to its low production cost and importance of its oil [2].

Seed dormancy is one of the biggest problems faced by researchers, as it is an adaptive mechanism that aims to protect the species so that the seed germinates, when environmental conditions are favorable for the survival of the seedlings [3]. The seeds of *H. annuus* L. are usually dormant at the time of harvest, and this state may per-sist for weeks or months. This behavior is a problem for the producer who needs homogeneous planting and high productivity to meet market demand [4].

Sunflower seeds presents physical dormancy, also known as tegumentary dormancy, a condition in which the germination process is inhibited by the protective structures of the seed [4].

In the literature, methods for overcoming dormancy ranges from mechanical or manual scarification, immersion in sulfuric acid, to the application of ultrasound (US).

Ultrasound is a mechanical wave, and its application can be innocuous, deleterious or induce development in living tissues, depending on its intensity, exposure time, wave frequency and distance from the transducer to the target [5].

looseness-1In animal cells, US can accelerate the process of cell division, whereas in plant cells, the vibration of particles can break the cell wall of a seed integument, overcoming physical dormancy. This fact makes it possible to start germination in a shorter period [6].

Some studies indicate that the application of ultra-sound to seeds of Watermelon (*Citrullus lanatus*), Caranaí (*Chelyocarpus chuco*), Pumpkin (*Cucurbita maxima*) and Mahogany (*Swietenia macrophylla*), soaked in liquid medium, increased their germination and growth rate by facilitating the absorption of water [7–11].

Thus, the objective of this work was to evaluate the effect of ultrasound as a treatment for germination and increase in the emergence performance of sunflower seeds *Helianthus annuus* L. in organic compost and clay substrate, at the proportion of 1:1.

2 Materials and Methods

The seeds were purchased commercially (brand Zaeli©) within the expiration date. The US application took place in the cell culture laboratory of the BioNorte Network laboratory complex of the Federal University of Acre, in the city Rio Branco, on October 23rd, 2021.

In this work, seven groups were used (with 75 seeds each), including the Control Group - CG, all immersed in water for 10 min. Ultrasound was applied to each of the six groups (tests). Groups T1, T2 and T3 were treated with a frequency of 1 MHz and an intensity of 0.5; 1.0 and 1.5 W·cm-2 respectively. Treatments T4, T5 and T6 used a frequency of 3 MHz at the same intensities. All protocols were applied for 1 min, with a Sonomed V therapeutic ultrasound device from Carci® properly calibrated by the ultrasound laboratory (LUS) of the Federal University of Rio de Janeiro (UFRJ). During the applications, 12 seeds were used at a time, in a disposable cup with 50 mL of water. Contact gel was added be-tween the face of the transducer and the bottom of the beaker to improve ultrasound transmission (Fig. 1).

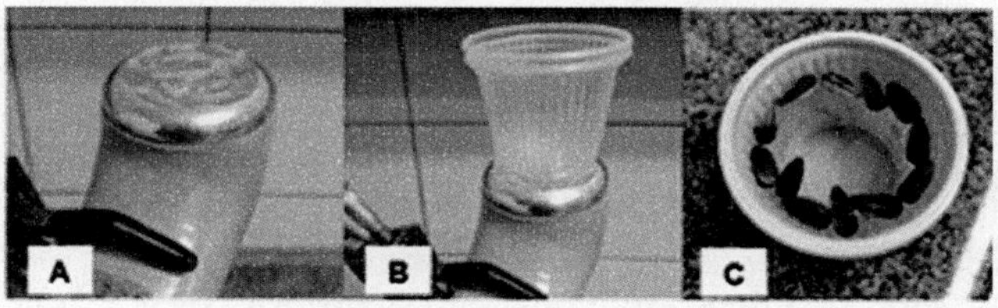

Fig. 1. A. - Transducer with gel. B - Transducer with disposable cup. C - Seeds around the entire disposable cup with 50 mL of water in the transducer.

After application, the seeds were immersed in an aqueous solution of sodium hypochlorite at 50% for 10 min for disinfection and, subsequently, washed in distilled water for the same period. The seeds were immersed in distilled water with 2.5 times their weight for 24 h. They were sown one day after the US application, conditioned in plastic bags with substrate of organic compost and clay, in the proportion of 1:1. They were monitored and watered daily in a nursery with a greenhouse at 50% insolation. The germination count was daily, from the individuals with cotyledons exposed above the substrate. Emergence, from seedlings with true leaves. The first count was carried out on October 26, 2021, three days after application, running for another 24 days after planting. The values were tabulated and, subsequently, the parameters Germination Percentage - %G, Germination Speed Index - GSI [12] and Average Germination Time - AGT [13] parameters, as well as Emergency Per-centage - %E, Emergency Speed Index - ESI [12] and Average Emergency Time - AET [13] were obtained.

The Germination Percentage (%G) was calculated by (1):

$$G = \mathrm{N}/75 \times 100 \tag{1}$$

where N is the number of seeds germinated at the end of the test, divided by the number of seeds sown.

The Germination Speed Index (GSI) and the Average Germination Time (AGT) were calculated by (2) e (3):

$$\mathrm{GSI} = (\mathrm{G1}/\mathrm{N1}) + (\mathrm{G2}/\mathrm{N2}) + (\mathrm{G3}/\mathrm{N3}) + \ldots + (\mathrm{Gn}/\mathrm{Nn}) \tag{2}$$

in which:

G1, G2, G3,..., Gn = number of seedlings computed in the first, second, third and last counts.

$$\mathrm{AGT} = \sum(ni.ti)/\sum ni \tag{3}$$

N1, N2, N3,..., Nn = number of days from sowing to the first, second, third and last count.

in which:

ni = number of seeds germinated per day;

ti = incubation time elapsed between the beginning of germination and the i-th count;

$i = 1, 2, 3\ldots 14$ days.

The Emergency Percentage (%E) was calculated by (4):

$$E = ((ni/N))/100 \quad (4)$$

where ni is the number of seedlings emerged at the end of the experiment and N is the total number of seeds sown.

The Emergency Speed Index (ESI) and the Average Emergency Time (AET) were calculated by (5) and (6):

$$ESI = (E1/N1) + (E2/N2) + \ldots (EN/Nn) \quad (5)$$

where E is the number of emerged seedlings recorded in the first, second and last counts, respectively. N1, N2 and Nn are the number of days after sowing in the first, second and last counts.

$$AET = (\Sigma(ni.ti))/(\Sigma ni) \quad (6)$$

where ni is the number of seedlings emerged in the i-th observation and ti is the number of days after sowing in the i-th observation.

3 Results and Discussion

3.1 Germination

The first germination count took place on October 26. It was observed that, in a 2-day interval from the sowing, on October 24, there were already germinated seeds, some with cotyledons exposed above the substrate. Throughout the experiment, a daily count was carried out, the last one on November 8th, 2021.

The control group had higher germination percentage - %G and Germination Speed Index - GSI with average values above the treatments, except for AGT, in which the best result was for T5 (Table 1).

Table 1. Seed germination (G, GSI e AGT).

Group		G (%)	GSI	AGT (days)
CG		**66.67**	**2.65**	18.98
1 MHz	**T1**	61.33	2.38	19.50
	T2	37.33	1.46	19.29
	T3	49.33	1.97	19.03
3 MHz	**T4**	33.33	1.28	19.64
	T5	50.67	2.05	**18.68**
	T6	56.00	2.23	18.90

Note that, for the germination percentage variable, the CG, on average, obtained the best result due to a greater number of germinated seeds (Fig. 2).

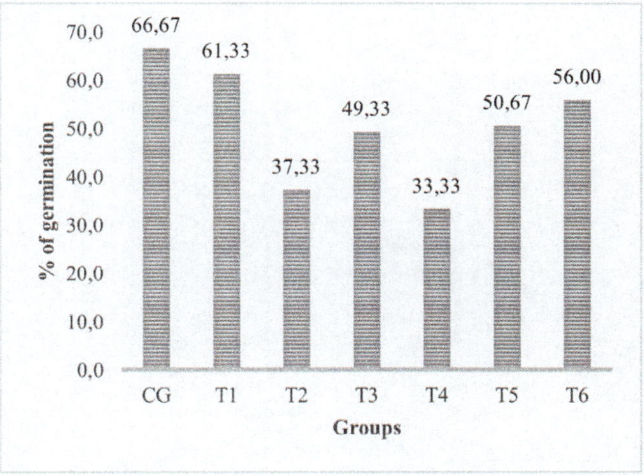

Fig. 2. Graph of Percentage of Germination (%G) according to the groups.

According to França-Neto et al. (1983) [14], seeds with high vigor have approximately 12% more plant height yield, as well as 24% productivity. Such data cor-roborate with those found in this study, with a variation in the results of the applications, depending on the vigor of the seeds.

According to Nazari and Eteghadipour (2017) [15], when analyzing several studies on the use of ultrasound, found that there is a favoring on the germination of several species. In a specific one, *Calanthe discolor*, treated with ultrasound, an increase of 50% was obtained in relation to the untreated ones. In another study, there was a 93% increase in the germination speed of pea seeds of the species *Pisum sativum.*

If the US intensity used is too high, germination inhibition can occur, given that the change in the physical state of the cell membrane from gel to crystalline can be very sudden and its restructuration cannot occur during the water absorption process, thus, causing serious damage to the plant [16]. This fact corroborates the results for %G, in which no treatment obtained a germination average higher than the control group, with T2 and T4 lower than the CG by 29.3% and 33.3% respectively (Fig. 2).

For GSI, the control group, on average, also obtained the best result, and the higher the index, the faster the germination, implying a faster emergence of seedlings (Fig. 3).

Bacaxixi et al. (2011) [17], when analyzing the germination percentage of *Helianthus annuus* in eight replications with 50 seeds, incubated in a B.O.D (Biochemical Oxygen Demand) chamber at 25 °C, obtained varied results due to the difference between the physiological characteristics of the seeds, such as age and period of numbness. Of these, only three repetitions (02, 07 and 08) presented %G above 70%. Therefore, with the study by Bacaxixi et al. (2011) [17], it is evident that for future studies, it is necessary to evaluate new doses of ultrasound in a greater number of seeds of this species, to find the best combination of time and intensity that favors the germination percentage.

As for AGT, the shorter the average germination time, the better (Fig. 4). T5 obtained the best result, with germination taking an average of 18.6 days.

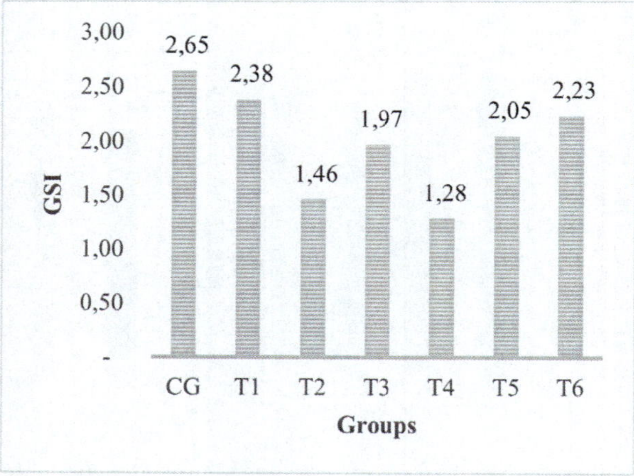

Fig. 3. Germination Speed Index as a function of groups. The control group had higher GSI.

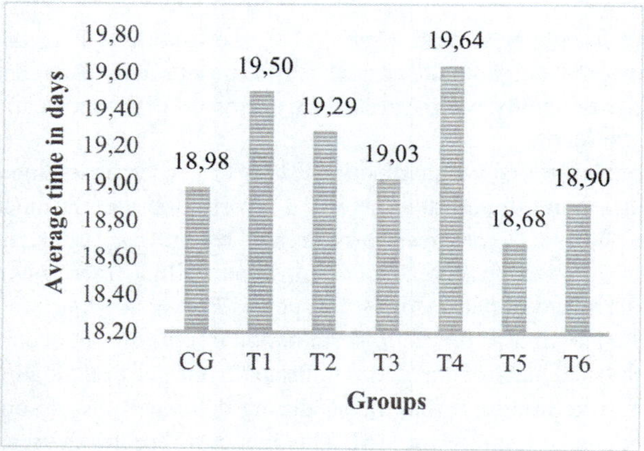

Fig. 4. Average Germination Time measured in days depending on the groups. The T5 group showed better AGT.

According to Ferreira et al. (2001) [18], the AGT is a good index to assess how quickly a species will establish itself in each environment. High ultrasound intensity can result in germination inhibition, impairing the water absorption process, causing serious damage to the plant [16]. This corroborates the result obtained for T4, which had the worst performance.

3.2 Emergency

The T1 treatment presented higher Emergency (%E) and Emergency Speed Index (ESI) values which means values above the ones for other treatments and for the control group, except for AET, in which the best result was for T2 (Table 2).

Table 2. Seed emergency (E, ESI e AET)

Group		E (%)	ESI	AET (days)
CG		33.33	2.73	9.52
1 MHz	**T1**	**36.00**	**2.88**	10.19
	T2	21.33	1.92	**8.38**
	T3	21.33	1.55	11.94
3 MHz	**T4**	18.67	1.33	11.86
	T5	25.33	2.08	10.05
	T6	30.67	2.54	9.39

The first emergency count took place on November 1, 2021. Throughout the experiment there was a daily count, with the last one on November 17, 2021. Treatment T1 had a higher Percentage of Emergency – %E (Fig. 5).

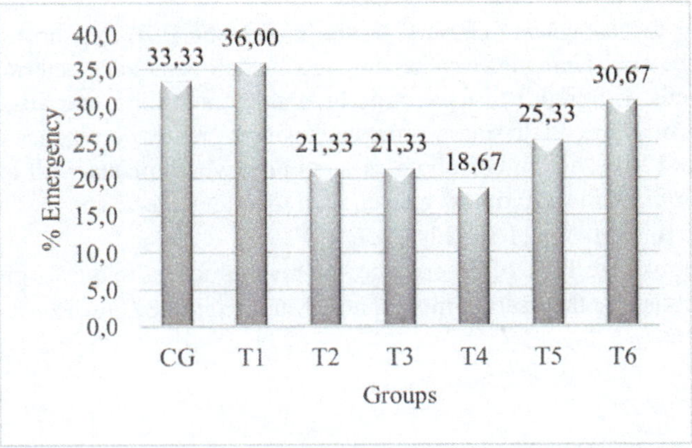

Fig. 5. Percentage of Emergencies by groups. T1 had the highest mean value.

High vigor results in an accelerated and uniform emergence in the medium, with normal seedling development under different environmental conditions [19]. The storage time of sunflower seeds influences their quality, and they can be stored for up to four months [20]. This fact corroborates the data of the present study, whose seeds were acquired commercially, not being sure under what storage conditions they were, having a direct impact on the results of the treatments, where germination and emergence were not standards regarding the best treatments of each stage.

Santos et al. (2020) [21] obtained better %E for T1 when using seeds of *Copaifera reticulata* D. under the same parameters of the current study, thus corroborating that ultrasonic waves of low frequency and intensity can break dormancy of seeds of different species.

T1 also showed a higher Emergency Speed Index - ESI with a mean value above the other treatments (Fig. 6).

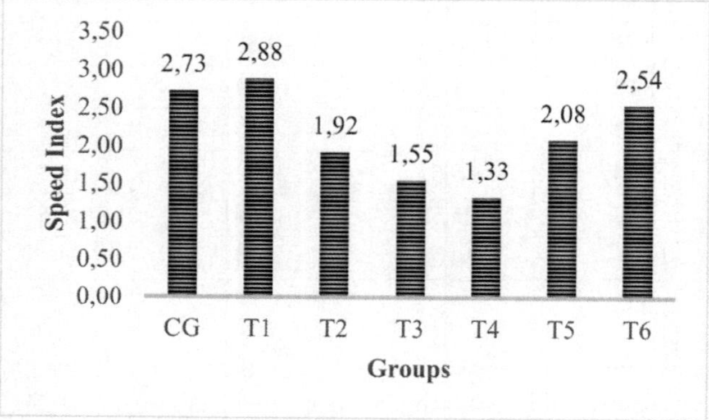

Fig. 6. Emergency Speed Index according to groups.

According to Nakagawa (1999) [21], the higher the ESI, the shorter the time of exposure to the environmental weather impacts, which can cause deterioration. Seedlings from seeds with high ESI have more favorable conditions, and this variable may be related to a high gain in dry mass, quickly causing a greater dominance of the species in the medium [20]. Thus, it corroborates our study, where plants with lower ESI had higher predation and lower biomass gain compared to the others. The best performances were obtained in seedlings of T1 (Fig. 6).

As for the Average Time of Emergence, the best value was to the T2 group, with the seedlings establishing themselves more quickly in the middle (Fig. 7).

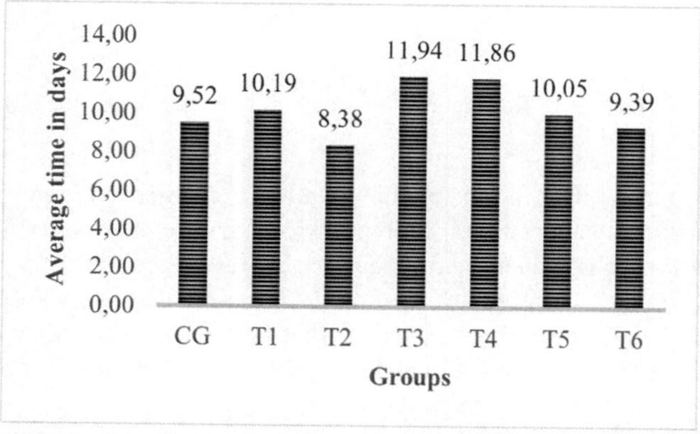

Fig. 7. Average Emergency Time measured in days depending on the groups. The T2 group showed better AET.

Although T2 presented better AET (Fig. 7), it does not mean that it had better performance throughout the experiment, considering that it had the second lowest percentage of germination (Fig. 2) and did not present the best result for AGT (Fig. 4). Such data explain the lower AET, since there were fewer germinated seedlings, di-rectly implying the result of emergence, leaving fewer seedlings to emerge over the days and consequently an apparent satisfactory result.

4 Conclusions

The results showed that, on average, the germination of the group of seeds treated with the ultrasonic dose applied was not superior to the control group, however, for emergence it presented satisfactory results, with T1 with better results for %E and ESI in relation to CG. New experiments should be carried out with the application of ultrasound at different doses and with a greater number of seeds per treatment, to minimize the weight of the physiological effects of the seeds on the uniform development of the seedlings.

Acknowledgment. The authors are grateful CAPES, CNPq, UFAC and FAPERJ.

Conflict of Interest. The authors declare that they have no conflict of interest.

References

1. Junqueira, A.H., Peetz, M.S.: Brazilian consumption of flowers and ornamental plants: habits, practices, and trends. Ornamental Hortic. **23**(2), 178–184 (2017)
2. Santo, G.R.E., Fernandes, D.A., Oliveira, G.P.: Vigor de sementes de girassol submetidas ao tratamento térmico. Biodiversidade **17**(3), 53–61 (2018)
3. Ferreira, A.G., Borghetti, F.: Germinação do básico ao aplicado. Artmed, Porto Alegre (2004)
4. Rodríguez, M., et al.: Effect of storage temperature on dormancy release of sunflower (Helianthus annuus) achenes. Seed Sci. Res. **28**(2), 101–111 (2018)
5. Hebling, S.A., Silva, W.R.: Efeitos do ultra-som de baixa intensidade na germinação de sementes de milho (*Zea mays* L.) sob diferentes condições de disponibilidade hídrica. Scientia Agricola **52**(3), 514–520 (1995)
6. Ter Haar, G.: Therapeutic ultrasound. Eur. J. Ultrasound **9**(1), 3–9 (1999)
7. Nascimento, C.C.F., Pinheiro, R.M., Santos, E.A. et al.: Análise do Efeito do Ultrassom no Tempo de Germinação de Sementes de Caranaí (*Chelyocarpus chuco*). In: 68ª Reunião Anual da SBPC, Porto Seguro/BA. (2016)
8. Costa, L.R.S., et al.: Efeito do Ultrassom na Germinação da Semente de Mogno (*Swietenia macrophylla King*) Meliaceae. In: 68ª Reunião anual da SBPC, pp. 1–1. SBPC, Porto Seguro (2016)
9. Nascimento, C.S., Maggi, L.E., Pereira, W.C.A.: Análise do efeito de Ultrassom em sementes de Abóbora *Curcubita maxima*. In: 68ª Reunião anual da SBPC, pp. 1–1. SBPC, Porto Seguro (2016)
10. Catarin, G.B., Maggi, L.E., Pereira, W.C.A.: Análise do Efeito do Ultrassom em Sementes de Melancia (*Citrullus lanatus*). In: 68ª Reunião anual da SBPC, pp. 1–1. SBPC, Porto Seguro (2016)

11. Mello, A.S., et al.: Análise do efeito do ultrassom em sementes de alface (*Lactuca sativa*). In: Congresso Regional de Pesquisa do Estado do Acre e XXIV Seminário de Iniciação Científica da Universidade Federal do Acre, pp. 21–22. UFAC, Rio Branco (2015)
12. Maguire, J.D.: Speed of germination aid in selection and evaluation for seedling emergence and vigor. Crop Sci. **2**(2), 176–177 (1962)
13. Labouriau, L.G.A.: Germinação das sementes. Monografia. Secretaria Geral da OEA (1983)
14. França-Neto, J.B., Krzyzanowski, F.C., Costa, N.P., et al.: Efeito de níveis de vigor das sementes sobre diversas características agronômicas da soja. In: EMBRAPA. Resultados de pesquisa de soja 1982/83. pp. 70–73. Centro Nacional de Pesquisa de Soja, Londrina, PR (1983)
15. Nazari, M., Eteghadipour, M.: Impacts of ultrasonic waves on seeds: A mini-review. Agric. Res. Technol. **6**(3) (2017)
16. Wang, Q., et al.: Modeling analysis on germination and seedling growth using ultrasound seed pretreatment in switchgrass. PLoS ONE **7**(10), 1–10 (2012)
17. Bacaxixi, P., et al.: Teste de germinação de girassol *Helianthus annuus* L. Revista científica eletrônica de agronomia **10**(20), 1–5 (2011)
18. Ferreira, A.G., Cassol, B., Rosa, S.G.T.D., et al.: Germinação de sementes de Asteraceae nativas no Rio Grande do Sul Brasil. Acta Botanica Brasilica **15**(2), 231–242 (2001)
19. ISTA. International Rules for Seed Testing. The International Seed Testing Association, Zurich, Switzerland. (2004)
20. Cabral, J.C.O.: Qualidade Fisiológica das Sementes de Girassol Submetidas à Secagem e ao Armazenamento. Instituto Federal Goiano, Campus Rio Verde. p. 48 (2022)
21. Nakagawa. J.: Testes de vigor baseados no desempenho das plântulas. In: Krzyzanowski, F.C.; Vieira, R.D.; França Neto, J.B. (Ed.). Vigor de sementes: conceitos e testes. Londrina: ABRATES p. 21 (1999)

Analysis of the Brazilian Job Market for Biomedical Engineering: A Pilot Study

Carolina Benetti(✉), Andrea Dorion Rodas, and Anderson Gariel Santiago

Center for Engineering, Modelling and Applied Social Sciences, Federal University of ABC, São Bernardo do Campo, Brazil
c.benetti@ufabc.edu.br

Abstract. The entry of the biomedical engineer into the Brazilian labor market is recent since the first undergraduate course in biomedical engineering dates from 2001 and the professional recognition of the biomedical engineer at the Federal Council of Engineering and Agronomy (CONFEA) and the Regional Council of Engineering and Agronomy (CREA) took place in 2018. This work is a pilot study that aims the analysis of job vacancies for graduates of biomedical engineering courses after three years of career regulation. An analysis of available job opportunities on the Indeed job search platform was conducted for biomedical engineering professionals and related fields. Vacancies with adherence to the profile of the biomedical engineer were selected and classified according to the area of activity and main activity. At the end of the analysis, 97 vacancies were analyzed, and the results suggest that the national market does not direct the search for a professional with a degree in biomedical engineering, expanding the search for professionals with other backgrounds.

Keywords: Biomedical engineering · Job vacancy · Carrier development · Job seekers

1 Introduction

The 21st century has undergone significant changes in the area of work, mainly in the creation and review of careers considered solid, especially in areas related to Engineering. One of these cases is Biomedical Engineering, which was a postgraduate specialization in courses such as Mechanical and Electrical Engineering. The first graduate course in Brazil in Bachelor of Biomedical Engineering began in 2001 at University of Paraíba Valley (UNIVAP) and its first professionals took over the market in 2005 [7]. The recognized professional took place on July 26th, 2018 by the Federal Council of Engineering and Agronomy (CONFEA) and the Regional Council of Engineering and Agronomy (CREA) resolution no. 1,103, in which the activities and skills of these professionals are detailed [1].

J. L. B. Marques et al. (Eds.): CLAIB 2022/CBEB 2022, IFMBE Proceedings 98, pp. 511–519, 2024.
https://doi.org/10.1007/978-3-031-49401-7_53

The Brazilian Society of Biomedical Engineering presents the Curriculum References for the Bachelor's Degree in Biomedical Engineering [7]. The topics covered in the formation of the Biomedical Engineer are interdisciplinary that go through basic training in engineering, administration, health sciences, and advanced technologies for diagnosis and therapies. The set of these themes promotes an interdisciplinary view of several areas of knowledge which allows its work in a large number of working environments, such as device industries, equipment, systems, materials, and dental-medical-hospital supplies; in clinics, medical laboratories, and hospitals; in companies, and scientific and technological research laboratories. Also, the biomedical engineers can work autonomously, in their own company, or by providing consultancy [7].

The Pedagogical Project of the Federal University of ABC systematically addresses the topics of Biomedical Engineering formation through compulsory and optional courses. The compulsory courses provide the essential career knowledge, and the elective subjects allow students to head the course to one of the specializations areas: Biomaterials and Implantable Devices; Biomechanics and Neuromotor Control; Clinical Engineering, Instrumentation and Processes for Diagnosis and Therapy; and Computational Systems Applied to Life Science [6].

The labor market has been increasingly looking for interdisciplinary professionals, that can view a problem from different perspectives and who can come up with innovative solutions, this desire has a direct impact on undergraduate courses that want their graduates to be successful professionals and the productively exercise their citizenship [5]. In this sense, this work aims to make a first overview after three years of recognition of the Biomedical Engineer by CONFEA/CREA in the Brazilian labor market.

2 Materials and Methods

The database chosen for the extraction of vacancies was the Indeed [4] web site, which is free, easy to register, and allows automated data extraction (web-scrapping). All accesses were performed from November 7 to 18, 2021 and for this, an Application Programming Interface (API) in Python 3.9 was implemented using the Requests-HTML library, with the result stored in a database.

For the search on the job portal, in addition to the term "Biomedical Engineering", 10 other professions that have an interface with Biomedical Engineering were defined: Chemical Engineering, Computer Engineering, Electrical Engineering, Mechanical Engineering, Mechatronics Engineering, Materials Engineering, Computer Science, Physical Engineering, and Data Science.

For each job, the following fields were created: "Identification Code" - a primary key that uniquely identifies the job within the Indeed portal database; "Position Name" - the first description of the vacancy; "Company Name" - if present, the company responsible for the vacancy; and "Position Description" - job description, the main field where the main analyzes presented in this work are carried out. Then, redundant entries were excluded along with those with a technician profile and those without "Position Name" or "Position Description".

An initial set of keywords was defined as a filter to identify job vacancies related to Biomedical Engineering profile, such as health, hospital, pharmacy, clinic, and biology. This filter was applied in the "Position Description" field.

The automatically selected job vacancies were evaluated by three reviewers, and the job vacancies incompatible with the biomedical engineering degree were excluded.

The selected job vacancies were again evaluated by three reviewers, and classified according to the knowledge area. Six knowledge areas were considered: 1. generic (no specific knowledge required), 2. computer science, 3. instrumentation, 4. biomechanics, 5. clinical engineering, and 6. biomaterials. This classification was based on the pedagogical project of Biomedical Engineering at UFABC [6].

In addition to the classification by the knowledge area, the job vacancies were also classified according to the main activity related to the job description. This classification was also performed manually by the three reviewers.

Finally, the job vacancy descriptions were used to create a word cloud. Natural Language Programming algorithms were applied to remove high-frequency terms and identify specific skills needed for each vacancy, resulting in the final analysis presented in this paper.

3 Results

The search using only the terms "Biomedical Engineer" or "Biomedical Engineering" without any additional filter resulted in a sample of 39 job vacancies. After using the filters to exclude unrelated vacancies and the reviewers' evaluation, this number was reduced to 14 job vacancies. Of this final sample, five vacancies were classified as clinical engineering, four as computer science, two as instrumentation, one as generic, and two job vacancies were classified as related to two areas: instrumentation and computer science and clinical engineering and computer science. Table 1 shows how the percentage of classification of the work areas was distributed.

Table 1. Area classification of the biomedical engineer job proposals.

Area	%
Clinical Engineering	37.4
Computer Science	28.6
Instrumentation	14.3
Clinical Engineering and Computer Science	7.1
Clinical Engineering and Instrumentation	7.1
Generic	7.1

Considering job vacancies in the areas of Biomedical Engineering, Electrical Engineering, Chemical Engineering, Production Engineering, Management Engineering, Mechanical Engineering, Mechatronics Engineering, Materials Engineering, Computer Science, Physical Engineering and Data Science, Pharmacy and Biological Sciences, 4597 were evaluated. After applying the job vacancy selection filters considering the exclusion criteria and terms related to Biomedical Engineering, 196 job vacancies were selected, and after the reviewers' evaluation, 96 job vacancies remained.

Of the 96 job selected positions, three were classified in more than one area of expertise: clinical engineering and computer science, clinical engineering and instrumentation, and clinical engineering and instrumentation and computing. For classification in the knowledge area, these job vacancies were doubled, totaling 99 vacancies. Figure 1 shows the result of the classification performed, 63.6% of the job vacancies were classified as generic (63 vacancies), 17.2% were classified as clinical engineering (17 vacancies), 14.1% as computer science (14 vacancies), and 5.1% as instrumentation (5 vacancies). No vacancies were classified in the areas of biomechanics and biomaterials. It is important to mention that the job vacancies classified as generic can be performed by a professional graduate of the biomedical engineering course, however, they are not specific to this professional profile.

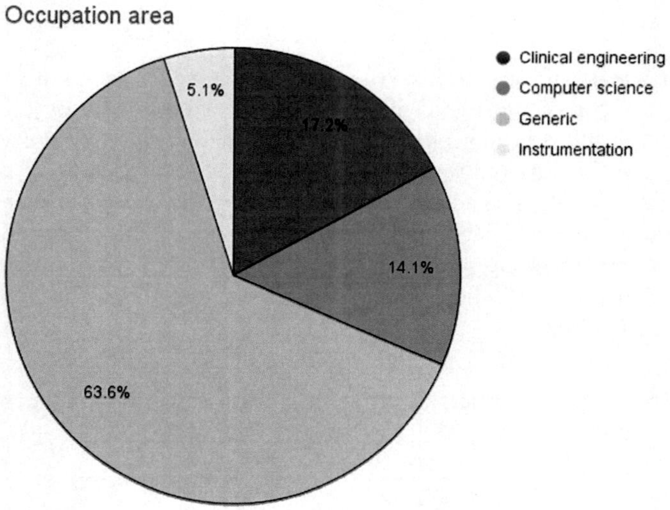

Fig. 1. Classification of the vacancies according to their area of expertise in biomedical engineering: generic, computer science, instrumentation, biomechanics, clinical engineering and biomaterials.

In the classification of the job vacancies in different knowledge areas, it was possible to classify 34 job vacancies in areas of biomedical engineering (computer science, instrumentation, biomechanics, clinical engineering, and biomaterials), which corresponds to an increase of 19 job vacancies in the number of vacancies found only searching for jobs related to biomedical engineering.

In the analysis of the 96 job vacancies, the main activity found was "management", which was the main function in 62 vacancies (64.6%). The other actions and counts are shown in Fig. 2. After management, the main actions are in clinical engineering (8.3%), programming (7.3%), regulation (4.2%), maintenance (3.1%), and quality control (3.1%).

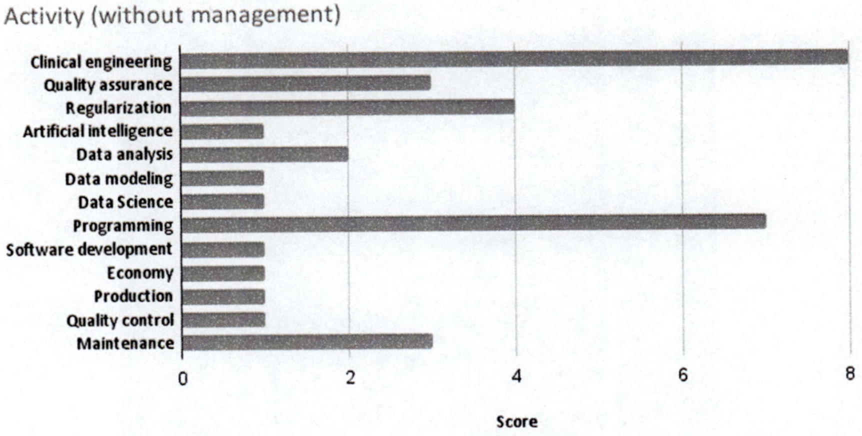

Fig. 2. Classification of the vacancies according to their main activity, disregarding the vacancies in which the classification was "management".

Figure 3A presents the word cloud created from the description of the 96 selected job vacancies. The most prominent words and their respective counts are: process (160), quality (125), project (124), and product (104). The other words had less than 100 counts. Figure 3B was made from the description of the 34 job vacancies whose main skill classification of the description was different from management. The most prominent words and their respective counts are: project (59), process (43), data (42), and quality (41). The other words had less than 40 counts.

4 Discussion

The results and discussions of this study must be considered in light of the limitations of the work. As it is a pilot study, just one career website was used to obtain job vacancies, Indeed. This fact limits the number and profile of the analyzed vacancies since there are other similar websites and also other means of advertising a job vacancy. For a global view of the biomedical engineer's job market in Brazil, it is necessary to expand the tools for obtaining vacancies, by using other search platforms and means of communication.

In this work, it was possible to perform the automation of the search for vacancies, which made it possible to select 197 out of a total of 4597, however, the

Fig. 3. Word cloud built from job description terms. A) Cloud of words considering all the vacancies.eps

elaborate routine needs to be improved, since of the 197 selected vacancies only 96 were classified as adherent to the profile of the engineer biomedical research after reviewers' evaluation. The main difficulty in automating job selection is the lack of standardization of the description of the ads, which in Indeed does not have a standard format, being at the discretion of the advertiser [4] and the wide spectrum of professional performance of the biomedical engineer [7], which makes it difficult to limit search terms.

The specific job search for biomedical engineers resulted in 14 vacancies and the inclusion of the careers of Electrical Engineering, Chemical Engineering, Production Engineering, Management Engineering, Mechanical Engineering, Mechatronics Engineering, Materials Engineering, Computer Science, Engineering Physics and Data Science, Pharmacy and Biological Sciences, resulted in the 96 job vacancies analyzed. These 96 jobs cannot be understood as the only job options available for a biomedical engineer on Indeed. It is possible that generic vacancies related to other careers were excluded in the automatic selection since this study proposal was to find vacancies with adherence to the biomedical engineer's career. In addition, considering the difficulty in the automatic selection of the aforementioned job positions, it is possible that vacancies with adherence to biomedical engineering registered in other careers may have been excluded in the automatic selection.

Considering that in Brazil, until 2022, there are 24 undergraduate courses in biomedical engineering [3] that together have the potential to release 4115 biomedical engineers per year [3], it is possible to conclude that the number of targeted job vacancies for biomedical engineers is low. One of the possible reasons for this is that the biomedical engineering career is relatively new [7] when compared to other engineering courses. The job market in Brazil may not search directly for biomedical engineers because it does not know their role and is used to looking for other professionals for job vacancies with adherence to the career. This hypothesis is reinforced by the 142% increase in the number of vacancies in the biomedical engineer's areas of activity when the search for a vacancy included other careers.

In Fig. 1, it is possible to observe that more than half of the job vacancies analyzed were classified as generic (63.6%). These vacancies seek professionals with non-specific training, often engineers or professionals with some exact sciences degree. Thus, it is possible for a graduate of the biomedical engineering course to apply for the vacancy, however, it is not possible to classify job vacancies in areas of biomedical engineering.

Still, in Fig. 1, clinical engineering, and computer science were highlighted areas of the vacancies classified in areas of biomedical engineering, reproducing the behavior of Table 1, where only job vacancies directed to biomedical engineering were analyzed. This result indicates that the two areas are the ones with the most opportunities in the national market, and also, that they are areas in which the job market most directs the search for a biomedical engineering professional.

It was found that "management" is the main function (64.6%) among the 96 job vacancies analyzed. This result is related to the number of generic vacancies found, given that most of these vacancies have this main activity related to the job description. Based on the results of Fig. 2, we can observe that after "management", the main activities in the analyzed job vacancies are related to clinical engineering and programming knowledge, corroborating the results found in the analysis of the vacancies' areas of activity. It is important to note that the analysis of the main function of a vacancy is based on its description, which in fact, is carried out by the employer, and therefore depends on the perception of the designer of the ad for the vacancy.

In addition to being subject to the perception of the designer of the job vacancy, the analysis of the main activity of the job descriptions is also subject to the impression of the reviewers. Thus, in order to reduce human inference in the analysis of job vacancies, a word cloud was constructed from the job description. The word cloud is formed by counting the words in the descriptions. A disadvantage of this process is that words with more than one meaning or that can be used in different contexts are more likely to be displayed than specific words or terms. However, even with this limitation, the word cloud tool is useful for providing an overview of the activities described in the job vacancies as a whole.

In the word cloud in Fig. 3A, the four highlighted words are: process, quality, project, and product. The National Curriculum Guidelines for the Undergraduate Course in Engineering [2] make direct reference to three of the four highlighted words of the cloud when establishing essential aspects in the formation of an engineer, they are: process, project, and product. This fact reinforces the generalist aspect of the vacancies found, which seek an engineer, and not necessarily a biomedical engineer.

Figure 3B shows the cloud without the job vacancies whose main activity was classified as "management". Regarding the cloud in Fig. 3A, three of the highlighted words remain to have a big display: project, process, and quality; The word "data" replaced "products" in the set of four words with greater emphasis, which reinforces the importance of the area of data analysis in the context of biomedical engineering.

The word clouds are a reflection of how the labor market looks for the Biomedical Engineer, the overlap of skills in the different specializations of Engineering is evident, perhaps for this reason the job applicant prefers to prepare less restricted ads and select different candidates according to their "vocation", and in the description of job vacancies, words from more generic profiles are used, as stated in Resolution number 2 of April 24th, 2019 of the National Education Council [2].

5 Conclusion

In this pilot study, it was possible to carry out an initial analysis of the Brazilian market for Biomedical Engineering. The results suggest that the national market is still not looking specifically for a professional trained in biomedical engineering, with the areas of clinical engineering and computer science being the most frequent among the job vacancies analyzed. It was observed that job vacancies on the Indeed platform tend to have generic ads, since there are overlapping profiles in different bachelor's degrees. However, for a complete analysis of the labor market, it is necessary to evaluate job vacancies from other platforms and means of dissemination.

Acknowledgements. The authors thank the Federal University of ABC.

Conflict of Interest. The authors declare that they have no conflict of interest.

References

1. D.O.U, 2018 D.O.U - Diário Oficial da União 08/08/2018, Edičão 152, Sečão 1, Página 137.at. https://www.in.gov.br/materia/-/assetpublisher/Kujrw0TZC2Mb/content/id/35900023/UCEQlTzKXPyVi6cWuD3q0ksQ. Accessed: 2022-05-30
2. D.O.U, 2019, D.O.U - Diário Oficial da União 26/04/2019, Número 80, Sečáo 1, Página 43. https://pesquisa.in.gov.br/imprensa/jsp/visualiza/index.jsp?data=26/04/2019&jornal=515&pagina=43&totalArquivos=94. Accessed 30 May 2022
3. eMEC at https://emec.mec.gov.br/. Accessed 30 May 2022
4. Indeed at https://br.indeed.com/
5. Fernanda Nunes da Rosa Mangini and Regina Célia Tamaso Mioto. "A interdisciplinaridade na sua interface com o mundo do trabalho". In: Revista katálysis 12 (2009), pp. 207–215
6. PPC-Engenharias, 2017 at https://prograd.ufabc.edu.br/cursos/eb. Accessed 30 May 2022
7. SBEB at https://www.sbeb.org.br/. Accessed 30 May 2022

Study of the Lethality Rate of Respiratory Diseases in Surrounding Cities Within 50 km of Cement Industries

F. J. H. Moraes[1(✉)], F. T. C. S. Balbina[1], E. Araujo[1,2], and E. Munin[1,2]

[1] Biomedical Engineering Center, Universidade Anhembi Morumbi (UAM), São Joé Dos Campos, São Paulo, Brazil
fernandojhmuerj@gmail.com

[2] Center for Innovation, Technology and Education (CITE), São José Dos Campos, São Paulo, Brazil

Abstract. Human activities promote significant environmental changes. Industrial plants emit pollutants of different types in the atmosphere, placing at risk the health of the population in the vicinities of manufacturing plants. The knowledge of the impact of industrial manufacturing processes of diverse products on public health is of major interest, as the first step for decision-making toward a cleaner and healthier world. In this work, information on the lethality rate of respiratory system diseases (RSD) is collected from the Brazilian Unified Health System (SUS) databases. This work makes a compilation of data for cities situated at distances up to 50 km of cement manufacturing plants and intends to evaluate the effect on public health of particulate pollutants produced by this type of industrial activity. Data are statistically analyzed and compared to the overall population of the southeastern region, as well as with the overall Brazilian population, considered here as background data. The results indicated that the surrounding cities within 50 km of the cement industries in the south-eastern region of Brazil have mean values of respiratory disease lethality higher than the values found in the southeastern region of the country, to which the sampled cities belong, and higher than the national average.

Keywords: Air quality · Air pollution · Atmospheric pollution · Cement industry · Particulate pollution

1 Introduction

The human respiratory system is exposed to potential hazards from the external environment. The air we breathe is subject to contamination such as viruses [1], bacteria [2], or even pollutants [3]. Diseases of the respiratory system (RSD) make it difficult to absorb oxygen, causing discomfort and limitation of physical activity [4, 5].

Chronic respiratory diseases (CRD) are aggravated by air pollution. Of which 4.2 million annual death, account for CRD, 3.7 million are directly linked to existing air pollution. Are CRD: asthma, allergic rhinitis, occupational lung diseases, pulmonary

J. L. B. Marques et al. (Eds.): CLAIB 2022/CBEB 2022, IFMBE Proceedings 98, pp. 520–529, 2024.
https://doi.org/10.1007/978-3-031-49401-7_54

hypertension, chronic obstructive pulmonary disease (COPD) [3, 6]. In addition to mortality, CRDs have caused bed occupancy and costs to the public health system [7]. For example, asthma was responsible for about 120,000 hospitalizations (2047 deaths) in 2013 [8]. In turn, COPD has a rate of 290,000 hospitalizations per year (25% of hospitalizations for respiratory problems), the fifth most frequent cause of death in Brazil [7].

The cement industry is responsible for producing 5% of total CO_2 emissions into the atmosphere. It's listed among the main sources of atmospheric pollution [9–11]. In addition, the chemical pollutants produced during the manufacturing process – Sulfur Oxide (SOx), Nitrogen Oxide (NOx), Carbon Dioxide (CO_2) and Chromium (Cr) [9, 12, 13] – cause a cloud of particulate material (PM) of differentiated granulometry. This particulate material is characterized by having diameters between 2.5 to 10 mm (MP10) or smaller than 2.5 mm (MP2;5) [12, 14].

The impact of pollutants on the respiratory health of the population surrounding cement factories has been studied, given the concern that such air pollution can cause deleterious effects on human health [9, 10, 12–14]. For example, in one study, the air quality in the vicinity of three cement plants in India was analyzed, pointing to a moderate risk for particulate matter and light (not harmful to health) for other pollutants [9]. Also in India, in Chandrapur, another study found a high incidence of health risk in the local population [13]. In a study by Nkhama et al., two communities were analyzed, one located 1 km and the other 18 km from the cement industry showing a greater number of residents with health problems in the nearest community [14]. In Brazil, in the community of Queima de Lençol, in the Federal District, located between two cement plants, a study evaluated the air quality, as a function of the concentration of total suspended particles, classifying the local air as inadequate [15]. In general, these studies indicate that the pollution and degradation of air quality caused by a cement factory reaches a radius of 1 to 4 km from its surroundings [13, 16].

The objective of this work is to evaluate the influence of the cement industry and the lethality of diseases of the respiratory system in the population of cities surrounding cement factories, which are located at a distance of up to 50 km, in the southeastern region of Brazil, during the period of 1998 to 2017. The identification and quantification of the impacts of environmental pollution generated by the cement industry can contribute to better planning of public health policies, with the identification and treatment of diseases [15, 17], as well as the adoption of preventive actions.

2 Methods

This is a retrospective, descriptive and cross-sectional study, with secondary analysis of data obtained from official Brazilian databases. The data of the present study were collected via official databases and online platforms, with a measurement tool and maps, in the public domain.

This study uses data from public domain platforms, not having direct interaction with living beings. Therefore, approval by the Research Ethics Committee is unnecessary.

This research comprised the location of cement units made available by cement.org, an organization that aims to disseminate information on the cement market [18]; location of cities through Google Maps [19]; the number of hospitalizations and deaths classified in Chapter ICD 10, item X for diseases of the respiratory system (Hospital Information System of the Brazilian Unified Health System - SIH/SUS - related to the Outpatient Information System - SIA) [20]; the radial distance from the city to the nearest cement plant (online CalcMaps measurement platform) [21].

The collected data refer to the number of deaths from respiratory diseases, by place of residence, in cities surrounding cement plants, located in the southeastern region of Brazil, from 1998 to 2017. Cities surrounding the cement industries are limited to 50 km radial distance from the nearest cement plants and with a population of 150,000 or less. Cities close to large industrial and urban centers are excluded, so there is no interference from particulates of another nature. The choice of the cities studied is made according to the availability of cities around the cement plant.

The number of deaths from diseases of the respiratory system is analyzed considering the number of hospital admissions also caused by diseases of the respiratory system in the cities of the study during the period from 1998 to 2017.

The number of deaths for each year was divided by the number of hospitalizations in the respective city, in the respective year, thus generating a percentage value of the lethality of RSD cases in a given city and in a given year, xRSD.

The average xRSD of cities located up to 50 km from the cement industry for a given year is denominated, in this work, as the population lethality coefficient of respiratory diseases (PLC).

Data analysis uses descriptive statistics, considering the population lethality coefficient of respiratory diseases. In turn, the inferential statistics uses ANOVA, followed by the posthoc multiple comparison Tukey test, to analyze the possibility of having a statistically significant difference (at a significance level $\alpha = 0.05$) in the PLC data among the ranges of radial distances.

3 Results

As a first result of the present research, 65 cities were identified as being located in the surroundings of the cement industry, therefore meeting the methodological requirments for selection. From those, a number of 10 cities are in the State of Rio de Janeiro:– Bom Jardim, Cantagalo, Carmo, Conceição de Macabu, Cordeiro, Itaocara, Macuco, Santo Antônio de Pádua, São Sebastião do Alto and Trajano de Morais. A number of 35 out of the 65 cities are in the State of Minas Gerais:– Alpinópolis, Arcos, Barbacena, Barroso, Bom Sucesso, Campo Belo, Carandaí, Carmo da Cachoeira, Congonhas, Conselheiro Lafaiete, Córrego Fundo, Entre Rio de Minas, Formiga, Iguatama, Ijaci, Itaú de Minas, Jaboticatubas, Lagoa da Prata, Lagoa Dourada, Matozinhos, Nazareno, Nepomuceno, Ouro Branco, Pains, Paraopeba, Passos, Pedro Leopoldo, Perdões, Piumhi, Prados, Pratápolis, Santo Antônio do Amparo, São João Del Rei, São Tomas do Aquino and Tiradentes. The remaining 20 cities are located in São Paulo State:– Apiaí, Cajatí, Cananeia, Capão Bonito, Eldorado, Guapiara, Ibiúna, Iporanga, Jacupiranga, Nova Campina, Pariquera-Açu, Piedade, Pilas do Sul, Registro, Ribeirão Branco, Ribeirão Grande, Salto de Pirapora, São Miguel Arcanjo, Sete Barras and Taquarivaí.

Before generating the definitive results of the data obtained for the cities studied during the period from 1998 to 2017, outliers were identified and removed from the final statistics. Thus, analyzing the distribution of the averages of each city during the period of time with aid of box plots, not shown here, the following cities and years were identified as outliers and excluded from the analysis: Macuco → 2000, 2011, 2012; Arcos → 2010, 2016; Ijaci → 2012, 2013, 2014, 2015, 2017; Itaú de Minas → 2017; Cajatí → 2004, 2016, 2017; Ribeirão Grande → 1998, 2007, 2009, 2013, 2015; Cordeiro → 2011, 2013, 2016, 2017; Prados → 2016; Guapiara → 2015, 2016; Conselheiro Lafaiete → 2016, 2017; Santo Antônio do Amparo → 2000, 2008, 2009, 2011, 2012; Tiradentes → 2017; Ibiúna → 2010, 2011; Iporanga → 1999, 2001, 2012, 2015; Carmo → 2004, 2008, 2013; Itaocara → 2017; Entre Rios de Minas → 2015, 2017; Lagoa da Prata → 2017; Lagoa Dourada → 2014; Nazareno → 2014; Registro → 2015; Santo Antônio de Pádua → 2014, 2015, 2016, 2017; Alpinópolis → 2014; Carmo da Cachoeira → 2017; Congonhas → 2017; Ouro Branco → 2005; Paraopeba → 2015; Nova Campina → 2017; Taquarivaí → 2016.

We computed the xRSD and PLC of the cities, for each year, by using the outlier-free data. Figure 1 presents a Pare-to diagram with the average xRSD from 1998 to 2017 for each city in the study. Also in Fig. 1, we show the averages of the Southeast Region of Brazil, where these cities are located, and the Brazilian national average, for the same period.

Figure 1 shows that only 12.3% of the cities studied are below the Brazilian national average and 36.9% are below the average for the Southeast region of Brazil. Within this universe of cities, the national average is below the 25th percentile, while the Southeast average is between the 25th percentile and the median. The cities that obtained a result above the 95th percentile are: Sete Barras-SP, Cananeia-SP and Jacupiranga-SP. Table 1 shows the number of cities in each state between each percentile described in Fig. 1.

Table 1. Number of cities in each state by percentile

Percentile (p)	São Paulo	Rio de Janeiro	Minas Gerais
$p \geq 95$	3	0	0
$95 > p \geq 75$	4	1	8
$75 > p \geq 50$	4	3	10
$50 > p \geq 25$	4	3	9
$p < 25$	5	3	8

For percentile p less than 95 and greater than 75, the result shows 1 city in the State of Rio de Janeiro (10% of the cities in the State), 4 in São Paulo (20%), 8 in Minas Gerais (22.9%). Between the 75th and 50th percentile (median) 3 cities in Rio de Janeiro (30%), 4 in São Paulo (20%), 10 in Minas Gerais. Below the median, p is less than 50 and greater than 25, in 3 cities in Rio de Janeiro (30%), 4 in the State of São Paulo (20%) and 9 in Minas Gerais (25.7%). Finally, below the 25th percentile, the result points to 3 cities in Rio de Janeiro (30%), 5 cities in São Paulo (25%) and 8 cities in Minas Gerais (22.9%).

Table 2 presents the results of xPLC during the period from 1998 to 2017. The table also shows the average for the Southeast region and the national average for each year

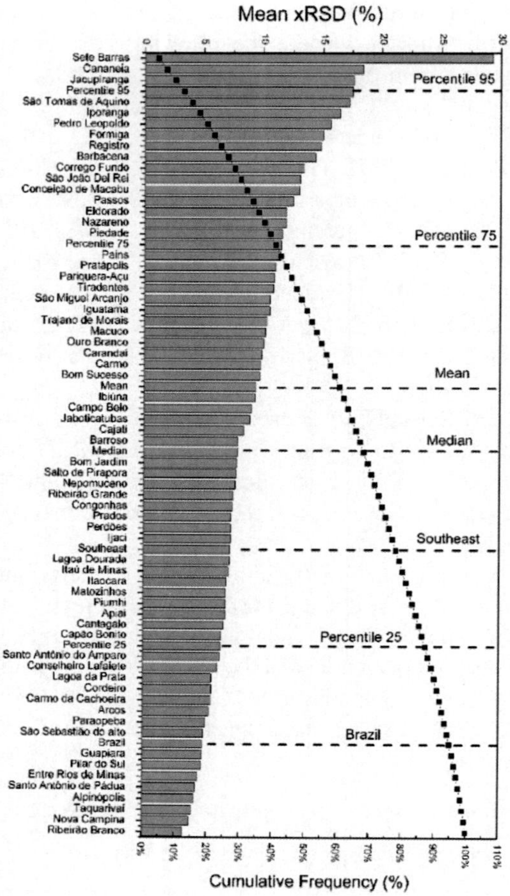

Fig. 1. Pareto diagram with the results of the average lethality rate of each city studied for the period from 1998 to 2017.

studied, in addition to the arithmetic average, standard deviation, average, maximum and minimum standard error of this time period. The results of Table 2 are also presented in graphical form (Fig. 2) comparing the variation in lethality xPLC over the period from 1998 to 2017. Through a linear extrapolation of the results of Fig. 2, the growth rate of each of the measured lines was identified for the time interval studied - from 1998 to 2017. The growth rate obtained for xPLC is 0.54% per year, while the lethality rate for the Southeast region and for Brazil are 0.283 and 0.289 a year, respectively.

The distribution of lethality in the period from 1998 to 2017 of the data stratified in xPLC, Southeast Region and Brazil can be seen in Fig. 3, through the box chart. The results regarding the distribution – minimum value, quartile 1, median, quartile 3 and maximum value – and the Skew-ness and Kurtosis values are presented in Table 2.

Table 2. Average values of xPLC during the period 1998 to 2017

Year	Studied cities (%)	Southeast (%)	Brazil (%)
1998	6.351	4.583	2.824
1999	5.433	4.982	2.965
2000	6.019	5.025	2.972
2001	6.1	5.307	3.221
2002	6.579	5.723	3.436
2003	7.379	6.11	3.726
2004	7.278	6.469	3.959
2005	8.301	6.75	4.268
2006	7.754	6.982	4.498
2007	8.847	7.547	4.762
2008	9.53	7.034	4.62
2009	9.258	7.324	4.816
2010	10.516	8.117	5.396
2011	11.012	8.483	5.832
2012	10.526	8.557	6.272
2013	12.539	8.903	6.711
2014	13.622	9.045	7.088
2015	14.653	9.522	7.532
2016	15.695	9.895	8.236
2017	15.833	9.759	7.991
Mean	9.661	7.306	5.056
SD	3.31	1.686	1.745
SEM	0.74	0.377	0.39
Mínimum	5.433	4.583	2.824
Quartile 1	6.9287	5.9166	3.5809
Median	9.0523	7.1789	4.6911
Quartile 3	11.7755	8.7304	6.4913
Máximum	15.833	9.895	8.236
Skewness	0.64297	-0.03162	0.48007
Kurtosis	- 0.73551	- 1.19892	- 0.98875

Table 3 shows the output of the post-hoc, multiple comparison statistical Tukey test, comparing the populacional lethality coefficient (xPLC) of the studied cities, Brazilian southeast region, and whole country along the studied time spam (from 1998 to 2017).

The results ponts out significant differences between the statistical distributions of each studied stratum.

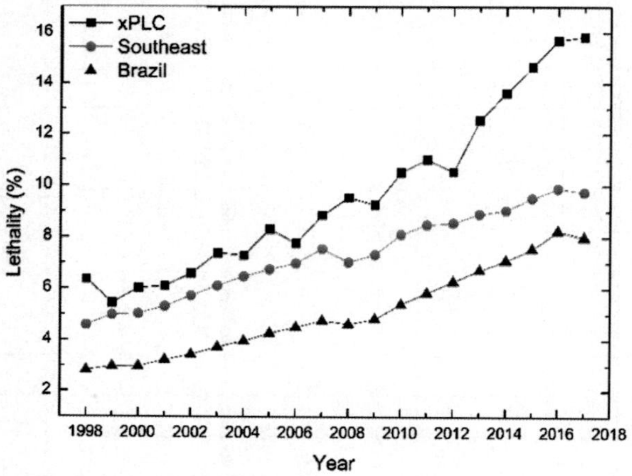

Fig. 2. History of average lethality during the period from 1998 to 2017 for the studied set of cities (square), for the Brazilian southeast region (circles) and for the whole country (trian-gles).

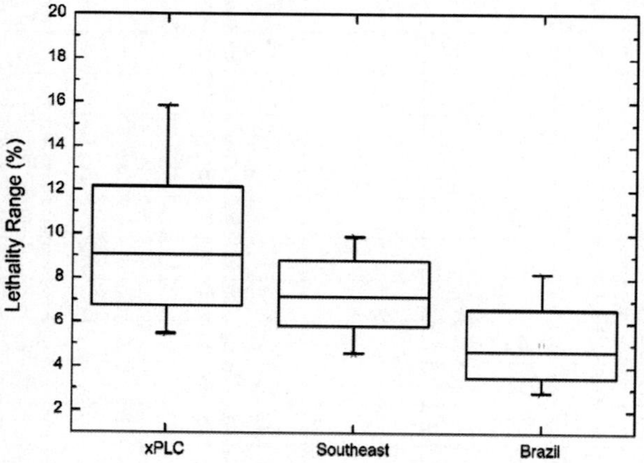

Fig. 3. Distribution of average lethality in the period from 1998 to 2017 for the studied set of cities, for the Brazilian southeast region and for the whole country

4 Discussion

The study presents a number of cities that satisfy the criterion of 50 km of radial distance of cement plants, with the state of Minas Gerais participating with 53.84% of the selected cities, while Rio de Janeiro and São Paulo participated with 15.38 and 30.76%,

Table 3. Results of the Tukey test for multiple comparisons between group mean (s stands to statistically significant difference between group means).

	Studied cities	Southeast	Brazil
Studied cities	–	s	s
Southeast	s	–	s
Brazil	s	s	–

respectively, of cities that met the selection criteria for this research. The predominance of the state of Minas Gerais is consistent with the large concentration of cement industry, due to the geological nature of that state, which increases the possibility of finding cities that satisfy the criteria established for the work.

The results of the distribution of cities for each state (Table 1) show a somewhat homogeneous distribution between the percentiles for the state of Minas Gerais, where 22.9% of the cities in the state are above the 75th percentile, 28.6% are between the 50th and 75th percentiles, 25.7% are between the 25th and 50th percentiles, and 22.9% are below the 25th percentile. The distribution result for the state of São Paulo reveals 3 cities above the 95th percentile, which represents 15% of the cities of São Paulo that were included in this study. São Paulo was the only state with cities above the 95th percentile. However, when looking at the number of cities above and below the median, the result shows a certain balance of 55% and 45%, respectively. Among the three states, Rio de Janeiro may have the cities with the lowest lethality rates in the study, since 90% of cities are below the 95th percentile, and 60% below the median.

Figure 1 shows that 50% of the cities that are above the measures correspond to 68% of the value of the summed averages and that the cities above the 75th percentile this value is approximately 40%.

Table 2 shows the results of xPLC for the sampled set of cities from 1998 to 2017, along with the average information for the southeast region and Brazil for the corresponding period. The values plotted in Fig. 1 indicate that the studied group of cities have mean lethality values higher than the means for the Southeast region and Brazil for all of the studied years. The growth rate of xPLC for the group of cities sampled in this study is higher than the growth rates of the Southeast region and the whole country, indicating a future trend of an even greater distancing in the values of the mean lethality of the studied cities in relation to the Southeast region. And the Brazilian national scenario.

The distribution of xPLC in cities up to 50 km away from the cement industry, in the Southeast region and in Brazil reveals a distribution with a low degree of skewness, with the exception of the sampled cities stratum that showed a moderate positive value (0.64). As for kurtosis, the distributions were slightly platykurtic ("flattened").

The analysis of data from the cities surrounding the cement industry, using a database and public domain tools, brought valuable informations about the lethality behaviour of respiratory diseases in these cities. The use of DATASUS proves to be a great research tool to map possible problems in different municipalities.

5 Conclusions

This research work indicated that the cities surrounding the cement industries within distances up to 50 km in the southeastern region of Brazil have mean values of the lethality of respiratory diseases higher than the values found in the southeastern region, to which the cities belong, and also, higher than the national average. This result indicates a possible public health problem linked to pollutants generated by the cement industry. Therefore, for future work, a more detailed study in these cities is necessary, in order to quantify the real level and type of pollutants found in the air of these cities.

References

1. La, R.G., Fratini, M., Libera, S.D., et al.: Viral infections acquired indoors through airborne, droplet or contact transmission. Ann Ist Super Sanità **49**, 124–132 (2013)
2. Pasanen, A.L., Korpi, A., Kasanen, J.A., et al.: Critical aspects on the significance of microbial volatile metabolites as indoor air pollutants. Environ. Int. **24**, 703–712 (1998)
3. da Saúde, M.: Secretaria de Vigilância em Saúde: Perfil da mortalidade por doenças respiratórias crônicas no Brasil, 2003 a 2013. Boletim Epidemiológico. **2016**(47), 1–9 (2016)
4. WHO. Frequently Asked Questions: Ambient and Household Air Pollution and Health. https://www.who.int/phe/healthtopics/outdoorair/databases/faqs-air-pollution.pdf?ua=1
5. Kurniati, A., Rizky, Z.P., Ramdhan, D.H.: Particulate matter 2.5 concentration and subjective acute respiratory effects among production workers at a cement factory. KnE Life Sci. **2017**:139–148 (2017)
6. Malta, D.C., L et al.: Mortalidade por doenças crônicas não transmissíveis no Brasil e suas regiões, 2000 a 2011 Epidemiol. Serv. Saúde **23**:599–608 (2014)
7. Godoy, I.L., et al.: Programa de cessação de tabagismo como ferramenta para o diagnóstico precoce de doença pulmonar obstrutiva crônica. J. BrasPneumol **33**:282–286 (2007)
8. Cardoso, T.A., L et al.: Impacto da asma no Brasil: análise longitudinal de dados extraídos de um banco de dados governamental brasileiro. J. BrasPneumol **43**:163—168 (2017)
9. Chaurasia, S., Tiwari, A.: Assessment of ambient air quality in the vicinity of cement industries. Int. J. Appl. Res. Technol. **1**:39–46 (2016)
10. Abu-Allaban, M., Qudais, H.: A impact assessment of ambient air quality by cement industry: a case study in Jordan. Aerosol Air Qual. Res. **11**, 802–810 (2011)
11. Lei, Y., et al.: An inventory of primary air pollutants & CO_2 emissions from cement production in China 1990–2020. Atmos. Environ. **45**:145–154 (2011)
12. Sánchez-Soberón, F.L., et al.: Main components and human health risks assessment of PM10, PM2.5, and PM1 in two areas influenced by cement plants. Atmos. Environ. **120**:109–116 (2015)
13. Kesarwani, S., James.: A Effect of air pollution on human health problems residents living around the cement plant, Chandrapur, Maharashtra, India. J. Pharmacognosy Phytochem. **6**, 507–510 (2017)
14. Nkhama, E., et al.: Effects of airborne particulate matter on respiratory health in a community near a cement factory in Chilanga, Zambia: results from a panel study. Int. J. Environ. Res. Publ. Health **14**, 1351 (2017)
15. Maury, M.B., Blumenschein, R.N.: Produção de cimento: Impactos à saúde e ao meio ambiente Sustentabilidade em. Debate **3**, 75–96 (2012)
16. Marcos, A.L., et al.: Association between PM10 concentrations and school absences in proximity of a cement plant in northern Italy. Int. J. Hygiene Environ. Health **217**, 386–391 (2014)

17. Junior, C.J.A., Filho, J.R.D.C., Sousa, A.L.: Prevalência de sinais e sintomas respiratórios em população residente próxima a uma fábrica de cimento, Cezarina. Goias Cad. Saúde Colet **22**, 120–126 (2011)
18. Cimento.org at: https://cimento.org/cimento-no-brasil/. Last accessed 21 Sept 2021
19. Google at https://www.google.com/maps/. Last accessed 02 Oct 2021
20. DATASUS Epidemiologia e Morbidade at: http://www2.datasus.gov.br/DATASUS/index.php?area=0203&id=6926. Last accessed 02 Oct 2021
21. Calcmaps.com at: https://www.calcmaps.com/pt/map-radius/. Last accessed 02 Oct 2021

Check for updates

Biomedical Document Management Analysis Methodology for Colombian Hospitals

M. J. Ardila, N. D. Mejía, H. A. Rojas(✉), and R. A. Espinosa

Universidad ECCI, 110311 Bogotá, Colombia
harolda.rojasd@ecci.edu.co

Abstract. Within the management of the technical area of a hospital, it is necessary to monitor compliance with current biomedical regulations on documentation, in order to identify adverse events and incidents that may occur due to errors in the document management of medical equipment. The objective of this article is to propose a methodology to evaluate the biomedical document management process in order to maintain order in the administrative area. As a methodology, a series of steps and characteristics that a hospital must comply with in order to be evaluated by the proposed methodology were established. From this, as relevant results, tables were constructed to facilitate the identification of variables necessary for a correct analysis of biomedical document management. Following this, a comparison is made with some similar international methodologies in order to identify the differences that may arise in biomedical document management in some countries. Finally, it is necessary to clarify that the methodology seeks to quantify the level of failures presented, which indicates whether biomedical document management is adequate and complies with Colombian regulations.

Keywords: Document management · Hospital · Methodology · Medical equipment

1 Introduction

A hospital is the space in which all types of services related to health are developed, therefore, diseases are diagnosed and different types of treatments are performed to restore the health of patients. Within a hospital there are several units and areas that work to fulfill all the tasks related to its operation [1]. The main areas into which a hospital is divided are: hotel area, administrative area, technical area, medical area, among others.

The technical area of the hospital covers the needs of the institution in terms of infrastructure and medical equipment. These include: producing efficient and effective health services that meet the quality standards established for this purpose; guaranteeing social and financial profitability through adequate management; providing second and third level medical services to the population that does not have social security; conducting medical research; and developing human resources in the medical and paramedical areas based on the standards and regulations in force [1].

J. L. B. Marques et al. (Eds.): CLAIB 2022/CBEB 2022, IFMBE Proceedings 98, pp. 530–538, 2024.
https://doi.org/10.1007/978-3-031-49401-7_55

In order to manage the technical area of a hospital, compliance with current biomedical regulations must be monitored. One of these, decree 4725 of 2005, regulates the control of adequate facilities and their specific areas, trained personnel, sanitary and hygienic conditions, equipment and current documentation of each process [2]. On the other hand, resolution 4816 of 2008, which regulates the National Technovigilance Program, allows the identification of adverse events and incidents that may occur in the use of medical devices, and establishes a protocol for reporting [3], in order to maintain the traceability of medical devices.

Hospitals in Colombia comply with regulations through methodologies that demanded the INVIMA (National Institute for Drug and Food Surveillance), which regulates the use of biomedical document management, must be followed. This encompasses activities aimed at ensuring the efficiency of the service provided and guaranteeing the safe operation, maximum performance and cost-effectiveness of all medical devices, through the collection and organization of information and compliance with different processes [4].

It has been evidenced that there is a poor handling of biomedical document management in the different IPS (Institution Providing Health Services) in Colombia, usually a large number of documents are generated, which can be lost if they are not well organized. It is important to highlight that these institutions are inspected through audits generally carried out by the departmental health secretariat, in order to ensure that the regulations are complied with. During these audits, documents are reviewed, and any faults in this area can lead to consequences ranging from sanctions to the closure of a service or the institution. Generally, this happens when there is no adequate methodology that integrates all the documentation required for the conservation and proper functioning of medical equipment.

According to national regulations, documentation must be kept in physical form (file), but there are digital alternatives (through google drive or shared work clouds) or more elaborate tools such as specialized software. This paper proposes a methodology to analyze the document management carried out in an IPS, in order to identify failures, evaluate their seriousness and establish possible improvement actions, taking into account the guidelines established in the current regulations (Decree 4725 of 2005, ABC of medical devices, Resolution 4816 of 2008).

2 Methodology

The current national regulations establish the necessary guidelines for the complete documentation related to medical equipment. Based on this, a methodology was proposed to identify the errors that may exist in the biomedical document management of a hospital in Colombia.

The objective of this methodology is to ensure that the biomedical document management process is kept in perfect condition, i.e., that all biomedical equipment has the documents required by the standard (life sheet, maintenance history, preventive and corrective maintenance reports, Invima registration, marketing permit, quick use guide, calibration certificates) and that they are properly filled out. In addition, all of them must have a place in the annual maintenance schedule and in the hospital's asset inventory report.

Therefore, the methodology is characterized by facilitating the identification of errors in the hospital's biomedical documentation. Among these errors are: incomplete completion of documents, inconsistencies in the preventive maintenance and calibration reports with respect to the proposed schedule and the absence of documents.

The application of this methodology in the different institutions is carried out through the development of the following steps:

1. Conducting a survey of the inventory and documentation of medical equipment to the responsible personnel.
2. Based on the results of the survey, the errors and documents in which there is a higher incidence of failures are identified.
3. By means of tables, the causes and effects corresponding to the errors found as a result of the poor management of biomedical documentation are identified.
4. Based on a table containing the criteria established according to the possible effects of poor biomedical document management, the seriousness of the errors is evaluated.
5. Corrective actions are assigned, taking into account the seriousness of the errors, based on the recommendations presented by this methodology.

On the other hand, in order to apply the methodology in a hospital, it must meet the following characteristics:

- The hospital should have a biomedical engineering department, which should be solely responsible for the documentation of biomedical equipment.
- The hospital must have at least one person in charge of the inventory of biomedical equipment.

3 Results

Based on the specifications made in the method regarding the purpose and general structure of the methodology for the analysis of a hospital's biomedical document management, it was proposed as follows:

1. A survey should be applied to the people who make up the biomedical administrative area of the hospital and another to those in charge of fixed assets. The purpose of this survey is to gather information about the failures that the hospital may be experiencing with regard to biomedical document management. Therefore, the survey has specific questions to clarify in which documents there are shortcomings. Below is a list of the questions that make up the survey addressed to those in charge of fixed assets:
   I. In the last month have any inconsistencies been found in the medical equipment inventory?
      a. Yes
      b. No.
   II. If yes, which inconsistency is the most frequent? (You can choose more than one answer)
      a. Lack of equipment in inventory.
      b. Equipment left over in inventory.
      c. Equipment location error.
      d. Error in the equipment identification data.

Since the personnel in charge of fixed assets is the only one in the hospital with knowledge of the alterations suffered by the inventory, and is also the one in charge of making the modifications to it, he is able to identify the problems that arise in this area.

On the other hand, the following questions make up the survey addressed to the people who make up the biomedical administrative area:

I. In the last month have you found any regulatory documents missing from a piece of equipment?
   a. Yes
   b. No.

II. If yes, which document is it? (You can choose more than one answer)
   a. Resume of the medical team
   b. Maintenance history
   c. Invima registration
   d. Commercialization permit
   e. Quick use guide.

III. Do preventive maintenance reports match the dates established in the maintenance schedule?
   a. Yes
   b. Sometimes
   c. No.

IV. Are the calibration certificates up to date according to the manufacturer's specifications?
   a. Yes
   b. Sometimes
   c. No.

   Taking into account that the biomedical administrative area contains all the regulatory documents for medical equipment, and that the personnel assigned to this area are in charge of maintaining control of these documents, this survey was conducted in order to identify the problems that may be occurring in biomedical document management.

2. After conducting the survey, a group knowledgeable about the process and its different aspects should be formed to objectively analyze the survey, so that the general premises given by the workers can be recognized and, based on these, the failures that occur in document management and the documents in which there is a higher incidence of failures can be identified.
3. After identifying the failures that the hospital has with regard to biomedical document management, the group in charge of the analysis must identify the possible causes of the failures through Table 1. Subsequently, the group must identify the effects that occur in the institution through Table 2. To fill in these tables, it is necessary to select whether or not the causes and effects shown are present or not.
4. Taking into account that the effects and causes of errors in document management have already been identified, it is necessary to evaluate the seriousness of the errors in order to prioritize the correction of the most serious errors. Table 3 lists 5 severity criteria and their respective values; these criteria were created based on the effects that could occur and the impact they could have on the operation of the equipment.

The errors identified in step 2 should be assigned a severity value according to the criteria.

Table 1. Identification of causes

Cause	Presented	Not presented
Poor control and storage of documents		
Failure to fill out maintenance forms immediately		
The equipment supplier does not provide all the documents that must be included		
The new team started the operation before completing their respective documentation		
Equipment that has reached the end of its life cycle is not derecognized from fixed assets		
Labor times do not coincide with the maintenance schedule		
The periodicity of calibration of the equipment according to the manufacturer's specifications is not clear		

Table 2. Identification of effects

Effect	Presented	Not presented
Medical equipment is constantly undergoing corrective maintenance		
Medical equipment is not within normal operating ranges		
The equipment suffers damage that cannot be repaired quickly		
The equipment is in a different location than the one established in the documents		
There is equipment that is not properly marked		
There are problems with audits conducted by regulatory agencies		
Failure in the services of a specific area due to lack of equipment in good condition		

If the severity is in the range levels above 3, the error should be considered as needing an immediate solution. On the other hand, if it turns out to be lower it is not necessary to prioritize the error. Then, the average of the severity values assigned to the different errors found should be calculated by means of the following equation:

$$\overline{x} = \frac{x_1 + x_2 + x_3 ... + x_n}{N} \tag{1}$$

where x refers to each of the severity values and N to the number of errors evaluated. The average severity found should be a value between 1 and 5.

Table 3. Error severity classification

Severity	Criteria	Value
Very low imperceptible impact	It is unreasonable to expect this minor bug to have any real effect on documentation performance	1
Low irrelevant repercussions barely perceptible	The type of failure would cause a slight inconvenience with document management. There will probably be a slight deterioration in the performance of the documentation. It is easily correctable	2
Moderate relatively important defects	Failure results in loss of documentation information, no record is kept. Deterioration in system performance will be observed	3
High	The failure can be critical, with several pieces of equipment showing evidence of failure to carry the required documentation. If not corrected, there will be serious repercussions on the condition of the equipment	4
Very high	Very critical potential failure mode, affects the operation of document management or seriously involves non-compliance with regulatory standards	5

5. After calculating the average severity of the errors, the status of the institution's biomedical document management should be found from Table 4.

Table 4. States of document management according to average severity.

Range	Value
Slight	1–2.5
Moderate	2.6–4
Severe	4.1–5

If the result is in the "slight" range, it will be considered that the institution maintains acceptable document management, which requires minor modifications that do not affect the performance of the equipment. On the other hand, if the result is in the "moderate" range, it will be considered necessary to continuously monitor the documentation, since errors cause alterations in the management of the equipment. On the other hand, if the result is in the "severe" range, immediate action must be taken, as there is a significant deterioration in biomedical document management and adverse events may occur.

Finally, taking into account the effects that are occurring in the institution, corrective actions should be sought to resolve the errors and prevent their recurrence.

4 Discussion

After having proposed the methodology, we inquired about other methodologies at the international level in order to make a brief comparison between the characteristics of each one, and thus be able to identify the changes that may occur according to the regulations under which they were established. This comparison can be seen in Table 5.

Table 5. Comparison of methodologies.

	Proposed methodology	Methodology for the implementation of Document Management Systems to support ISO 9001:2008 quality management systems	The importance of the methodology for the success of the document management system
Authors	M. J. Ardila, N. D. Mejía, H. A. Rojas y R. A. Espinosa	José Manuel Conde Hernada, Cristina González Gaya	Company QFlow Systems
Type	Research paper	Research paper	Software (Management System)
Features	The objective of this methodology is to ensure that the biomedical document management process is maintained in perfect conditions, i.e., that all biomedical equipment has the documents required by the standard "Decree 4725 of 2005, ABC of medical devices, Resolution 4816 of 2008"	Based on the analysis of the document requirements established in "ISO 90001:2008" and the document processes associated with the management system and taking advantage of the existing tools, six steps have been proposed	A document management system is a secure repository where an organization's documents are centralized, organized and accessible

(continued)

It is expected that the methodology proposed in this article will be implemented in the future as a standard evaluation method for biomedical document management in hospitals nationwide.

Table 5. (*continued*)

	Proposed methodology	Methodology for the implementation of Document Management Systems to support ISO 9001:2008 quality management systems	The importance of the methodology for the success of the document management system
Similarities	Improvement actions for biomedical document management to create workflows that optimize and automate identified problems while complying with legislative standards and mitigating patient risk		
Step by step	Step 1. Inventory and Documentation Survey Step 2. Identifying errors and failures in documentation Step 3. Identification of causes from an error evaluation table Step 4. Evaluation of the seriousness of errors according to the criteria of the effects of poor document management Step 5. Assignment of corrective actions based on recommendations	Step 1. Definition of documentary requirements Step 2. Evaluation of existing systems Step 3. Identification of document management strategies in the organization Step 4. Design of the Document Management System Step 5. Document Management System Implementation Step 6. Maintenance and continuous improvement of the Document Management Systems	Step 1. Consistent documentation standards Step 2. Document storage and retrieval Step 3. Security Protocols

5 Conclusions

This study proposes a methodology for the document management of biomedical equipment in a hospital, which allows quantifying the level of failures presented during the process of collection, organization and storage of information and identifying failures in biomedical document management in order to apply actions that help to comply with the regulations and avoid sanctions or adverse events to patients or medical personnel. Based on this methodology, a series of steps were defined that lead to a quantitative value as a final result, which seeks to identify whether the biomedical document management is adequate and complies with the regulations, or on the contrary, if it shows a deterioration that directly affects the operation of biomedical equipment, seeking to reduce adverse effects that may occur in the biomedical area. In addition, its effectiveness will depend on the administrative conditions of the institution.

Additionally, it involves the participation of personnel from some areas of the institution related to the documentation of the equipment, for example, those in charge of fixed assets, those in charge of the administrative area of the biomedical department and those that the institution considers pertinent to carry out the evaluation of the methodology. This methodology, in turn, was designed based on the regulations in force in Colombia with the objective of benefiting institutions at the national level.

References

1. Miranda, F.: Hospitales. SlideShare, a Scribd Company. https://es.slideshare.net/Marianamirandacarrasco/Hospitales-23629800. Accessed 28 March 2022
2. Alejandra, A.A.: Repository Uniminuto (2017). Retrieved from: https://Repository.Uniminuto.edu/bitstream/10656/4920/1/t.ind_ArizaEstradaAngie_2017.pdf
3. Ministerio de salud y protección social: Resolución n.° 4816, Programa nacional de tecnovigilancia (2008)
4. Sánchez, M.C., Rodríguez D., Ernesto, B.: Gestión Tecnológica clínico-hospitalaria". Springerlink. https://doi.org/10.1007/978-3-540-74471-9_264. Accessed 8 March 2022
5. Ministerio de Salud y Protección Social: ABC de Dispositivos Médicos. Bogotá D.C. (2013)
6. Colombia: Ministerio de salud y protección social. (2005, 26 de noviembre). Decreto número 4725 de 2005, Decreto número 4725 de 2005
7. Ministerio de Salud y Protección Social: Documento propuesta para lineamientos a seguir en la gestión de mantenimiento de equipos biomédicos. Bogotá D.C., Version 009 (2014)
8. International Organization for Standardization: ISO 9001:2015 Quality Management Systems-Requirements. ISO Online Browsing Platform (OBP). https://www.iso.org/obp/ui/#iso:std:iso:9001:ed-5:v1:es. Accessed 5 April 2022
9. International Organization for Standardization: ISO 30300:2013 Management Systems for Records (2013)
10. Ministerio de Trabajo y Asuntos Sociales España: NTP 679: Análisis modal de fallos y efectos. AMFE. Barcelona (2004). Accessed 2 April 2022. [OnEnline]. Available: https://www.insst.es/documents/94886/326775/ntp_679.pdf/3f2a81e3-531c-4daa-bfc2-2abd3aaba4ba
11. Lossio-Ventura, J.A., Jonquet, C., Roche, M., Teisseire, M.: Biomedical term extraction: overview and a new methodology. Inf. Retrieval J., 2016, Med. Inf. Retrieval **19**(1), 59–99 (2016). https://doi.org/10.1007/s10791-015-9262-2, lirmm-01274539
12. QFlow Systems: The Importance of Methodology for Document Management System Success | QFlow. Content Services Platform for Enterprises | QFlow Systems. https://www.qflow.com/the-importance-of-methodology-for-document-management-system-success/. Accessed 26 Aug 2022
13. Conde, J.M., González, C.: https://www.dbc.wroc.pl/dlibra/publication/44195/edition/39674. ScienceDirect **13**(17), 13–16, (2016). Accessed 26 Aug 2022 [Online]. Available: https://doi.org/10.15611/dm.2016.13.02
14. Peréz, M., Ferreira, T., Lourenco, A., Igrejas, G.: Boosting biomedical document classification through the use of domain entity recognizers and semantic ontologies for document representation: the case of gluten bibliome. ScienceDirect, **139**, 100 (2013). Accessed 27 Aug 2022. [Online]. Available: https://doi.org/10.1016/j.hydromet.2013.07.001

Estimation of Cardiac Output in Insect Hearts Paced at Different Rates

F. H. K. Inazumi[1,2], N. C. B. Calderon[1,2(✉)], A. R. Santana[1,2], R. A. Bassani[1,2], and J. W. M. Bassani[1,2]

[1] Department of Electronics and Biomedical Engineering, School of Electrical and Computer Engineering, Universidade Estadual de Campinas (UNICAMP), Campinas, SP, Brazil
n160942@dac.unicamp.br

[2] LabNECC, Center for Biomedical Engineering, UNICAMP, Campinas, SP, Brazil

Abstract. Insects have been often used as experimental animal models because of their considerable genetic homology with humans, short lifespan and low-cost maintenance. This study aims to estimate hemodynamic parameters (both systolic and the cardiac output) in response to increase in heart rate in the perfused in situ cardiac tube of the mealworm beetle *Tenebrio molitor* using a software developed in our laboratory, in which the heart is modeled as a prolate spheroid. A negative relationship between rate and the systolic parameters (e.g., stroke volume and ejection fraction) was observed, but the cardiac output was not statistically affected by heart rate. Considering that the end-diastolic volume was not significantly changed by rate, the negative staircase cannot be attributed mostly to the heterometric (Frank-Starling) regulation of cardiac systolic force.

Keywords: Insect · *T. molitor* · Hemodynamic parameters · Stroke volume · Cardiac output · Heart rate

1 Introduction

Research on insect physiology is important because this is the dominant taxonomic group of metazoans on Earth, which has marked impact on the environment [1]. Moreover, insects have been used as alternative experimental models for studying mammalian physiology because of the considerable degree of genetic homology of insects and mammals [2], and also due to the significantly lower cost of their breeding and maintenance when compared to more morphologically complex species.

Due to the significant genetic homology with humans, *Drosophila melanogaster* (fruit fly) is often used in molecular, genetic, behavioral and physiological studies [3–6]. Other insects are also used for these purposes, such as *Galleria mellonella* (honeycomb moth), *Bombyx mori* (silk worm), *Periplaneta americana* (American cockroach), *Locusta migratoria* (grasshopper) and *Tenebrio molitor* (mealworm beetle) [7–12]. Some of these species have been used for cardiovascular studies, with the development of cardiomyopathy models [2, 13].

The most common parameters to assess the function of the insect cardiac tube are the heart rate and the systolic variation in the tube diameter, e.g., [14–17]. However,

J. L. B. Marques et al. (Eds.): CLAIB 2022/CBEB 2022, IFMBE Proceedings 98, pp. 539–547, 2024.
https://doi.org/10.1007/978-3-031-49401-7_56

chronotropic and inotropic functions, although interdependent, may undergo separate regulation by neurotransmitters and other factors. A better, more direct indicator of the cardiac pumping activity would be the cardiac output (i.e., the blood volume pumped per minute), which is defined as the product of the heart rate and the average blood volume ejected by each contraction (stroke volume).

To better understand how the cardiac output in insects is affected by the sole variation in heart rate, without change of the hemodynamic load, the aim of the present work was estimate the cardiac output in the *T. molitor in situ* heart preparation stimulated at different frequencies. It is expected that this approach may be helpful for better quantitative characterization of the cardiac activity in insect experimental models, and that the results may shed light on how the heart rate affects the overall pumping function in this model.

2 Materials and Methods

2.1 Animals and the In Situ Heart Preparation

Adult beetles (*T. molitor*; 1–2 weeks old) of both sexes were kept in aerated plastic containers at 24 ± 1 °C, 65 ± 5% relative humidity, 12-h photoperiod, where they were fed potatoes, carrots, cornmeal and wheat bran.

Under cold anesthesia, the insect was decapitated and its wings and legs were removed. An incision was made in the ventral midline from the thorax to the abdomen, and the dorsal cuticle was fixed on a rubber surface with pins. Digestive organs and other viscera were removed to expose the dorsal vessel (cardiac tube). The preparation was then fixed to the coverslip bottom of a perfusion chamber, which was placed on the stage of a stereomicroscope (Carl Zeiss, Germany) under halogen lamp illumination. The dorsal vessel was then perfused at 1 ml• min^{-1} with the specially designed *T. molitor* solution (TS) with the following composition (mM): 274 NaCl, 19 KCl, 4 $CaCl_2$, 4 glucose, and 5 N-2 hydroxyethylpiperazine-N”-2 ethane-sulfonic acid (HEPES); pH adjusted to 7.0 [15]. Experiments were carried out at room temperature (24 ± 1 °C).

2.2 Measurement of the Cardiac Luminal Diameter and Estimation of the Cardiac Output

Videos of the cardiac tube located in the insect abdominal region (2.5× amplification) were recorded at 30 frames per second with a webcam (Philips, mod. SPC1030NC/00) connected to a computer. From the videos, it was possible to measure the average heart rate and the luminal diameter of the cardiac tube (based on a micrometric calibration) with a program written in Python that detects motion and the contour of the dorsal vessel lumen (FHK Inazumi & JWM Bassani, unpublished).

The software allows estimation of the hemodynamic parameters by adjusting a prolate spheroid to the heart tube lumen (Fig. 1).

The volume of the prolate spheroid (V_{PS}) is described by Eq. 1:

$$V_{PS} = \frac{4}{3}\pi\frac{a}{2}\left(\frac{b}{2}\right)^2 \tag{1}$$

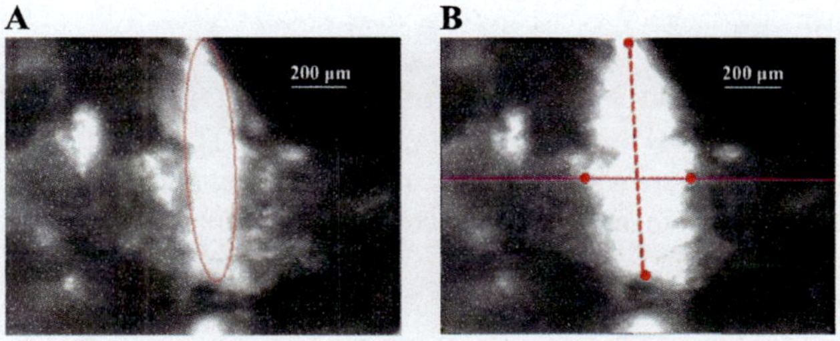

Fig. 1. Heart tube video image during systole (A) and diastole (B). In (A), the spheroid initially generated by the software is shown. (B) shows the luminal axes: the major axis (dotted line) is automatically set by the software, and the minor axis (continuous line) can be adjusted by the user.

where a and b are the length of the major and minor axes, respectively.

The fractional shortening of the cardiac tube during systole (*FS*) is calculated according to Eq. 2, where *EDD* is the end diastolic luminal diameter (i.e., at the end of the diastole), and *ESD* is the end systolic diameter.

$$FS = \frac{EDD - ESD}{EDD} \times 100 \quad (2)$$

The stroke volume (*SV*) is the difference between the end-diastolic (*EDV*) and the end-systolic (*ESV*) volumes (i.e., volumes at the end of diastole and systole, respectively were calculated as V_{PS} using Eq. 1), expressed in nl (Eq. 3):

$$SV = EDV - ESV \quad (3)$$

The ejection fraction (*EF*) is calculated from *SV* and *EDV*, as shown in Eq. 4.

$$EF = \frac{SV}{EDV} \times 100 \quad (4)$$

Finally, the cardiac output (*CO*) was obtained as the product of the average heart rate and *SV* (Eq. 3). The CO was expressed in ml/min/kg, considering an average body weight of 120.5 mg.

$$CO = HR \times SV \quad (5)$$

2.3 Experimental Protocol

Voltage pulses (15–25 V, 4 ms duration) were delivered by an electric stimulator through a pair of stainless steel electrodes placed transversally to the heart vessel. The spontaneously beating preparation was allowed to stabilize for at least 5 min under perfusion with TS. Then, electrical pacing at 1.5 Hz was started, and the rate was subsequently increased to 2 Hz. Videos of the heart tube were recorded at all frequencies. The stimulation period at each rate was that sufficient for the stabilization of contraction amplitude (1–5 min).

2.4 Statistical Analysis

Data are expressed as mean ± standard error of the mean (N = 8 hearts). D'Agostino e Pearson normality test was used to test data distribution. The effect of stimulation rate on the estimated parameters was assessed using one-way analysis of variance for paired samples. Statistical significance was considered to occur when $p < 0.05$.

3 Results

As the frequency was increased from the spontaneous value of 1.3 ± 0.2 Hz to 2 Hz, FS was reduced ($p = 0.032$), in the negative staircase pattern fashion (Fig. 2). A similar effect of pacing rate was observed for SV and EF ($p = 0.028$ and 0.038, respectively; Fig. 3).

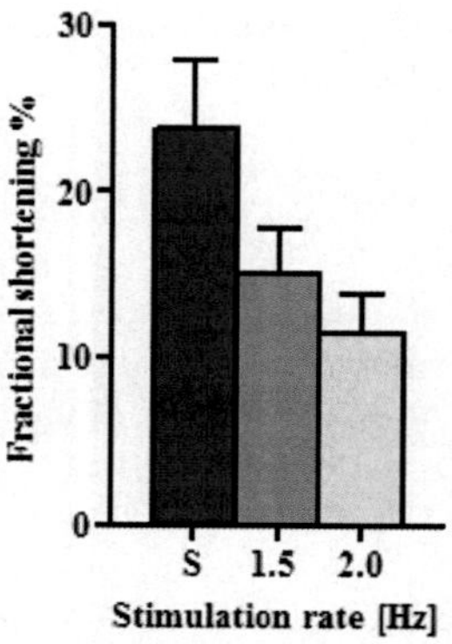

Fig. 2. Negative staircase of fractional shortening of the perfused *T. molitor* heart tube in response to increase in heart rate by electric pacing. S = spontaneous frequency (1.3 ± 0.2 Hz). N = 8. $p < 0.05$, one-way analysis of variance for paired samples.

On the other hand, increasing heart rate did not significantly alter CO ($p = 0.743$; Fig. 4). Likewise, varying heart rate between ~ 1.3 and 2.0 Hz did not produce a significant effect on EDV ($p = 0.837$; Fig. 5).

4 Discussion

In the present study, we observed depressed systolic function, characterized mainly by decreased SV and EF, with increase in the beating rate of the in situ, perfused *T. molitor* heart tube. Nevertheless, CO, which describes the time-averaged cardiac pumping activity, varied little and was not significantly affected by the heart rate.

Considering the isolated systolic function, SV depression might be attributable either to a rate-dependent influence on excitation-contraction coupling and/or the contractile apparatus, or to the heterometric mechanism of regulation of cardiac force (Frank-Starling mechanism), characteristic of vertebrate striated muscle. This is an essentially

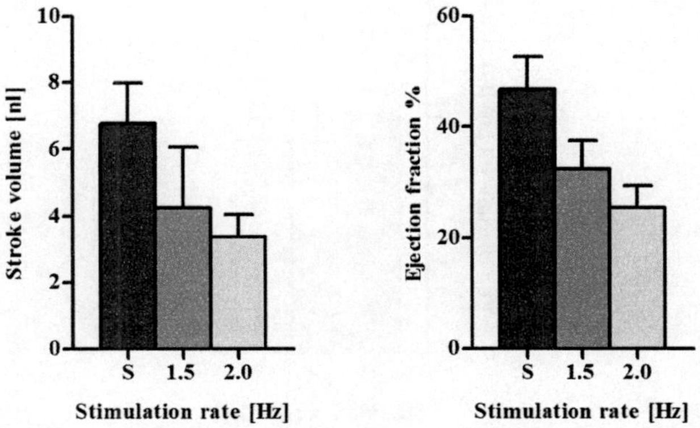

Fig. 3. Negative staircase of stroke volume (left panel) and ejection fraction (right panel) of the perfused *T. molitor* heart tube in response to increase in heart rate by electric pacing. S = spontaneous frequency (1.3 ± 0.2 Hz). N = 8. $p < 0.05$ for both, one-way analysis of variance for paired samples.

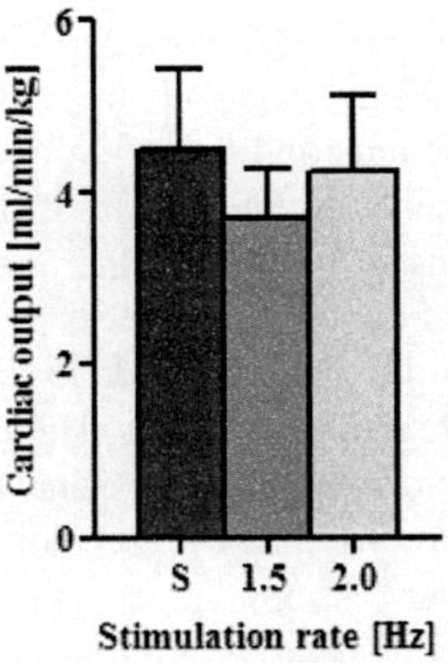

Fig. 4. Cardiac output values estimated in perfused *T. molitor* heart tube in response to increase in heart rate by electric pacing. S = spontaneous frequency (1.3 ± 0.2 Hz). N = 8. $p > 0.70$, one-way analysis of variance for paired samples.

adaptative cardiovascular mechanism, in which the force developed at systole depends on the length of the myocyte sarcomeres determined by the stretch imposed on the chamber wall by the preload, i.e., amount of blood it contained prior to systole (i.e., EDV). This mechanism allows the heart to match its output (CO) to the input (venous return) under a variety of situations, from resting to increased circulatory demand (e.g., exercise) [18].

If the heterometric regulation of cardiac force was the main mechanism responsible for the negative relationship between heart rate and the systolic variables, it should be accompanied by a rate-dependent decrease in hemodynamic preload (i.e., EDV) due to abbreviation of the inter-beat interval, and thus, of the time available for cardiac filling. However, this was not observed: EDV was little affected by contraction frequency, thus

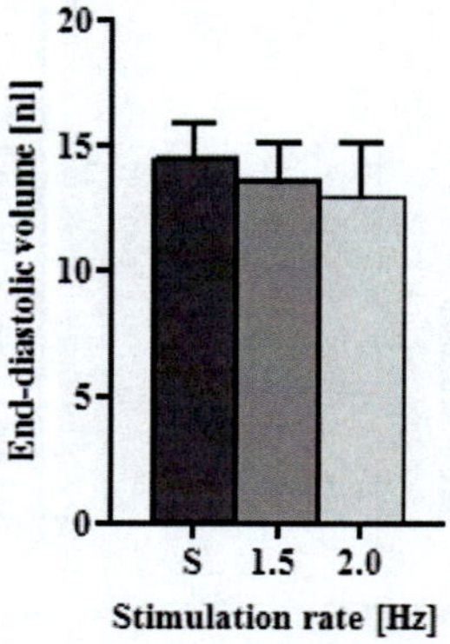

Fig. 5. Values of end-diastolic volume estimated in perfused *T. molitor* heart tube in response to increase in heart rate by electric pacing. S = spontaneous frequency (1.3 ± 0.2 Hz). N = 8. $p > 0.80$, one-way analysis of variance for paired samples.

indicating that fairly similar diastolic cardiac filling could be achieved even after a ~ 35% reduction of the interbeat interval. The insensitivity of EDV and CO to heart rate is thus consistent with the absence of known changes in the input of hemolymph to the heart under the present experimental conditions.

On the other hand, the present results indicate that the depression of the systolic function with higher rates seems to be due to alteration of steps associated with the triggering and/or development of contractile process.

The negative staircase of contraction amplitude in response to increase in the rate of contraction was reported for the first time in 1871 by Henry Bowditch [19], who described the relationship between the heart rate and the cardiac contractile force. Nevertheless, the pattern of the so-called "force-frequency" relationship, observed also in isolated myocardial preparations, may vary among species and functional conditions. In large mammals, including humans, it has been reported as positive in the absence of cardiac disease and negative under heart failure, whereas in healthy rodents it is usually negative [20, 21], although it may less often be flat or positive [22–24]. This relationship was shown to be negative in a variety of fish species [25], and also in the heart tube of insects, such as *D. melanogaster* [26] and *T. molitor* [14].

Evidence obtained from mammalian myocardium has supported a role of decreased Ca^{2+} fluxes across the sarcolemma and/or the membrane of the main intracellular Ca^{2+} store (sarcoplasmic reticulum), which result in diminished availability of the ion to the contractile apparatus as contraction rate is increased [22–24]. Future studies should establish if this is also the main mechanism for the negative SV and EF staircase in the insect heart. It is noteworthy, however, that in the whole, pumping heart, CO is not diminished by the rate-dependent systolic depression. The adaptive advantage of this behavior is possibly related to myocardial energy economy, but still remains to be determined.

5 Compliance with Ethical Requirements

5.1 Conflict of Interest

No conflicts of interest, financial or otherwise, are declared by the author(s).

5.2 Statement of Human and Animal Rights

Although Brazilian law does not require ethical approval for experimentation with invertebrates, care was taken to follow standard international ethical procedures concerning the adequate use of experimental animals.

6 Conclusion

The present results show that the depression of systolic contraction previously observed in *T. molitor* heart is also manifested in hemodynamic parameters, such as SV and EF. However, under conditions of unchanged hemodynamic load, such a depression is not reflected on the CO, which showed to be independent of the heart rate. The lack of variation in cardiac preload (EDV) with rate argues against a major role of length x force relationship in this phenomenon. The main underlying mechanism might be related to a change in fluxes and availability of Ca^{2+} to the myofilaments, a possibility that should be investigated in future experiments.

Acknowledgment. The technical support of Ms. Elizângela S. Oliveira is greatly appreciated. Financial support: Brazilian National Council for Scientific and Technological Development - CNPq (304010/2016-2 grant to JWMB and scholarship, for FHKI, 127197/2021-4); Brazilian Ministry of Health/FINEP (grant# 01-13-0214-00); and Coordination for the Improvement of Higher Education Personnel (CAPES, scholarships for NCBC and ARS).

References

1. Stork, N.E.: How many species of insects and other terrestrial arthropods are there on Earth? Annu. Rev. Entomol. **63**(1), 31–45 (2018). https://doi.org/10.1146/annurev-ento-020117-043348
2. Wolf, M.J.: Modeling dilated cardiomyopathies in Drosophila. Trends Cardiovasc. Med. **22**(3), 55–61 (2012). https://doi.org/10.1016/j.tcm.2012.06.012
3. Johnson, E., Sherry, T., Ringo, J., Dowse, H.: Modulation of the cardiac pacemaker of Drosophila: cellular mechanisms. J. Comp. Physiol. B Biochem. Syst. Environ. Physiol. **172**(3), 227–236 (2002). https://doi.org/10.1007/s00360-001-0246-8
4. Pauls, D., Blechschmidt, C., Frantzmann, F., el Jundi, B., Selcho, M.: A comprehensive anatomical map of the peripheral octopaminergic/tyraminergic system of Drosophila melanogaster. Sci. Rep. **8**(1), 11–14 (2018). https://doi.org/10.1038/s41598-018-33686-3
5. Ranson, D.C., Ayoub, S.S., Corcoran, O., Casalotti, S.O.: Pharmacological targeting of the GABAB receptor alters Drosophila's behavioural responses to alcohol. Addict. Biol. **25**(2), 1–9 (2019). https://doi.org/10.1111/adb.12725

6. Bushnell, P.J., et al.: Genetic targets of acute toluene inhalation in Drosophila melanogaster. Toxicol. Sci. **156**(1), 230–239 (2017). https://doi.org/10.1093/toxsci/kfw243
7. Mikulak, E., Gliniewicz, A., Przygodzka, M., Solecka, J.: Galleria mellonella L. as model organism used in biomedical and other studies. Przegl. Epidemiol. **72**(1), 57–73 (2018)
8. Soumya, M., Reddy, H.A., Nageswari, G., Venkatappa, B.: Silkworm (Bombyx mori) and its constituents: a fascinating insect in science and research. J. Entomol. Zool. Stud. **5**(5), 1701–1705 (2017)
9. Stankiewicz, M., Dąbrowski, M., Lima, M.E.: Nervous system of Periplaneta americana cockroach as a model in toxinological studies: a short historical and actual view. J. Toxicol. **2012**, 1–11 (2012). https://doi.org/10.1155/2012/143740
10. Chowanski, S., Rosinski, G.: Myotropic effects of cholinergic muscarinic agonists and antagonists in the beetle. Curr. Pharmac. Biotechnol. **18**(3), 1088–1097 (2017). https://doi.org/10.2174/1389201019666180124233440
11. Wang, X., et al.: The locust genome provides insight into swarm formation and long-distance flight. Nat. Commun. **5**(1), 2957 (2014). https://doi.org/10.1038/ncomms3957
12. Adamski, Z., et al.: Beetles as model organisms in physiological, biomedical and environmental studies: a review. Front. Physiol. **10**, 319 (2019). https://doi.org/10.3389/fphys.2019.00319
13. Viswanathan, M.C., Kaushik, G., Engler, A.J., Lehman, W., Cammarato, A.: A Drosophila melanogaster model of diastolic dysfunction and cardiomyopathy based on impaired troponin-T function. Circ. Res. **114**(2), 6–17 (2014). https://doi.org/10.1161/CIRCRESAHA.114.302028
14. Fim Neto, A., Bassani, R.A., Oliveira, P.X., Bassani, J.W.M.: Sources of Ca^{2+} for contraction of the heart tube of Tenebrio molitor (Coleoptera: Tenebrionidae). J. Comp. Physiol. B Biochem. Syst. Environ. Physiol. **188**(6), 929–937 (2018). https://doi.org/10.1007/s00360-018-1183-0
15. Feliciano, DF., Bassani, RA., Oliveira, PX., Bassani, JWM.: Pacemaker activity in the insect (T. molitor) heart: role of the sarcoplasmic reticulum. Am. J. Physiol.—Regul. Integr. Comp. Physiol. **301**(6), 1838–184 (2011). https://doi.org/10.1152/ajpregu.00089.2011
16. Markou, T., Theophilidis, G.: The pacemaker activity generating the intrinsic myogenic contraction of the dorsal vessel of Tenebrio molitor (Coleoptera). J. Exp. Biol. **203**(22), 3471–3483 (2000)
17. Pacholska-Bogalska, J., Szymczak, M., Marciniak, P., Walkowiak-Nowicka, K., Rosiński, G.: Heart mechanical and hemodynamic parameters of a beetle, Tenebrio molitor, at selected ages. Arch. Insect Biochem. Physiol. **99**(1), 1 (2018). https://doi.org/10.1002/arch.21474
18. Lakatta, E.G.: Starling's law of the heart is explained by an intimate interaction of muscle length and myofilament calcium activation. J. Am. Coll. Cardiol. **10**(5), 1157–1164 (1987). https://doi.org/10.1016/S0735-1097(87)80361-3
19. Endoh, M.: Force-frequency relationship in intact mammalian ventricular myocardium: physiological and pathophysiological relevance. Eur. J. Pharmacol. **500**(1–3), 73–86 (2004). https://doi.org/10.1016/j.ejphar.2004.07.013
20. Koch-Weser, J., Blinks, J.R.: The influence of the interval between beats on myocardial contractility. Pharmacol. Rev. **15**, 601–652 (1963)
21. Mulieri, L.A., Hasenfuss, G., Leavitt, B., Allen, P.D., Alpert, N.R.: Altered myocardial force-frequency relation in human heart failure. Circulation **85**(5), 1743–1750 (1992). https://doi.org/10.1161/01.CIR.85.5.1743
22. Antoons, G., Mubagwa, K., Nevelsteen, I., Sipido, K.R.: Mechanisms underlying the frequency dependence of contraction and $[Ca^{2+}]i$ transients in mouse ventricular myocytes. J. Physiol. **543**(3), 889–898 (2002). https://doi.org/10.1113/jphysiol.2002.025619

23. Dibb, K.M., Eisner, D.A., Trafford, A.W.: Regulation of systolic [Ca^{2+}]i and cellular Ca^{2+} flux balance in rat ventricular myocytes by SR Ca^{2+}, L-type Ca^{2+} current and diastolic [Ca^{2+}]i. J. Physiol. **585**(2), 579–592 (2007). https://doi.org/10.1113/jphysiol.2007.141473
24. Bassani, R.A., Ricardo, R.A., Bassani, J.W.M.: The mechanism of negative staircase in rodent myocardium. In: 16th Symposium on Ca^{2+}-Binding Proteins and Ca^{2+} Function in Health and Disease, p. 1. Annals, Pucón, Chile (2009)
25. Shiels, H.A., Vornanen, M., Farrell, A.P.: The force–frequency relationship in fish hearts: a review. Comp. Biochem. Physiol. Part A Mol. Integr. Physiol. **132**(4), 811–826 (2002). https://doi.org/10.1016/S1095-6433(02)00050-8
26. Balcazar, D., et al.: SERCA is critical to control the Bowditch effect in the heart. Sci. Rep. **8**(1), 1–8 (2018). https://doi.org/10.1038/s41598-018-30638-9

Development of a Low-Cost Video Laryngoscope

Roberto C. Ceccato[1](✉), Gabriel O. Alves[2], Denilson Storck Fomin[3], Luiz Eduardo Flório Junior[3], Thiago Nouer Frederico[3], and Henrique Takachi Moriya[1]

[1] Biomedical Engineering Laboratory, Universidade de São Paulo, 05508-010 São Paulo, Brazil
roberto.ceccato@usp.br
[2] Instituto de Física da Universidade de São Paulo, 05314-970 São Paulo, Brazil
[3] Hospital Israelita Albert Einstein, 05652-900 São Paulo, Brazil

Abstract. Tracheal intubation is a widely needed procedure in medicine and veterinary. Ranging from surgeries to intensive care unit, intubation must be performed in order to provide mechanical ventilation to patients. Hence, a laryngoscope is normally used to grant the visibility of the superior airways and to guide the endotracheal tube. However, conventional laringoscopy usually has a low rate of success when compared to its video-based counterpart. Notwithstanding, commercial video laryngoscopes are extremely expensive, limiting its receptiveness. In this work we present the proof of concept of a 3D-printed low-cost video laryngoscope, putting forth a feasible approach and a detailed description of its manufacture.

Keywords: Laryngoscopy · Intubation · Airway management · Equipment and supplies · Low-resource settings

1 Introduction

The use of video laryngoscopes improves the rate of success of intubation procedures [1–3]. The method provides clear-cut clinical improvements, such as the possibility to avoid problems like esophageal intubation or yet non intubation, which can lead to non ventilation scenarios, requiring the realization of a tracheostomy procedure. Likewise, further benefits such as a reduction of inpatient costs and length of hospital stay have also been observed [4]. However, the costs associated with such devices are still a great obstacle to their adoption in low and middle-income countries. For that reason, intratracheal intubation is mostly performed with direct laryngoscopes even nowadays.

Surveys have shown that medical technologies and minimally invasive surgery tools in low-resource settings (LRS) are some of the priorities among bioengineering professionals. Furthermore, 3D printed medical devices for personalized medicine appear as relevant emerging technologies in this context [5].

J. L. B. Marques et al. (Eds.): CLAIB 2022/CBEB 2022, IFMBE Proceedings 98, pp. 548–559, 2024.
https://doi.org/10.1007/978-3-031-49401-7_57

Moreover, relevant criteria for designing medical devices under low-resource settings have been subject to discussion in [6]. In particular, criteria such as cost and equipment lifetime have been highlighted in the aforementioned work.

Under these considerations, low-cost video laryngoscopes arise as a potential technology in LRS [7–12] (for a review see [13]). Among the prevalent proposals, the video is broadcast to a cellphone device through a borescope, which is attached to a standard laryngoscope blade (see e.g. [14]).

In this work we show how a video laryngoscope can be built with 3D-printed materials and a single-board computer (SBC). We believe our construction allows for a robust and systematic assembly. The advantages of such construction are twofold: while standard 3D-printed materials are readily available in the market at low cost, the design and composition of the blade, such as bio-compatible resins, can be easily modified in order to cater a specific need or constraint. Furthermore, the use of a SBC eases the implementation of desirable software features, such as video recording and remote video streaming.

The paper is structured as follows: in Sect. 2 we present details on the construction of the laryngoscope. First we discuss the software implementation, afterwards we present the 3D-model in detail. Results and further considerations can be found in Sect. 3. Concluding remarks can be found in Sect. 4.

2 Materials and Methods

The construction of the video-laryngoscope can be divided into two main segments: the manufacturing of the physical parts of the equipment and the software development. Details of their implementations will be dealt with in Sects. 2.1 and 2.2, respectively. The components used in the prototype, shown in Fig. 1, are:

- A Raspberry Pi Zero, which was our chosen SBC
- A SD card
- A 3.5" LCD display (125 MHz SPI interface and 320×480 resolution)
- A 7.3 mm-diameter endoscope camera
- 3D-printed body (Details are given in Sects. 2.1. and 3)
- A battery/a power source.

The manufacturing process that we adopted was the 3D-printing method based on fused deposition modeling (FDM), with acrylonitrile butadiene styrene (ABS) as the chosen material. The Raspberry Pi Zero, a SBC broadly available in the market, was the software processing unit used in this project. Moreover, the chosen display also provides-touch screen capabilities, which allows for direct interaction with the device, dispelling the need for physical buttons in the body of the laryngoscope, which greatly simplifies the project design. The latency and the hardware performance are discussed in more detail in Sect. 3. Furthermore, the endoscope shown in Fig. 1 has a IP67 certification [15] and embedded LEDs, with controllable light levels, features also suited for the intended use. Finally, we have used *Python*, with the aid of some widespread open-source libraries, to interface some of the components and to implement the desired functions. Details are described in the following subsections.

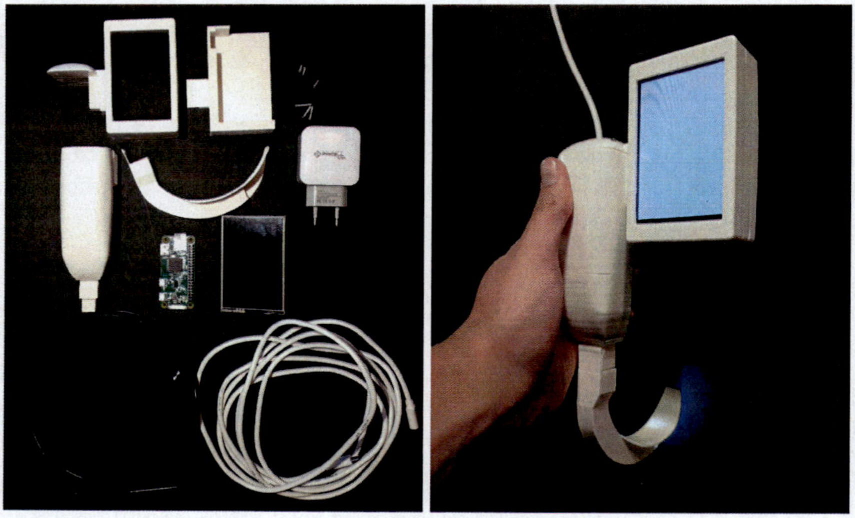

Fig. 1. (Left) Components of the video laryngoscope. The body, blade, borescope, screen and Raspberry Pi can all be seen in the figure. (Right) Fully assembled video laryngoscope.

2.1 Design

The starting point of the hardware design was the analysis of a commercial video laryngoscope curved blade (*McGrath® X blade*™, *Aircraft Medical, USA*). By basing our design on its curvature profile, the blade shown in Fig. 1 was assembled. Once the geometry of the the blade was established, the remaining parts, namely the main body and screen frame, were developed.

Our group encountered a few challenges when using ABS-printed components, specially in regard to elongation at failure. This is a desirable property for the equipment, since it allows both for a better accommodation of the blade at superior airways and also for a better mechanical interference between the blade and the body. Thus, in order to increase the 3D-printing elongation at failure and surface smoothness, every 3D-printed component used in the equipment body went through an acetone vapor bath at a temperature around 25°C during 50 min. The acetone vapor bath acts on the ABS printed layers, partially dissolving some of the material in the surface. This induces a material flow which fills up the spacing between different layers in the printed material, consequently increasing elongation at failure [16,17]. In the same spirit, some of the challenges regarding the appropriate design of the blade have been previously investigated in [18].

2.2 Software

The software was devised to execute three main functions: i) capture the video, ii) transmit it locally through the LCD screen and iii) stream and store the captured data in a host device, such as a desktop computer, if desired by the user. As depicted in Fig. 2, the open-source library *OpenCV* was used to read, write and stream video files [19]. Meanwhile, we used the *Flask* framework to build a web application which can be used to download and stream the captured video in a remote device [20]. The video can be displayed either through *OpenCV* or through *MPlayer*, an open-source cross-platform movie player with several embedded functionalities [21]. All of the routines used here were implemented either as *Python* code or *Shell* scripts.

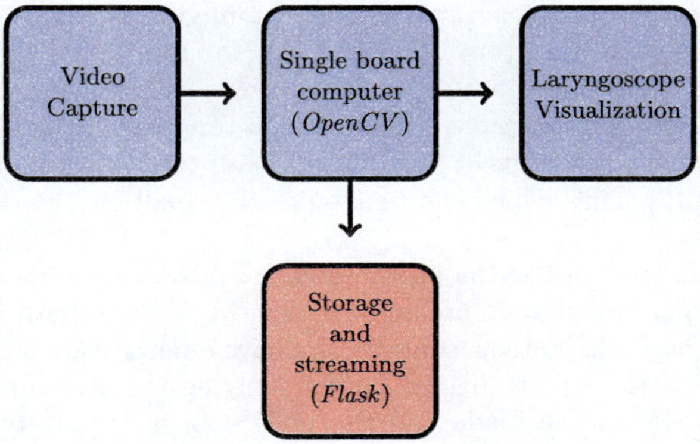

Fig. 2. Schematic diagram of software components and functions.

3 Results

The mains result achieved in this work were the 3D-printed physical parts, their design and manufacture method, as well as the software development. We opted for a modular construction where different parts are sufficiently independent of each other. This way they can be changed, removed or modified as needed, without compromising the rest of the project. Certain critical components, such as the blade, can be adapted as needed, be it due to any possible financial, medical or resource constraints. The use of the laryngoscope in a pediatric setting is an instance where this may come at hand: it is possible to integrate a blade design suited for pediatric use with minimal alterations in the rest of the project. Moreover, the software used is also highly customizable, and further features can be implemented as desired, which can greatly enhance its potential for educational uses.

The laryngoscope body itself is comprised of four different parts, which are printed separately:

- Main body
- Upper frame
- Lower frame
- Blade.

An overall view of the main body and both frames is given in Fig. 3 (a), while the blade is shown in Fig. 3 (b). Detailed dimensions and views are shown throughout Fig. 3 (c) - (j). In Fig. 3 (c) and (d) we show the front and the back view of the main body, respectively. The hole in the main body, shown in Fig. 3 (c), is used as a passage for the endoscope camera. We also opted for a hollow construction, this not only reduces the quantity of ABS needed for the manufacturing process, but it also makes up space for the wires which should go inside the body. The holes in the back view in Fig. 3 (d) are used to screw the case in the main body, as illustrated in the bottom view in Fig. 3 (e). The upper frame, shown in Fig. 3 (f) (front view) and (h) (top view), and the lower frame, depicted in Fig. 3 (h), fit together, forming a single case which encapsulates the Raspberry-pi and the LCD screen. A side and a top view of the blade are given in Fig. 3 (i) and (j), respectively. The blade is thus connected to the main body through a snap fit joint, which can be seen as the small gray part in Fig. 3 (a), (c) and (e).

The snap fit joint in the the blade and the body was the most fragile part in our prototype, and that is precisely where the acetone bath has shown to been the most useful in our construction. We have qualitatively observed a great increase in the resistance in this area after applying the acetone bath, making the interface between the blade and the body robust for the typical use. As discussed in [16], the effects of the acetone bath and the resulting increase in the elongation at failure are the most pronounced for lower thickness, possibly due to the smaller area over volume ratio. This possibly explains why the snap fit joint, the smallest in the project, with thickness of order of $1mm$ in its design, has seen such an increase in durability. We have also observed some improvement in other areas, such as the in the frame borders, but they far less noticeable. Thus, while the acetone is vital for a good functioning of the snap fit, and also possibly useful for the whole blade itself in order to decrease the surface roughness, it is not strictly necessary for other components of the laryngoscope. A more robust design to interlock the blade and the main body may then eliminate the need for the acetone bath and simultaneously may also make the prototype more robust.

In regard to the software development, the *OpenCV* not only makes video capture available, but it can also be used to implement additional features in the video interface. A few of these features can be seen in Fig. 4(a): we were able to implement a drawing feature using built-in functions in the library itself. Moreover, we could also add basic usage components, such as a button for erasing the annotations and a button to close the program, since the application is run in full-screen. A *Shell* script can also be used to initialize the program on startup if desired. While the drawing feature can be useful even in a clinical setting, we believe that it is even more useful in educational applications. Other uses, such as a screen capture button for instance, can also be easily implemented.

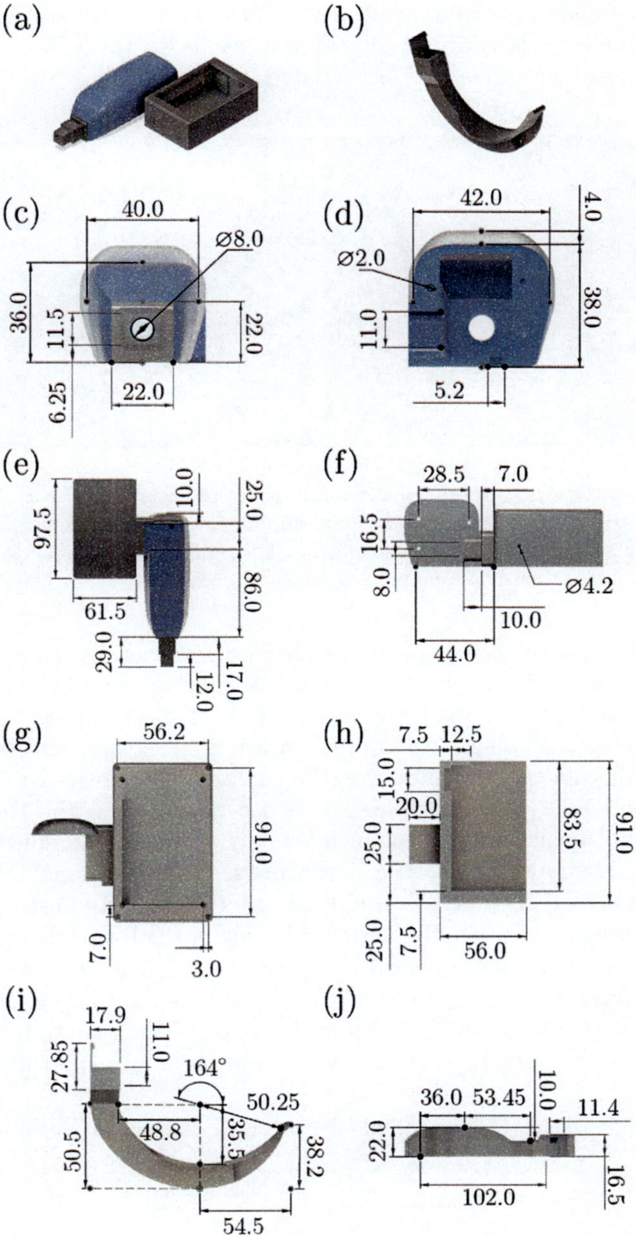

Fig. 3. Illustration of the 3D-model for (a) the main body and the Raspberry-pi case and (b) the blade. In (c) - (j) we show the dimensions of the model. All units are in displayed *mm*.

It suffices to make the proper modification in the *Python* script which runs the *OpenCV*-based application. Moreover, since all the libraries used are free and open-source, collaborations and improvements to the video laryngoscope software development are greatly stimulated in this scenario.

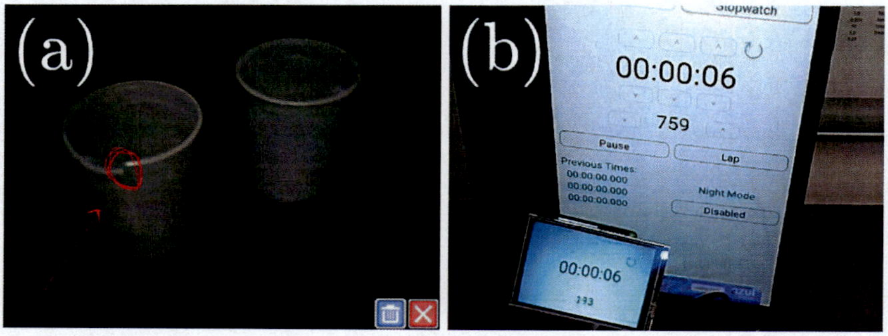

Fig. 4. (a) A screen capture of the video-laryngoscope interface. (b) A picture of the setup using the Raspberry-Pi and the endoscope camera. Here a digital stopwatch in a screen being recorded by the endoscope. The time difference between the screen and the recorded video is used to quantify the video latency.

Furthermore, as previously mentioned in Sect. 2.2, we were also able to implement remote streaming capabilities. *Flask*, a light-weight web framework for Python, can be used to stream the video shown in the Raspberry pi to a host computer. For prototyping purposes, it is possible to set up a simple *html* page and a builtin local development server. We qualitatively observed a considerably latency in this case, when compared to video displayed in the Raspberry-pi. Thus, this implementation is recommended only when considerable delay is not a concern, e.g. remote storage. And that is precisely an additional feature which these libraries make possible. We could also use these scripts to store the video. While the video can very simply be stored locally in the Raspberry-pi itself, and be retrieved either by the SD card or remotely trough a SSH connection, for instance, another interesting use would be to save the video remotely. We were able to use *Flask* and *OpenCV* to store the video in the remote host machine. This feature could be integrated with some database routine, for instance.

Table 1. Delay observed in different software framework conceptions.

Framework	Delay $[ms]$
MPlayer	625
Python with *OpenCV*	568

In addition, we have also investigated the latency of the video shown in the laryngoscope. The basic setup is shown in Fig. 4 (b). We used the laryngoscope camera to record a digital stopwatch, displayed on a desktop computer. Finally, we took a picture of the whole setup and from the resulting image we could determine the latency of the program. Results are shown in Table 1. As we can see, displaying the video through *MPlayer* introduces a slightly larger latency. On the other hand, executing the scripts and displaying the video diretcly in *Python* using *OpenCV* displays some slight improvement. We also observed a qualitatively larger frame drop in the video capture with the *MPlayer* framework in comparison to the *Python* one. The latency and the frame drop should be taken into account depending on the intended use for the video laryngoscope presented here. Applications which require minimal latency, such as clinical usage, should either optimize the hardware or the scripts themselves. Since *OpenCV* is also available in C++, it would also be possible to use it to implement it to some of our routines. It is however unkown to us whether this would result in a significant improvement.

Table 2. Typical price of the components used in the project.

Component	Price (US$)
Raspberry Pi Zero	20
3.5" display	55
Endoscope camera	10
Cellphone charger	5
SD-card	5
ABS ($\sim 125g$)	2.5

Finally, average costs are shown in Table 2, amounting to roughly $US\$100$ at current prices. As we mentioned before, certain components or materials can be chosen and substituted depending on their availability and their prices due to regional differences and market conditions. The table shows that the equipment can be built for prices much lower than the typical cost of usual commercial video-laryngoscopes. Moreover, the costs can be reduced even further when one is interested in simpler implementations where a local SCB with a screen is not needed and the camera is directly connected to a cellphone or a computer, for example.

4 Discussion

Intratracheal intubation is often associated with complications, which can be severe, irreversible and fatal. These can be avoided through a better visualisation of the upper airways, which is made possible by a video laryngoscope.

Despite the previously discussed benefits of the equipment, the ones currently available commercially are sold at prices between $US\$1000$ and $US\$15000$. Such high cost withholds a more widespread adoption, specially among lower income countries, groups and facilities. In this sense, this work provides a proof of concept and a feasible approach for the implementation and the manufacture of a low-cost video laryngoscope. Our proposal provides a cheap, functional and potentially versatile prototype of the equipment. The implementation presented here had a unity cost of approximately $US\$100$. Its lower price can also be useful to promote academic and educational uses, encompassing applications in medical training and in research involving intratracheal intubation.

However, such low price does not take the very expensive homologation process into account. Video laryngoscopes have to satisfy a large number of standards, which can be seen in [22]. The compliance of the equipment with the technical specifications on each of the standards is essential to ensure patient safety and well-being. While this is not an obstacle to educational applications, one should bear these costs in mind when medical use is considered.

Finally, we briefly compare our proposal with the existing literature on LRS video laryngoscopes. One of the original and most common implementations consists in a setup where the endoscope camera is attached to a standard laryngoscope blade and the video is displayed on, e.g., a cellphone or a computer. This proposal is one of the cheapest alternatives in the literature [9] and bears small video latency. Previous works provide some evidence of its clinical [7] and educational [8] effectiveness. Other proposals provide a middle-ground where either the blade [10] or the handle [8] is 3D-printed. Meanwhile, a custom printed project, as ours, allows for modifications and fine-tuning depending on the purpose of the equipment, which might be useful in research applications and maintenance, for instance. Facilities where 3D-printing equipment may be readily available, such as universities, eliminates big part of the overhead associated with our construction, both in terms of cost and material availability. Our equipment weighs 325g, which is similar to commercial video laryngoscopes [23, 24] and, as seen in Fig. 1, we could assemble it in a very similar fashion to the commercial ones, where the SBC and the screen are attached to the body itself. We believe that this tighter construction allows for better handling, opposed to several LRS implementations in which screen and the handle are detached. Furthermore, by removing the dependence on external peripherals such as cellphones, our construction avoids some sanitary risks associated with their use, simplifies the process of disinfection and makes the project modular and self-contained. However, as it is common to LRS alternatives, suitability under a regulatory framework, due to e.g. safety constraints and proved effectiveness might be one of the biggest obstacles in our proposal. Nevertheless, we believe that our work provides a starting pathway in this sense, where more robust hardware, bio-compatible resins and clinical trials could be used to further improve the equipment in a clinically suited sense.

Future works may focus on reviewing the software conception with the objective of reducing the latency and frame drop. Additionally, further investigations regarding the hardware development are also welcome, such as strategies to increase the resistance of critical regions of the body (like the junction between the blade and the body) and a quantitative analysis of the effect of the acetone bath in the the increase of traction strength. Also, bio-compatible resins will be used in the manufacture of the equipment, aiming to perform *in vivo* tests. Once in possession of a bio-compatible and design revised version (with veterinary aimed versions of the blade), animal tests will be done in order to quantitatively evaluate the benefits of the video laryngoscope usage and it resistance analysis in field. Additionally, a *GitHub* repository is being developed in order to turn the work into a totally open source project. The code used here and detailed tutorials will be added to the repository.

5 Conclusion

In this paper, a proof of concept of a low-cost video laryngoscope was put forth. The minutia presented were a proposal for the manufacturing process, discussing details of a concrete 3D-model for the equipment, the hardware specifications, the software development and its manufacturing. The possibility of assembling a low-cost video laryngoscope overcomes the high cost barrier associated with the equipment and opens up new opportunities of usage in fields such as educational, veterinary and human respiratory research. Furthermore, the equipment will be subject to clinical tests with animals and all the relevant code and models will be shared through a *GitHub* repository.

Acknowledgments. The authors would like to thank the endowment fund Amigos da Poli for the financial support with the 2020 call notice and the Hospital Israelita Albert Einstein for providing medical technical support. G.O.A. acknowledges the financial support from the São Paulo founding agency FAPESP (Grant No. 2020/16050-0). This study was financed in part by the Coordenacao de Aperfeicoamento de Pessoal de Nivel Superior - Brazil (CAPES) - Finance Code 001 and the Conselho Nacional de Pesquisa e Desenvolvimento Científico e Tecnologico - Brazil (CNPq) (308280/2019-9 to HTM).
Conflict of interest The authors declare that they have no conflict of interest.

References

1. Kilicaslan, A., Topal, A., Tavlan, A., Erol, A., Otelcioglu, S.: Effectiveness of the C-MAC video laryngoscope in the management of unexpected failed intubations. Braz. J. Anesthesiol. (English Edition) **64**(1), 62–65 (2014). https://doi.org/10.1016/j.bjane.2013.03.001
2. Lewis, S., Butler, A., Parker, J., Cook, T., Schofield-Robinson, O., Smith, A.: Videolaryngoscopy versus direct laryngoscopy for adult patients requiring tracheal intubation: a Cochrane systematic review. Braz. J. Anaesth. **119**(3), 369–383 (2017). https://doi.org/10.1093/bja/aex228,https://linkinghub.elsevier.com/retrieve/pii/S0007091217537502
3. Liu, D.X., Ye, Y., Zhu, Y.H., Li, J., He, H.Y., Dong, L., Zhu, Z.Q.: Intubation of non-difficult airways using video laryngoscope versus direct laryngoscope: a randomized, parallel-group study. BMC Anesthesiol. **19**(1) (2019). https://doi.org/10.1186/s12871-019-0737-3, https://bmcanesthesiol.biomedcentral.com/articles/10.1186/s12871-019-0737-3

4. Zhang, J., Jiang, W., Urdaneta, F.: Economic analysis of the use of video laryngoscopy versus direct laryngoscopy in the surgical setting. J. Comp. Effectiveness Res. **10**(10), 831–844 (2021), https://doi.org/10.2217/cer-2021-0068,https://www.futuremedicine.com/doi/10.2217/cer-2021-0068
5. De Maria, C., Díaz Lantada, A., Jämsä, T., Pecchia, L., Ahluwalia, A.: Biomedical engineering in low- and middle-income settings: analysis of current state, challenges and best practices. Heal. Technol. (2022). https://doi.org/10.1007/s12553-022-00657-8
6. Piaggio, D., Castaldo, R., Cinelli, M., Cinelli, S., Maccaro, A., Pecchia, L.: A framework for designing medical devices resilient to low-resource settings. Globalization Health **17**(1) (2021). https://doi.org/10.1186/s12992-021-00718-z. https://globalizationandhealth.biomedcentral.com/articles/10.1186/s12992-021-00718-z
7. Karippacheril, J.G., Umesh, G., Ramkumar, V.: Inexpensive video-laryngoscopy guided intubation using a personal computer: initial experience of a novel technique. J. Clin. Monit. Comput. **28**(3), 261–264 (2014). https://doi.org/10.1007/s10877-013-9522-x
8. Nyrhinen, E., Nyrhinen, A.: 3D-printed, $20 video laryngoscope designed for resource-constrained settings. Ann. Glob. Health **82**(3), 593 (2016). https://doi.org/10.1016/j.aogh.2016.04.559
9. Cohen, Y., Rubinstein, R.M., Berkenstadt, H.: Abstract PR536: A Do-It-Yourself Videolaryngoscope for Under 25$—a solution for lower income countries? Anesth. Analg. **123**(3), 680–681 (2016) https://doi.org/10.1213/01.ane.0000492919.95559.72. http://journals.lww.com/00000539-201609002-00512
10. John, A., John, S., Lambert, C.: Development and testing of a low cost videolaryngoscope in a resource limited setting. Ann. Glob. Health **83**(1), 4 (2017). https://doi.org/10.1016/j.aogh.2017.03.009
11. Hasija, N., Kale, S., Girdhar, K.K.: Digitizing the direct laryngoscopy experience: the economic way!! Korean J. Anesthesiol. **71**(6), 486–487 (2018), https://doi.org/10.4097/kja.d.18.00093. http://ekja.org/journal/view.php?doi=10.4097/kja.d.18.00093
12. Saoraya, J., Musikatavorn, K., Sereeyotin, J.: Low-cost videolaryngoscope in response to COVID-19 pandemic. W. J. Emerg. Med. **21**(4), 817–818 (2020). https://doi.org/10.5811/westjem.2020.5.47831
13. Hamal, P.K., Yadav, R.K., Malla, P.: Performance of custom made videolaryngoscope for endotracheal intubation: a systematic review. PLoS ONE **17**(1 January) (2022). https://doi.org/10.1371/journal.pone.0261863
14. Uribe, V.F., Andrés, D., Ríos, V., Jiménez, L.C., Diego, J., Felipe, A., Reina, P., Camilo, J., Salgado, G., Calderón, M.: Borescope versus laryngoscope in difficult airway management by non-expert personnel : a non-randomized pilot study in a simulated environment. JSM Clin. Anesthesiol. (2019)
15. International Electrotechnical Commission: IP ratings. Available at: https://www.iec.ch/ip-ratings
16. Neff, C., Trapuzzano, M., Crane, N.B.: Impact of vapor polishing on surface roughness and mechanical properties for 3D printed abs. In: Solid Freeform Fabrication 2016: Proceedings of the 27th Annual International Solid Freeform Fabrication Symposium—An Additive Manufacturing Conference, SFF 2016 **i**, pp. 2295–2304 (2016)
17. Gao, H., Kaweesa, D.V., Moore, J., Meisel, N.A.: Investigating the impact of acetone vapor smoothing on the strength and elongation of printed ABS parts. JOM **69**(3), 580–585 (2017). https://doi.org/10.1007/s11837-016-2214-5

18. de Almeida, V.S., de Almeida, V.S., Campos, G.O.: Challenges of prototyping, developing, and using video laryngoscopes produced by inhouse manufacturing on 3D printers. Braz. J. Anesthesiol. (English Edition) **71**(3), 309–311 (2021). https://doi.org/10.1016/j.bjane.2021.01.012, https://linkinghub.elsevier.com/retrieve/pii/S0104001421001731
19. Bradski, G.: The OpenCV Library. Dr. Dobb's Journal of Software Tools (2000)
20. Pallets Organization: Flask. Available at: https://github.com/pallets/flask
21. MPlayer Team: Mplayer. Available at: https://mplayerhq.hu/design7/documentation.html
22. Aircraft Medical, USA: McGrath® MAC EMS Video Laryngoscope Operator's Manual. Available at: https://www.strykeremergencycare.com/globalassets/assets/oi-ifu/mcgrath-mac-ems-oi.pdf
23. Medtronic, USA: McGrath® MAC Video Laryngoscope Product Brochure. Available at: https://www.medtronic.com/covidien/en-us/products/intubation/mcgrath-mac-video-laryngoscope.html
24. Infinium Medical, Inc., USA: Infinium ClearVue® VL3D Product Brochure. Available at: https://www.strykeremergencycare.com/globalassets/assets/oi-ifu/mcgrath-mac-ems-oi.pdf

Demystification of Colors in Kirlian Photography as a Clinical Diagnostic Method

Wilson Picler[1,2](✉) and Eduardo Tavares Costa[1]

[1] FEEC/State University of Campinas (Unicamp), Campinas, SP, Brazil
Prof.wilsonpicler@gmail.com

[2] Uninter International University Center, Curitiba, PR, Brazil

Abstract. Kirlian Photography is the photography of the Luminous Halo, known as the Corona effect resulting from the ionization of the air due to the application of high-voltage electric fields in objects, fingertips, or plant leaves. Many studies have disseminated this technique, in the context of alternative therapies, as applicable to the diagnosis of various diseases and mental and bioenergetic states of human beings. Although digital photography techniques have reached higher quality levels than analog films, we are experiencing a phenomenon of the resurgence of analog techniques. This inspires the need of reviewing Kirlian Photography approaches with photographic films as an alternative method of diagnosis.

The present work focuses on the evaluation of this technique to demystify some patterns in these photographs as indicators for clinical diagnoses. Kirlionic Corona emission spectroscopy revealed a large amount of ultraviolet radiation with a smaller portion of light in the visible spectrum with a predominance of violet/blue. Analyzing Kirlian Photography's technique, it was observed that Corona can occur both at the top and at the bottom of the photographic film. Experiments have shown that when Corona occurs only at the top of the film, Kirliangraphy is revealed in blue color. However, it was also observed that when Corona occurs under the photographic film, the radiation generated sensitizes the film from the opposite side, producing a severe chromatic aberration that many unsuspecting practitioners, including medical doctors, psychologists, and therapists, mistakenly believe to be related to the biological states of the patient.

Keywords: Kirlian · Bioelectrography · Alternative Therapies

1 Introduction

Kirlian photography, Kirliangraphy or Bioelectrography (Fig. 1), is a technique for photographing the luminous effect, called Corona (Light Crown), resulting from the ionization of the air around biological objects (fingers and plants) energized with High Voltage fields.

A simplified representation of the Kirlian Photography techniques is shown in Fig. 1.

Figure 2 exhibits a typical Kirliangraphy (KG) obtained with analog (chemical) film.

Kirliangraphy has been the object of study in several areas in universities from several countries, and it has even been the subject of theses and dissertations in public and private

J. L. B. Marques et al. (Eds.): CLAIB 2022/CBEB 2022, IFMBE Proceedings 98, pp. 560–571, 2024.
https://doi.org/10.1007/978-3-031-49401-7_58

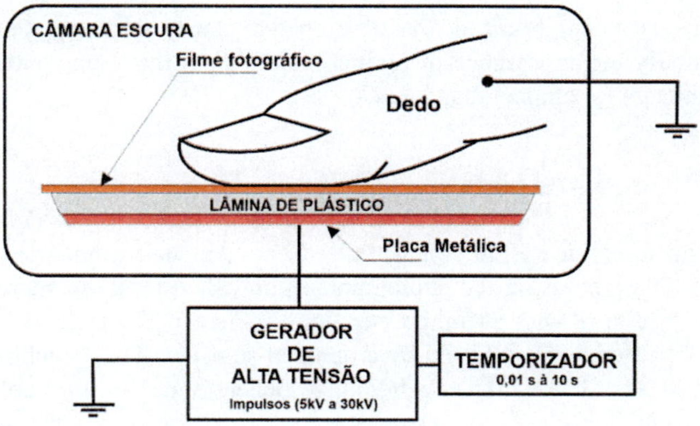

Fig. 1. Representation of the Kirlian Photography technique

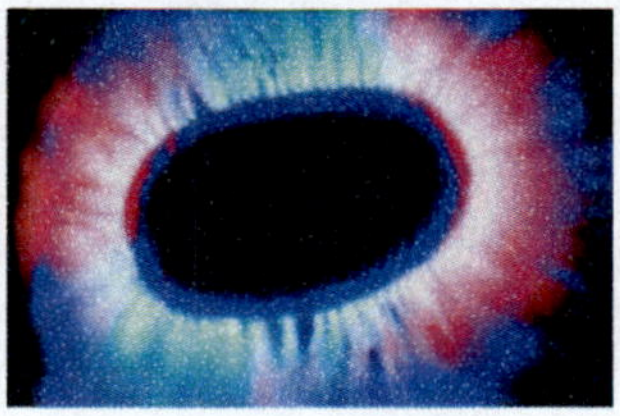

Fig. 2. Typical commercial standard KG

universities. This technique has been widespread since the end of the 19th century. It received the dedication of famous researchers such as the German mathematician and philosopher Georg Christoph Lichtenberg in 1777; Nikolas Tesla, with his experiments with High Voltage, and even the Brazilian Jesuit Priest Roberto Landell de Moura, who in 1904 in Rio Grande do Sul made hundreds of Bioelectrographies. In 1939, in Krasnodar, Russia, Semyon, and Valentina Kirlian began a series of experiments with Bioelectrophotography that culminated in publications in 1959 and 1961, in which they reported that electrographs not only showed changes from person to person but also changed with age. Their emotional state. Thus, a true odyssey began with the photo of the Human Aura. Since then, many works have been published concerning different patterns of these Bioelectrographies with the most diverse aspects of human physical and mental health. From the late 1990s onwards, bioelectrography evolved to the use of CCD cameras instead of photographic films.

In the age of digital technology, Researcher and Professor Dr. Kosntantin Korotkov [1], from the Federal Technological University of Saint-Petersburg, Russia, developed a technique he called *GDV* (*Gas Discharge Visualization*) and later *Bio-Well*, with which he processed Corona images, acquired through a CCD camera, and identifies patterns that, according to him, are related to various pathologies.

Once that was stated, this work was developed to clarify some aspects of this technique, particularly the appearance of multicolored shapes that some authors claim to serve as parameters for clinical diagnoses.

2 Materials and Methods

To carry out this research, a High Voltage Pulse Generator, an instrument to measure the force that the finger exerts on the photographic film called Dactylodynamometer, and a Film Holder equipped with a Compressor Frame (Fig. 3). The High Voltage value is about 10 to 30kV transient followed by a damped sinusoid. The switching frequency ranges from 100 Hz to 10 kHz. The High Voltage Generator has an adjustable timer from 0.1 to 10 s.

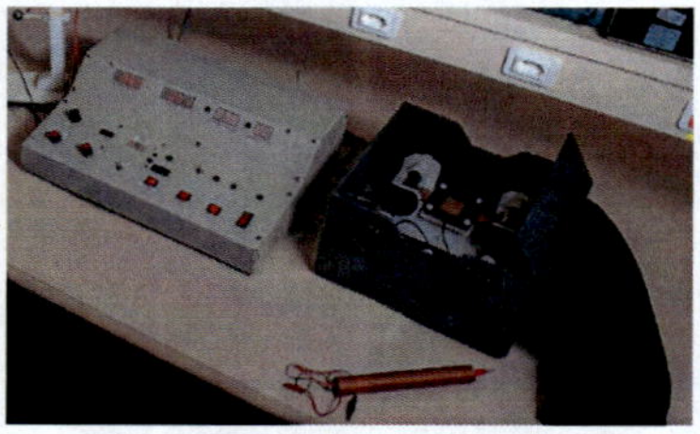

Fig. 3. Equipment developed for Kirliangraphy

Using the High Voltage pulse generator, the spectroscopy of the Corona generated under the same conditions as the Bioelectrographs was performed. For this purpose, an experimental arrangement was developed in which it was possible to collect a sample of the Corona radiation and perform the spectroscopy.

The spectrum obtained in this experiment (Fig. 4) revealed a large amount of strong spectral lines in the ultraviolet range.

The most used color photographic film in Kirliangraphy is *Fujicolor Superia* [2] which has 3 light-sensitive layers in the RGB *(Red, Green, and Blue*) standard. All emulsions of a colored photographic film are sensitive to UV radiation and violet light (Fig. 5). To prevent the UV/Violet radiation from reaching the layers responsible for the green and red there is a filter between the blue layer and the others, which prevents ultraviolet and blue/violet light from passing to the lower layers.

Analyzing how a Newton Milhomens' Standard Film Holder [3] is built (Fig. 6) shows a misalignment of the spools in relation to the polarizing/dielectric plate that causes the existence of air gaps under the Film.

When high voltage pulses are applied to the polarizing plate, the electric field ionizes the air and produces the Corona around the finger and in the mentioned air gaps.

Thus, the ultraviolet radiation will sensitize the film in three manners (Fig. 7). (A) By the correct upside, sensitizing just the Blue Layer due to the action of the Filter layer. (B) On the opposite side, sensitizing the Red and Green Layers. (C) Both sides, sensitizing the Red, Green, and Blue Layers in different intensities and regions.

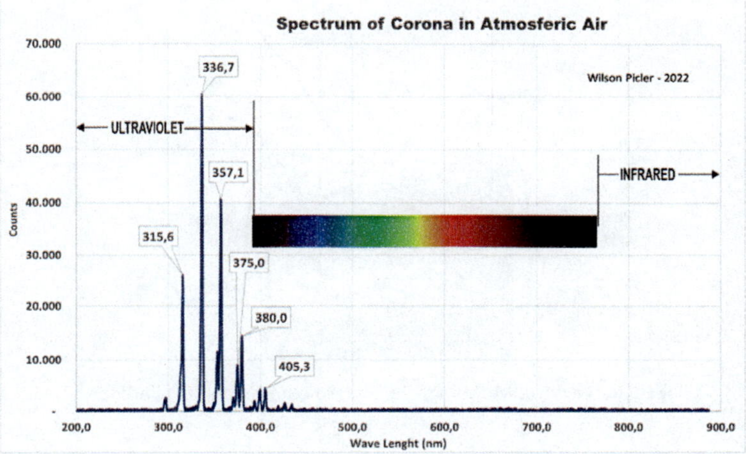

Fig. 4. Corona spectrum

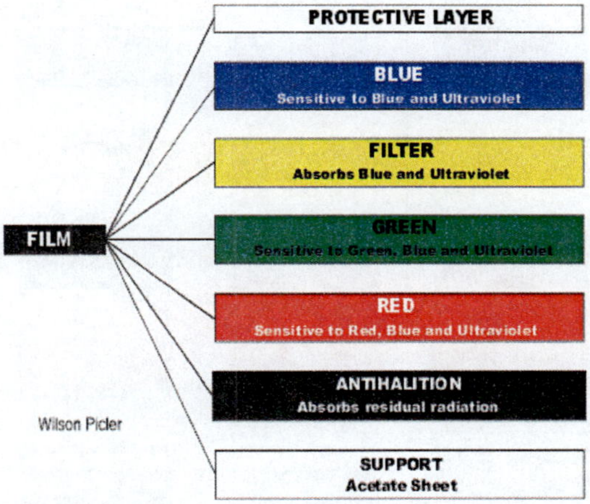

Fig. 5. Simplified structure of a colored photographic film

In the RGB process, mixing red and green generates different orange and yellow tones. Where the finger firmly touches the film, there is no air underneath it, and the Corona occurs only on the upper face of the photographic film, and as expected by the spectral characteristic of the Corona, the photograph results in a blue/violet color.

It can be hypothesized that the red and blue colors in Fig. 8A and the yellow and red colors of Fig. 8B, which Milhomens [4] claims are related to the patient's bioenergetic and emotional states, are chromatic aberrations resulting from the ionization of the air underneath the photographic film [4].

The following experiments were designed to demonstrate the veracity of the mentioned hypothesis.

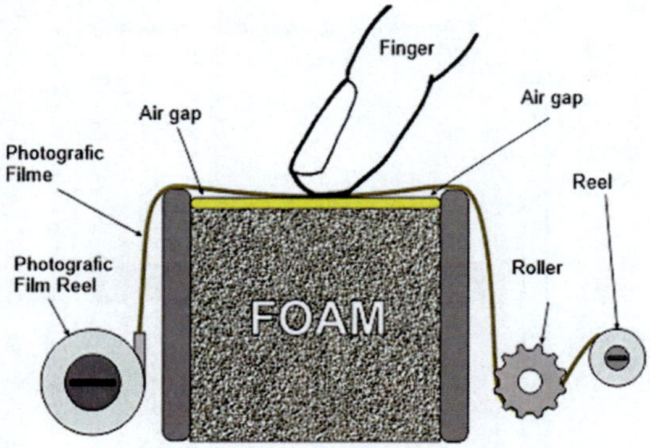

Fig. 6. Newton Milhomens' standard commercial film holder

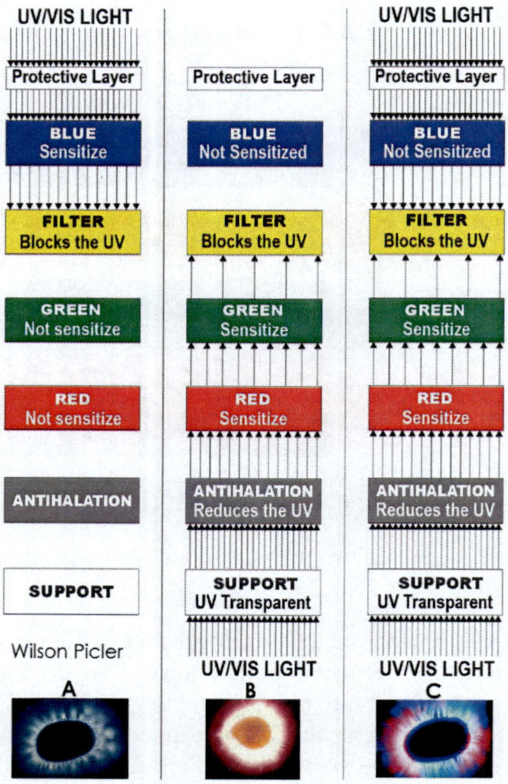

Fig. 7. Possible ways of sensitizing the film by UV radiation

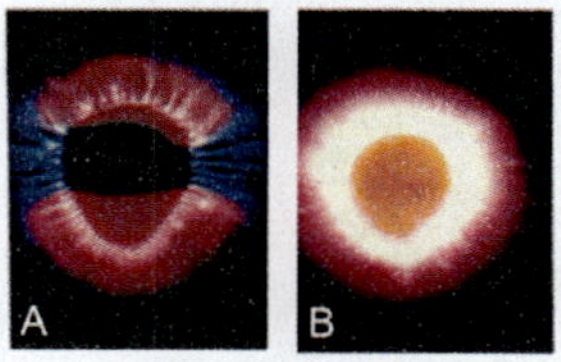

Fig. 8. Photos obtained with the standard Newton Milhomens film holder

2.1 Plastic Sheet Cut into Strips

A 0.3 mm thick plastic acetate sheet cut into stripes in the form of a flattened comb was introduced (Fig. 9) to force the existence of air in well-defined regions with known geometry, the fillets, and the channels.

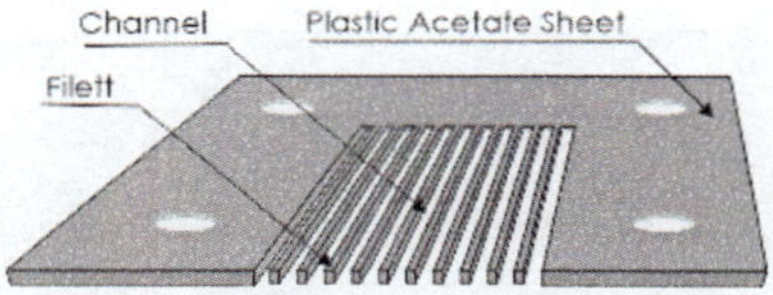

Fig. 9. Cut plastic acetate sheet

With this sheet, when applying high voltage to the set, the air in the channels under the film will be ionized, and consequently, the Corona will occur in these regions. In the fillet's regions, only the air on the film around the finger will be ionized (Fig. 10).

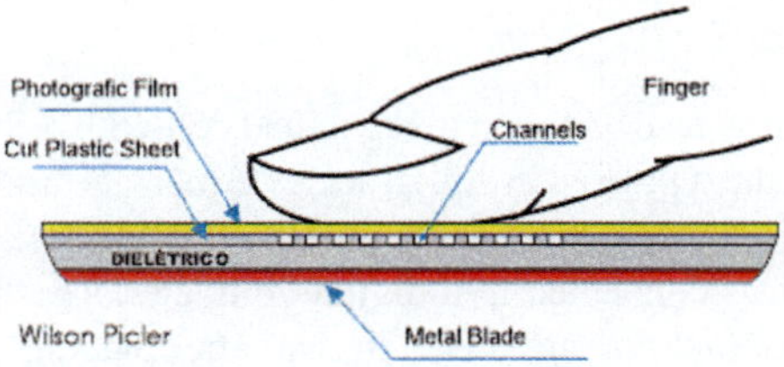

Fig. 10. Cut sheet inserted under the film

2.2 Compressor Frame

To avoid the formation of air gaps under the photographic film, a Compressor Frame was built, fixed by four Polyacetal screws. To prevent friction in the displacement of the film from scratching, the underside of the Frame was covered with felt (Fig. 11).

Figure 12 shows the Film Holder with the Compressor Frame attached using the white Polyacetal screws.

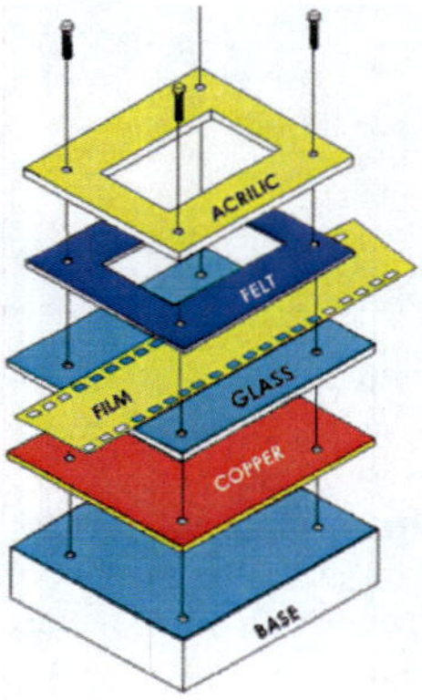

Fig. 11. Compressor frame

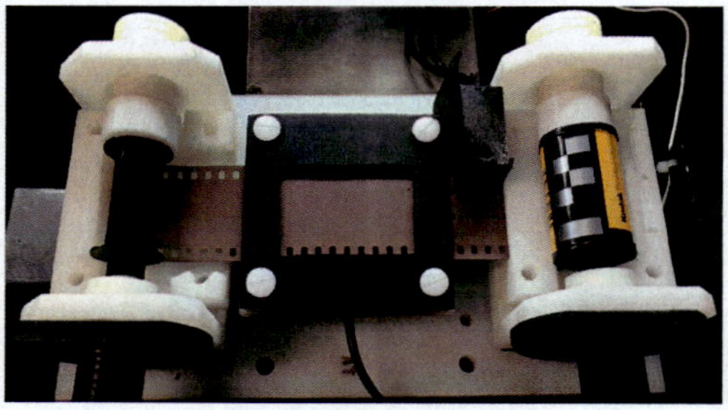

Fig. 12. Film holder with compressor frame installed

2.3 Dactylodynamometer

A Dactylodynamometer was developed using a load cell with 4 Strain Gages forming a Wheatstone Bridge, of the type used in digital scales to measure and monitor the force that the finger exerts on the photographic film. To amplify the cell signal was used an INA101 differential amplifier was connected to a digital voltmeter. The Dactylo-dynamometer is adjusted using an external potentiometer so that, after placing the photographic film, it displays a value of 0.00 V. To calibrate the scale, the amplifier gain was adjusted to display the value of 1.00 V when a weight of 1.00 kg was placed.

2.4 Digital Bioelectrography

To obtain digital Bioelectrographs, a Kirlionic Chamber was built in which the polarizing plate consists of a glass slide with a thin film deposit of ITO that makes the glass electrically conductive and transparent at the same time (Fig. 13).

The leading face is positioned downwards. The finger touches the top surface which is not conductive. The thickness of the glass is 1.5 mm which supports an insulation voltage of more than 30 kV. Special care is taken to insulate the sides of the glass with epoxy resins within a black acrylic frame.

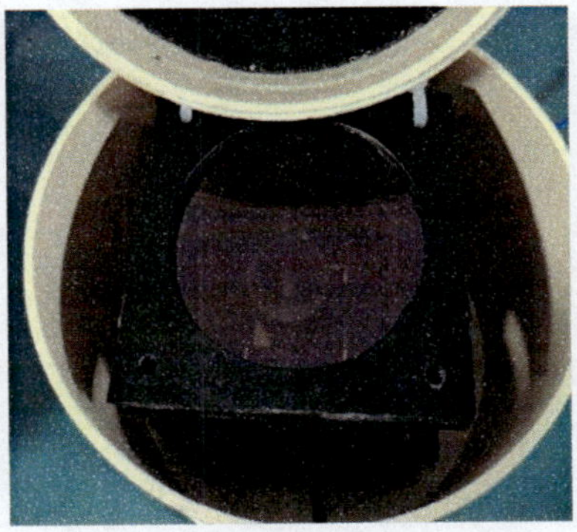

Fig. 13. Device for digital Kirliangraphy with CCD camera

To capture the Corona image, a CCD camera with a focus adjustment lens is positioned under the transparent polarizing plate (Fig. 14).

Fig. 14. Color USB CCD camera

3 Results

The experiments with the cut plastic sheet and the Compressor Frame were surprisingly successful.

3.1 Results with the Cut Plastic Sheet

The use of the Cut Plastic Sheet under the film resulted in Kirliangraphs with stripes in red (Fig. 15) coinciding where there was air only under the film, blue where there was air only on top of the film, and pink where there was air on both sides, constituting itself in an indisputable proof of the opposite sensitization hypothesis of the film.

The exposure time was increased to saturate the red layer and also to cause sensitization of the layer responsible for the green, causing the mixture of red and green. In this condition, the central stripes appeared in yellow/orange color (Fig. 16).

3.2 Result of Application of Compressor Frame

By applying the Compressor Frame, the air gaps under the film were eliminated resulting in a blue photograph (Fig. 17).

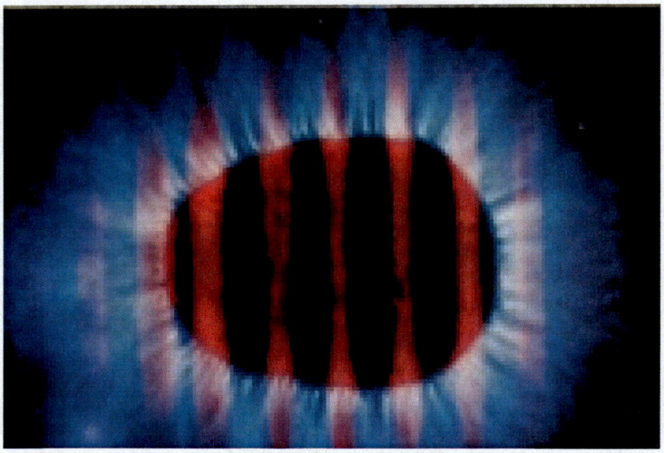

Fig. 15. KG with the cut plastic sheet inserted. 50 pulses of 10kV

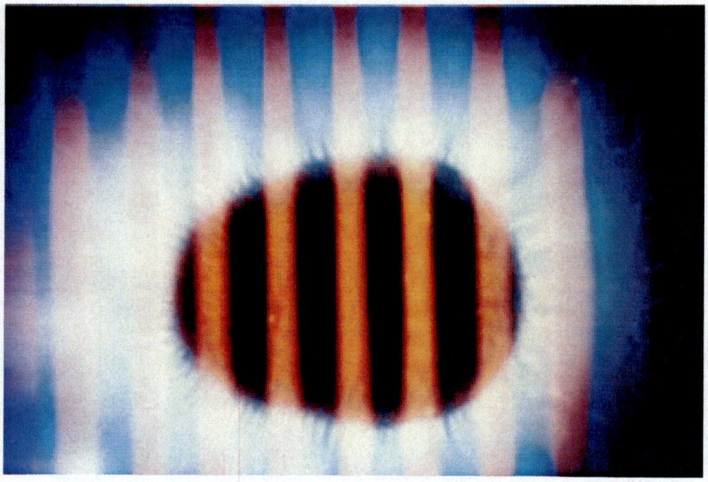

Fig. 16. KG with the cut plastic sheet inserted. 250 pulses of 10kV

3.3 Results with Different Strengths

Using the developed *Dactylodynamometer*, it was possible to experience different forces that the finger exerts on the photographic film with and without the Compressor Frame, applying a very light force with the finger just touching the film softly, Fig. 18 A and B and applying forces of 50 g-force (gf), Fig. 18 C and D.

3.4 Result with CCD Digital Camera

Using the chamber of Fig. 13, a Digital Bioelectrography was obtained (Fig. 19) by applying pulses to the polarizing plate with a repetition frequency of 2.7 kHz - Duty

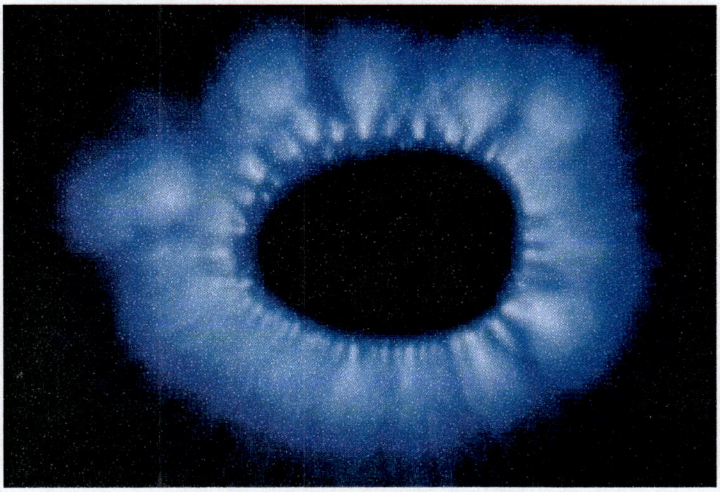

Fig. 17. KG with the compressor frame applied

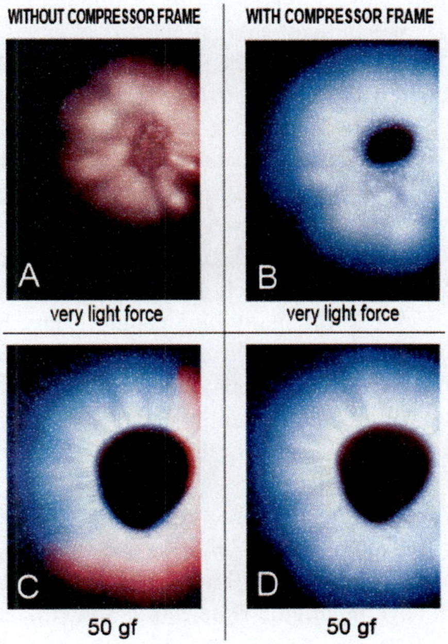

Fig. 18. KGs with and without compressor frame with different strengths.

Cycle 25%. The CCD Camera was programmed for an exposure time of 3 s, which resulted in a completely blue KG.

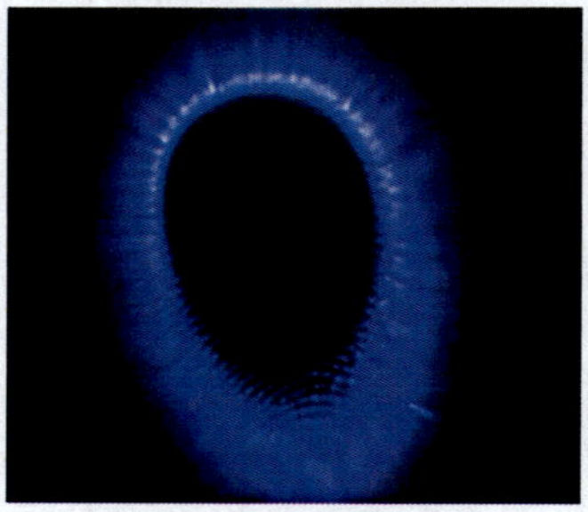

Fig. 19. Digital bioelectrography obtained with the device in Fig. 12.

4 Discussions

The experiments showed that when the Kirliangraphy are carried out with technical conditions that allow the formation of air gaps under the film, they are presented with red, pink, orange, and even yellow colors. This occurs because the electric field formed by the high voltage pulses permeates the entire dielectric and film set assembly and ionizes the air in the vicinity of the polarizing plate, producing the Corona effect, with strong emission of ultraviolet and violet radiation.

This emitted radiation is recorded by the photographic film, and when coming from the upper face of the film, it results in a violet/blue photo. When the radiation originates from Corona in the air gaps under the film it sensitizes the film from the opposite side, generating photos with red, orange, yellow, and pink colors. The experiments also showed that the force exerted by the finger on the film changes the shapes of Kirliangraphs and combined with the formation of air gaps under the film gives rise to a diversity of patterns of shapes and colors. When the air gaps under the film are eliminated using a Compressor Frame, the Kirliangraphs appear completely blue. The Digital Bioelectrography using the CCD Camera also resulted in a totally blue pattern. Thus, both results were plain compatible with the Corona emission spectrum.

Korotkov's GVD method does not analyze colors, it relates different regions of the fingers with other organs based on Traditional Chinese Medicine and measures the intensity of Corona at these points and with these attempts to identify patterns that can serve as parameters for clinical diagnoses.

There is a vast literature on the internet about the GVD technique invented by Korotkov [1], however, the author cannot infer any statement regarding the real applicability of GVD as a diagnostic method, however, states that it is necessary to have more academic research covering digital Bioelectrography to verify the veracity of the referred technique.

5 Conclusions

Based on the results and discussions, it can be said that the practice of clinical diagnosis using color analog photographic films with film holders devoid of resources to eliminate the air under the film constitutes a serious mistake and can be understood as quackery, pseudoscience, etc.

Unfortunately, many alternative professionals and their patients have been deceived for decades by this practice, precisely because they present these fascinating multicolored aspects that, as demonstrated in this work, are artifacts resulting from technical failures.

The results also showed that the GVD Technique that uses CCD or CMOS image sensors does not present the described chromatic aberrations, however, the GVD technique by Korotkov [1], should receive more interest in future academic research, with critical approaches, to confirm or refute claims of its applicability in clinical diagnosis.

Conflict of Interests. The author declares that there may be a conflict of interest with the IUMAB - International Union of Medical and Applied Bioelectrography [5] for having adopted Newton Milhomens' standard [5] as a reference in Brazil.

He also declares that he is a supporting partner at Uninter Educational Corporation and finances this research with his resources and in his private laboratory with the consent of Prof. Eduardo Tavares Costa, his Ph.D. advisor at FEEC/Unicamp.

Research Ethics Committees. The Project on which this work was based was approved by the Research Ethics Committees of Uninter and Unicamp under the numbers:

CEP/Uninter/CAAE 40277820.4.0000.5573
CEP/Unicamp/CAAE 40277820.4.3001.5404.

References

1. Korotkov, K.: Bio-well—Revolutionary Instrument to Reveal Energy Fields of Human and Nature. Bio-Well Company, Saint-Petersburg, Russia (2018). www.bio-well.com
2. Fujifilm Data Sheet AF3-008E: (s.d.). Fujicolor Superia 200. Fuji Photo Film Co. Ltd., Tokyo, Japan
3. Milhomens, N.: Fotos Kirlian: Como Interpretar. IBRASA, São Paulo (1988)
4. Boyer, D.G., Tiller, W.A.: The colors in Kirlian photography-fact or artifact? Funct. Photogr. **1**(1), 24–27, 38 (1976)
5. IUMAB: International Union of Medical and Applied Bioelectrography I. Protcolo do V Congresso Mundial da IUMAB. IUMAB, Curitiba (2000)

Influence of Exposure to Ultrasonic Waves for 5 min on Seeds of *Helianthus annuus* L.

R. A. N. Souza[1(✉)], I. O. Santos[1], L. E. Maggi[1], Z. P. V. Navarro[2], R. C. A. Franco[1], and M. C. Silva[1]

[1] Federal University of Acre, CCBN/Biophysics Laboratory, Rio Branco, AC, Brazil
souzaronald930@gmail.com

[2] Federal University of Acre, BIONORTE/Laboratory of Nanobiotechnology, Rio Branco, AC, Brazil

Abstract. This work evaluated the influence of ultrasound application on the germination and emergence of seeds of the species *Helianthus annuus* L., (sunflower), adopting some parameters such as Germination Percentage (%G), Emergence Percentage (%E), Germination Speed Index (GSI), Emergence Speed Index (ESI), Average Germination Time (AGT) and Average Emergence Time (AET). The following treatments were used: T1, T2 and T3 with 75 seeds each by 1 MHz frequency, and 0.5; 1.0 and 1.5 W/cm^2 of intensity respectively. In T4, T5 and T6 the same intensities were applied at a frequency of 3 MHz in 75 seeds in each treatment and Control Group - GC did not receive application. All applications followed 5 min in duration. After application, the seeds were immersed in aqueous solution for 10 min for disinfection and, later, washed in distilled water for the same period. They were immersed in distilled water with 2.5 times its weight for 24 h. They were sown one day after ultrasound application, conditioned in plastic bags with a 1:1 ratio of sand and clay. T1 performed better than the control group in G% = 60 and E% = 40. The 3 MHz groups were below CG for germination, except for AGT. As for emergence, they obtained better values in everything in relation to the CG. T2 was the worst treatment in all parameters and could be discarded in further studies. More experiments are needed with different doses and a greater number of seeds to obtain greater accuracy in the technique, which is widely explored.

Keywords: Ultrasound · Seed dormancy · Seedling production

1 Introduction

The species *Helianthus annuus*, popularly known as sunflower, is the fourth oilseed in the production of vegetable oils in the world, after soybean, palm, and canola. It stands out for its quality due to its physicochemical and nutritional characteristics [1]. This plant presents a series of biologically active compounds responsible for its pharmacological potential, due to its chemical composition for the defense against abiotic stresses such as light, temperature, water availability, nutritional stresses, defense against biotic agents such as predators and parasites [2].

J. L. B. Marques et al. (Eds.): CLAIB 2022/CBEB 2022, IFMBE Proceedings 98, pp. 572–579, 2024.
https://doi.org/10.1007/978-3-031-49401-7_59

Sunflower cultivation in Brazil was brought by the first european expeditions at the end of the 19th century to stimulate and expand its cultivation in different regions of the country. However, only in 2003, with the National Biodiesel Program, the sunflower returned to the list of oilseeds. Intended for human consumption and vehicular energy, their demand grew by a global average of 1.8% per year. The internal demand grows 13%, forcing Brazil to import from other countries, mainly from Argentina, demanding research and development actions aimed at the organization and expansion of the sunflower culture [3].

According to Venâncio (2019), dormancy is a problem in plant cultivation because it is a natural mechanism that helps its propagation in a natural environment, however, it prevents or delays germination, occurring in all seeds. There are seeds that are ready to germinate, but when they fall into the ground and encounter unfavorable conditions, they do not emerge, remaining viable, germinating only when they find favorable conditions [4, 5].

Among some physical and chemical methods for breaking seed dormancy, the application of ultrasound is more practical, safer and does not generate chemical residues [5]. Ultrasound waves act to break dormancy, applying stimuli in the form of mechanical pressure on the seeds, in addition to being free from contamination with chemical products [6].

This research evaluated the effect of therapeutic ultrasound application on germination and emergence of *Helianthus annuus* L. seedlings at different intensities and frequencies in sand and clay substrate in the proportion 1:1.

2 Materials and Methods

The seeds of *H. annuus* L. used in this work were from the brand Zaeli©, which are used to feed parrots. The application took place at the Federal University of Acre, Rio Branco campus, in the Laboratory Complex of the BIONORTE Network. A 50% aqueous solution of sodium hypochlorite was used for 10 min for disinfection and subsequently washed in distilled water for the same time. Seven groups with 75 seeds each were used, one of them being the Control Group - CG without ultrasound application (only immersion in distilled water for 10 min). The Test Group (TG) received a specific protocol with a Sonomed V therapeutic ultrasound device from Carci® for 5 min, frequency of 1 MHz, divided into T1, T2 and T3 treated at intensities of 0.5: 1.0 and 1.5 W/cm^2 respectively. The other treatments, T4, T5 and T6, were submitted to the same intensities mentioned above at a frequency of 3 MHz.

The application consisted of 12 seeds per application in a disposable cup with 100 mL of distilled water. Between the surfaces of the transducer and the bottom of the cup, gel was inserted for coupling, for better ultrasound transmission. See Fig. 1.

After application, the seeds were immersed in an aqueous solution of sodium hypochlorite at 50% for 10 min for disinfection and, subsequently, washed in distilled water for the same period. The seeds were immersed in distilled water with 2.5 times their weight for 24 h.

The seeds were sown one day after the application of ultrasound in the Zoobotanical Park of the Federal University of Acre in plastic packages in sand and clay substrate,

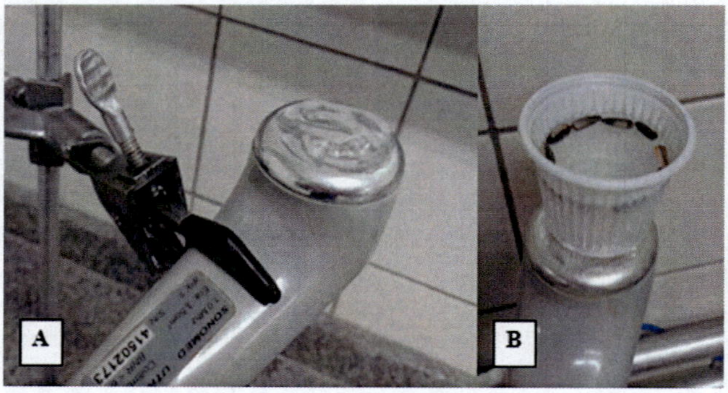

Fig. 1. A-Therapeutic ultrasound transducer with gel for contact coupling. B-Transducer and plastic container for seed application.

proportion of 1:1. They were monitored and watered daily in a nursery with a greenhouse at 50% insolation. The germination count was daily, from the individuals with cotyledons exposed above the substrate. Emergence, from seedlings with true leaves.

The first count was carried out on October 26th, 2021, three days after application, running for another 24 days after planting. The values were tabulated and then the parameters were calculated: Germination Percentage (%G), Germination Speed Index (GSI) [7] and Average Germination Time measured in days (AGT) [8].

%G is calculated using the Eq. (1).

$$G = (\mathrm{N}/75) \times 100 \quad (1)$$

where: N is the number of seeds germinated at the end of the test, divided by the number of seeds sown.

The Germination Speed Index (GSI) and the Average Germination Time (AGT) were calculated by (2) and (3).

$$\mathrm{GSI} = (\mathrm{G1}/\mathrm{N1}) + (\mathrm{G2}/\mathrm{N2}) + (\mathrm{G3}/\mathrm{N3}) + \ldots + (\mathrm{Gn}/\mathrm{Nn}) \quad (2)$$

G1, G2, G3,..., Gn = number of seedlings computed in the first, second, third and last counts.

N1, N2, N3,..., Nn = number of days from sowing to the first, second, third and last counts.

$$\mathrm{AGT} = \sum (ni.ti) \,/\, \sum ni \quad (3)$$

In which:

ni represents the number of seeds germinated in the interval between each count.

ti = incubation time elapsed between the beginning of germination and the i-th count in days divided by the sum of germinated seeds per day.

Emergency percentage (%E), Emergency Speed Index (ESI) [7] and Average Emergency Time (AET) [8] were calculated by (4), (5) and (6).

$$\mathrm{E} = ((\mathrm{ni}/\mathrm{N}))/100 \quad (4)$$

where ni represents the number of seeds emerged at the end of the experiment and N represents the total number of seeds sown.

$$ESI = (E_1/N_1) + (E_2/N_2) + \ldots (E_N/Nn) \quad (5)$$

where E is the number of emerged seeds recorded in the first, second and last counts, respectively. N1, N2 and Nn are the number of days after sowing in the first, second and last counts.

$$AET = \left(\sum [(ni.ti)]\right)/\left(\sum ni\right) \quad (6)$$

where i is the number of seeds emerged per day, ni is the number of seeds emerged in the i-th observation and ti is the number of days after sowing in the i-th observation.

3 Results and Discussion

3.1 Germination

The interval between sowing and first germination was two days. Germinations were counted daily in a period that started on October 26th, 2021, until November 8th, 2021.

As for the Germination Percentage (%G), there were sharp differences between treatments. The treatment that presented the highest %F was T1, with approximately 5% of germinations above the control group. T2 showed the lowest percentage of germination, presenting 40% less germination in reference to the control group.

Regarding the GSI, no group presented a better result than the control group. For AGT, T2 and T3 showed a small difference, with T2 being higher and T3 being lower.

In Fig. 2, T1 presented a higher performance in relation to the control group. T2 showed the lowest values. The %G result, in general, did not present satisfactory data for the use of the ultrasound technique, except for T1, which exceeded the %G of the control group with an increase of 5.33%. This fact corroborates the work of Machikowa et al. (2013) [9] where effects of ultrasonic treatment, at different intensities and in a time of 5 min with *Helianthus annuus* L. seeds, resulted in an increase of 5.88% in the germination percentage.

In relation to GSI (Fig. 3), the control group expressed the highest velocity and T1 was, among the treatments, the values closest to the CG, and the highest among the treatments.

GSI and AGT results can be related to Cipriani et al. (2018) [10], who used several seeds of forest importance in their work and none of them, with the influence of an ultrasonic bath, positively or significantly altered the results of GSI and AGT.

Regarding the AGT (Fig. 4), there was little variation within all the treated groups, with T5 showing the best results, obtaining the lowest AGT value. T2, on the other hand, presented the worst performance, taking an average of 20, 36 days to germinate.

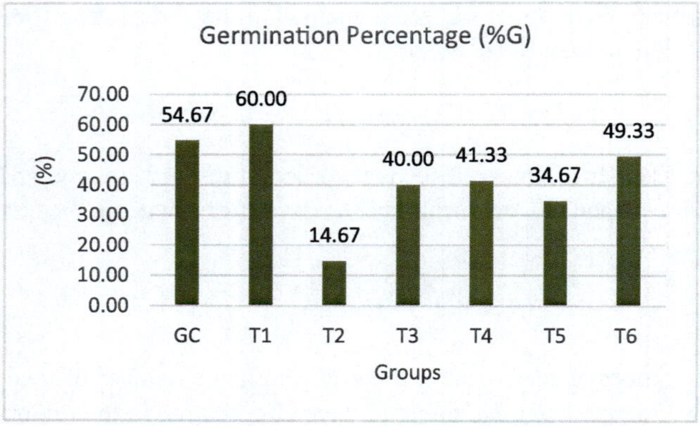

Fig. 2. Germination percentage (%G) depending on the groups.

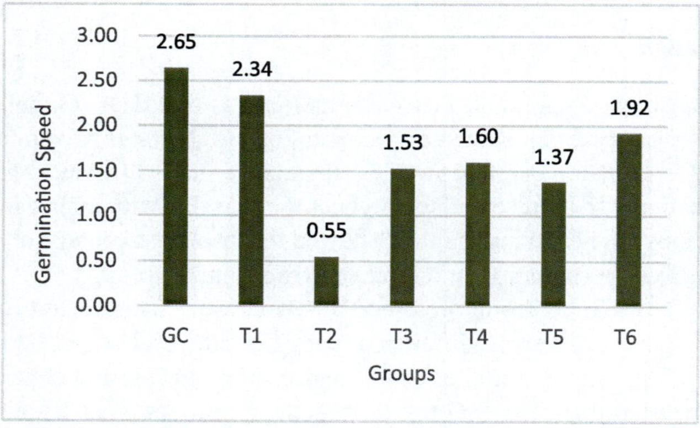

Fig. 3. Germination speed index depending on the groups.

3.2 Emergency

AET showed average variation within treatments. There was a very similar pattern of %E (Fig. 5) with %F (Fig. 2), however, in %E, T6 was better in relation to CG. T1 was better again, using the same parameters presented by Santos et al. [11] of 1 MHz, 0.5 W/cm^2 for 5 min and having the highest %E of the group. However, using seeds of *Copaifera reticulata* D. (Fabaceae). This time T1, with a higher difference in relation to the CG and the other treatments, presented 12% more in emergencies compared to the control group.

For emergence speed (Fig. 6), again we noticed the same pattern seen before, T1 presented the best treatment, while T2 had very low patterns, T6 repeats the pattern seen in %E, showing a small difference regarding CG. The ESI could result in more accurate data, similar to those of Neto et al. (2019) [12], if the experiment were carried out in a more controlled environment (in terms of temperature, predation, humidity etc.).

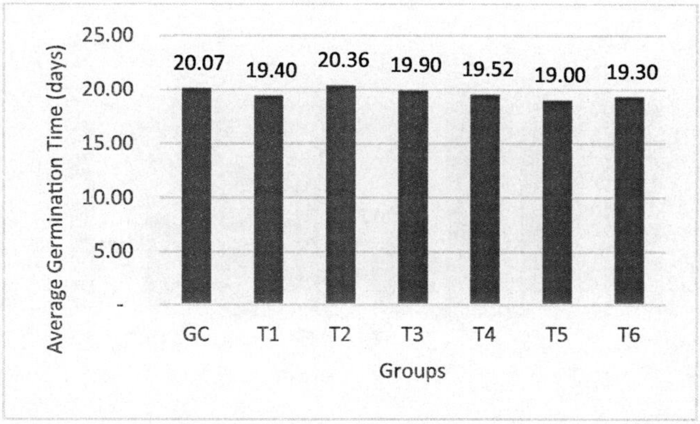

Fig. 4. Average Germination Time depending on the groups.

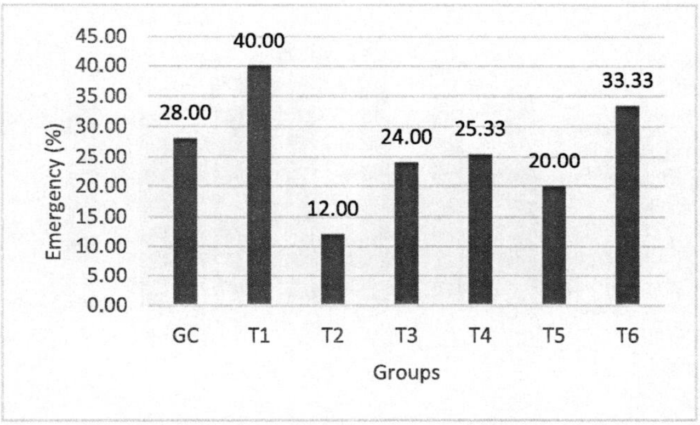

Fig. 5. Percentage of Emergencies according to groups.

As for AET (Fig. 7), as well as in AGT (Fig. 4), there were minimal variations, and for this variable, T5 presented better results in relation to CG, with the worst treatments being T3, T2 when compared to CG.

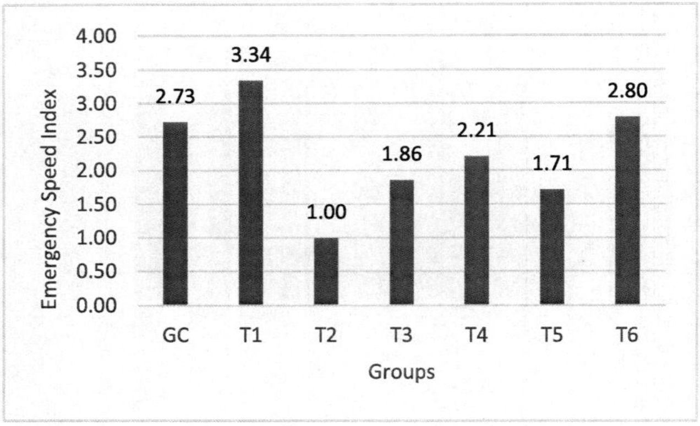

Fig. 6. Emergency speed index depending on the groups.

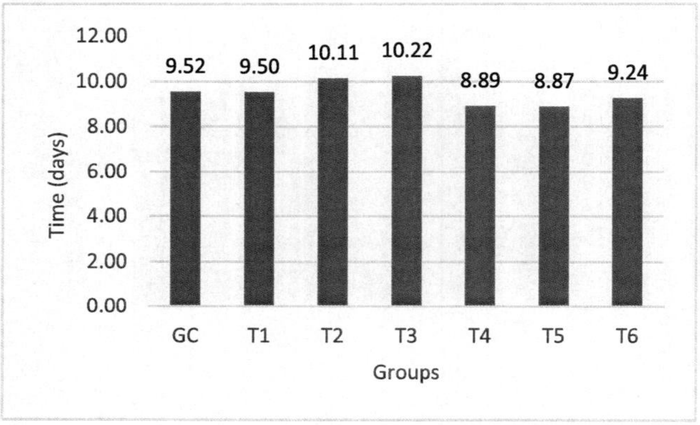

Fig. 7. Mean time to emergence depending on the groups.

4 Conclusion

At the end of this work, it became possible to state that it is not recommended to use 1 MHz, with an intensity of 1 W/cm^2. The use of 3 MHz waves did not show good results in 5 min for germination, except in AGT, but one treatment in particular, (1 MHz–0.5 W/cm^2–5 min) - T1 was better than CG in all variables, except in GSI, and in AGT. For emergence, better values were obtained in all parameters in relation to CG. It was not possible to obtain mean, standard deviation or implement any statistical test with the current experiment as it was performed only once. For this, it would be necessary to carry out the experiment in triplicate. As this depends on the amount of viable seeds. Biochemical and physiological analyzes should be performed to explain the biological mechanisms that support the use of ultrasound (in the amplitude and frequency range used in this study) to break the dormancy of sunflower seeds.

Acknowledgement. The authors thank the Brazilian agencies CAPES, CNPq.

Conflict of Interest. The authors declare that they have no conflict of interest.

References

1. Campo, C.B.H., Saraiva, O.F.: resultados de pesquisa da EMBRAPA soja - 2001 Girassol e Trigo. 1st edn. Embrapa, Londrina-PR (2002)
2. Bashir, T., Mashwani, Z.U.R., Zahara, K., Haider, S., Mudrikah, S.T.: Chemistry, pharmacology and ethnomedicinal uses of *Helianthus annuus* (Sunflower): a review. Pure Appl. Biol. **4**(2), 226–235 (2015)
3. Ungaro, M.G.R., Castro, C., Farias, J. B.R.: GIRASSOL. In: Girassol http://www.alice.cnptia.embrapa.br/alice/handle/doc/579641. Last accessed 2022/05/28
4. Venâncio, R.S.S., Martins, A.C.G.: Overcoming dormancy of *Senna multijuga* seeds with an ultrasonic probe the comparison with ultrasound and sulfuric acid baths. Ciencia Rural **49**(9), 1–7 (2019)
5. Mori, E.S., Rodrigues, F.C.M.P., Freitas, N.P.: Sementes florestais: Guia para germinação de 100 espécies nativas. Instituto Refloresta. https://www.researchgate.net/publication/236596314_Sementes_florestais_guia_para_a_germinacao_de_100_especies_nativas. Last accessed 2022/05/28
6. Nazari, M., Eteghadipour, M.: Impacts of ultrasonic waves on seeds: a mini-review. Agric. Res. Technol. **6**(3), 1–5 (2017)
7. Maguire, J.D.: Speed of germination-aid in selection and evaluation for seedling emergence and vigor. Crop Science. https://doi.org/10.2135/cropsci1962.0011183X000200020033x. Last accessed 2022/05/29
8. Labouriau, L.G.: A germinação das sementes. Secretaria Geral da OEA (1983)
9. Machikowa, T., Kulrattanarak, T., Wonprasaid, S.: MACHIKOWA. Word Acad. Sci., Eng. Technol (2022). https://medicalexposedownloads.com/PDF/effects-of-ultrasonic-treatment-on-germination-of-synthetic-sunflower-seeds.pdf. Last accessed 2022/05/29
10. Cipriani, H.N., Vieira, A.H., Passos, A.M.A., Baptista, I.S., Júnior, G.M. Anais 4° Encontro Brasileiro de Silvicultura, in Anais 4° Encontro Brasileiro de Silvicultura (2022). https://www.alice.cnptia.embrapa.br/handle/doc/1092957. Last accessed 2022/05/29
11. Santos, I.O., et al.: Use of ultrasound in the emergency and initial growth of *Copaifera reticulata Ducke* (Fabaceae). IFMBE Proc. **83**, 2385–2389 (2020)
12. Neto, S.M., Correia, E.C.S.S., Ferreira, J.C.A., et al.: Germinação e crescimento inicial do girassol (*Helianthus annuus* L.) sob concentração salina," in 8ª Jornada Científica e Tecnológica da Fatec de Botocatu (2022). http://www.jornacitec.fatecbt.edu.br/index.php/VIIIJTC/VIIIJTC/paper/view/1956. Last accessed 2022/06/09

Impact of Social Reaction on the Dynamics of Epidemics

Helio Schechtman[1](✉) and Max O. Souza[2]

[1] Programa de Computação Cientfica, Fundação Oswaldo Cruz, Rio de Janeiro, Rio de Janeiro, Brazil
helio.schechtman@fiocruz.br

[2] Instituto de Matemática e Estatística, Universidade Federal Fluminense, Niterói, Rio de Janeiro, Brazil

Abstract. The effects of social reaction in an epidemic were investigated by means of in-silico experiments using the classical S(usceptible), I(nfected), R(emoved) mathematical model, but assuming a periodic function for the infection rate. This choice is known to be prone to counter-intuitive results: There is an increase in the total number of infected people over the course of the epidemic—the so-called final size of the epidemic. On further investigation, it was observed that this effect was associated with a decrease in the maximum number of infected individuals at any given time. For a disease that draw on resources of the public health sector, the latter behaviour is a desired one since it prevents the health system to succumb to an excess in diseased individuals due to the epidemic. Moreover, it was also noticed that the magnitude of both effects was seasonally dependent on the moment of the initial infection. Overall, these results suggest that modelling non-pharmaceutical interventions requires a more consistent linking between theory and epidemiological data.

Keywords: Epidemics · Non-pharmaceutical interventions · SIR

1 Introduction

Mathematical modelling of the dynamics of diseases is one of the main tools to help with the decisions made by public health authorities [1,2]. This is the case for human pathogens [2–5] and also for veterinary diseases—particularly those of economic importance such as foot and mouth disease [6]. In this vein, the use of dynamical systems are ubiquitous and the subject of textbooks on Mathematical Biology, for example [7,8], and research monographs [9]. More complex models might include stochastic effects, but in this work we shall restrict ourselves to determinist models.

Modelling an infectious disease in a given population demands many assumptions, some explicit and many others implicit. Three of such assumptions are that

J. L. B. Marques et al. (Eds.): CLAIB 2022/CBEB 2022, IFMBE Proceedings 98, pp. 580–589, 2024.
https://doi.org/10.1007/978-3-031-49401-7_60

there are no births nor deaths nor migration, incoming or outgoing, of individuals in that given well-mixed population (which by itself is a fourth assumption). Probably the simplest way for representing the time evolution of an infectious disease with acquired immunity in that population is known as the SIR model. This model states that a susceptible individual, when encountering an infectious individual might become infected with a certain given probability, and this now infectious individual will also evolve to a lifelong recuperated state impervious to that infection.

Therefore, the population might be considered of comprising three compartments. A compartment made up of S(usceptible) individuals, a second one containing the I(nfectious) individuals and a third compartment consisting of the R(ecovered) individuals, and the total number of individuals in the population is N.

Considering the fraction of individuals of each compartment, $s = S/N$, $i = I/N$, and $r = R/N$, the basic model is then given by the following equations:

$$\begin{cases} \dfrac{ds}{dt} = -\beta si \\ \dfrac{di}{dt} = \beta si - \gamma i \\ \dfrac{dr}{dt} = \gamma i \end{cases} \tag{1}$$

where ds/dt is time derivative of the fraction of susceptible individuals; di/dt is the time derivative for the fraction of infectious individuals; dr/dt is the time derivative for the fraction of recovered individuals; β (in fact $\beta(t)$) is the infection rate; and γ (again, in fact $\gamma(t)$) is the recovery rate (or the inverse of the infectious period) [10]. Obviously, if we multiply both sides of the equations by N we obtain the number of individuals in each compartment.

Even though the model is a relatively simple one, the term βsi is quite complex in its meaning. The term si might be understood as the probability of encounters of S(usceptible) individuals with I(infectious) individuals, assuming that the population is well mixed. The term si multiplied by N then yields the expected value of potentially infective encounters, *i.e.* encounters of S(usceptible) with I(infectious) individuals.

Having thus explained in this modelling effort the term si, the parameter β remains to be described in its effects. β in reality comprises two modulating factors. As explained by Weiss [11], suppose that each infectious individual has k contacts (each sufficient for transmission of the disease) per unit time and that k is independent of the population size. Then kS/N of these contacts are with susceptible individuals. If the fraction τ of adequate contacts results in transmission of the disease, then each infectious individual infects $k\tau S/N$ susceptible individuals per unit time. The parameter τ is, naturally, called the transmissibility of the infectious disease. In summary, one factor reflects the strength of the individual's social interaction whilst the other incorporates the transmissibility of the disease, and thus $\beta = k\tau$.

1.1 General Coefficients

For any given initial condition at $t = 0$, denoted by $(s_0, i_0, r_0) \geq 0$ and subjected to $s_0 + i_0 + r_0 = 1$, it is a standard result in the theory of ordinary differential equations [12] that there exists a unique solution for all positive time satisfying $(s(t), i(t), r(t)) \geq 0$ and $s(t) + i(t) + r(t) = 1$.

From this result, we can easily conclude that *every disease governed by Equation* (1) *will die out* after sometime.

More precisely, we have that $\lim_{t\to\infty}(s(t), i(t), r(t)) = (s^*, 0, r^*)$. The quantity $r^* - r_0$ is named the *final size of the epidemic*, since it comprises all individuals that have being infectious at some positive time. Notice that the results above do not take into account any information on the function β and γ.

1.2 Basic Reproduction Number

An epidemic will occur if the number of infected individuals increases, *i.e.*, $di/dt > 0$. However for that to happen, $\beta si - \gamma i > 0$ or $(\beta/\gamma)si > i$. At the outset of an epidemic, nearly everyone (except the index case) is susceptible. So it can be assumed that $s \approx 1$, and the following inequality $\mathcal{R}_0 = (\beta/\gamma) > 1$ holds. The term $\mathcal{R}_0$ is the expected number of secondary infections in a population with initially a single infected individual and is named the *basic reproduction number*—cf. [13–15].

The formal mathematical approach would be to look at the next generation matrix. *i.e.* the square matrix G in which the jk^{th} element of G, g_{jk} is the expected number of secondary infections of type j caused by a single infectious individual of type k, again assuming that the population of type j is entirely susceptible. That is, each element of the matrix G is a reproduction number, but one where who infects whom is accounted for. The basic reproduction number is given by the spectral radius of G, or the dominant eigenvalue of G. The next generation matrix has a number of desirable properties from a mathematical standpoint. In particular, it is a non-negative matrix and, as such, it is guaranteed that there will be a single, unique eigenvalue which is positive, real, and strictly greater than all the others; this is the $\mathcal{R}_0$.

1.3 Constant Infectivity

This is the case of constant β, studied by Ross and Hudson [16] and Kerman and McKendrick [17], which is by now classical textbook material on mathematical epidemiology, e.g. [7–9].

For this brief analysis, we will follow Weiss [11]:

- The transcendental formula for r^*:

$$(1 - r^*) \exp(\mathcal{R}_0 \frac{r^* - r_0}{1 - r_0}) = 1 - i_0 - r_0.$$

- The epidemic threshold theorem:
  Let $\mathcal{R}_e = s_0\mathcal{R}_0$—this the so called effective basic reproductive number. If $\mathcal{R}_e \leq 1$, then $i(t) \downarrow 0$ at $t \to \infty$. If $\mathcal{R}_e > 1$, then $i(t)$ is initially monotonically increasing, reaches a maximum, and then it is monotonically decreasing. The scenario when $\mathcal{R}_e > 1$ is called an epidemic.

These two results are the basis of public health management since the beginning of the twentieth century up to now. The threshold theorem shows that an epidemic can be avoided (or at least mitigated) if appropriate actions are taken to reduce $\mathcal{R}_e$. These include:

- Reduce the susceptible base s_0—vaccination.
- Reduce the infection rate $\beta(t)$ by either reducing the disease transmissibility (use of masks for airborne diseases) or by reducing the contact rate (suspension of some activities or more severe lockdowns).
- Increase the removal rate—efficient medical care with quick identification and isolation of infected individuals.

Furthermore, the transcendental formula implies that r^* is monotonically increasing as a function of $\mathcal{R}_0$—and it is also monotonically increasing as a function of i_0. Thus, for a given initial infection, the fraction of the population that will be eventually infected will decrease if $\mathcal{R}_0$ is reduced. Moreover, it can also be shown that the peak size is a monotonic function of $\mathcal{R}_0$. These two observations combined show that decreasing $\mathcal{R}_0$ will reduce the overall burden of the disease in a given population.

1.4 Periodic

As soon as $\beta(t)$ is no longer constant the disease dynamics becomes richer and far less intuitive. For the periodic case, the corresponding results are more recent—twenty first century—and are due to Bacaer and Gomes [18].

In order to provide a clear formulation, we first state the weak threshold theorem for the constant $\beta(t)$ case.

- If $\mathcal{R}_0 < 1$, then
$$r^* - r_0 \to 0 \text{ as } i_0 \to 0$$
- If $\mathcal{R}_0 > 1$, then
$$r^* - r_0 \geq (1 - r_0)\left(1 - \frac{1}{\mathcal{R}_0}\right) \quad \text{for all } 0 < i_0 < 1 - r_0$$

The above result can be interpreted in the following sense: when $\mathcal{R}_0 < 1$ then an effective policy of epidemiological surveillance and proper quarantine procedures will maintain the disease controlled. On the other hand, when $\mathcal{R}_0 > 1$, a single infected that slip into the population will give rise to a significant number of cases bounded below by the right hand side of the inequality above.

Consider, as stated by Thieme [19], $\beta(t)$ to be a periodic function of period τ and let

$$\mathcal{R}_0 = \frac{1}{\gamma(t)\tau} \int_0^\tau \beta(t)\,\mathrm{d}t.$$

The above expression for $\mathcal{R}_0$ turns out to be the correct one for the periodic case inasmuch as it represents the expected number of secondary infections in a population with initially a single infected individual. Notice also that it reduces to the classical case when $\beta(t) = \beta$.

So far, the only threshold result available for the periodic case [18] is akin to the weak form discussed above for constant β. Notice, however, that no monotonicity result is now available. Indeed, numerical simulations by those authors show that r^* is no longer a monotonic function of $\mathcal{R}_0$ and, moreover, that its behaviour can be quite sensitive to this value. It should be also pointed out that an analytical verification of these results is yet to be obtained.

Bearing in mind that monotonicity is inexistent, the epidemiological parameters r^* and the maximum value of $i(t)$ are quite sensitive to small disturbances in the model parameters, and might display complex and counter-intuitive behaviour. The authors revisited the results of Bacaer and Gomes [18] and numerically investigated the effects of small disturbances on the initial parameters t_0/τ and R_0.

1.5 Aims and Objectives

Models with non constant β can exhibit dynamics that are quite counter-intuitive. Indeed this is the case of several models for non-pharmaceutical intervention - e.g. [5,20].

Therefore, the aim of this work is to revisit the very basic SIR model with time-periodic coefficients and bring attention to the fact that its dynamics are even more complex and poorly understood than some of the results in current literature might suggest. Furthermore, the authors wish to highlight that extreme care should be exerted when attempting to infer from the results of such a model.

2 Materials and Methods

The **SIR** model was implemented and simulated computationally using the Python programming language version 3.8.2. For these *in silico* experiments, we used the same choice of parameters as in [18], namely $\beta(t) = \bar{\beta}(1 + \frac{1}{2}\sin 2\pi t)$ and $\gamma^{-1} = \frac{1}{52}$, with time measured in years. Such a choice for β can ben understood as a truncation of the Fourier series of a possibly more complex periodic function. The initial condition was also always chosen as $(1 - 10^{-3}, 10^{-3}, 0)$ — with $r_0 = 0$, which are appropriate for modelling an emerging disease as it was the case of the initial epidemic of COVID-19, for instance. This initial condition was enforced at different choices of $\frac{t_0}{\tau}$.

3 Results and Discussion

One of the striking results in [18] is the lack of monotonicity of r^* as a function of $\mathcal{R}_0$. In Fig. 1 we display one of the simulations in [18] with additional data on the infected individuals. These results confirm the absence of monotonicity of r^* as a function of $\mathcal{R}_0$, but also shown the surprising result that the peak dynamics of infectious individuals can be more subtle: in this example, while the final size increases with a decreasing $\mathcal{R}_0$, the single peak that appears with $\mathcal{R}_0 = 2.5$ case splits into two significantly smaller ones, when $\mathcal{R}_0 = 2$. Thus, while there is a larger fraction of the population that gets infected over time in the former case, for a disease that might likely lead to hospitalisation or even the need of ICU beds, the pressure on the health system is much smaller in the latter case.

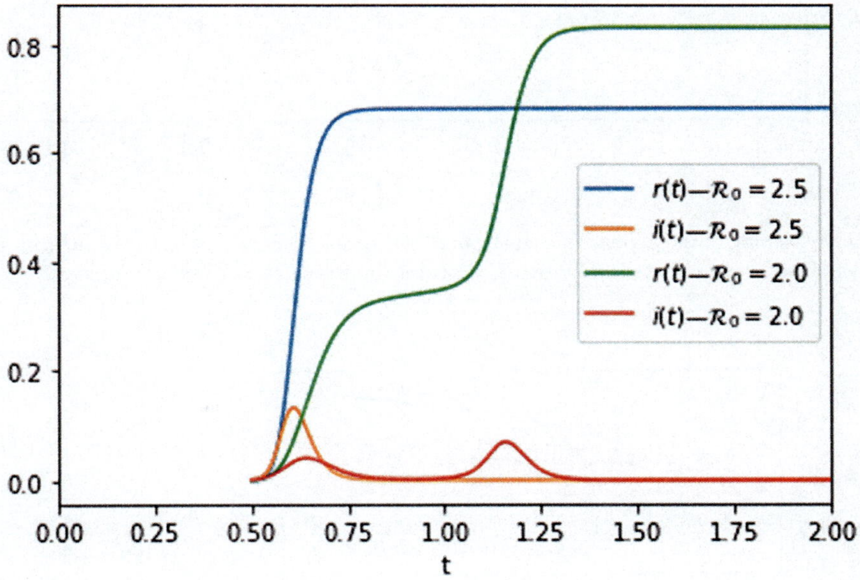

Fig. 1. Display of the time evolution of the fraction of infected, $i(t)$, and recovered individuals, $r(t)$ for two values of the basic reproduction number, $\mathcal{R}_0$—2.0 and 2.5.

The previous results raise the question whether this behaviour is a prevalent one. To improve our understanding we broad our investigation and study the effect of the initial time of infection, *i.e.* appearance of the first infected individual, in these two quantities of interest to public health management: the final size of the epidemic and the maximum peak of infected individuals. Figure 2 displays the behaviour of these quantities as a function of $\frac{t_0}{\tau}$ for $\mathcal{R}_0 \in \{2, 2.5\}$: these results show that there are two different windows where the expected monotonicity behaviour is inverted. Interesting enough, these windows overlap around $\frac{t_0}{\tau} = \frac{1}{2}$.

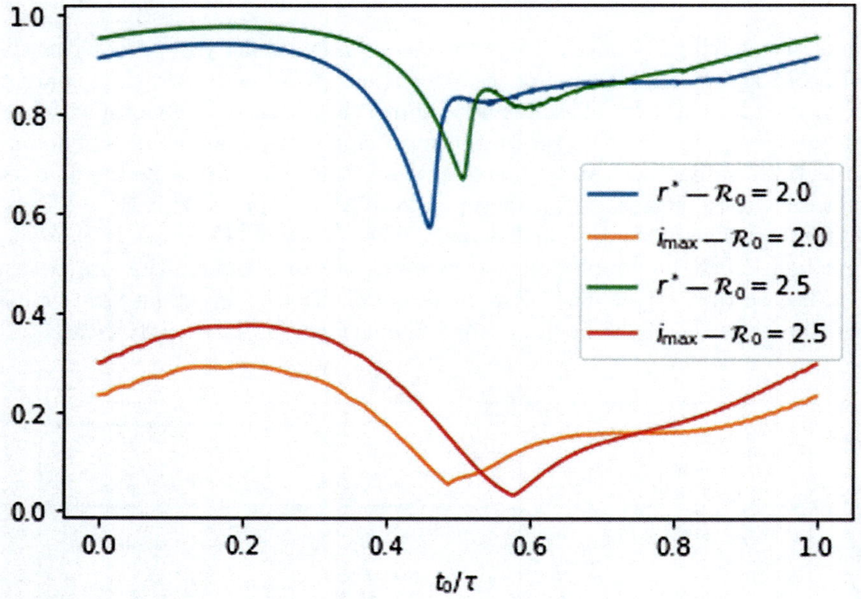

Fig. 2. Display of the dependence of the final size of the epidemic, r^*, and the maximum of the fraction of infected individuals, $i_{\max}$, on the initial time of infection, $\frac{t_0}{\tau}$.

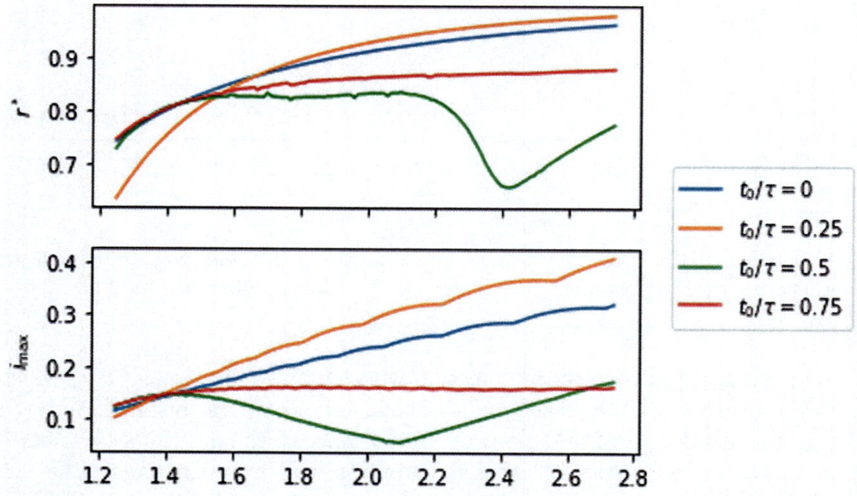

Fig. 3. Display of the dependence of the final size of the epidemic, r^*, and the maximum of the fraction of infected individuals, $i_{\max}$, on the basic reproduction number, $\mathcal{R}_0$, for four different values of $\frac{t_0}{\tau}$.

Finally, Fig. 3 studies the dependence on R_0 for four different values of $\frac{t_0}{\tau}$ that can be understood as the four different seasons in the year. These results show an interesting behaviour: for the seasons that might be considered favourable to the disease, *i.e.* autumn, spring and winter, the behaviour of these quantities retain the expected monotonic behaviour. On the other hand, for the season that is unfavourable to the disease, *i.e.* summer, this behaviour is more complex. Notice that for r^*, there is a strongly non-monotonic behaviour when $\frac{t_0}{\tau} = \frac{1}{2}$, whereas for $\frac{t_0}{\tau} = \frac{3}{4}$ monotonicity is retained although very weakly. For $i_{\max}$ this dependence is essentially constant in the case of $\frac{t_0}{\tau} = 0.75$, where once again there is a strongly non-monotonic behaviour in the case of $\frac{t_0}{\tau} = 0.5$.

4 Conclusion

We have numerically investigated the dynamics of two important quantities for public health management: r^* and $i_{\max}$, the former being the final size of the epidemic and the latter the maximum peak of infected individuals. In the classical case the monotonic dependence on $\mathcal{R}_0$ of both quantities suggest a clear course of action in order to reduce the burden of an epidemic. However, the dynamics of a seasonal disease can be much more complex, involving non-monotonic behaviour occurring for both quantities, and typically in different parameter windows. Moreover, the time where the infection starts can significantly impact the development of a particular epidemic.

The results presented above indicate that enforcing measures that reduce the infectivity rate—social distancing and use of masks for instance—thus reducing the $\mathcal{R}_0$ might induce contradictory effects, *i.e.* an increase in the final size of the epidemic—r^*, vis-a-vis a reduction on the maximum fraction of the population that gets infected over time, *i.e.* multiple peaks for the fraction of infected individuals $i(t)$. The latter is a desired effect as it reduces the pressure on the health system as less infected individuals will need hospitalisation or even ICU beds at any specific time throughout the epidemic.

Additionally, it might be concluded that small errors on the estimate of the initial time of infection (t_0) will affect quite significantly the outcome of the final size of the epidemic, r^*. Therefore, extra care has to be taken when modelling an infectious disease with a time dependent infection rate $\beta(t)$.

These results suggest that additional care should be exerted when modelling non-pharmaceutical interventions. Also, a more consistent link of the modelling and epidemiological data should be established. In addition, the possible dynamical outcomes should be further explored in order to yield a comprehensive understanding of the influence of such interventions.

Acknowledgments. HS and MOS were partially financed by Coordenação de Aperfeiçoamento de Pessoal de Nível Superior - Brasil (CAPES) - Finance code 001. MOS was also partially financed by CNPq (grant # 310293/2018-9) and by FAPERJ (grant # E-26/210.440/2019).

Conflict of Interest. The authors declare that they have no conflict of interest.

References

1. Lo, N.C., Andrejko, K., Shukla, P., Baker, T., Sawin, V.I., Norris, S.L., Lewnard, J.A.: Contribution and quality of mathematical modeling evidence in World Health Organization guidelines: A systematic review. Epidemics **39**, 100570 (2022). https://doi.org/10.1016/j.epidem.2022.100570
2. McKenzie, F.E., Samba, E.M.: The role of mathematical modeling in evidence-based malaria control. Am. J. Trop. Med. Hyg. **71**(2 Suppl), 94–96 (2004). https://doi.org/10.4269/ajtmh.2004.71.94
3. Chretien, J.P., Riley, S., George, D.B.: Mathematical modeling of the West Africa Ebola epidemic. eLife **4**, e09186 (2015). https://doi.org/10.7554/eLife.09186
4. Poletti, P., Ajelli, M., Merler, S.: The effect of risk perception on the 2009 H1N1 pandemic influenza dynamics. PLOS ONE **6**(2), e16460 (2011). https://doi.org/10.1371/journal.pone.0016460
5. Liu, Z., Magal, P., Seydi, O., Webb, G., Liu, Z., Magal, P., Seydi, O., Webb, G.: Predicting the cumulative number of cases for the COVID-19 epidemic in China from early data. Math. Biosci. Eng. **17**(4), 3040–3051 (2020). https://doi.org/10.3934/mbe.2020172
6. Keeling, M.J.: Models of foot-and-mouth disease. Proc. R. Soc. B **272**(1569), 1195–1202 (2005). https://doi.org/10.1098/rspb.2004.3046
7. Murray, J.D.: Mathematical Biology: I. An Introduction, 3rd edn. Springer, New York (2007)
8. Britton, N.F.: Essential Mathematical Biology. Springer, London (2003)
9. Diekmann, O., Heesterbeek, J.A.P.: Mathematical Epidemiology of Infectious Diseases: Model Building, Analysis and Interpretation. Wiley (2000)
10. Ridenhour, B., Kowalik, J.M., Shay, D.K.: Unraveling R0: considerations for public health applications. Am. J. Public Health **104**(2), e32–e41 (2014). https://doi.org/10.2105/AJPH.2013.301704
11. Weiss, H.: The SIR model and the foundations of public health. Mater. Matematics **2013**(3) (2013)
12. Teschl, G.: Ordinary Differential Equations and Dynamical Systems, Graduate Studies in Mathematics, vol. 140. American Mathematical Society, Providence (2012)
13. van den Driessche, P., Watmough, J.: Reproduction numbers and sub-threshold endemic equilibria for compartmental models of disease transmission. Math. Biosci. **180**(1), 29–48 (2002). https://doi.org/10.1016/S0025-5564(02)00108-6
14. Diekmann, O., Heesterbeek, J.A.P., Metz, J.A.J.: On the definition and the computation of the basic reproduction ratio R0 in models for infectious diseases in heterogeneous populations. J. Math. Biol. **28**(4), 365–382 (1990). https://doi.org/10.1007/BF00178324
15. Diekmann, O., Heesterbeek, J.A.P., Roberts, M.G.: The construction of next-generation matrices for compartmental epidemic models. J. R. Soc. Interface **7**(47), 873–885 (2010). https://doi.org/10.1098/rsif.2009.0386
16. Ross, R., Hudson, H.P.: An application of the theory of probabilities to the study of a priori pathometry.–Part II. Proc. R. Soc. Lond. A **93**(650), 212–225 (1917). https://doi.org/10.1098/rspa.1917.0014
17. Kermack, W.O., McKendrick, A.G.: A contribution to the mathematical theory of epidemics. Proc. R. Soc. Lond. **115**(772), 700–721 (1927). http://www.jstor.org/stable/94815

18. Bacaër, N., Gomes, M.G.M.: On the final size of epidemics with seasonality. Bull. Math. Biol. **71**(8), 1954 (2009). https://doi.org/10.1007/s11538-009-9433-7
19. Thieme, H.R.: Mathematics in Population Biology. Princeton University Press (2003)
20. Wynants, L., Calster, B.V., Collins, G.S., Riley, R.D., Heinze, G., Schuit, E., Bonten, M.M.J., Dahly, D.L., Damen, J.A., Debray, T.P.A., Jong, V.M.T.d., Vos, M.D., Dhiman, P., Haller, M.C., Harhay, M.O., Henckaerts, L., Heus, P., Kammer, M., Kreuzberger, N., Lohmann, A., Luijken, K., Ma, J., Martin, G.P., McLernon, D.J., Navarro, C.L.A., Reitsma, J.B., Sergeant, J.C., Shi, C., Skoetz, N., Smits, L.J.M., Snell, K.I.E., Sperrin, M., Spijker, R., Steyerberg, E.W., Takada, T., Tzoulaki, I., Kuijk, S.M.J.v., Bussel, B.C.T.v., Horst, I.C.C.v.d., Royen, F.S.v., Verbakel, J.Y., Wallisch, C., Wilkinson, J., Wolff, R., Hooft, L., Moons, K.G.M., Smeden, M.v.: Prediction models for diagnosis and prognosis of covid-19: systematic review and critical appraisal. BMJ **369**, m1328 (2020). https://doi.org/10.1136/bmj.m1328

High Frequency Device as a Therapeutic Resource in the Treatment of Skin Aging: Evaluation of Female Self-esteem

A. F. Santos[1], L. Assis[1], C. C. S. Martignago[2], L. dos Santos[1], J. C. Tarocco[1], R. S. Pessoa[1,3], and C. R. Tim[1](✉)

[1] Biomedical Engineering Department, Scientific and Technological Institute of Brazil University, Carolina Fonseca, 235, São Paulo, Brazil
carla.tim@universidadebrasil.edu.br
[2] Department of Bioscience, Federal University of São Paulo, Santos, Brazil
[3] Department of Physics, Technological Institute of Aeronautics, São José Dos Campos, Brazil

Abstract. The aging of the integumentary system, makes the skin flaccid, wrinkled, dehydrated and irregularity in pigmentation. The increase in the life expectancy of the population has increased the concern with aging, good appearance, quality of life, thus providing greater self-esteem and acceptance in social life. The high frequency generator, a low-cost and widely used resource, has relevant impacts on the repair process and provides anti-inflammatory and bactericidal effect. Objective, to evaluate the self-esteem of women after the use of the high frequency generator in the treatment of signs of skin aging. The study followed ethical precepts, according to Resolution 466/12 of the National Health Council and was approved by the CEP (4,599,360). The participants received the application of a high frequency generator in the periorbital and nasolabial regions, for 30 min, 3 times a week, for 8 weeks. Twenty-six women with a mean age of 54.4 years participated in the study, most of them had dry and oily skin, sometimes wearing sunscreen, and 35% had phototype III skin according to the Fitzpatrick scale, with grade II and III photoaging, according to Glogau. Improved self-esteem was obtained during the sessions, and good satisfaction with the treatment. Overall cosmetic improvement was significant, with 50% of participants rating "much better", 42% "much better" and 8% "better" using the global scale. Treatment with the high frequency generator had satisfactory effects on the skin, allowing good appearance and increasing self-esteem. Noninvasive procedure, easy to apply, which increases the patient's treatment.

Keywords: High frequency · Skin aging · Photoaging

1 Introduction

In order to clarify the aging process, many studies have been carried out to understand how all the dynamics actually occur. Aging is of a dynamic, gradual, inevitable process, which occurs in all body systems, caused by intrinsic factors, (also called genetic), chronological factors, and extrinsic factors, (called photoaging), which are related to sun exposure, pollution, alcohol, drugs etc. [1].

J. L. B. Marques et al. (Eds.): CLAIB 2022/CBEB 2022, IFMBE Proceedings 98, pp. 590–599, 2024.
https://doi.org/10.1007/978-3-031-49401-7_61

The color of the skin depends mainly upon the density and distribution of melanin, the pigment which protects the skin from ultraviolet radiation of the sun and the skin is a clinically classified as phototype. The gold standard of classification systems is Fitzpatrick scale, which is based on self-reported erythema sensitivity and tanning ability. This phototype scale is graded on a number (from I to VI) which, to some extent, reflects the degree of color in the skin and its level of sensitivity to the damage generated by ultraviolet radiation [2]. Individuals with skin phototype I to III are the most sensitive to ultraviolet radiation, because they have lower melanin production, the main physiological mechanism photoprotective skin, being exposed to aggression stemming from sun exposure [3].

As a result of the aging of the integumentary system, the skin becomes flabby, wrinkled, dehydrated and with irregular pigmentation. These weaknesses become evident due to the structural and functional disorders caused in the skin. The skin is a primordial organ. It has several essential functions in the life of the living being, and it is presented in layers [4, 5].

The epidermis is the outer layer of the skin, subdivided into four to five layers (depending on thickness), exercising all its protective function against external agents. Closely linked to the epidermis, the next layer, called the dermis, is made up of connective tissue, which has capillaries, fibroblasts and several other fundamental components for homeostasis in its structure [6].

The fibroblasts contained in the dermis provide the skin with firmness and resistance, as they are responsible for the synthesis of the extracellular matrix, collagen, elastin, among other fundamental components [7]. As the days go by, the cells lose their capacity for cell renewal, bringing a reduction in the number of fibroblasts, this reduction can be affected by chronological factors, and potentiated by photoaging factors, thus evidencing skin aging [8].

The increase in population life expectancy has lead people to be concerned about aging with a good appearance, with quality of life, providing an increase in self-esteem and acceptance in social life, in addition to favoring bene-fits to physical and mental well-being. Although aging is an inevitable process, there are certain characteristics that can be minimized and even postponed, because of this, there has been much research done on treating the effects brought about by aging. Many treatments rely on invasive, costly procedures, making access difficult for all people who go through such aging events [9].

Through analysis of the weakness of the skin experienced by the physiological process of aging, and the difficulty of access for everyone to have a treatment that minimizes such signs of time, it was identified that the equipment that generates high frequency (HF) plasma, has the possibility of softening wrinkles, restoring firmer skin and improving pigmentation. The HF is a device widely used in aesthetics for various therapeutic purposes, which has significant physiological effects that can favor tissue repair [10].

The HF equipment tied to electric field generation, works with alternating currents, high voltage and low intensity, and glass electrodes whose interior contains vacuum rarefied air, and neon gas, argon or xenon. With the passage of the electromagnetic current through the air or gas occurs the ionization of the molecules, which, suffers the strong energetic impact, become fluorescent and cause the formation of ozone that can act in the balance of oxide reduction [10]. Thus, it can be said that the aesthetic and therapeutic benefits using HF are vast, as it adds the effects of electrotherapy with those of ozone [11].

Thus, the study in question aimed to assess the self-esteem of women aged 50 to 60 after using the high-frequency device in the treatment of signs of skin aging.

2 Material and Methods

The study was a single-blind pre- and post-treatment clinical trial. First, patients were evaluated and then received treatment. After 4 and 8 weeks the patients were evaluated again, thus the patients knew of treatment and only the evaluators were blind. Following all ethical precepts, approved under the opinion number 4,599,360. The research participants were oriented as to the objective and purpose of the procedure that would be submitted, and after the guidelines, they signed the Free Informed Consent Term, in two copies, one for the patient and the other for the re-searcher, and only after the signature started the procedure. The dynamics of recruitment were based on spontaneous demand, dissemination of a post on social networks, and indication of the volunteers themselves. The research took place in the city of Rondonópolis, Mato Grosso - Brazil. The participants were selected by dissemination of the project to the community by pamphlets and social net-works. The inclusion criteria, the following criteria were used: women aged between 50 and 60 years, with photo-type I, II, III and IV according to the Fitzpatrick scale, presence of signs of facial photoaging grade II to V, according to the Glogau scale, and Free and Informed Consent by signing. The exclusion criteria the following criteria were followed: Pregnant women; history of neoplasm; use of some type of cosmetic in the facial region in the last six months; rhytidectomy and/or blepharoplasty surgery, and ablative chemical or physical peeling in the last 12 months; and, use of botulinum toxin or fillers with collagen implants in the last six months; carriers of diseases aggravated by light; patients using photosensitizing drugs; patients with a history of keloid or hypertrophic scarring; patient with inability to perform photoprotection during the entire treatment period; patients with active skin diseases in the facial region; presence of implanted electronic or metallic devices; refusal to sign the free and informed consent form and discontinuation: withdrawal of participants; three consecutive fouls; present any disorder/infection that implies the use of antibiotics throughout the study. As for the study design, the participants were interviewed, filled out an anamnesis form, evaluated for skin phototype using the Fitzpatrick scale and facial photoaging using the Glogau aging scale. The instrument for measuring self-esteem was applied with the Brazilian version of the Rosenberg self-esteem scale, and the instrument was also used to assess the measurement of perception and satisfaction with treatment, and finally the instrument to assess global aesthetic improvement, using a scale to assess the improvement of the general aesthetics – Global Aesthetic Improvement Scale (GAIS). Participants had their faces photographed at baseline (T0) and after 8 weeks (T8). The care protocol was performed in 8 weeks, with 3 sessions per week, totaling 24 sessions. For the application of the high frequency generator, the participant's skin was sanitized with gauze soaked in 0.9% saline and dried with gauze. After the dry face, the HF equipment (Ibramed) was applied, using the larger spherical glass electrode. The direct application or effluviation was applied, in which the electrode is passed slowly and regularly in direct contact with the skin, causing the formation of sparks. The HF device was turned on and tested in the forearm region and then placed in the region to be treated so that it slid over the region maintaining constant movement, with complete and uniform contact on the skin surface during therapy. For treatment with the HF device, the participant's face was divided into three regions: right, left periorbital and nasolabial periorbital, always following this order for application. Each region received the treatment for 10 min, totaling 30 min of treatment. After the interventions, the participants received sunscreen in the region, Sun Protection Factor 30 (SPF), and guidance to maintain daily use, for which the SPF 30 protector was provided to each participant. At the end of each session, asepsis of the electrode was performed with gauze soaked in 70% alcohol.

3 Results

The study recruited 98 women; 61 of these were ineligible by the exclusion criteria, 38 candidates were younger than 50 years old or older than 60 years old, and 23 women had some type of cosmetic procedure in the last six months; Thus, 37 women started the protocol, but 11 of them did not show assiduity to receive the treatment, and for this reason they did not finish, and these were omitted from the sample, 26 participants completed the research (Fig. 1).

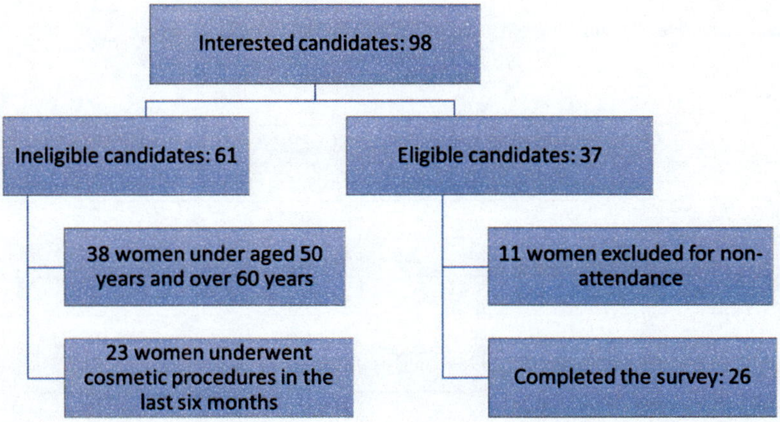

Fig. 1. Selection of patients. *Source* Own authorship

The mean age of the participants was 54.4 ± 3.3 years. In the dermatological history, 31% of the participants have mixed skin type, 27% have oily skin, 23% have dry skin and 19% represent women with normal skin. Of the 26 participants, 35% use sunscreen daily, 42% use sunscreen sometimes, 8% said they use it often, 12% do not use it and a minority represented by 4% only use it when exposed to the sun. When asked about the sun protection factor (SPF) used, the most used represented by 42% of the participants use SPF 30, 23% use SPF 50, 15% use SPF 70, 4% represent those who use 40 and 60 SPF. And 12% don't wear SPF. As for the use of makeup, 77% said they use makeup sometimes, 12% do not adopt this practice, 8% use makeup frequently and 4% use it daily, of those who use makeup, 73% remove it to sleep, 15% do not withdraw, and 12% don't wear makeup (Table 1).

Table 1. Table captions should be placed above the tables.

Variable	Participant (n = 26)
Age	54.4 (± 3.3)
Skin type	
Normal	5 (19%)
Dry	6 (23%)
Mixed	8 (31%)
Oily	7 (27%)
Use sunscreen	
Always/daily	9 (35%)
Often	2 (8%)
Sometimes	11 (42%)
On Sunny days	0 (0%)
When exposed to the sun	1 (4%)
Never	3 (12%)
FPS	
30	11 (42%)
40	1 (4%)
50	6 (23%)
60	1 (4%)
70	4 (15%)
None	3 (12%)
Use makeup	
Always/daily	1 (4%)
Often	2 (8%)
Sometimes	20 (77%)
Never	3 (12%)
Remove makeup to sleep	
Yes	19 (73%)
No	4 (15%)
Don't wear makeup	3 (12%)

Skin phototype assessment was performed according to the Fitzpatrick classification (FZ). Of the 26 candidates evaluated, 12% are skin phototype FZ I, 23% are skin phototype (FZ) II, 35% are skin phototype FZ III, 23% are skin phototype FZ IV and 8% are skin phototype FZ V (Table 2).

Table 2. Fitzpatrick scale.

Variable	Participant n = 26
FZ-I	3 (12%)
FZ-II	6 (23%)
FZ-III	9 (35%)
FZ-IV	6 (23%)
FZ-V	2 (8%)
FZ-VI	0 (0%)

To assess photoaging, the Glogau Scale was used, so 50% characterize the participants with photoaging G2, 43% represent those with photoaging G3 and G4 represented by 4% (Table 3).

Table 3. Glogau scale.

Variable	Participant n = 26
G1	0 (0%)
G2	13 (50%)
G3	12 (46%)
G4	1 (4%)

The results obtained in relation to self-esteem, using the Brazilian version of the Rosenberg self-esteem scale, applied to 26 participants, it was possible to observe that the scale scores had an average of 5.8 ± 3.4 before treatment and in the mean last session of 2.2 ± 2.5 (Table 4).

Table 4. Rosenberg self-esteem scale.

Variable	Score
Self-esteem (T0)	5.8 (± 3.4)
Self-esteem (T8)	2.2 (± 2.5)

Regarding the perception of treatment, all candidates, that is 100%, stated that their appearance and self-esteem improved after undergoing treatment (Table 5).

Table 5. Perception of treatment.

Variable	Participant (n = 26)
Appearance after treatment	
Improved	26 (100%)
Remained the same	0 (0%)
Worsened	0 (0%)
Self-esteem after treatment	
Improved	26 (100%)
Remained the same	0 (0%)
Worsened	0 (0%)

Regarding satisfaction with the treatment, 50% say maximum satisfaction, 38% high satisfaction and 12% say moderate satisfaction (Table 6).

Table 6. Satisfaction with the treatment.

Variable	Participant (n = 26)
Maximum satisfaction	13 (50%)
High satisfaction	10 (38%)
Moderate satisfaction	3 (12%)
Little satisfaction	0 (0%)
No satisfaction	0 (0%)

The overall aesthetic improvement was considered significant, with 50% of the candidates rated "better, 42% "best" and 8% "good", the evaluation was performed using the Global Aesthetic Improvement Scale) (GAIS) (Table 7).

4 Discussion

Aged skin affects quality of life, self-esteem, changes in self-image, and can have psychological impact. Thus, the research showed that the proposed treatment offered the participants significant improvements in the general context. During the treatment, it was observed that the care and attention of the participants with their skin changed, with each session they reported differences that were perceived throughout the treatment. The average age of the participants in this study was 54.4 years old, those participants exhibited some of the aesthetic dysfunctions brought about by aging, such as age, hormonal decrease, and cell life span [12].

Table 7. Global aesthetic improvement scale (GAIS).

Variable	Participant (n = 26)
Best	11 (42%)
Better	13 (50%)
Good	2 (8%)
No change	0 (0%)
Poor	0 (0%)
Worse	0 (0%)
Much worse	0 (0%)

The care given to the skin is directly linked to aging, as photoaging is a consequence of exposure and mistreatment of the skin tissue [13]. It is important to observe that only 35% participants of this study, used sunscreen daily, and that the others used it sometimes, frequently and some even did not use it. Solar radiation causes significant skin lesions that can be avoided with the use of sun protection [14]. During the treatment, all participants used sunscreen daily.

Skin phototype was also evaluated, using the Fitzpatrick scale, which is based on skin sensitivity to ultraviolet radiation and degree of aging [15]. Most of the candidates, (35%), had skin classified as type III, had a sensitive ultra-violet reaction, burned moderately and gradually, and the skin tone is light to medium. There are some reports that correlates wrinkles and skin phototype, thus deep wrinkles are more severe in skin phototypes I, which has a low tolerance to sunlight [16, 17].

The increase in self-esteem during the treatment process was perceptible, both in the participants' reports and in the responses obtained in the application of the Rosenberg scale. According to Avelar and Veiga [18], self-esteem is equivalent to wanting yourself well and that when reduced, this can reflect as an inferiority complex, feeling of incapacity, apathy, discouragement, indifference to your values, personal appearance, among others, which are usually presented as a state of sadness.

The improvement of the results is evaluated by the maximum satisfaction of the volunteers in relation to the procedure, by the perception of the appearance which all of them affirmed that they obtained improvement, both in the appearance and in the self-esteem after the procedure and also by the application of the global aesthetic improvement scale of the procedure. Gais global aesthetic improvement scale, where it is evaluated whether the resource used is applicable to such aesthetic dysfunction. The evaluation shows that the application of HF as a wrinkle treatment showed significant improvements, with evidence that 50% of the participants presented results classified as much better, 42% as much better and 8% as better.

In view of the above information, it is evident that treatment with HF can be a promising resource in the treatment of skin aging, It was possible to identify improvements considered positive in the participants, increasing self-esteem and still showing good satisfaction with the treatment used. This finding may be related to the fact that the HF is an equipment that has different types of glass electrodes, filled with gases (argon, neon and xenon) or rarefied air, producing an alternating current of low intensity and high voltage that, in contact with the skin promotes sparking that allows the generation of ozone species. The electrode commonly used for direct application or effluviation is the mushroom or onion type. Ozone can intervene in the redox balance, promoting the formation of reactive oxygen molecules, which favors biochemical events of cellular metabolism [19–21].

Its physiological effects can be described as increased cellular oxygenation, bactericidal and antiseptic action, improvement of dermal trophism, in addition to the thermal effect, hyperemia and vasodilation [19]. These mentioned effects justify the tissue improvement obtained in the skin of the participants, as the local thermal action increases vasodilation with the opening of the capillaries, improving tissue trophism, thus obtaining nutritional gain of oxygen, nutrients for the tissue, and still improving the drainage system of cellular waste, thus providing the production of quality elastic fibers, acting on fibroblasts and other cells [22].

Despite the knowledge of such physiological effects obtained by HF, it is necessary to understand the effectiveness of the equipment in the revitalization of the skin, in special wrinkles treatment.

5 Conclusion

The manuscript aims to use the high frequency device as a stimulus for anti-aging. It is easy to apply and relatively low cost, which can bring a solution compared to other methods of high cost and low accessibility. It is important to consider that the aging process is not only linked to changes in the joints, muscular or endocrine system, but mainly reflects on the perception and body image of the individual. However, it is necessary to expand the methods of evaluation of the cutaneous effects of high frequency so that we can better validate the results.

Acknowledgment. We would like to acknowledge the contributions of the funding agency CAPES for the financial support of the present research.

Conflict of Interest The authors declare that they have no conflict of interest.

References

1. Addor, F.: Influence of a nutritionl suplemente containing collagen peptides on the properties of the dermis. Surg. Cosmet. Dermatol. **2**, 116–121 (2015). https://doi.org/10.5935/scd1984-8773.201572636
2. Fitzpatrick, T.B.: The validity and practicality of sun-reactive skin types I through VI. Arch. Dematol. **6**, 869–871 (1988)
3. Videira, I.F.S., Moura, D., Magina, S.: Mechanisms regulating melanogenesis. An. Bras. Dermatol. **1**, 76–83 (2013). https://doi.org/10.1590/s0365-05962013000100009
4. Bravo, B.S.F., Azulay, D.R., Luiz, R.R., et al.: Oral isotretinoin in photoaging: objective histological evidence of efficacy and durability. An. Bras. Dermatol. **4**, 478–486 (2015). https://doi.org/10.1590/abd1806-4841.20153703

5. Arango, A.C.M., Munoz, S.V.F., Sanclemente, G.: Mechanisms of skin aging. Iatreia Abr-Jun 2:160–170 (2017)
6. Barbara, E.C.O.S., Brito, V.S.C., Rocha, W.R.S., et al.: Uso da micropuntura no tratamento de rugas. Fisioterapia Bras. **4**, 481–488 (2017)
7. Tobin, D.J.: Introduction to skin aging. J. Tissue Viability **26**, 37–46 (2016)
8. Zhang, S., Duan, E.: Fighting against skin aging: the way from bench to bedside. Transplante Cell. **5**, 729–738 (2018). https://doi.org/10.1177/0963689717725755
9. Pilkington, S.J., Belden, S., Miller, R.A.: The tricky tear trough: a review of topical cosmeceuticals for periorbital skin rejuvenation. J. Clin. Aesthet. Dermatol. **9**, 39–47 (2015)
10. Oliveira, L.M.N.: Utilização do ozônio através do aparelho de alta frequência no tratamento da úlcera por pressão. Rev. Bras. Ciên. Saúde Paraíba **30**, 41–46 (2011)
11. Pereira Sá, H., et al.: Estudo comparativo da ação do laser GaAlInP e do gerador de alta frequência no tratamento de feridas cutâneas em ratos: estudo experimental. Conscientiae saúde **3**, 360–366 (2010)
12. Silva, A.R.S., Santos, A.C.O., Gonçalvez, V.M., et al.: Radiofrequência no tratamento de rugas faciais. Rev. Universidade Ibirapuera São Paulo **7**, 38–42 (2014)
13. Feitosa, B.V., Palma, A.L.R.: Estudo dos efeitos da radiofrequência no tratamento facial em mulheres entre 30 e 50 anos. Pesquisa, Soc Desenvolvimento **2**, e28811225696 (2022). https://doi.org/10.33448/rsd-v11i2.25696
14. Renno, F.C., Renno, R.C., Nassif, P.W.: Atualização em Fotoprotetores. RevUningá 3:56 (2014)
15. Mota, J.P., Barja, P.R.: Classificação dos fototipos de pele: análise fotoacústica vesus análise clínica. UNIVAP/IP&D/FASBio, 2561–2564 (2002)
16. Nagashima, H., Hanada, K., Hashimoto, I.: Correlation of skin phototype with facial wrinkle formation. Photodermatol. Photoimmunol. Photomed. I **15**, 2–6 (1999)
17. Kligman, A.M., Zheng, P., Lavker, R.M.: The anatomy and pathogenesis of wrinkles. Br. J. Dermatol. **113**, 37–42 (1985)
18. Avelar, C.F.P., Veiga, R.T.: Como entender a vaidade feminina utilizando a autoestima e a personalidade. Rev adm empres Ago **4**, 338–349 (2013)
19. Martins, A., Silva, J.T., Fréz, A.R., et al.: Efeito bactericida do gerador de alta frequência na cultura de Staphylococcus aureus. Fisioterapia e Pesquisa **19**, 153–157 (2012)
20. Korelo, R.I.G.: Gerador de alta frequência como recurso para tratamento de úlceras por pressão: estudo piloto. Fisioter Mov Curitiba **4**, 715–724 (2013)
21. Traina, A.A.: Efeitos biológicos do ozônio diluído em água na reparação tecidual de feridas dérmicas em ratos. Tese de Doutorado Faculdade de Odontologia Universidade de São Paulo (2008). https://doi.org/10.11606/T.23.2008.tde-08042009-150340
22. Carvalho, G.F., Silva, R.M.V., Mesquita, F.J.J.T., et al.: Avaliação dos efeitos da radio freqüência no tecido conjuntivo. Especial Dermatologia **68**, 10–25 (2011)

Author Index

J. L. B. Marques et al. (Eds.): CLAIB 2022/CBEB 2022, IFMBE Proceedings 98, pp. 601–603, 2024.
https://doi.org/10.1007/978-3-031-49401-7